Chemistry of
Natural Products

Volume I

Chemistry of
Natural Products

Volume I

Ashutosh Kar

Formerly

Professor, School of Pharmacy, Addis Ababa University,
Addis Ababa **(Ethiopia)**

Dean, Chairman & Professor, Faculty of Pharmaceutical Sciences,
Guru Jambheshwar University, Hisar **(India)**

Professor, School of Pharmacy, Al Arab Medical University,
Benghazi **(Libya)**

Professor, College of Pharmacy, University of Delhi,
Delhi **(India)**

Professor & Head, Department of Pharmaceutical Chemistry,
Faculty of Pharmaceutical Sciences, University of Nigeria,
Nsukka **(Nigeria)**

CBSPD

CBS Publishers & Distributors Pvt Ltd

New Delhi • Bengaluru • Chennai • Kochi • Kolkata • Lucknow • Mumbai
Gujarat • Hyderabad • Jharkhand • Nagpur • Patna • Pune • Uttarakhand

Chemistry of Natural Products

Volume I

ISBN: 978-81-239-1874-7

Copyright © Author and Publisher

First Edition: 2010
Reprint: 2014, 2018, **2025**

Published by **Satish Kumar Jain** and produced by **Varun Jain** for

CBS Publishers & Distributors Pvt Ltd
4819/XI Prahlad Street, 24 Ansari Road, Daryaganj, New Delhi 110 002, India
Ph: 011-23289259, 23266838   Website: www.cbspd.com
                                                          e-mail: delhi@cbspd.com
Corporate Office: 204 FIE, Industrial Area, Patparganj, Delhi 110 092
Ph: 011-4934 4934   Fax: 011-4934 4935   e-mail: publishing@cbspd.com; publicity@cbspd.com

Branches

- **Bengaluru:** Seema House 2975, 17th Cross, K.R. Road, Banasankari 2nd Stage, Bengaluru 560 070, Karnataka, India
  Ph: +91-80-26771678/79   Fax: +91-80-26771680   e-mail: bangalore@cbspd.com

- **Chennai:** 18/8B, Subbarayan Street, Shenoy Nagar, Chennai 600 030, Tamil Nadu, India
  Ph: +91-44-42032115, 26681266   e-mail: chennai@cbspd.com

- **Kochi:** 42/1325, 1326, Power House Road, Opp KSEB, Power House, Ernakulam 682 018, Kerala, India
  Ph: +91-484-4059061-65   Fax: +91-484-4059065   e-mail: kochi@cbspd.com

- **Kolkata:** 147, Hind Ceramics Compound, 1st Floor, Nilgunj Road, Belghoria, Kolkata-700056, West Bengal, India
  Ph: 033-25633055, 033-25633056   e-mail: kolkata@cbspd.com

- **Lucknow:** Basement, Khushnuma Complex, 7-Meerabai Marg (Behind Jawahar Bhawan), Lucknow 226001, India
  Ph: 0522-4000032   e-mail: tiwari.lucknow@cbspd.com

- **Mumbai:** PWD Shed. Gala no. 25/26, Ramchandra Bhatt Marg, Next to JJ Hospital Gate no. 2 Opp. Union Bank of India Noorbaug
  Mumbai-400009, Maharashtra, India
  Ph: 022-66661880/89   e-mail: mumbai@cbspd.com

Representatives

• **Gujarat**	0-9879558667	• **Hyderabad**	0-9885175004	• **Jharkhand**	0-9811541605	• **Nagpur**	0-8692091830
• **Patna**	0-9334159340	• **Pune**	0-9664372571	• **Uttarakhand**	0-9716462459		

Printed at Rashtriya Printers, Dilshad Garden, Delhi, India

To Dearest Grandchildren
Aditi and *Aryan*
My source of Inspiration
and Drive

> "Always Look At What You Have Left,
> Never Look At What You Have Lost."
>
> — Robert H Schuller

> "Everything Has Its Limit—
> Iron Ore Cannot Be Enriched Into Gold."
>
> — Mark Twain

> "We All Live in a *World guided by Certain Laws*,
> Just there is a *Universal Law of Gravity*.
> In case one falls off the *Leaning Tower of Pisa*,
> it hardly matters, if one is a *Noble Human Being* or a
> *Crooked Person*, one is surely Going to Hit the Ground."
>
> — Anonymous

Preface

The specific objective of *Chemistry of Natural Products: Volume I* is to provide fundamental knowledge of various class of natural products *viz*. **Carbohydrates; Amino Acids, Peptides and Proteins; and Alkaloids** to the postgraduate students of Chemistry in Indian universities besides other professional programmes of study. It is also intended to encourage the interests of those students who are actively engaged in PG courses in natural products; and other health professionals.

An array of biosynthetic processes which have evolved specifically in *living organisms* have resulted in the unique generation of *diverse chemical structures* that essentially possess an amazing range of *biological activity profiles*. The natural product chemist shoulders the herculian task for stimulating an intensive search of nature to find newer leads for further scope in future.

The author has taken into consideration the New Syllabi of University Grants Commission (UGC) and developed the entire text matter accordingly so that the PG students in all Indian universities will benefit from the textbook enormously.

The book deals with the above cited three topics in a methodical manner. Most structures have been taken from the latest *Merck Index*, proper explanations have been included to help the student grasp the subject matter. The relevant matter has been adequately *expanded-explained-expatiated* for the satisfaction of the reader.

Modern spectroscopic methods of analysis, used for characterization of **natural products**, have been included and explained adequately.

The subject matter has been duly substantiated with appropriate references as a *Footnote* wherever necessary. Each chapter ends with *Selected Bibliography* and *Review Questions* to enable the students to prepare for the examination.

The author is confident that this book will prove to be a pathfinder to undergraduates **B.Sc. (Hons.)** and postgraduates *M.Sc.* in **Chemistry, Biotechnology, Biochemistry, Pharmaceutical Chemistry, Molecular Biology, Microbiology** and other allied fields.

The author warmly acknowledges the overwhelming support extended by Shri Satish Kumar Jain, Managing Director; and Shri Vinod Kumar Jain, Production Director, CBS Publishers & Distributors Pvt. Ltd., New Delhi, for bringing out this book in a record time-frame.

Gurgaon
March 2010

ASHUTOSH KAR

Contents

Chemistry of
Natural Products

1. Carbohydrates

2. Amino Acids, Peptides and Proteins

3. Alkaloids

Contents at a Glance

1 Carbohydrates

1. INTRODUCTION

'Carbohydrates' predominantly represent one of the most abundantly occurring categories of 'Natural Products. Any living organism belonging to either *plant* or *animal kingdom* do essentially involve in the so called **Carbohydrate Metabolism'.** The actual genesis of the terminology **carbohydrate** was duely derived for the inherent presence of *three* important elements from the **Mendeleeve's Periodic Table** *viz.,* **Carbon, Hydrogen,** and **Oxygen** in it. Importantly, the latter *two elements i.e.;* **H** and **O** are critically present in the same ratio (2 : 1) as in water (H_2O). In other words, one may visualize the **carbohydrates** as the *'hydrates of carbon* [Greek : *carbo* means-carbon; *hydrates* means-hydrates]; and, therefore, they intimately correspond to the following *generalized formula:*

$$C_a (H_2O)_b$$

Based on the above **generalized formula,** let us look at the emperical formula of a few carbohydrates, namely:

Glucose	:	$C_6 H_{12} O_6$	in which the prevalent
Sugar	:	$C_{12} H_{22} O_{11}$	*'Ratio'* of 'H' and 'O' is
Amylose, Amylopectin	:	$(C_6 H_{10} O_5)_n$	evidently 2 : 1

Nevertheless, one may also come across of a host of such specific **'carbohydrates'** that do not correspond to the aforecited **generalized formula,** such as:

Rhamnose $C_6H_{12}O_5$ 1 : 2 : 1

Besides, there exists quite a few **chemical organic compounds** that essentially bear the **ratio of 'H' and 'O' as 2 : 1,** but they do not fall into the class of **'carbohydrates'** for instance:

Inositol	:	$C_6H_{12}O_6$	
Lactic Acid	:	$C_3H_6O_3$	they essentially have the
Acetic Acid	:	$C_2H_4O_2$	'Ratio of 'H' and 'O' as 2 : 1
Formalin	:	CH_2O	

Based on the already discussed wide variance in the overall statutory requirements and qualifications for a chemical entity to enjoy the status of being called a 'carbohydrate', a latest and up-to-date definition has been duly forwarded:

'an optically active polyhydroxyaldehydes, or ketones, or such chemical entities that may be duly 'hydrolyzed' to either of them'.

Medical sciences have amply proved that the 'carbohydrates' do designate as:

- **source of energy** : *viz.,* sugars, glucose, fructose etc.,
- **store of energy** *viz.,* starch, glycogen etc.,
- **biochemical pathways** *in vivo* : *viz.,* Kreb's cycle, conversion to alcohol etc.,

In addition to the above the **carbohydrates** also play a variety of vital and important roles, such as:

- **major support to 'plant tissues'** : *viz.,* cellulose fibres,
- **main constituent of shells of sea-foods** : *viz.,* chitin, (*e.g.,* **crabs, shrimps, lobsters etc.**)
- **major constituents of 'Nucleic Acids'** : *viz.,* ribose, deoxyribose or aldopentoses, (*e.g.* DNA-Deoxyribonucleic acid, RNA-Ribonucleic acid)
- **natural sweetness in 'fruits'** : *viz.,* fructose, levulose etc., (*e.g.,* grapes, oranges, pineapple, mangoes, banana, melon, strawberry etc.,)
- **mother's milk, cow's milk and dairy products** *e.g.,* **cheese, whey powder, skimmed-milk powder, whole-milk powder etc.,**: *viz.,* lactose
- **fermentation of 'carbohydrates' yield potable alcohols** : *viz.,* wines, Beer, Scotch whiskey, Vodka, Champene, Gin, Rum, Cognac, Brandy, Liquors, (*e.g.,* **grape juice, potatoes, malt wort, molasses, sweet-beat**)
- **energy drinks** (*e.g.,* **instant energy for 'athletes'**) : *viz.,* carbonated 'Glucose Drinks' with fruity flavours,
- **baby foods (weaning foods)** : *viz.,* Farex [R],

It is, however, pertinent to state here that the 'primary source of natural carbohydrates' is the 'plant kingdom', that ultimately cause their formation from *carbon dioxide* (CO_2) and *water* (H_2O) in UV-light:

$$x\ CO_2 + x\ H_2O \xrightarrow[\text{Chlorophyll}]{\text{UV ray}} C_n(H_2O)_n + O_2\uparrow$$

Carbon dioxide   Water   Carbohydrate   Oxygen

2. NOMENCLATURE

In general, the nomenclature of the 'carbohydrates' is not so simple and easy but is rather **complex in nature;** and, therefore, each individual compound needs to be designated duly with a 'specific name'. In usual conventional manner, the *suffix* '-ose' is invariably used for naming the various **sugars** or **carbohydrates.** Following are some of the typical instances wherein the portion of the *particular name* just preceeding the *suffix* sometimes distinctly represents the actual origin, history or source of the chemical entity, for instance:

Lactose	:	Milk sugar ('*lactis*' in Latin means 'milk');
Maltose	:	Malt* sugar;

---

* **Malt:** It is obtained by the germination process of moist barley in an aerated and perforated beds at a temperature of 18±2°C for 4-5 days, when starch bundles in barley get converted to **maltose,** and the process is called 'Malting' and product as **Malt.**

Sucrose	:	Sugar derived from *'Sugarcanes'*;
Fructose	:	Sugar obtained from *'Fruits'*;

3. CLASSIFICATION

The **Carbohydrates** are broadly classified into *two* major categories, namely:

- **Sugars**, and
- **Non-Sugars (or Polysaccharides)**.

These *two* categories of **carbohydrates** shall now be classified individually in the sections that follows:

3.1. Sugars

Generally the **sugars'** are sweet in taste, and found to be mostly crystalline in nature, as well as water-soluble. Besides, the molecular weights of sugars are well defined for a specific pure compound.

The **'sugars'** are further subdivided into *two* sub-classes, such as:

- **Monosaccharides**, and
- **Oligosaccharides**.

3.1.1. Monosaccharides

In a broader perspective, the **'monosaccharides'** are considered to be the simplest form of sugars. These may be defined as –**'polyhydroxyketones or polyhydroxyaldehydes that may not be further hydrolyzed to simpler forms of sugars'**.

Nevertheless, the **'monosaccharides'** may be further classified into *two* sub-classes usually based upon :

- **nature of carbonyl** ($-\overset{O}{\underset{||}{C}}-$) **moiety, and**
- **number of C-atoms in a molecule.**

(*a*) **Nature of Carbonyl** ($-\overset{O}{\underset{||}{C}}-$) **Moiety :** If the **carbonyl moiety** belongs to an **aldehyde functional group** ($-\overset{O}{\underset{||}{C}}-H$) – they are known as **'aldoses'**, such as : **glucose**. In case, the **carbonyl moiety** represent a **ketone functional group** ($-\overset{O}{\underset{||}{C}}-$) -they are termed as **'Ketoses'**, for instance: **fructose.**

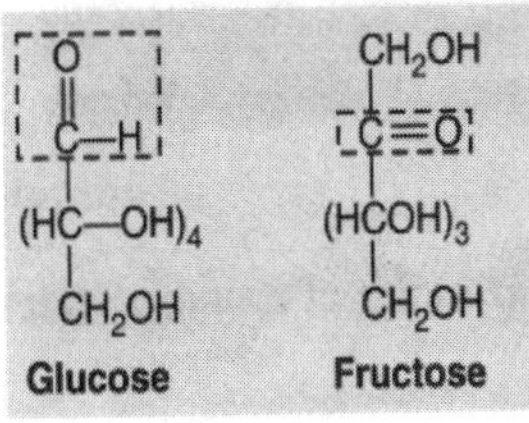

$$
\begin{array}{ccc}
\begin{array}{c} \overset{O}{\underset{||}{C}}-H \\ (HC-OH)_4 \\ CH_2OH \\ \textbf{Glucose} \end{array}
&
\begin{array}{c} CH_2OH \\ C=O \\ (HCOH)_3 \\ CH_2OH \\ \textbf{Fructose} \end{array}
&
\begin{array}{c} \textbf{Emperical Formula}=C_6H_{12}O_6 \\ \textbf{Molecular Wt.}=180 \end{array}
\end{array}
$$

(*b*) **Number of C-Atoms in a Molecule :** Various compounds bearing different C-atoms do have a specific name according to the **number of C-atoms** duly present in the molecule, such as :

Compounds with **3–C atoms** : Trioses;

Compounds with **4–C atoms** : Tetroses;

Compounds with **5–C atoms** : Pentoses;

Compounds with **6–C atoms** : Hexoses;

Compounds with **7–C atoms** : Heptoses;

Compounds with **8–C atoms** : Octoses;

Compounds with **9–C atoms** : Nonoses;

Compounds with **10–C atoms** : Decoses;

3.1.2. Oligosaccharides

The **'oligo saccharides'** upon careful **hydrolysis** give rise the **'monosaccharide molecules'** ranging between **2 to 10 in number.** Therefore, based on the aforesaid perception the **'oligosaccharides'** may be further classified into different categories depending exclusively upon the precise and exact number of the emanated *'monosaccharide units'* obtained duly after performing the **hydrolysis.** Thus, we may have :

3.1.2.1. Disaccharides

The **'disaccharides'** may be defined as **'sugars that yield upon hydrolysis usually two molecules either of the same or altogether different monosaccharides'.**

Examples : Following are *two* **typical examples :**

(a) **Maltose yields two moles of 'glucose' :**

$$C_{12}H_{22}O_{11} + H_2O \xrightarrow{\text{Hydrolysis}} 2, C_6H_{12}O_6$$

Maltose (Malt Sugar)   **Glucose**

(b) **Sucrose yields one mole each of Glucose and Fructose:**

$$C_{12}H_{22}O_{11} + H_2O \xrightarrow{\text{Hydrolysis}} C_6H_{12}O_6 + C_6H_{12}O_6$$

Sucrose (Cane Sugar)   **Glucose   Fructose**

3.12.2. Trisaccharides

Generally, the **'trisaccharides'** upon hydrolysis produce *one mole* each of *three* **monosaccharide variants** as exemplified under:

$$C_{18}H_{32}O_{16} + 2H_2O \xrightarrow{\text{Hydrolysis}} C_6H_{12}O_6 + C_6H_{12}O_6 + C_6H_{12}O_6$$

Raffinose   **Glucose   Galactose Fructose**

3.1.2.3. Tetrasaccharides

The **'Tetrasaccharides'** invariably give rise to *four* moles of **monosaccharides** upon hydrolysis, such as:

$$C_{24}H_{42}O_{21} + 3H_2O \xrightarrow{\text{Hydrolysis}} 4C_6H_{12}O_6$$

Stachyose   **Monosaccharides**

3.2. Non-Sugars [or Polysaccharides]

It has been observed that the **non-sugars** (or **polysaccharides**) are invariably found to be **taste-less, solids, amorphous in nature and** usually occur as:

- **water-insoluble compounds,** and/or
- **form colloidal suspensions.**

Besides, the characteristic features of '**polysaccharides**' resemble very much akin to the so called '**high polymers**' *viz.,* the critical **molecular weight** of a '**polysaccharide**' solely depends upon the number of small units it actually holds as may be observed with **Dextran–50** and **Dextran-75** (*i.e.,* these have *molecular weights* of **50,000** and **75,000** respectively).

Another common features usually observed amongst the '**polysaccharides**' being that they are tasteless, and upon careful **hydrolysis** in the presence of **dilute mineral acids** produce an array of **sweet monosaccharide moles.** Indeed the most abundantly distributed and usually noticeable '**polysac-charides**' predominantly being represented by the following **generalized formula** of **starch, cellu-lose,** and the like.

$$[C_6H_{10}O_5] + nH_2O \xrightarrow[\text{[H}^+\text{]}]{\text{Hydrolysis}} n[C_6H_{12}O_6]$$

Starch or
Cellulose

Glucose

Importantly, the **pentosans,** having the molecular formula $(C_5H_8O_4)_n$, and distributed in nature rather rarely do also belong to the class of '**polysaccharides**'.

There are **two polysaccharide** variants available, namely:

(a) **Homopolysaccharides-** are those which upon acid hydrolysis yield the **same (or single) type of monosaccharide units** *viz.,* **starch, insulin, and cellulose,** and

(b) **Heteropolysaccharides-** are those that on being subjected to acid hydrolysis give rise to **different kinds of monosaccharide units** *viz.,* **chondroitin sulphate.**

Highlights of Carbohydrates Classification : Bearing in mind the foregoing statement of facts, examples, and explanations one may draw the following inferences that:

- **monosaccharides** actually designate the most simple and ultimate carbohydrates; whereas, the others are duly regarded to be the **genuine condensation products of the monosaccharides.**
- '**pentoses**' and **hexoses**' are the most *vital* and *important* amongst the **monosaccharides** based upon their abundant availability in **natural products,** and
- the '**hexoses**' do show an edge over the '*pentoses*' in terms of their occurrence and overall reactivity in a host of critical biochemical processes in the living organisms.

It would be worthwhile to study the '**monosaccharides**' more exhaustively and comprehensively in order to understand the most authentic and fundamental principles of the '**carbohydrate chemistry**'.

4. CHEMISTRY OF MONOSACCHARIDES

As we have already discussed earlier that the '**monosaccharides**' are being designated as the **polyhydroxyaldehydes** or **polyhydroxyketones;** and hence, they mostly respond to the critical char-acteristic generalized reactions pertaining to:

- **carbonyl functional moiety,**
- **alcoholic functional moiety, and**

- both carbonyl and alcoholic moieties present together, in a plethora of compound.

> **Note:** As 'aldoses' are represented duly by 'glucose', and 'ketoses' by 'fructose', the various chemi-
> cal reactions (*i.e.*, their *chemistry*) shall be specifically studied pertaining to their references.

The **'chemistry of monosaccharides'** will be discussed with particular reference to the following typical reactions and supplemented duly by appropriate examples whenever necessary; together with plausible explanations to understand the mechanism(s) of reactions:

- Oxidation,
- Reduction,
- Acetylation,
- Formation of Glucosides,
- Phenylhydrazine Reaction,
- Epimerization and
- Reaction with alkali.

4.1. Oxidation

Monosaccharides undergo *'oxidation'* to yield a good number of *products of reaction* which solely depends upon the exact nature of the oxidizing agent being employed. Following are an array of **oxidizing agents** together with their respective **products of reaction** duly obtained from *'glucose'* and *'fructose'*:

4.1.1. Bromine Water

It is usually regarded to be a **'mild oxidizing agent'**, which critically oxidizes the **aldose** (*viz.*, Glucose) to an *alcohol* known as **alditol** (or **sorbitol**) through *three* **sequential steps** as indicated below:

Explanations: The above course of reactions are explained as under:

Step-I: The *Oxidation* of **aldose** (Glucose) with Br_2–water helps in the conversion of the terminal *aldehyde function* into the corresponding **carboxylic acid function** thereby producing **aldonic acid (gluconic acid).**

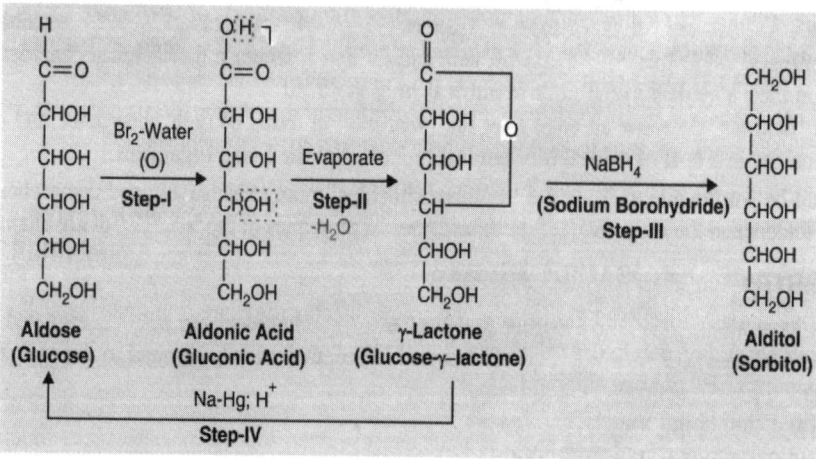

Step-II: Careful evaporation of the resulting product (*i.e.*, **Gluconic acid**), preferably under vacuum, gives rise to the formation of **γ-lactone** due to the abstraction of a mole of water (as shown by dotted line).

Step-III: The **γ-lactone** on being subjected to reduction in the presence of **sodium borohydride** (a strong **'reducing agent'**) converts the terminal carbonyl function into a **primary alcohol** ($-CH_2OH$); also causes the cleavage of the **'epoxide bridge'** between C–1 and C–4 to form a **secondary alcohol** (–CHOH), thereby yielding **alditol (sorbitol).**

Step-IV: The **γ-lactone (Glucose-γ-lactoue)** may be gainfully reduced to the original starting **material glucose** by reaction with sodium-amalgam (Na–Hg) with a trace of an acid (H^+) *i.e.*, another *mild reducing agent.*

Note: Ketoses do not get oxidized by bromine water.

4.1.2. Nitric Acid

The *strong oxidizing agent* like *nitric acid* (HNO_3) critically oxidizes **aldose (Glucose)** almost completely to the ultimate corresponding **'dicarboxylic acid'** known as **aldaric acid (Saccharic acid)** as given under:

$$
\begin{array}{ccc}
H & & H \\
| & & | \\
C=O & & C=O \\
| & \xrightarrow[\substack{\text{(Oxidation)} \\ -H_2O}]{HNO_3 \text{ Conc.}} & | \\
(CHOH)_4 & & (CHOH)_4 \\
| & & | \\
CH_2OH & & C=O \\
& & | \\
& & OH
\end{array}
$$

Aldose   **Aldaric Acid**
(Glucose)   **(Saccharic Acid)**

Explanations: The generously strong oxidation capability of HNO_3 renders the terminal **aldehyde function** into the **carboxylic acid function;** besides the other end of the terminal **primary alcoholic function** into the second **carboxylic acid moiety** due to the elimination of a mole of water (*i.e.* in all **'three'** *O-atoms* are actually used up in the said conversion).

Interestingly, the **'ketoses'** in particular *viz.,* **fructose,** on being subjected to oxidation with HNO_3 invariably gives rise to a mixture of acids, **glycolic acid** (2 C-atoms) and **tartaric acid** (4C-atoms) on account of the strategic cleavage occurring between C–2 and C–3 carbon chain, as stated under:

$$
\begin{array}{cccc}
CH_2OH & & & \\
| & & CH_2OH & COOH \\
C=O & & | & | \\
\text{- - -}| \text{- - -} & \xrightarrow{HNO_3} & COOH \quad + & (COOH)_2 \\
(CHOH)_3 & \text{Oxidation and} & & | \\
| & \text{Cleanage bet.} & & COOH \\
CH_2OH & \text{C–2 And C–3} & & \\
& \text{carbon chain} & &
\end{array}
$$

Ketose   **Glycolic**   **Tartaric Acid**
(Fructose)   **Acid**

4.1.3. Lead Tetra-Acetate [Pb (CH$_3$CO)$_4$] and Periodic Acid [H$_5$ IO$_6$]

It has been amply demonstrated that these *two* reagents *i.e.,* **lead tetra-acetate** and **periodic acid** do categorically help in the **oxidation** of the *1, 2-glycolic residues* duly embedded in the

monosaccharide molecules by virtue of the rupturing of the C–C linkage to yield *two* **aldehyde compounds** separately as indicated in the generalized reaction given below:

$$
\begin{array}{c}
\text{1,2-Glucolic} \\
\text{residue}
\end{array}
\left\{
\begin{array}{c}
|\\
\text{CHOH} \\
| \\
\text{CHOH} \\
|
\end{array}
\right.
\xrightarrow[\text{H}_5\text{IO}_6]{[\text{Pb(CH}_3\text{CO)}_4]}
\text{or}
\left.
\begin{array}{c}
|\\
\text{CHO} \\
+ \\
\text{CHO} \\
|
\end{array}
\right\}
\text{Two aldehyde compounds}
$$

Important Points: These essentially include:

(1) The above cited reagents fail to attack such compounds that do not contain *two hydroxyl* (–OH) *moieties* duly located on the **two adjacent C–atoms** as indicated in **'compound X'** below:

$$
\begin{array}{ll}
| & \\
\text{CHOH} & \text{Hydroxyl group} \\
| & \\
\text{CHOH}_3 & \text{Methoxy group} \\
| & \\
\text{CHOH} & \text{Hydroxyl group} \\
|
\end{array}
$$

Compound X

(2) Interestingly, between **each two 1, 2-glycolic moieties** duly present in the *reacting compound* at least **'one molecule of the said oxidizing reagent'** gets consumed precisely as depicted in the oxidation of **'compound Y'** given under:

$$
\begin{array}{c}
|\\
\text{I} \left(\begin{array}{c} \text{CHOH} \\ \text{---}\!\!+\!\!\text{---} \end{array}\right. \\
\text{II} \left(\begin{array}{c} \text{CHOH} \\ \text{---}\!\!+\!\!\text{---} \end{array}\right. \\
\text{III} \left(\begin{array}{c} \text{CHOH} \\ \text{---}\!\!+\!\!\text{---} \\ \text{CH}_2\text{OH} \end{array}\right.
\end{array}
\xrightarrow[\substack{\textbf{Periodic Acid} \\ \text{[One mole each for} \\ \text{one pair of glycolic moiety]}}]{3\text{H}_5\text{IO}_6}
\begin{array}{ll}
| & \\
\text{CHO} & \text{An aldehyde compound} \\
+ & \\
2\text{HCOOH} & \\
+ & \\
\text{HCHO} & \text{An aldehyde compound}
\end{array}
$$

Compound Y

Notes: (1) **The total number of 'dotted lines' depicts clearly the exact number of the reagent molecules taken up during oxidation.**

(2) **The 'dotted lines' explicitly indicate the point of attack.**

(3) **Due to the ease of convenience in handling the H_5IO_6 is always preferred to [Pb(CH$_3$CO)$_4$].**

(4) **The oxidation by periodic acid [H_5IO_6] is proved to be the most authentic procedure for determining the size of the ring in *'aldoses'* exclusively.**

4.1.4. Tollen's Reagent [or Benedict's Reagent]

Based on the fact that both **Tollen's Reagent** (*i.e., ammoniacal silver nitrate solution*) and **Benedict's Reagent** (*i.e., alkaline solution of copper sulphate complexed with citrate ion*) being exclusively mild oxidizing agents, they are capable of **oxidizing** the **'aldoses'** to the corresponding **aldonic acid (Gluconic acid),** and the *respective reagent* (s) get duly reduced.

Reducing Sugars: Importantly, all such **sugars** that get ultimately oxidized by these specific reagents are invariably termed as **'reducing sugars'**, such as: **glucose, fructose.**

Examples: (*a*) **Reaction with Tollen's Reagent:**

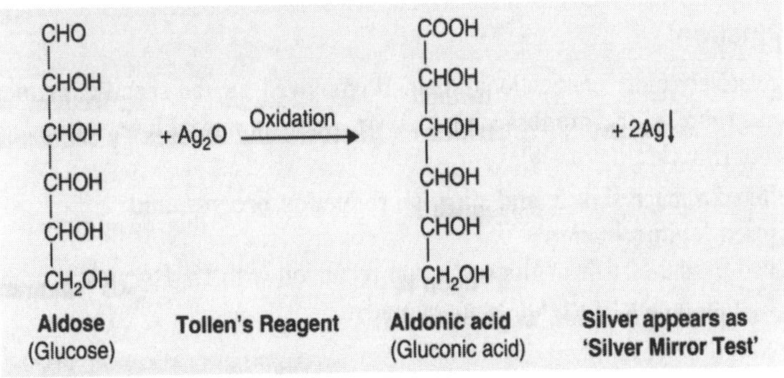

| Aldose | Tollen's Reagent | Aldonic acid | Silver appears as |
| (Glucose) | | (Gluconic acid) | 'Silver Mirror Test' |

(*b*) **Reaction with Benedict's Reagent:**

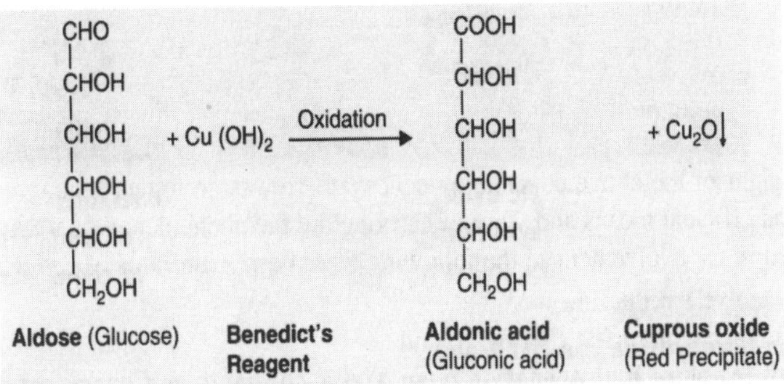

| Aldose (Glucose) | Benedict's Reagent | Aldonic acid (Gluconic acid) | Cuprous oxide (Red Precipitate) |

Note: In '*Qualitative Analysis*' of organic compounds both *Totten's reagent* and *Benedict's's Reagent* are profusely used for the detection of reducing sugars viz., Glucose, and Fructose.

Reaction with 'Ketoses' (or a-Hydroxy Ketones) : It has been duly observed that the **a-hydroxy ketones** do also get *oxidised* effectively and promptly by these *two aforesaid reagents* to the corresponding **acids,** obviously the '**ketoses**' critically bearing the **hydroxy ketonic moiety** also undergoes **oxidation.**

Explanations: The above reaction may be explained by assuming that there is a cleavage between C–2 (*ketonic moiety*) and C-3 (*secondary alcoholic function*) by the **Benedict's reagent** thereby giving rise to **glycolic acid** from the upper cleaved segment, and **tartaric acid** from the lower cleaved segment. Thus, the reagent containing **cupric hydroxide** gets duly reduced to **cuprous oxide** as the **red precipitate** settling at the bottom of test tube.

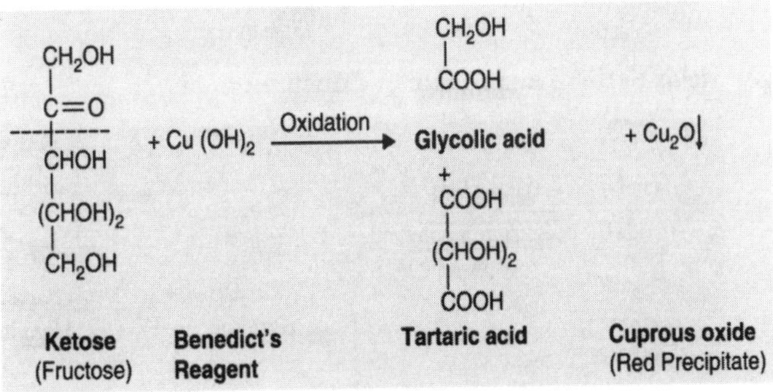

| Ketose | Benedict's | Tartaric acid | Cuprous oxide |
| (Fructose) | Reagent | | (Red Precipitate) |

Note: The above statement of fact adequately explains why *'fructose'* not having an aldehydic functional moiety, gets duly *reduced* by Tollen's reagent, Benedict's Reagent, and Fehling's solution.

4.2. Reduction

The relatively older conventional method employed for the specific **'reduction'** of the **'ketonic functional group'** in the **monosaccharides** *viz.*, **ketoses** and **aldoses** by *sodium-amalgam* (Na-Hg) posed *two* drawbacks:

- offered a much **slower and sluggish reduction process, and**
- caused **'epimerization'**

as evidenced by the fact that **'glucose'** upon reduction with Na-Hg yields **'mannitol'** instead of the expected normal product *sorbitol,* as given under:

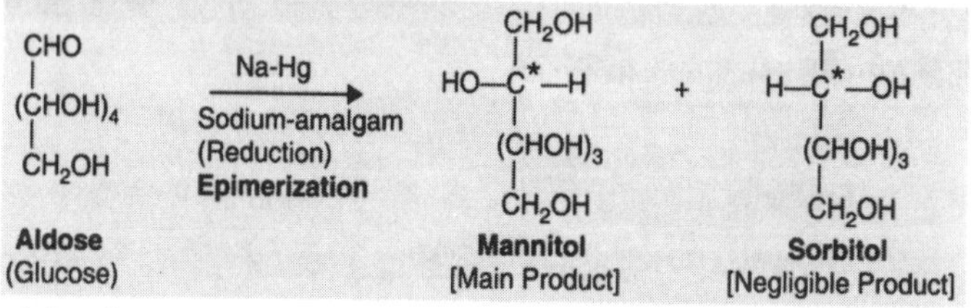

In the light of the above concrete evidences, there was an urgent requirement to look for better, specific, and efficient means and ways of carrying out the phenomenon of **'reduction'**, which eventually resulted in the evolvement of the following *three* **newer methods of reduction,** namely :

- **electrolytic reduction,**
- **sodium borohydride (Na BH$_4$), and**
- **high pressure hydrogenation over** *Nickel chromate* or *Copper chromate* **as most viable catalysts.**

Amazingly, the aforesaid **newer methods of 'reductions'** predominantly and definitely produce always the desired products, such as:

Sorbitol (or Alditol)- an alcohol : From **'aldose'** (*viz;*, **glucose**);

Sorbitol and Mannitol-mixture of alcohols : from **'ketose'** (*viz.,* **fructose**)

(*a*) **Aldose yields Sorbitol:**

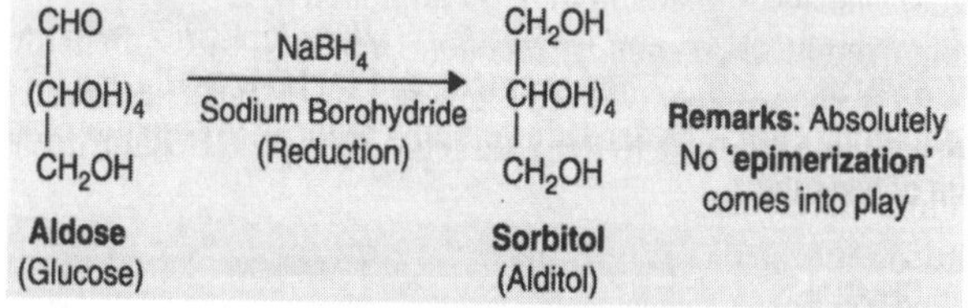

(*b*) **Ketose yields Sorbitol and Mannitol (Mixture):**

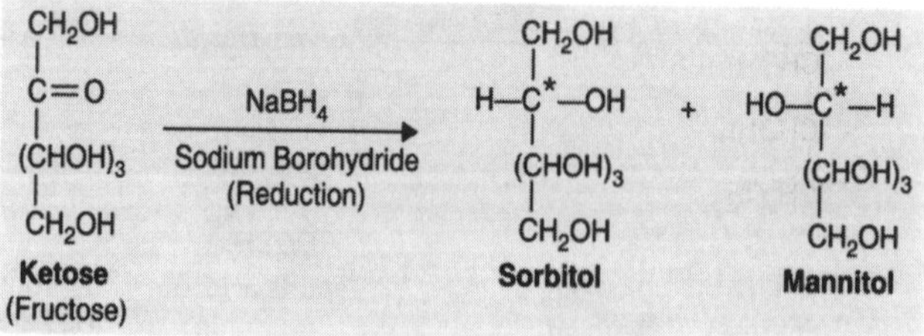

Note: The C–atom marked (*) is asymmetric in nature *i.e.,* all the *'four attachments or substituents'* to this C–atom are different. Such compounds are also termed as 'chiral compounds', and the C-atoms as *'chiral centres'.*

Explanation: In the first instance (*i.e.,* **aldose**) there exists practically *no chiral centre;* and, therefore, it always produce only one alcohol (*i.e.,* **sorbitol**) upon reduction. However, in the second instance (*i.e., Ketoses*) there exists **'one chiral centre';** and, hence, it gives rise to two altogether different configurations (with regard to the positions of H and OH at C–2 atom) thereby producing *two different alcohols,* namely : **sorbitol,** and **mannitol.**

4.3. Acetylation

The monosaccharides *viz.,* **aldoses** and **ketoses,** on being subjected to treatment with **acctic anhydride [(CH$_3$CO)$_2$O]** to yield the respective **polysaccharide acetates** as given under.

CHO		CHO			CH$_2$OH		CH$_2$OAC	
(CHOH)$_4$	5AC$_2$O	(CHOAc)$_4$			C=O	5AC$_2$O	C=O	
CH$_2$OH	Acetylation	CH$_2$OAC	+5CH$_3$COOH		(CHOH)$_3$	Acetylation	(CHOAc)$_3$ + 5CH$_3$COOH	
Aldose (Glucose)		**GLUCOSE** Pentaacetate	**Acetic Acid**		CH$_2$OH		CH$_2$OH	
					Ketose (Fructose)		**Fructose** Pentaacetate	**Acetic Acid**

(a) Accetylation of Aldose (Glucose):   **(b) Accetylation of Ketose (Fructose):**

Explanations: In **'acetylation'** the H-atom attached to the *alcoholic groups* both in *glucose* and *fructose* is duly replaced by the **acetyl moiety** $\overset{O}{\underset{}{\text{(CH}_3-\overset{\|}{\text{C}}-)}}$ to form the corresponding **esterification** *viz.*, $-\text{O}-\overset{\overset{O}{\|}}{\text{C}}-\text{CH}_3$ or **acetylation.** As both **aldose** and **ketose** do have in them **five distinct -OH groups,** producing **glucose pentaectate** and **fructose pentaectate** respectively together with *five moles* each of **acetic acid.**

Note: Importantly, *'acetylation'* determines critically the exact *'number of hydroxyl groups'* present in an organic compound *viz.,* formation of *'glucose pentaacetate'* evidently shows/confirms that glucose essentially possesses *'five hydroxyl moieties'.*

4.4. Formation of Glucosides

The **'monosaccharide'** on being specifically treated with **methanolic hydrogen chloride** gives rise to the formation of *two isomeric compounds,* usually termed as **methyl glucosides.** The **aldehydic moiety (HC=O)** present in *glucose* gets duly converted into an **acetal moiety** $\text{H}-\overset{\overset{O}{\|}}{\text{C}}-\text{OCH}_3$ Therefore, the *acetals* (and hence the *glucosides*) are found to be *fairly stable* in an **alkaline medium,** and *labile* in an **acidic medium.**

Example: Formation of Glucosides from Glucose:

Explanations: The carbonyl group $(-\overset{\overset{}{|}}{\underset{\underset{O}{\|}}{\text{C}}}-)$ present in the aldehydic group in **glucose** get converted into a **methoxy moiety (–OCH₃)**; whereas, the **lactone ring** gets duly formed between C–1 and C–5 with the elimination of a mole of water. The **C-atom marked (*) is a chiral atom,** hence one may get *two* different configurations α and β-**methyl glucoside** as shown below:

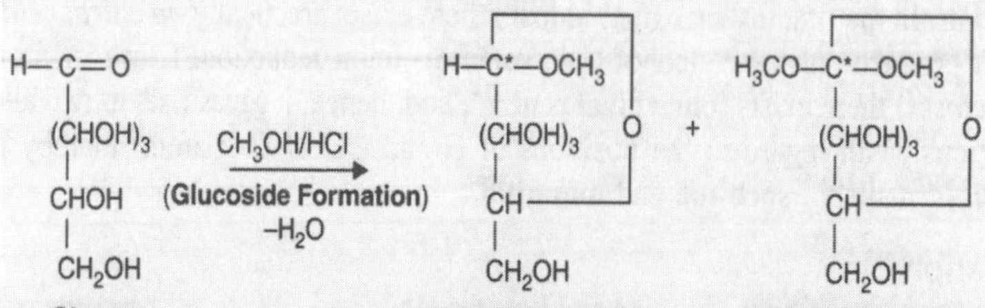

Glucose	**α-Methylglucoside**	**β-Methylglucoside**

Dimethyl Sulphate [(CH₃)₂SO₄]/Methyl Iodide [CH₃–I] The appropriate methylation of the '**remaining four hydroxyl moieties**' *i.e., from C–2 to C–4 and C–6 critically present in the 'methyl glucoside* (α *and* β) may be conveniently and easily '**methylated**' using either **dimethyl sulphate** or **methyl iodide** as given below to yield **methyl tetra–methyl glucoside (I).** Consequently, when **(I)** is carefully treated with a mineral acid **(H⁺)** the former undergoes crucial hydrolysis exclusively at the '**hemiacetal linkage' at C–1 atom,** and the remaining *four* linkages remain very much '**intact**' to obtain **tetramethyl glucose (II).**

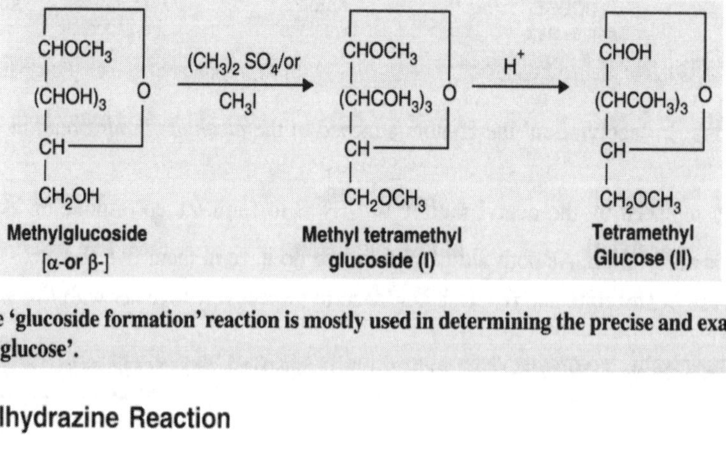

Methylglucoside [α-or β-]	**Methyl tetramethyl glucoside (I)**	**Tetramethyl Glucose (II)**

Note: The 'glucoside formation' reaction is mostly used in determining the precise and exact structure of 'glucose'.

4.5. Phenylhydrazine Reaction

It is a known fact that '**osazone formation**' designates typical reaction given by almost all **α-hydroxy aldehydes and α–hydroxyketones.** Nevertheless, the normal carbonyl moiety ($-\overset{O}{\overset{\|}{C}}-$) invariably present in **aldoses** (*glucose*) and **ketoses** (*fructose*) particularly react with **phenylhydrazine** [C₆H₅.NH.NH₂] to produce '**phenylhydrazone**' at *only one molecular ratio.* Importantly, unlike the **normal carbonyl moiety,** the *sugars* usually undergo farther reaction with an *excess of phenylhydrazine* to form the respective '**osazones**' at *three molecular ratio* through the **phenylhydrazones.**

Ever since various theories have been put forward in order to explain the possible '**mechanism of osazone formation**', but none of them proved to be acceptable reasonably. Keeping in mind the different reaction parameters *two* well recognized and broadly accepted **mechanisms** shall now be discussed briefly in the sections that follows:

4.5.1. Fischer's Mechanism

Fisher mechanism judiciously suggest that the *osazone formation either* from **aldoses** or **ketoses** actually takes place in *three* stages in a sequential profile:

Stage-I: Conversion of Aldose/Ketose to Corresponding Phenylhydrazone:

It takes up the very *first molecule of phenylhydrazine* with the elimination of one mole of water.

Stage-II: Conversion of Glucose/Fructose Phenylhydrazone to Corresponding Glucosone/Fructosone: In this case, the *second molecule of phenylhydrazine* specifically oxidizes:

- the C-2 hydroxyl (-OH) moiety in *'glucose'* to a carbonyl group, or
- the C-1 hydroxyl (-OH) moiety in *'fructose'* to a carbonyl group.

Besides, phenylhydrazine gets reduced to one mole each of **aniline** and **ammonia**.

Stage-III: Conversion of Newly Formed Carbonyl Moiety (Stage-II) Glucosone/ Fructosone to Corresponding Glucosazone/Fructosazone : In the final stage, the third molecule of phenylhydrazine reacts duly to produce **Glucosazone/Fructosazone.**

These reactions are provided as given under:

(*a*) **Phenylhydrazine Reaction with Glucose:**

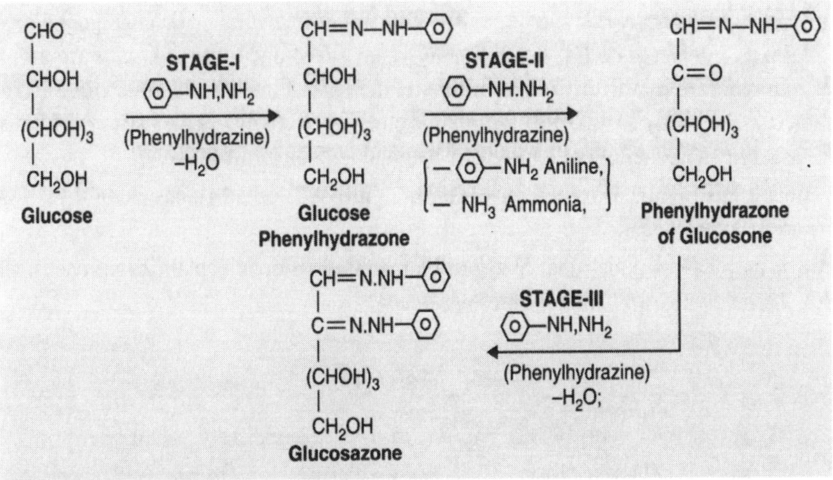

(*b*) **Phenylhydrazine Reaction with Fructose:**

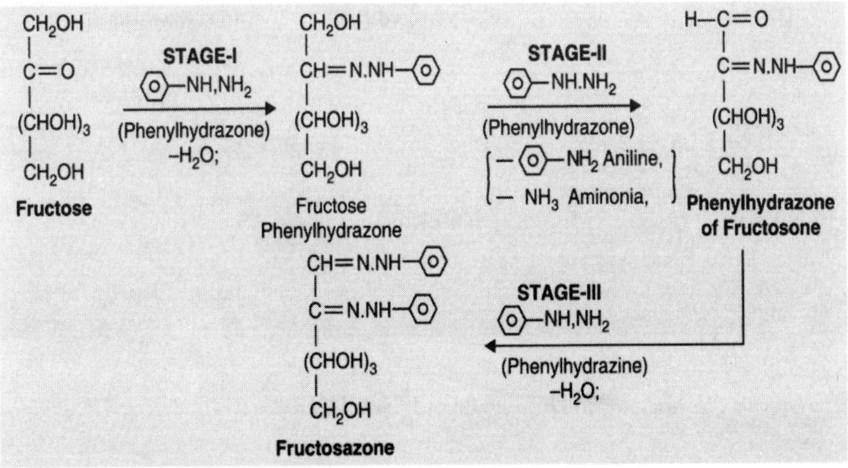

Note: (1) Glucose and Fructose though are definitely different 'sugars', but amazingly both produce the *similar osazones*.

(2) It also ascertains and proves that both *'glucose'* and *'fructose'* differ only in the *two* C-atoms (*viz;*, C–1 and C–2) that are intimately engaged in the 'osazone formation'

Inexplicability in Fischer's Mechanism: *Fischer's mechanism* obviously fails to explain the following *two* aspects miserably:

(*i*) **Phenylhydrazine** being a strong **reducing agent does behave in these reactions as an oxidizing agent.**

(*ii*) It fails to explain logically as to why the reaction comes to a hault after the **stage-III** *i.e.,* soon after the two C-atoms (*viz.,* **C-1** and **C-2**) get duly introduced with **'two phenylhydrazine residues'.**

4.5.2 Weygand's Mechanisms

Weygand *et al.* (1940)* first and foremost suggested the critical **'osazone formation'** that specifically makes use of the **Amadori Rearrangement** with the *phenylhydrazone* (I) obtained initially. Later on after 25 years, Shemyakin *et al.* (1965)* reaffirmed that *phenylhydrazine* (I) undergoes a reversible reaction to form an **intermediate (II),** that eventually gets rid of one mole of *aniline via* an intricate **intramolecular rearrangement** to yield an **iminoketone (III).** Finally the *iminoketone* (III) reacts with two molecules of **phenylhydrazine** one after the other sequentially to give rise to the formation of the desired **'osazone'** with the loss of a mole of water and ammonia respectively.

The various steps involved in the conversion of **'glucose'** to the **'glucosazone'** by *Weygand's mechanism* are as stated under:

In short, it may be concluded that **Weygand's mechanism** does explain comprehensively and satisfactorily the fallowing *two* cardinal aspects, namely :

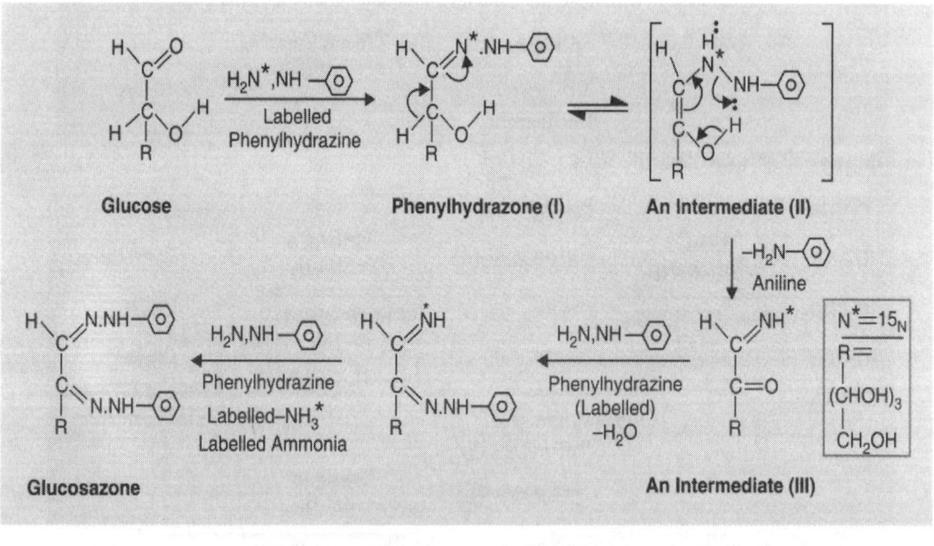

Glucose   Phenylhydrazone (I)   An Intermediate (II)

Glucosazone   An Intermediate (III)

---

* Finar IL : **Organic Chemistry,** Vol.1., Longmans Green & Co., Ltd., London (UK), 5th ed., p.477, 1967.

- shortcomings and inexplicabilities of **Fischer's mechanism,** and.
- substantiates adequately the critical observation that when **'phenylhydrazone'** duly prepared by the reaction of **glucose** with ^{15}N **labelled phenylhydrazine** is treated alternatively with unlabelled phenylhydrazine, it would yield specifically **'unlabelled glucosazone'** accompanied particularly by the elimination of **'labelled ammonia'.**

Likewise, the various steps involved in the conversion of *'fructose'* to the *'fructosazone'* by **Weygand's mechanism** are as given below:

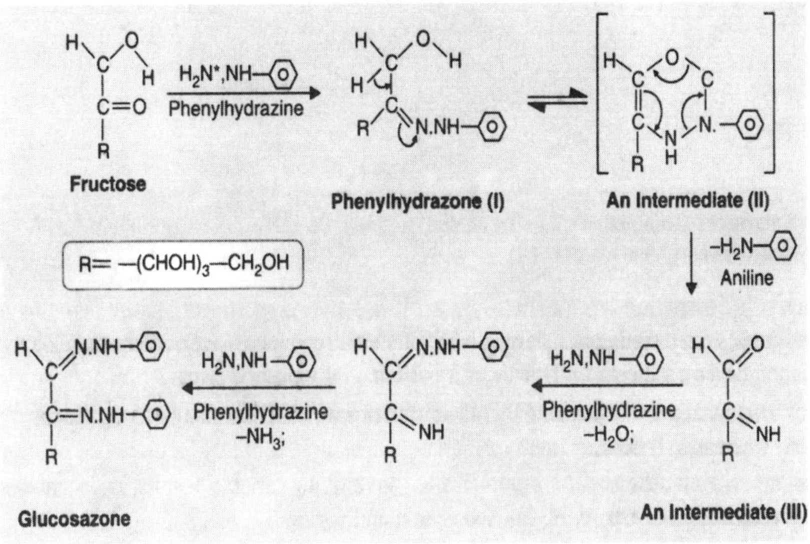

Fructose   **Phenylhydrazone (I)**   **An Intermediate (II)**

$$R = -(CHOH)_3 - CH_2OH$$

Glucosazone   **An Intermediate (III)**

Interestingly, the **Weygand's mechanism,** very much akin to the **Fischer's mechanism,** does not fully justifies the particular reason for the critical involvement of the **first two C-atoms** *viz.,* C-1 and C-2) present duly in *glucose* and *fructose* for the **'osazone formation'.**

Stability of the 'Osazones': It has been amply documented that no sooner the **'osazones'** are duly obtained, they are properly and securedly stabilized by **'internal hydrogen bonding'** or **'chelation'.** Therefore, probably no other **carbon atoms,** except **C–1** and **C–2,** do take part in the osazone formation as shown under:

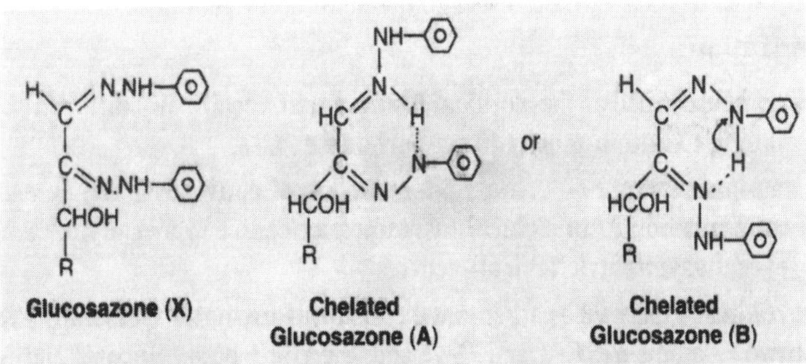

Glucosazone (X)   **Chelated Glucosazone (A)**   or   **Chelated Glucosazone (B)**

The above *two* chelated structures of **'glucosazone', A** and **B,** have been adequately confirmed by sophisticated physico–chemical methods, such as : **X–ray diffraction analysis** (Bjamer *et al.,* 1963), and **NMR–spectral analysis** (Mester *et al.,* 1965).

Later on, Blair *et al.* (1969) rejected the *two chelated structures* 'A' and 'B' solely based on their elaborative **IR and UV–spectral studies,** and suggested confidently the following '**phenylazo structure**' strategically located **at C-2 atom** for the designated **glucosazone compound** obtained duly in the '*solid form*'.

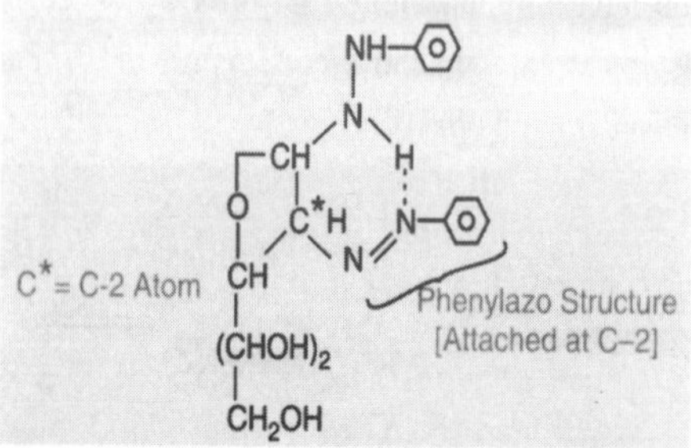

$C^* =$ C-2 Atom

Phenylazo Structure [Attached at C–2]

Note: Amazingly, the above structure in the '*solution state*' attains an *equilibrium* with open-chain structure of 'glucosazone' (X)

Ostriazole Formation from Fructosazone: It is, however, pertinent to state here that the **characteristic osazones** from the **sugars,** having a bright crystalline yellow solid forms go a long way in their identification based on their **mp.** However, it has been duly observed that:

- exact time taken for '**osazone formation**' varies with different sugars *viz.*, **glucosazone** takes *2 mts.*, whereas, **fructosazone** *5 mts,* and
- more critically their *melting points* often give quite variable results; and hence, renders the characterization of '**sugars**' not so reliable and correct.

Alternatively, the formation of '**ostriazoles**' by the interaction of *osazones* upon heating with aqueous **cupric sulphate solution,** thereby producing beautiful crystalline compounds having both **sharp and dependable characteristic melting points,** and with the elimination of one mole of **aniline.**

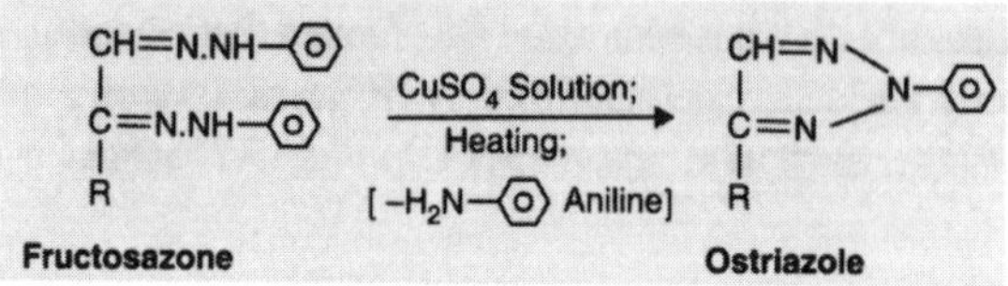

Fructosazone   **Ostriazole**

4.6. Epimerization

Epimers may be defined as the- '**optical isomers that specifically differ in the configuration at only one asymmetric C-atom mostly the α-carbon** *i.e.*, **C-2.**

Epimerization refers to the- '**critical phenomenon of converting one epimer into another** *i.e. inversion of configuration at one chiral (asymmetric) centre (generally α-carbon) in a molecule comprising several asymmetric (chiral) centres**'.

As we have already examined earlier that the '**osazone formation**' essentially causes to participate only the first two C–atoms *i.e.* **C–1** and **C–2**; and, therefore, both '**glucose**' and '**fructose**' predominantly give rise to the formation of exactly the '**same osazone**' (see section 4.5.2), which categorically suggests that the *two sugars* must hold the **same configurations** at C-3, C-4, and C-5. Besides, it has been observed critcally that another *'aldohexose'* per se **mannose** also yields precisely the '**same** osazone' which indeed offers an enormous solid evidence that its configuration at **C–3 through C–5** happen to be absolutely identical to those of **'glucose'** and **'fructose'** Hence, it may be overwhelmigly concluded that- **'such pair of diasteromeric aldoses'** *viz., glucose and* **mannose which only differs in the configuration at C-2 are known as epimers.**

At the fag end of the nineteenth ceutury **Fischer** keenly observed that **'epimerzation'** gets accomplished effectively *via* **aldonic acid** (or **gluconic acid**). Importantly, first of all, **aldonic acid** is carefully heated to about **150°C** in an aqueous medium containing **pyridine or quinoline,** in order to check and prevent the formation of *'lactone'* of the starting material (*i.e.,* **aldonic acid**) to give rise to an **'equilibrium mixture'** comprising the **original acid** (*aldonic acid*) and its corresponding **'epimer'**. In other words, the precise underlying **mechanism of epimerization** resembles to a great extent to that of **'racemization'**, such as : the classical example of **epimerization of 'Glucose'** to **'Mannose'** as expatiated under:

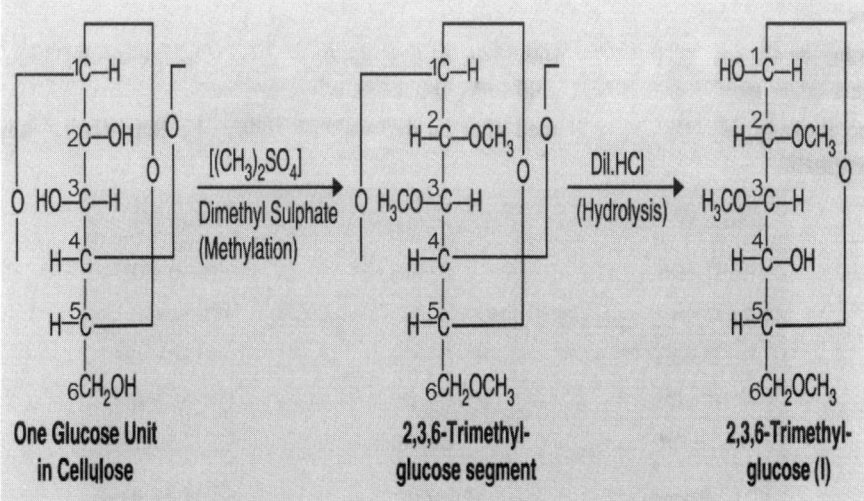

One Glucose Unit in Cellulose   **2,3,6-Trimethyl-glucose segment**   **2,3,6-Trimethyl-glucose (I)**

Explanations: The various sequential steps may be explained as under:

(1) **D (+)–Glucose (I)** is oxidized with Br_2-water to produce the **aldonic acid** (or **gluconic acid**) due to the conversion of–CHO group to -COOH group.

(2) The resulting **aldonic acid** is treated with aqueous quinoline or pyridine *i.e.,* slightly *basic environment* so as *deprotonate* the carboxyl moiety to give rise to the corresponding **carboxylation.**

(3) The end product obtained in (2) above is subjected to an obstruction of proton (H^+) at C-2 atom to generate the **carbanion** (*i.e.,* C-2 atom bearing a negative charge on it).

(4) The resulting **carbanion** due to the **intramolecular electronic effect** shifts the **double bond** located in the **carbonyl function** at C-1 to the **new position between C-1 and C-2.**

(5) The product obtained in (4) having C–1 loaded with *two negatively charged O-atoms* on being duly **protonated** yields the corresponding **mannoic acid.**

(6) The ultimate step involves first careful **evaporation** followed by usual reduction of **carboxyl moiety** at C-2 to the corresponding **aldehyde moiety** to result into the formation of **D(+)– mannose (II).**

> **Note:** It may be explicitly noticed that in **D (+) glucose (I) and D (+)-mannose (II)** the critical difference being the orientation of H and OH at C–2 only.

Obviously, one may easily lay hand on a quite convenient practical means to identify *'a pair of epimers'* by the careful preparation of **'osazones'** only. Since, in the critical formation of **osazones'** the prevailing asymmetry at C-2 atom of the **glucose (aldose)** is lost completely; and, therefore, the **'epimers'** *viz.*, **glucose** and **mannose** invariably produce **identical osazones.**

4.7. Reaction with Alkali [or Lobry de Bruyn-van Ekenstein Rearrangment]

It may be observed commonly that **sugars** on being warmed with concentrated alkali (NaOH) solutions first changes into yellow-brown-and turn into a resinous product finally; however, the probable mechanism still remains a mystery. Importantly, the *sugars* on being treated with *amines* or *dilute alkalis* undergo spectacular rearrangment to produce rather a mixture comprising **more than one sugar entities.**

Example: Glucose when reacted with *dilute alkali (e.g., 0.2 M NaOH)* gets duly converted into a mixture of *three* **monosaccharides** *viz.*, **glucose, mannose, and fructose.**

The above typical reaction is named after its discoverer as **Lobry de Bruyn-van Ekenstein rearrangment.**

D (+)-Glucose   'Enediol'   D (+)-Mannose

D (–)-Fructose

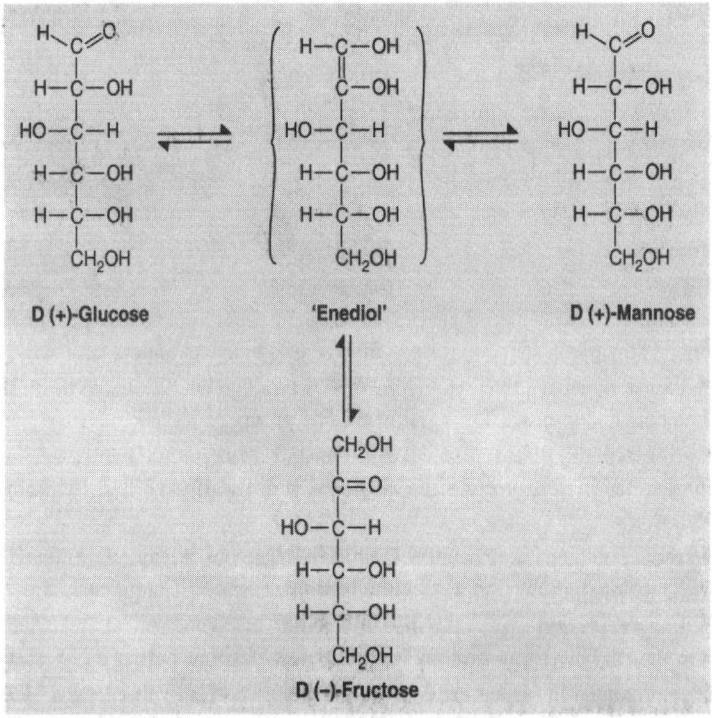

Later on, Wohl (1900) duly postulated that a common **'enediol'** *intermediate* is obtained in the course of the **actual transformation** *via* the **1, 2–enolisation** of three **monosaccharides,** namely:

D (+)-Glucose, D (+)-Mannose, and D (–)-Fructose, as given under:

Topper *et al* (1951) reconfirmed the **1,2–enolisation** process during the **transformation(s)** amongst the *monosaccharides* by employing **deuterium oxide.** They, in fact, further substantiated that **fructose** as well as **mannose** may not be actually produced from a *'single enediol residue';* nevertheless, they are duly formed *via* two **geometrical isomeric enediol intermediates** *e.g., trans*–diol, and *cis*–diol as depicted vividly under:

| D (+)-Glucose | *trans*–Diol | D (-)-Fructose | *cis*–Diol | D (+)-Mannose |

tran-**Diol** : Here, the hydroxyl (–OH) groups at C-1 and C-2 are located diagonally opposite to each other.

eis-**Diol** : In this instance, the hydroxyl (–OH) moieties at C–1 and C–2 are positioned on the *same side* of the **double bond.**

5. INTERCONVERSIONS IN MONOSACCHARIDES

The highly specific reactions pertaining to the **interconversions in monosaccharides** are of paramount interest *vis-a-vis* significant applications not only in the critical synthesis but also in the precise and exact determination of the configuration of both *aldoses* and *ketoses.*

The **interconversions** are of *four* different types, namely :

(*i*) Conversion of 'Aldose' into 'Ketose',

(*ii*) Conversion of 'Aldose' into next higher 'Ketose',

(*iii*) Conversion of 'Aldose' into 'Ketose' with two additional carbon chain, and

(*iv*) Conversion of 'Ketose' into 'Aldose'.

5.1 Conversion of 'Aldose' into 'Ketose'

The conversion of **'aldose'** into **'ketose'** is most easily accomplished by first preparing its **osazone,** which on being treated with *benzaldehyde* or *concentrated HCl* yields **ozone** due to the removal of *'phenylhydrazine residues'.* Based on the theoretical concept that an **aldehydic moiety** gets reduced faster and promptly *vis-a-vis* a **ketonic functional moiety,** the resulting **'osone'** is effectively reduced using *Zn/CH$_3$ COOH mixture* to the desired **'ketose',** as shown under:

| Aldose (Glucose) | An Osazone | An Osone | Ketose (Fructose) |

Nevertheless, an *alternative method* is also available that makes use of the simple process of warming the *aldose* in an agueous medium to produce an admixture of the *'epimerization.'* Thus, from the admixture of *aldose* and *ketose,* the former is being removed specifically by the following *two means*:

• careful crystallization, and

• critical oxidation to **aldonic acid (gluconic acid)** duly followed by precipitation as the Ba-salt, and the *ketose* remains in the residual product.

5.2 Conversion of 'Aldose' into Next Higher 'Ketose'

Wolform *et al.* (1956)* used the **Arndt-Eistert reaction** involving critically *six* sequential steps adopted for the conversion of **D-arabinose** (an *aldopentose*) into **D-fructose** (a *ketohexose*) as shown under:

Explanations: These essentially include:

(1) The *five-membered aldopentose i.e.,* **D-arabinose** is oxidized with Br_2-water to produce **arbinoic acid** in **Step-I** to convert C-1 aldehyde to C-1 carboxyl moiety.

(2) Acetylation in **Step-II** protects the *three* alcoholic moieties to produce tetracetyl arabinose acid plus four moles AcOH eliminated.

(3) Treatment with thionyl chloride converts the hydroxyl group of COOH at C-1 into a **chloride group** to yield corresponding **acid chloride (X)** in **Step-III.**

(4) Treatment of the resulting **acid chloride (X)** with diazomethane in **Step-IV,** replaces the Cl group with **diazomethane function (Y)** quantitatively.

> Note: The 'acetylation' in Step-II is of paramount importance in order to critically get-rid of the possibility of 'methylation' of the three available alcoholic moieties by diazomethane.

(5) Hydroxylation of the **diazomethane function** at C-1 in (y) by **aqucous acetic acid (Step-V)** produces *tetracetyl fructose*(**Z**).

(6) Careful treatment with dilute NaOH solution hydrolyses the *four acetyl moieties* into corresponding *hydroxyl groups* to yield **'fructose'** in **Step-VI** *i.e.,* to get a **6-membered Ketose.**

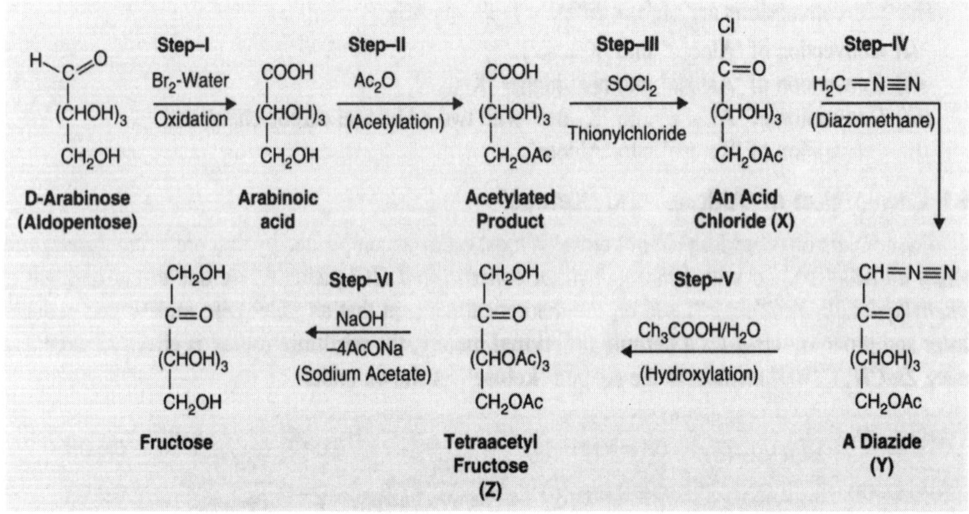

It is, therefore, quite possible to afford the conversion of a **membered 'aldose'** to the next higher ketose *i.e.,* **6-membered 'ketose'** by the application of **Arndt–Eistert reaction.**

5.3. Conversion of 'Aldose' into 'Ketose' with Two Additional Carbon Chain

Sowden (1950) adopted the *'nitromethane synthesis'* for the conversion of **aldose** into the corresponding **ketose** having two additional carbon atom chain as given under:

---

* Wolform *et al: J Am Chem Soc.,* 78: 2489,1956

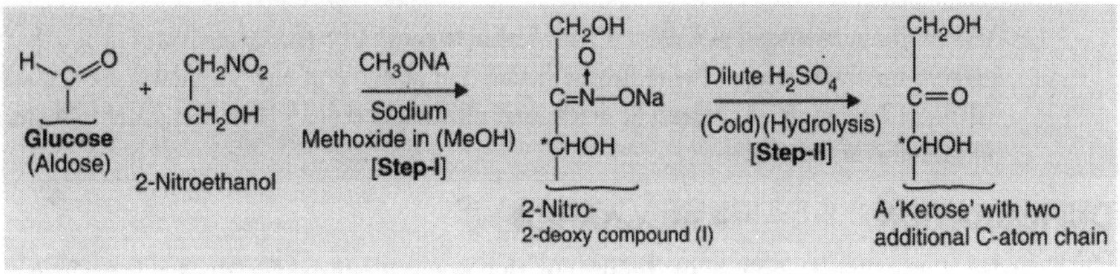

Explanations : Various steps involved may be expatiated as under:

Step-I: Glucose, a 6-membered aldose when treated duly with 2–nitroethanol in the presence of methanolic sodium methoxide (freshly prepared), **Step-I** comes into play with the production of the corresponding **2–nitro–2deoxy compound (I)**. It may be observed that 2 additional C-atoms chain has already been introduced to the original **glucose molecule.**

Step-II: The resulting **higher C-atom chain compound (I)** is subjected to hydrolysis with dilute H_2SO_4, whereupon a **'Ketose' with two additional, C-atom chain** is accomplished.

5.4. Conversion of 'Ketose' into 'Aldose'

The actual conversion of **'ketose' (fructose)** into an **'adlose' (glucose)** may be accomplished by carrying out the following *four* steps in a sequential order:

Step-1: Catalytic reduction using H_2-Ni (Raney **Nickle**) helps to reduce the **'ketose'** to the corresponding **alcohol (I).**

Step-2: The resulting alcohol when gets oxidized carefully with **Br_2-water the terminal CH_2OH moiety is duly oxidized to a corresponding monocarboxylic acid (II)** *i.e.,* **gluconic acid.**

Step-3: Subjecting compound **(II)** to gentle heating causes **lactonization** to form a oxygen-bridge between C–1 and C–4 to **γ-lactone (or glucose-γ-lactone) (III)**

Step-4: The **lactonized compound (III)** when subjected to **reduction** with *sodium-amalgam (Na-Hg)* and exposed to an acidic environment gives rise to the formation of the desired **'aldose' (glucose) (IV).**

The various reactions involved are as given under:

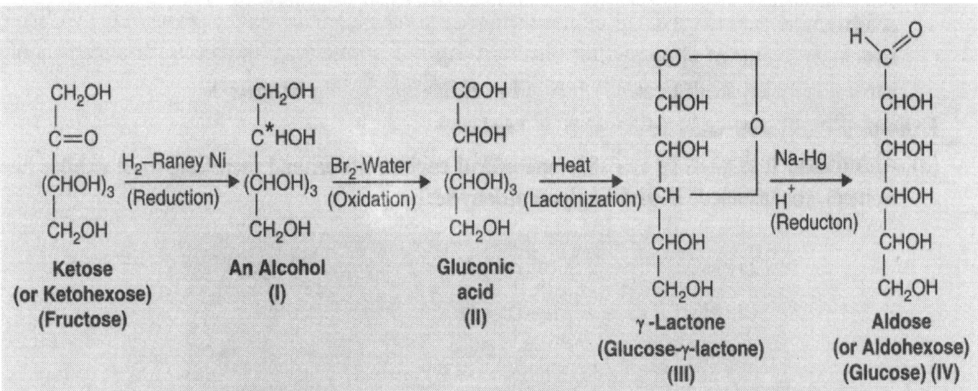

Notes: (1) Importantly, the very first step (1) generates one additional *chiral carbon atom* marked (*), which would give rise to *two isomeric alcohols* (upon reduction).

(2) Besides, the subsequent *oxidation* of the alcohol to result into the formation of *two carboxylic moieties* on either side of the molecule (instead of at one end only as shown above), which will form 2 alcohols –4 carboxylic acids, and ultimately there will be a mixture of 4 aldoses (in unequal amounts).

6. CONFIGURATION OF MONOSACCHARIDES

In a broader perspective, one may designate the *monosaccharides* as the largely existing **polyhydroxyaldehydes** or **polyhydroxyketones,** which may easily and conveniently be hydrolyzed into further smaller chemical entities. Based on the several critical evidences from the survey of litera- ture, it has been adequately observed that the **'monosaccharides'** could be classified strictly according to the following *two* criteria, namely:

- **precise and exact nature of the carbonyl** $(-\overset{\overset{O}{\|}}{C}-)$ **functional moiety, and**

- **exact number of carbon atoms.**

| Aldodiose | Aldotriose | Aldotetrose | Aldopentose | Aldohexose | Ketohexose | Ketoheptose |

Important Points: These **important points** must be noted carefully:

(1) **Aldodiose** is classified under **'sugars'** only on *theoretical basis,* since it does not contain any *asymmetric* (or **'chiral'**) carbon atom; and, therefore, fails to display any **'optical activity'** at all, whereas **'sugars'** in general show **optical activity.**

(2) A critical glance at the above **monosaccharides** (both *aldoses* and *ketoses*) viz., **'aldotriose'** through **'ketoheptose'**, one may *notice* the presence of **one or more than one 'chiral carbon atoms'** (as vividly indicated by **asterik * sign**).

(3) The critical presence of **number of chiral centres** in the **'aldoses'** and **'ketoses'** actually determine the *exact type of* **monosaccharide** possessing either two or *more than two* 'stereoi- somers' by virtue of **altogether different configuration(s)** *i.e.,* the precise arrangement of the various moieties strategically positioned at **chiral C-atom or atoms:**

Examples: Following are a few *typical examples:*

(a) **Aldotriose:** It essentially has only **one chiral carbon atom;** and therefore, must exhibit *two* isomers, for instance: **D-and L-glyceraldehyde:**

| Aldotriose | D-Glyceraldehyde | L-Glyceraldehyde |

Explanations: Rosanoff (1906) specifically selected the *'simplest sugar' glyceraldehyde* (derived from *aldotriose*) as a **randomly selected standard** so as to carry out a comprehensive investigative studies pertaining to the **configuration of carbohydrates.**

According to Rosanoff:

- When the **hydroxyl moiety** in *glyceraldehyde molecule* is located to **'right'** when –CHO group is at the top or when the direction from the top to the right exists between H → OH (as shown above) – the glyceraldehyde is said to be belonging to the *D-series* or **'Dextrorotatory'** (and marked usually by a **+ sign**), and

- When the **hydroxyl moiety** in, *glyceraldehyde molecule* is critically located to **'left'** when –CHO group is at the top or when the direction from the top to the left exists between H → OH (as shown above) – the glyceraldehyde is regarded to be belonging to the *L-Series* or **'Levorotatory'** (and marked normally by a **– sign**).

(4) In fact, the actual total number of **optically active isomers,** also referred to as, **'stereoisomers'** with regard to different variants in commonly available **'aldoses'** and **'ketoses' is invariably given by the following generalized formula '2^n',**

where, n = number of **'dissimilar asymmetric C-atoms'.**

Enantiomeric Pairs: There are in all 2^{n-1} enantiomeric pairs, also known as the **'mirror images'.**

Table: 1.1 records the total number of **'optical isomers'** and **'enantiomeric pairs'** pertaining to the spectrum of **monosaccharides** *vis-a-vis* their number of **chiral C-atoms** present.

Table: 1.1 Monosaccharide Variants *vis-a-vis* Number of Chiral C-atoms, Optical Isomers and Enantiomeric Pairs.

		NUMBER OF		
S.No.	Monosaccharide Variants	Chiral C-Atoms	Optical Isomers	Enantiomeric Pairs
1	Aldotriose	1	2	1
2	Aldotetroses	2	4	2
3	Aldopentoses	3	8	4
4	Aldohexoses	4	16	8
5	Ketopentoses	2	4	2
6	Ketohexoses	3	8	4
7	Ketoheptoses	4	12	6

Keeping in view the above *factual data and figures* it is quite possible to adjudge these *isomeric forms of different kinds of monosaccharide variants* which may be studied in details both traditionally and conventionally within the domain of **'configuration of monosaccharides'**

At this point in time, it is pertinent to examine and study the **'configuration'** of certain vital and cardinal **monosaccharide variants,** namely:

- Aldotrioses,
- Aldotetroses,
- Aldopentoses,
- Aldohexoses, and
- Ketohexoses,

that will be discussed individually in the sections that follows:

6.1. Aldotrioses

Aldotrioses i.e., glyceraldehydes [H–$\overset{\overset{\displaystyle O}{\|}}{C}$–CHOH–CH$_2$OH], happens to be the only member of the so called **'aldotriose series'.** It has been duly observed that the **glyceraldehyde** predominantly occurs in *two typical forms* that differ critically in these *two* cardinal aspects:

- configurational series of compounds, and
- specific sign of rotation.

D-Glyceraldehyde: It particularly rotates the **'plane of polarized light'** to the **right**, and is also known as **dextro – (d) – rotatory.**

L-Glyceraldehyde: It specifically rotates, the **'plane of polarized light'** to the **'left'**, and is also called as **laevo – (l) – rotatory.**

The actual proclaimed **'configuration of other carbohydrates'** are duly originated right from the **D-and L-glyceraldehydes** supported adequately by the glaring factual evidence(s) that the – **'aldoses and ketoses with the critical and specific configuration of their penultimate C-atom very much akin to that of either D- or L-glyceraldehyde are duly assigned to the respective D- and L-series of the sugars'.**

Thus, we may have: ·

$$H—\overset{|}{\underset{|}{C}}—OH \qquad\qquad HO—\overset{|}{\underset{|}{C}}—H$$
$$CH_2OH \qquad\qquad\qquad CH_2OH$$
$$\textbf{D-Series} \qquad\qquad\qquad \textbf{L-Series}$$

Interestingly, the rather, older way of using the **D–** and **L–ssymbols,** for actual representation of the **series,** have been outrightly replaced with *d–(dextro)* and *l–(laevo)* respectively; whereas, the **'signs of rotations'** are duly designated by **(+) and (–) signs** accordingly.

> **Note: (1) Configuration is absolutely unrelated to the actual** *signs of rotation i.e.,* **a specific 'sugar' with a** *designated configuration* **may be able to rotate the** *plane of polarized light* **either to 'right' or to 'left'.**
>
> **Example: Let us consider** *two sugar entities* **that are particularly derived from** *d*-glyceraldehyde **(i.e., they belong to either D– or d– series), it never indicates/suggests that they should critically rotate the** *plane of polarized light to the right.* **However, in actual practice they may or may not rotate the** *plane of polarized light to the right,* **that may exclusively be ascertained by actually determining their 'rotation by a sensitive and callibrated** *polarimeter.'*
>
> **(2) In order to eliminate even an iota of doubt and element of confusion, the old symbols D– and L– are solely used for 'families'; whereas, the symbols** *d–* **and** *l–* **are employed exclusively to designate the 'signs of rotations'.**

6.1.1. Origination of Higher Aldoses

It has been duly observed that **'higher aldoses'** may be originated from the *two* divergent series of **D–** and **L– glyceraldehydes** by the *simple addition of on C-atoms at a time* by means of the **Kiliani Synthesis** with the critical formation of an **'epimeric pair of sugars'** belonging to the next **'higher homologue'** at each step forward.

Importantly, the determination of the **'configuration of a monosaccharide'** essentially comprises in ascending the lower homologue of known configuration by one C-atom to give rise to the formation of two higher isomeric monosaccharides, the actual nature of which is duly established by their respective oxidation to the corresponding acids'.

Examples:

(1) **Representation of Aldotrioses (or Glyceraldehydes):**

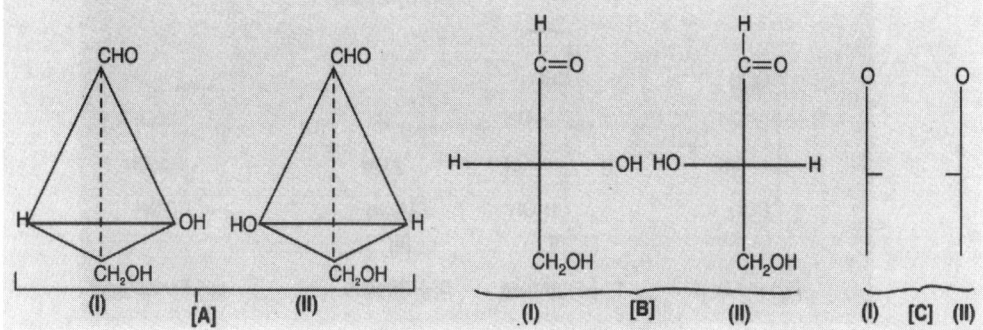

$$(I) = D (+)-Glyceraldehyde$$
$$(II) = L (-)-Glyceraldehyde$$

[A] = **Tetrahedron structure of Glyceraldehyde:**

$$1 = D (+)-Glyceraldehyde$$
$$II = L (-)-Glyceraldehyde$$

[B] = **Line diagrams of Glyceraldehydes:**

$$I = D(+)-Glyceraldehyde$$
$$II = L (-)-Glyceraldehyde$$

[C] = **Conventional Planar Diagram of Glyceraldehyde:**

Head Represents: with symbols **'O'** indicates specifically the *aldehydic moiety* (–CHO) present

$$\overset{O}{\underset{\|}{}}$$

in **'aldoses'** or *ketonic moiety* $(-\overset{O}{\underset{\|}{C}}-)$ present in **'ketoses',**

Tail Represents: critically the **–CH$_2$OH (hydroxy methyl) group,** and

Straight Line: projecting sideways indicates explicitly the **hydroxyl** (–OH) moiety (while the *H-atom* positioned opposite to *–OH group* is usually *not indicated*).

(2) **Representation of Aldotetroses:** It is, however, pertinent to state here that since **'aldotetroses'** [CHO – (CHOH)$_2$ – CH$_2$OH] essentially possess *two chiral centres* (*i.e.,* **asymmetric C-atoms**), the *four optically active states* are *actually possible* or *two pairs of enantiomorphs*). Alternatively, all of these are invariably termed as: **D– and L– Erythrose,** and **D– and L– Threose** as given under:

Explanations: From the above reaction(s) one may critically observe that D(+)–glyceraldehyde on being subjected to **Kiliani Synthesis** give rise to one mole each of **D(+)–Erythrose (I)** and **D (–) – Threose (II)** respectively; however, at this particular stage it may not be ascertained definitely which one is **'erythrose'** and **'threose'.**

Nevertheless, it has been established beyond any reasonable doubt that one will critically obtain the *mesotartaric acid* from **D(–) erythrose (I)** exclusively thereby affirming the following *two* cardinal facts:

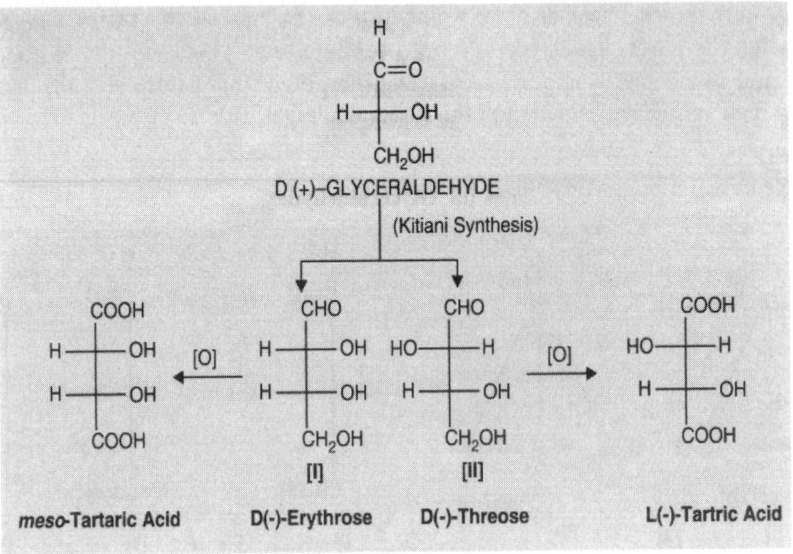

D (+)–GLYCERALDEHYDE

(Kitiani Synthesis)

meso-Tartaric Acid   D(-)-Erythrose   D(-)-Threose   L(-)-Tartric Acid

- **Structure [I] is D (–)–Erythrose,** and
- **Structure [II] is D (–) Threose.**

(3) **Representation of Aldopentoses:** Aldopentoses essentially have the following generalized structure:

$$H-\overset{\overset{\displaystyle O}{\|}}{C}-CHOH-CHOH-CHOH-CH_2OH$$

and hence, possesses *three* **chiral centres** or *three* **asymmetric C-atoms.** Based on the formula 2^n, where 'n' being the number of **chiral centres** present, we have 2^3 or **8 optically active forms** *i.e.,* **four pairs of enantiomorphs.** Nevertheless, all of these forms are known actually, and do emphatically correspond to the respective **D– and L– forms of** the following sugars:

D (–)–Ribose (III),

D (–)–Arabinose (IV),

D (+)–Xylose (V), and

D (–)–Lyxose.

Importantly, the strategic configuration of the aforesaid *four* **'enantiomeric pairs'** could be established explicitly by carefully adopting either of the *two* procedures sequentially, as described under:

Method I: In this particular instance, the **'aldopentoses'** are duly obtained by stepping up the **two aldotetroses,** namely:

D– (–) –Erythrose and D–(–) –Threose, by carrying out these *two* steps sequentially:

- stepping up of the *two aldotetroses* with one additional C-atom *via* **Kiliani synthesis,** and
- resulting products *viz.,* **D–(–) Ribose and D–(–)–Arabinose; D–(+)–Xylose and D–(–)–Lyxose,** are duly *examined optically* by using a calibrated polarimeter.

Explanations: The various reactions included in **Method I** are expatiated below:

(1) As seen earlier that **D–(–)–Erythrose (I)** yields **D–(–)–Ribose (III)** *via* the **Kiliani Synthesis** (with *one additional C-atom*), which upon due *oxidation gives rise to the* **'dibasic acid'** [III

A] and is *optically inactive.*

(2) Likewise, **D–(–)-Arabinose [IV]** obtained from **D–(–)–Erythrose (I)** upon due oxidation produces a 'dibasic acid' **[IV A]** that proves to be **optically active.**

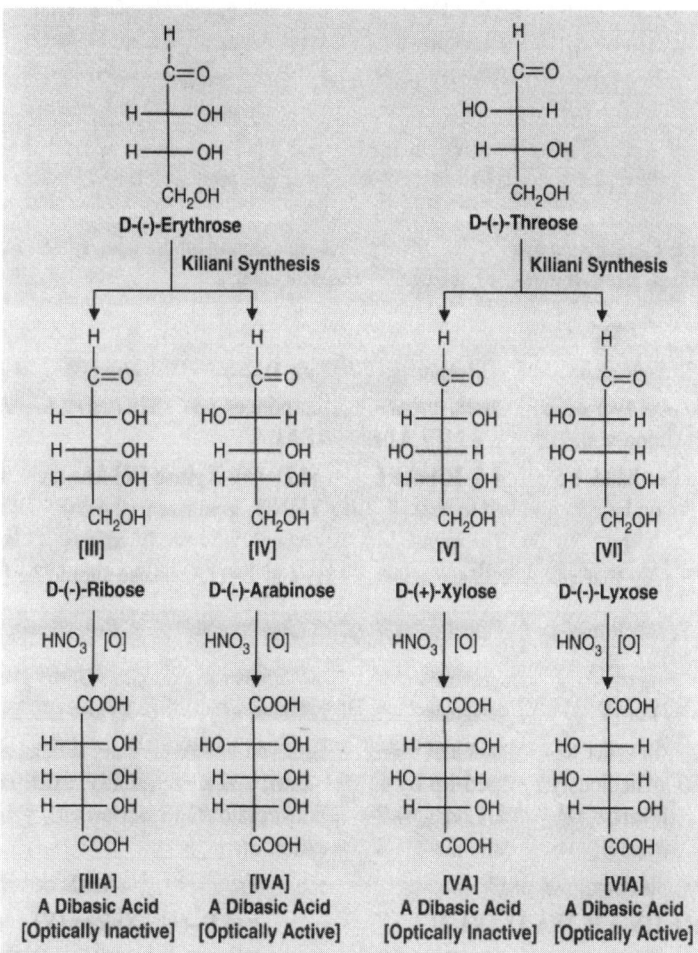

(3) In the same vein, **D–(–) –Threose [II]** on being subjected to **Kiliani Synthesis** (add on *one additional C-atom*) produces **D–(+) –Xylose [V],** that on oxidation gives rise to a 'dibasic acid'**[VA]** and proves to be optically inactive.

(4) Likewise, **D–(–)–Lyxose [VI]** upon oxidation gives rise to a 'dibasic acid' **[VIA]** which is found to be 'optically active'

(5) Thus, it may be inferred that the **compound [III]** *i.e.,* **D–(–) Ribose** duly derived from **D– (–)– Erythrose [I],** and producing **optically inactive dicarboxylic acid [III A]** should be 'Ribose'; and the compound **[IV]** *i.e.,* **D–(–)–Arabinose.** Likewise, the corresponding **compound [V]** *i.e.,* **D–(+) –Xylose,** and compound **[VI]** *i.e.,* **D–(–)–Lyxose** are duly confirmed.

Method II. Importantly, in this particular instance one may not necessarily have a pre-determined, knowledge with respect to the critical **configuration** of the 'aldotetroses' *i.e.,* the *parent compounds.* In fact, the *four* **possible enantiomeric pairs** in the *D– series of aldopentoses* may be given as under:

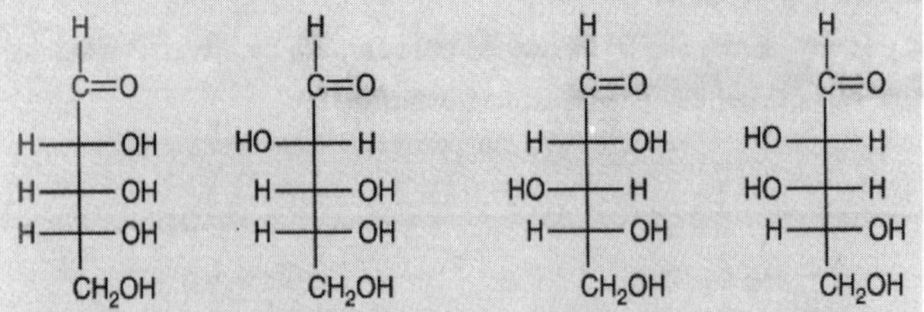

| [III]
D-(–)-Ribose | [IV]
D-(–)-Arabinose | [V]
D-(+)-Xylose | [VI]
D-(–)-Lyxose |

Note: It should be observed critically that the aforesaid configurations have been duly derived and based upon the well defined 'principles' described earlier.

IMPORTANT POINTS

(1) We have since observed in 'Method I' that both **D–(–)–Arabinose [IV]** and **D–(–)–Lyxose [VI]** *i.e.*, the **two aldopentoses**, prominently produces *optically active* dicarboxylic acids upon oxidation with HNO_3 *e.g.*, [IV A] and [VI A].

(2) On the other hand, both **D–(–)–Ribose [III]** and **D–(+)–Xylose [V]** *i.e.*, the **aldopentoses** (in *Method I*), on being subjected to oxidation with HNO_3 give rise to the corresponding *optically inactive* acid by virtue of the presence of a critical **'plane of symmetry'**, which suggests predominantly that **D–(–)–Ribose** must be **[III]** and **D–(+)–Xylose** should be **[V]**.

Note: Nevertheless, it may be observed prominently at this 'crucial stage' to ascertain emphatically:

- whether [IV] or [VI] could be assigned as **D–(–)–Arabinose** and **D–(–)–Lyxose** respectively, and
- whether [III] or [V] may be designated as **D–(–)–Ribose** and **D–(+)–Xylose** respectively.

(3) Therefore, in order to ascertain and their precise/exact nature all these **aldopentoses** are carefully and meticulously **stepped-up by one C-atom;** and subsequently, **oxidized** to the corresponding **dicarboxylic acid.** Consequently, the aforesaid *two* **reaction** *viz.*, **Kiliani Synthesis** and **Oxidation** (HNO_3) results into the formulation of:

Two 'active dicarboxylic acids'   :   From **D– (–) –Arabinose [IV]**,
(*e.g.*, [IV A1], [IV A2], [VA 1], [VA 2]   and **D–(+) –Xylose [V]** ; and
One 'inactive (or meso) dicarboxylic acid' +   :   From **D–(–) –Ribose [III]**,
One 'active dicarboxylic acid')   and **D–(–) –Lyxose [VI]**.
(*e.g.*, [III A1], [III A2], [VI A1], and [VI A2]

| [III]
D-(–)-Ribose | [IV]
D-(–)-Arabinose |

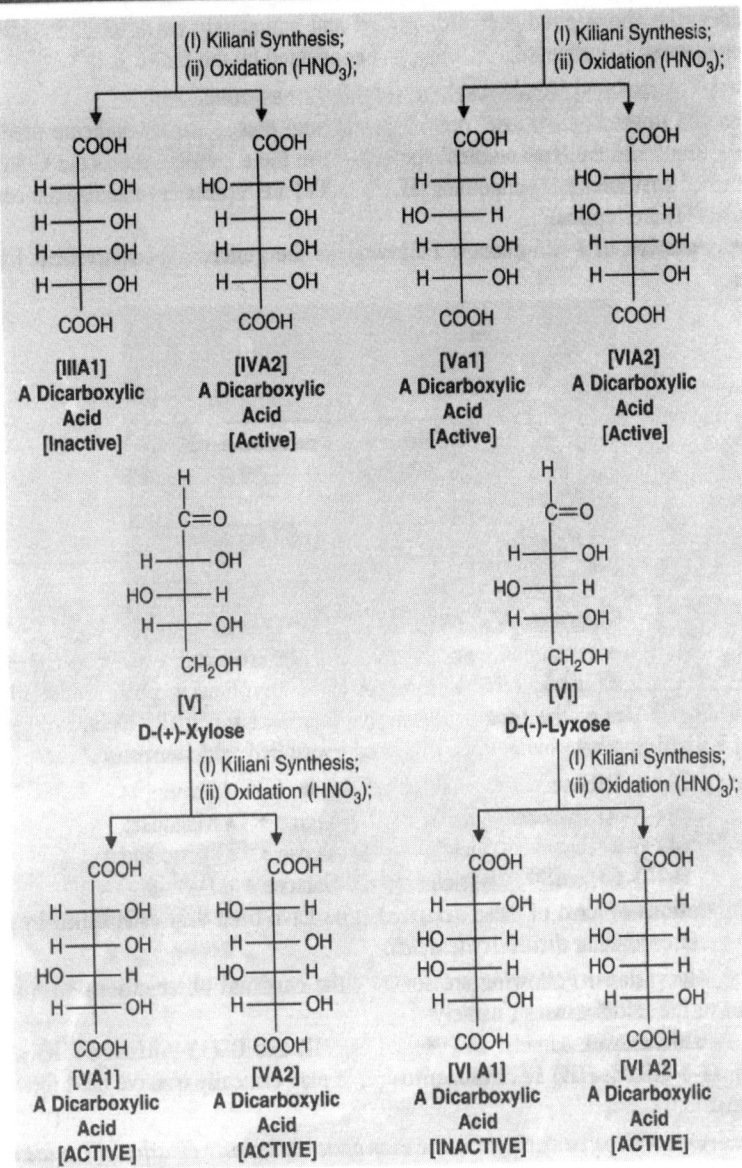

Important Inferences: These essentially comprise:

(i) From the above reactions it is quite evident that the **compounds [III]** and **[VI]** are **D–(–)–Ribose** and **D–(–)–Lyxose** respectively.

(ii) In the same vein, compounds **[IV]** and **[V]** represent **D–(–)–Arabinose** and **D–(+)–Xylose** respectively.

(iii) However, the *oxidation* of the 'aldopentoses' give rise to the formation of **[IV]** and **[VI]** *i.e,* **Arabinose** and **Lyxose**; whereas, **[III]** and **[V]** designate duly **Ribose** and **Xylose.**

(iv) Therefore, one may affirmatively confirm that: **[III] = Ribose; [IV] = Arabinose; [V] = Xylose;** and **[VI] = Lyxose** as adequately supported by the *two* factual statements submitted above.

(v) Besides, the above cited duly determined and adequately established 'configurations of aldopentoses' is further substantiated and confirmed by the fact that:

- both 'Ribose' and 'Arabinose' yield the 'same osazone', and
- both 'Xylose' and 'Lyxose' produce the 'same osazone', that indicate predominantly in both instances the 'two sugars' specifically in the configuration of the C-atom located at second position *i.e.*, compounds [III] and [IV] are 'epimers', and also the compounds [V] and [VI] are 'epimers'.

(4) Representation of Aldohexoses: Following is the **generalized structural** formula of the 'aldohexoses':

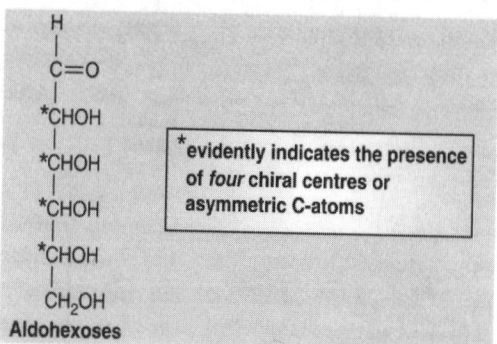

Aldohexoses

Since, the **aldohexoses** essentially possess *four* **chiral centres;** and, therefore, they would exist in sixteen (*i.e.*, $2^4 = 16$) **optically active isomers** or **eight enantiomeric pairs.** In actual practice, all these 'enantiomeric pairs or the 'optically active isomers' are actually obtainable by the critical application of **Kiliani Synthesis** only one time to the respective 'aldopentoses'.

Examples: **D–(–)–Ribose**   yields : D–Allose + D–Altrose;
      D–(–)–Arabinose yields : D–Glucose + D–Mannose;
      D–(+)–Xylose   yields : D–Glucose + D–Idose; and
      D–(–)–Lyxose   yields : D–Galactose + D–Talose.

The **configurations** of most of these *derived sugars* have been duly established by their specific oxidation to the corresponding *dicarboxylic acids*.

Cardinal Observations: Following are some of the **cardinal observations** with respect to the configurations of the 'aldohexoses', namely:

(1) The *two* aldohexoses, namely: **D–(+)–Allose (VII)** and **D–(+)–Altrose (VIII)**, duly obtained from **D–(–)–Ribose (III)** *i.e.*, aldopentose, one may critically observe the following **configurations:**

- the sugar which particularly gives rise to an *optically inactive dicarboxylic acid i.e.*, Allomucic Acid upon oxidation is found to be **D–(+)–Allose (VII)** and
- the sugar which specifically produces an *optically active* dicarboxylic acid *i.e.*, Talomucic Acid on being oxidized happens to be **D–(+)–Altrose (VIII)**.

(2) Likewise, the other *two* aldohexoses, namely: **D–(+)–Galactoside (IX)** and **D–(+)–Talose (X)**, duly produced from **D–(–)–Lyxose (VI)** *i.e.*, aldopentose, one would specifically see the following **configurations:**

- the sugar that categorically produces an *optically inactive* dicarboxylic acid *i.e.*, Mucic Acid upon oxidation with HNO_3 is found to be **D–(+)–Galactose (IX)**, and
- the sugar which particularly gives rise to an *optically active* dicarboxylic acid *i.e.*, Talomucic Acid on being oxidized establishes to be **D–(+)–Talose (X)**.

(3) It has been amply established that **oxidation** of **D–(+)–Glucose (XI)** and **D–(–)–Mannose (XII)**, which are both obtained from **D–(–)–Arabinose (IV)**; and likewise, **D–(–)–Gulose (XIII)** and **D–(–)–Idose (XIV)**, that are both derived from **D–(+)–Xylose (V)** do provide *optically active* **dicarboxylic acids**. Nevertheless, their **configurations** are actually determined and established by means of some modified method.

(4) From the above observations one may observe that both:

D–(+)–Glucose (XI) and **D–(–)–Gulose (XIII)** critically give rise to the same 'dicarboxylic acid' i.e., **D– or L–Saccharic Acid**.

Besides, it is a well-known fact that in the course of usual oxidation with HNO_3, the 'terminal groups' viz, **aldehydic (CHO) moiety** and **hydroxy methyl (CH_2OH) moiety** are converted exclusively into the respective **carboxyl (COOH) moieties** (to yield *dicarboxylic acid*); whereas, the **rest of the molecule do not get affected at all.**

(5) Based upon the above observations, one may critically take cognizance of the fact that the two compounds **D–(+)–Glucose (XI)** and **D–(–)–Gulose (XIII)** essentially differ solely in the strategical locations of these **two terminal moieties** (viz.,–CHO and –CH_2OH); and, therefore, if one may be able to **'interchange the position of one of these moieties, one would get the other'**.

(6) Consequently, an intimate and closer look at the *structural formulae: XI, XII, XIII* and **XIV** evidently reveals that the aforesaid **'possibility'** does prevail in the compounds **XI** and **XIII** specifically. Hence, it may be stated that **compound XI** should be **D–(+)–Glucose** which may be further substantiated by the fact that it is duly obtained from **D–(–)–Arabinose (IV)**; and in the same vein, the **compound (XIII)** should be **D–(–)–Gulose**.

(7) It may be concluded that **compound (XII)** is **D–(–)–Mannose**, and **compound (XIV)** happens to be **D–(–)–Idose**.

Following are the various reactions emanating from **D–(–)–Ribose (III)** and **D–(–)–Lyxose (VI)**:

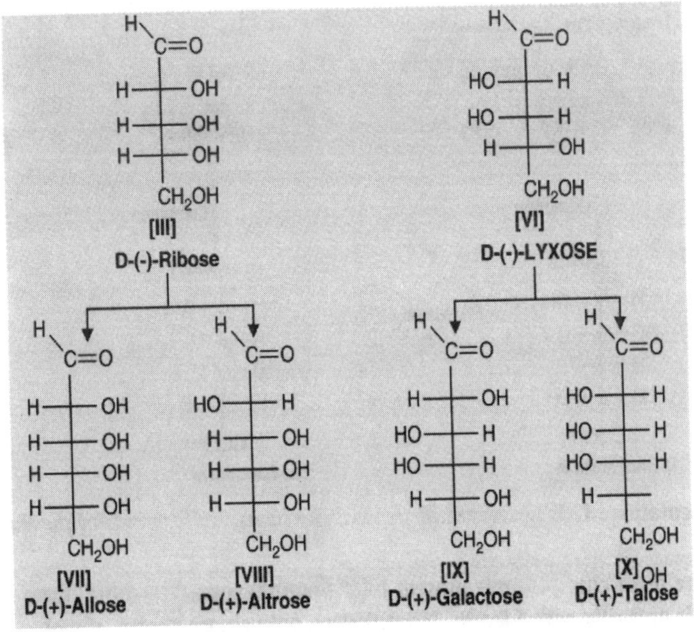

(Contd.)

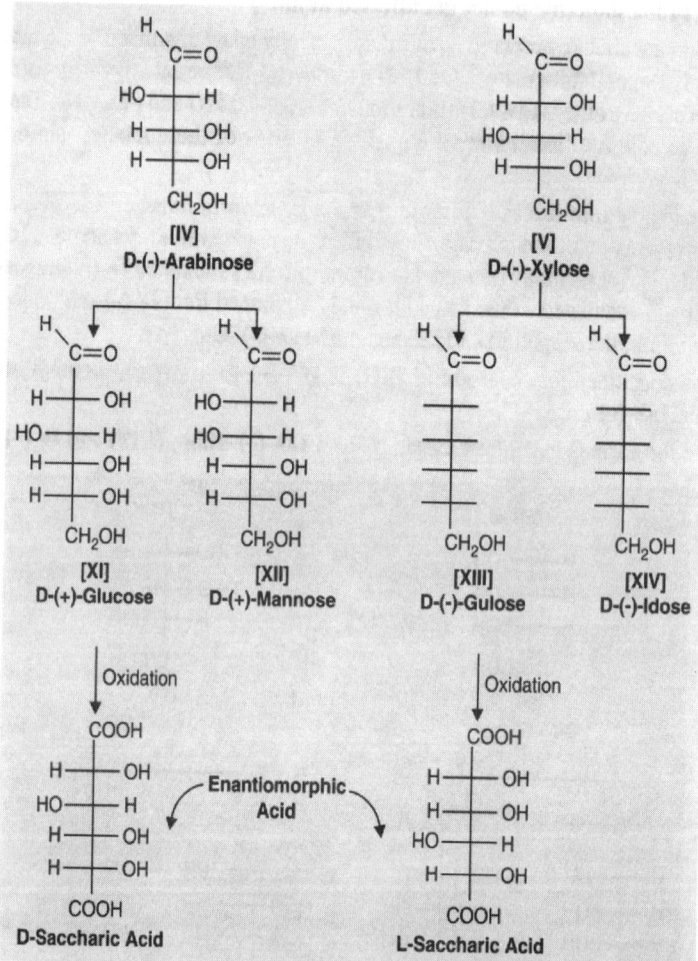

Following are the different reactions emanating from **D–(–)–Arabinose (IV)** and **D–(+)–Xylose (V).**

(5) Representation of Ketohexoses: The *generalized structural formula of the 'ketohexoses'* is as given below:

As there are *three* chiral centres present in the 'ketohexoses; therefore, there may exist **eight** [$2^3 = 8$] **possible optically active forms.** Nevertheless, only *six* are known, namely:

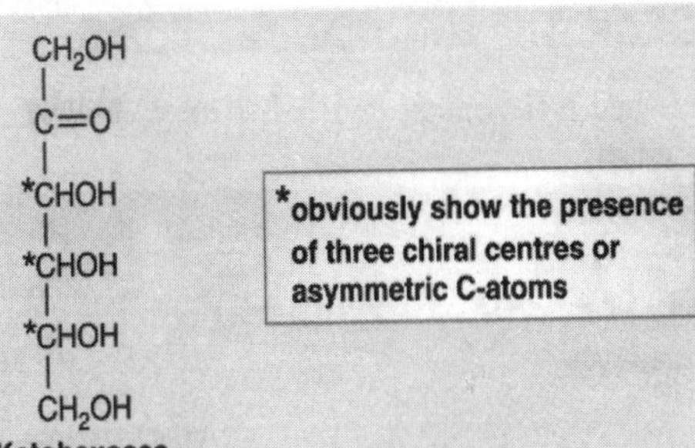

$$
\begin{array}{l}
CH_2OH \\
| \\
C=O \\
| \\
*CHOH \\
| \\
*CHOH \\
| \\
*CHOH \\
| \\
CH_2OH
\end{array}
$$

Ketohexoses

> *obviously show the presence of three chiral centres or asymmetric C-atoms

D–(–)–Fructose; D–(+)–Sorbose; D–(+)–Tagatose;

L–(+)–Fructose; L–(–)–Sorbose; L–(–)–Psicose;

Interestingly, out of these *six* known optically active forms of 'ketohexoses' only *three* are usually found in nature, such as: **D–(–)–Fructose; L–(–)–Sorbose; D–(+)–Tagotose;**

Configuration: Importantly, the specific 'configuration of ketohexoses' is broadly established by the preparation of its corresponding 'osazone'; and subsequently, comparing it with the 'osazones obtained duly from the aldohexoses'. It is, however, worthwhile to mention here that the critical 'osazone formation' invariably and predominantly engages exclusively the **first two C-atoms**. Therefore, either a 'ketose' or an 'aldose' essentially forming the *same osazone'* shall have the 'same

configuration with an exception in the *first two C-atoms* **that are always** $-\overset{\overset{\displaystyle O}{\|}}{C}-CH_2OH$

Nevertheless, the aforesaid 'method' may be further expatiated by considering the following example of **D–(–)–Fructose (I)** that essentially gives rise to the formation of the **same osazone as D–(+)–Glucose (II)'.** Evidently, the configuration of the **last three asymmetric C-atoms** (*viz.,* C-3 to C-5) pertaining to **D–(–)–Fructose (I)** is virtually the same as that of **D–(+)–Glucose (II)**, as exemplified under:

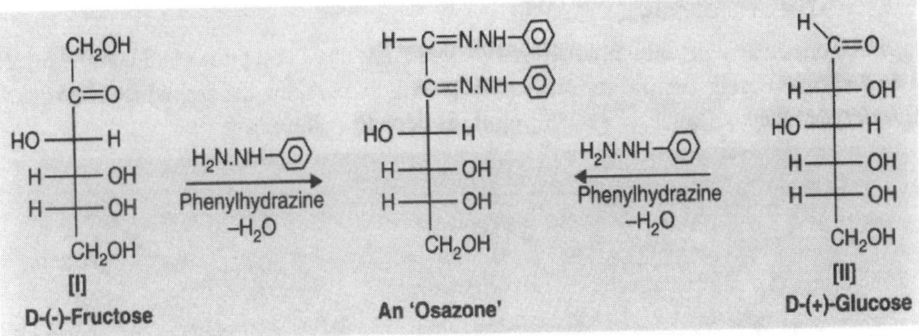

Likewise, based upon the above cited analogy the configurations of other 'ketohexoses' have been established duly as given under:

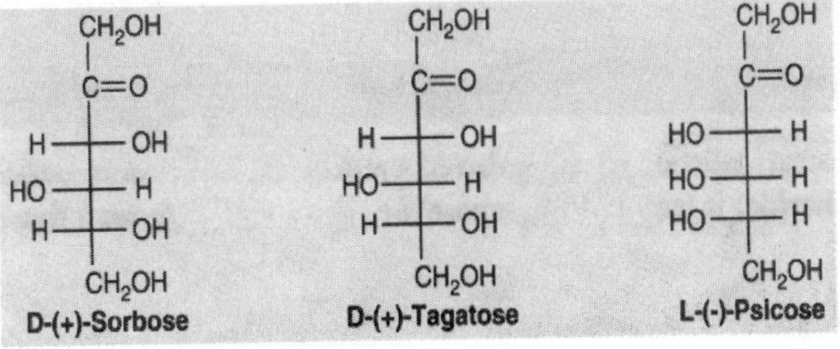

7. CYCLIC STRUCTURES OF THE MONOSACCHARIDES

It has already been established that both γ– and δ–hydroxy aldehydes invariably exist as *cyclic hemiacetals,* as exemplified under:

(i) $H_3C-CH-CH_2 CH_2 CH_2-C=O$

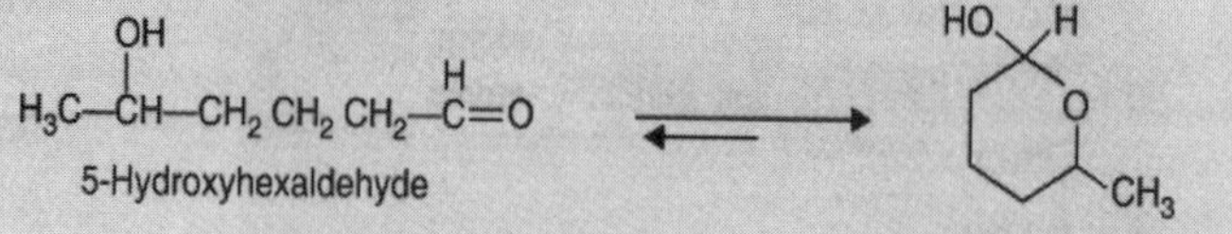

5-Hydroxyhexaldehyde

A Cyclic Hemiacetal
[Six-Membered 'lactone ring'] ...... (a)

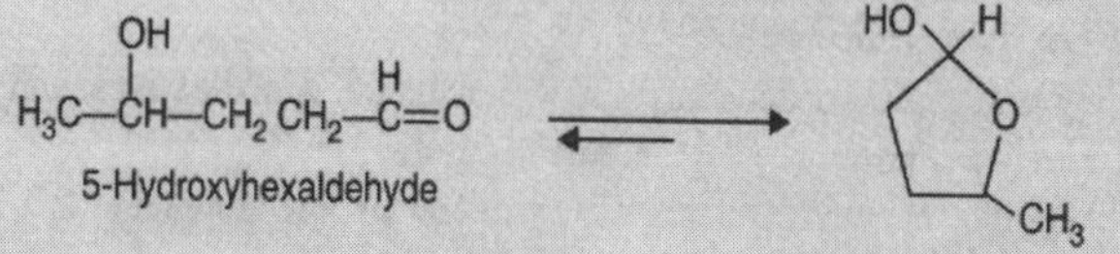

(ii) $H_3C-CH-CH_2 CH_2-C=O$

5-Hydroxyhexaldehyde

A Cyclic Hemiacetal
[Five-Membered 'lactone ring'] ...... (b)

That is, the Eqns. (a) and (b) above illustrate the formation of the 'cyclic hemiacetals' from the respective hydroxylated **hexaldehyde** and **pentaldehyde**.

Obviously, the same observation is found to be 'true' of the 'aldoses', and 'ketoses'. Nevertheless, in actual practice the 'monosaccharides' are usually written by an established convention as the 'acyclic carbonyl compounds'; however, they do exist predominantly as the 'cyclic acetals'.

Example: In an aqueous medium **glucose** mostly comprise:

Free 'aldehyde' : ~ 0.003%,
Hydrate : traces only,
Cyclic hemiacetals : > 99.99%

It has been established beyond any reasonable doubt that in a good number of carbohydrates both **5– and 6– membered cyclic hemiacetals** are most possibly based upon the fact which **hydroxy (–OH) moiety** *undergoes the phenomenon of cyclization,* as depicted below:

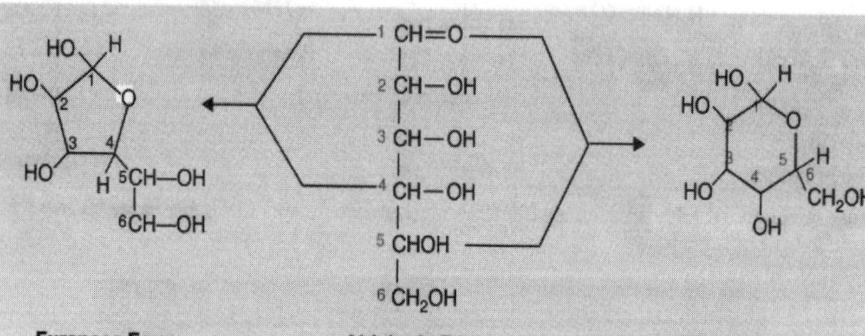

Furanose Form   **Aldehyde Form**   **Pyranose Form**

Furanose: In actual practice, a **5–membered cyclic acetal form** of a *carbohydrate* (more precisely a **monosaccharide**) is termed as 'furanose' *i.e.,* derived after *furan* a five-membered oxygen heterocycle .

Pyranose: A six-membered cyclic acetal form of a *carbohydrate* (*viz,* a **monosaccharide**) is usually known as a **'pyranose'**, which is duly derived from *pyran*–a **6-membered oxygen hetero-**

cycle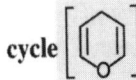

It has been duly observed that the **aldopentoses** as well as the **aldohexoses** invariably do exist as **'pyranoses'** in aqueous solution; however, the so called *'furanose'* forms of certain **carbohydrates** are noticeably vital and important.

7.1. Nomenclature

The various ways and means usually adopted in assigning the appropriate *names* of *carbohydrates* are as exemplified under:

- **Glucose:** usually given when referring to any or all forms of the **carbohydrate,**
- **'Prefix' derived from name of carbohydrate + 'Suffix' derived from type of hemiacetal ring***: normally a **6-membered cyclic hemiacetal from** of *D-glucose* is known as **D-glucopyranose,** which may be further expatiated as under:

D–Glucopyranose

a   b

    'a'-designates the *prefix* from **glucose,** and

    'b'- represents the *suffix* from **pyranose** (*i.e.,* **6-membered cyclic hemiacetal form**).

Likewise, the **5-membered cyclic hemiacetal form of D-mannose** shall be termed as **D-mannofuranose.**

7.2. NMR Spectroscopy in Revelation of Cyclic Structure of Aldoses

Inspite of the fact that the *cyclic structures of aldoses* were proved and established by specific **chemical degradative studies,** these *cyclic structures (viz.,* **pyranoses** and **furanoses***)* are meticulously and rapidly revealed by the help of **NMR spectroscopic studies.**

Examples: The following typical examples will essentially elaborate these spectacular revelations as given below:

(*a*) One may observe that the *aldehydic proton resonance of glucose* falling very much within the **range of δ 9-10** pertaining to its NMR spectrum is found to be *very weak* so as to have them **detected under the prevailing ordinary circumstances.** Nevertheless, there exists a **'doublet'** at **δ 5.2** corresponding to a strategically located **proton α to two oxygen atoms,** namely:

- Proton at **C-1 of the pyranose structure.**

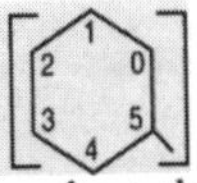

(*b*) **CMR-Spectrum** reveals that the carbonyl carbon absorption in the **δ 200 region** of the spectrum fails to discern in an *ordinary spectra.*

7.3. Anomers

It is, however, pertinent to observe that- **'furanose and pyranose form of a carbohydrate essentially possesses one more asymmetric carbon (or chiral centre)** *vis-a-vis* **the open-chain form** *viz.* **C-1 in the particular instance of the 'aldoses'.**

---

* *i.e.,* **Pyranose** for a **6-membered ring;** and **Furanose** for a **5-membered ring**

Based on the above specific comparison one may observe *two* **possible diastereoisomers of D-glucopyranose** as given under:

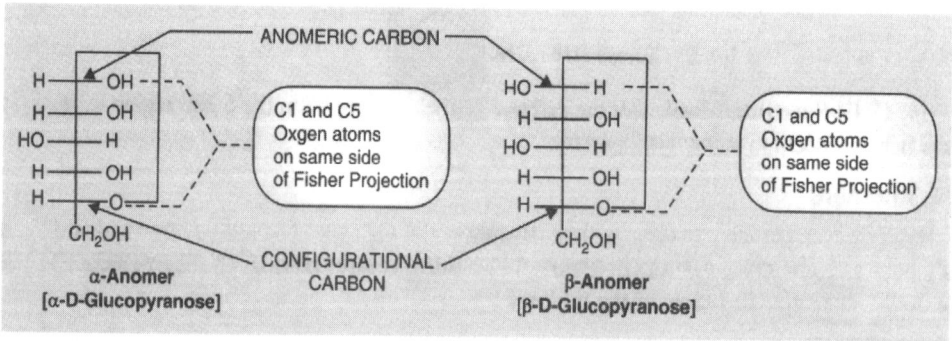

7.3.1 Fischer Projections

In true sense, both of the above drawn compounds *i.e.,* α-*anomer* and β-**anomer** represent different forms of **D-glucopyranose;** and in actual practice, one may lay hand on a mixture of both **anomers** from a **glucose solution.** Since, they designate as diastereoisomers; and, therefore, may be separated conveniently having altogether different characteristic features.

Thus, when two cyclic forms of a carbohydrate essentially their configurations exclusively at the specific 'hemiacetal carbons' they are termed as **anomers.** Alternatively, the **anomers** designate the so called *cyclic forms of the carbohydrates* which are particularly 'epimeric' at the *hemiacetal carbon atom.* Therefore, the two forms of **D-glucopyranose** are nothing but the 'anomers of glucose'. In other words, the *hemiacetal carbon (i.e.,* C-1 of an 'aldose') is commonly known as the **anomeric carbon.**

In the aforesaid structures, the 'anomers' are usually referred to as the Greek letters α and β. However, this particular **nomenclature** turns to for information the 'cyclic form of a carbohydrate', having *'all C-atoms'* expressed in a **straight vertical line.**

The α-Anomer: In the α-anomer– 'the hemiacetal hydroxyl (–OH) moiety is located on the same side of the Fischer projection as the oxygen atom at the *configurational carbon'*.*

The β-Anomer: In the β-anomer– 'the hemiacetal hydroxyl (-OH) moiety is located on the side of the Fischer projection just opposite to the oxygen atom at the *configurational carbon'.**

Interestingly, the actual application of the aforesaid definitions to the nomenclature of the *D-glucopyranose anomers* is as stated under:

```
            ANOMERIC CARBON
H——▲ OH ----.                            HO ——▲ H ---.
H —— OH      :  ╭──────────────╮          H —— OH    :  ╭──────────────╮
HO —— H      :  │ C1 and C5    │          HO —— H    :  │ C1 and C5    │
H —— OH      :  │ Oxgen atoms  │          H —— OH    :  │ Oxgen atoms  │
H ——▼ O====.'   │ on same side │          H ——▼ O===.'  │ on same side │
CH₂OH           │ of Fisher    │          CH₂OH         │ of Fisher    │
                │ Projection   │                        │ Projection   │
α-Anomer        ╰──────────────╯          β-Anomer       ╰──────────────╯
[α-D-Glucopyranose]   CONFIGURATIDNAL     [β-D-Glucopyranose]
                        CARBON
```

C1 and C5 Oxgen atoms on same side of Fisher Projection

CONFIGURATIDNAL CARBON

α-Anomer
[α-D-Glucopyranose]

β-Anomer
[β-D-Glucopyranose]

---

* **Configurational Carbon:** It refers to the C which is used for specifying the D, L–designation; *i.e.,* C-5 for the aldohexoses.

7.3.2. Conformational Representations of Pyranoses

It has been reasonably established that the **Fischer projections** of the **monosaccharides** *in particular* and the carbohydrates *in general* are found to be quite convenient for specifying their ensuing **configurations** at each **asymmetric carbon** (or **chiral centre**); however, the **Fischer projections** provide virtually *no information* with respect to the **conformations of carbohydrates.** At this critical juncture, it becomes almost vital and important to justify as well as relate the **Fischer projections** to the *conformational representations* of the **carbohydrates.**

Example: Following represents a *typical example* for the meticulous conversion of the **Fischer projection** of β-D-glucopyranose into the corresponding **'chair conformation'.** However, it may be expatiated adequately by following these steps in a sequential manner:

(1) First and foremost, it is required to redraw the **Fischer projection** for β-D-glucopyranose in an equivalent **Fischer projection** wherein the **'ring oxygen'** is duly projected in the **downward position.** Interestingly, it may be accomplished by making use of a **'cyclic permutation and combination of the various moieties strategically located on C-5',** which falls very much within the *'permitted manipulation(s) of the Fischer projections'.* The redrawn structure of β-D-glucopyranose based on *Fischer projection* is as follows:

β-D-Glucopyranose   β-D-Glucopyranose

Explanations: In order to understand the **conformational representations of pyranoses** in a better way let us assume that the *C-backbone* of a **Fischer projection** of this type is hypothetically supposed to be artistically folded around a *cylindrical drum* or *barrel* or *cask* as illustrated below in Fig. 1.1

Fig. 1.1: Illustrates D-Glucose Molecule with Four Chiral Carbons. [Groups that are far away from the 'eye of an observer are 'Vertical' in the Fischer projection, and moieties near the observer are 'Horizontal'.]

---

* **Haworth Projection:** It is named after the noted British Carbohydrate chemist Sir Walter Norman Haworth (1883-1950), who actually performed critical and excellent research specifically pertaining to the **'cyclic structures of carbohydrates** Sir Haworth was gracefully knighted in the year 1947, and bagged the prestigious *Nobel Prize in Chemistry in 1937.*

An intelligent and logistic explanation of this kind would perhaps lead to an appropriate interpretation of the **Fischer projection** pertaining to **β-D-glucopyranose** that essentially gives rise to the following structure wherein the ring critically lies in a plane absolutely perpendicular to the flat page as depicted in Fig. 1.2.

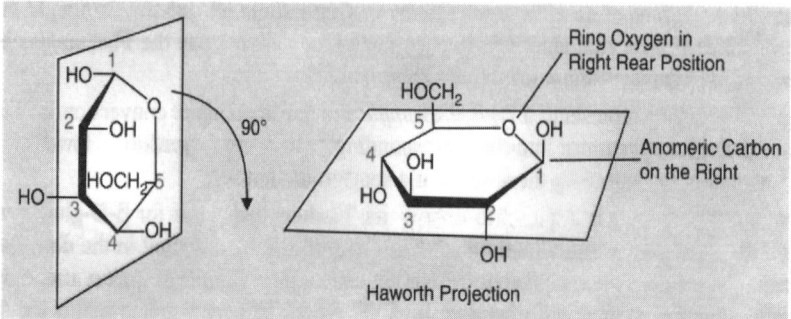

Fig. 1.2: Haworth Projection of β-D-glucopyranose [In Fig. 1.2., the H-atoms are not shown.]

(2) **Haworth Projection:** In a situation when the plane of the ring is **rotated 90°** (see LHS of Fig. 1.2.) in such a manner that the *anomeric carbon takes its position on the RHS, and the* **ring** oxygen is positioned in the rear, the moieties present in **'up positions'** are those which are located on the **LHS** in the **Fischer Projection;** whereas, the moieties in **'down positions'** are those that are present on the **RHS** in the **Fischer projection**.

Thus, a *planar structure* of this type is termed as the **Haworth projection**. Nevertheless, it is worthwhile to state here that in a **Haworth projection**, the *'ring'* is usually drawn in a plane at **right angles** to the page, and the **positions of the substituents are indicated with up or down bonds**. Besides, the *shaded bonds* are invariably shown *in front of the page* (*i.e.,* projected towards the 'observer'), and the *remaining bonds* are shown *in back of the page* (*i.e.,* projected away from the 'observer').

(3) **Limitations of Haworth Projection:** It has been duly observed that a **Haworth projection** fails to specify the actual ensuing conformation of the ring. Interestingly, the **6-membered carbohydrate rings** do closely resemble (or match) the **substituted cyclohexanes;** besides, the former essentially exist and prevail in *'chair conformations'* very much akin to the **substituted cyclohexanes.**

Conformational Representation of 3-D-Glucopyranose: It may be accomplished completely by carefully drawing either one of the *two chair conformation,* wherein the *'anomeric carbon'* plus the **ring oxygen** are exclusively present in the **same relative positions** as they are observed in the preceding **Haworth projection.** Subsequently, allocated the various moieties present in *up* and *down* in both *axial* and *equatorial positions,* as deemed to be appropriate*.

* (1) Despite the *'chair-flip'* affords change in the equatorial moieties to axial ones, and *vice versa,* it fails to change whether a group is either up or down.

  (2) It, therefore, hardly matters which of the two feasible and possible **chair conformations** one would draw first.

(4) Furanoses (5-Membered Rings) *Vs* Haworth Projections: Inspite of the basic facts that the 5-membered rings of **'furanoses'** are totally **nonplanar,** they appear to be **close enough to planarity** which categorically place the **Haworth formulae** in good and reasonable approximations to their *actual structures.* Therefore, the so called **Haworth projections** are abundantly employed for the **'furanoses'** for this cardinal reason. Hence, one may arrive at the **Haworth projection** pertaining to the **B-D-Ribofuranose** as given under in a sequential manner:

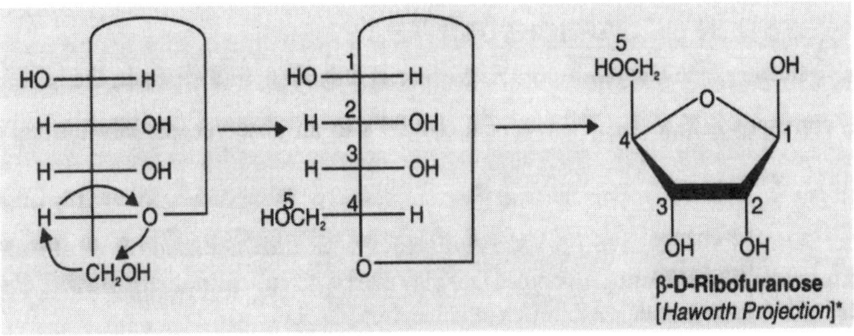

β-D-Ribofuranose
*[Haworth Projection]**

Since, we have already seen earlier that the structure of **(3-D-glucopyranose** is relatively easy to remember due to the underlying fact that it occurs in a **more stable chair conformation** having its all **ring substituents** located **equitorially.**

Conformation of (3-D-Galaetopyranose): As **D-galactose** and **D-glucose** designated as *epimers* at C-4; subsequently, the **B-D-glucopyranose** may be represented promptly by affording an *interchange* of the **–H and –OH moieties** at C-4 in the said compound (*i.e.,* **(3-D-glucopyranose)** as given under:

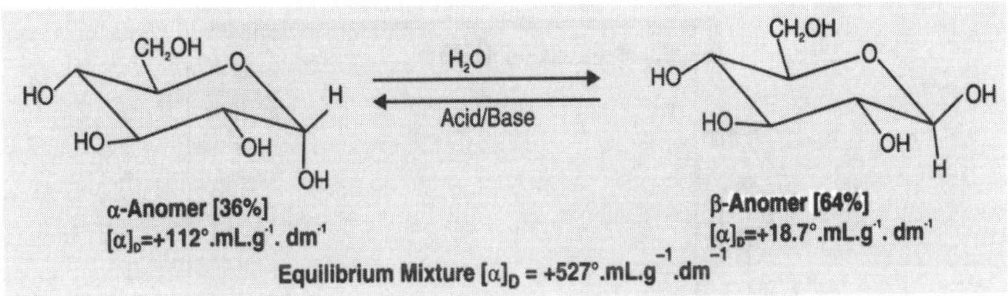

α-Anomer [36%]
$[\alpha]_D = +112°.mL.g^{-1}. dm^{-1}$

β-Anomer [64%]
$[\alpha]_D = +18.7°.mL.g^{-1}. dm^{-1}$

Equilibrium Mixture $[\alpha]_D = +527°.mL.g^{-1}.dm^{-1}$

$$[\alpha]_D = + 112°.mL.g^{-1}.dm^{-1} \qquad [\alpha]_D = +18.7°.mL.g^{-1}.dm^{-1}$$

Equilibrium mixture $[\alpha]_D = +52.7°.mL.g^{-1}.dm^{-1}$.

Exactly, in the same vein, since both **mannose** and **glucose** are found to be *epimeric* at C-2, the structure of **D-mannopyranose** may be simply obtained by the interchangeability of **–H and –OH moieties** at C-2 of the aforesaid structure (*i.e.,* **D-mannopyranose**).

Squiggly Bond: In certain cases, it is virtually necessary to draw the proper conformation of a given *'carbohydrate'* which may be present as:

- a mixture of **'anomers',** or
- an uncertain mixture of **'anomeric composition'.**

Therefore, in these particular instances the **'configuration at the anomeric carbon'** is duly represented by a typical **'squiggly bond'.**

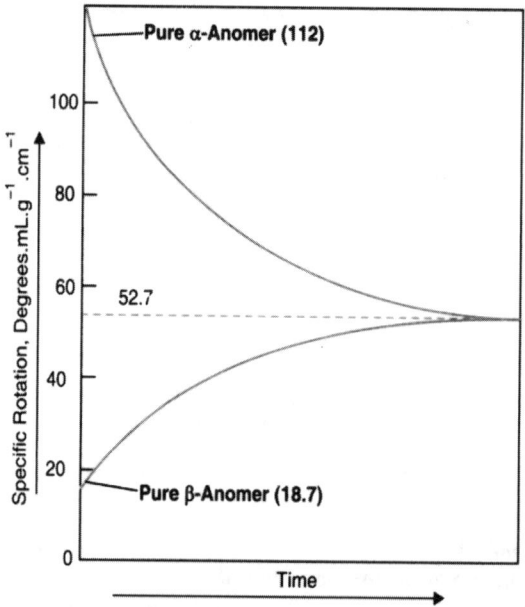

D-Glucopyranose of Mixed of Uncertain Anomeric Composition.

8. MUTAROTATION OF CARBOHYDRATES

In a typical situation when crystalline **α-D-glucose** is duly dissolved in water, the initial rotation of the solution, corresponding to $[\alpha]_D^{20} = 120°$, usually falls to an observed equilibrium value of **52.7°** (the β form has $[\alpha]_D^{20} = 18.7$). This particular phenomenon is commonly termed as **'mutarotation'**.*

In other words, the change of **optical rotation** *vis-a-vis* **time** is known as **mutarotation** (*muta* = change). Importantly, **'mutarotation'** comes into play even when **pure anomers of other carbohydrates'** are duly dissolved in an aqueous solution as depicted in Fig. 1.3.

Mechanism: It has been duly observed that the **'mutarotation of glucose'** is invariably caused by the ensuing conversion of the **α- and β-glucopyranose anomers** into an *equilibrium mixture of both* **preferentially.** Amazingly, one may also observe the same 'equilibrium mixture', from either pure:

- **α-D-Glucopyranose,** or
- **β-D-Glucopyranose.**

Note: 'Mutarotation' is normally catalyzed by both *acid* and *base* but also takes even in pure aqueous phase.

Figure: 1.3: Observed Mutarotation of D-Glucose.

_____

* Norman R *et al.*: **Principles of Organic Synthesis,** 3rd ed., Nelson Thornes, New York, 2005.

The mechanism of conversion of **α-anomer** into **β-anomer** is given as under:

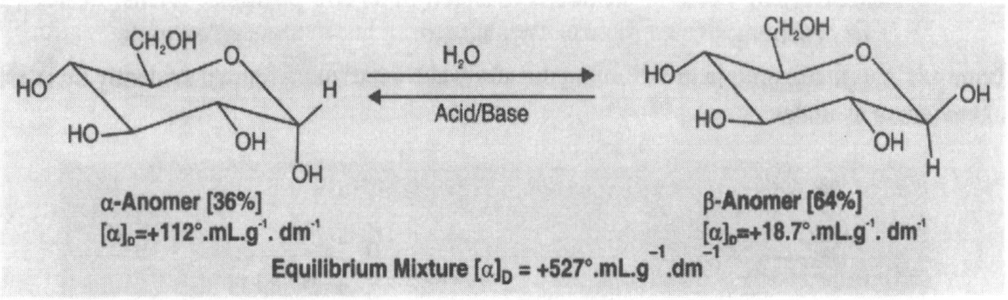

α-Anomer [36%]
$[\alpha]_o = +112°.mL.g^{-1}.dm^{-1}$

β-Anomer [64%]
$[\alpha]_o = +18.7°.mL.g^{-1}.dm^{-1}$

Equilibrium Mixture $[\alpha]_D = +527°.mL.g^{-1}.dm^{-1}$

Cardinal Observations: Following are some of the **cardinal** observations with regard to 'mutarotation' in the *cyclic hemiacetal* forms of 'glucose'.

(1) First and foremost, it may be noted that the phenomenon of **'mutarotation'** is a typical characteristic feature pertaining to the *cyclic hemiacetals* forms of **glucose** *i.e.*, an **'aldehydic function'** **cannot undergo mutarotation.** Nevertheless, the earlier **carbohydrate chemists** well conceived and assumed that **'mutarotation'** happened to be one of the phenomena which categorically put forward the idea that **'aldoses'** might exist as **cyclic hemiacetals.**

(2) **'Mutarotation'** normally comes into play in *two* ways:

• **'opening of pyranose ring'** to the **'free aldehyde form'** which, in fact, is actually the **reverse of hemiacetal formation,** and

• **rotation of 180° at the carbon-carbon double bond** to the **carbonyl** $(-\overset{O}{\underset{||}{C}}-)$ **function** allows the critical **'reclosure'** of the ensuing **cyclic hemiacetal** by virtue of the attack of the **hydroxyl**

(–OH) function strategically located on the *opposite face* of the **carbonyl** $(-\overset{O}{\underset{||}{C}}-)$ **carbon** as depicted under:

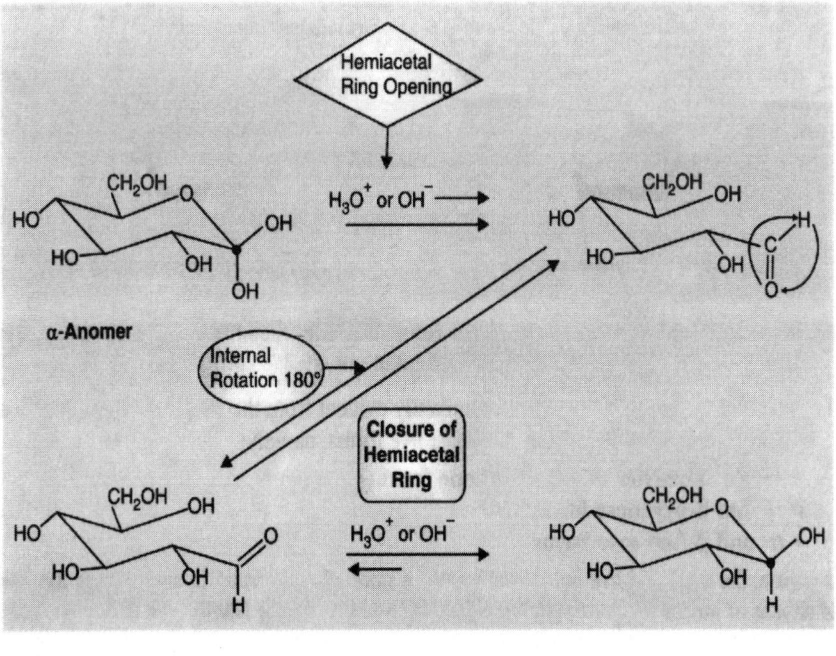

(3) **Mutarotation of Glucose:** In reality the **'mutarotation of glucose'** is almost caused exclusively by virtue of the *interconversion* of its **two puranose forms.** Besides, certain **'other carbohydrates'** do usually undergo rather more complex mutarotations.

Example: A typical example in exhibiting the aforesaid behaviour is shown evidently by **D-fructose** *i.e.,* a *ketohexose* as under:

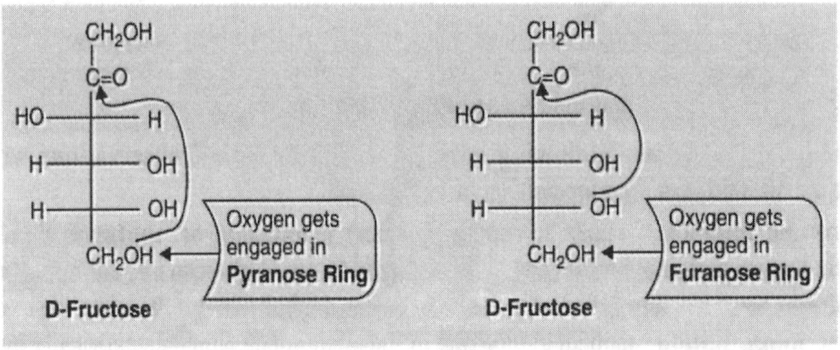

Explanations: Based on the above structures pertaining to the **hemiacetal states of D-fructose** may be obtained from its **carbonyl** ($-\overset{\text{O}}{\underset{}{\overset{\|}{\text{C}}}}-$) **form** as the corresponding **D-fructose** in the **β-D-fructopyranose** form usually in crystalline states. Besides, the *crystals* of this particular form on being dissolved in water gets duly **equilibrated** to both **'pyranose'** and **'furanose'** forms.

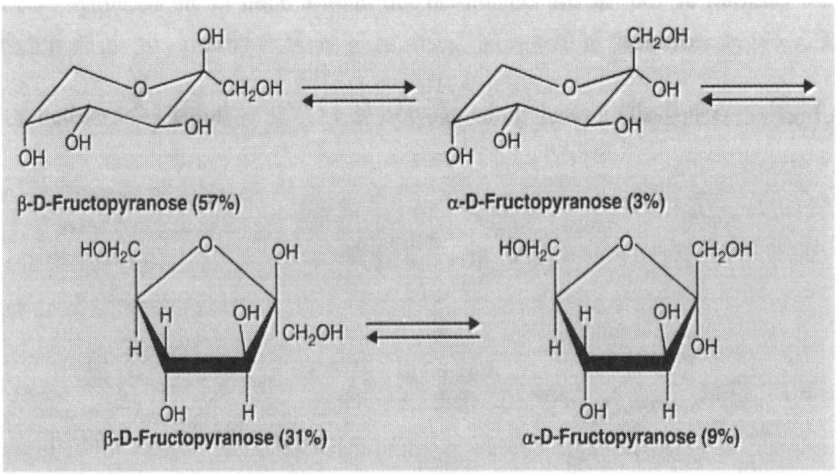

Note: Incidentally, glucose in solution also consists of the 'furanose form'; however, there are invariably present in exceptionally small quantum (~0.2% each).

(4) **Variants of Single Hexose:** It is abundantly evident from the above discussions that a **'single hexose'** may invariably occur in at least *five* forms, namely:

- **acyclic aldehyde or acyclic ketone form,**
- **α- and β-pyranose form, and**
- **α- and β-furanose forms.**

In addition, the aforesaid *five* forms accomplish a *state of equilibrium* in an **aqueous medium.** As to date, the advent of modern physico chemical techniques, specifically **NMR spectroscopy,** have enabled chemists to- **'determine for several carbohydrates the exact and precise quantum of the different forms that are eventually present at equilibrium.'**

Table 1.1. Records the compositions of **'monosaccharides'** at equilibrium in an aqueous solution at 40°C:

Table 1.1 : Precise Compositions of Monosaccharides at Equilibrium in Aqueous Solution at 40°C.

S.No.	Sugar	Percent at Equilibrium				Aldelyde or Ketone
		Pyranose Form		Furanose Form		
		α	β	α	β	
1	D-Glucose	36	64	Traces*	traces*	0.003
2	D-Mannose	68	32	trances*	0	traces
3	D-Allose	18	70	5	7	–
4	D-Altrose	27	40	20	13	–
5	D-Idose**	39	36	11	14	–
6	D-Talose	40	29	20	11	–
7	D-Xulose	37	63	–	–	–
8.	D-Ribose	20	56	6	18	0.02
9.	D-Fructose	0-3	57-75	4-9	21-31	0.25

Salient Features: The *three* vital and important **salient features** which may be concluded in a broader perspective are as follows:

1. It has been duly observed that largely the **'aldopentoses'** and **'aldohexoses'** do occur importantly as the **'pyranoses',** although a few of them possess appreciable quantum of the respective **'furanose forms'.**

2. A large segment of the **monosaccharides** do essentially comprise comparatively small quantum of their **'noncyclic carbonyl forms'.**

3. Innariably the mixtures of **α- and β-anomers** are observed, however, the precise and exact quantum of each does vary from one particular instance to another.

Note: Interestingly, the exact fraction of any form in solution at equilibrium is usually determined by its *inherent stability vis-a-vis* all other corresponding forms.

9. BASE-CATALYZED ISOMERIZATION *vis-a-vis* EQUILIBRATION OF ALDOSES AND KETOSES

It has been established beyond any reasonable doubt that in base-catalyzed isomerization both **'aldoses'** and **'ketoses'** undergo critical and rapid equilibration to certain decisive turns of *other aldoses and ketoses* as given below:

In fact, the above cited **'transformations'** give a glaring example pertaining to the *Lobry de Bruyn-Alberd van Ekenstein Reaction* duly named after two well-known **Dutch Chemists.*****

---

* *i.e.,* in an aqueous dioxane (10% w/v) glucose contains 0.1-0.2% of each **'Furanose'.**

** At 25°C.

*** Cornelius Abriaan van Trooostenbery Lobry de Bruyn (1857-1904); and Willem Alberda van Ekenstin (1858-1907).

9.1 Formation of Enolate Ion

In addition, **glucose** in aqueous solution usually occurs in its **cyclic hemiacetal forms,** it also exists in equilibrium with a comparatively small quantum of its **'acyclic aldehyde form'.** However, the latter form *i.e.,* the **'aldehyde',** very much akin to other **carbonyl** $(-\overset{O}{\overset{\|}{C}}-)$**compounds** having essentially the **α-hydrogens,** gets critically ionized to yield small quantum of its respective **enolate ion in base.** Amazingly, the consequent **'protonation'** of the resulting **'enolation'** specifically taking place at one face of the double bond yields **'glucose';** whereas, *protonation* at the other face produces **'mannose'** as depicted below:

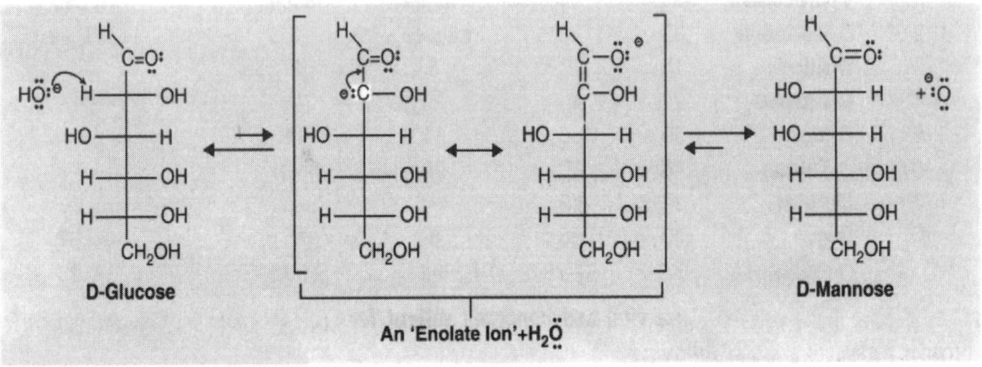

D-Glucose

An 'Enolate Ion'+H₂O̤

D-Mannose

9.2. Formation of Enediol

Importantly, the aforesaid **'enolate ion'** may be specifically protonated on the O-atom to yield an altogether **'new enol'** usually known as an **'enediol'.** Besides, an **enediol** essentially comprises a **hydroxy (–OH) moiety** at **each end of a double bond.** Thus, the **'enediol'** truly obtained from **glucose** being represented simultaneously by *enol* pertaining to:

- *aldoses* **glucose** and **mannose,** and
- *ketose* **fructose.**

The actual conversions of **glucose or mannose,** and **fructose** into their respective **'enediol'** are shown adequately as given under:

(a) Conversion of Glucose or Mannose to Enediol

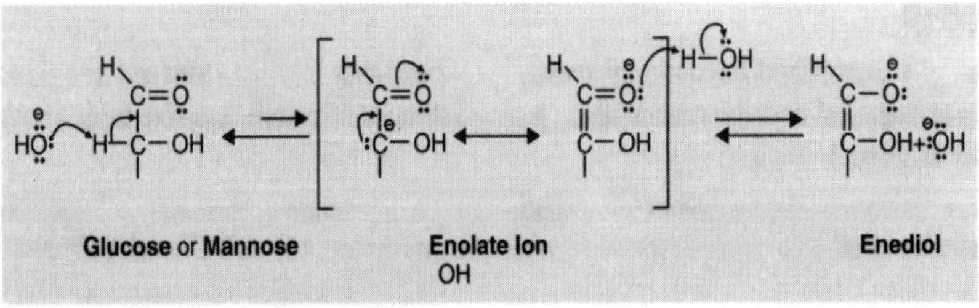

Glucose or **Mannose**   **Enolate Ion**   **Enediol**
OH

Explanations: These essentially comprise:

(1) The negatively charged OH⁻ ion donates a pair of electrons to the H-atom at C-2 of glucose or mannose which eventually shifts the electrons from C–H bond at C-2 to the C–C bond between C-1 and C-2.

(2) Consequently, an 'enolate ion' is formed as an intermediate (as shown on the LHS of square bracket), which in turn undergoes rearrangement to establish a 'double-bond' between C-1 and C-2 and a negative charge upon O-atom.

(3) Further, this negatively charge on O-atom is shared with the water molecule electronically as shown on the RHS of the square bracket.

(4) The enolate ion (in the square bracket) accepts one proton and gets converted duly to the desired 'enediol' plus a hydroxy (–OH) ion. This reaction is reversible.

(b) Conversion of Fructose to Enediol:

Fructose   Enolate Ion   Enediol

Explanations: These invariably include:

(1) The enolate ion (LHS) on reaction with a mole of water retains one H-atom from it to yield largely 'fructose' plus a hydroxyl (–OH) ion.

(2) The enolate ion (RHS) shifts a pair of electrons from the carbonyl O-atom to the C-O linkage; besides, there is also a shift of electron from C-2 to C-1 double bond (as an intermediate).

(3) The intermediate enolate ion, as obtained above, undergoes a reversible reaction to produce the desired 'enediol'.

Therefore, one may safely infer that this kind of a base catalyzed epimerization and aldose-ketose equilibration may not essentially cease at C-2 only.

Example: D-Fructose epimerizes strategically at C-3 upon the prolonged treatment with an appropriate base.

Advantageous Transformations in Metabolism: It has been duly established that quite a few typical transformations of this kind are extremely vital and important in metabolism.

Example: The particular conversion of 'glucose-6-phosphate' into 'fructose-6-phosphate', takes place predominately in the breakdown of glucose i.e., glucolysis involving the series of reactions by which glucose gets adequately utilized by a well-known source of food. Because, most 'biochemical reactions' invariably take place at pH 7; therefore, extremely low quantum of the hydroxide (OH) ion is normally present to catalyze the reaction effectively. Evidently, the said reaction gets duly catalyzed by an enzyme termed as, glucose-6-phosphate isomerase. The reaction is as given below:

Glucose-6-phosphate   Fructose-6-phosphate

10. ETHER AND ESTER DERIVATIVES OF CARBOHYDRATES

Since the 'carbohydrates', in general, do comprise several –OH moieties, it may be duly apprehended that carbohydrates should ordinarily undergo a host of reactions pertaining to alcohols. Thus, one of these reactions is directly concerned with the ether formation. It has been observed that in the presence of a *concentrated base, carbohydrates* invariably get converted into the *ethers* by such *reactive alkylating agents* as:

- dimettryl sulphate [(CH,),SO,],
- benzyl chloride [–Ⓞ–CH,–Cl], and
- methyl iodide [CH,–I].

Following is the reaction between **D-glucose** and **dimethyl sulphate** in the presence of **NaOH** (*i.e.,* a *strong concentrated base*) to yield **methyl 2, 3, 4, 6-tetra-*O*-Methyl-D-glucopyranoside (I)** and *methyl sulphate anion* as given below:

D-Glucose   Dimethyl Sulphate

Methyl 2,3,4,6-tetra-O-
methyl-D-glucopypanoside (I)   Methyl sulphate ion

Note: It may be noticed that the '*ethers*' are usually termed as '*O*-alkyl derivatives of the carates.bohydr

Explanations: The aforesaid examples, in fact, do refer to the **Williamson ether synthesis**, which with a large spectrum of alcohols needs predominantly a '**base**' that should be essentially stronger than the **hydroxyl ion (OH$^{\ominus}$)** to give rise to the formation of the ultimate **conjugate-base alkoxide**.

Besides, they observed to be distinctly more acidic (pKa = 12) *vis-a-vis* those of the **ordinary alcohols.** *

As a result, appreciable concentrations of their **conjugate-base alkoxide ions** are duly generated in **concentrated NaOH**. To accomplish this it is always recommended beneficially to use a **large excess of the reactive alkylating agents** *viz.*, **dimethyl sulphate, benzyl chloride**, and **methyl iodide**; since, the **hydroxide ions [OH$^-$]** itself, present in surmountable quantum, also undergoes reaction with the **alkylating agents** at large.

However, it is pertinent to state here that:

- **practically insignificant**, or
- **absolutely base-catalyzed epimerization free,**

reaction comes into play despite the extremely strong basic environment employed.

---

* One may observe the **higher acidity** by virtue of the carbohydrate hydroxy moieties is indeed attributable to the polar effect of several neighbouring oxygens present in the molecule.

Obviously, under such conditions the *critical alkylation* of the **hydroxyl (–OH) moiety** strategically located at the **anomeric carbon*** is observed to be *much faster* in comparison to **'epimerization'**. Importantly, once this particular **O-atom** gets duly alkylated, evidently **epimerization** may not take place at all.

Reagents Used for Methylethers of Carbohydrates: A few other typical reagents that are commonly used preparing the respective **methylethers of carbohydrates** are:

- combination of **methyl iodide** (CH_3I) with **silver oxide** (Ag_2O), and
- strongly basic **sodium amide** ($NaNH_2$) in liquid **ammonia** (NH_3) followed immediately by CH_3I.

10.1 Alkoxy Moiety at Anomeric Carbon

It is a known and established analogy that the **'alkoxy moiety'** strategically positioned at the **anomeric carbon atom** (*viz.,* **C-1)** is altogether different from the other **alkoxy moieties** in an **'alkylated carbohydrate** by virtue of the fact it essentially forms an integral part of the **'glycosidic linkage'** in an aqueous acidic environment under relatively mild conditions as indicated below:

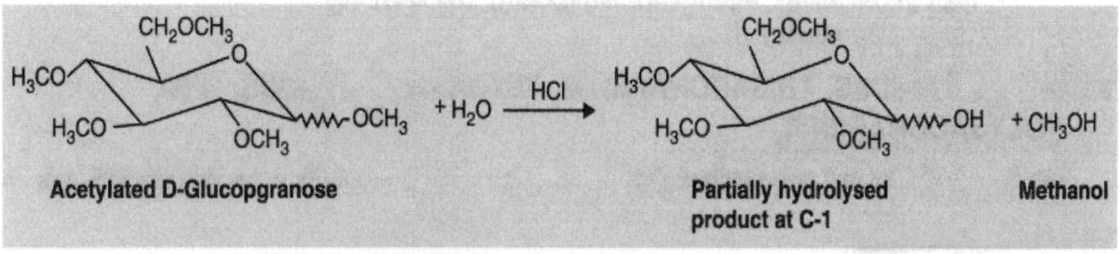

CH2OCH3
H3CO
H3CO
OCH3
OCH3
+ H2O
HCl
CH2OCH3
H3CO
H3CO
OCH3
OH
+ CH3OH
Acetylated D-Glucopgranose
Partially hydrolysed product at C-1
Methanol

Note: The other 'alkoxy moieties' are found to be *ordinary ethers;* and, therefore, fail to hydrolyze under the aforesaid experimental parameters. Actually, they do require *much stronger* experimental conditions for causing the necessary cleavage.

10.2. Esterification of Alcohols

Importantly, another prominent reaction of **alcohols** is *esterification*, whereby the **'hydroxy moieties'** of the *carbohydrates* may be duly *esterified* very much akin to other **alcohols** as illustrated under:

10.3. Saponification of Esterified Carbohydrates

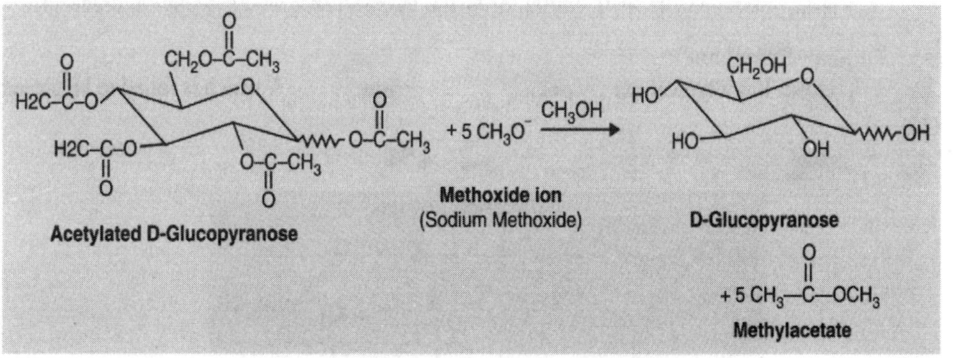

CH2O-C-CH3
H2C-C-O
H2C-C-O
O-C-CH3
O-C-CH3
+ 5 CH3O⁻
CH3OH
Methoxide ion
(Sodium Methoxide)
Acetylated D-Glucopyranose
CH2OH
HO
HO
OH
OH
D-Glucopyranose
+ 5 CH3—C—OCH3
Methylacetate

* The **cyclization** of sugar obviously gives rise to an altogether **'new chiral centre at C-1'** in the **cyclic form**; and, therefore, this carbon is known as the **'anomeric carbon'**.

The **ester derivatives of the carbohydrates** may be subjected to *saponification* in *two* ways:

- **treatment performed in a 'base' (*viz.*, NaOH),** or
- **removed carefully by '*transesterification*' in the presence of an** *alkoxide e.g.*, **methoxide (CH₃ONa).**

Important Points: Following *two* points are fairly important:

(1) One may critically observe that both **ethers** and **esters** are invariably employed for the strategic **protection of various moieties** that are intimately involved in the **carbohydrates.**

(2) Since, the derivatized forms of **ethers** and **esters of carbohydrates** do categorically inherit and possess the following *two* characteristic features, namely:

- **distinctly higher solubility,** and
- **exceptionally greater volatility,**

When compared with the **parent carbohydrates.**

(3) In addition, they find critical usage in the much needed **'characterization of carbohydrates'** by either of the *two* following recognized analytical methods, such as:

- **high-performance liquid chromatography (HPLC),** and
- **mass spectrometry (MS).**

11. DISACCHARIDES, TRISACCHARIDES, TETRASACCHARIDES, AND POLYSACCHARIDES

The **disaccharides** and **polysaccharides** shall be discussed individually in the sections that follows:

11.1. Disaccharides

The *disaccharides* actually represent the most *vital* and the *simplest* **oligosaccharides.** In actual practice, the **'disaccharides'** essentially comprise *two* **monosaccharide units** duly convected by a **glucosidic bondage.** Generally, the **disaccharides** do possess the following remarkable characteristic features, namely:

- sweet palatable in nature,
- water-soluble substances,
- hydrolyzed by enzymes easily,
- hydrolyzed by diluted mineral acids, and
- causes intermolecular dehydration occurring between two monosaccharide molecules.

11.1.1. Typical Examples

[A] (+) Lactose: It is a typical example belonging to a **'disaccharide'**, which is found to be present in both:

- human milk: upto 6-7%, and cow's milk: ~ 4.5%.

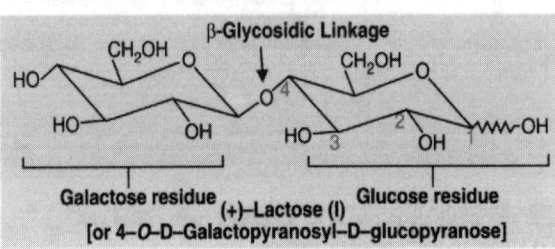

Observations: These essentially include:

(1) In **(+) -lactose**, one may observe that a **D-glucopyranose molecule** is strategically linked by its **O-atom at C-4** to **C-1** of **D-galactopyranose**.

(2) In true sense, **(+)-lactose** happens to be a *'glycoside'*. wherein the *carbohydrate* is **'galactose'**), and the **'alcohol is 'glucose'**.

> **Note:** It may be remembered that the 'glycosidic linkage' proves to be an *'acetal'*; and, therefore, the *'acetals'* usually get hydrolyzed in an *'acidic environment'*.

(3) Interestingly, **(+)-lactose** may be hydrolyzed in an **aqueous acidic medium** to yield one equivalent mole each of **D-glucose** and **D-galactose**, which very much resembles to that of a **'methyl glycoside'** being hydrolyzed to give a respective *carbohydrate* and *methanol*.

Thus, we may have the following reactions:

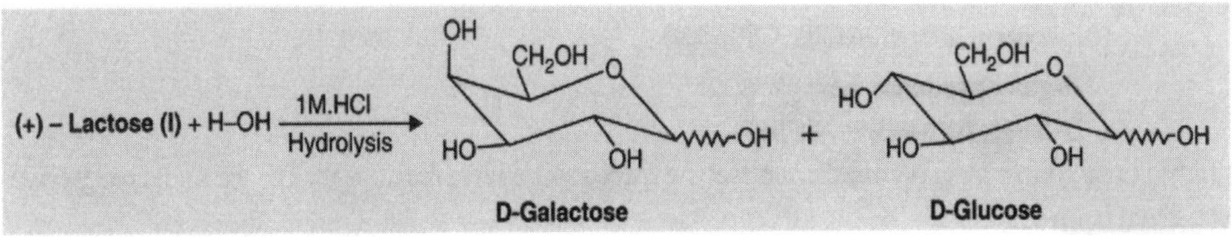

$$(+) - \text{Lactose (I)} + \text{H–OH} \xrightarrow[\text{Hydrolysis}]{\text{1M.HCl}}$$

D-Galactose   **D-Glucose**

Compared to Methyl Glycoside

Compared to Methyl Glycoside

$$+ \text{H–OH} \xrightarrow[\text{Hydrolysis}]{\text{1M.HCl}}$$

Methyl Glycoside   **D-Galactose**   **Methanol**

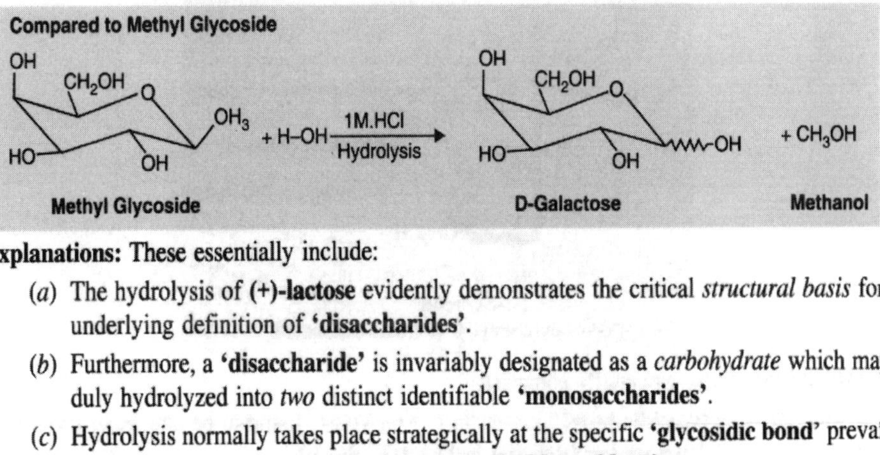

Explanations: These essentially include:

(a) The hydrolysis of **(+)-lactose** evidently demonstrates the critical *structural basis* for the underlying definition of **'disaccharides'**.

(b) Furthermore, a **'disaccharide'** is invariably designated as a *carbohydrate* which may be duly hydrolyzed into *two* distinct identifiable **'monosaccharides'**.

(c) Hydrolysis normally takes place strategically at the specific **'glycosidic bond'** prevailing predominantly between the *two* **'monosaccharide residues'**.

(4) **Streochemistry* of Glycosidic Bond (+)-Lactose:** It has been duly established that the ensuing **'stereochemistry of glycosidic bond'** present in **(+)-lactose** has a β-*orientation*. In other words, the *stereochemistry* of **O-atom** joining critically the **two monosaccharide residues** in the **glycosidic linkage** very much corresponds to that in the **β-anomer of D-galactopyranose**.

(5) **Concept of Reducing Sugar:** Since, **C-1** of the *galactose residue* in **(+)-lactose** is intimately engaged in the formation of a **glycosidic bondage'**, it *cannot be oxidized* at all. Nevertheless,

---

* **Stereochemistry:** In fact, the *'stereochemistry'* exerts a great importance in *biology, due* to the reason that the **higher animals** do possess an *enzyme*, β-**galaetosidase**, which categorically helps in the catalysis of hydrolysis of the ensuing β-glycosidic linkage very close to the neutral pH. Besides, it permits **(+)-lactose** to serve as a **'source of glucose'**.

C-1 of the *glucose residue* essentially forms a part of a **'hemiacetal moiety'***, is observed to be in *equilibrium* with the **'free aldehydic function'**; and, therefore, capable of undergoing typical characteristic **aldehydic reactions.**

In this manner, the treatment of the **carbohydrates** *viz.,* **(+)-lactose** which may be subjected to oxidation with Br_2-water affords, reaction with **'glucose residue'**; in this manner are usually known as the **'reducing sugars'.** Thus, the **'glucose residue'**; of the ensuing **disaccharide**; whereas, the ensuing **'galactose residue'** at the **'non reducing terminal'.**

> Note: Since, *(+)-lactose* does possess a *'hemiacetal moiety'*, it also undergoes a host of other reactions pertaining to the *'aldose hemiacetals'*, for instance: *maturational.*

[B] (+)-Sucrose: The so called *'table sugar' i.e.,* **(+)-sucrose** is widely recognized as another most important **disaccharide.** An approximately **120 million tonnes of sucrose** is produced duly across the globe annually.

The **(+)-sucrose** predominantly comprises:

- **D-glucopyranose residue**, and
- **D-fructofuranose residue**,

linked to each other by a *'glycosidic bond'* at the specific **anomeric carbons** of both the **monosaccharides,** as given below:

(+)-Sucrose
[or α–DGlucopyranosyl–β–D–fructofuranocide

Explanations: These essentially comprise:

(1) The critical **'glycosidic bond'** present in *(+)-sucrose* is found to be altogether different from the one in **(+)-fructose** [as discussed in 11.1.1(1) above].

(2) In this particular instance *viz.,* **(+)-lactose**, only one of the *lactose residues viz.,* the **'galactose residue'** contains specifically an **'acetal carbon'** (*i.e.,* a *glycosidic carbon*).

(3) Contrarily, in **(+)-sucrose**, the *glycosidic bond* crucially forms a **'bridge'** between **C-2** belonging to the *'fructofuranose residue'*, and **C-1** of the **'glucopyranose residue'**. In fact, they are the **carbonyl carbons (-C-)** duly present in the **noncyclic forms** of the *individual monosaccharides.***

---

* It resembles very much akin to the **hemiaetal moiety** of the **monosaccharides.**

** It may be recalled that the **carbonyl ($\overset{O}{\overset{\|}{C}}$) carbons** invariably give rise to the **acetal** or **hemiacetal carbons** in the cyclic forms.

Thus, we may have:

From the above explicit structure it may be inferred that the **'free hemiacetal moiety'** present in **sucrose** is:

- *neither* in **'fructose'**, and
- *nor* in **'glucose'**.

Therefore, **(+)-sucrose** has an *advantageous natural features*, namely:

- fails to undergo oxidation by **'Bromine-water'**, and
- fails to be subjected to **'Mutarotation'**.

Importantly, all such carbohydrates that critically meet the above *two* requirements usually fall into the category of **'nonreducing sugars'**.

(4) **Invert Sugar**: It has been duly observed that very much akin to several other **'glycosides'**, the *(+)-sucrose* may be effectively hydrolyzed to its various recognized **monosaccharides**. Nevertheless, in a broader perspective the **(+)-sucrose** gets hydrolyzed by either:

- an enzyme known as **'invertase'**, or
- an aqueous **mineral acid** (*viz.*, **HCl**),

produces an equimolar mixture of **D-glucose** as well as **D-fructose**. The resulting mixture is occasionally termed as **invert sugar**.

Example: Following are *two* typical examples:

(a) *Levulose:* It refers to the *negative rotation* which is duly caused by the **strongly -ve rotation of fructose** that stands at $-92°.mL.g^{-1}.dm^{-1}$.

(b) *Dextrose:* This refers to the *positive rotation* which is duly caused by the corresponding +ve **rotation of glucose** that comes out to be $+52.7°.mL.g^{-1}.dm^{-1}$.

Notes: (1) It may be carefully observed that the -ve rotation (of *'levulose'*) shows an apparent greater magnitude *vis-a-vis* the +ve rotation (of *'dextrose'*) *i.e.*, a difference of almost *40°*.

(2) Fructose is found to be *twice* as sweet as sucrose, which is why *'natural honey'* has a distinct intense sweetness due to the presence of *'invert sugar'*.

Constitution

The various steps involved in establishing the constitution of **(+)-sucrose** are as enumerated under:

(1) The molecular formula of **sucrose** is $C_{12}H_{22}O_{11}$.

(2) **Hydrolysis of (+)-Sucrose:** The **(+)-sucrose** upon critical hydrolysis with *enzymes* or *dilute mineral acids* (*e.g.*, **HCl**) yields almost equimolar quantum of **D-glucose** and **D-fructose,** that perhaps is solely responsible in constituting the *two* said **monosaccharide** units of *sucrose*.

(3) **Reactions with Fehling's Solution or Phenylhydrazine:** **(+)-Sucrose** neither causes reduction with **Fehling's solution*** nor reacts with **phenylhydrazine** which indicate evidently that

---

* Morrison RT and Boyd RN : **Organic Chemistry**, Prentice-Hall of India, New Delhi, 6th edn., 1997.

both the **monosaccharides** (*viz.*, **D-Glucose** and **D-Fructose**) are duly engaged in establishing the **glycosidic linkage**. It also ascertains the fact that the said linkage comes into being duly to:

- the *C-atom* C-1 (*i.e.*,) present in **D-Glucose**, and
- the *C-atom* C-2 (*i.e.*,) present in **D-Fructose**.

(4) Evidential Proof of α-D-Glucose unit and β-D-Fructofuranose

The **evidential proof** for the formation of the aforesaid *two* compounds may be expatiated as given under:

(*a*) **α-D-Glucose Unit:** It is obtained by the hydrolysis of **(+)-sucrose** by the enzyme *maltase*, and not by *emulsion* that obviously shows the particular existence of an **α-D-glucose unit.**

(*b*) **β-D-Fructofuranose Unit:** Contrarily, **(+)-sucrose** gets hydrolyzed by an enzyme *takainvertase* which seems to be critically specific for the glycoside **β-fructo furanosides**, thereby suggesting that a **β-D-fructofuranose** unit is duly present in **(+)-sucrose**, as shown in the structure of **(+)-sucrose** earlier.

It may be observed that the aforesaid *stereochemistry* aspects of *(+)-sucrose* have been duly ascertained by **Hudson's Isoprotation Rule.**

In 1909, Hudson CS (US Public Health Service) put forward the following remarkable proposals, namely:

- **In the *D-series*, the more dextrorotatory member of an α, β pair of anomers is to be named α-D; whereas, the other one being named as β-D, and**
- **In the *L-series*, the more levorotatory member of such a pair is assigned the name α-L, and the other being termed as β-L.**

Therefore, the 'enantiomer' of a-D-(+)-glucose is a-L-(–)-glucose.

In addition, the hydroxyl (–OH) or methoxy (–OCH$_3$) moiety at C-1 is strategically located on the right in an **α-D-anomer**, and on the left in a **β-D-anomer**, as depicted under for the *'aldohexoses'* (*viz.*, one may not that the *"right"* indicates *"down"* in the structure given below.*

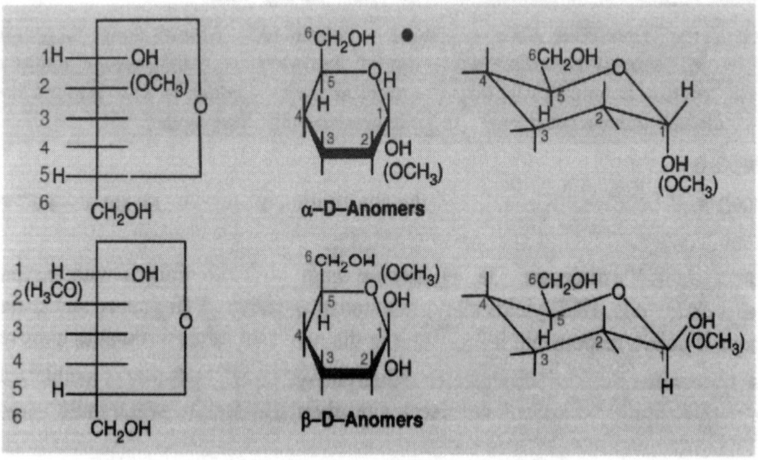

α–D–Anomers

β–D–Anomers

Observation of Hudson's Proposals: These essentially comprise:

(1) They are actually based upon certain apparent but *unestablished relationships* existing be-tween **ensuing configuration** and **optical rotation**. Incidentally, all the evidences critically indicates that the *assigned configurations* opt out to be the **correct ones**.

Example: It has been proved beyond any reasonable doubt that both α-D-glucose and methyl α-D-glucoside essentially do have the **same configuration**, as occurs in **β-D-glucose** and **methyl β-D-glucoside**.

Evidence: The *enzymatic hydrolysis* of **methyl α-D-glucoside** liberates predominantly:

- an excess quantum of the more highly rotating **α-D-glucose**, and
- an initial liberation of **β-D-glucose** from the hydrolysis of **methyl β-D-glucoside**.

(2) Importantly, the configuration occurring at **C-1** is virtually the same in the **methyl α-glyco-sides** of all the D-aldohexoses.

Evidence: In actual practice, they all more or less yield the same compound on being subjected to oxidation by the **hydroiodic acid [HIO₄]**.

(5) **Determination of the size of Glucose and Fructose Units:** The crucial determination of the size of 'glucose unit' and 'fructose unit' is usually ascertained by carrying out:

- **complete methylation of (+)-sucrose using dimethyl sulphate in an alkaline environ-ment (NaOH solution), and**
- **hydrolysis in an aqueous acidic medium to obtain 2,3,4,6-tetra O-methyl-D-glucose (I), and 1,3,4,6-tetra-O-methyl-D-fructose (II).**

Amazingly, the structure of the above *two* resulting products *i.e.,* (*I*) and (*II*) suggests explicitly that the so called *'glucose residue'* duly present in (+)-**sucrose** happens to be a **'glucopyranose moiety'**; while, the *'fructose residue'* is normally present in the form of **'fructopyranose moiety'** as given under:

Interestingly, (+)-**sucrose** may be represented by the following *three* different types of duly recognized and proposed structures, such as:

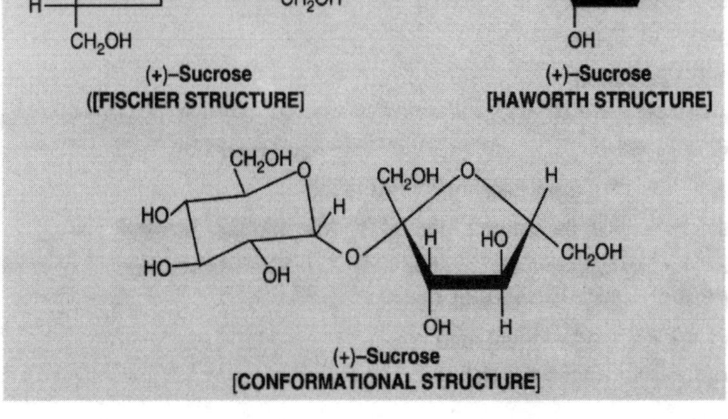

(+)–Sucrose
([FISCHER STRUCTURE]

(+)–Sucrose
[HAWORTH STRUCTURE]

(+)–Sucrose
[CONFORMATIONAL STRUCTURE]

- **Fischer structure,**
- **Haworth structure, and**
- **Conformational structure.**

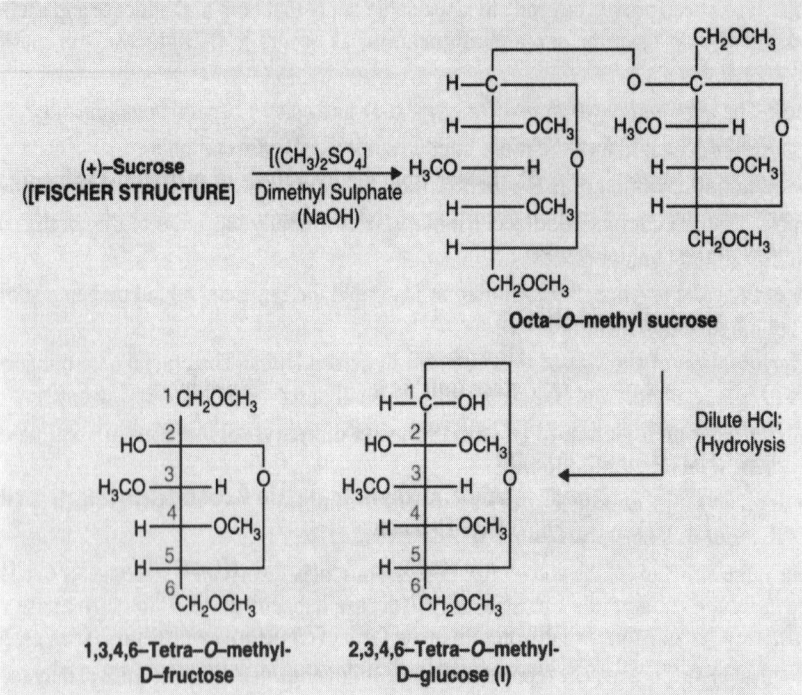

1,3,4,6–Tetra–O–methyl-D–fructose

2,3,4,6–Tetra–O–methyl-D–glucose (I)

The following reactions vividly show the conversions of:

- **(+)-sucrose** into octa-O-methyl sucrose, and
- **hydrolysis** with dilute mineral acid to yield the products (I) and (II) respectively.

(6) Confirmation of (+)-Sucrose Structure

The aforesaid *Fischer structure* for (+)-sucrose has been duly confirmed by a plethora of physical and chemical logistic evidences and supportive acceptable reasonings, namely:

- periodic acid [H_5IO_6] oxidation, and
- X-Ray Diffraction analysis.

Periodic Acid Oxidation

In fact, the **periodic acid (H_5IO_6) oxidation** actually ascertains the precise and exact structure of (+)-sucrose with particular references to the following vital aspects solely:

- size of the **'two monosaccharide unit'**, and
- fails to establish the nature of the **'glycosidic bondage'** exceptionally.

Thus, (+)-**sucrose (A)** reacts gainfully with five moles of H_5IO_6 and gives rise to the formation of one mole each of a **tetraaldehyde (B)** and a **formic acid**. Consequently, the latter upon treatment with:

- bromine-water (oxidation), and
- hydrolysis with aqueous HCl,

yields *one mole* of **glyoxylic acid (C)**, *two moles* of **glyceric acid (D)**, and *one mole* of **hydroxy pyruvic acid (E)** respectively as shown under:

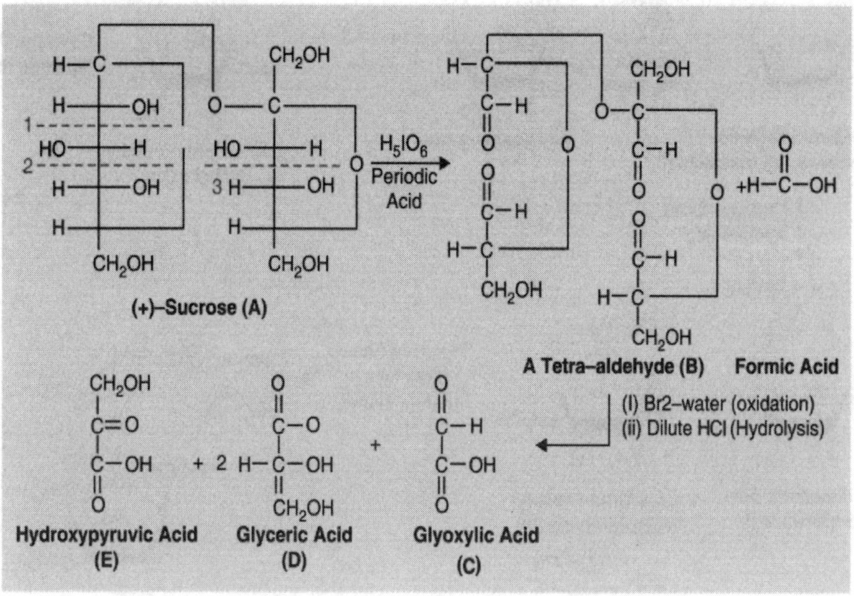

(+)-Sucrose (A)

A Tetra–aldehyde (B)   Formic Acid

(I) Br2–water (oxidation)
(ii) Dilute HCl (Hydrolysis)

Hydroxypyruvic Acid (E)   Glyceric Acid (D)   Glyoxylic Acid (C)

X-Ray Diffraction Analysis

Beavers *et al.* (1947)* first and foremost carried out the **X-ray diffraction analysis** of *sucrose sodium bromide dihydrate*; and eventually confirmed the stereochemical configuration determined earlier chemically along with the **5-membered ring structure of fructose**.

Synthesis

Lemieux (1956)* was pioneer in synthesizing **sucrose** by adopting the following steps in a sequential manner as stated under:

(1) The intermediate **1, 2-anhydro-α-D-glucopyroanose-3, 4, 6-triacetate (I)** is prepared by the interaction of **2,3,4, 6-tetra-O-acetyl-β-D-glucose** with PCl$_5$ to obtain **1-chloro-2,3,4,6-tetra-O-acetyl-β-D-glucose**. The latter product (*i.e.,* the **chloro derivative**) is now treated carefully with **ammonia in ether** followed by **ammonia in benzene** to obtain (**I**).

(2) The **epoxide ring** between C-1 and C-2 in (I) is cleaved meticulously to form an **acetonim ion** linking C-6 and C-1, thereby resulting into the formation of **α-D-glucopyronose-3,4,6-triacetate (II)**.**

(3) The resulting compound (II) on being reacted with **1,3,4,6-tetra-O-acetyl-D-fructofuranose (III)** in a *sealed tube* and heated upto **100°C** for a long duration of **104 hrs** to obtain the desired **sucrose hepta-acetate (IV)**.

* Beavers *et al.* (1947) : In : **Sucrose-Properties and Applications** (eds) Mathlonthi M and Reiser P [Good bookshooping.com]; & *Proc.Roy Soc., 190A;* 257, 1947.

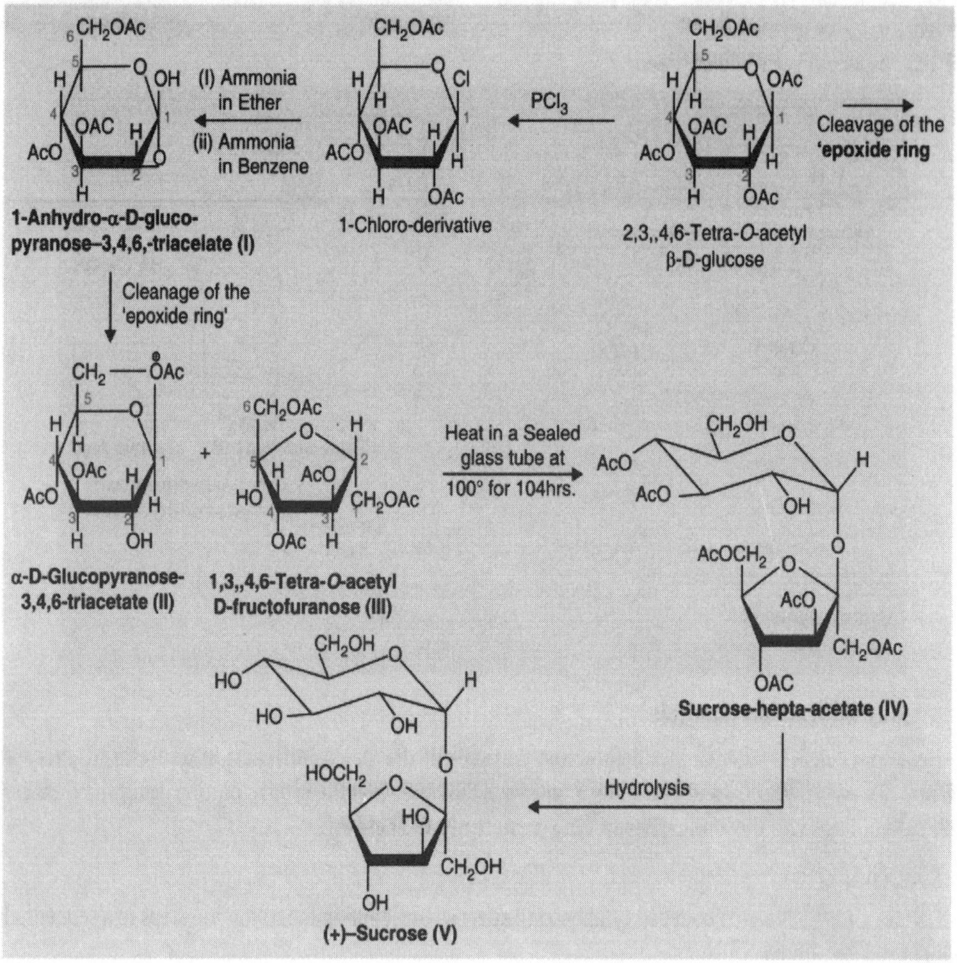

1-Anhydro-α-D-gluco-pyranose–3,4,6,-triacelate (I)

1-Chloro-derivative

2,3,,4,6-Tetra-O-acetyl β-D-glucose

α-D-Glucopyranose-3,4,6-triacetate (II)

1,3,,4,6-Tetra-O-acetyl D-fructofuranose (III)

Sucrose-hepta-acetate (IV)

(+)–Sucrose (V)

(4) The above **ester (IV)** on being subjected to *critical hydrolysis* gives rise to the **(+)-sucrose (V)**. All the reactions cited above from (1) through (4) are as stated above.

Inversion of Sucrose

It is well-known that **(+)-sucrose** is a **disaccharide** which essentially comprise:

- a *'unit of glucose'* (usually in the **acetal form**), and
- a *'unit of fructose'* (invariably in the **ketal form**),

that is strategically, linked *via* **C-1 of glucose**, and **C-2 of fructose** *viz., a 1-2′ linkage.*

Importantly, in **(+)-sucrose**, one may crucially take cognizance of the fact that neither **'glucose'** nor **'fructose'** can actually exist in an **'open-chain form'** due to the obvious formation of both **'acetal'** and **'ketal'** as depicted under:

---

* Lemieux H, *J Am Chem Soc.*, **78**:4117, 1956..

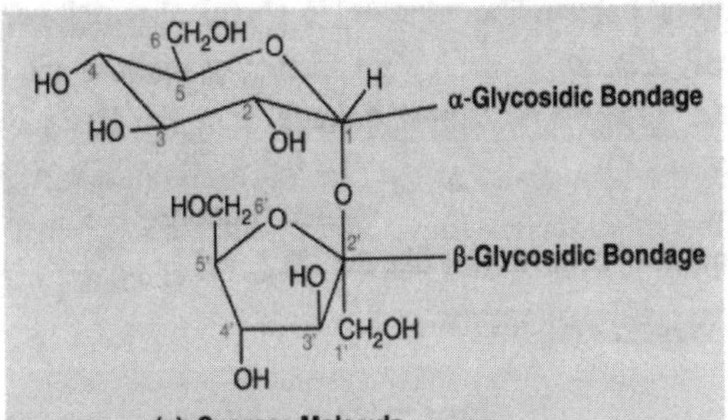

(+)-Sucrose Molecule

Nevertheless, it may be observed that:

- (+)-**sucrose** never represent a '**reducing sugar**',
- (+)-**sucrose** fails to exhibit '**mutarotation**',
- *Specific rotation* $[\alpha]_D$ **of sucrose** stands at **+66°**, and
- ultimate hydrolysis of (+)-**sucrose** gives **glucose** and **fructose** having **specific rotations** $[\alpha]_D$ to be **+52.5°** and **–92°** respectively, thereby rendering the overall resulting mixture to be **laevorotatory (–)**.

Based on the above critical **phenomenon of sucrose** is usually termed as the '**inversion of sugar**'; whereas, the ensuing **mixture of monosaccharides** is largely known as '*invert sugar*'*

[C] Maltose

Maltose (I) is a *disaccharide*, composed of *two* separate units of '**glucose**' which are strategically linked with **a-linkage** between **C-1 of one unit** and **C-4 of the other unit** through an **O-atom**.

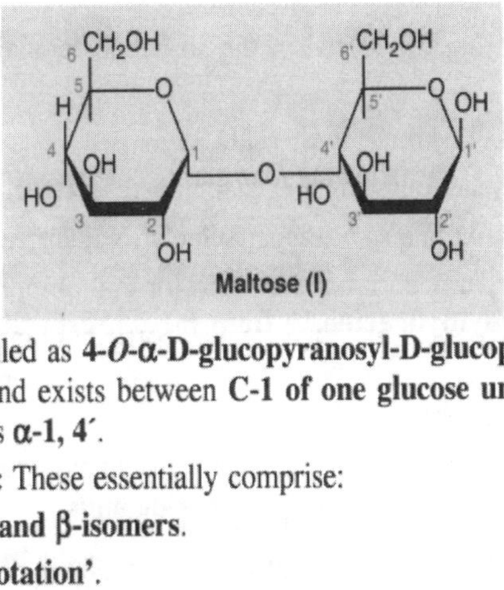

Maltose (I)

It may be **chemically** called as **4-*O*-α-D-glucopyranosyl-D-glucopyranose**; in '**maltose**', as the prevailing '*linkage*' is '**α**', and exists between **C-1 of one glucose unit** and **C-4' of the other**, the ensuing '**linkage**' is termed as **α-1, 4'**.

Characteristic Features: These essentially comprise:

(1) **Matose** occurs as **α-and β-isomers**.

(2) It undergoes '**mutarotation**'.

(3) **Maltose** reacts with **Benedict's** and **Fehling's Reagents** to form *red precipitate* of **Cu_2O** showing the characteristics of reducing sugars.

(4) It also reacts with **phenylhydrazine** to give rise to the production of the *typical* '**phenyl osazone**'.

---

* It is the essential component of '**natural honey**'; and hence, it is always **sweeter** in comparison to '**sucrose**'

(5) Why does 'Maltose' behave like 'glucose' in chemical reactions?

One may have a closer look at the following structure of **maltose** and observe critically that the **LHS-glucose** prominently possesses an **'acetal linkage'** (*i.e.,* the **glycosidic linkage**); whereas, the **RHS-glucose** still possesses the **hemiacetal at C-1′**. It has been duly ascertained that **glucose** may invariably exist in an **equilibrium of the ensuing α-and β-anomers**, as well as the **open-chain form**. Therefore, this is why **'maltose'** behaves just like the **'glucose'** in chemical reactions.

Thus, we may have **'maltose'** as follows:

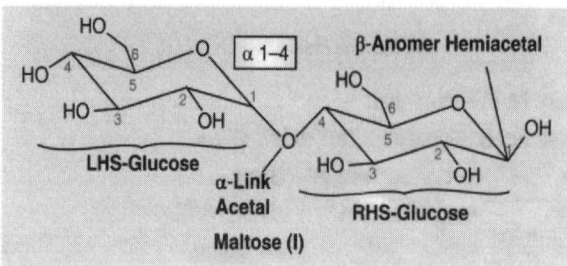

Maltose (I)

Occurrence

Malt essentially consists of the grain of the cereal **'Barley'**, *Hordeum distichon (Family : Gramineae)*, which as been **partially germinated*** and dried. **Maltose** is the *major carbohydrate* of **malt** and **malt extract**, which is employed profusely in making **'malted-milk food'** (*viz.* **Ovaltine, Bournvita, Milo**) and in the pharmaceutical preparations for the administration of **'cod-liver oil'**, and **'iron-vitamin tones'**.

Maltose is also known as a **'malt-sugar'** duly obtained by the action of the *enzyme* **'diastase'** upon the starch content (in *Barley*).

Constitution: The various steps involved in the so called **'constitution'** of **Maltose** are as stated under:

(1) The molecular formula of **Maltose** is $C_{12}H_{22}O_{11}$.

(2) Since, **maltose** exhibits the following *four* glaring and spectacular properties, namely:

- causes reduction of **Fehling's solution, Tollen's Reagent**, and **Benedict's solution,**
- gets oxidized by **Br$_2$-water** to **maltonic acid (or maltobionic acid),**
- reacts with **phenylhydrazine** to form the corresponding **osazone analogue (mp, 206°C)**, and
- exhibits **'mutarotation'**.

Thus, all these aforesaid facts suggest predominantly, that maltose must possess a **'latent aldehydic (–CHO) moiety'** duly present in one of its **monosaccharide units.**

(3) **Hydrolysis of Maltose:** Interestingly, **'maltose'** gets usually **hydrolyzed** either by the *enzyme maltose* or *emulsin* thereby indicating that the ensuing bondage joining the **reducing half of the maltose** to the corresponding **non-reducing half of the maltose** is linked by **α 1-4′** (as shown in **'Maltose'** earlier).

---

* The **'germination'** of water-soaked barley grains is carried out at a temperature of 20±2°C with a circulation of air, maintenance of humidity, and in the presence of **giberellic acid** (a *plant-growth hoimone*) upto 96 hrs.

Amazingly, **maltose** when hydrolyzed in an **aqueous acidic environment**, it categorically gives rise to the formation of **D-glucose** exclusively thereby ascertaining the fact that the former exclusively contains **'glucose'** as the **monosaccharide present**.

(4) **Determination of 'Ring Sizes of Monosaccharide Units Present in Maltose:** The meticulous determination of the precise and exact *ring sizes* of the various **monosaccharide units** that are essentially present in the **'maltose'** molecule may be accomplished gainfully by adopting the following *three* steps in a sequential manner:

(*a*) **Oxidation with Br$_2$-water:** It necessarily helps **maltose** (I) to undergo oxidation to **maltonic acid (II)** (or **malto bionic acid**), thereby suggesting that the so called **'latent aldehydic function** gets duly oxidized to the corresponding **carboxylic (–COOH) function**.

(*b*) **Methylation of (II) with Dimethyl Sulphate to Methyl Ester of Octa-*O*-Methyl Maltonic Acid (III):** The intermediate compound (III) is invariably accomplished the **complete methylation** of **maltonic acid (II)** using *dimethyl sulphate via* the formation of these *two* intermediary chemical entities, namely:

- **2,3,4,6-tetra-*O*-methyl-D-glucopyranose (IV)** duly obtained from the **'glucoside segment'** in (II), and

- **2, 3, 5, 6-tetra-*O*-methyl gluconic acid (V)** duly derived from the **'alcohol segment'** in (II).

(*c*) The strategical presence of **'free hydroxyl moieties'** in the aforesaid products at positions **C-1** and **C-4** respectively suggest overwhelmingly, that the desired **'glycosidic bondage'** present critically in **'maltose'** actually involved these *two* **specified C-atoms;** besides, the **'original ring sizes'** in **maltose** were those of **'pyranose'** only.

Based on the above logical and self-explanatory statements one may safely arrive at the structure of **maltose (I)** that eventually expatiates all the above degradations:

(5) **C-4′ of the Reducing Half (in RHS-Glucose) of Maltose (I) Intimately Involved in Glycosidic Linkage:** The earlier statement of facts affirms legitimately that:

- **maltose** is a **'reducing sugar';** and hence, **C-1** is *free*, and

- **Maltose** also forms an **'osazone'** (with *phenylhydrazine*); and, therefore, **C-2** remains *free*.

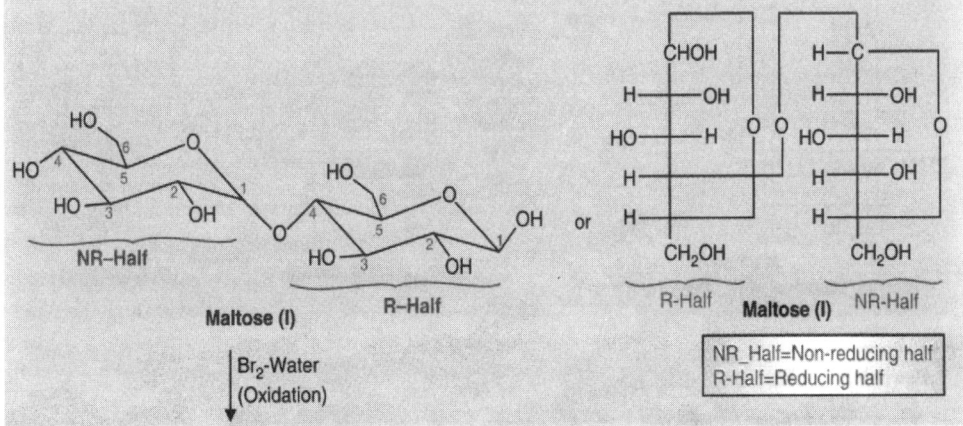

Maltose (I)

Br$_2$-Water (Oxidation)

NR_Half=Non-reducing half
R-Half=Reducing half

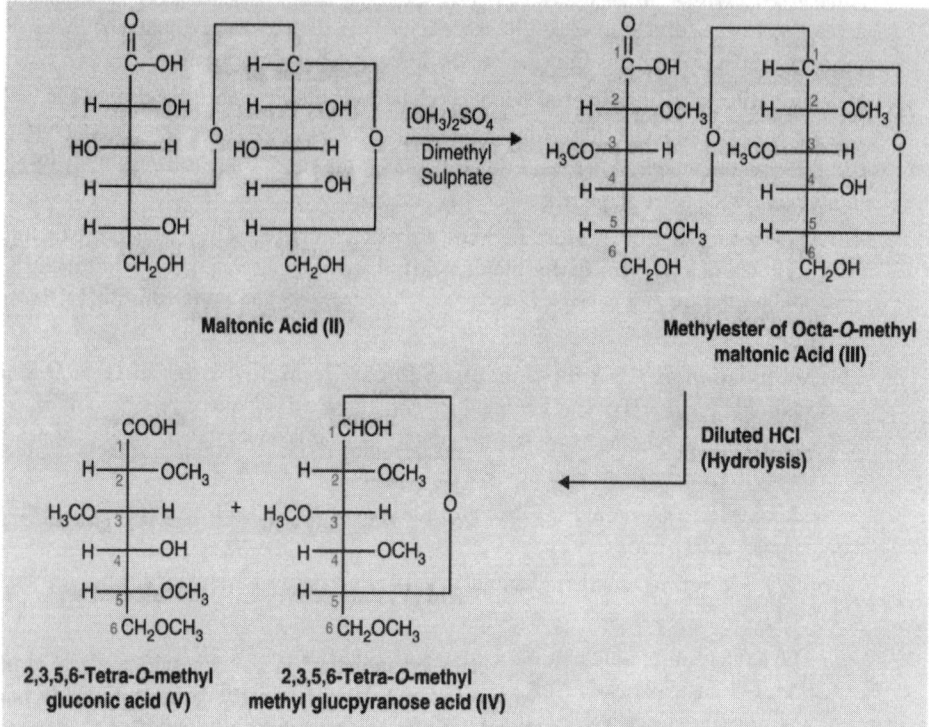

Maltonic Acid (II)

Methylester of Octa-O-methyl maltonic Acid (III)

Diluted HCl (Hydrolysis)

2,3,5,6-Tetra-O-methyl gluconic acid (V)   2,3,5,6-Tetra-O-methyl methyl glucpyranose acid (IV)

Haworth (1926)* carefully carried out the degradation of **maltose** by only one **C-atom** by the help of **Wohl's method,**** and observed critically that the 'resulting product' could again produce an 'osazone', which suggests prominently that **C-3** is *free* distinctly.

Obviously, Haworth again made a futile attempt to carry out the degradation of the above 'resulting product' but *never* obtained an **osazone** that promulgates categorically **C-4 in maltose** is *not free*; and, therefore, suggests **C-4** being engaged absolutely in the so called 'glycosidic linkage'.

Thus, we may have the following reactions:

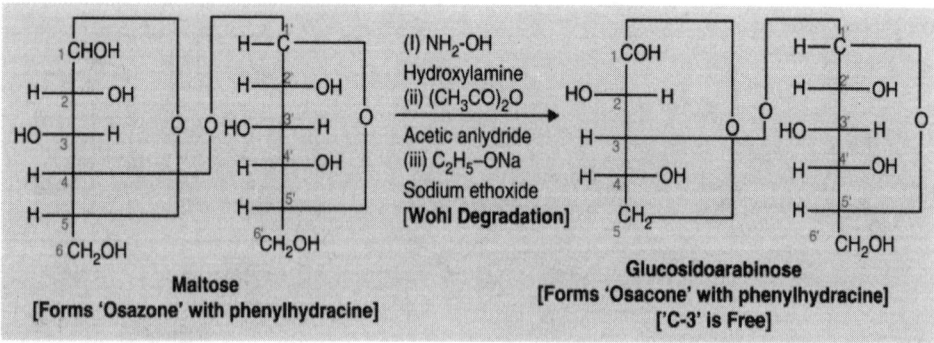

Maltose
[Forms 'Osazone' with phenylhydracine]

Glucosidoarabinose
[Forms 'Osacone' with phenylhydracine]
['C-3' is Free]

---

* Haworth P, : *J Chem Soc.*, 3094, 1926.

** Loudon GM : *Organic Chemistry*, Oxford University Press, New York, 4th ed. 1350, 2002.

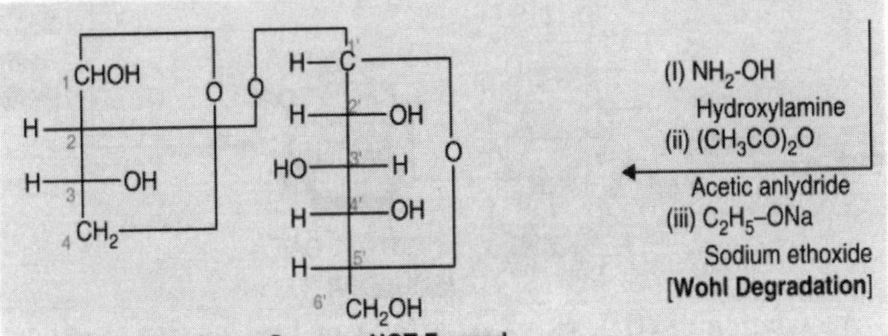

Glucosidoerythrose Osazone NOT Formed
[C-4 is NOT Free]

(I) NH$_2$-OH
Hydroxylamine
(ii) (CH$_3$CO)$_2$O
Acetic anlydride
(iii) C$_2$H$_5$-ONa
Sodium ethoxide
[Wohl Degradation]

(6) **Maltose Exists in α- and β-Anomers:** Conclusively, it may be observed that the critical presence of a 'free –CHOH moiety at C-1' in **maltose (I)** enables eridently to exist in the present *two* distinct forms, namely:

- **α-anomer of maltose**, and
- **β-anomer of maltose**,

as depicted under (C) Maltose (section-4).

Importantly, the very existence of *two* **isomeric maltose** explains eventually the crucial *phenomenon of mutarotation* whose observed values are as stated under:

Specific Rotation [α]$_D$		Values
α-Maltose	:	+168°,
α-Maltose	:	+112°, and
Equilibrium Mixture	:	+136°.

(7) **Synthesis:** Khan (1981)* established the crystal and molecular structure of **maltose**. However, Lemient *et al.* (1953) first and foremost carried out the synthesis of **maltose** almost in the same manner as of **sucrose** [under section (B)].

The various steps that essentially involved in the synthesis of '**maltose**' are as enumerated under:

(*a*) The starting materials for the synthesis are, namely:

- **1,2-Anhydro-3, 4, 6-tri-*O*-acetyl-α-D-glucopyronose (I)**, and
- **1, 2, 3, 6-Tetra-*O*-acetyl-β-D-glucose (II)**,

which are carefully heated together at 120°C for a duration of 12-14 hrs. To obtain the **Heptaacetyl-β-D-maltose (III)**.

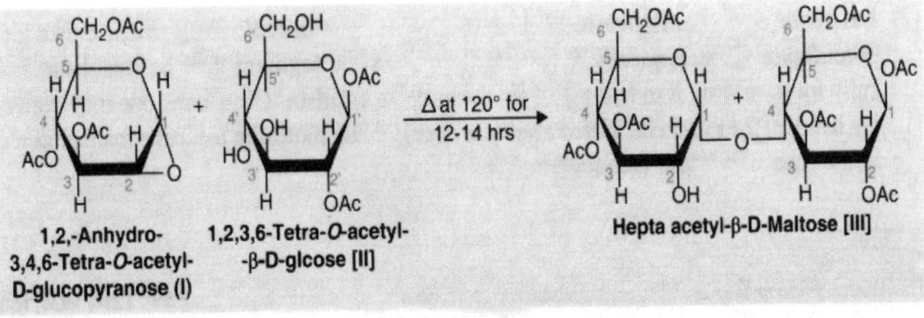

1,2,-Anhydro-
3,4,6-Tetra-*O*-acetyl-
D-glucopyranose (I)

1,2,3,6-Tetra-*O*-acetyl-
-β-D-glcose [II]

Δ at 120° for
12-14 hrs

Hepta acetyl-β-D-Maltose [III]

---

* Khan R : *Carbohyder Chem Biochem.*, **39** : 213-278, 1981.

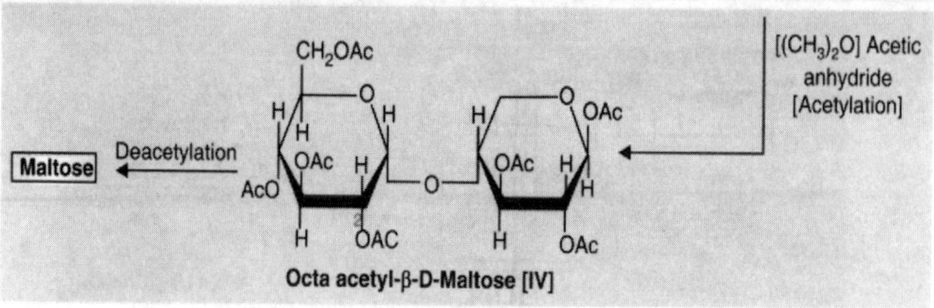

Octa acetyl-β-D-Maltose [IV]

(b) **Acetylation of (III) with Acetic Anhydride:** This step specifically acetylates the hydroxyl (–OH) moiety at C-2 to yield **octa-acetyl-β-D-maltose (IV)**.

(c) **Deacetylation of (IV) yields Maltose (I):**

The various reactions stated above from (a) through (c) may be expressed shown above.

[D] Cellobiose

It has been observed that 'cellobiose', just like *maltose*, is composed of *two* 'units of glucose'; however, the prevailing **1-4´-linkage** is found to be β, instead of α. Hence, **cellobiose** can be chemically termed as **4-O-β-D-glucopyranosyl-D-glucopyranose** as given below:

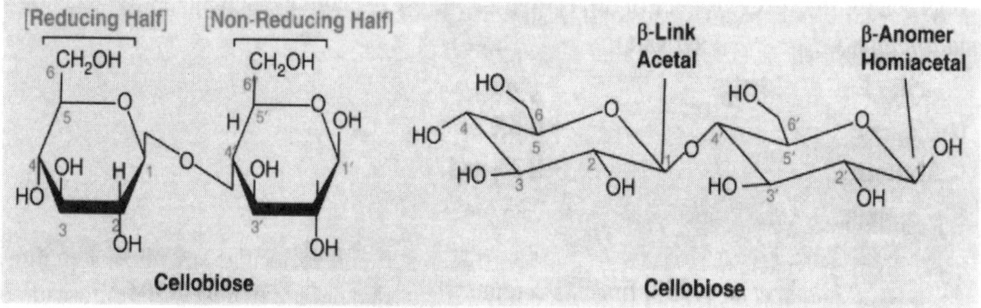

Cellobiose   Cellobiose

Interestingly, **cellobiose** is regarded to be the well recognized '**building block of cellulose**' *i.e.*, a *polysaccharide*. It is, therefore, worthwhile to state here that **cellobiose** does not occur as such in **nature or as glucoside**, but it is duly formed as an intermediate product in the course of *hydrolysis* of the former in an *acidic* environment. Besides, **cellobiose** may not be obtained in perfect pure *crystalline form* by the careful hydrolysis of *cellulose*.

Constitution: The various cardinal steps that are essentially involved in the **constitution of cellobiose** are as stated below:

(1) Its molecular formula is $C_{12}H_{22}O_{11}$.

(2) **Existence of β-Configuration at the 'Glycosidic Segment':** The critical existence of the β-configuration strategically located at the **glycosidic segment** may be established emphatically by subjecting it to *hydrolysis* by the enzyme **emulsin**. Obviously, one may conveniently lay hand on the *exact elucidated structure of cellobiose* almost in the same manner as was done in the case of **maltose** [see *section (C)*].*

---

* Haworth H: *J Chem Soc,* **119** : 193, 1923; Charlton *et al.* : *ibid* : 89, 1926; Zemplen, *Ber,* **59** : 1254, 1926; Haworth *et al.* : *J Chem. Soc.,* 2809, 1927; Peterson, Spencer, *J Am Chem Soc.,* **49**, 2822, 1927; Helferich *et al. Ber.* **63** : 992, 1930.

(3) A close look at the structure of **cellobiose** reveals that the C-1 of the **reducing half-portion** of the same predominately gives rise to critical existence in the following *two* forms:

- **anomeric α,** and
- **anomeric β,**

which essentially has the values of the specific rotation $[\alpha]_D$ as:

α-**Anomeric Form** : **+72°;**
β-**Anomeric Form** : **+16°;** and
Equilibrium Mixture : **+35°.**

(4) **Synthesis**: The synthesis of **cellobiose** was duly put forward by Haskins *et al.* (1942)**. However, a rather simple and unambiguous synthesis essentially comprise the following steps, namely:

- Interaction between **2,3,4,6-Tetra-*O*-acetyl-α-D-glucopyranosyl bromide (I)** [also known as '**acetobromoglucose**'], and **1,2,3,6-tetra-*O*-acetyl-β-D-glucopyranoside (II)** in the presence of Na-metal gives rise to the formation of the corresponding **cellobiose acetate (III)**.
- Hydrolysis of the resulting product (III) *i.e.*, **deacetylation** yields the desired product **cellobiose (IV)**.

Thus, we may have these reactions:

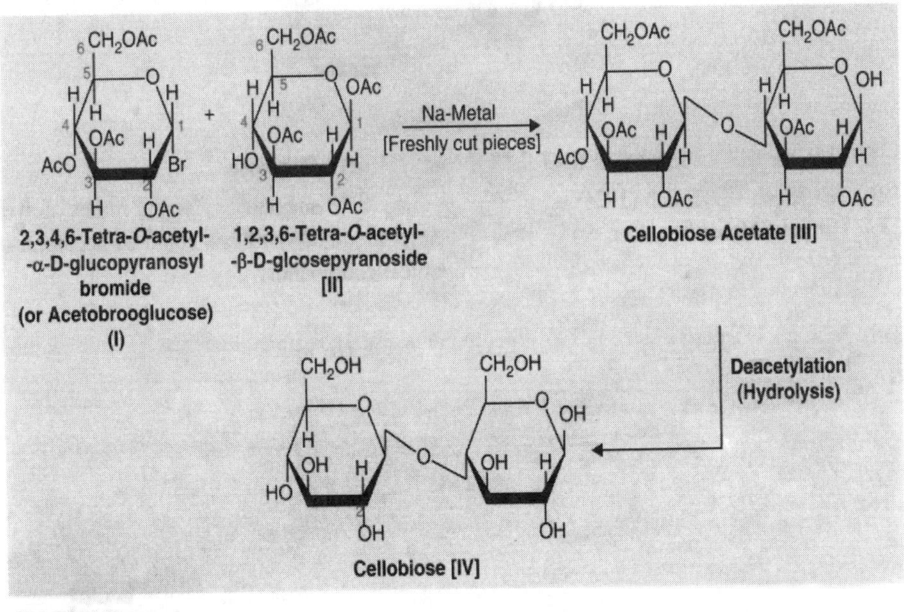

2,3,4,6-Tetra-*O*-acetyl--α-D-glucopyranosyl bromide (or Acetobrooglucose) (I)

1,2,3,6-tetra-*O*-acetyl--β-D-glcosepyranoside [II]

Cellobiose Acetate [III]

Na-Metal [Freshly cut pieces]

Deacetylation (Hydrolysis)

Cellobiose [IV]

[E] Gentiobiose

Gentiobiose, the **disaccharide** has been isolated originally by carrying out the careful hydrolysis of a **trisaccharide** known as '*gentianose*'** by partial hydrolysis with H_2SO_4 (0.2%) or with the enzyme *invertin*. It may also be obtained from **D-glucose** by **enzymatic synthesis** with the enzyme *emulsin*. Nevertheless, it was duly and meticulously isolated from the *glycoside* **amygdalin**.

---

* Haskins *et al.* : *J Am Chem Soc*, **64** : 1289, 1942.

** It is a **disaccharide** obtained from '**gentian**' normally used as a **bitter tonic** due to presence of **bitter glycosides**.

Constitution: The various aspects of the **constitution** of *gentiobiose* are as stated under:

(1) Its molecular formula is $C_{12}H_{22}O_{11}$.

(2) **Presence of Glucose Units**: **Gentiobiose** gets hydrolyzed in dilute aqueous mineral acids to produce two moles of **'glucose'** that suggests specifically that the former comprises two **glucose units**.

(3) **Presence of Reducing Sugars**: **Gentibiose** usually responds to the so called various specific well-known tests for the presence of **'reducing sugars'** namely:

- **Tollen's Reagent Test,**
- **Barfoed's Reagent Test, and**
- **Benedict's Solution Test,**

thereby indicating, the critical existence of at least **'one latent aldehydic moiety** in it.

(4) **Hydrolysis by Almond Emulsin**: The enzymatic hydrolysis of **'gentibiose'** by *almond emulsin* predominantly indicates the presence of β-**glycosidic linkage**.

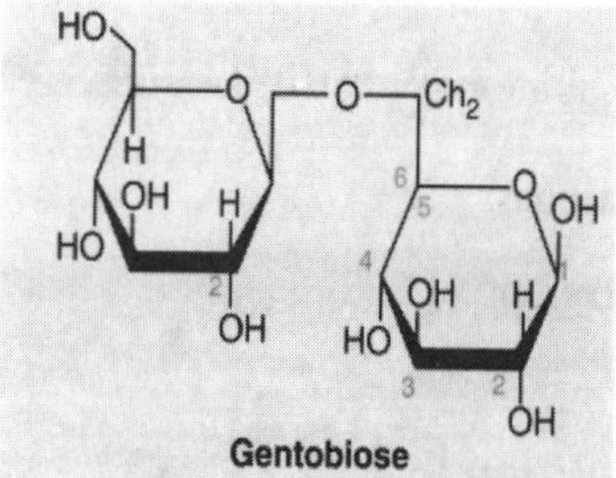

Gentobiose

(5) It is chemically known as **6–O–β-D-glucopyranosyl–D–glcose.**

(6) **Synthesis:** Helferich (1926)* and Reynolds (1938)** suggested different routes of **synthesis** for **gentibiose;** however, the relatively simpler mode of **synthesis** could be accomplished by the *condensation* of the following *two* chemical compounds:

- 2,3,4,6,–Tetra–O–acetyl–D–glucosyl bromide (I)
- 1,2,3,4–Tetra–O–acetyl–β-D–glucopyranoside (II) as given under:

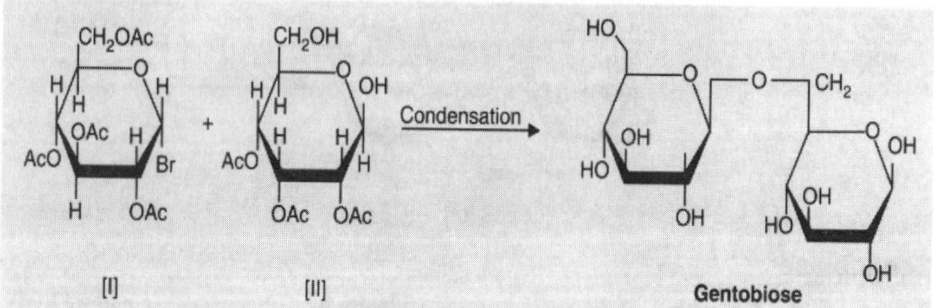

[I]   [II]   **Gentobiose**

(F) **Melibiose:** *Melibiose* is duly prepared form **raffinose*** by direct fermentation with the aid of *top yeast* that particularly removes the **fructose.**

** Helferich K:*Ann*, **450**, 219, 1926.

** Reynolds E: *J Am Chem soe*, **60** : 2559, 1938.

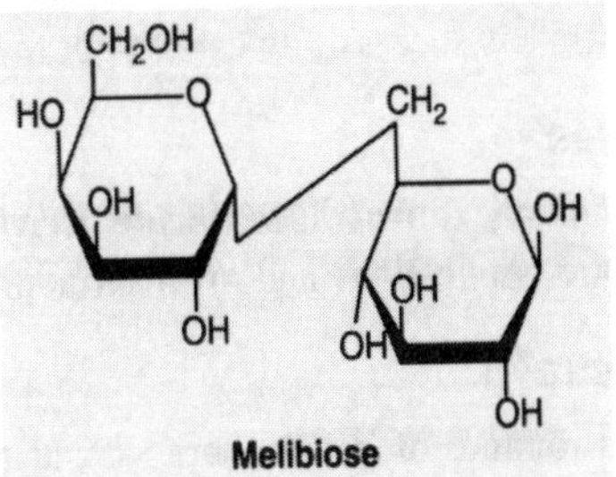

Melibiose

Constitution: The various cardinal aspects of the constitution are as enumerated under:

(1) Its molecular formula is $C_{12}H_{22}O_{11}$.

(2) It is chemically known as **6–O–α–D–galactopyramosyl–D–glucose.**

(3) The presence of reducing sugars is duly characterised by such reactions a mentioned under section (E) item 3.

(4) **Melibiose** undergoes *mutarotation.*

(5) **Melibiose** also forms **'osazone'** with *phenylhydrarine.*

(6) **Hydrolysis** with dilute mineral acids gives rise to the production of one mole each of **D–glucose** and **D–galactose.**

(7) **Methylation followed by hydrolysis:** It has been adequately established beyond any reasonable doubt that **melibiose** upon complete **methylation** followed by **hydrolysis** yields *two* distinet products of reaction, namely:

- 2,3,4,–Trimethyl–D–glcose (I), and
- 2,3,4,6–Tetramethyl–D–galactose (II),

as given under, which indicated cirdently that the *'latent aldehydic moiety'* is located definitely in the **'glucose unit':**

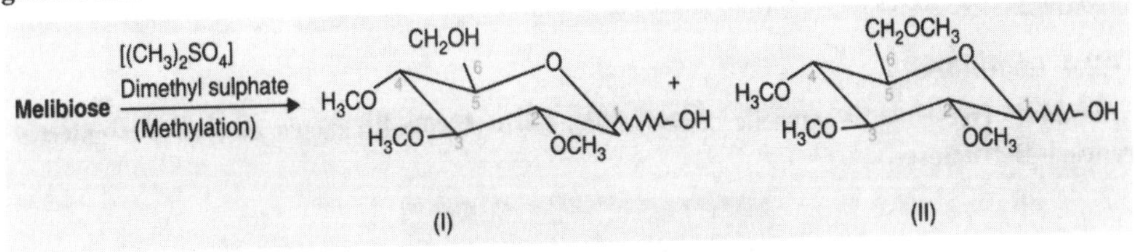

(8) **Nature of Glucosidic Linkage:** In order to ascertain the exact and precise **nature of the glucosidic linkage** present in **melibiose** one has to depend upon its **specific rotation** $[\alpha]_D = +113°$, which suggests prevalently that the **glycosidic linkage** is difinitely 'α'.

Furthermore, one may critically observe the following sequential steps of reaction that ultimately do provide a reasonably strong and solid with regard to the underlying fact that the C–6 present in the **glucose moiety** (*i.e.,* the *'reducing segment'*) are intimately and prominently involved in the creation of the so called **'glycosidic linkage'** in **melibiose.**

---

* **Raffinose:** It is a **trisaccharide** invariably found in **eucalyptus manna** and **sugar beet** (*i.e., white beet* from which *sugar is obtained*).

(9) Heleferich and Bredereck (1928)* has successfully carried out the synthesis of **melibiose.**

11.1.2 Disaccharide Derivatives

Following are the *five* most commonly known **disaccharide derivatives** which find their vital and important utilities both in therapeutic applications and artificial approved sweeteners, namely:

11.1.2.1 Sucrose Esters [Olestra®]

It is invariably recognized as a mixture of **hexa–, hepta–,** and **octaesters** of *sucrose,* and also of **fatty acids** derived specifically from the 'edible oils' is normally regarded to be a 'fat substitute'. The **as–FDA** in 1995 have duly approved the **sucrose esters** based on the following *two* good *reasons:*

- **not absorbed** systematically,
- **not digested** systematically,

their restricted used in certain food products.

11.1.2.2 Maltitol [Maltitol Syrup (Lycasin)®]

Importantly, **maltitol** which being a 'polyalcohol' never occurs naturally; and is duly produced by synthesis from **pure glucose syrups** rich in 'maltose'. **Lycasin** (R) is an artificial approved 'sweetener' in France.

11.1.2.3 Isomalt

Isomalt designates the products of the **catalytic hydrogenation of isomaltulose****. In fact, it is an admixture of the following *two* products:

- α-D-glucopyranosyl–(1 → 6)–sorbitol, and
- α-D–glucopyranosyl–(1 → 6)–mannitol,

combinedly available as **Palatirit** (R).

> **Note:** It is also used as an authorized sweetener in France.

11.1.2.4 Lactulose

Lactulose represents the synthetic disaccharide, and is chemically known as **4–O–β–D–galacto pyranosyl–D–fructose.**

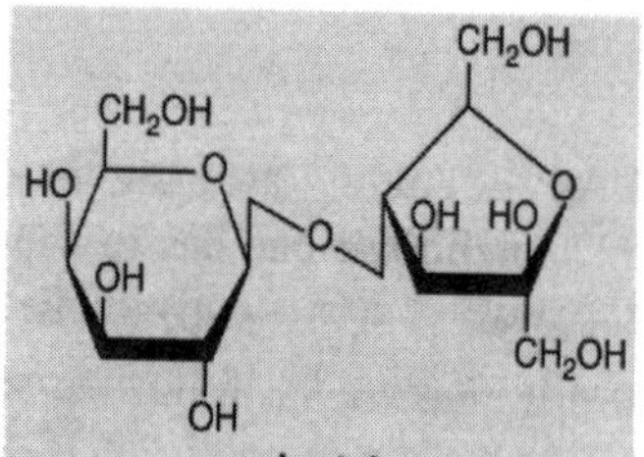

Lactulose

It is also commercially known as **Bifeteral; Cephulac; Duphalac; Generlac; Lactuflor; Laevilac;** and **Normase**. Importantly, **lactulose** critically serves as:

---

* Helferich and Bredereck: *Ann;,* **465**:166, 1928

** **Isomaltulose:** It refers to the product of the enymatic tranformation of (+)–**sucrose** by **Protaminobacter rubrum.**

- osmotic laxative,
- lowers blood ammonia, and
- stimulated intestinal peristalsis.

Lactulose exerts its action due to the lowering of the **'colon pH'** by the acids arising by virtue of its crucial degradation by the ensuing microflora which decreases the **intestinal absorption** of **ammonia (as NH$_3$)**; and, therefore, enhances its diffusion out of the blood followed by its **critical trapping,** and ultimately gets eliminated as **NH$^+_4$ion.**

The products is used in *constipation, hepatic encephalopathies via* the **nasogastric administration** or by **enema.**

11.1.2.5 Lactilol

Lactilol represents the catalytic hydrogenation product of **lactulose (seetion 11.1.2.4).** Definitely,

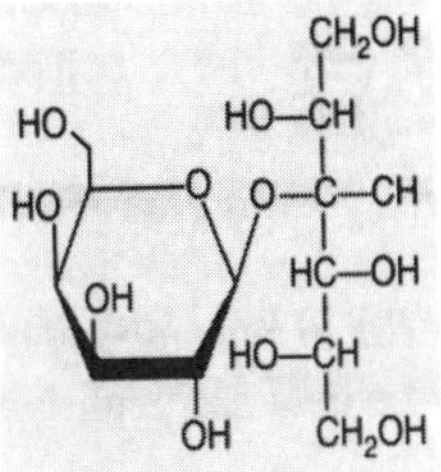

Lactilol

it is *not a* **sugar** of the **natural plant origin.** However, it does exhibit almost the *same indications* as well as the same *potential side effects viz;,* **abdominal cramps,** and **diarrhoea.**

11.2 Trisaccharides

In general, the **trisaccharides** are obtained from *two* divergent sources, namely

- naturally occurring plant sources, and
- partially hydrolyzed products of polysaccharides.

11.2.1 Natural Sources

The various documented and recognized sources of the **trisaccharides** are as stated under:

S.No.	Name	Products of Hydrolysis	Occurrence
1	Gentianose	Glucose, glucose, fructose	Gentiana spp.
2	Melezitose	Glucose, fructose, glucose	Manna from Larix
3	Plantease	Glucose, fructose, galacotose,	Seeds of *Psyllium* spp.
4	Raffinose	Galactose, glucose, fructose	Many seeds (*viz,* cotton–seed)
5	Manneotriose	Galactose, galactose, glucose	Manna of ash, *Fraxinus ornus.*
6	Rhaminose	Rhamnose, rhamnose, galactose.	*Rhamnus infectoria*
7	Scillatriose	Rhamnose, glucose, glucose	Glycoside of squill.

Note: The other typical examples of **'trisaccharides'** are usually found amongst the **glyosides of Digitalis.**

11.2.2 Hydrolyzed Products of Polysaccharides

In fact, the various typical examples of the *trisaccharides* that are obtained by the partial hydrolysis of the **polysaccharides** essentially comprise:

- **Cellotriose:** obtained from **cellulose;**
- **Maltotriose:** obtained from **starch.**

Note: The 'trisaccharides' invariably turn out to be either reducing type or non-reducing type.

11.2.3 Gentianose

Gentian (**Gentian Root, BP, EP, BHP**) is reported to be *rich in sugars,* which essentially comprise **gentianose** *i.e.,* the **trisaccharide.** During the *phenomenon of fermentation* the sugar gets partially hydrolysed into **glucose** and **fructose.** In a situation when the ensuing *fermentation process* is allowed to proceed a little too far, one may observe that the **hexose sugars** are subsequently converted into **carbon dioxide (CO_2)** and ethanol. Menkovic *et al.* (2000)* reported the chemical composition *vis-a-vis* the seasonal variations observed duly in the content of specific **secondary metabolites** present prominently in the aerial portions of *G. lutea.*

Characteristic Features: Following are the various **characteristic features** of 'gentianose':
  (1) It is a **non-reducing sugar.**
  (2) Upon hydrolysis **gentianose** gives rise to *two moles of* **D–glucose** and *one* mole of **D–fructose.**
  (3) **Structure of Gentianose:** It may be duly accomplished by adopting the following steps in a sequential manner:
    - **hydrolysis** with the enzyme *invertase* results into the formation of one molecule each of **Gentiobiose** and **D–Fructose,** and
    - **hydrolysis** with the enzyme *emulsin* yields **(+)–Sucrose** and **D–Glucose.**

Thus, from the above *two* critical findings and observations it may be inferred that the *'structures of gentiobiose'* essentially comprise:

- *Two* units of **'glucose',** and
- *One* unit of **'fructose',**

that are *intimately combined* so as to designate **'gentiobiose'** and **'sucrose'** *viz;,* the *three* aforesaid **monosaccharide** molecules, namely: **Glucose–Glucose–Fructose** are arranged in this way we may have the structure of **'gentianose'** as given under:

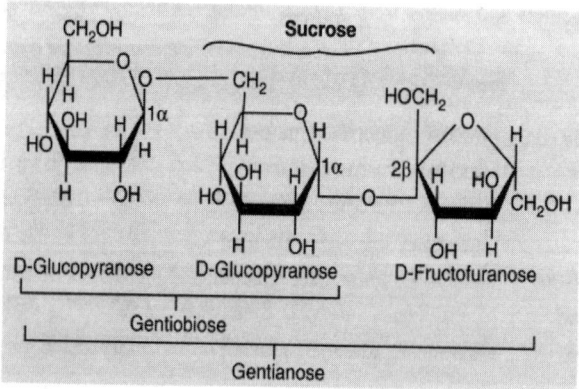

*Menkovic N *et al.* : *Planta Medica,* **66**:178, 2000

11.2.4 Raffinose

Raffinose is regarded to be the most abundantly found **trisaccharide** in nature, such as:

- **Australian manna** (from *Eucalyptus spp. Myrtaceac),*
- Cottonseed meal,
- Seeds, leaves, branches, and roots of several plant species,
- Mother liquor obtained from the sugar–beet crystallisation, and
- Crystalline exudation of certain **'eucalyptus tree'.**

Characteristic Features: The **characteristic features** of *raffinose* are as stated under:

(1) Its molecular formula is $C_{18}H_{32}O_{16}$.

(2) **Raffinose** responds to the characteristics of **non-reducing sugars** thereby indicating that all the **'anomeric C–atoms'** duly present in it are totally engaged in the **'glucosidic linkages'.**

(3) **Acid Hydrolysis:** The *acid hydrolysis of* **raffinose** yields 1 mole each of **D–galactose, D–glucose,** and **D–Fructose,** which categorically suggests that each of them actually constitute **raffinose** molecule.

(4) **Action with Invertase:** In reality, the enzyme *invertase* splits **raffinose** into *two* known **disaccharides,** namely:

- **melibiose** [see (F)], and
- **saccharose** (or **sucrose**) [see (B)].

Furthermore, the **trisaccharide** *raffinose* on being subjected to hydrolysis by the enzyme *invertase* affords:

- **Melibiose and D–Fructose;**

whereas, the particular enzyme for the α–glycosidic linkage *i.e.,* α–galactosidase gives rise to the formation of:

- **Sucrose** (or **saccharose**) and **D–galactose.**

Therefore, the aforesaid **hydrolysis products** do suggest explicitly that:

- In **raffinose** mole the **D–glucose** units are intimately linked together as could be seen in **'melibiose',** and
- In **raffinose** mole the **D–glucose** and **D–fructose** units are duly joined together as may be observed in **'sucrose'.**

In other words, the *three* aforesaid **monosaccharide molecules** are usually linked to one another in the following sequence:

Galactose–Glucose–Fructose.

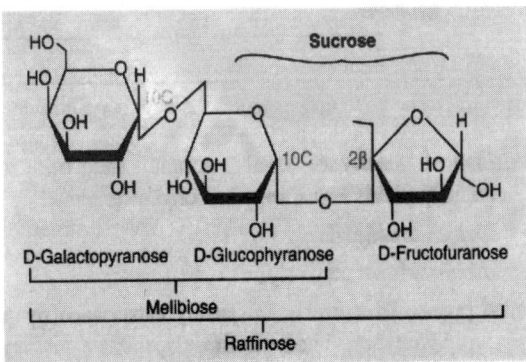

Thus, one may designate duly the *structure of* **raffinose** as stated under:

(5) Confirmation of Raffinose Structure: The confirmation of the aforesaid **structure of raffinose** has been duly confirmed by the following sequential steps:

- Complete **'methylation'**, and
- subsequent careful **hydrolysis,**

to an admixture of the following derivatised products:

❏ 2,3,4,6–Tetra–*O*–mettyl–D–galactopyranose [I],

❏ 2,3,4–Tri–*O*–mettyl–D–galactopyranose [II], and

❏ 1,3,4,6–Tetra–*O*–mettyl–D–galactopyranose [III], as given under:

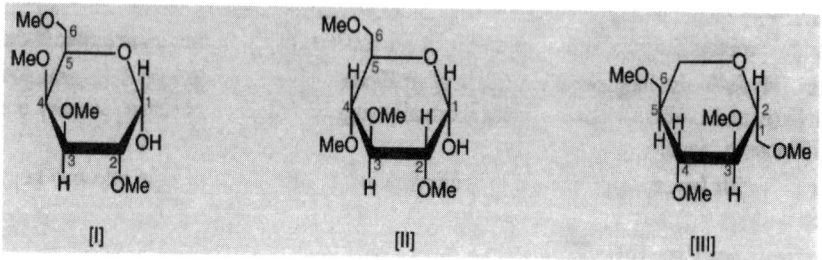

11.3 Tetrasaccharides

The partial hydrolysis of the **'polysaccharides'** afford several critical examples of the **tetrasaccharide.** However, **stachyose** is regarded to be the most important example of this *particular category* that occurs essentially in the roots of a good number of **plant species.**

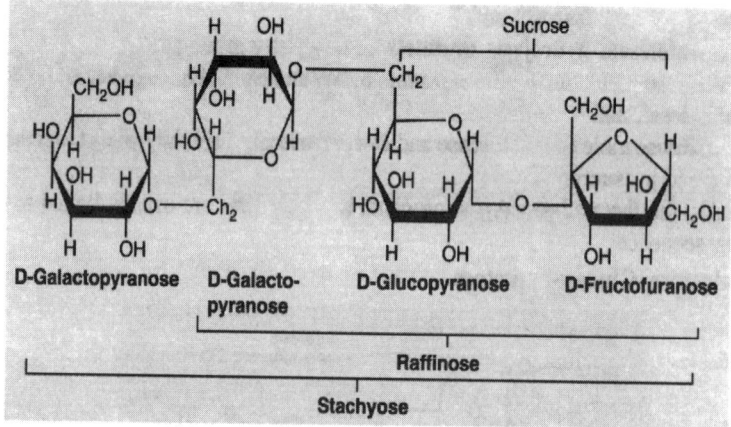

Constitution: The **Constitution** of **'stachyose'** is duly accomplished by carrying out carefully its *enzymatic hydrolysis* that eventually yields *two known* carbohydrates, namely:

- **(+)–sucrose** – a *disaccharide,* and
- **raffinose** – a *trisaccharide* respectively.

Importantly, its exact and precise structure is adequately established by adopting the following *two* reactions sequentially viz., • *methylation;* and • **hydrolysis,** thereby giving rise to the formation of the following *four* derivatized products of reactions, such as:

- 2,3,4,6–Tetra–*O*–mettyl–D–galactose,
- 2,3,4–Tri–*O*–mettyl–D–galactose,
- 2,3,4–Tri–*O*–mettyl–D–glucose, and
- 1,3,4,6–Tetra–*O*–mettyl–D–fructose.

Hence, one may arrive at the following structure of **'Stachyose':**

12. POLYSACCHARIDES [or GLYCANS]

Polysaccharides (or Glycans), based on a random choice, are defined as–**'high–molecular weight* polymers that are solely resulting from the due condensation of a good number of monosaccharide molecules'.**

However, rather more explicitly each and every **'sugar moiety'** is *directly* and *intimately* linked to its *'neighbour' via* a **glucosidic bondage** created arbitrarily by a so called **theoretical elimination** of a **'mole of water'** existing between:

- **the hemiacetal** *hydroxyl (–OH) moiety* **located strategically located strategically at C–1 of one sugar entity, and**
- **any of the hydroxyl (–OH) groups positioned on the other 'sugar molecule'.**

In a broader perspective, the terminology **'polysaccharide'** is applied invariably to such **'polymers'** having predominantly more than **'ten monomer units';** and, therefore, the **'polysaccharides'** do comprise **100 to 90,000 monosaccharide units.**

12.1 Nomenclature

In general, the *'prefix'* **–an** is largely employed to designate the **'polysaccharides'** which are otherwise also termed as **glycans.**

Example: Following are a few *typical examples:*

- **Glucans : polysaccharides with only 'Glucose Units'** *e.g.,* **cellulose;**
- **Fructans : designate the 'Fructose Polymers';**
- **Pentans : polysaccharides with only** 'Pentose units';

Note: It may be noted categorically that his specific nomenclature is not most applied to various known *'polysaccharides'.*

12.2 Structure of Polysaccharides

The critical structure education of **Polysaccharides** embraces prominently the following *five* cardinal characteristic features just similar to the structure of **disaccharides,** namely:

- **Nature of** *'monosaccharide'* **becoming the part of the** polysaccharide **along with their sequential arrangement,**
- **Ascertaining the precise and exact presence of either a 'pyranose' or a 'furanose' structure in the** *monosaccharides,*
- **Critical presence of the 'glycosidic linkage' as** α–or β–**or a mixture of the** *two,*
- **Critical presence of either a 'linear' or a 'branched' structure in the** *polysaccharide,* **and**
- **Exact determination of the 'molecular weight'* to ascertain the approximate total quantum**

---

* Ranging between 16,000 to 14,000,000.

of the so called **monosaccharide units** present actually in the *parent molecule.*

Based on the recent galloping advances accomplished in the field of **'chemistry of natural products'** there are *three* classically distinguished means and ways to establish the **structure of polysaccharides,** namely:

(*a*) **Periodic-Sequence Polysaccharides:** In this particular instance, the *sugars* usually come into being along with the chain in a pattern which distinctly repeats itself at a regular interval (*e.g., **amylose, cellulose*** etc.,). Importantly, the conformation of this specific kind of polymer is invariably determined solely by the *conformation of the* **glycosidic linkage** present in it:

- **Case–1:** When β–(1 → 4), the shape adopts a reasonably **'elongated ribbon'** such as : cellulose;

- **Case–II** When α–(1 → 4), the ensuing *polymer* may eventually adopt a vivid **'helical shape'** for instance: **Amylose;** and

- **Case–III** In certain highly specific instances the resulting **'conformation'** turns out to be:

  - **loose in structure, and**
  - **flexible in shape,**

perhaps due to a reasonably *large extent of freedom* vis-a-vis the prevailing **rotation involved** such as : the particular **(1 → 6)** bonds.

(*b*) **Interrupted-Sequence Polysaccharides:** It refers to the *specific segments* haring **regular periodicity** that critically alternate with the *heterogeneous ones.* Nevertheless, the observed potential **polymer-polymer interactions** invariably render the phenomenon of **'gel formation'** both feasible and possible.

(*c*) **Completely Heterogeneous Polysaccharides:** It has been duly established that the ensuing **'potential interactions'** do belong to the **polymer-solvent type.**

A comprehensive survey of literature has revealed that there are several widely accepted and duly recognized **'generalized techniques'** which may find their abundant utility and applicability for elucidating the **'structure of polysaccharides'.** A few of them shall now be discussed at a greater length in the sections that follows:

(1) **Hydrolysis in Acidic Environment: Polysaccharide** may undergo hydrolysis in an **acidic medium (or environment)** using dilute mineral acids (*viz.,* HCL), which categorically converts it into its **respective constituent monosaccharides** that may be ultimately identified by:

- physical characteristic features, and
- preparation of appropriate derivatives.

Important Points: These essentially comprise:

(*a*) The **quantitative determination(s)** of *polysaccharide* is usually accomplished by a variety of sophisticated analytical procedures:

- GLC,
- FT-IR, and
- Zone-Electrophoresis.

---

* **Molecular Weight:** It is duly determined by the **mass spectroscopic method.**

(b) Both controlled and careful hydrolysis results into the formation of **di-, tri-, and higher oligosaccharides,** that will consequently ascertain the following cardinal aspects, namely:

- **pattern of the 'glycosidic linkage', and**
- **ring size of individual monosaccharide entities.**

Note: However, it may be noted that the particular linkages (1→4) found to be the most resistant to acid hydrolysis.

(c)**Furanose/Pyranose Form of Monosaccharide:** *Acid hydrolysis* provides meticulously the direct information with regard to the exact **'ring-size'** of the *monosaccharide units.* Thus, one may have these *two* forms, namely: **furanose** and **pyranose** duly indicated as stated below:

Furanose Form: When the **'polysaccharide'** undergoes *acid hydrolysis* quite *rapidly,* the monosaccharide is present duly in the **'furanose form'.**

Pyranose Form: When the **'polysaccharide'** undergoes *acid hydrolysis* with *great difficulty,* the monosaccharides is definitely present in the **'pyranose form'.**

Note: Nevertheless, these findings/observations need to be reaffirmed by other known methodologies as well.

(2) Enzymatic Hydrolysis: It has been duly established that the most reliable, useful, and dependable hydrolyzed products obtained with a series of **'enzymes'** are indeed capable of removing either one or more than one **'sugar entities'** particularly from the **'reducing-end of the actual polysaccharide used'.**

(3) Simultaneous Acetylation Followed by Hydrolysis (or Acetolysis): This specific procedure, in fact, comprises in the meticulous and very careful treatment of the **polysaccharide** with an admixture of **acetic anhydrite** and **dilute H_2SO_4 (3 to 5%).** However, it has been observed that besides attacking *other existing linkages* present in the **polysaccharide** it also attacks the **(1→6 linkages)** rather more rapidly with great ease and fervour.

(4) Methylation: Srivastava *et al.* (1989)* proposed a *quantitative method* pertaining to the methylation of specifically the **'lower molecular weight polysaccharides'.** The procedure essentially includes:

- **Methylation** performed by the careful incorporation of **barium oxide (BaO_2)** and **methyl iodide (CH_3I)** into a solution of **polysaccharide** in freshly distilled **dimethyl sulfoxide [DMSO, $(CH_3)_2SO_4$]** at $20 \pm 2°C$.
- **Completion of Methylation:** It is usually tested by:
  - **Ziesel Method** *i.e.,* to ascertain precisely whether the total number of **methyl moieties gets either** *increased* or *decreased* by **further methylation;** or
  - **IR-Spectrum** *i.e.,* to notice critically the absence of the **hydroxyl (–OH) bond.**

Furthermore, **IR-spectrum** also reveals and expatiates such other vital and important clues in the resulting **methylated products,** such as:

---

*Srivastave R *et al.* : **Bioactive Polysaccharides from Plants,** *Phytochemistry,* **28** : 2877-2883, 1989.

- nature of functional moieties present duly in the ensuing **polysaccharide,** for instance:
  - ❑ **Carboxyl (–COOH) group, and 1**
  - ❑ **acetamido (–CH$_2$C̈–NH$_2$) group,**

which may help predominantly as the **crucial decisive factor** for the α- or β–glycosidic linkages.

- **Hydrolysis of Methylated Product(s):** Finally, the hydrolysis of the **completely methylated polysaccharides** is carried out, and subsequently the **'products of reactions'** are duly *characterized* and *identified* by known **analytical techniques.**

- **Point of Attachment of 'Sugar Units' :** This is confirmed duly by the aforesaid procedures. It is, however, pertinent to state here that these ensuing *'sugar units'* may be judiciously subjected to:

  - **further hydrolysis,** and/or
  - **oxidized appropriately,**

so as to ascertain the *ring size* as well as the *point of glucosidic linkage.*

(5) Periodic Oxidation [or Smith's Degradation]: In a broader perspective, a good number of *carbohydrate* do possess essentially an array of **'vicinal glycol units';** and, therefore, very much akin to other **1,2-glycols** are usually **oxidized by periodic acid [H$_5$IO$_6$].** Interestingly, an obvious complication comes into being when, as in several carbohydrates, **more than two adjacent C–atoms** prominently bear the **hydroxy (–OH) moieties.** Thus, one of the oxidation products turns out to be an α–hydroxy aldehyde:

Example: A carbohydrate under the **periodic oxidation** gives:

- α–hydroxy aldehyde plus **another simple aldehyde** (which never gets oxidized further anymore), and
- α–hydroxy aldehyde on being subjected to further **periodic oxidation** yields **formic acid** plus **another simple aldehyde [Eqn. (b)].**

Thus, we may have the following reactions:

Eqn: (a):   A Carbohydrate   Periodic Acid [Oxidation] → α-Hydroxy-aldehyde + A Simple aldehyde [Not oxidizable anymore]

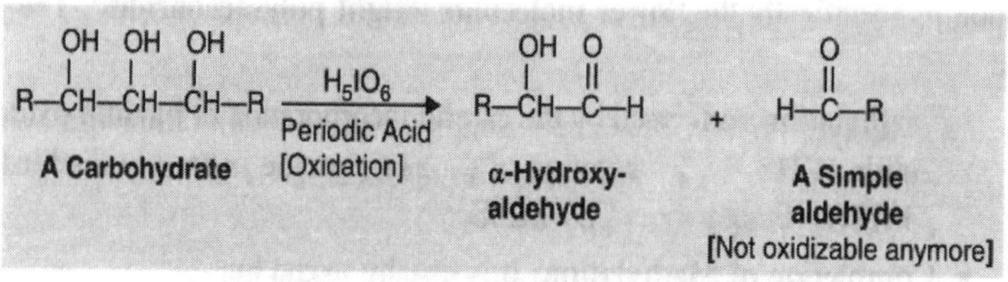

Eqn: (b):   α-Hydroxy-Aldehyde   Periodic Acid [Oxidation] → A Simple Aldehyde + Formic Acid

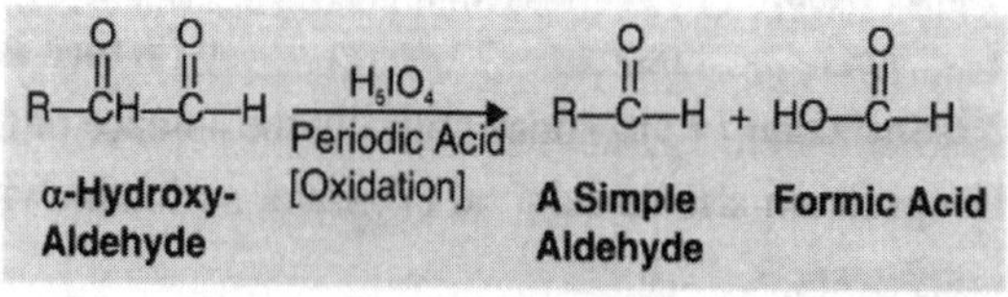

Note: By *'analogy'* an *a-hydroxy Ketone* gets oxidized to an *aldehyde* plus a *carboxylic acid.*

Importantly, based on the underlying fact that it is now quite possible as well as feasible to estimate most precisely and accurately both the quantum of the **'periodic acid (H_5IO_6)'** actually *utilized vis-a-vis* the quantum of **'formic acid (HCOOH)'** actually *generated;* hence, the **periodic oxidation** may be justifiably employed to accomplish a **clear-cut differentiation** between the prevailing *'pyranose'* and *'furanose'* structures of the so called **saccharide structural analogues** (or **derivatives**).

Example: Following represents a typical example of the **periodate oxidation:**

- **Methyl–α–D–glucopyranoside** yields *one* equivalent of *formic acid* as given under:

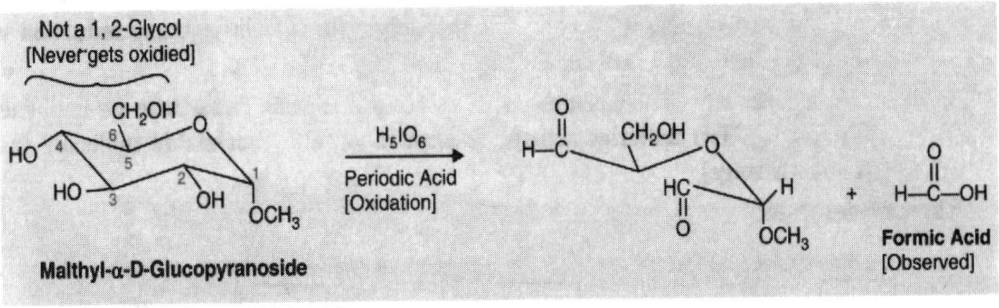

Malthyl-α-D-Glucopyranoside

Methyl–α–D–glucopyranocide

That is, a **'pyranose form'** of the *glycoside* yields **formic acid** solely

- **Methyl α–D–glucopyranocide** gives rise to *one* equivalent of *formaldehyde* as stated below:

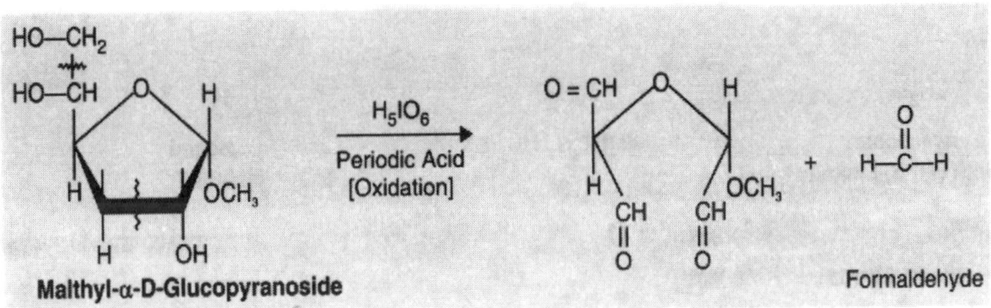

Malthyl-α-D-Glucopyranoside

Methyl–α–D–glucofuranoside

That is, **'furanose form'** of the *glycoside* yields **formaldehyde** exclusively.

Note: The **'periodic oxidation'** pertaining to the *carbohydrate* was first and foremost developed by Hudson CS (1881-1952), a well-known *American Carbohydrate Chemist*. In fact, it has been exploited both extensively and intensively to corelate the so called *'anomeric coufigurations'* of a host of *carbohydrate structural analogues* (or *derivatives*).

Interestingly, another school of thoughts vehemently believes that the **'periodic oxidation'** designates the most vital and important means for establishing the prevalent **'mode of linkages'** exclusively in the *disaccharides* and *polysaccharides* as well. Nevertheless, it remarkably gives rise to altogether *different product variants* in:

- (1 → 3) linkages,
- (1 → 4) linkages,
- (1 → 6) linkages, and
- (1 → 2) linkages.

The generalized accomplished method essentially comprises the following cardinal steps, such as:

- critical **oxidation of the polysaccharide molecule** immediately followed up by **reduction***, and

- hydrolysis in an acidic environment to obviously proving the prevailing '**mode of glycosidic linkage**' predominantly present in the **parent polysaccharide molecule.**

(a) **Polysaccharide Possessing (1 → 3) Linkage:** In this particular instance, one may critically take cognizance of the following facts:

- practically '**No Oxidation**' occurs; and, therefore, fails to undergo any **reduction** with **sodium borohydride (NaBH$_4$),** and

- aforesaid observations may be duly expatiated by virtue of the fact there are **no adjacent hydroxyl (–OH) moieties** actually present in the **polysaccharide molecule [as in (1 → 3) linkage]**

Thus, we may have:

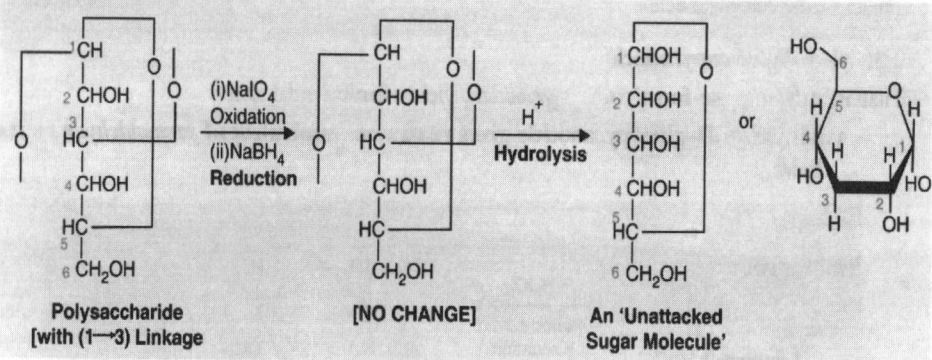

Polysaccharide   **[NO CHANGE]**   **An 'Unattacked**
[with (1→3) Linkage   **Sugar Molecule'**

(b) **Polysaccharides Possessing (1 → 4) Linkage:** In this specific case, the following critical observations may be noted:

- at first stage of **oxidation with NaIO$_4$,** a cleavage occurs between C–2 and C–3 atoms to yield an '**open-chain aldehyde**' (I),

- secondly, the **reduction of (I) with NaBH$_4$** to convert the *open-chain aldehyde* into the corresponding **alcoholic functions (II)** (both at **C–2** and **C–3**), and

- thirdly, the **hydrolysis with mineral acid (HCl) splits (II)** at C–2 and C–3 to yield **glycolaldehyde (III)** plus a *four* **carbon-chain alcohol** known as **crythritol (IV)**

The various reaction may be shown as under:

(c) **Polysaccharide Possessing (1 → 6) Linkage:** In this specific example, the **polysaccharide with (1 → 6) linkage** on being subjected to **oxidation** with *two* moles of **NaIO$_4$,** helps to abstract a mole of **formic acid** thereby removing **C–3** from the *polysaccharide,* and affording *two* **aldehydic moieties at C–2 and C–4** with an **open–chain configuration (I).** The resulting

---

* Reduction: It is primarily carried out so as to minimize the **aldehydic function (–CHO)** because in the course of hydrolysis they may exhibit a tendency to undergo both **degradation** and **condensation.**

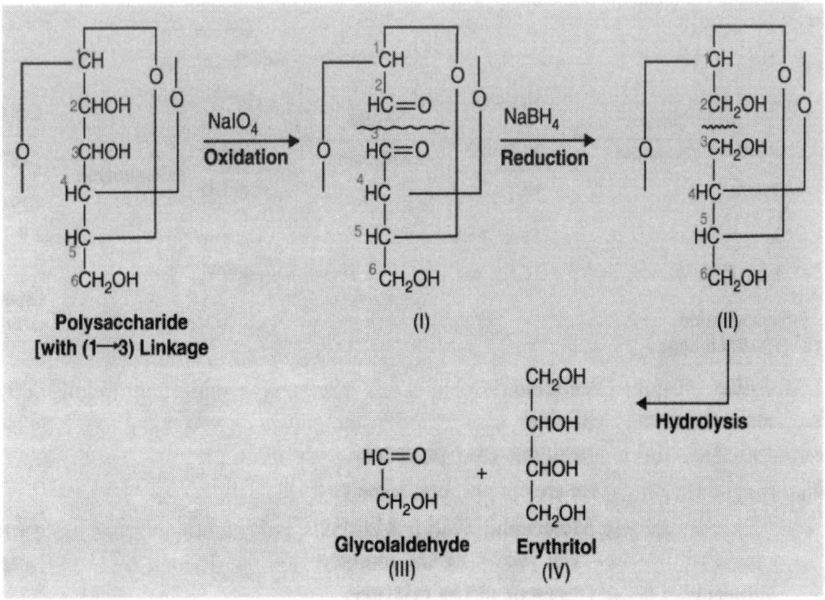

product (I) upon reduction with **NaBH$_4$** renders the *aldehydic functions* in (I) into respective **alcoholic functions** (II). Finally, the **alcoholic product (II)** upon hydrolysis in the **acidic medium** causes a distinet cleavage to yield a **tritydric alcohol** (*i.e.,* **Glycerol**) (III) plus an aldehyde known as **glycolaldehyde (IV).**

Thus, we may have the following reactions:

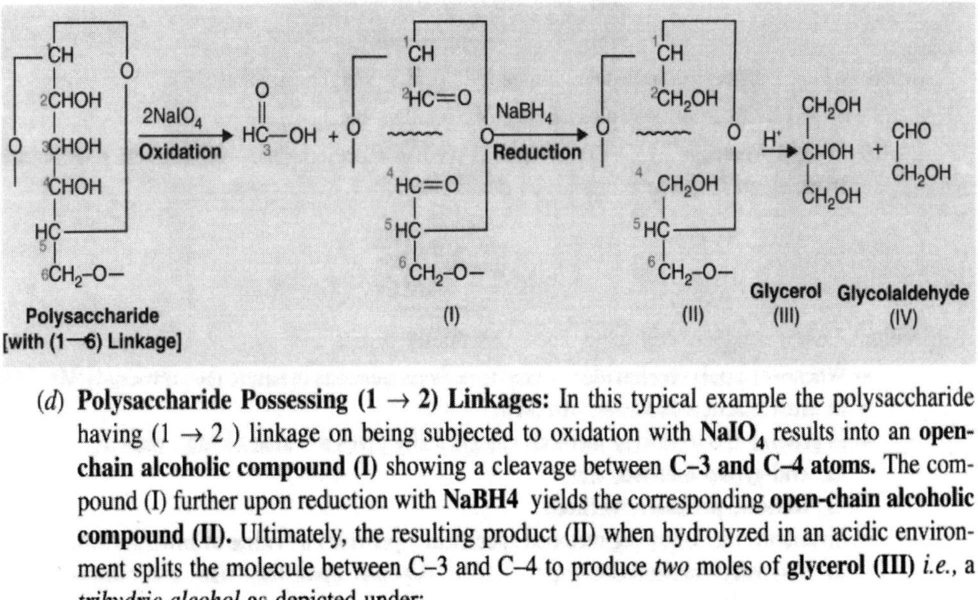

(d) **Polysaccharide Possessing (1 → 2) Linkages:** In this typical example the polysaccharide having (1 → 2) linkage on being subjected to oxidation with **NaIO$_4$** results into an **open-chain alcoholic compound (I)** showing a cleavage between C–3 and C–4 atoms. The compound (I) further upon reduction with NaBH4 yields the corresponding **open-chain alcoholic compound (II)**. Ultimately, the resulting product (II) when hydrolyzed in an acidic environment splits the molecule between C–3 and C–4 to produce *two* moles of **glycerol (III)** *i.e.,* a *trihydric alcohol* as depicted under:

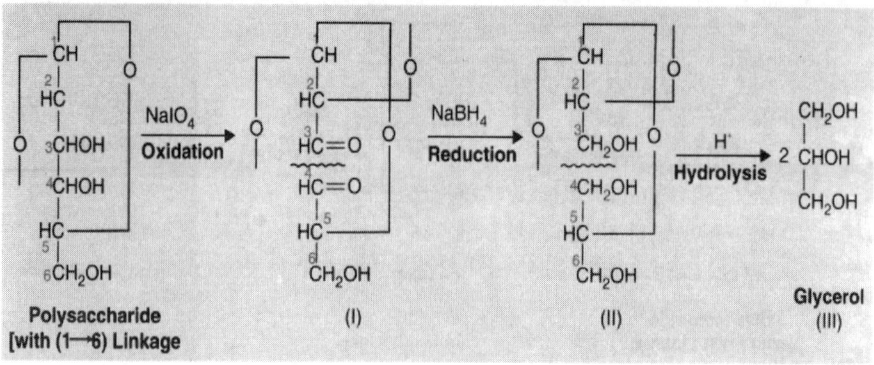

Polysaccharide
[with (1→6) Linkage

(I)   (II)   **Glycerol**
                      (III)

(6) **Molecular Weight Determination:** In actual practice, the **molecular weight determination** provides a most critical and vital clue with regard to the *indication* pertaining to the total number of the **monosaccharide units** constituting the **polysaccharide molecule.**

Molecular Weight Variants: There are of *two* types, namely:

(a) **Number Average Molecular Weight (Mn):** It designates an **average molecular weight,** and is defined as– **'the weight of the analyte sample divided by the total number of molecules (n) duly present in the mixture.**

It may be expressed as:

$$Mn = \frac{Weight}{n}$$

However, in general it may be designated as:

$$M_n = \frac{n_1 M_1 + n_2 M_2 + ...}{n_1 + n_2 + ...} = \frac{n_1 M}{n_1}$$

where, n_1 = Number of molecules having Molecular Weight M_1, and

n_2 = Number of molecules having Molecular Weight M_2

(b) **Weight Average Molecular Weight (Mw):** It represents the **weight average molecular weight,** and may be expressed as given below:

$$Mw = \frac{n_1 M_i^2}{n_1 M_1}$$

Salient Feature of Mn and Mw: These essentially comprise:

- Whenever a **polysaccharide** is found to be homogeneous in nature then obviously Mn=Mw,
- In usual practice, however, Mn>Mw,
- In general, the following methods categorically gives a 'value of Mn', such as:
  ❑ **end-group method, and**
  ❑ **osmotic-pressure method,**
- Invariably, the following methods specifically provides a **'value of mw',** for instance:
  ❑ **viscosity method** (perhaps it solely depends upon the *'size'* and *'shape'* of the molecule,
  ❑ **rate of sedimentation, and**
  ❑ **sedimentation equilibrium.**

12.3 Behaviour of Polysaccharides: Gel Formation

Gel refers to a semisolid condition of precipitated or coagulated colloid which essentially contain a large quantum of water.

It has been duly established that a plethora of **polysaccharides** are duly characterized by their inherent potential and ability to form gels *i.e.,* solid **3D-macromolecular arrays** which predominantly retain the so called **'liquid phase'** particularly confined within their **lattice.***

In other words, **'gel formation'** is, in a manner, designates explicitly the **passage from disorder** (*viz;, a* **true solution**) to a specific order duly brought into existence by the partial association of either of **chains** or of **segments of a chain.**

Important Features of Gel formation: These essentially comprise:

(1) The greater the *chains* or *chain segments* **associate,** the greater would be the **rigidity of the gel,** which may rise to *partial syneresis* (*i.e.,* **gel contraction).**

(2) In case, the ensuing organization of **'gel formation'** tends to be too great, the resulting structure actually approaches that of a **precipitate.**

(3) Importantly, the possible observed reversibility of the **'gel formation'** suggests overwhelmingly that the **'inter-polymer bonds'** are rather **weak in nature** *viz;,* **H-bonds, coordination bonds.**

(4) Interestingly, the corollary of the aforesaid **'bond weakness'** is perhaps due to the fact that the said bonds should be able to produce in relatively sufficient quantum; and, therefore, solely proves to be the most **critical determining influential factor** related to the **polymer structure** upon its ability to produce **'gels',** namely:

- **Regular Homogeneous Polymers:** These possess the following characteristic functionalities, such as:
  - ❑ produce extremely **sensitive junction segments,**
  - ❑ possess **high structural organization,** and
  - ❑ mostly and predominantly represented by precipitates.

- **Heterogeneous Polymers:** They do not have **'regular sequences',** and get invariably *dispersed* in the solvent thereby forming **viscous solutions** ultimately.

- **Polymers Dotted with Regular Sequences Interrupted by Irregular Patterns:** These essentially produce critical **'punctual functional segments';** and, therefore, form the ultimate *elsastic gels.* It is pertinent to state here that the **junction segments** (or **zones**) may involve the following *two* types of characteristic structures, namely:
  - ❑ **Helical Structures:** *e.g.,* **agarose, carrageenans,** and
  - ❑ **Chain Pile Ups:** *e.g.,* **pectins, alginic acid salts.**

The following diagramatic representations illustrate vividly the precise and exact mode of **'get formation'** *vis-a-vis* the formation of the so called **'punctual junction segments'.**

---

* **Lattice:** A **network** or **framework** formed by structures interwined usually at right angles with each other.

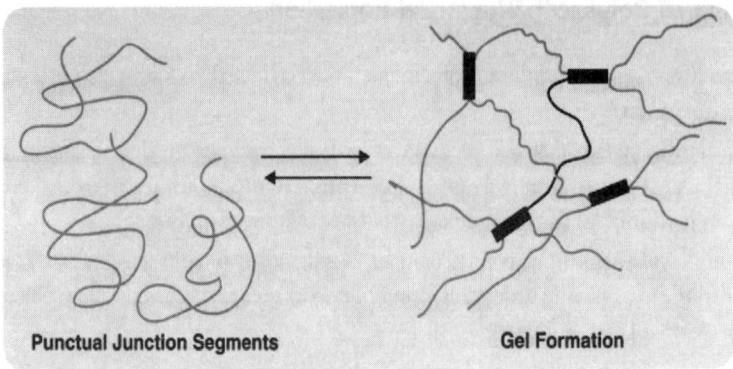

Punctual Junction Segments   **Gel Formation**

12.4 Isolation and Structural Analysis

The vital and important aspects, namely : **isolation** and **structural analysis** in the *polysaccharides* do occupy a pivotal status; and, hence, would be discussed briefly and separately in the sections that follows:

12.4.1 Isolation

First and foremost the **polysaccharides** are dissolved carefully in water in *two different environments:*

- in the presence of *dilute mineral acids** (*e.g.,* HCl), and
- in the presence of *various salts***.

Besides, one is competent enough to make use of the following means and ways in a **small–laboratory-scale set-up:**

- effective use of **'aprotic dipolar solvents'** *viz;,* **ether, hexane, methylenc chloride;**
- effective use of **'dialysis'** to get rid of ensuing *salts* and *low-molecular weight molecules;*
- effective usage of **ion-exchange resins** and/or **molecular gel filtration;** and
- effective usuage of **extraction procedures** such as:
  - ❏ **plant pigments** by *acetone* or *ethanol,* and
  - ❏ **oligosaccharides** by *acetone* or *ethanol.*

Polysaccharide Isolation (or Fractionation): It is indeed a rather delicate procedure that essentially involves:

- by using various known **'precipitation techniques'** *i.e.,* **salt addition, immiscible solvents,** and **alteration in pH;**
- **'chromatographic techniques'** – usually are of wide and broad application upon an array of adsorbent variants, such as : **activated charcoal** (chromatographic grade), **ion-exchangers,** and **native or substituted reticulated polyglucan gels;** and
- **'specialized fractionation techniques'** – these normally comprise such methods:
  - ❏ formation of **'boric-acid complexes',**

---

* In order to augment and facilitate the crucial extraction of **'Pectin'.**

** **Carbonates**– usually exmployed in the particular instance of **'Algin'.**

❑ formation of **'inclusion derivatives'** and

❑ usage of **'quaternary ammonium salts'**.

Importantly, in all the aforesaid *three* **polysaccharide fractionation** procedures the following steps are common:

⇒ purification,

⇒ physical and chemical estimation,

⇒ optical rotation,

⇒ molecular weight determination,

⇒ elemental composition, and

⇒ electrophoresis assays.

12.4.2 Structural Analysis

In a broader perspective, the *structural analysis* of the **polysaccharides** seems to be rather an exceedingly *complex* as well as *complicated* problem, which essentially requires the *judicious* and *meticulous* combined application of such prominent methodologies, namely:

- **physical techniques** *viz, spectrophotometer method* (**UV, FT-IR**), and *spectroscopic methods* (**NMR, MS**) and

- chemical methods *viz;,* **partial hydrolysis, hydrolysis, methanolysis, derivatization, controlled degradation of the respective polymer,** and **its subsequent derivations.**

12.5 Classification of Polysaccharides

The advent of modern research and the immense contributions made by famous **'Carbohydrate Scientists'** across the globe have put forward the following recognized and accepted **classification of polysaccharides** as given under:

- Polysaccharides from Lower Plants (or Polysaccharides from Microorganisms and Fungi),
- Polysaccharides of Lower Plants (or Algal Polysaccharides),
- Homogeneous Polysaccharides (or Homopolysaccharides),
- Heterogeneous Polysaccharides (or Heteropolysaccharides)

The aforesaid *four* distinct class of the **'polysaccharides'** shall now be treated individually in the sections that follows:

12.5.1 Polysaccharides from Lower Plants (or Polysaccharides from Microorganisms and Fungi)

Since the past couple of centuries upto the twenty-first century the human being have used profusely the various *'sugar polymers'* largely derived from:

- **higher plants** *(viz.,* cereals, tubes etc.,) or
- **semisynthesized** duly from **'natural polymers'**.

However, the **polysaccharides** derived from different *plant origins* do come across with a good number of **disadvantages** and **drawbacks** caused mainly due to:

- irregular supply due to **'inconsistent climatic conditions'**,
- exhorbitant fluctuations in **'cost effectiveness'**,

- enormous variations in **'product quality'**, and
- absolute lack in the **reproducibility** of the physical characteristic features on account of the *observed variability* inherent in the **living matter.**

Remarkable and superb progress in the field of **'biotechnology'** usually **combat** as well as **alleviate** most of the aforesaid *disadvantages* and *drawbacks:*

- **polysaccharides** are produced with great precaution meticulously, and
- products obtained with spectacular **constant quality,** and having excellent physical characteristic properties.

Following are the *three* typical **'polysaccharides'** duly produced by the microorganism which have been *authorized for sale* by the relevant competent authority:

12.5.1.1 Dextrans [Dextran (INN)*]

Dextran designates a *polysaccharide* produced by the microbial action of *Leuconostoc mesenteroides* upon **sucrose.** It is available in various *molecular weights,* and is mostly employed as a **'plasma volume expander'.**

In general, **dextrans** are *glucose polymers* or *glucans* duly made up of **α–D–glucopyranosyl residues** having **(1 → 6) linkage.** The **dextran molecules** are invariably found to be in the **'branched state'** having:

- high molecular weight ranging between **40×10^6 to 50×10^6,** and
- synthesized by the aid of an **'extracellular enzyme'** that are present in various microorganisms belonging to the **genera** *leuconostoc, lactobacillus,* and *streptococcus.*

Nevertheless, the enzyme **'dextransucrase'** is observed to cause the **polymerization** of the α–glucopyranosyl groups by affording due **transfers from sucrose.**

It has been established that each and every **'dextran polymer'** is highly specific to the *particular strain* which generates it; and, therefore, it may usually comprise either **(1 → 2)** or **(1 → 3)** or **(1 → 4) linkages,** but the *most predominant* one being the **(1 → 6) linkage.** Importantly, the actual **extent of branching** varies specifically within a range of 5 to 33%, and in majority of cases one may find the *lateral chains* to be:

- **extremely short dimension** (*viz;,* **1 or 2 glucose units),** and
- critically linked to the **principal chain** either by a **(1 → 3)** or **(1 → 2) linkage.**

Parenteral** Preparations: For the **parenteral preparation,** the *dextrans* **40, or 60, or 70** which are duly provided in the **Official Compendia** *viz.,* **European Pharmacopoea, United States Pharmacopoea, British Pharmacopea, International Pharmacopea, Indian Pharmacopoea.** Further more, it has been duly specified that these *low molecular weight* **'dextrans'** do represent **"an unique blend of polysaccharides' that are obtained from Leuconostoc mesenteroides strain (NRRL–B–512) or substrains obtained therefrom (*e.g., L.mesenteroides* B–512F=NCTC 10817)."**

---

* **Dextran (INN):** *Dextran* is the **International Nonproprietory Name.**

** **Parenteral :** Denoting any medication route other than the *alimentary cangal viz.,* **intravenous (IV), intramuscular (IM), subcutaneous (SC),** or **mucosal.**

Production: The commercially available **'dextran'** categorically represents a *polymer* comprising essentially;

- ~ **95% of a–D–(1 → 6) linkages, and**
- ~ **5% of a–D– (1 → 3) linkages,**

that are involved in the **'lateral branching'** exclusively.

However, the various steps engaged in its *production* include:

(1) use of only selected strains of *L.mesenteroides* carefully cultivated upon the **sucrose-rich media.**

(2) soon after completion of the *culture, enough* **ethanol** is added so as to cause precipitation of the **polymer** *i.e.,* **polysaccharide).**

(3) Bearing in mind that the **molecular weight of the ensuing polymer** (or depolymerization) still remains on the *higher-side,* a critical **partial-hydrolysis** follows immediately to *dispose of polymers* of **40K to 75 K molecular weight,** which may be adequately accomplished by means of :

- hydrolysis (partial) in an **acidic medium,**
- hydrolysis (partial) in **fungal enzymes**), and
- hydrolysis (partial) in **ultrasonic treatment.**

(4) After **depolymerization** (or **deionization**) the following *two* steps are to be adopted sequentially:

- **precipitation with acetone,** and
- **recrystallization,**

to obtain the **'medicinal dextran.**

Tests for 'Selected Dextrans': The following crucial **'tests'** are being conducted, namely:

- determination of the **'residual solvents'** by **Gas Liquid Chromatography (GLC),**
- presence of traces of **'heary metals'** (e.g., Hg, Pb)
- any possible **'contaminants',**
- presence of **'bacterial endotoxins',** and
- molecular mass distribution by **'Size–Exclusion Chromatography' (SEC)*.**

Application: These largely include:

(1) **Intravenous Infusion:** Dextrans having an average molecular weight 60,000 **[Dextran 60]** in *6 % (w/v) solution;* and alternatively of molecular weight 40,000 **[Dextran 40]** in *3.5 % or 10% (w/v) solution.*

> **Notes: 1.** Both the *'osmolarity'* and *'viscosity'* of these *two* aforesaid solutions are quite similar to those of *'plasma'.*
> **2.** Dextran is absolutely 'non-toxic' in nature.
> **3.** It is also neutral serologically.
> **4.** It exhibits a prolonged action and gets eliminated almost completely.

(2) **Plasma Substitute: Dextran** finds its enormous applications as a **'plasma substitute'** specifically for the following indications:

---

* Kar A.: **Pharmacatical Drug Analysis,** New Age International, New Delhi, 3nd. ed 2010.

- in severe **'shock'** caused due to **haemorrhage,**
- in acute condition of **'shock'** due to **haemorrhage,**
- in typical cases of **'toxicinfection,** and
- in critical **preoperative hemodilution** agent.

(3) Since, **Dextran** indulges in causing apparent **interferences** with particular reference to **'hemostasis'** the **maximum/optimum dosage regimen** is usually set at $1.5.kg^{-1}.day^{-1}$ or 20mL. kg^{-1}. **Dextran 40** also recommended for **similar indications.**

(4) **Dextran** is also employed for the formulation of **'Eye Drops'** indicated invariably for;

- symptomatic treatment of **'lacrymal insufficiency',** and
- prominently improving the comfort of **'contact–lens users'** by maintaining a **lubricating film crated upon the cornea.**

(5) **Dextran Sulfate** has gainfully entered into the array of such **Formulations** having **anti-inflammatory combination** utilized among other use, namely:

- **traumatology** (*viz,*. **contusions, dislocations,** and **sprains),**
- **phlebology** *(mild phlebitis),* and
- **rheumatology** *(viz;,* **small joint arthropathy** and **tendinitis).**

(6) **Dextranomer [INN]:** It is employed exclusively as an external agent for:

- **mechanical cleansing'** of wounds *via* absorption of both *tissue debris* and *exudateds:*

Examples: These essentially include:

- *Wet wounds*–both with/without infection *viz;,* *leg ulcers* and *leg ulcers due to venous stasis.*

12.5.1.2 Xanthun Gum

The **xanthum gum** designates a *polysaccharide gum* duly produced by the bacterium *xanthomonas* **Examples***. It is essentially composed of *two* vital segments, such as:

- **D–glycosyl, D–mannosyl, and D–glucosyluronic acid residues,** and
- **different proportions of** *O*–acetyl and pyruvic acid acetal.

In fact, the **xanthum gum** represents the *primary structure* comprising a **cellulose backbone** having **trisaccharide side-chains;**

whereas, the repeating unit being a **pentasaccharide** predominantly.

Preparation: The preparation of **xanthan gum** is initiated from a bacterium *X.campestris*, that develops very commonly upon certain species of ***Brassicaceal.*** In this manner, the *vegetative substrate* is being duly exploited in producing a **'gummy exudate'** termed as the **'xanthan gum',** which designates a relatively **'high-molecular-mass of the anionic polysaccharide',** having nearly 1×10^6. As per the **European Pharmacopoea (Addendum to 3rd edn.)** - 1998 the **xanthan gum** comprises:

- pyruvinyl moieties not less than 1.5%, and
- as their respective Na,K or Ca-Salts.

---

* Jeans *et al.* : *J Polymer Sci,* **5** : 519 (1961).

However, on a **Commercial scale,** the *xanthan gum* is produced by the **bacterial culture** consisting of:

- an appropriately **'buffered'** and **'aerated'** media containing essentially :
- carbohydrates,
- source of N, and
- minerals.

Further on the completion of the on-going fermentation process the *generated polymer (i.e.,* the **polysaccharide)** is duly recovered by precipitation carefully with **isopropanol,** *filtered, dried,* and powered to the desired particle size.

Critical Tests: These essentially include:

(1) Absence of residual solvents by **GLC.**

(2) Absence of other contamination as polysaccharides by carrying out **TLC** of the respective hydrolysate.

(3) Absence of total microbial contamination in the **'final product'.**

(4) Spectrophotometric (UV) determination of **pyruvic acid** using *dinitro phenylhydrazine.*

Structure: The actual structure* of **xanthan gum** is based upon a backbone very much similar to that of:

- **cellulose** [*i.e.,* D–glucopyranose lined to (β–(1$\rightarrow$4)], and
- **trisaccharides** forming branches right from C–3 of the glucose units.

Salient Features: These essentially comprise:

(1) Each **'trisaccharide molecule'** is made up of:

- **one mole of D–glucuronic acid salt,** and
- **two moles of D–mannose** of which:
  - ➢ one is attached to the **'main chain',** and is duly acetylated at C–6, and
  - ➢ the other being the **'terminal'** gets attached to a mole of **pyruvic acid** through an **acetal** involving its **hydroxyl (–OH) moieties** at **C–4 and C–6 positions.**

(2) Approximately 50% of these **'terminal mannose units'** do form a *cyclic ketal* having **pyruvic acid** [*i.e.* 4,6– *O*–[1–carboxyethylidene]–D–mannose].

Note: However, their precise and exact distribution across the entire polymer is not yet fully known.

(3) **Branching Point:** In reality, the branching point vividly determines a highly specific conformation, which eventually explains the observed utmost resistance offered duly to:

- the prevailing *'enzymes',* and
- the **physical characteristic features** specific to **xanthan gum,**

The structure of the **xanthan gum** is given as under:

---

* Sloneker *et al., Can J chem.,* **40:** 2066, 2188, 1962; Jansson PE *et al. Carbolyed Res,* **45** : 275, 1975; Morris Er *et al. J Mol Biol,* **110** : 1,1977.

$$\cdots\cdots\!\!\rightarrow\!)\beta\text{--}D\text{--}Glcp\text{--}(1\!\rightarrow\!4)\text{--}\beta\text{--}D\text{--}Glcp\text{--}(1\!\rightarrow\!4)\text{--}\beta\text{--}D\text{--}Glcp\text{--}(\!\rightarrow\!\cdots\cdots$$
$$1\!\!-\!\!\!+\!\!\!\rightarrow\!3$$
$$\alpha\text{--}D\text{--}Manp\text{--}6\text{--}O\text{--}Ac$$
$$1\!\!-\!\!\!+\!\!\!\rightarrow\!2$$
$$\beta\text{--}D\text{--}Glcp\text{--}A$$
$$1\!\!-\!\!\!+\!\!\!\rightarrow\!4$$
$$\beta\text{--}D\text{--}Manp$$
$$4\ (|)\ 6$$
$$H_3C\text{--}C\text{--}COOH$$

Xanthan Gum

Characteristic Properties: The various **characteristic properties** invariably include:

(1) **Xanthan gum** is both soluble in cold and hot water.

(2) It essentially forms aqueous solutions rapidly the **viscosity** of which almost remains unchanged by:

- variation in temperatures, and
- alternation in pH range.

(3) **Pseudoplastic–Type Solutions:** The **xanthan gum** solution behaves as **pseudoplastic-type solutions** due to the following facts:

- **reduction in viscosity directly proportional to shearing, and**
- **instantaneous recovery of the initial viscosity on the discontinuation of shearing.**

Note: Therefore, the 'xanthan gum' enjoys a host of allied non pharmaccentical utilities, such as:
- 'emulsions' which flow out of the bottle, but do stick onto the surface of 'salads', and
- 'cleansing products' which tend to become fluid rapidly upon application; however, specifically adhere to the surface *'without dripping' e.g.,* ceiling paints.

(4) In general, the incidence of the **'incompatibilities'** are observed to be rare *e.g.,* **borates, free-radical generator, hypochlorites,** and **peroxides.**

(5) **Compatibility Profile:** The **compatibility profile** of the **xanthan gum** is quite commendable:

- no adverse reactions with most salts,
- fairly good with moderate surfactant concentrations,
- enough tolerance with most preservatives,
- withstands up to 50% (*v/v*) alcohol concentrations,
- compatible with a good number of **'vegetable hydrocolloids'**,
- fails to form **'gels'** by itself,
- gives rise to **'thermally-reversible'** in the critical presence of **'galactomannans'** belonging to the family *Fabaceae* **(carob),** and
- it fails to exhibit the least toxicity profile.

Applications of Xanthan Gum: These predominantly comprise:

(1) It is regarded to be the **'first-choice stabilizer'** meant for the formulation of **emulsion and suspensions.**

(2) **Xanthan gum** is highly prized *natural gum* for attributing **pseudoplasticity of its solutions,** which has eventually boosted up its global market requirement in such a short span.

(3) **Food processing industries** *i.e.,* based on latest developments in **'Food Technology'** has made an intensive and extensive use of it, because of the **'authorized application'*** as given under:

- **For Instant Soups :** 0.1% (w/v); and
- **For Dessert Puddings/Custards :** 0.5 % (w/v)

(4) **Stabilizing and Gelating Agent:** It prominently serves as **a reliable stabilizing and gelatin agent** for being an integral part of various types of *'sauces'* namely:

- salad dressings,
- seasonings,
- gravies,
- soups (instant-type),
- jellies,
- milk–based jellied desserts,
- bakery products (*viz.,* sandwich loaf-bread)
- fruit-based preparations,** and
- certain canned products.

(5) **Multifarious Utilities–** include: paints, cleansing products, explosives, polishes, pesticides, photographic emulsions, textiles, and printing aids.

12.5.1.3 Lentinan

Lentinan is a **natural polysaccharide** isolated primarily from the edible fungus or mushroom *lentinus edodes* (Berk.) Sing. Importantly, its structure is a **glucan** essentially comprising a principal chain having β–(1→ 3) **linkage,** duly substituted by (1→ 6) **linked glucoses,** and of molecular weight ranging between 400,00 to 800,000.

Chihara *et al.* (1969)*** and chihara *et al.* (1970)**** put forward the primary structure of **'lentinan'** as a : β–1,3 D–glucan having *two* β–1, 6–glucopyranoside branchings for each and every 5 β–1, 3–linear linkages.

Sasaki *et al.* (1976)***** carried out a detailed structural study of **lentinan** and proposed the following **chemical structure:**

---

* **European ID code E 415;**

** Usually combined with **'pectin'** so as to check and prevent *'syneresis'.*

*** Chihara G'*et al.: Nature,* **22:** 637, 1969.

**** Chihara G'*et al.: Cancer Res.,* **30 :** 2776, 1970.

***** Sasaki *et al., Carbolydrate Res.,* **47:** 99, 1976.

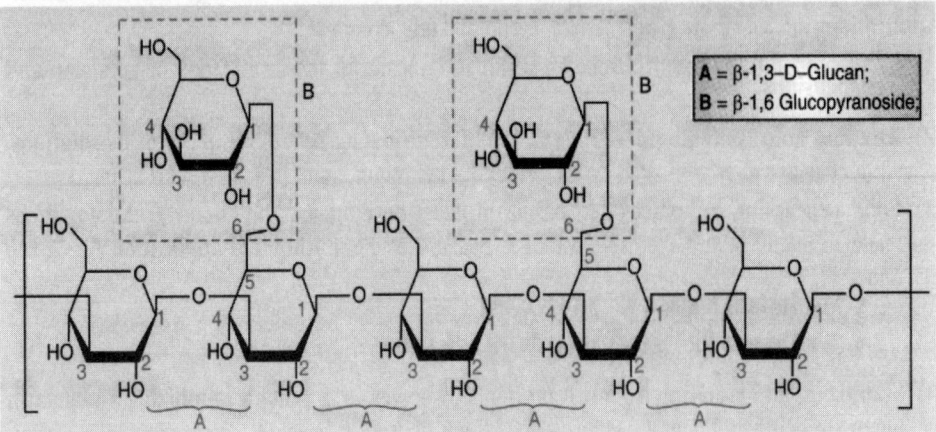

Lentinan

Characteristic Properties: These essentially consists of:

(1) The **antitumour** characteristic properties of *lentinan* has been duly established to be caused not due to any **cytotoxic properties,** but exclusively on account of an **immunogenic activity.**

(2) **Lentinan** particularly stimulates the proliferation of:
- *T–lymphocytes* in the presence of **Interleukin–2,**
- **macrophage activity,** and
- critical production of **Indtrleukin–1**

(3) **Immunostimulant Activity**: Maeda and ehiharea (1971)* studied the *immunostimulant activity* profile of **lentinan** and observed encouraging results.

(4) **Gastric and Colorectal Cancer:** Chihara (1982)** reviewed the clinical trials carried out with **lentinan** upon the **gastric** and **colorectal cancer** on human subject indicating fruitful results.

> **Note:** **There are a good number of 'fungi', specifically *Basidiomycetes*, which meticulously manufacture typical 'polysaccharides' having significant characteristic features very much akin to those of 'lentinan'.**

12.5.2 Polysaccharides of Lower Plants [or Algal Polysaccharides]

A survey of literature has revealed that one of the most characteristic elements pertaining to the various spectrum of **phyla** which essentially comprise the **'algae'** being that they include predominantly:
- next to **'unicellular organisms',** and
- **multicellular organisms;**

which categorically gives rise to the formation of *'complex thalluses' i.e.,* **cell aggregates** that are invariably found to be **differentiable, flexible,** and **free-from lignin.**

Nevertheless, the **'algal polysaccharides'** do have the following *exceptions,* namely:

---

* Maeda YY and Chihara G, *Nature,* **229** : 634,1971
** Chihara G, *Int.J.Tissue Res.* **4** : 207-225, 1982

(a) 'matrix engulfing the *algal cells* remains acidic in nature, and

(b) 'constituent polysaccharides' designate as polymers capable of producing 'gels'.

Interestingly, the genuine and critical adaptation to the specific 'marine environment' necessarily demands more *flexibility vis-a-vis rigidity*, which is logically and scientifically based upon the fact that it never exerts its effects on these **marine plants,** as it normally does on the **terrestrial plants.**

Following are some of the recognized, well-accepted, and largely used **polysaccharides of lower plants** (or **algal polysaccharides**), namely:

12.5.2.1. Alginic Acid-Alginates

Alginic acid is a *hydrophilic colloidal polysaccharide* duly obtained from **sea weeds,** which in the form of mixed salts of Ca, Mg, and other bases, usually occurs as a structural component of the cell wall.

European Pharmacopea (1998) designates **alginic acid'** as an admixture of **polyuronic acids** chiefly obtained from **algae** belonging to the family *phacophyceae*. Besides, it consists of not <19% and not more than 25% of the total carboxyl (–COOH) moieties, calculated with reference to the dried substance.

It may be worthwhile to state here that **'alginic acid'** is also produced by certain specific microorganisms. With the advent of *biotechnological advanced approaches* rendering it to be appreciably **cost-effective,** it could be feasible and possible to exploit these organisms *viz., Azotobacter, Pseudomonas,* to secrete **'exopolysaccharides'** having critical **modified and uniform structural characteristic features.**

Structure: Alginic acid is a linear polymer of *two distinct* entities, namely:

- B-(1 → 4)–D–mannosyluronic acid (**I**) and
- α–(1 → 4)–L–gulosyluronic acid residues (**II**), the relative proportions of which vary remarkably upon:
  - **botanical source of the plant,** and
  - **state of maturation of the plant.**

Importantly, the linkage observed between the aforesaid *two* **monomers (I) and (II)** is found to be of the type β–(1 → 4). In fact, these *two acids* are duly present in the polymer in the form of the following structures, such as:

- **Poly–M homogeneous block (I),** and
- **Poly–G homogeneous block (II),**

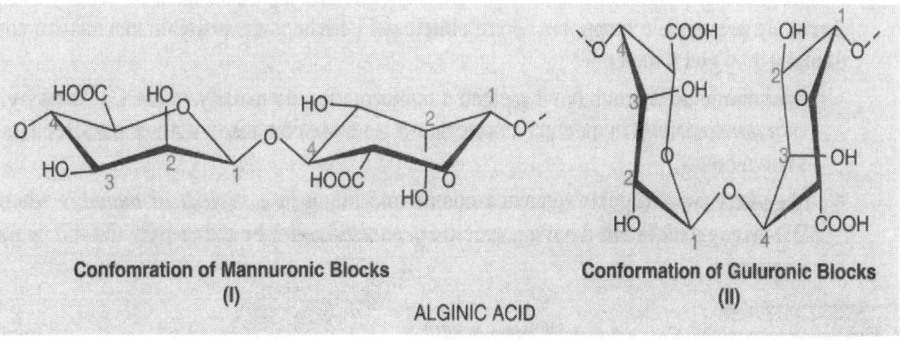

Conformation of Mannuronic Blocks
(I)

ALGINIC ACID

Conformation of Guluronic Blocks
(II)

which are duly separated by regions, wherein they may appear as **alternate positions** *viz.,* [G–M–G M...] as given below:

Preparations: The preparation of **alginic acid** and **alginates** is vehemently marked by a marked and pronounced **polyanionic inherent character** that eventually renders **'alginic acid'** to:

- remain **insoluble in water** and enable it to form **'salts',**
- **water-soluble** Na, K or NH_4 salts, and
- **water-insoluble Ca salts.**

However, the extraction of the duly *'fragmented'* or *'crushed'* **thalluses** usually commences with a critical and careful treatment with **deionized acidified water** that eliminated specifically:

- **soluble mineral salts,** and
- **sugars.**

In actual practice, it commences with the following steps in a sequential manner:

- maceration and stirring of the **'thallamus fragments'** in *hot aqueous alkaline solution**, that helps to solubilize the **alginic acid** completely, and
- filtration and elimination of **calcium alginate** as the *subsequent residues* is adequately prepared by the addition of $CaCl_2$ **solution** (2%. w/v) to the filtrate dropwise,
- careful recovery of the *decolorized* and *deodorized* **precipitate** is accomplished, and
- purification is done by **redissolution** and **representation** as the pure **alginic acid.**

Alternative Method: Alginic acid may also be isolated easily and conveniently by the direct *acidification of the alkaline solution;* however, this procedure categorically gives rise to *two serious drawbacks,* namely:

- **alginic acid (polymer)** loses its solubility profile to an appreciable extent, and
- evolution of CO_2(as a gas from the ensuing reaction normally carries the product (*i.e.,* **alginic acid**) almost near to the above surface.

Note: In the *two* procedural variants one may be able to produce different types of salts (*viz.,* Na, K, NH_4, or Ca), or alternatively an ester i.e., propylene glycol alginate.

Characteristic Properties: The various cardinal **characteristic properties** of **alginic acid** are as stated under:

(1) The **'alginates'** of the so called *monovalent cations (viz.,* Na$^+$, K$^+$, and NH_4+) and also of the *bivalent cation (viz,* Mg $_2$+) distinctly produce **viscous colloidal solutions** exhibiting explicitly **'pseudoplastic behaviour'** at relatively *lower concentrations.*

(2) Besides, further progressive addition of **divalent cations** (*eg.,* Ca^{2+}ions) invariably causes the thermally irreversible formation of an **'elastic gel'.** Perhaps the possible mechanism could be explained as given under:

- **'guluronic acid'** units having pleated conformation do usually retain Ca^{2+}ions by virtue of *coordination,* in perfect cooperation and coordination with a parallel chain, as shown below,
- **'egg-box type'** regular sequence comes into being in a *periodical manner;* whereby, a **3D–array gets formed** having specific *organized zones* by either **poly (M–G)** or **poly–M** segments.

---

* A sodium carbonate (Na_2Co_3) solution 5% (w/v) at 50°C.

Therefore, in the light of the above statement of facts, the ensuing structure of 'alginic acid' (polymer) is considered to be the– 'critical determining factor of the rheological behavioral pattern of the alginic acid gels. Importantly, the precise and exact proportion of the poly G-blocks, and their subsequent length actually control the *crucial generation* as well as the *critical strength* of the gels that are duly accomplished in the presence of Ca^{2+} ions*.

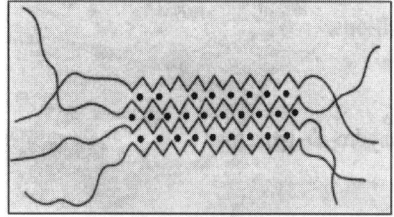

Showing wordination of Ca^{2+} ions ( ) by Poly–G units; and *Creation of function iones* pertaining to the **Egg–Box Type**

Alginate Gel Formation

Applications of Alginic Acid: Though **alginic acid** is being profusely used in **pharmaceutical formulations** in many ways, such as:

- **Gastric Antacids–** for the *pathogenic disturbances viz., reflux, oesophagitis, and pyrosis.*
- **Treatment of Obesity–***Ca–salt of β–poly–D–mannuronic acid.*
- **Pharmaceutical Technology–** is valued extremely for their **thickening, binding,** and **disintegration properties.**

Other application of **'alginic acid'** essentially comprise:

(1) In France, the *'Official Fucus'* is being strongly used as an **plant-based medicament** for the treatment of **weight-loss** problems**.

(2) The Germans, recommend it profusely for **thyroid disorders** and **obesity.*****

> **Note:** The French Official *Fucus* should have a total iodine content ranging between 0.03 to 0.2%. Besides, the I2 concentration of the *sea-weed-based products* happens to vary widely, and gets lowered distinctly during storage.

12.5.2.2 Carrageenans [*Syn:* Carageen; Carrageenin;]

The **'carrageenans'** designate a mixture of *sulphated polysaccharides* duly extracted from **red seaweed** belonging to the family *Rhodophyceae.* Besides, there are also several **major sources,** namely: *Chondrus crispus* (L.) Stackhouse, and *Gigartina stellata (Gmammillosa)* (Goodenough and Woodward) J.Aghardh, belonging to the family *Gigartinaceae.* They are mostly found abundantly in *North Atlantic Coastal Regions* extending right from **Norway to North Atlantic coastal regions** extending right from **Norway** to **North Africa.** The **seeweed** is also termed as **'Irish Moss.**

Nomenclature: Importantly, the very name **'carrageenan'** has been duly derived from the Irish coastal town of *Carragheen.*

Structure: The **'carrageenans'** are **galactans** or **polymers** of **D–galactose,** which are found to be **heavily sulphated;** and do represent as 'specific anions' having multiple electrolytes with *molecular weights* ranging from 10^5 to 10^6, It is, however, pertinent to state here that almost **'all carrageenans'** essentially have a *linear structure* pertaining to the **(AB)n type** and possess $(1 \rightarrow 3)$ and $(1 \rightarrow 4)$

---

* It has been duly established that *in vivo*, the **oldest** and the **hardest** tissues are found to be the richest in **L–guluronic acid.**

** *French Expl. Note, 1998.*

*** Mongraph of **German Commission E, 1990.**

bonds at *alternate positions,* where **'A'** and **'B'** represent the **galactopyranosyl residues,** as given under:

$$\ldots \ -3_A1\beta{-\!\!-}4_B1\alpha{-\!\!-}4_B1\alpha{-\!\!-}3_A1\beta{-\!\!-} \ \ldots$$

Importantly, both **'A'** and **'B'** units are *invariably sulphated* as stated under:

Unit A : particularly at **2–or 4–positions; and**

Unit B : specifically at **2–or 6–positions.**

Interestingly, the **Unit B** may be either:

- **D–galactose** or
- its **'internal ether'** *viz.,* 3, 6–anhydro–D–galactose

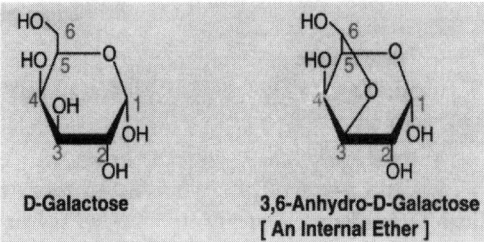

D-Galactose   **3,6-Anhydro-D-Galactose**
                         [An Internal Ether]

Classification: The **carrageenans** may be classified into **seven distinct variants** which are duly distinguished as a *critical function of the nature of the sequence.* However, the structures of these **repeating units** (*eg., K, U, V, Q and E*) together with each individual **A Units** and **B Units** have been summarized meticulously as given below:

A Units	B Units	Carrageenans
D–Galactose–4–sulphate	D–Galactose–6–sulphate	μ
	D–Galactose–2, 6–disulphate	ν
	3, 6–Anhydro–D–galactose	κ
	3, 6–Anhydro–D–galactose–2–sulphate	ι
D–Galactose–2–sulphate	D–Galactose–2–sulphate	ε
	D–Galactose–2, 6–disulphate	λ
	2, 6–Anhydro–D–galactose–2–sulphate	θ

Structure: The structural variance as observed in the **carrageenans** are namely due to several critical governing factors, such as:

- **source of the carrageenan,**
- **the specific species of the plant origin,**
- **'alternaing generations',**
- **'internal either formation'** particularly in the **B units** (*i.e.,* 3,6–hydro–D–galactose) and the formation of **sulphuric acid half–esters** essentially engaging the **4–hydroxyl (OH) moieties** of the **A units** that are found to be highly specific to the **haploid gametophyte,** and
- *sulphation* taking place at the C–2 position of the **'residue A'** exclusively in the **deploid tetrasporophyte.**

The *content variability* as observed in **3, 6–anlydro–D–galactose** is found to be:

- In **'haploid gametophyte'** – the *carrageenan seerus* to be linked to a *varying extent* prevalently between the structures of the **u and v types** (*ie.*, they are regarded to be the **'precursors'**.

- However, it may be observed crtically that the *two cardinal characteristic features, namely:*
  - ➢ propostion of **diploid tetra sporophytes,** and
  - ➢ proportion of **haploid gametophytes,**

which are duly present in a *'given population'* depends solely upon such factors as:

- variability associated with **'geographical origin'** of the *species*, and
- enzymatic conversion rate duly influcneed by the **environmental parameters.**

Following are *two* typical structures of **κ–Carrageenan** and **λ–Carrageenan:**

κ-Carrageenan   λ-Carrageenan

Characteristic Features: These predominantly comprise:

(1) **K–and L–carrageenans:** These **Carrageenans** gets dissolved in water rapidly, whereby the **macromolecular chain** undergoes proper distribution **statistically** (*ie.*, **random coi**). Besides, at normal room temperatures, one may see the **'regular segments'** of the ensuing molecules being intimately associated to form **'double helices'**. The resulting product are **stabilized** adequately by the aid of **weak-inter–chain bonds,** thereby yielding an ultimate **thermally reversible gel.**

3D–Structures: In reality, the crucial presence of observed irregularities within the polymer apparently forms *linkages*, which in turn exerts force upon each chain to associate with a host of **'neighbouring chains'** to produce a **3D–structure** duly responsible for the overall **cohesion of the gel.**

(a) **L–carrageenans:** In this particular instance, the **sulphate (SO_4^{2-}) moieties** that are strategically located on the *outer zones* of the **'double helices'** to undergo any sort of **'association'** perhaps due to the ensuing **electrostatic repulsive forces**; and, therefore, the ultimate **'gel'** formed remains very much *elastic in nature*, and also fails to *contract*.

(b) **K–Carrageenans:** They essentially give rise to the formation **'double helices'**, which eventually in the critical absence of the **sulphuric acid half-ester groups** located at **C–2** of the **B– unit** may tend to *aggregate* and the resulting **'gels'** found to be:

- brittle in nature,
- rigid in nature, and
- more capable of undergoing synthesis

(2) **λ–Carrageenans:** Interestingly, the solution of **λ–Carrageenans:** do not **'gel'** at all. It is perhaps due to the crucial perusal of the **sulphuric acid hall-ester moieties** which prevent and check ultimately the formation of the **helical structures.** In fact, the **λ–Carrageenans** are soluble in **cold water** to yield relatively **extremely viscous solutions.**

Test for Carrageenans: The various characteristic **'tests'** for the **carrageenans,** as recommended in the **French Pharmacopoea*** are as stated under:

- **Identification and characterization of galactose** by *TLC* after careful hydrolysis of the polymer (carrageenan) with dilute H_2SO_4.
- **Apparent viscosity–** assay of a **15%** (w/v) aqueous solution at 75°C.
- **Limit test–** for heary metals *viz'*. Pb, Hg etc.,
- **Residual level in McOH** and **Propanol–**by **GLC** method (<0.1%)
- **Quantitative determinations of** SO_4^{2-} – by *barium perchlorate* **[Ba(CO₄)₂]** after appropriate mineralization in O_2.

Application: A few vital and important application of **carrageenans** are as given under:

(1) In *'pharmaceutical industry'*, in the formulation of **pastes, creams, and emulsions.**

(2) **Carrageenans** have successfully gained entry into the various commonly used formulations related to **'hygiene'** and **'cosmetic'** products, such as: **toothpastes, shaving creams, shampoos, skin–lotions, gels, ointments** and **creams.**

(3) The **non–absorbable, non–digestible,** and **non–toxic K–and τ–carrageenans** are being employed as additives** in *low concentrations* in **specialized food products,** for instance:

- inhibition of ice–cream crystallization, in milk–based (dairy) products, and as stabilizers/gel-forming agents,
- **λ–Carrageenans** find their enormous usuage as the prevalent *stabilizers* and *thickners* of **'emulsions'.**

12.5.2.3. Agar [or Gelose]

Agar is also commonly known as: Bengal Isinglass; Ceylon Isinglass; Chinese Isinglass; and Japan Isinglass.

Agar refers to a *polysaccharide complex* duly extracted from the agarocytes of algal *Rhodophyceac. Nevertheless, the major and predominant* **agar–producing** genera are namely: *G'elidium, Gracilaria, acanthopeltis, ceramium,* and *Pterocladia* generously found in the pacific and Indian Oceans, and Japan Sea.

However, **agar** may be adequately sperated into *two* distinctly separate fractions, namely:

- **'neutral gelling fraction'** *e.g;, agarose,* and
- **'sulphated non–gelling fraction** *e.g;, agaropectin****.

Structure: The **structure of agar** is believed to be a *complex range of polysaccharide chains* with strategically positioned alternating α–(1 → 3) and β–(1 → 4) linkages; and also varying in **'total charge content'.** In fact, *three* **extremes of structures** haven been duly noted, for instance:

---

* French Pharmacopoea, 10th, ed, 1998
** European ID Code E407.
***Araki : *J Chem Soc Japan*, **58:** 1338, 1937.

- **neutral agarose,**
- **pyruvated agarose with little sulphation,** and
- **sulphated galactan*** *i.e.,* **poor in 'internal ethers'.**

Sequence:
-(1→ 3)-β-D-gal-(1→ 4)-α-L-gal---

Agar

Neutral Agarose – designates a barely sulphated linear polymer, duly formed as a **'linear structure'** of the *(AB) n–type* haring distinct **alternate 1 → 3 and 1 → 4 bonds,** wherein:

- the **A units** are mettylated partially to D–galactoses, and
- the **B units** are essentially L–enantiomers of galactose (mostly of the 3,6–ablydro–L– galactose type**).

Pyruvated Agarose (or *Pyruvyl Agarose*)– seems to be hardly sulphated and essentially consists of a relatively large segment of **internal 3, 6–anhydrides;** whereas, a small portion of its **A units** are **4, 6–)–(1–carboxyethylidene)–D–galactoses.** That is, the **hydroxyl (–OH)** moieties present duly at C–1 and C–6 positions are actively engaged in the formation of a **'cyclic petal'** by subsequent interaction with pyruvic acid.

Note: The actual quantum and proportions of the *'two forms',* namely : 'neutral agarose' and 'pyruvated agarose' do vary significantly depending upon the source of the species.

Characteristic Properties: These invariably comprise:

(1) **Agar** consists of colourless of pale–yellow, translucent, and occurs as resilient ribbons or flakes.

(2) **Agar** is usually characterized by:
  - critical flecting colour reaction with iodine solution,
  - precipitation of SO_4^{2-} ions with dilute mineral acid and slight warming,
  - solubility in hot water (1%, w/v, solution),
  - specifically gives rise to the formation of a **'gel'** between a temperature ranging **30–35°C,** that liquefies finally only above **80°C,** and
  - dissolution in hot water and duly forms **thick gels** upon cooling.

(3) **Agarose** prominently forms **'double–helical structures'** (*ie;,* just similar to **DNA**) which indulges into *actual aggregation* to result into a *3D–network* capable of **retaining enough water molecules.** Amazingly, the **3D-network of agarose** has the following *three* **cardinal features,** namely:

---

*   Duck worth *et al.: Carbolyd.* Res., **16** : 189, 435, 446, (1971)
**   Please see seetion 12.5.2.2.

- undergoes the phenomenon of **'assimilation'** *i.e., absorption*),
- fails to undergo **'fermentation'**, and
- exhibits absolute **'non-toxic profile'**.

> **Note:** Based on the aforesaid exceptional features, *agarose* is used as : (a) *mechanical laxative;* and (b) *gastrointestinal (GI) protective* agents.

(4) **In Microbiology: Agar** essentially and widely serves as a **'classic culture medium.**

(5) **In Tissue-Culture Studies: Agar** caters for the *in vitro* production of *plants.*

(6) **In Chemical and Biochemical Studies: Agar** largely forms **'gels with great resilience** and **multiple utilities** either with other **polysaccharides** or as a **single entity,** such as:

- **stationary phase** in *Size Exclusion Chromatography* (SEC)*,
- an effective medium in **'Gel Electrophoresis' (GE)****, and several known **'Immunological Methods'**, and
- in **'Affinity Chromatography'** after due substitution with various additives.

12.5.3 Homogeneous Polysaccharides [or Homopolysaccharides]

The **homogeneous polysaccharides** essentially comprises: *starch, cellulose,* and *fructans,* which shall now be discussed individually in the section that follows:

12.5.3.1 Starch [Amylum]

Starch represent the **'carbohydrate polymer'** duly stored by *plants,* very much analogous to *storage of fats by the animal species.* It usually occurs as **discrete granules** in the mature grain of:

- **corn,** *Zea mays* Linné (Family: G'raminaceae),
- **wheat,** *Triticum activum* Linne (Family : G'raminaceae),
- **tubers of potato,** *Solanum tuberosum* Linne (Family : Solanaceae), and
- **rice,** *Oriza sativa* Linné (Family : G'raminaceae).

In a rather broader perspective, the starches are found to be an admixture of *two* distinct **polymers,** namely:

- **Amylose** *ie.,* a **linear (1 → 4)–a–D–glucan,** and
- **Amylopectin** *ie.,* a **branched D–glucan with mostly:**
  - ➢ α–D–(1 → 4) linkage, and
  - ➢ α–D–(1 → 6) linkage (approx. 4%).

The **'corn starch'** critically contains: **amylose** *(approx. 27%),* and **amylopectin** *(approx. 73%);* and furthermore, these **two polymers** are so intimately associated very much in the **'crystal lattice'** that they almost cease to become soluble in *water* and alcohol.***

---

*   Kar Ashutosh, **'Pharmaceutical Drug Analysis'**, New Age International, New Delhi, 3rd, ed, 2010.

**   Kar Ashutosh, **'Pharmaceutical Analysis'** : Vol.II, CBS Publishers & Distributors, New Delhi, 2009.

*** Bemiller JN : **'Starch Amylose'** In *Industrial Gums* 'RL Whistler (ed.), Academic Press, New york, 2nd ed. pp: 545-566, 1973.; and Powell El : **'Starch Amylopeclin,** *ibid,* pp 567-576.

Production: In general, **starch** is commercially produced by extraction *primarily* from : **corn** and **tubes of potato**; and secondarily from: **wheat** and **cassava**.

The **'corn starch'** is invariably prepared by a **'wet process'***, that eventually yields **63 kg of starch** from **one quintal of corn.**

Structure and Composition:

It is, however, pertinent to state here that **starch** (*ie.*, the *saccharide fraction*) designates a mixture of **two polymers**, namely:

(*a*) **Amylose**–being *essentially linear;* and (b) **amylopectin**–being a *ramified molecule.* Besides, the **starches** are broadly and explicitly differentiated by their respective *amylose content,* for instance:

- **Rice:** amylose content 16 to 17%,
- **Potato:** amylose content 20%,
- **Barley:** amylose content 23–24%,
- **Wheat:** amylose content 25–28%,
- **Smooth Peas:** amylose content 35%, and
- **Amylomaize:** amylose content 65–70%.

It has been duly observed that **amylose** predominantly comprises **D–glucose units** in the most preferred **$4C_1$ conformation**, which being *most stable;* and **linked critically and quasi exclusively** by **a–(1 → 4) glycosidic bonds.** Nevertheless, one may notice the crucial existence of a *small quantum of short* a–(1 → 6) glycosidic branched chains also. The **partial structure of amylose is as given under:**

A Partial Structure of Amylose

Interestingly, **amylose** account for almost 20% by weight of starch, and has an average molecular weight of >10^6. The hydrolysis of **amylose** gives rise to *maltose.* It reacts with **iodine** to result into the formation of a *colour complex* (**bluish-black in appearance**). In reality, it represents the **colour reaction of starch** which is a **confirmatory test.**

(*b*) **Amylopectin**– designates the major constituent of *starches* that accounts for nearly **80%**. It essentially consists of hundreds of **glucose units** strategically linked together by **a–(1→4)** and **a–(1→6)** *glycosidic bonds.* **Amylopectin** approximately comprises **nonlinear branches having 1 in every 20–25 glucose units.** Its hydrolysis gives rise to the formation **'maltose'** (discussed earlier in this chapter).

---

*   Brueton Jean: **Pharmacognosy: Phytochemistry of Medicinal Plants,** 2nd ed., Intercept Ltd., New york, 1999.

A **partial structure of amylopectin** is as given below:

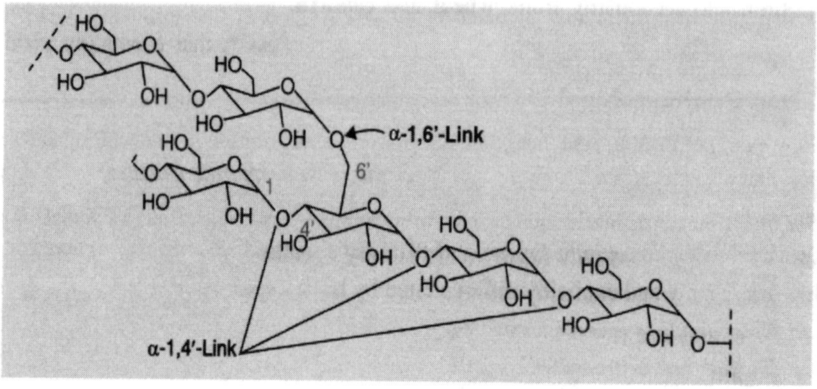

Characteristic Properties: The *characteristic properties* of **starch** are as enumerated under:

(1) Based upon two exceptionally typical properties, namely:

- essentially **'linear characteristic feature'**, and
- *homogeneity* of its **'glycosidic linkages'**,

the *amylose* may invariably adopt a unique **helical conformation** (as in **DNA**); and, therefore, it is readily capable of forming:

❑ **hydrophobic molecules** *eg.,* I_2, *fatty acids, alcohol, emulsifiers, and lipids.*

Thus, we may critically take cognizance of these facts, such as:

- analytical characterization of **starch** is exclusively based upon the reaction of **amylose** with I_2 to give bluish black colour.
- the most peculiar and highly specific formation of *typical complexes* by meticulous insertion of the **aliphatic alcohols** very much into the available **hydrophobic helicoidal cavity** present duly in the **'amylose molecule'** may allow in certain parameters the fractionation of the *two* vital components *viz;*, *amylose* and *amylopectin**.
- the presence of **amylopectin** is solely responsible for attributing the inherent **'crystallinity in starch'**.

Examples: A few typical examples are as given under:

❑ **Starch Source:** cereal starch of Type A, and

❑ **Tuber Starch or Retrograded Starch:** starch of Type B.

Note: However, the extent and nature of *'crystallinity in starch'* solely depends on the *stacking pattern of the double helices ie.,* monoclinic or hexagonal symmetry; besides, the *extent of hydration.*

- The **'retrogradation profile of starch'** may be modified suitably and conveniently by virtue of meticulous combinations with other **lipids** and **polysaccharides**.

---

* The same characteristic property is being fully utilized to keep the **'bread'** from becoming **'rancid'**; and, therefore, the use of the **trapped molecules of fatty acid monoglycouide** renders the mechanism still a mystery.

Modified Starches

Intelligently, a considerable degree of active and progressive researches have been carried out in order to accomplish the following *two* excellent **physical properties** of *starch*, namely:

- modification in the **rheological properties** of the '**starch gels**', and
- expansion of the '**possible uses**' of starch.

However, these '**modified starches**' may be prepared duly by adopting several means and ways as given below:

(1) Alterations in the respective proportions of **amylopectin** and **amylose,** which actually augments the '**varietal selection objectives**'.

(2) **Simple physical treatment** specifically confined to:

- **pregelatinized starch** *ie.,* obtained duly by dehydration and also by subjecting to pre-cooking,
- **extruded starch** *ie.,* prepared by the method of **extrusion,** and
- **compacted starch** *ie.,* prepared by mere compaction by simple mechanical means.

(3) **Chemical modifications** *ie.,* taking due advantage pertaining to the reactivity of the inherent *primary–* and *secondary*–**alcoholic (–OH) functions present in starch,** for instance:

- esterification either by **acetic anhydride** to obtain respective **starch acetates** or by **phosphoric acid** to obtain respective **starch phosphates,**
- **production of nonionic starches** by esterification to the corresponding **hydroxyalkyl starches**–to **carboxymethyl starches** (*ie., anionic starches*), and ultimately subjecting to '**colonization**' due to the following *two* modalities:
  - ❑ careful '**grafting**' of **tertiary amines,** and
  - ❑ skilful '**grafting**' of **quaternary ammonium salts,**
- **oxidation** of starch by **sodium hypochlorite [Na ClO],** and
- **hydrogenation**–as applicable to the **oligosaccharides** obtained duly from **depolymerization.**

(4) **Reticulation:** *Reticulation* refers to a phenomenon whereby **starch** is treated carefully at a temperature *just below* to that of its '**gelatinization stage**' by several *chemical reagents'* such as:

- **epichlorohydrin** ,[△ Cl]
- **phosphorus oxychloride [POCl$_3$],**
- **formaldehyde [HCHO],** and
- **acetic anhydride [(CH$_3$CO)$_2$O],**

that meticulously inducts the critical formation of a rather *small percentage* of the '**intramolecular bridges**'.

(5) **Controlled Depolymerization:** The *controlled depolymerization i.e.,* **partial hydrolysis of starch,** may be accomplished effectively in an *acidic environment*, but as to date it is invariably carried out **enzymatically.** Importantly, it makes use of such *enzyme variants as:*

- ❑ '**debranching enzymes**' *e.g., isoamylase* or *pullulanase* **type,** which help in bringing out the cleavage specifically at the α–(1 → 6) **bonds,**

❑ 'amylases' *e.g.*, α–amylase, that eventually gives rise to the **oligosaccharides in particular; and** β–amylase, which causes induction of 'recovering hydrolysis profile' right from the *non-reducing end of the linear chain* and ultimately yields 'maltose', and

❑ 'amyloglucosidases' *e.g.*, **exoenzymes** which *hydrolyzes repetitively* the following *two* **bonds** critically:

- α–(1 → 4) **bonds**, and
- α–(1 → 6) **bonds**,

   thereby producing ultimately 'glucose'.

In short, it may be added that the overall applicability of these **enzymatic techniques** do help in the large-scale production of such products:

❑ **maltodextrins** (*i.e.*, 'dextrinization of starch), of the *glucose syrup*, and *hydrolysates*,

❑ **saccharification** of *maltodextrins*, and

❑ **isomerization** of *fructose*.

Application of Starches: The various *application of starches vis-a-vis* their **product variants** and **composition** are summarized as stated under :

S.No	Product Variants	Composition	Application Profiles
1	Amylum Iodisatum	Iodized Starch	• Administered internally in *syphilis*, and *allied cachexias* along with milk, arrowroot, gruel, and water.
2	Cataplasma Amyli	Starch Poultice	• As a substitute for the household *'bread poultice'* for the household *'bread poultice'* for usage in small *superficial ulceration*.
3	Catalplasma Amyli Acidi Borici	Starch/Boric Acid Poultice	• As an antiseptic poultice for application to ulcerated wounds: [*Composition:* Starch; 10; Boric Acid: 6; and Water (sufficient) : upto **100 mL.**
4	Glycerium Amyli	Glycerin of Starch	• As a soothing emollient for use in *chilblains*, and *chapped hands in winter.*
5	Mucilago Amyli	Mucilage of Starch	• *As a commonly used bases for* **enemata.**

12.5.3.2 Cellulose

The 'cellulose' designates predominately the **polysaccharide** having several *glucose units* duly linked just as may be seen in **cellobiose** (described earlier). In a broader perspective, **cellulose** is regarded overwhelmingly as the *chief constituent* pertaining to the 'fibre of plants'. However, 'cotton' represents the 'purest natural form' containing almost **90% of cellulose.***

Cellulose essentially possesses a 'linear-chain structure'. Perhaps this characteristic feature enables **different cellulose molecules** to interact to generate *large aggregate structure* that are held together by means of the **hydrogen-bonds.** It first yields on *hydrolysis* **cellobiose,** which on farther hydrolysis breaks down to **'glucose residues'.**

Cellulose Variants: Cellulose is regarded to be the most *universal biological polymer* found in the **nature.** It usually occurs in *two* well known forms, namely:

(a) **Cellulose from Cotton Plant and Cotton** [*Gossypium* spp.; **Family:** *Malvaceae*] The different *breeds* and *varieties of cotton plants* usually are of *four* species, such as:

- **Two Asian Diploid Species:** with *thick and short fibres viz., Garboreum* L., and G *herbaceum* L.; and

- **Two American Tetraploid:**   with *medium fibres (G'.hirsutum* L.,), and with *long fibres* **(or Amphidiploid species)**   (*Gbarbedense* L.,).

Preparation of Cellulose: The **cotton cellulose** essentially comprises approximately 2 % of impurities embodying chiefly such substances as :

❏ **'wax'** as the protective film coating,

❏ **'oil'** present in the seeds,

❏ **'pectic'** substances

❏ **'minerals'** *viz;,* sand [SiO_2] and

❏ **'colouring matters',**

which are removed meticulously so as to refrain from any sort of *degradation* of the **'cellulose molecule.**

Besides, the other relatively *'heavy impurities e.g;,* stones particulate matters, and dried leaves are duly discarded by suitable **'mechanical means'.** Soon after removal of the aforesaid *'heavy impurities'* the **cotton fibres** are boiled gently with dilute aqueous sodium hydroxide (NaOH) preferably in an **'inert environment'** so as to get rid of **oil, wax,** and **pectin** as far as possible. The small **cotton stubs** or **fibres** are subsequently bleached with dilute **sodium hypochlorite [NaClO]** to obtain ultimately the **pure cellulose** (~ 99.7%).

(b) **Cellulose from Wood [or Wood Pulp]:** The **cellulose** currently being used originated from *delignifying wood* either in an alkaline or in an **acidic** medium. Wood find its abundant application in the **'paper industry',** and also used in the **production of fibres.** The processes that are involved in the development as *'pilot experiment'* do allow the critical and need-based recovery of **cellulose** plus other **wood constituents,** such as : **hemicellulose** and **lignin.**

Preparation of Cellulose: The various steps included are as given below:

(1) The **fatty** and **waxy impurities** are duly removed by treatment with *methanol.*

(2) **Hemicellulose** and **lignin** are carefully eliminated by treating the *resulting wood pulp* in *two* different manners as stated under:

❏ **Sulphite Process:** Treatment with dilute mineral acid (HCl) or with **calcium or sodium sulphite,** and

---

* Ott E *et al.* **Cellulose and Cellulose Derivatives, Vols. 1–3,** Interscience, New York, 1954, 1955; and Shafizadch *Pure Appl. Chem,* **35**; 195-208, 1973.

❑ **Sulphate Process:** Treatment with dilute sodium hydroxide (NaOH) solution or with **sodium sulphate** at a temperature ranging between **200–250°C** under **35–40 bar pressure**, and followed by *brutal return* to the normal atmospheric pressure.

(3) **Bleaching Process:** The resulting treated pulp containing the fibres are subsequently bleached either with **chlorine (Cl$_2$) gas** or **sodium hypochlorite [NaOCl]**.

(4) Finally, the bleached **cellulose fibres** are subjected to ultimate *purification* by using NaOH solution (15–20% (w/v)]

> **Note:** The two processes engaged in the purification steps are found to be *mild* relatively, and in doing so a portion of the *cellulose gets hydrolyzed.* Hence, by altering the various steps undertaken for the *purification process* one may obtain altogether *'different forms of cellulose'* that would certainly possess *varying average size of the molecule.* The *three* vital and important cellulose forms are α, β and γ.

Structure: The **structure of cellulose** essentially designates a **linear polymer** which is made up of **β–(1 → 6) linked D–glucose units.** Importantly, the prevailing **D–glucopyransoe molecules** are duly present in the **4C$_1$ chair conformation** *i.e.,* the *hydroxyl (–OH) moieties,* as well as the *glycosidic linkage* are strategically located all in the **'equatorial position'.** However, the H-atoms are present on the **axial position.**

β–*Nature of Glycosidic Linkage:* Interestingly, the **β–nature of the glycosidic linkage** essentially causes a **180° rotation** to every other *glucose unit (i.e.,* the **fundamental pattern seen in 'cellobiose').** This, however, inherits the ultimate cellulose molecule a typical **ribbon-like structure** duly consolidated by specific **intramolecular hydrogen bonds,** more critically located between the:

- **C–3 hydroxyl (OH) moiety,** and
- **intracyclic O–atom of the neighbouring glucose unit.**

The **X-ray diffraction spectrometric studies** do reveal the presence of *intermolecular hydrogen bonds* intimately associated with chains right into the **'microfibrils'** showing **explicit crystallinity,** thereby ascertaining the presence of the crucial **'amorphous zones'** in the *cellulose molecule.* Nevertheless, the molecular weight of **cellulose varies** predominantly from **5 × 10^5 to 2.5 × 10^6,** which depends solely upon the particular **'botanical origin',** besides the *nature of the tissue,* and the *exact process used for its preparation.*

The structure of **'cellulose'** is as given under:

X: β-1,4-Glycosidic Linkage

Partial Structure of Cellulose
[Showing β-(1→4) linked D-glucose Units]

Constitution of Cellulose: The various steps involved in establishing the **constitution of cellulose** are as stated under:

(1) The elementary analysis indicated that the molecular formula of **cellulose** is $(C_6H_{10}O_5)n$.

(2) **Acid Hydrolysis:** Careful and controlled hydrolysis of **cellulose** critically gives a quantitative yield of approximately 96% of the **crystalline pure D–glucose** which evidently indicates that the 'cellulose' is composed of the 'D–glucose units' exclusively.

(3) **Acetylation, Nitration, and Methylation:** These *three* different *chemical reactions* obviously determine that the *three* prominent **hydroxyl (–OH) moieties** located strategically at **C-2, C-3, and C-6 remain absolutely free** in each *glucose unit*.

(4) **Enzymatic Hydrolysis:** The complete and exhaustive enzymatic hydrolysis of cellulose gives rise to the production of the following *two* products:

- 2, 3, 6–Trimettyl glucose (I) nearly **86%**, and
- 2, 3, 4, 6–Tetra methyl glucose (II) approximately **0.6%**,

which evidently provides a strong *indicative suggestions* and *possible results* as stated under:

❏ **Formation of 2, 3, 4–Trimethyl glucose (I)–** It vividy suggests that in **cellulose molecule** the *hydroxyl (–OH) moieties* are duly located at **C-2, C-3, and C-6** positions. Therefore, one may rightly draw a conclusion that:

➤ **two glucose units** are linked *via* **C-1** and **C-4:** provided the 'glucose units' are duly present in the form of 'pyranose ring', or

➤ **two glucose units** are linked *via* **C-1** and **C-5:** if the 'glucose units' are duly present in the form of 'furanose ring'.

❏ It is, however, pertinent to state here that the *two* said 'glucose units' present in the 'cellulose molecule' are definitely linked *via* **C-1** and **C-4.**

Therefore, it is feasible and possible to assign the following probable structure to *cellulose*:

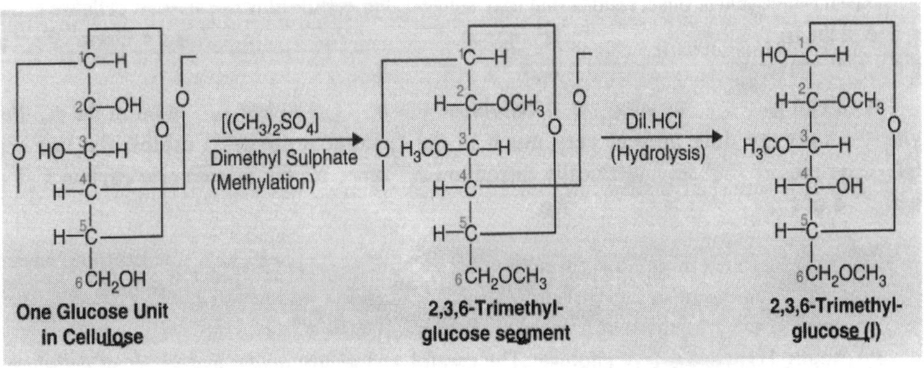

Explanations: The above cited course of reactions may be explained as under:

Step I: First of all one single glucose unit present in cellulose is duly methylated with dimethyl sulphate, whereby **C-2, C-3, and C-4** undergo methylation having open-ended oxygen linkages at **C-1** and **C-4** respectively.

Step II: Secondly, the resulting product obtained from **Step-I** is duly hydrolyzed in the presence of dilute HCl to produce **2, 3, 6-trimethyl glucose (I)**.

> **Note:** In case, the linkages were adequately interchanged specifically at *C-4* and *C-5*, we would have obtained the same 2, 3, 6-trimethyl glucose (I), as shown above, under identical treatment of the starting material (*maltose*).

- **Percentage Yield of (I):** The actual yield of compound **2, 3, 6-trimethyl glucose (I)** stands at **0.6%** only that indicates explicitly that **cellulose** essentially has a **chain-length of approximately 100 to 200 glucose units**.
- **Total Absence of Dimethyl D-Glucose:** The total absence in the *formation* (and hence the *isolation*) of **dimethyl D-glucose** clearly suggests that **cellulose** possesses a **linear polymeric structure**.

(5) **Acetolysis:** *Acetolysis* refers to the– 'simultaneous acetylation duly followed up by hydrolysis'.

Therefore, the careful *'acetolysis of cellulose'* carried out with an admixture of **acetic anhydride** and **dilute sulphuric acid** gives rise to the formation of **cellobiose octaacetate (I)** upto 50% as shown below.

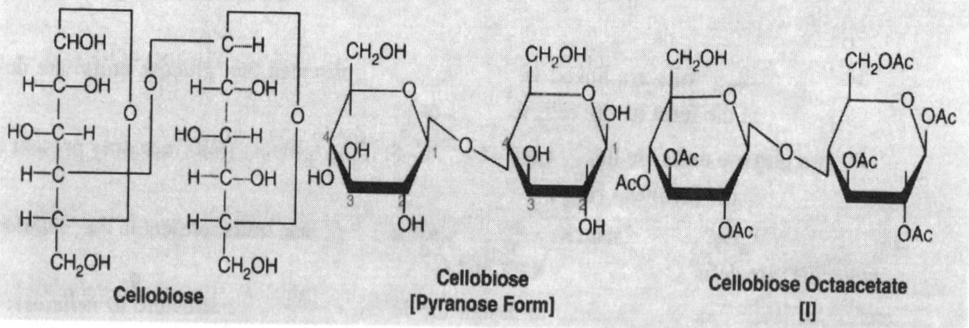

| Cellobiose | Cellobiose [Pyranose Form] | Cellobiose Octaacetate [I] |

Based on the above observations one may suggest that **cellulose** is made up of **cellobiose units** by virtue of the fact that under the **acetolysis parameters'** the *backward reaction* is not possible at all *i.e.,* the actual conversion of *glucose* to *cellobiose octaacetate* (I).

A careful glance at the structure of **cellobiose** one may get a clear indication in the *cellulose* the *glucose units* are duly present very much in the **'pyranose form'**. Besides, the *two* separate 'glucopyranose units' are strategically linked to each other *via* the β-anomeric carbon C-1 of one with C-4 of the other as shown above.

> **Note:** It is pertinent to observe here that the actual formation of the *'cellobiose structural analogues'* fails to indicate and clarify whether each and every *'pair of cellobiose units'* are critically linked to each other either by α-or β-glycosidic linkage.

(6) **Acidic Hydrolysis of Cellulose:** The careful and gently *acidic hydrolysis of cellulose* and a detailed study on the **'hydrolyzate'*** has revealed that in addition to *cellobiose* it also contains essentially: **cellotriose, cellotetraose, cellopentaose, cellohexose,** and **celloheptaose**. Interestingly, the **optical**

---

* Helferich *et al.* : *Ber,* **63:** 992, 1930; Peterson, Spencer, *J. Am. Chem. Soc.,* **49;** 2822, 1927.

rotation calculations further established that in all the above *hydrolyzate entities the* **C-1 linkage** is in the **β-configuration.**

(7) Cellulose on being dissolved in water forms usually a **colloidal solution** thereby indicating that the substance is indeed a very **large molecule.** Besides, it invariably gives rise to fibres (*viz.,* **rayon**) it clearly ascertains that the **molecule is linear. However, both these characteristic features have been duly established by X-ray diffraction analysis.**

(8) **Probable Structure of Cellulose:** Therefore, based upon the above mentioned evidences and statement of facts from (1) through (7), one may assign the following **probable structure of cellulose:**

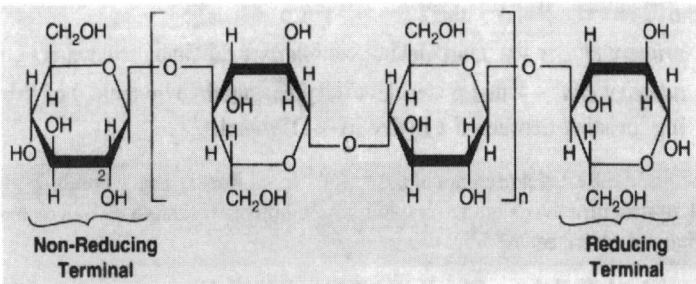

(9) **Characteristic Features of Cellulose Molecule:** The various **characteristic features of cellulose molecule** are as enumerated under:

- Primarily the **cellulose molecule** does not have a *'planar structure',* instead it possesses a **'screw-axis'** *i.e.,* each and every *glucose unit present in it is located strategically at* **right angles (90°)** to the previous *glucose unit.*
- **Rigid chain molecular status**–*Cellulose* clearly shows an absence of **'free rotation'** in the vicinity of the **C-O-C linkage** on account of the existence of visible **steric effect;** besides, the *compact packing of atoms* provide a status of the **rigid-chain molecule.**
- The **long rigid chains in cellulose** are held together by the **inherent hydrogen bonding** that provides categorically a *3D-brick work* structure to **cellulose.**
- **Rigidity** *vis-a-vis* **Flexibility**–Normally, the *long rigid chains* in **cellulose** shall render its *fibres* more strong with **rigidity,** but with little **flexibility.** As a result, though the **fibres of cellulose** may exhibit *great tensile strength,* yet they could not be either *knotted* or *tied up* without giving a shape to, thereby indicating that these should essentially possess **flexibility.**

(10) **Value of 'n'***: The precise and exact determination of the **value of 'n'** in assigning and establishing the structure of **cellulose** may be accomplished by finding out the **molecular weight.**

Nevertheless, the **molecular weight** of *cellulose* can be determined by *several methods,* namely:

- **Chemical Methods** *i.e.,* determination of the **terminal-groups** which could be **reducing** or **non-reducing** or **both,** by any one of the following procedures, such as:
  - ❏ **Haworth methylation process,** and
  - ❏ Periodic oxidation method (or **Hirst Method**),

---

* That is, the **chain length** or **molecular size** of *cellulose molecule.*

- **Physical Methods** *i.e.,* **viscosity, osmotic pressure,** and **ultracentrifugal sedimentation measurements** indicate explicitly that the *unreacted live cellulose* possesses **molecular weights** ranging between 3×10^5 to 5×10^5, which otherwise is almost equivalent to an approximate **chain-length of 2000 to 3000 glucose units.**

(11) **X-Ray Diffraction studies of cellulose:** The meticulous and critical *X-ray diffraction studies of cellulose* reveal the following glaring facts and informations, such as:

- Length of *each unit cell* **(10.25 Å) which corresponds to the equivalent length of** *two* **D-glucopyranose units (6.15 Å)** each, thereby indicating specifically that in the **cellulose fibre** the respective *'glucose units'* are strategically arranged in chains almost parallel to the **fibre axis.** Besides, there are *two* types of bonds:

  ❑ *primary chains* – that are **held in position** by the help of **'covalent bonds'**, and

  ❑ *adjacent chains* – that are critically **linked together** by the aid of **'hydrogen bonding'** existing between the **hydroxyl (–CH)** moieties.

Note: Importantly, the crucial presence of the covalent as well as hydrogen bonds duly present in the cellulose structure overwhelmingly explains the inherent strength as well as the overall non-reactivity of native cellulose fibres.

- **Mercerized Cellulose:** The *X-ray diffraction analysis* revealed that the **mercerized cellulose** apparently displayed the unit cell, length to be the same; whereas, the *adjacent chains* do become more distinctly separated. Therefore, the **'mercerized cellulose'** turns out to be much more reactive chemically *vis-a-vis* the **'ordinary cellulose'.**

Note: The augmented *'chemical reactivity'* profile explains vividly the enhanced capability of water absorption plus the ease and fervour with which it may be *'dyed'*. Besides, the *regenerated cellulases* are found to be very much akin to the *mercerized cellulose.*

- **Crystalline Nature of Cellulose:** The observed *X-ray diffraction* pattern predominantly shows that the **cellulose** is *not crystalline* throughout, but interestingly the *crystalline segments* are strategically located and dispersed in the relatively **smaller amorphous regions exclusively.** In addition to the above observations the **crystallize zones** are particularly oriented almost parallel to the **'fibre axis'**, and also are duly surrounded by the ensuing **amorphous zones.**

Hemicellulose: *Hemicellulose* refers to one of a **group of polysaccharides** that essentially differ from *cellulose* in that they may be hydrolyzed by dilute mineral acids, and from other poly saccharides in that they are not readily digested by the **amylases.** In fact, the group predominantly includes: **pentosans, galactosans, (agar-agar),** and **pectins.**

Xylan

In other words, **hemicelluloses** belong to an *integral group of polysaccharides* that usually occur in association with the **cellulose** very much in **plant cell-walls. Interestingly, the pentosans** *viz.,* specifically *xylan* comprise several **D-xylose units** which are meticulously linked *via* **C-1** and **C-4** linkages having a β-**configuration** as given under:

Semisynthetic Derivatives of Cellulose

Cellulose essentially represents a **polyhydroxylated polymer** which can be readily subjected to 'esterification' and 'etherification'.

Esterification of Cellulose: The *cellulose esters* predominantly contain *two* important derivatives, which shall be discussed briefly as under:

(*a*) **Nitrocellulose or [Cellulose Nitrates]:** These are obtained easily by carrying out the *nitration of cellulose* comprising a mixture of **nitric acid** and **sulphuric acid** in an aqueous medium *importantly, the* **'degree of nitration'** is critically monitored and adjusted by:

- *actual concentration* of the **nitrating mixture** employed,
- *time of nitration* and
- critical *proportion of* **'nitrating mixture' to cellulose.**

In this manner, one may obtain the **'cellulose trinitrate'** (or **guncotton**), which is an *'explosive'*, by carrying out the reaction to *almost completion*, whereby precisely **three nitro functional moieties** actually gain entry to the *cellulose molecule*, as given under.

No₂ Group-is duly attached to C-2,C-3 and C-6 atoms of the glucopyranose strycture

Cellulose Trinitrate
[or Gun Cotton]

(*b*) **Cellulose Acetates:** In actual practice, the **cellulose acetates** are duly obtained by acetylating the 'cellulose' in the form of *cotton linters* with an admixture of **glacial acetic acid** and **acetic anhydride.** There are *three* important **cellulose acetate variants**, namely: **tri-, di-, and mono cellulose acetates.** The commercial applications of the aforesaid *three variants of cellulose acetates* are as enumerated under:

- **Cellulose triacetate:** It has recently accomplished an enormous commercial utility and importance by virtue of its excellent *solubility profile* in **methylene chloride [CH₂Cl₂]** exclusively. Besides, it has been employed largely in:
  - ❑ **photographic film**, and
  - ❑ **textile fibres** (usually known as **'tricel'** or **'aranel'**).
- **Cellulose diacetate:** It has proved to be much more **valuable** as well as useful in comparison to the **cellulose triacetate.** In addition, it has an added advantage over the **cellulose dinitrate** in terms of the following *two* distinct features, namely:

- ❑ definitely more stable, and
- ❑ not so inflammable in nature.

- • Cellulose diacetate finds its enormous applications in the critical commercial production of:
  - ➤ plastics,
  - ➤ safety film basses, and
  - ➤ rayon acetate

- • Cellulose monoacetate: It finds its abundant usage in the large-scale production of a variety of **utility products,** such as:
  - ❑ toilet accessories,
  - ❑ handle of knife,
  - ❑ toys for children, and
  - ❑ automobile components.

(c) **Miscellaneous Cellulose Esters:** The **cellulose acetate-butyrate ester** is being employed overwhelmingly in *two* most vital and important industries, such as:

- • plastic industries, and
- • textile industries

Interestingly, the **cellulose acetate-butyrate ester** is regarded to be having an edge to the corresponding **cellulose acetate** due to the following good reasons, namely:

- • gets plasticised easily,
- • shows better water resistances,
- • exhibits improved weathering properties, and
- • proves to be tougher in strength.

Etherification of Cellulose: The **etherification of cellulose** specifically yields **water-soluble polymers** having a plethora of *technological applications*, such as: **methyl** ($-CH_3$), **ethyl** ($-C_2H_5$), **propyl** ($-C_3H_7$), and **carboxymethyl celluloses.**

The **etherified cellulose molecules** are duly prepared by adopting the following steps in a sequential manner:

- • action of an **alkyl halide** upon **cellulose** *i.e.*, treating first of all by an *alkaline substance,*
- • *frequent incorporation of either* **ethylene oxide** or **propylene oxide** to the resulting medium that ultimately leads to the formation of '**mixed ethers**' *e.g.,*
  - ❑ methylhydroxyethyl cellulose, and
  - ❑ methylhydroxypropyl cellulose.

Examples: Hydroxylate celluloses are in variably used in the formulation of **sustained release tablets** and **suspensions.**

Natrosol® (*i.e.*, **hydroxyettyl cellulose**) represents solely a **nonionic water-soluble polymer** which is employed both *intensively* and *extensively** as: **thickening agent, binder, stabilizer, protective colloid,** and **suspending agent.** Besides, **Nitrosol**(R) also finds its enormous usage in '**cosmological**

---

* Specifically in such utilities in which the need for a **nonionic material** is a prime necessity and requirement.

preparations' to serve as a **thickner** for *shampoos, conditions liquid soaps, sizes, shaving creams, textile printing inks,* and *a variety* food products (**ice-creams, beverages, ready-to-drink milk-shape shake preparations**).

Highlights of Semisynthetic Cellulose Derivatives: These essentially comprise certain typical features, namely:

(1) *Solubility in water–which solely depends upon the actual* **degree of substitution** pertaining to the **hydroxyl (–OH) moieties** of the *'parent polymer'.*

(2) It has been duly observed that a majority of them do dissolve in water to result into the formation of **extremely viscous solutions.**

(3) **Highly viscous solutions**–in turn render them to be used in cosmetological and pharmaceutical preparations abundantly; since, they critically serve as:

- **film-coating or film-forming agents,**
- **thickeners, binders, stabilizers, and lubricants** in the most preferred formulations in *gels, creams, lotions, toothpastes, make-up products, moisturisers* etc., and
- **'pharmaceutical technologists'** usually make a *judicious choice of polymer* with respect to the perfect **design of specific coatings,** such as:
  - ❑ **enteric coatings,**
  - ❑ **sustained-release microgranules,** and
  - ❑ **accomplish microencapsulation.**

Following are *two* typical examples of semisynthetic cellulose derivatives, which will be discussed briefly along with their applications:

(a) **Carmellose [***Syn:* **Carboxymethyl ether cellulose sodium salt; Carmethose; Carboxym-ethylcellulose sodium]:** *Carmellose* is prepared by treating **alkaline cellulose** with **sodium dichloroacetate.** It is a *water-soluble ionic ether* (*i.e.,* **carboxymethyl cellulose**). Importantly, its actual **degree of substitution (DS)** invariably ranges between **0.5 to 1.2*.**

Though there are certain **inherent incompatibilities,** such as:

- **alkaloids,**
- **antibiotics,** and
- **trivalent cations** (*viz.,* Al^{3+}, Fe^{3+}, Cr^{3+} etc.),

it finds an enormous application as a **pharmaceutical aid** for:

- ❑ **stabilizer for suspensions,** and
- ❑ **direct and wet compression.**

Note: **Several countries have duly approved the usage of** *'Carmellose'* **as a component of 'Low-Calorie Diets'** (*i.e.,* Appetite Suppressants).

(b) **Hypromellose [***Syn:* **Hydroxypropyl methyl cellulose; Geniosol; Lacril; HPMC]:** *Hypromellose* designates a nonionic water-soluble ether of methyl cellulose which critically gives rise to solutions having a **wide spectrum of viscosity (400-15,000 cp).**

It essentially possesses a variety of diversified applications, namely:

- film former,
- protective colloid,

---

* In the specific instance of **'cellulose esters',** the DS ranges between 0 to 3, due to the presence of **substitutable hydroxyl (–OH) moieties** located at **C-2, C-3, and C-6.**

- stabilizer,
- suspending agent,
- thickner in food products,
- ophthalmic solution to improve upon comfort with **hard and soft contact lenses** (but not the **'hydrophilic lenses'**),
- in *severe cases* of **'dry eye syndrome'**–as a tiny thin cylinders to be inserted right inside the **'inferior conjunctival canal'** so as to *stabilize the lachrymal film,*
- **pharmaceutical aid**–as *suspending agent, tablet excipient, viscosity enhancer, demulcent,* and *ophthalmic lubricant,*
- in **'adhesives'**–as *asphalt emulsions, tile mortars, caulking compounds, plastic mixes cements,* and *paints,*
- as sticker for **'agricultural dusts and sprays'**, and
- as **'eye-drops'** for topical protection of the *corneal epithelium* in the course of **functional ocular explorations.**

12.5.3.3 Fructans

Fructans represent the **'fructose polymers'** that are essentially liked by a β-(2 → 1) bond to a *terminal glucose molecule.* In fact, they may invariably regarded to he the **'higher homologues of sucrose'**. Besides, the **fructans** usually constitute an indispensable form of carbon duly fixed up by **photosynthesis.**

Types of Fructans: There are *two known types of naturally occurring* **'fructans'**, namely:

- **Insulin-type fructans, and]**
- **Phlein-type frutans,**

which will be discussed briefly in the sections that follows:

(*a*) **Inulin-Type Fructans:** They belong to the natural order *Asteraceae* and **Boraginaceae**. Importantly, the basic unit predominantly being a **β-(2 → 1) D-fructo furanosyl pattern;** and

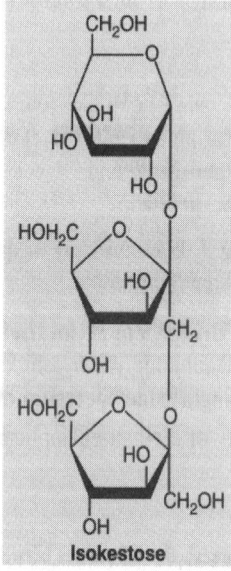

Isokestose

the very first recognized member of this series is **'isokestose'** which is actually a **trisaccharide** as given under:

Inulin, when administered *via* intravenous route, never gets metabolized; and, therefore, is not bound by the *plasma proteins in vivo*. However, it may be of **immense interest** for the ensuing **exploration phenomenon** pertaining to the **renal function.** It may be duly, explained due top its ability to undergo **'glomerular filtration,** thereby enhancing the osmotic pressure of the **tubular fluid.**

(b) **Phlein-Type Fructans (*Poaceae*):** They essentially possess a *basic unit,* β-(2 → 6)-D-**fructofuranosyl;** and the very first member of this series happens to the **'Kestose'** and an isomer of it known as **'isokestose'** as given under:

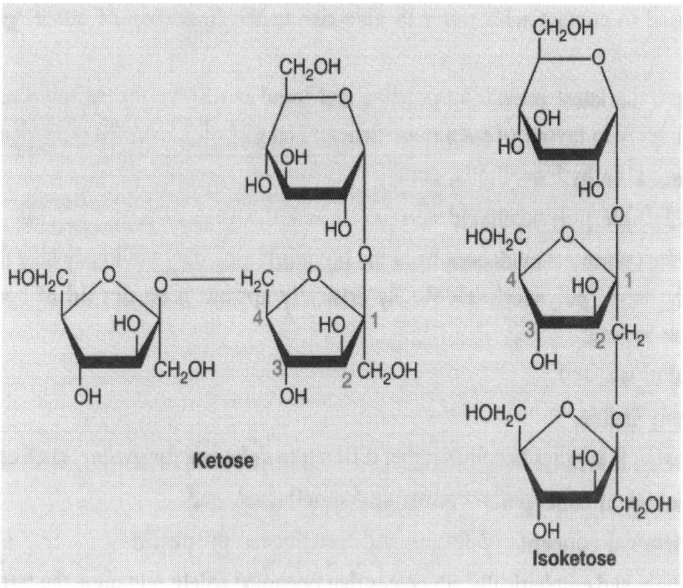

Ketose

Isoketose

Note: In fact, the 'branched fructans' *i.e.*, *'neoketose'* and even the higher homologous without the *'terminal glucose moiety'* are invariably found to be rather more rarely seen (*viz.*, *Asparagus officinalis* L.,).

Characteristic Features: These essentially comprise:

(1) **Fructans,** in a broader perspective, are very **flexible polymers'** due to the glaring fact that they do possess *three* **prominent bonds** that are strategically located at **C–C–O–C** instead of **C–D–O–C** as could be seen in several polysaccharides.

(2) **Fructans,** in general, are observed to be:

- **Levorotatory,**
- **Non reducing in nature,**
- **Readily soluble in water,** and
- **Extremely sensitive to hydrolysis in an** *'acidic environment'*.

(3) *Extent of polymerization*–is observed to be rather prevailing at a *'low ebb'* e.g., in **garlic:** upto **10;** and in **onion** upto **250.** Besides, the **degree of polymerization** duly changes both with the **type of species,** and the **physiological status.**

(4) It has been duly observed that particular appellations (*viz,* **inulin, tritein,** and **asparagosan**) invariably represent exclusively an admixture of homologous members of various **'degree of substitution' (DP)** as may be seen in a given series of **kestose** and **isokestose.**

12.5.4 Heterogeneous Polysaccharides [or Heteropolysaccharides]

Preamble: The *two commonly employed terminologies viz.,* **'gums'** or **'mucilages'** invariably designate the so called **'polysaccharide macromolecules'** which critically undergoes dissolution virtually on being exposed to contact with water to give rise to the formation of either *'gels'* or *'colloidal solutions'.*

Interestingly, the latest prevailing practice and trend is to give up entirely these orthodox-type terminologies largely in favour of such more general and globally accepted ones, namely:

- **vegetable hydrocolloids,** and
- **vegetable polysaccharides.**

Based ion the enormous evidences from the literatures one may obviously take due cognizance of the fact that the term **'polysaccharide'** may critically appear to be **devoid of specificity,** since it generally applies to both:

- ❏ **cellulose,** and
- ❏ **gum arabic.**

Nevertheless, it is at times becomes rather difficult to *delineate* (or *outline*) such cardinal aspects as:

- **biological concepts of 'gums and mucilages',** and
- **chemical concepts of 'gums and mucilages and pectins',**

which strongly and emphatically advocates that one must solely rely upon the **'structural criteria'** in order to accomplish the proper **classification of these polymers,** wherever seem to be possible.

Thus, having formally categorized the term of **'gum'**-for all **'plant exudates',** one may distinctly distinguish:

- **'neutral heterogenous polysaccharides'** *e.g.,* **mannens and derivatives,** and
- **'acidic heterogeneous polysaccharides'** *e.g.,* **galacturonans** (or **pectins).**

In addition, the **heterogenous polysaccharides** (or **hetero polysaccharides**) are of immense **industrial utilities.** Nevertheless, they usually encounter a firce neck-to-neck competition, *based upon the solely economical reasons,* with the **polymers of :**

- **bacterial origin,** or
- **semisynthetic origin** (*e.g., cellulose derivatives*).

Following are a few **important criteria** which vividly distinguish the **'gums'** and the **'mucilages'**:

S.No.	Gums	S.No	Mucilages
1.	**Gums** are complex molecules, and are invariably **branched, heterogeneous,** and countain **uronic acids.**	1.	**Mucilages** invariably designate the **normall cell constituents** that may prexist in *specialized histological formation,* which are fairly common in the **external legnment of seeds.**
2.	They usually flower on the **'outer surface'** of plant, and are generally considered to result from a **trauma*.**	2.	**Acidic mucilages**–present in the **Malvales;** whereas, **neutral mucerlages**–present in **Fables.**
3.	They normally *concrete* by **desiccation,** and are **insoluble in organic solvents.** In fact, this specific criterion critically *differentiates* from the respective **'resins'.**	3.	They exert in **'germination process'** due to their *water-retention criteria.*

12.5.4.1 Gum Arabic [*Acacia Spp.: Mimosaceae*] [*Syn:* Acacia]

Gum arabic designates the **air-hardened gummy exudate** that flows almost naturally or upon **'tapping'** from the respective *stem, trunk* and *branches* of *Acacia senegal* (L.) Willed; and other Acacia species belonging to the **African origin**.**

The precise and exact estimation of the **molecular weight** ranges from **2,40,000***** to **5,80,000****.**

Chemical Composition: The **'gum arabic'** in its raw form essentially comprises the following constituents, namely:

- **water** (10-15%),
- **tannins** (as in *'coloured drugs'* specifically),
- **pexoxidase enzyme,** and
- practically **absence of 'starch'.**

Acacia was originally thought to be composed exclusively of the *four* **chemical constituents*****,** such as : **(–)–arabinose; (+)–galactose; (–)–rhamnose,** and **(+)–glucuronic acid,** as given under:

(–)–Arabinose   (+)–Galactose   (–)–Rhamnose   (+)–Glucuronic Acid

---

* Although, the **tragacanth gum** is found to be duly *'stored prior'* to any **external aggression.**

** **European Pharmacopoeia,** 3rd ed., 1998.

*** Oakley, *Trans, Faraday Soc;,* **31** : 136, 1935.

**** Anderson *et al., Carbolyed Res,* **3** : 308, 1967.

***** Kar, A: **Pharmacognosy and Pharmacobitechnology,** New Age International, New Delhi, 2nd ed,. 2007.

Hydrolysis with 0.1 M H₂ SO₄: **Gum arabic** (or **acacia**) on being subjected to hydrolysis with 0.1 M H_2SO_4 does help in the removal of the **'combined product (I) of (–)–arabinose and (+)–galactose;** whereas, the residue comprises the **'combined product (II) of (+)–glactose and (+)–glucuronic acid'.** However, the said two products (I) and II are invariably formed in the **ratio of 3:1** as given under:

(I)   (II)

Applications of Acacia: These essentially comprise:

(1) Commercial production of **spray-drited** *'fixed'* flavours to *remain stable.*

(2) Manufacture of **'powdered flavours'** usually used in **packaged dry-mix products,** such as: **cake mixes, puddings,** and **desserts,** where both **stability of flavour** and **prolonged shelf-life** are primarily required.

(3) It is used as **colloidal stabilizer;** and in **paper-coating.**

(4) It is skilfully used in conjunction with *'gelatin'* to produce typical **conservates** for the **'microcneapsulation* of drugs'.**

(5) **'Mucilage of acacia'** is mostly used as a **demulcent.**

(6) It is extensively used in making **candy** and other **food products.**

(7) **Gum arabic** is found to be *quite compatible* with a host of other **plant hydrocolloids, starches, carbohydrate,** and proteins.

(8) **Pectin** is used as **'latex thickness',** and as *substitutes* for **agar** in the **microbiological media.**

Specific Test for Acacia: It has been observed that a 10% aqueous solution of acacia fails to form any precipitate with a dilute solution of **lead acctate** (a clear distinction from **Tragacanth** and **Agar**). Besides, **acacia** fails to produce any colour change with **iodine solution** (a marked distinction from **starch** and **dextrin).** Finally, **acacia** does not give rise to a **bluish-black colouration** with a **ferric chloride (FeCl₃)** solution (an apparent distinction from **Tannins).**

12.5.4.2 Pectin

Pectin designates a group of **polysaccharide materials** usually found in the cell walls of all seed bearing plant tissues that essentially functions as an **'intercellular cementing substance'.** In fact, **pectin** acts in combination with both *cellulose* and *hemicellulose* to angment the aforesaid **intercellular cementing agent.** It has been duly observed that **lemon and orange rind,** which comprises nearly **30% of peetin** do cater for as the *richest sources of pectin.*

---

* **Microcncapsulation :** It refers to the formation of very small capsules that essentially coutain an **'active ingredient'** (*ie.,* **drug**). The coating of the capsule is so desigued so as to enable its disintegration very much within the body in order to release the active ingredient.

Pectin occurs naturally as the *'partial methyl ester'* of α–(1 → 4) linked D–polygalacturonase sequences adequately interrupted with (1 → 2)–L–rhamnose residues. However, the **natural sugars** *e.g.;*, **D–galactose, L–arabinose, D–xylose,** and **L–fucose** do form the side chains upon the **'pectin molecule'**.

Constitution of Pectin: The *constitution of pectin*, which predominantly consists of *three* major **polysaccharide** products, namely:

- **pectic acid** (as the *'methyl ester'*),
- **araban,** (as the *'methyl ester'*),
- **araban,** and
- **galactan.**

Sehneider and Bock (1938) put forward the following probable structure of **'pectin galacturonan':**

Pectin Galacturonan

Now, the *structure* of each of the *three* major **polysaccharides,** as stated above, shall be treated in the sections that follows:

(*a*) **Structure of Pectic Acid:** The procedures involved in establishing the **constitution of pectic acid** has proved to be a difficult task mainly due to the following *two* important reasons, namely:

- **High molecular weight** and the **colloidal nature of the polymers** actually pose a serious and complicated problem pertaining to their complete separation of **'pectic acid'** from the so called **'adsorbed arban and galactan',** and
- **Pectic acid** commonly found in various *plant species* is found critically to be present in an **'insoluble form',** which may be duly explained by the inherent **insoluble** characteristic feature of **'pectic acid'** by virtue of:
  - ❑ combination of its **carboxyl (–COOH) moieties** with other *polymers,* for instance: **araban, galactan,** and **cellulose,** and
  - ❑ crucial presence of **'insoluble Mg and Ca salts of pectic acid'.**

Besides, **pectic acid** also shows the following *typical reaction,* namely:

(1) **Hydrolysis with dilute HCl/enzyme pectinase:** It results in the formation of *d*–galactouronic acid*.

(2) **Oxidation with periodic acid (H_5IO_6) followed by hydrolysis:** It eventually forms 'tartaric acid' which shows evidently that the critical and specific **C–2 and C–3 hydroxy (–OH) moieties** present in *d*–galactouronic acid unit are **absolutely free.**

(3) **Hydrolysis of methylated pectic acid:** The completion of this reaction yields the respective **2, 3–dimethyl–d–galactouronic acid** thereby further ascertaining the fact that the **hydroxyl**

---

* Since, it possesses a **higher molecular weight,** it was regarded to be as **polygalactouronic acid.**

(–OH)' moieties strategically positioned at **C–2 and C–3** are not actually involved in the *d*–galactouronic acid linkages.

(4) The aforesaid *two* **degradation reaction** [*viz.,* (2) and (3)] clearly suggest that the 'pectic acid' molecule exclusively comprises *d*–galactouronic acid residues duly joined *via* the respective *hydroxyl moieties* either located on **C–1 and C–4** or **C–1 and C–5** to result into the formation of an ultimate 'linear polymer'.

However, as the **pectic acid molecule** is *fairly stable* to the usual *acidic hydrolysis*, it may be rightly inferred to have a 'pyranose ring structure'. Therefore, one may arrive at an evidential proof that **pectic acid** happens to be a *poly*–galacturonic acid *via* the **C–1 and C–4 linkages** only; and, hence justifiably explains the above cited reactions from (1) through (4).

Thus, we may have following reactions:

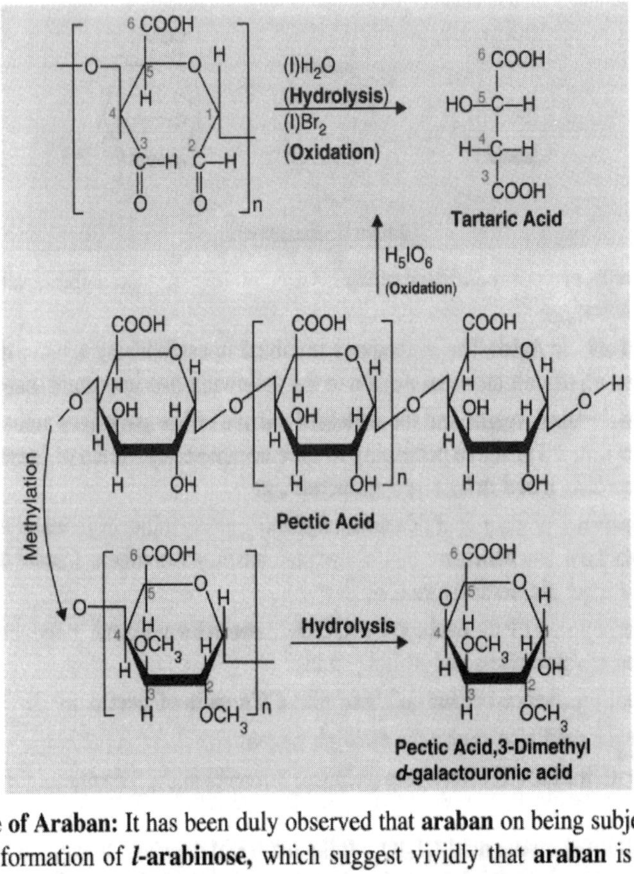

(b) **Structure of Araban:** It has been duly observed that **araban** on being subjected to hydrolysis gives rise to the formation of *l*-arabinose, which suggest vividly that **araban** is essentially a *poly* arabinose.

The Complete **structure of araban** has been adequately established by carrying out:

- **methylation,** and
- **followed by methanolysis,**

to yield *three* following **intermediates** as well as the **probable compound :**

❑ **2, 3, 5–trimethyl methyl–*l*–arabinoside (I),**

❑ **2, 3–dimethyl methyl–*l*–arabinoside (II),** and

❑ **2–methyl methyl–l–arabinoside (III),**

the structures of which are as given under:

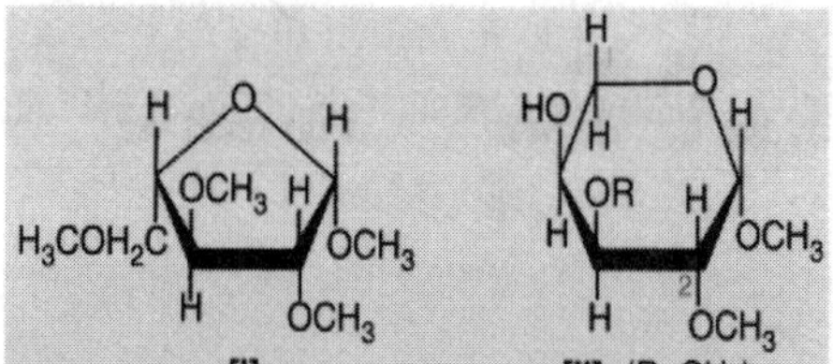

[I]

[II]: (R=CH₃)
[III]: (R=H)

The ultimate formation of the aforesaid *three* products (*viz.,* **I, II, and III**); and the remarkable *ease of hydrolysis* overwhelmingly suggest that in **'araban'** the various **'arabinose units'** are duly present in the **'furanose form' only'** Thus, the **'araban'** designates a **branched polymer (IV)** as given below:

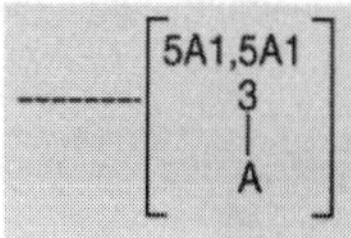

[IV]

A=*i*-Arabino furanose unit duly linked *via* the numbered C-atoms.

(*c*) **Structure of Galactan:** The various steps that are involved essentially in the establishment of the **structure of galactan** are as enumerated under:

(1) **Hydrolysis of galactan** yields only *d*–galactose thereby ascertaining the fact that it is a *poly-galactose.*

(2) **Methylation–Hydrolysis** sequential treatment gives rise to the formation of **2, 3, 6–trimethyl–** *d*–galactose (a) plus a small guantum of **2,3,4,6–tetramethyl–*d*–galactose (B).**

(3) Evidenced by the fact that **'galactan'** undergoes acidic hydrolysis with great difficulty, clearly indicates that the *d*–galactose residues duly present very much in the *'pyranose form'.*

(4) Based on the fact that the extent of rotation of the **polysaccharide** (*i.e.,* **galactan**) is relatively small, which explicitly suggest that the in-built **'d–galactose residues'** mostly present are linked *via* the **β–glycosidic bondages.**

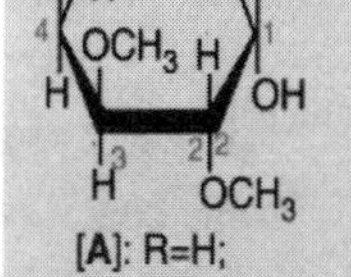

--------[4G1,4G1,4G1,-----]------

G=*d*-Galactopyranose linked *via* the numbered C-atons.*d*

[A]: R=H;
[B]: R=CH₃;

(5) The aforesaid factual evidences may ultimately lead to the following assigned structure of **'galactan' (C).**

13. GLYCOSIDES

13.1 Introduction

Glycosides are usually regarded to be such compounds that are essentially derived by the replacement of the H–atom attached to the **squiggly bond** of the **'cyclic hemiacetal'** by an **organic radical** *viz;*, **methyl (–CH₃) group.** Hence, the realistic formation of a **'glycoside'** is fairly comparable to the formation of **'acetals'**, whereby the **simple hemiacetals** do react with another molecule of alcohol as given under:

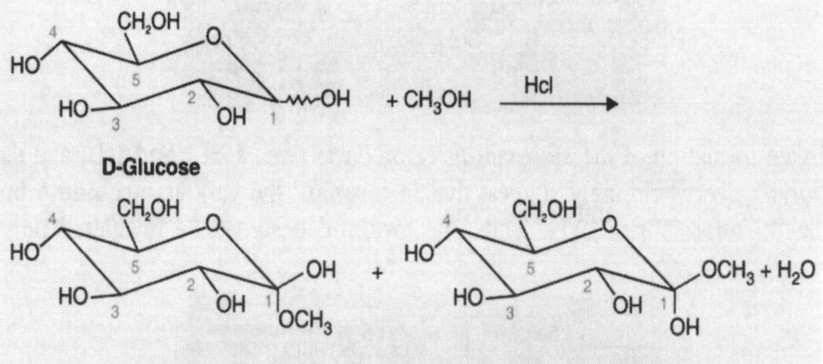

D-Glucose

Methyl-α-D-glucopyranoside (I)   **Methyl-β-D-glucopyranoside (II)**

> **Note:** The actual yield of the above *two* isomers (I) and (II) vary between 83 to 85%, which may eventually be separated effectively by means of *'fractional crystallization.*

In other words, the H–atom at C–1 joined with a *squiggly bond* by the organic radical methyl (–CH₃) duly obtained from **methanol** thereby releasing a mole of water. Further, the orientation of –H and –OCH₃ at C–1 could be providing **α–or β–glucopyranosides** respectively. In fact, these types of compounds are invariably termed as **'glycosides'.**

Alternatively, a **'glycoside'** refers to a substance **derived from plants** which, upon hydrolysis, gives rise to (knows as **'aglycones'**). However, depending upon the critical nature of the **'sugar formed'**, the **glycosides** are usually designated as: *'glucosides'* or *galactosides'* or *'fructosides'.*

Glycosidic Hydroxyl Moiety: However, it is pertinent to state here that the hydroxyl moiety

derived exclusively from carbonyl group ($-\overset{\overset{\text{O}}{\|}}{\text{C}}-$)due to the ring formation *viz.,* C–1 in **'aldoses'** and C–2 in **'ketoses'** plays crucial role in the **glycoside formation;** and, therefore, the **hydroxyl moiety** is usually called as **'glycosidic hydroxyl moiety'.**

It may be observed that the reaction of a **cyclic hemiacetal** (*e.g.;*, **glucopyranose**) with the corresponding reaction of on usual ordinary aldehyde under the same conditions, such as:

Glycoside Formation: In this reaction, **D–glucose** on being reacted with an alcohol (R–OH) in an acidic environment incorporated duly **one–OR group** at the *squiggly bond* plus a mole of water is eliminated as given under:

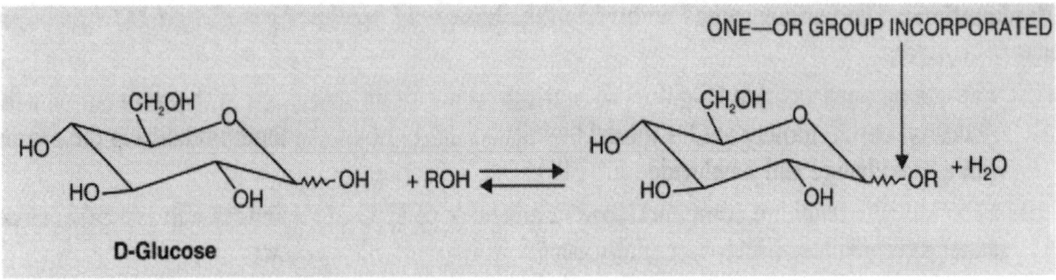

ONE—OR GROUP INCORPORATED

D-Glucose

$+ ROH \rightleftharpoons$ $+ H_2O$

Aldehyde Acetal Formation: Importantly, the aldehyde acetal formation designates a **cyclic hemiacetal,** for instance:

- **'glucopyranose'** incorporates actually **one alcoholic–OR moiety,** and
- **'ordinary aldehyde function'** critically incorporate **two –'OR' moieties.**

Note: **Perhaps this distinct difference between *'aldoses'* and *'ordinary aldehydes'* provides one of the solid reasons which the earlier *carbohydrate chemists* firmly believed, and suspected that the *'aldoses'* usually come into being as the so called 'cyclic hemiacetals'.**

However, the aforesaid **aldehyde–actual formation reaction** may be represented as given under:

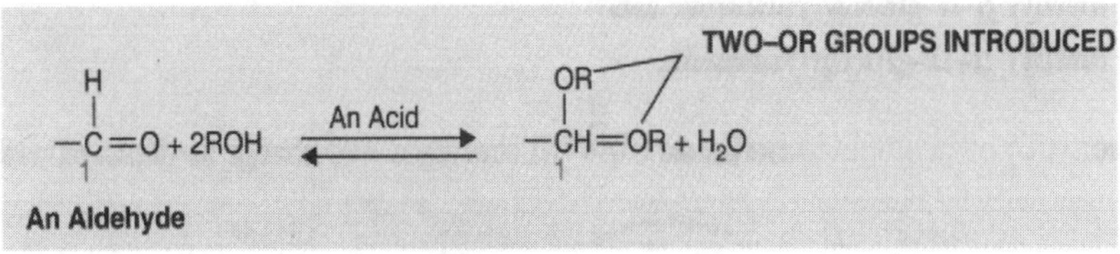

TWO–OR GROUPS INTRODUCED

$$-\overset{H}{\underset{1}{C}}=O + 2ROH \xrightleftharpoons[\text{An Acid}]{} -\overset{OR}{\underset{1}{CH}}=OR + H_2O$$

An Aldehyde

In a broader perspective, the **'glycosides'** are termed as *derivatives* of the **parent carbohydrase,** such as:

'Pyranoside'–shows that the *glycoside ring* is **6–membered,** and

'Furanoside'–indicates that the *glycoside ring* is **5–membered.**

just like the **'acetal formation'** as shown below :

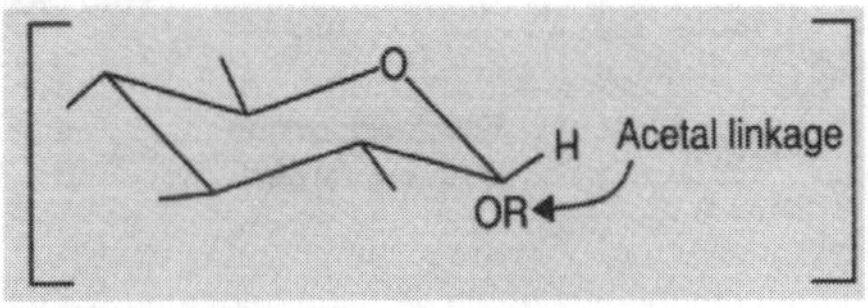

H   Acetal linkage

OR

the ensuing **'glycoside formation'** is eventually *catalyzed by acid,* and critically involves an α–alkoxy **carbocation intermediate** as given under:

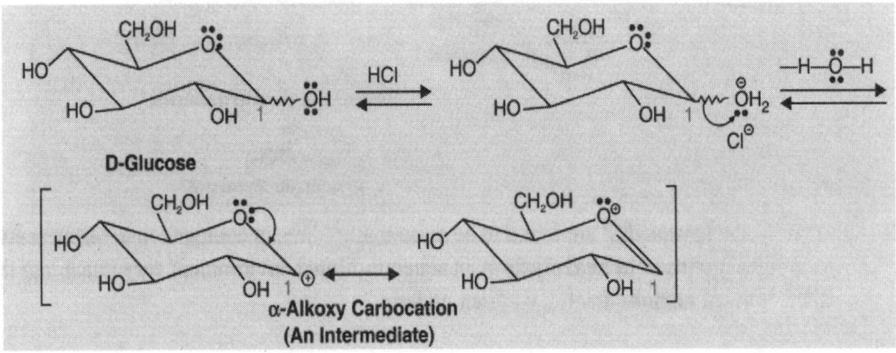

D-Glucose

α-Alkoxy Carbocation
(An Intermediate)

Explanations: The various steps involved in the above *acid catalyzed reaction* of **D–glucose** are explained as under:

(1) First, the monosaccharide **D–glucose** in the presence of an acid helps in the protonation of the hydruoxyl (–OH) moiety at C–1 joined with a **squiggly bond (--)** thereby rendering the former with a +ve change and a chloride ion (Cl⁻.)

(2) Secondly, the resulting compound loses a molecule of H_2O to yield an **α–alkoxy carbocation** as an intermediate, which essentially occurs in *two notable variants:*

   • **α–alkoxy carbocation** where in C–1 bears a **+ve charge** besides, a pair of electrons from the **O–atom** shifts them from the former to the covalent bond between C–1 and O–otom, and

   • **α–alkoxy carbocation** where in the O–atom essentially bears a **+ve charge,** and a double bond is duly established between the **O–atom** and the **C–1.**

Salient features: These essentially comprise:

(1) **Glycosides** are found to be **fairly stable to base** as exemplified below by citing the typical example of:

   • **methyl β–d–glucopyranoside,** and

   • **methyl α–D–glucopyranoside,**

in the presence of **protonated methanol** $[H_2\overset{\oplus}{\ddot{O}}CH_3]$ bearing a +ve charge as depicted under:

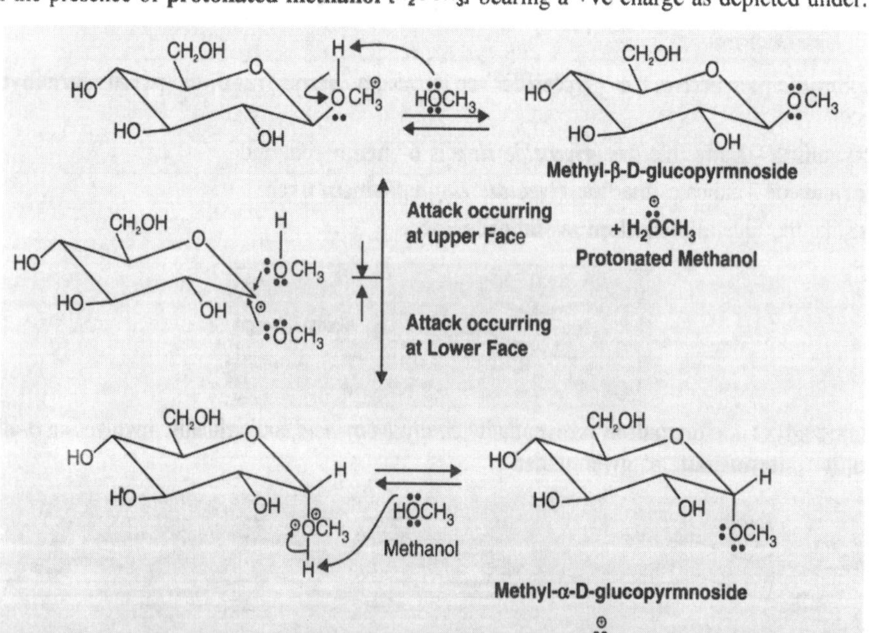

Methyl-β-D-glucopyrmnoside

Methyl-α-D-glucopyrmnoside

(2) Importantly, the 'glycosides' are found to be reasonably stable, in comparison to **other acetals;** however, they are **prone to hydrolysis** in an aqueous diluted environment very much into their so called **'parent carbohydrate',** as given under:

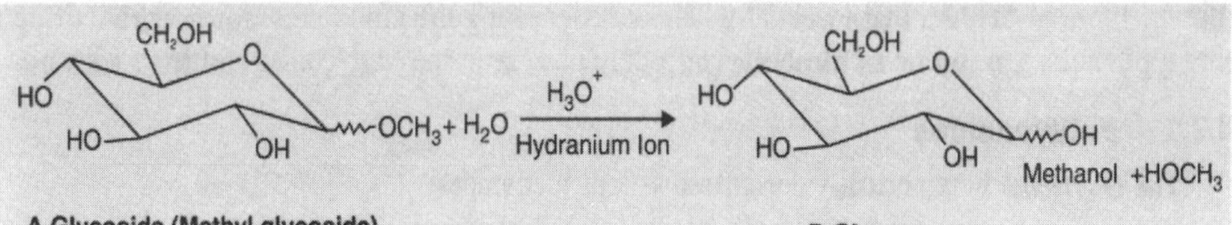

A Glycoside (Methyl glycoside) **D-Glucose**

(3) The glycoside of a **natural product** may be subjected to hydrolysis to give rise to its corresponding products:

- **alcohol,**
- **phenol,** and
- **carbohydrate,**

very much akin to the **simple methyl glucosides.**

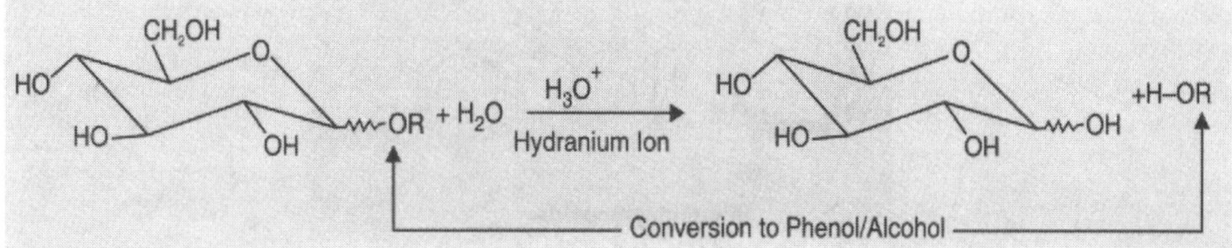

(4) Following are two typical examples of the **naturally occurring glycosides** which find their abundant usage in therapeutic applications, namely: **Doxorubicin** and **Salicin,** wherein the *carbohydrate segment* of each **glycosidic portion** is vividly indicated in the *dotted/coloured box.*

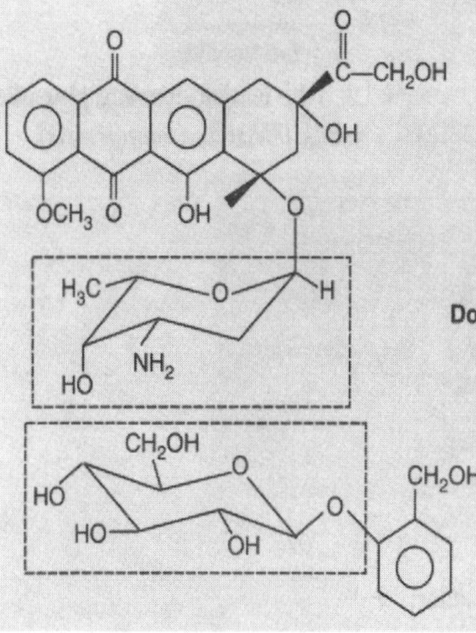

Doxorubiein–an 'antitumour drug'

Salicin–obtained from the extract of 'Willow Bark'

13.2 Naturally Occurring Glycosides

In general, the **glycosides** are most commonly formed, in nature by the interaction of the so called **'nucleoside glycosides'.**

Example: There exists actually *four* **critical point of linkages** between the *carbohydrate* (*i.e.,* the **'sugar moiety'**) and the *non-carbohydrate* (*i.e.;,* **'aglycone moiety'**) which could form *four*

different types of bridges that essentially helps in connecting the group present in **carbohydrate** to either a **phenolic group** or an **alcoholic group** duly present in the **'aglycone residue'**, such as:

13.2.1 O–Glycosides

The *O–glycosides* are normally designated as indicated under:

$$-OH \quad HO-C_6H_{11}O_5 \xrightarrow[-H_2O]{} \underbrace{-O-C_6H_{11}O_5}_{\text{O-Glycoside}}$$

It has been observed that the **O–glycosides** are profusely found in nature particularly in the *higher plants, namely:* **senna, rhubarb,** and **frangula.** A few typical examples of the **O-glycosides** are:

Rhein–8–glycoside; Amygadalin; Arbutin; Indican; Salicin etc.,

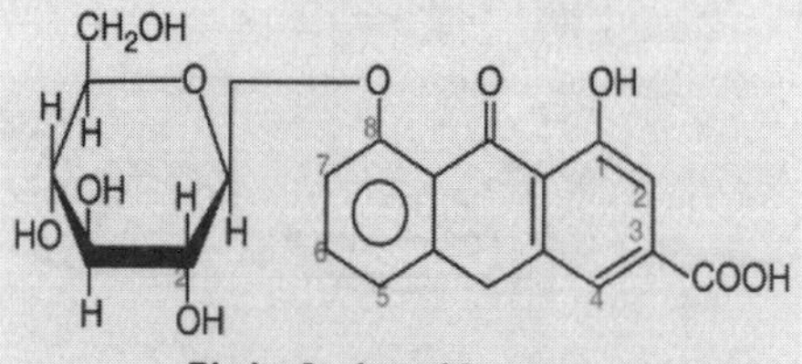

Rhein–8–glycoside

13.2.2 S–Glycosides

The *S–glycosides* are usually represented as stated below:

$$-SH \quad HS-C_6H_{11}O_5 \xrightarrow[-H_2S]{} \underbrace{-S-C_6H_{11}O_5}_{\text{S-Glycoside}}$$

Importantly, the **S–glycosides** are invariably confined to the **isothiocyanate glycosides**, for instance: **Sinigrin**, which is obtained from the **black mustard seeds** (*Brassica compestris*).

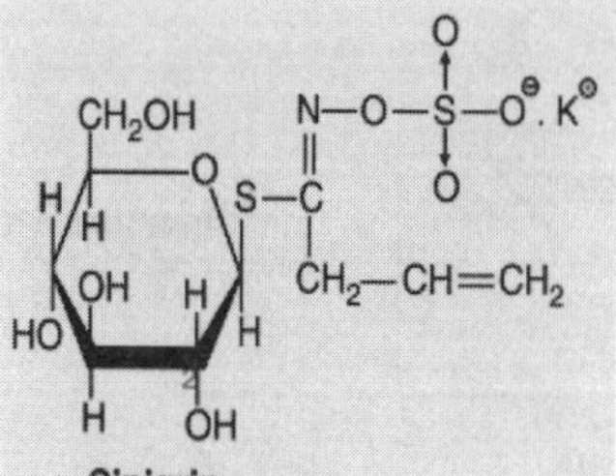

Sinigrin

13.2.3 N–Glycosides

The *N–glycosides* may be designated as under:

$$=N-H \quad HO-C_6H_{11}O_5 \xrightarrow[-H_2S]{} \underbrace{=N-C_6H_{11}O_5}_{\text{N-Glycoside}}$$

One may critically examine the **'muclosides'** as the most relevant and typical example of the **N-glycosides,** where in the respective *amino function* of the *'base'* reacts ultimately with the hydroxyl (–OH) group present in **ribose** or **deoxyribose.**

Adenosine represents the typical example of **N–glycosides,** which being abundantly distributed in nature *viz.,* available in **yeast nucleic acid:**

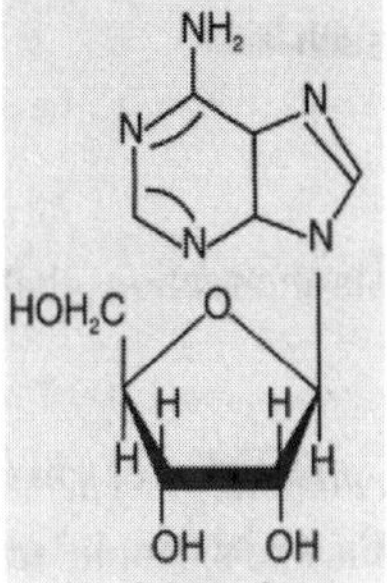

Adenosine

13.2.4 C–Glycosides

The *C–glycosides* may be represented as given below:

$$\geq CH \quad HO-C_6H_{11}O_5 \longrightarrow \geq C-C_6H_{11}O_5$$

C-Glycoside

It has been duly observed that the **C–glycosides** are generously present in a variety of plant products, namely: **Aloin** (in **'aloe'**), and **Cascarosides** (in **'cascara'**) :

(*i*) **Aloin:**

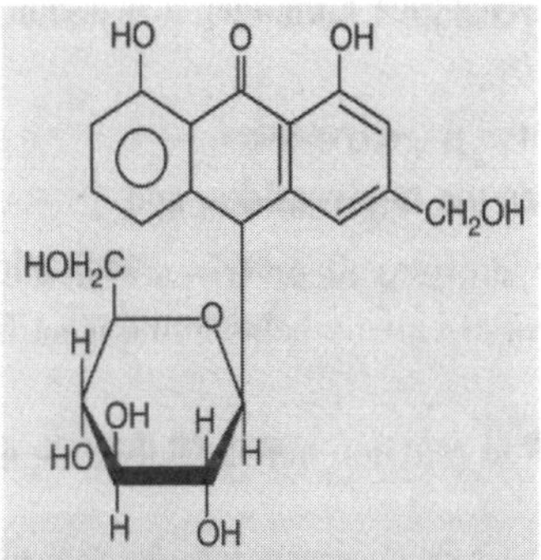

Aloin [or Barbaloin]

(*ii*) **Cascarosides (A, B, C, and D):**

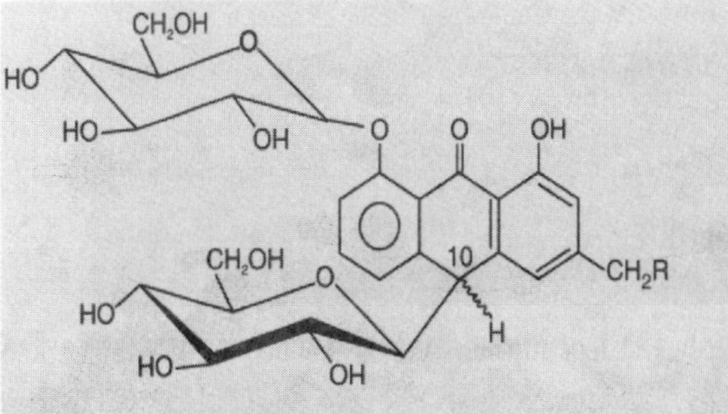

Cascaroside A: R=OH,(12S);
Cascaroside B: R=OH,(10R);
Cascaroside C: R=H,(10S);
Cascaroside D: R=H,(10R);

Cascarosides

13.3 Structure Determination of Glycosides

Generally, the structure of the **glycosides** is duly established and determined by the following *two* well-known and recognized procedures, such as:

- **Degradation,** and
- **Synthesis,**

which shall now be discussed briefly in the sections that follows:

13.3.1 Degradation

Degradation refers to the *physical, metabolic,* or *chemical* change to a relatively **'less complex form.** The phenomenon of **'degradation'** may be exemplified by the ingestion of *food products* that are physically degraded during chewing; and subsequently chemically degraded from *complete compounds,* for instance: **starches** and **proteins** to yield *carbohydrate* and *amino acids* respectively.

Thus, the **glycosides** are invariably subjected to *hydrolysis* in the presence of diluted mineral acids, and the resulting **hydrolyzed components** *viz.,* **carbohydrate(s)** and **aglycone** residues are carefully segregated by means of **paper chromatography.** Importantly, the following characteristic features are duly ascertained by specific chemical means as elaborated under:

(*a*) **Size of the Carbohydrate Ring:** The exact and precise *size of the carbohydrate ring* may be established by commonly available method(s) *e.g.,* **complete methylation immediately followed by hydrolysis,** and

(*b*) **Configuration of the Glycosidic Linkage:** It is usually established by the aid of crucial **'enzymatic hydrolysis'** *viz.,*

Emulsin – hydrolyse the **β–glycosides,**

Maltase – hydrolyses the **α–glycosides,** and

Specific Enzyme – hydrolyses **di–or tri–saccharide** only upto a certain extent (*i.e.,* **partial hydrolysis)** that essentially helps in the identification of the resulting **'oligosaccharides'.**

The various types of **degradation reaction** may be duly designated as stated under:

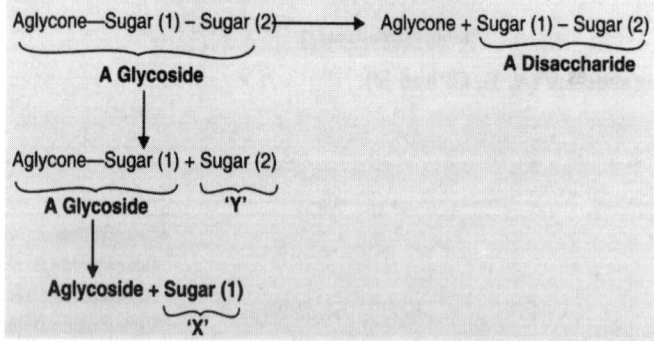

Hence, the **'glycoside;** upon hydrolysis (**degradation**) yields the *two* carbohydrate **'X'** and **'Y'** respectively as indicated above.

13.3.2 Synthesis

In actual practice, the structure of a 'glycoside' is duly established by the help of the aforesaid 'degradation reactions', which is eventually confirmed by a separate synthesis. In a broader perspective, the synthesis of a glycoside essentially involves several reactions as stated under:

(a) Reaction taking place between an O-acetyl-α-glycosyl bromide with a phenol or an alcohol (viz., the aglycone residue) in the critical presence of silver carbonate [Ag (CO$_3$)$_2$] in perfectly anhydrous benzene.

(b) Walden Inversion usually comes into play thereby giving rise to β–linked glycosides mainly.

(c) D–Glycosides are prepared invariably by carrying out the reaction particularly in the presence of either ferric chloride [FeCl$_3$] or mercuric acetate [Hg(CH$_3$CO)$_2$] instead of silver carbonate.

(d) O-Acetylaglycosyl bromide is actually prepared from the carbohydrate viz;, an aldohexose, which is duly acetylated in the presence of zinc chloride [Zn Cl$_2$] to yield the desired α–or β–penta acetate. The resulting product is treated subsequently with glacial acetic acid adequately saturated with HBr (hydrogen bromide) to form α–acctobromohexose in either instances i.e., whether the starting material is α–or β–penta acetate.

Therefore, commencing from aldohexose the various reactions involved in the 'synthesis' of α–or β–glycosides are as given under:

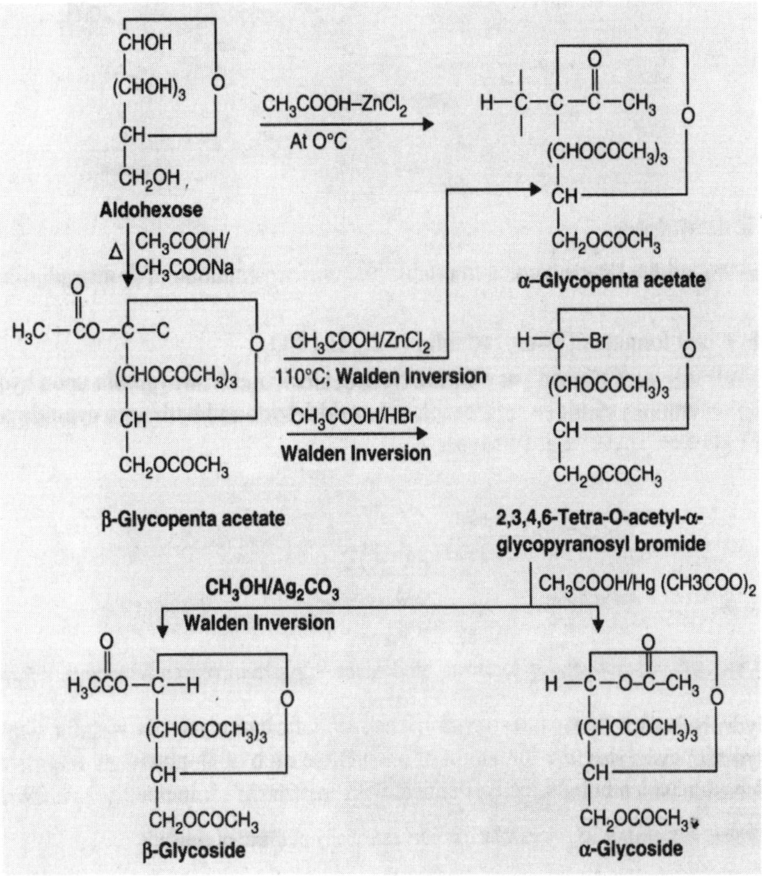

As we know that the 'glycosidic moiety' is observed to be fairly stable in an *alkaline environment*, the **acetyl moiety** may be removed most easily and conveniently by an *alkaline hydrolysis* to yield the desired 'glycoside'.

13.4 Important Glycosides

Following are some of the **vital and important 'glycosides'** which essentially are obtained from the natural products, such as:

13.4.1 β–Amygdalin

The **β–amygdalin** is known to be one of the earliest **glycosides,** which particularly occurs in **bitter almonds,** and derived from the pit or other seed parts of several plants *viz,* **apricots.** The **amygadlin** from which the *poisonous* **hydrocyanic acid (HCN)** can be obtained by a crucial **enzymatic action,** is the substance known in the **US** as **'Laetrile'.** Interestingly, it possesses little nutritional or therapeutic efficacy at all

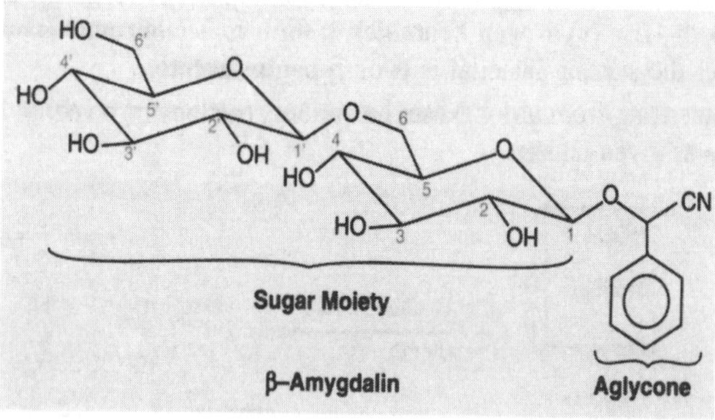

Sugar Moiety

β–Amygdalin **Aglycone**

13.4.1.1 Constitution

The various cardinal steps involved in establishing the constitution of **β–amygadalin** are as stated under:

(1) Molecular formula of **β–amygadalin** is $C_{20}H_{27}NO_{11}$.

(2) **Hydrolysis with Emulsin* or Dilute Hydrochloric Acid: Amygdalin** upon hydrolysis under these conditions yields *one mole* each of **benzaldehyde** and **hydrogen cyanide** plus *two moles* of **D–glucose** Thus, we may have:

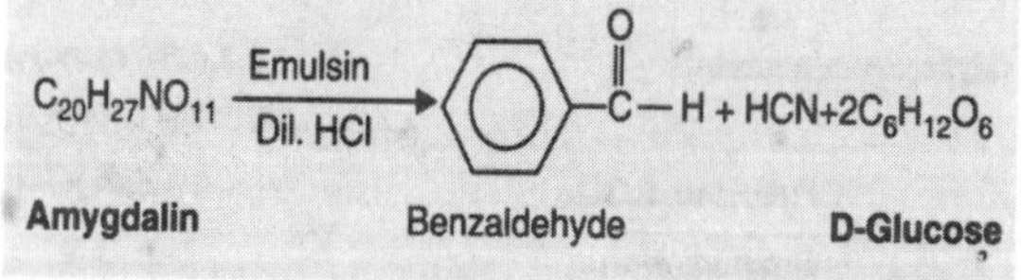

Amygdalin **Benzaldehyde** **D-Glucose**

Note: The *'glycosidic linkage'* is found to be β, since it gets hydrolyzed by *emulsin** (*an enzyme*).

(3) **Hydrolysis with Zymase (an enzyme):** The specific hydrolysis of **amygdalin** with the enzyme 'zymase' gives rise to the formation of one mole each of **D–glucose** and **mandelonitrile–D–glucoside** which being perfectly identical with **'prunasin'.** Importantly, **prunasin** designates

* **Emulsin:** Refers to a mixture of enzymes which occurs essentially in **bitter almonds.**

a naturally occurring 'glucoside' that evidently affirms the 'sugar entity' duly present in the glycoside is, in fact, a glycoside; and, therefore, the *aglycone residue* is nothing but **mandelonitrile.** Thus, we may have the following reaction:

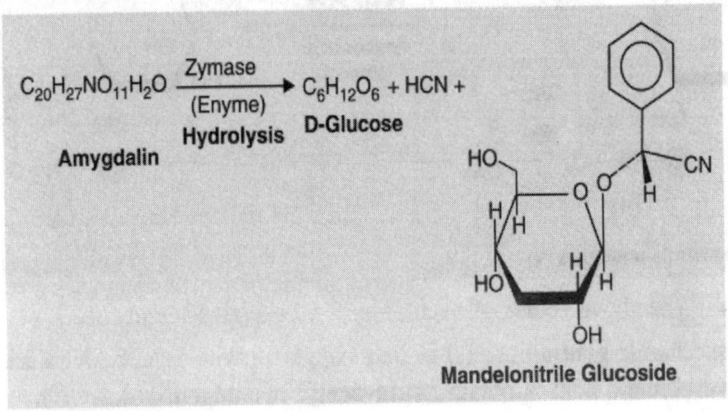

Mandelonitrile Glucoside

> **Note:** *'Prunasin'* is the *d–Form* of mandelonitrile glucoside, which on being treated with alkalies gets converted to 'prulaurasin'.

(4) **Synthesis of (+)–Amygdalin:** The synthesis of (+)–amygdalin may be accomplished using the disaccharide 'gentiobiose' as given under:

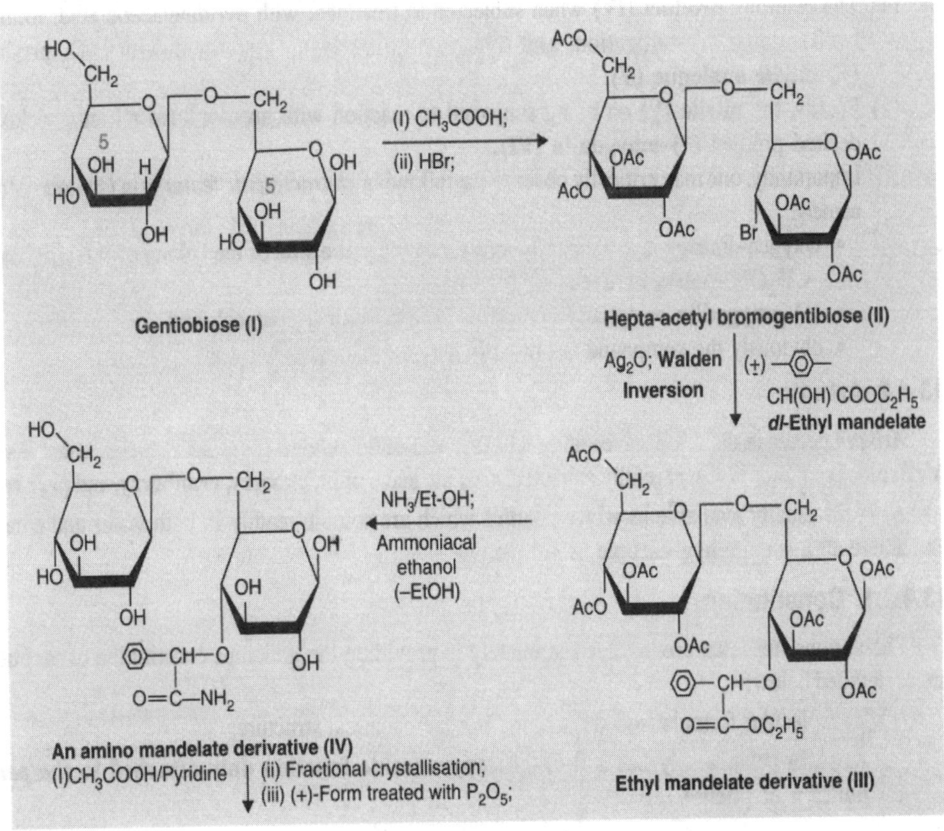

Gentiobiose (I)

Hepta-acetyl bromogentibiose (II)

An amino mandelate derivative (IV)

Ethyl mandelate derivative (III)

(+)-Nitrile analogue (V)   **(+)-Amygdalin (VI)**

Explanation: The above course of reaction may be explained as under:

(1) The disaccharide **gentiobiose (I)** is first acetylated with glacial acetic acid and then treated with hydrobromic acid to obtain **hepta–acetyl bromogentiobiose (II).**

(2) The resulting product **(II)** yields the corresponding **ethyl mandelate derivative** by treatment with *dl*–**ethyl mandelate (III)** in the presence of **silver oxide** when *Walden Inversion* comes into play.

(3) The **dervatized product (III)** in the presence of ammoniacal ethanol gives rise to the formation of an **amino mandelate derivative (IV)** with the loss of a mole of ethanol.

(4) The resulting **product (IV)** when subjected to treatment with pyridine/acetic acid, followed by fractional crystallization, and treating ultimately the **(+)–form** with P_2O_5 produces **(+)–nitrile analogue (V).**

(5) Finally, the **nitrile (V)** on being subjected to reaction with **ammoniacal ethanol** yields the desired product **(+)–amygdalin (VI).**

Importantly, one may critically observe the following *characteristic features* in (+)–amygadlin, namely:

- Oxygen–linkage is strategically located on the **same side** of the **'plane of the ring'** as the **–CH₂OH moiety at C–5.**
- **(+)–amygadlin** essentially comprises *several* **'chiral centres'**, and
- obviously the compound is **optically active** *i.e., dextrorotatory.*

13.4.2 Arbutin

Arbutin occurs in the dried leaves of **bearberry, and other related plant sources** *viz.*, **coactylis** and **McBride.** Besides, it is also carefully extracted from the leave of **blackberry, cranberry,** and **pear trees.**

Arbutin usually available as **white needles** which are soluble readily in both water and ethanol, and above all it is quite **hygroscopic in nature.**

13.4.2.1. Constitution

The various steps that are involved intimately in providing the judicious constitution of **'arbutin'** are as detailed below:

(1) Its molecular formula is $C_{12}H_{16}O_7$ and has the chemical structure:

Arbutin prominently has a **β–D–glucopyranoside function** duly attached to the *para*-position of phenol.

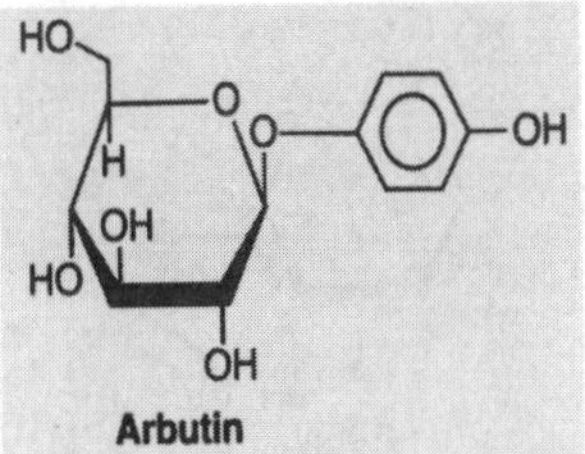

Arbutin

(2) **Ferric Chloride Test:** It gives rise to *blue coloration* with $FeCl_3$ solution thereby indicating the presence of **phenolic function.**

(3) **Hydrolysis with Emulsin or Dilute HCl:** The hydrolysis of **arbutin** with either **emulsin** or **dilute HCl** yields one mole each of **α–D–glucose** and **hydroquinone** as shown under:

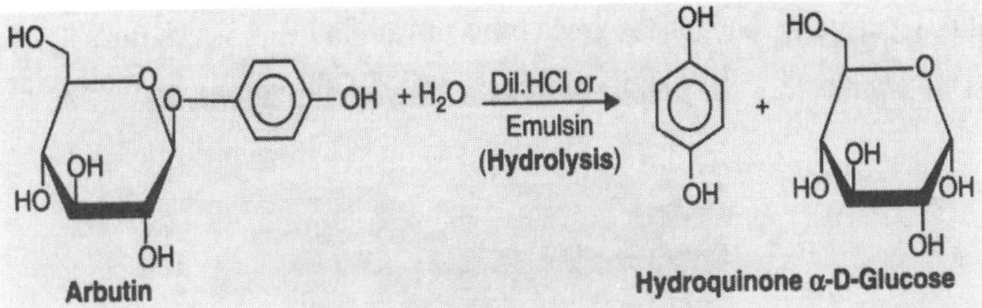

Arbutin   **Hydroquinone α-D-Glucose**

The above reaction ascertains the fact that in the structure of **arbutin** the **α–D–glucose** and the **hydroquinione** are hooked on to each other by the aid of a **β–linkage.**

(4) **Synthesis of Arbutin:** The various steps involved in the **synthase of arbutin** are as enumerated under:

    (a) The interaction between **2,3,4,6–letra–O–acetyl–α–D–glucopyranosyl bromide (I)** and **hydroquinone monobenzoate** in the presence of *silver oxide* and *quinoline* undergoes **Walden Inversion** to yield the corresponding **hydroquinone monobenzoate derivative (II).**

    (b) The resulting **product (II)** on being treat with *ammonia (NH₃)* undergoes critical and specific:

        • **deacetylation ,** and
        • **debenzoylation,**

to give rise to the formation of :

    ❏ **Arbutin,** and
    ❏ **Methylarbutin.**

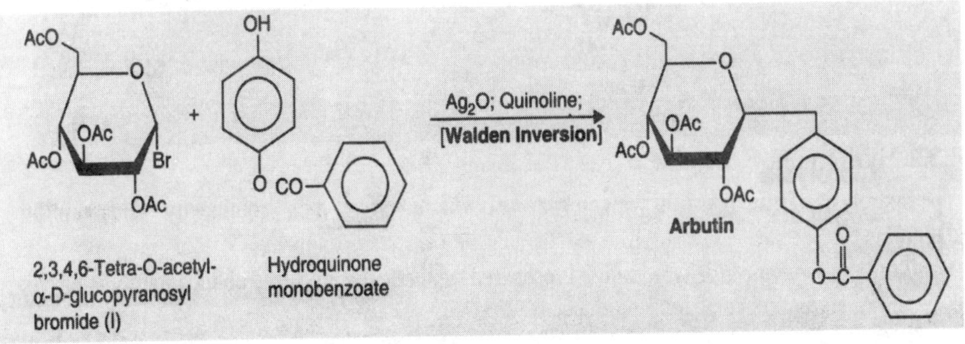

2,3,4,6-Tetra-O-acetyl-   Hydroquinone
α-D-glucopyranosyl   monobenzoate
bromide (I)

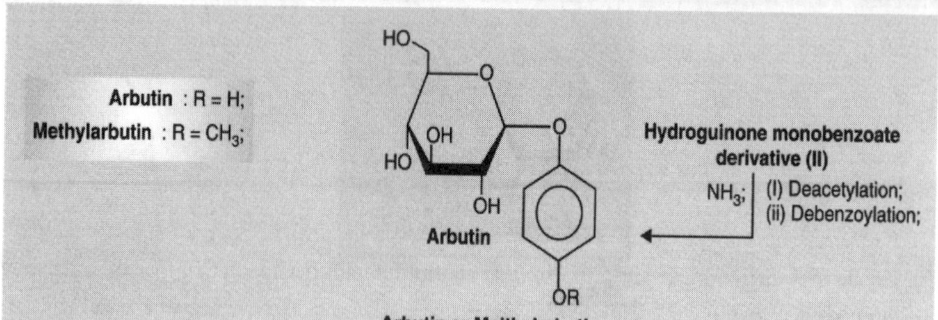

Arbutin : R = H;
Methylarbutin : R = CH$_3$;

Arbutin

Hydroguinone monobenzoate derivative (II)

NH$_3$; (I) Deacetylation;
(ii) Debenzoylation;

Arbutin or Melthylarbutin

13.4.3. Glucovanillin

Glucovanillin is g enerally found in the green fruits of **vanilla bean,** which finds its overwhelming usage in a host of *pharmaceutical preparation* as a **'flavouring agent.** Its molecular formula is $C_{14}H_{18}O_8$.

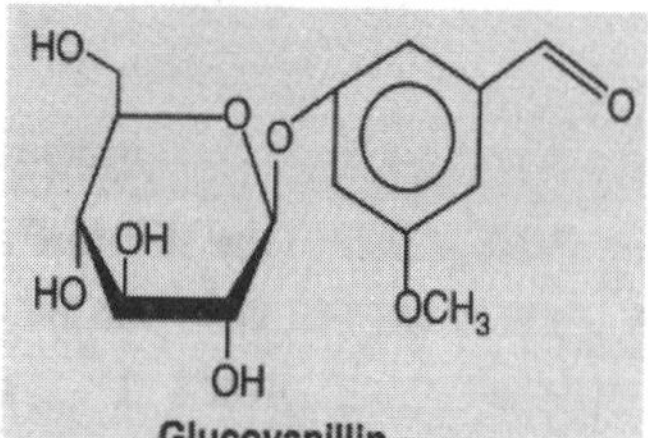

Glucovanillin

13.4.3.1 Synthesis of Glucovanillin

Glucovanillin may be synthesized by the following method:

- **'coniferin'** by oxidation with chromium–6–oxide [CrO$_3$].

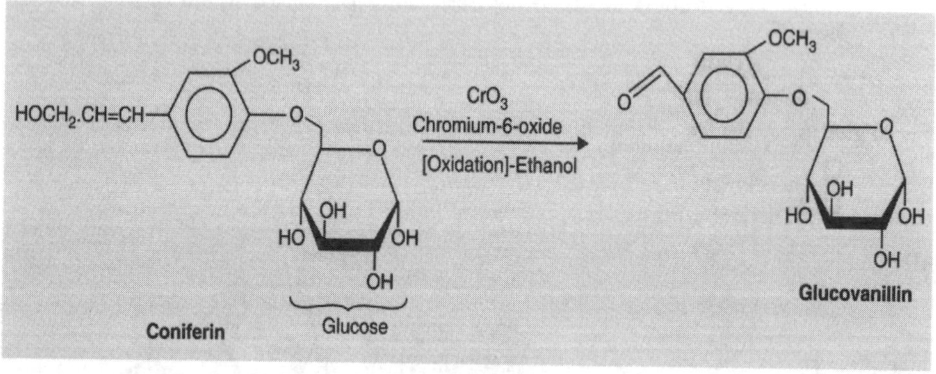

HOCH$_2$.CH=CH–

Coniferin   Glucose

CrO$_3$
Chromium-6-oxide
[Oxidation]-Ethanol

Glucovanillin

13.4.3.2 Hydrolysis

The hydrolysis of **glucovanillin** with an *enzyme* or *dilute mineral acid* prominently yields **vanillin** and **D–glucose.**

Vanillin is an *aromatic aldehyde*, which is prepared synthetically, and used abundantly in biscuit industry and a large number of food products.

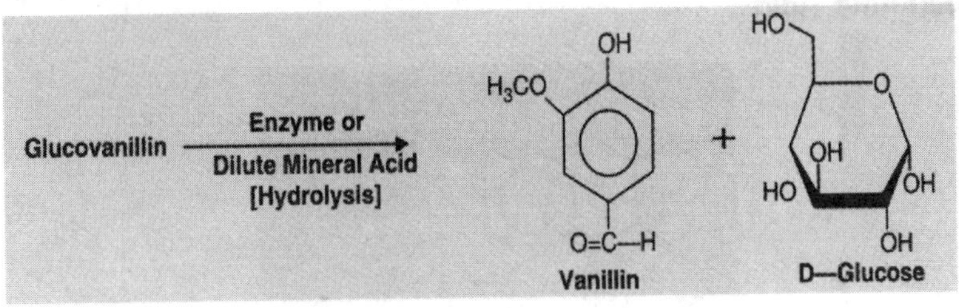

13.4.4 Indican

Indican is obtained from the leaves of the **indigo plant** and **wood plant.**

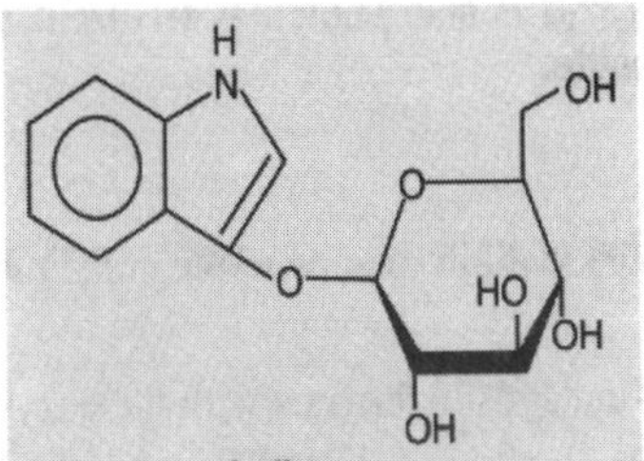

Indican

Its molecular formula is $C_{14}H_{17}NO_6$.

Indican upon hydrolysis with **emulsin** or **dilute mineral acid** yields one mole each of **2–hydroxy indole** (or **indoxyl**) and **D–glucose.**

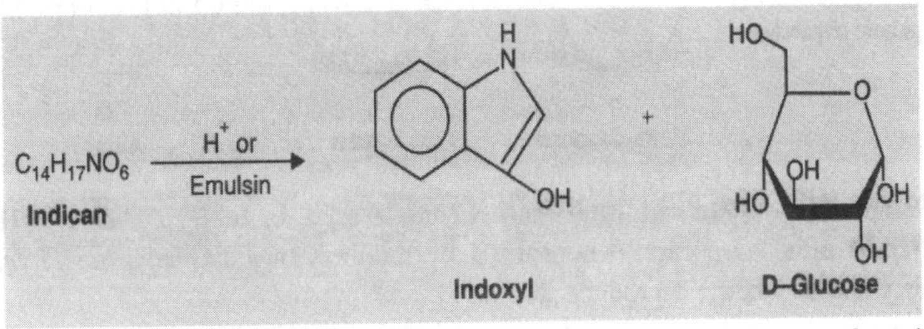

It has been observed that the indoxyl and D–glucose molecules are joing together *via* β–linkage.

13.4.4.1 Synthesis

Indican may be synthesized by the interaction of **indoxyl** and α–**acetobromoglucose** under *two* separate treatments in a sequential manner, namely:

- action with silver carbonate (Ag_2CO_3), and
- reaction with ammonia (NH_3).

13.4.5 Ruberythric Acid

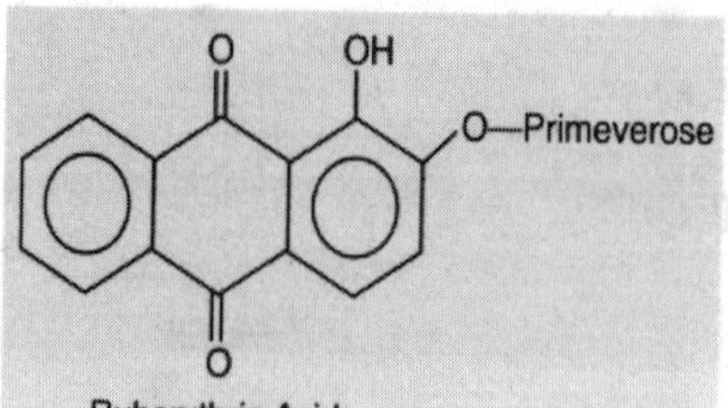

Ruberythric Acid

Ruberythric acid is a widely known **glycoside** obtained from the **madder roots.**

Its molecular formula is $C_{25}H_{26}O_{13}$. It is also known as : **rubian, rubianic acid, ruberythrinic acid,** and **β–2–alizarin primeveroside.**

13.4.5.1 Constitution

The different important steps that essentially comprise the constitution of **ruberhythric acid** are as stated under:

(1) **Hydrolysis with Emulsin:** It undegos hydrolysis with the enzyme **emulsin** to give rise to the formation of one mole each of **D–glucose, D–xylose,** and **alizarin** as given under:

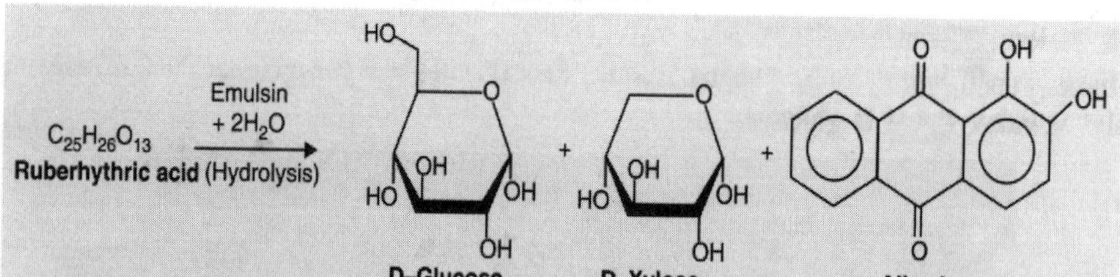

(2) **Hydrolysis with Enzyme obtained from** *Primula veris* **L. (or** *P. Officinalis* **(L.) Hill:** **Ruberhythric acid** upon careful hydrolysis by the **enzyme** obtained from *P. veris* or *P. officinalis* yields *two* distinct products:

• **alizarin,** and

• **disaccharide.**

Nevertheless, the disaccharide upon very critical *mild hydrolysis* products one mole each of **D–glucose, D–xylose,** and **primeverose (or 6–O–b–D–Xylopyranosyl–D–glucose)** as shown under:

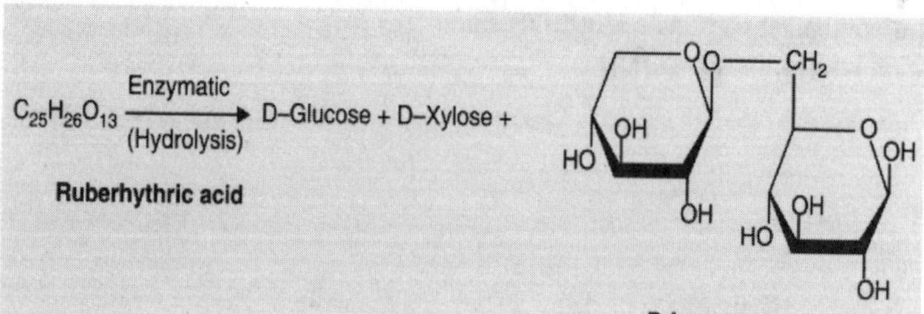

Primeverose

Note: The crucial formation of *'primeverose'* upon hydrolysis clearly indicates that the *alizarin* is strategically linked to the glucose structure of the *primeverose.*

(3) **Determination of specific Hydroxyl (–OH) Moiety in Alizarin Engaged in** β–Glycoside Linkage of Ruberythric acid: It can be critically established by carrying out the careful **methylation** followed by hydrolysis of **ruberythric acid** that gives critically the product, **alizarin–1–methyl ether,** which show vividly that the **C–1 of alizarin** *remains free* absolutely.

Therefore, it may be inferred obviously that the **primeverose molecule** is linked strategically *via* C–2 of the **alizarin molecule** to give rise to the formation of **ruberythric acid** as given below:

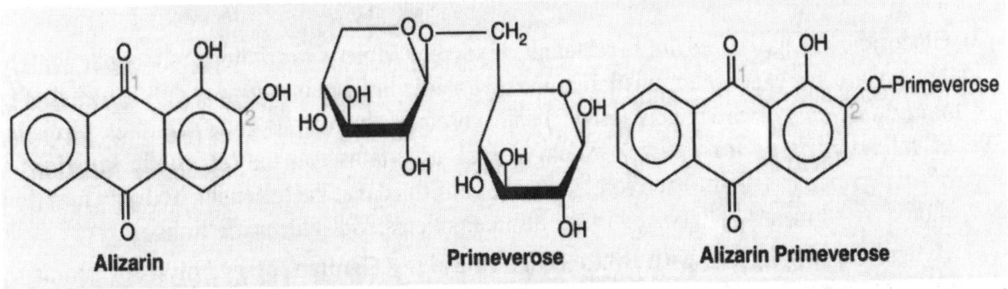

| Alizarin | Primeverose | Alizarin Primeverose |

(4) **Synthesis: Ruberythric acid** may be synthesized from **alizarin** (I) by the interaction with **hexa–O–acetylprimeverosyl bromide (II)** by treatment with **silver carbonate (Ag₂CO₃)** and **ammonia (NH₃)** to yield **ruberythric acid** with the elimination of **HBr** as given below:

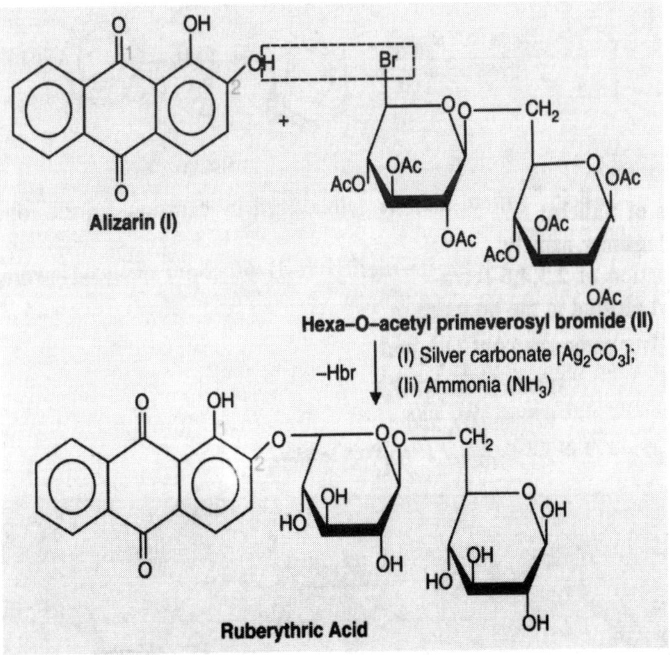

Alizarin (I)

Hexa–O–acetyl primeverosyl bromide (II)

–Hbr (I) Silver carbonate [Ag₂CO₃];
(Ii) Ammonia (NH₃)

Ruberythric Acid

13.4.6 Salicin

Salicin is obtained from the barks of **poplar** and **willow**, and also isolated from the leaves of *female flowers* of the **willow.**

(1) **Hydrolysis with Emulsin:** Salicin upon hydrolysis with the enzyme **emulsin** yields one mole each of **saligenin (an 'aglycone')** and **D-glucose** as stated under:

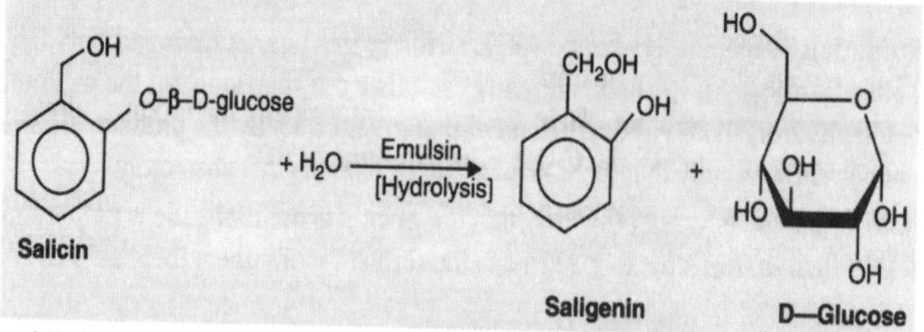

(2) Glucosidic linkage caused due to **'Phenolic Hydroxyl Moiety' or 'Salicylic Alcoholic Moiety:** The precise and exact determination of the **glyuosidic linkage** in *salicin* is duly established by *mild oxidation* with **nitric acid (HNO$_3$)** and subsequent hydrolysis gives rise to the formation of **salicylaldehyde (or** *helicin***)** which, in fact, ascertains that the **'alcoholic function'** is *absolutely free* in **salicin.** Therefore, it may be confirmed that the **'phenolic hydroxy function'** should by all means be involved in the desired/anticipated β-glucosidic linkage.

(3) **Hydrolysis of Saligenin with Dilute HCl and Boiling Gently:** The hydrolysis of **2 moles of saligenin** when carried out in an acidic environment by gentle boiling for a prolonged duration usually gets combined together to provide one mole each of **saliration** and **water** as stated under:

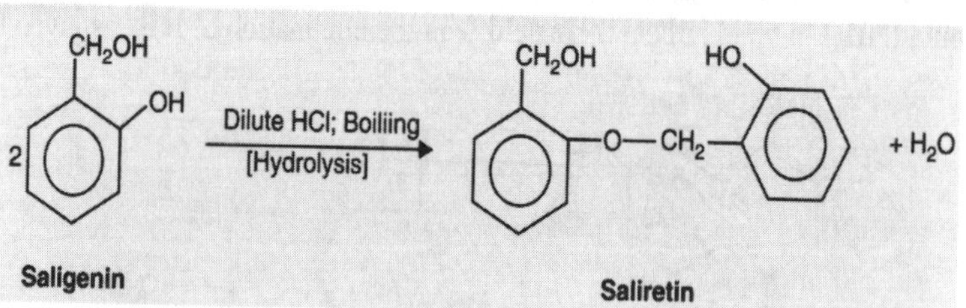

(4) **Synthesis of Salicin:** *Salicin* may be synthesized by carrying out the following steps in sequential manner, namely:

- interaction of **2,3,4,6-tetra-*O*-methyl-α-D-glucopyranosyl-1-bromindes** (I) with **salicyl alcohol** in the presence of:
  - ❑ silver carbonate (AgCO$_3$), and
  - ❑ dimethyl sulphate [(CH$_3$)$_2$SO$_4$],

to produce penta methyl salicin (II), and

- demethylation of the *product (II)* gives **salicin**

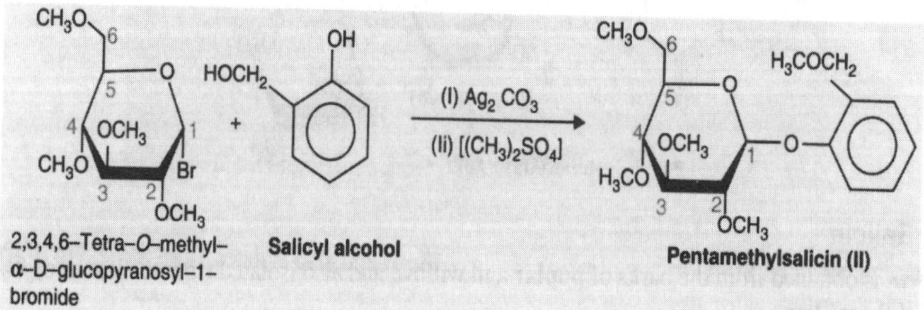

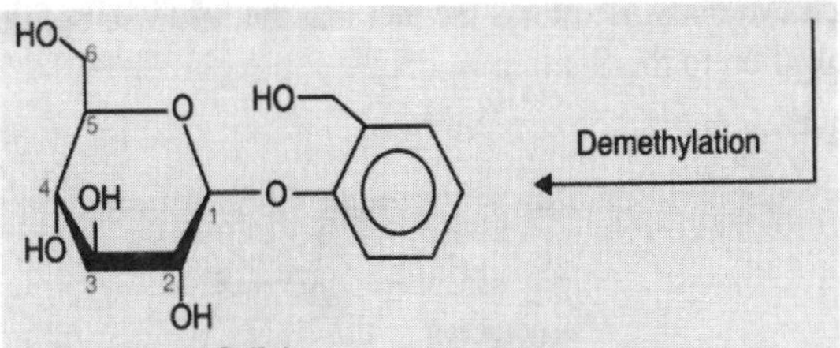

Salicin

13.4.7 Sinigrin

Sinigrin, a **thioglycoside** is obtained from the **black mustard seed.** It is also termed as: **allyl glucosinolate** or **potassium myronate.**

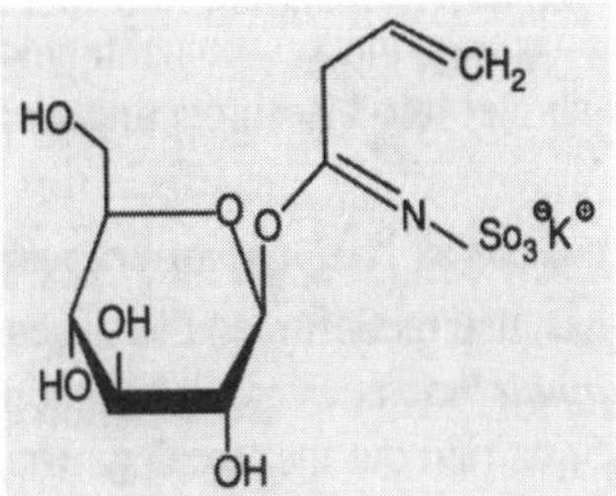

Sinigrin

Besides, **sinigrin** is also obtained from the **horseradish root** (*Alliaria officinalis* Andrz;, *Cruciferale*).]

13.4.7.1 Constitution

The various vital and important steps involved in the constitution of **'sinigrin'** are as follows:

(1) Its molecular formula is $C_{10}H_{16}NO_9S_2K$.

(2) **Hydrolysis of Sinigrin by the Enzyme Myrosin: Sinigrin** critically undergoes hydrolysis in the presence of an *enzyme* **myrosin** to produce one mole each of **allyl-isothiocyanate, D-glucose,** and **potassium hydrogen sulphate** as given under:

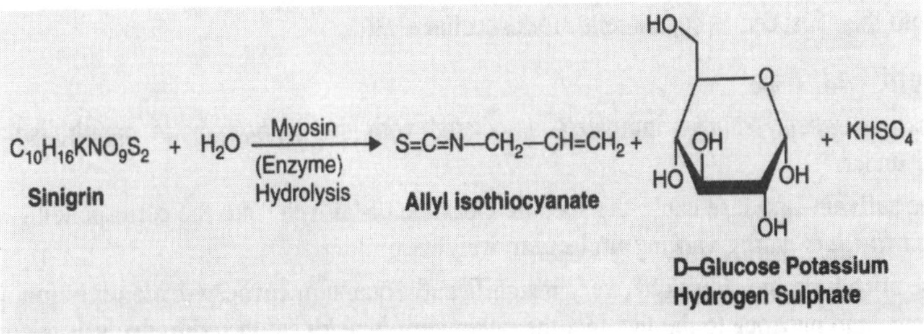

$$C_{10}H_{16}KNO_9S_2 + H_2O \xrightarrow[\text{(Enzyme)}]{\text{Myosin}} S=C=N-CH_2-CH=CH_2 +$$

Sinigrin · Hydrolysis · **Allyl isothiocyanate** · **D–Glucose Potassium Hydrogen Sulphate** · $+ KHSO_4$

Allyl isothiocyanate duly present in the **mustard oil** as the **'volatile oil component'** is responsible exclusively for the typical characteristic odour of the oil.

(3) **Degradation of Sinigrin with Sodium Methoxide [CH₃ONa]: Sinigrin** on being subjected to degradation with *freshly prepared* **sodium methoxide** gives rise to the formation of

thioglucose, which evidently ascertains the fact that the **'glucose reside'** strategically present in **sinigrin** is hooked on to the **S–atom** in it.

Thus, we may have:

$$C_{10}H_{16}KNO_9S_2 \ + \ CH_3ONa \ \xrightarrow{\text{Degradation}} \qquad + \text{ Other Products}$$

Sinigrin   Sodium Meltioxide

5–Thio–D–glucose

(4) Ettlinger *et al.* (1956) were pioneer in assigning the first proposed structure of **sinigrin** as shown above. In fact, they have duly made a concrete and providential suggestion that the hydrolyzed product [as in (2) above] **'allyl isothiocyanate'** is actually produced by **molecular rearrangement.**

(5) Besides, **D–glucose** is duly present as **β–D–glucopyranoside.**

(6) **X-Ray diffraction analysis** has also reconfirmed the structure of **sinigrin** by suggesting that the *'thioglucose'* and the *'sulphate'* moieties are definitely present in the *sym–*configuration. Later on, Eltinger *et al.* (1965) carried out the **'total synthesis'** of **sinigrin** and confirmed its assigned chemical structure.

14. CARBOHYDRATE METABOLISM

Carbohydrates designate one of the major **sources of energy** and a chief **component of diet** particularly for the *animal kingdom.* In fact, the **dietary carbohydrates** largely comprise: **starch, sucrose, lactose,** and **monosaccharides** [mainly **'glucose'** and **'fructose').** Importantly' **glycogen** is invariably found in meat products. In human beings, the **carbohydrates** are usually digested in **gastrointestinal (GI) tract** by means of several *typical enzymes,* namely:

- **salivary amylase,**
- **panercatic amylase,**
- **oligo-1,6-glucosidase, and**
- **disaccharidases** (*e.g.,* **lactase, maltase,** and **sucrase**),

right into their respective constituent **monosaccharides.**

14.1 Salient Features

The various *salient features* intimately associated with the **carbohydrate metabolism** are as enumerated under:

(1) The **salivary amylase** cause the specific cleavage of **'starch'** into the corresponding *smaller dextrin units* haring varying **molecular weights.**

(2) Specifically in the **stomach*,** very insignificantly quantum **carbohydrate digestion** actually comes into play due to the fact that the rather **simple acid–catalyzed hydrolysis** taking place at the *acetyl linkages* is found to be extremely slow in pace at the prevailing **body temperature** (*i.e.,* 37°C).

---

* Salvary amylase is transient in nature beacuse it gets duly deactivated.

(3) Consequently, the **carbohydrates** chiefly and solely undergo the phenomenon of **'digestion'** strategically at the **small intestine,** wherein they are duly attacked by the **pancreatic enzymes,** together with certain other enzymes as well that are critically secreted by the **mucosal cells of the intestine.**

(4) The array of **carbohydrates** duly present in the small intestine are converted meticulously into **galactose, fructose,** and **glucose** that get eventually transported to the *liver via* the **portal vein.**

(5) The **'liver'** helps to convert the **galactose** and **fructose to** *'glucose'* by the suitable enzymes **'isomerases'.**

(6) **'Liver glycogen'** is the store-house of systemic *glucose* till such time it is required to maintain the **'appropriate blood-sugar level',** or it is adequately despatched right into the *criculatory system* of the body and duly transported to the **respective cells.**

(7) **Metabolism:** It has been duly established beyond any reasonable doubt that the **'monosaccharides'** [see (4) above] are duly absorbed from the intestine, and subsequently, transported *via* the **hepatic portal blood** to the *'liver'* and finally to the *'extrahepatic tissues',* wherein they get **metabolized appropriately,** as illustrated in Fig.1.4:

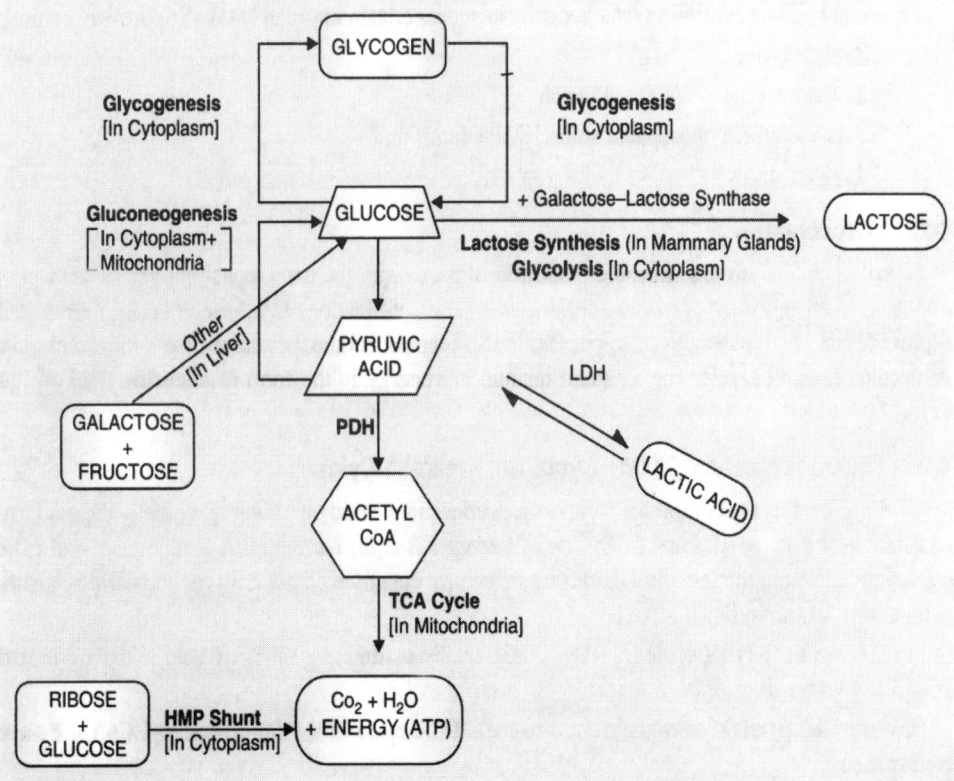

Figure: 1.4: Diagramatic Representation of Carbohydrate Metabolism [Anabolism and Catabolism]

HMP–Shunt	:	Hexose monophosphate shunt;
TCA–Cycle	:	Tricarboxylic acid cycle;
LDH	:	Lactate dehydrogenase (an enzyme);
PDH	:	Pyruvate dehydrogenize (a complex enzyme).

In a broader perspective, **carbohydrate metabolism** essentially comprises *two* cardinal **pathways,** namely:

- **Catabolic Pathways (or Carbohydrate Catabolism):** It is solely responsible for carrying out the essential **cleavage of saccharides,** for instance: **glucose, fructose, glycogen;** and is also commonly termed as **carbohydrate catabolism.**

- **Anabolic Pathways (or Carbohydrate Anabolism):** It is exclusively responsible for the **synthesis of saccharides,** such as: **glycogen, glucose** and **lactose;** and invariably known as **carbohydrate anabolism.**

It is, however, pertinent to state that the aforesaid *two* critical phenomena need to be discussed briefly in the sections that follows.

14.2 Carbohydrate Catabolism

The **carbohydrate catabolism** categorically includes certain important **catabolic pathways,** namely:

- ❑ glycolysis,
- ❑ tricarboxylic acid (TCA) cycle,
- ❑ hexose monophosphate shunt (HMP-shunt), and
- ❑ glycogenolysis.

14.2.1 Glycolysis

Glycolysis refers to the **anaerobic oxidation of glucose** to the corresponding **pyruvic acid** in the *cytoplasm.* In other words, *glycolysis* designates a series of reactions that usually convert a molecule of 'glucose' into *two* molecules of **pyruvate.** It also represents the first stage of the cell respiration of a molecule of **glucose,** releasing a small quantum of energy in the form of **adenosine triphosphate (ATP).**

14.2.2 Tricarboxylic Acid (TCA) Cycle [or Kreb's* Cycle]

TCA cycle refers to complicated series of reactions in the body involving specifically the **oxidative metabolism** of *pyruvic acid* and *liberation of energy.* It is, in fact, the most predominant and major *pathway* pertaining to the **terminal oxidation** in the process of which not only **carbohydrates** but also **proteins** and **fats** are utilized as well.

In other words, **Kreb's cycle** represents the **aerobic oxidation** of pyruvic acid to carbon dioxide $(CO_2\uparrow)$ and **water** in the *mitochondria.*

However, the **pyruvic acid** gains entry to the **TCA-cycle** in the form of **acctyl CoA** (*i.e.,* **acctyl coenzymeA).**

---

* Sir Hans Krebs–a *German* **biochemist** (1900–1981), co–winner of a **Nobel Prize** in 1953, who lived and worked in Great Britain.

14.2.3 HMP Shunt*

HMP shunt refers to an *alternative pathway* meant for the **aerobic oxidation of glucose.** Nevertheless, a few **pentoses** and **tetrose** may also be oxidized *via the* **HMP shunt** pathway.

14.2.4 Glycogenolysis

Glycogenolysis refers to the critical and specific conversion of **glycogen** into **glucose** particularly in the *liver* and *muscles.* Alternatively, **glycogenolysis** mentions the *break down of glycogen* in the cytoplasm of **hepatic** and **muscle fibre cell.**

14.3 Carbohydrate Anabolism

Importantly, the **carbohydrate anabolism** entails the **anabolic pathways,** such as:

- ❑ gluconeogensis,
- ❑ glycogenesis, and
- ❑ lactose synthesis.

14.3.1 Gluconeogenesis

Gluconeogenesis refers to the meticulons **'synthesis of glucose'** particularly from the **non-carbohydrate substances,** for instance: **glycogenic amino acids, lactic acid, glycerol,** and the **tricarboxylic acid (TCA) cycle intermediates.**

14.3.2 Glycogenesis

Glycogenesis refers to the critical formation of **glycogen** from **glucose** specifically in the *liver* and *muscles.*

14.3.3 Lactose Synthesis

Lactose synthesis mentions the **'synthesis of lactose'** (*ie., milk sugar*) from **glucose** and **galactose** specifically in the **mammary glands.**

14.4 Miscellaneous Pathways of Carbohydrate Metabolism

In addition to the **carbohydrate catabolism** and **carbohydrate anabolism** there are *four* other **miscellaneous pathways of the carbohydrate metabolism,** such as:

- ❑ Galactose metabolism,
- ❑ Fructose metabolism,
- ❑ Uronic acid pathway, and
- ❑ Cori cycle,

which shall now be discussed individually as under:

14.4.1 Galactose** Metabolism

Galactose metabolism refers to the actual conversion of **galactose** to **glucose** in the *liver.*

---

* HMP Shunt: Hexose Monophosphate Shunt.

** **Galactose: Galactose** is an isomer of **glucose** and is formed, along with *glucose,* in the **hydrolysis of lactose** (i.e., the **milk sugar**).

14.4.2 Fructose Metabolism

Fructose metabolism mentions the conversion of **fructose** to **glucose** or subsequently metabolized in the *liver.* **Fructose** is duly obtained from **cane sugar** (sucrose), **honey,** and **fruits.**

14.4.3 Uronic Acid Pathway

Uronic acid pathway refers to the critical oxidation of **glucose** to glucuronic acid *via* this pathway However, it may be mentioned at this point in time that **glucuronic acid** is essentially required for the meticulous *in vivo* synthesis of:

- **mucopolysaccharides** (*e.g.,* **heparin), and**
- **detoxification reactions.**

14.4 Cori Cycle

The *Cori Cycle* refers to the particular cycle in **carbohydrate metabolism** in which *muscle glycogen* breaks down, forms *lactic acid,* that eventually enters the blood-stream, and is converted to **liver glycogen.** Consequently, **liver glycogen** breaks down into *glucose,* which is meticulously carried to *muscles* where it gets reconverted to the *'muscle glycogen'.*

14.5 Glucose Origin to Living Organisms

In reality, there are *two* well recognized **glucose origins,** namely:

- **carbohydrates** originated from the **'diet'** *ie., food-intake,* which also refers to the **'exogenous source'** and
- **carbohydrates** originated from the **'liver glycogen'** *ie.,* **proteins,** that refers to the **'endogenous source'.**

These *two* entirely distinct and diversified **origins of glucose** shall now be discussed briefly in the sections that follows:

14.5.1 Exogenous Origin

It has been duly established that the direct **exogenous origin of carbohydrate** (*viz.,* **glucose** happens to be the **'dietary carbohydrates'** that becomes readily available to the different tissues present in the *living organisms* by *two* vital and important phenomena:

- **digestion,** and
- **absorption.**

Thus, it makes it rather necessary and even mandatory to possess and adequate knowledge of both **digestion** and **absorption** rertaining to the host of **dictary carbohydrates.**

Besides, the specifically *digestible* **polysaccharides** and **disaccharides** get duly hydrolyzed into the **intestinal lumen** and *inside the* **epithelial ells** ultimately to form various **monosaccharide components,** such as:

Starch, Glycogen, Maltose	:	converted to **glucose;**
Lactose	:	converted to **glucose** and **galactose;**
Sucrose	:	converted to **glucose** and **fructose.**

In addition to the above largely available carbohydrates, small quantum of **pentose** and **mannose** do also exist in **food products**. Following is the **rate of absorptions** of a number of commonly available **monosaccharides:**

Galactose>Glucose>Fructose>Mannose>Xylose>Arabinose

Mechanism of Absorption: There exists *two* established **mechanism of absorption** of *carbohydrates (sugars)* into the living organism, such as:

- **simple diffusion** — being solely *dependent* upon the **concentration gradient of sugars** prevailing between the *intestinal lumen* and *mucosal cells,* and

- **transport system** — being solely *independent* with respect to the **concentration gradient of sugars.**

Interestingly, before gaining a legitimate entry of **carbohydrate** in the **systemic circulation** eventually pass on to the **'liver'** *via* **adequate absorption** right into the **'portal blood'**, whereby both **fructose** and **galactose** get duly converted to *'glucose'* ultimately.

14.5.2 Endogenous Origin

Importantly, the spectacular **endogenous origin of glucose** being the *liver* **'glycogen'** and *muscle 'glycogen' respectively.*

Liver Glycogen:	It gets duly hydrolyzed ultimately to **'glucose on demand'**, otherwise known as **'glycogenolysis'.**
Muscle Glycogen:	It fails to undergo *direct conversion* into **'blood glucose'**. However, it usually gets converted into **glucose** *via* **lactic acid formation** in the muscles.

Besides, **'glucose'** is adequately synthesized *in vivo* (*ie.,* in **liver**) starting from a plethora of **non-carbohydrate,** such as:

- *amino acids* : derived from the **protein hydrolysates;**
- *glycerol* : obtained from the **hydrolysis of fats amidols;**
- *lactic acid* : formed duly during the **anaerobic oxidation of glucose.**

> Note: *Amino acids* **that specifically serve as the** *'precursors of carbohydrates'* **are termed as** *glycogenic substances;* **and, therefore, the critical phenomenon of** *'glucose synthesis'* **is usually known as** *'gluconeogenesis'.*

14.6 CARDINAL ASPECTS IN CARBOHYDRATE METABOLISM

Preamble: The various vital and important **aspects in carbohydrate metabolism** has been squarely treated under sections 14.1 through 14.5. However, an indepth-knowledge, understanding, and comprehensive details with respect to the following topics is an absolute necessity, namely:

(*a*) Embden–Meyerhof–Parnas pathway [or Glycolysis],

(*b*) Tricarboxylic Acid (TCA) Cycle [or Kreb's Cycle],

(*i*) Hexose Monophosphate (HMP) Shunt,

(*d*) Gluconeogenesis,

(*e*) Glycongenolysis,

(i) Glycogenesis, and

(I) Miscellaneous pathways of carbohydrate metabolism.

The above **carbohydrate metabolism** variants and aspects shall now be treated individually in the section that follows:

14.6.1 Embden-Meyerhof–Parnas (EMP) Pathway [or Glycolysis]

Glycolysis (*Greek* :**glyks** = sweet; and lysis = breakdown) refers to the ***anaerobic oxidation of glucose*** to **pyruvic acid. The Embden Meyerhof Parnas (EMP) pathway** is regard to be an **universal metabolic pathway** solely restricted to **catabolism of glucose,** which comes into play solely in **prokaryotes** and **eukaryotes.**

It has been duly observed that in **glycolysis** one mole of **glucose** gets *oxidized* under **anacrobic condition** *via* an array of reaction to produce *two* **moles of pyruvic acid** as depicted in Fig.1.5.

In fact, **glycolysis** virtually takes place in **cytosol** in all *human tissues*. Importantly, the *glycolytic pathway* is critically catalyzed by the specific **extramitochondrial enzymes** that are present in **cytosol.**

> **Note:** *'Glycolysis'* was, in fact, discovered by Embden, Meyerhof and Parnas; and, therefore, it is invariably known as **Embden-Meyerhof-Parnas (EMP) pathway.**

The **'glycolytic pathway'** *ie.,* the sequential cleavage of **'glucose'** to **'pyruvic acid'** categorically comprises **ten** *vital steps,* namely:

Step–1: Glucose gets phosphorylated by **ATP** and **Mg^{2+}** to **glucose–6–phosphate.** Then said reaction is duly catalyzed by the **particular enzyme** *glucokinase* in liver, and by a **non-specific enzyme** *hexokinase* in *hepatic* and /or *extrahepatic tissues.*

Important Points: These essentially include:

- **Glucokinase** and **hexokinase** designate the *'Regulatory Enzymes'* of the glycolytic pathway, and

- **Step–1** is essentially an **'irreversible biochemical reaction'.**

Step–2: The resulting **glucose-6-phosphate** in the presence of the enzyme **phosphohexose isomerase** (also known as **'phosphogluco isomerase')**

Thus, the reaction emphatically promotas *isomerization* of an **aldohexose** to the corresponding to **ketohexose.**

> **Note: Step–2 is a** *'reversible reaction'.*

Step–3: Fructose-6-phosphate gets **phosphorylated** in the presence of **ATP** and **Mg^{2+}** to yield **fructose-1, 6-**diphosphate, which is duly catalyzed by enzyme **phosphofructo kinase–1 (PFK-1).**

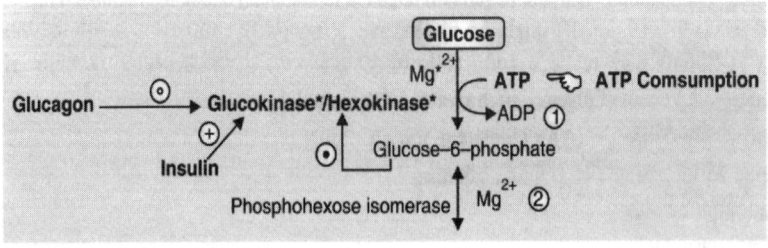

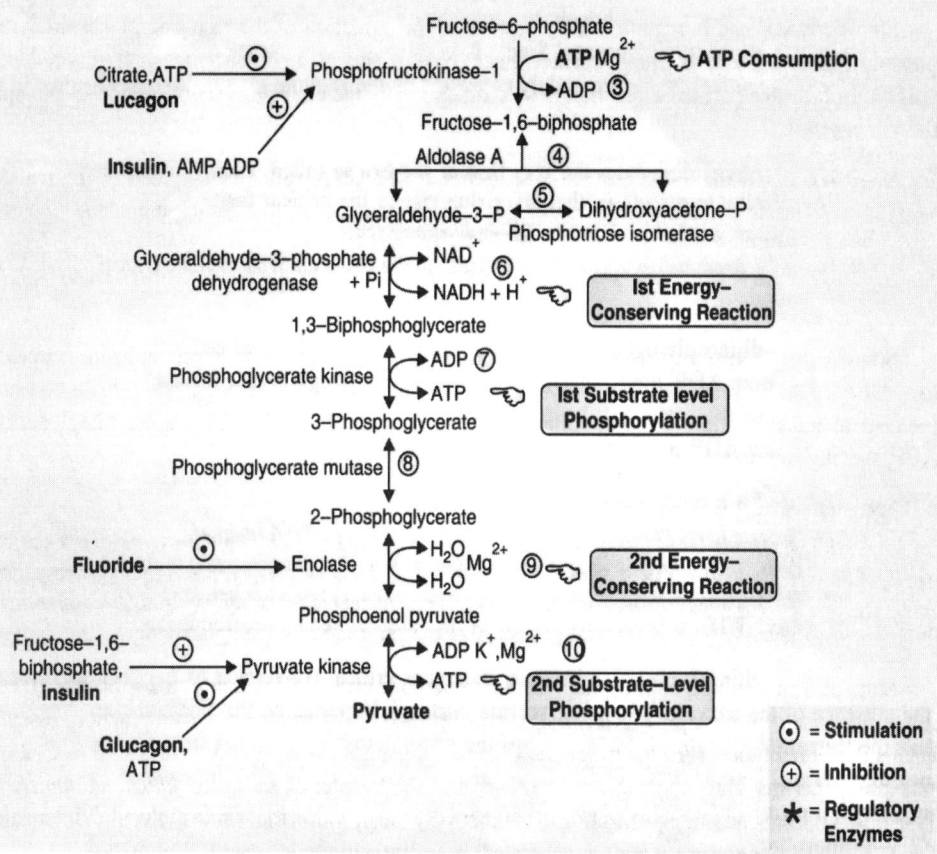

Fig: 1.5 : Glycolytic Pathway Involving Various Regulatory Steps 1 to 10.

Note: (1) *'PFK-1'* is a regulatory enzyme of glycolysis.
(2) *Step-3* is an irreversible reaction.

Step–4: The enzyme **'aldolase A',** found in most of the tissues, does help in the split up **fructose-1,6–diphosphate into** *two* distinct and different **triose phosphates,** namely:

- **glyceraldelyde-3-phosphate (exphospho–aldotriose),** and
- **dihydroxyacetone phosphate** (*or phospho–ketotriose*)

Note: *Step–4* happens to be a *reversible reaction.*

Step–5: The **dihydro acetone phosphate** duly formed in **Step–4** gets isomerized to **glyceraldelyde–3–phosphate** in the presence of the enzyme **phosphotriose isomerase** very much prior to its critical metabolism in **glycolysis.**

Note: (1) *Step–5* is also a *reversible reaction.*
(2) Up to *Step–5* two moles of glyceraldelyde–3–phosphate are duly produced right from only one molecule of glucose (see Fig. 1:5)

Step–6: NAD⁺ and **inorganic phosphate** cause the oxidation of the resulting **glyceraldelyde–3–phosphate** into **1,3–diphosphoglycerate.** Since, NAD⁺ acts as a *'co-enzyme'*, and gets reduced to **NADH.** In fact, this particular reaction is duly catalyzed by the enzyme **glyceraldelyde–3–phosphate dehydrogenase.**

Note: (1) It essentially designates the very first of the two so called 'energy conserving reaction' pertaining to *glycolysis,* thereby giving rise to the critical formation of a–*high-energy phosphate bon'* at C–1 in the 1,3–diphosphoglycerate.

     (2) NADH produced gets crucially oxidized *via* the mitochondrial electron transport chain to generate 3–ATP,

Step–7: In **1, 3–diphosphyoglycerate** the *high-energy phosphate bond* gets utilized meticulously to **synthesize ATP** from **ADP.** Interestingly, the enzyme **'phosphoglycerate kinase'** helps to catalyze the critical transfer of the ensuing **'phosphoryl moiety'** from *C–1* of **1,3–dephosphoglycerate** to **ADP,** thereby forming **ADP** and **3–phosphoglycerate** respectively.

Note: (1) Step–7 is a reversible reaction.

     (2) *Substrate Level Phosphorylation:* In fact, this type of *ATP formation* by critical *transfer of a phosphoryl moiety* from a *'substrate'* (*ie.,* 1, 3–dephosphoglycerate) to the corresponding *ADP* without the least implication and involvement of the mitochondrial *'electron transport chain'* (ETC) is invariably known as the *substrate level phosphorylation.*

Step–8: The resulting 3–phosphoglycerate undergoes critical conversion to **2–phosphoglycerate** in the presence of the enzyme *'phosphoglycerate mutase'*. Importantly, the said reaction, essentially entails the **'intramolecular rearrangement'** of the **phosphoryl moiety** right from **C–3** to **C–2** of the **phosphoglycerate.** Thus, the enzyme *'mestase'* aids the transfer of a specific *functional moiety* (*viz,*. **Phosphoryl**) from one strategic position to another very much within the **'same molecule'**. Interestingly, the **2, 3–diphosphoglycerate** is duly generated as an **'intermediate product'** in *Step–8.*

Step–9: The product obtained from *Step–8,* **2–phosphoglycerate,** gets duly dehydrated into a relatively **high–energy compound** *'phosphoenol pyruvate'*. The ensuing reaction undergoes catalysis by the enzyme *enolase.* Thus the dehydration of **2–phosphoglycerate** critically catalyzed by **enolase** prominently ensures the **'redistribution of energy'** very much within the molecule thereby affording the *conversion* of *low-energy phosphate bond* of **2–phosphoglycerate** into the corresponding *high-energy phosphate bond* of **phosphophenol pyruvate.**

Note: (1) It designates the *'second'* of the two prominent *energy–converting reaction* of 'glycolysis'.

     (2) The *'fluoride ion'* (F⁻) is duly recognized to be the *'strong inhibitor of enzyme enolase'*.

Step–10: The **'phosphophenol pyruvate'** having *high–energy phosphate bond* is fully used up to synthesize **ADP.** For this, the presence of enzyme **'pyruvate kinase'** helps in the catalyzation with regard to the critical transfer of the **'phosphoryl moiety'** from *phosphophenol pyruvate* to **ADP** thereby giving rise to **pyruvate** and **ATP** respectively, as shown in Fig.1.5.

Note: (1) *Step–10* remains and irreversible reaction.

     (2) *ATP* generation due to the critical *transfer* of a *phosphoryl moiety* from a *'substrate'* (*ie.,* phosphophenol pyruvate) to *ADP* without the actual usage of the mitochondrial *electron transport chain (ETC)* represents duly a second befitting example of the so called, *'substrate level phophorylation'* in glycolytic pathway (or *glycolysis*).

14.6.2 Tricarboxylic Acid Cycle [or TCA Cycle or Kreb's Cycle or Citric Acid Cycle]

Nomenclature: The 'tricarboxylic acid cycle' (TCA cycle) is commonly known as the **Kreb's Cycle** since, it was first and foremost put forward by **Sir Hans Adolf Krebs**– a German Biochemist, who bagged the **Nobel Prize** in 1953. Nevertheless, the **TCA cycle** is invariably termed as the **citric acid cycle,** which is based on the fact that the very first pioneer chemical entity duly formed in this cycle is *'citric acid'*. Interestingly, the **TCA cycle** has been appropriately christined on the obvious basic fact that **'citric acid'** *ie.,* the first ever compound produced happens to be a *tricarboxylic acid.*

Fig.1.6. illustrates the TCA cycle and its various regulatory steps involved in a sequential manner:

It is, however, pertinent to state here that *in humans'* the TCA Cycle is solely responsible for **nearly 70% of the total ATP production** *in vivo.* Besides, most of the reaction of 'TCA Cycle' actually come into play exclusively in the **mitochondrial* matrix,** which are *catalyzed* eventually to

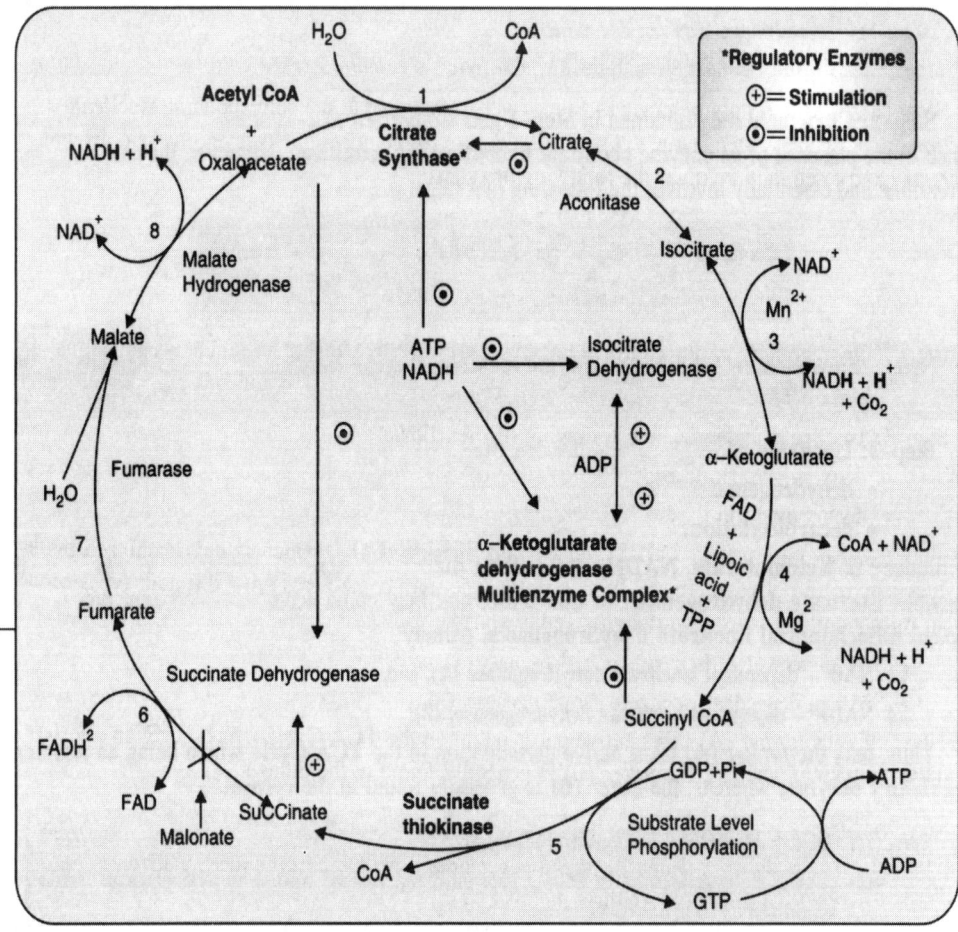

Figure: 1.6. TCA cycle and Various Regulatory Steps Involved

---

* **Mitochondria:** They refer to the *cell organelles* of rod or oval shape 0.5 μm in diameter. Further, these may be observed by the help of a **phase-constrast** or **electron microscope.** They essentially comprise the *enzymes* for the **aerobic stages** of cell respiration, and serve as the sites of most **ATP synthesis.**

the so called **mitochondrial enzymes.** Nevertheless, the *mitochondrial sites* do designate the critical site for *two* important **biochemical processes,** namely:

- **Electron transportation,** and
- **Oxidative phosphorylation.**

The *cyclic pathway* of the **TCA Cycle** predominantly comprise *eight* cardinal sequential steps as enumerated under:

Step–1: Both the enzymes **citrate synthase** and **acetyl CoA** help in the condensation with **oxaloacetate** to produce **citric acid** (a *tricarboxylic acid*). Amazingly, only a *small quantum* of **oxaloacetate** is actually needed for carrying out the **oxidation of a relatively** *large quantum* of **acetyl CoA** *via* the **TCA Cycle.** In reality, oxaloacetate plays a crucial *'catalytic role'* since it fails to be consumed in the course of the **TCA Cycle.**

Note: (1) *Step–1* is an *'irreversible reaction.*
 (2) *Citrate synthase* designates an *'allosteric* regulatory enzyme'.*

Step–2: Citric acid duly obtained in **Step–1** gets *isomerized* to the corresponding **isocitrate** very much in the presence of an enzyme **aconitate hydratase** (or **aconitase**). However, the said reaction is *reversible,* and essentially involves the following *two* steps:

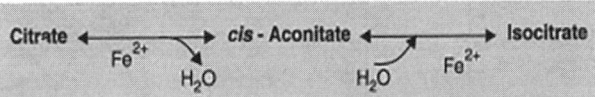

Note: *'Fluoroacetate'* inhibits the enzyme *aconitase* thereby causing an undue accumulation of *citric acid.*

Step–3: Isocitrate undergoes the following *two* reactions:

- **dehydrogenation,** and
- **decarboxylation,**

to produce α–**Ketoglutarate, NADH,** and **carbon dioxide (CO_2),** which is duly catalyzed by the enzymes **isocitrate dehydrogenase.** In true sense, one may come across *two* different types of the typical **mitochondrial isocitrate dehydrogenases,** namely:

❑ NAD^+ – **dependent isocitrate dehydrogenase (A),** and

❑ $NADP^+$ – **dependent isocitrate dehydrogenase. (B).**

Thus, only the *former* (A) takes active participation in the **TCA Cycle** which being an allosteric regulatory enzyme; whereas, the *latter* (B) is generally found in the **cytosol.**

Note: (1) *Step–3* represents a reversible reaction.
 (2) *NADH* duly generated in *Step–3* gets oxidized *via* the mitochondrial *electron transport chain (ETC)* to produce three moles of ATP.

---

* **Allosteric:** Allosteric interactions wherein **drugs, hormones** or **other therapeutic agents** attach to a site other than the *active site of an enzyme* either increasing or decreasing the overall effect.

Step–4: The resulting product α–Ketoglutarate (from *Step–3*) gets critically **decarboxylated** *via* **oxidative mode** to *succinyl CoA* (*ie.,* a substance having a *high–energy* **thioester linkage**), **NADH,** α–Ketoglutarate dehydrogenase multienzyme complex (I) *i.e.,* **regulatory enzyme.**

Just like the **PDH enzyme complex,** the **multienzyme complex (I)** is composed of *five enzymes,* such as : **FAD*, TPP**, lipoic acid, CoA, and NAD⁺.**

The α–Ketoglutarate dehydrogenase complex is nothing but a conglomeration of the following *three* **enzyme variants,** namely:

- **α–Ketoglutarate dehydrogenase,**
- **dihydrolipoyl transacetylase,** and
- **dihydrolipoyl dehydrogenase.**

Importantly, the **α–Ketoglutarate dehydrogenase complex duly represents** an *allosteric regulatory enzyme* of the TCA cycle.

> **Note: (1)** *Step–4* designates an *irreversible reaction.*
>
> **(2)** *NADH* yielded in *Step–4* gets duly oxidized *via* the mitochondrial ETC to produce three moles of *ATP.*

Step–5: Succinyl CoA processing a **high–energy thioester bond** gets converted to **succinate** by the enzyme *succinyl CoA synthase* (or **succinate thiokinase**). Importantly, the utilization of the **high energy thioester bond** of *succinyl CoA* takes place the corresponding **phosphorylate GDP,** and finally to GTP.

Thus, *'substrate level phosphorylation'* comes into play *i.e.,* the critical *phosphorylation* of **GDP to GTP** making use of the **high energy thioester bond** of a *'substrate'* (*viz.,* **succinyl CoA**) without any engagement of the **mitochondrial ETC***.**

> **Note: (1) The above biochemical transformation represents the only** *substrate level phophorylation in TCA* **cycle.**
>
> **(2) GTP formed in this manner specifically transfers its** *high energy phosphate* **to ADP thereby producing** *ATP* **and** *GDP* **respectively in the crucial presence of the enzyme** *nuclcoside diphosphate kinase.* **Hence, prevails an overall net gain of** *1 ATP* **in** *step–5.*

Step–6: Succinate, thus obtained from *step-5,* gets duly oxidized to **fumarate** in the presence of **succinate dehydrogenase,** which belongs to the class of *'flavoenzyme'* essentially comprising the *covalently bound* **coenzyme FAD****.** Importantly, **FAD** caters as a **hydrogen acceptor;** and hence, the *enzyme–linked* **FAD** gets duly reduced to **FADH** thereby oxidizing the *succinate* to *fumarate.* However, **malonate** (*ie.,* the structural analogue of **succinate**) predominantly poses as a **'competitive inhibitor'** of the enzyme **succinate dehydrogenase.**

> **Note: (1)** *Step–5* designates a reversible reaction'.
>
> **(2)** *FADH* generated in this step gets invariably oxidized to produce two moles of *ATP via* the *mitochondrial ETC.*

---

* **FAD :** Flavin adenine dinucleotide.

** **TPP :** Thiamine pyrophosphate.

*** **ETC :** Electron transport chain.

**** **FAD :** Flavine adenine dinucleotide.

Step–7: Fumarate is hydrated to yield **malate** by the enzyme **fumarase** (or *fumarate hydratase*). Evidently, **fumarase** cause the catalyzation of the specific hydration of the *trans* **double bond** to give rise to the formation of **malate.**

Step–8: In the last step of the **TCA cycle,** the resulting **malate** gets duly oxidized by **NAD$^+$–dependent malate dehydrogenase** so as to obtain **oxaloacetate** and **NADH** respectively. In this way, the 'oxaloacetate' gets regenerated; and, therefore, proceed ahead with the **TCA cycle** continuously *in vivo.*

> **Note: (1)** *Step–8* designates a *reversible reaction.*
>
> **(2)** *NADH* obtained in this particular step gets oxidized *via the mitochondrial ETC* to produce *three moles of ATP.*

14.6.3 Hexose Monophosphate Shunt [HMP Shunt]

The **Hexose Monophosphate Shunt (HMP Shunt)** is also commonly known as:

- **Warburg–Lipman Pathway (WL–Pathway),** or
- **Pentose Phosphate Pathway (PPP),** or
- **Phosphogluconate Pathway (PG–Pathway).**

It is worthwhile to state here that the **HMP shunt** is considered to be an *'alternative pathway'* to **glycolysis** as well as **Kreb's Cycle** (or **TCA Cycle**) meant for the aerobic oxidation of **'glucose'.** Most importantly one may understand the intricacies involved in the **HMP Shunt** which not necessarily cater for the source of **energy;** however, its major functionality remains confined to the production of the much desired **'reducing equivalents'** *viz.,* **NADPH** for carrying out the **critical synthesis** of a host of important substances:

- **cholesterol,**
- **fatty acids,**
- **hormones,** and
- **steroids.**

Besides, **HMP shunt** prominently provides **'ribose'** for the essential and crucial **synthesis of nucleic acids** *e.g.,* **DNA and RNA.**

Modus operandi: **HMP Shunt** invariably comes into play in certain typical *organs* and *tissues* in humans, for instance: **adipose tissues, adrenal cortex, lactating mammary glands, liver, ovary,** and **testes,** that are both actively and over whelimingly engaged in the overall synthesis of **steroids** and **fatty acids.**

Interestingly the **HMP Shunt** usually occurs in **'Cytosol';** and, therefore, almost all the *biochemical reactions* involved in this pathway are duly catalyzed by the **extramitochondrial enzymes** exclusively.

HMP Shunt Variants: The entire range of **HMP Shunt** *reaction* may be categorized into *two* broad **phases,** namely:

(a) **Oxidative Phase – that critically takes care of the oxidation of glucose–6–phosphate to pentose phosphate,** and

(b) **Non Oxidative Phase–that essentially deals with the crucial regeneration of glucose–6–phosphate from pentose phosphate.**

14.6.3.1 Oxidative Phase

In a broader perspective the **'oxidative phase'** of the **HMP Shunt** predominantly comprises *three* vital and important steps in a sequential manner (*viz.*, (1), (2), and (3) as given explicitly in Fig. 1.7, which ultimately result in the critical oxidation of **glucose–6–phosphate** to ribulose **5–phosphate.**

Step–1: Glucose–6–phosphate undergoes oxidation to yield **6–phosphogluconolactone,** which is duly catalyzed by the enzyme **NADP⁺–dependent glucose–6–phosphate dehydrogenase (G6PD).** Besides, the aforesaid oxidative reaction is usually accompanied by the critical reduction of **NADP⁺** to **NADPH.** In fact, this represents the very *first example* of NADPH generation specifically in the **HMP Shunt.**

Step–1 represents the *reversible reaction.*

Step–2: 6–Phosphogluconolactone obtained above gets hydrolyzed to **6–phosphogluconate** by the enzyme **lactones** (or **gluconolactone hydrolase).**

Stpe–3: 6–Phosphogluconate on being subjected to:

- **dehydrogenation,** and
- **decarboxylation,**

gives rise to the production of **ribulose–5–phosphate, NADH,** and **carbon dioxide (CO₂).** Nevertheless, the said reaction is catalyzed by **NADP⁺–dependent–6–phosphogluconate dehydrogenase,** which records the *second example* of the **NADPH generation** in the **HMP Shunt.**

Overall Reaction of the Oxidative Phase in HMP Shunt: These essentially comprise the following aspects:

(1) **Glucose–6–phosphate (6 moles)** get duly oxidized to give an even number of **ribulose–5–phosphate** (or *pentose phosphate*), **six moles of carbon dioxide,** and **twelve NADPH** (see Fig: 1.7).

Thus, we may have the following expression:

6 [Glucose–6–phosphate] + 12 NADP + 6 H_2O————6 [Ribulose–phosphate] + 12 NADPH + 12H⁺ + 6CO₂↑

or Glucose–6–phosphate] + 2 NADP + H_2O————Ribulose–5–phosphate + 2 NADPH + 2H⁺ + CO₂↑

(2) Therefore, in the course of the *oxidative phase* of the **HMP Shunt** we may lay hand on the following resultant products by the *oxidation* of **one mole of glucose–6–phosphate:**

- **Ribulate–5–phosphate** (or **Pentose phosphate):** 1 mole;
- **Carbon dioxide:** 1 mole ; and
- **NADH:** 2 moles.

14.6.3.2 Non–Oxidative Phase

The *non–oxidate-phase* of the **HMP Shunt** predominantly comprise the following *seven* cardinal **sequential steps,** namely:

Step–4: Ribulose–5–phosphate undergoes the *catalytic epimerzation,* in the presence of the enzyme **ribulose–5–phosphate epimerase,** to produce **xylulose–5–phosphate.**

Step-4 designates a reversible reaction.

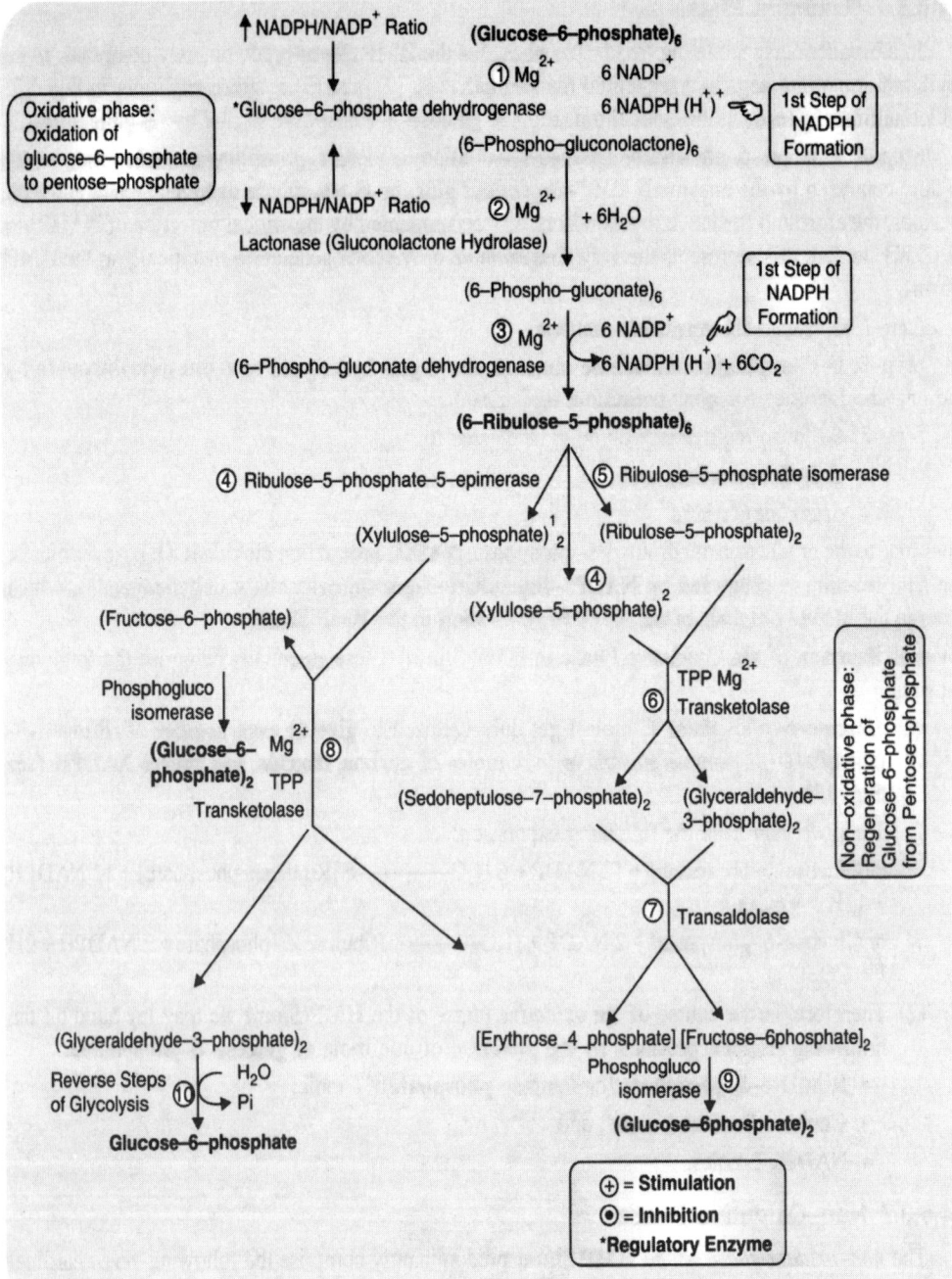

Fig. 1.7: HMP Shunt Showing Various Regulatory Steps.

Step-5: Ribulose–5–phosphate on being subjected to catalytic isomerization, with the *enzyme ribulose–5–phosphate isomerase,* gives rise to the formation of **ribose–5–phosphate.**

Step-5 represents a **reversible reaction.**

Step-6: Ribose–5–phosphate can aldose xylulose–5–phosphate (a **Ketose**) undergoes a catalytic conversion by the *enzyme transseudoheptulose–7–phosphate* respectively.

Note: (1) **TPP acts as the critical and specific coenzyme in this reaction.**

(2) *Step-6* designates a *reversible reaction.*

Step-7: Glyceraldelyde–3–phosphate and sedoheptulose–7–**phosphate** on being subjected to catalytic conversion in the presence of the enzyme **transaldolase** produces **fructose-6-phosphate** and **erythrose–4–phosphate** respectively.

Mechanism: The *enzyme* **transaldolase** enable the catalyzation of a specific **'3-carbon segment'** *ie.,* a *'dihydroxyacetone moiety'* right from the **sedoheptulose–7–phosphate** onto the **glyceraldehyde–3–phosphate** almost precisely.

Step-8: *Transketoketolase* strategically helps in the catalysis towards the critical conversion of **xylulose-5-phosphate** (a *ketose*) and **erythrose–4–phosphate** (an *aldose*) into **glyceraldyelyde–3–phosphate** and **fructose–6–phosphate** respectively.

Note: (1) **TPP serves as a** *'coenzyme'* **in this reaction.**

(2) *Step–8* is a *reversible reaction.*

Step-9: Glucose–6–phosphate is duly converted by the catalytic isomerization of **fructose–6–phosphate** (formed in *Step–7* and *Step–8)* in the presence of the *enzyme* **phospho glucoisomerase.**

Step-9 being essentially a *reversible reaction.*

Step-10: Glyeraldehyde–3–phosphate (2 moles) are meticulously converted to one mole of *glucose-6-phosphate via* the specific reversible steps incurred in the **glycolytic pathway** (or **glycolysis**) as given under:

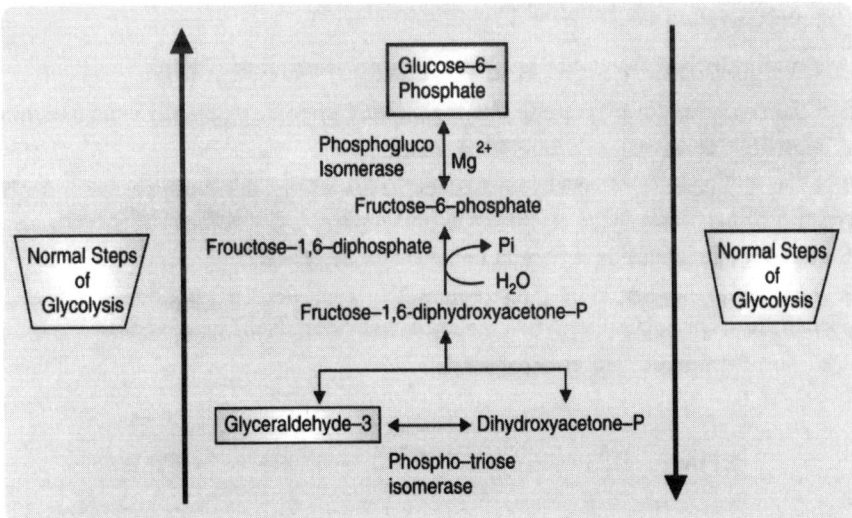

Overall Reaction of the Non-Oxidative Phase in HMP Shunt

It has been duly observed that in the course of the **non-oxidative phase of the HMP shunt,** only *six molecules* of *ribulose 5–phosphate* (critically produced in the previous 'oxidative phase') are *via* an array of biochemical reactions as discussed earlier in various *steps* from **4** through **10**.

Thus, we may have the following expression:

$$\text{6-Ribulose – 5-phosphate} + H_2O \longrightarrow \text{5 Glucose–6-phosphate} + Pi$$

14.6.4. Gluconeogenesis [or Neoglucogenesis]

Gluconeogensis (or Neoglucogenesis) **may be defined as– "synthesis of altogether** *'never'* **glucose (or glycogen) entities from the non-carbohydrate residues** *e.g.,* **pyruvate, lactate, propionate, glycerol, glycogenic amino acids, TCA cycle intermediates, and intermediates derived from glycolysis".**

Gluconeogenic or Glycogenic Precursors-refer to the particular *non-carbohydrate residues (molecules)* that may be converted essentially to **'glucose'** *via* the **gluconeogenic pathway.** Nevertheless, the phenomenon of **gluconeogensis** both *largely and preferentially* comes into play in *mammals* particularly in the *'liver'*, and to a relatively *smaller degree* in the *'kidneys'* (*ie.*, the **renal cortex).**

Degree of **gluconeogenesis** taking place at:

Liver	:	**90 %** ;
Kidneys	:	**10 %** ;

with respect to the formation of **'newly synthesized glucose'** *in vivo*. Both *brain* and *muscles* carry out **gluconeogensis** to a miserably poor extent only.

Both **mitochondria** and **cytosol** are the *crucial sites* where the **gluconeogenic pathway** usually takes place, but **cytosol** accounts for the **major spot.**

The **gluconcogenic pathway** *ie.,* the critical *synthesis of glucose* from **pyruvate** may be conveniently categorized into *two* important **phases,** namely:

- **conversion of pyruvate to phosphoenol pyruvate (PEP),** and
- **conversion of phosphoenol pyruvate to glucose.**

14.6.4.1 Conversion of Pyruvate to Phosphoenol Pyruvate (PEP)

In fact, the conversion of *pyruvate* to the *phosphoenol pyruvate* essentially comprises *two most important* steps (or **reactions)** as discussed under:

Step-1: The very first step essentially involves the formation of **oxaloacetate by the** *carboxylation of* pyruvate in the presence of → **pyruvate carboxylase** *ie.,* a *mitochondrial enzyme.*

Biotin serves as the critical *coenzyme* in the above stated conversion.

It is, however, pertinent to state here that the very first **'regulatory enzyme'** in the *gluconeogenesis pathway* being the **pyruvate carboxylase;** and, therefore, represents one of the *key steps* in the phenomenon of **'gluconeogenesis':**

$$\text{2, Pyruvate} + \text{2 ATP} + \text{2 CO}_2 \xrightarrow[\text{Mg}^{2+};\ \text{Biotin}]{\substack{\text{Pyruvate} \\ \text{Carboxylase}}} \text{2 Oxaloacetate} + \text{2 ADP} + \text{2 Pi}$$

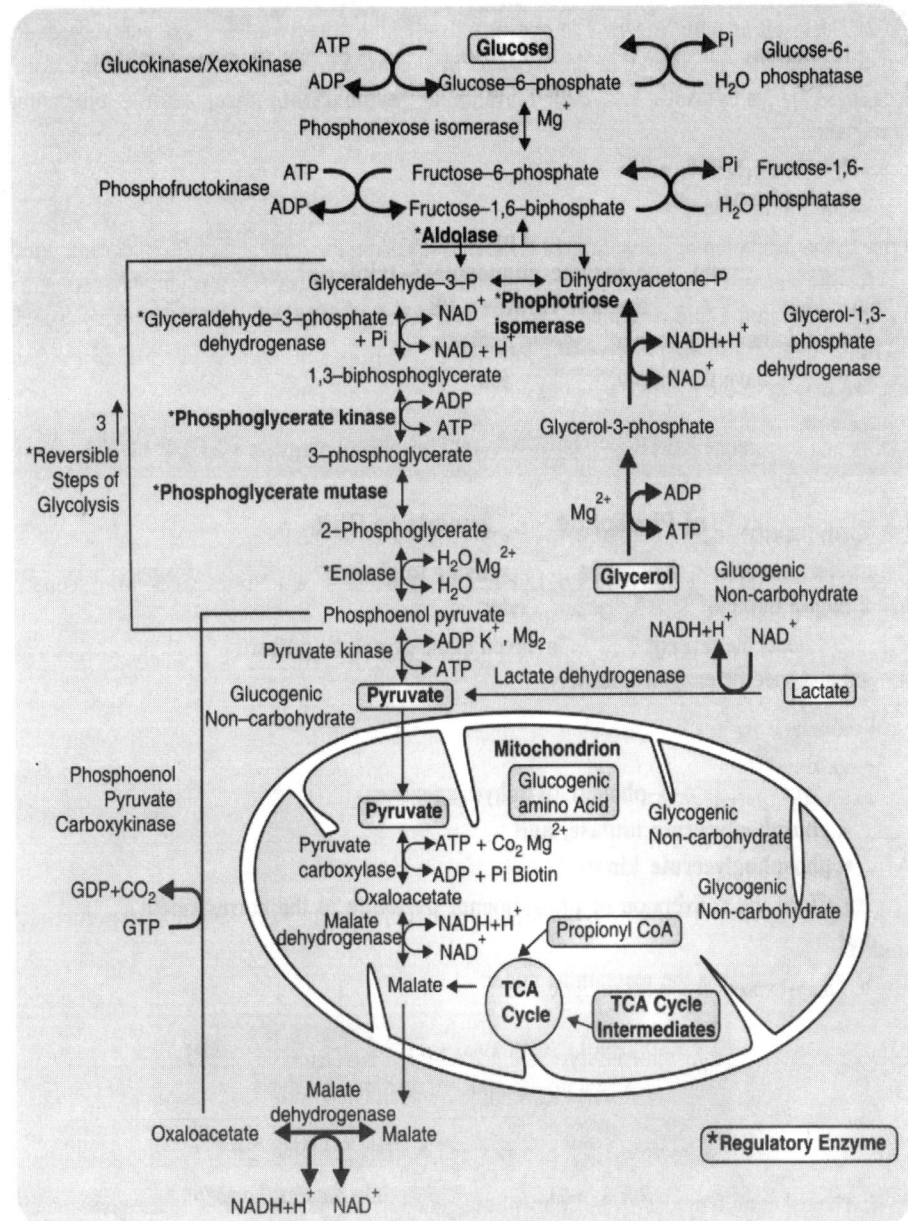

Figure: 1.8: Gluconcogenic Pathway

Importantly, the **oxaloacetate (OAA)** obtained in this way is duly reduced to malate in the presence of **NAD⁺–dependent malate dehydrogenase** (*i.e.*, a **mitochondrial enzyme of the TCA cycle**).

Hence, **Step-1** designates a highly critical and specific reaction pertaining to the **gluconcogenic pathway**; whereas, the subsequent reduction of *oxaloacetate* to the corresponding *malate* represents a reaction of the **TCA Cycle** (**discussed earlier**)– that also makes a positive contribution to the process of **gluconeogenesis**.

Step-2: The resulting **malate** (*Step-1*) gets diffused out of the **mitochondrion,** and gets subsequently oxidized to **oxaloacetate (OPA)** in the **cytosol** in the presence of **NAD$^+$–dependent malate dehydrogenase** (*ie., a* **cytosolic enzyme**). Furthermore, **oxaloacetate** undergoes *two* biochemical reaction sequentially:

- **decarboxylation, and**
- **phophorylation,**

to give rise to the formation of **phosphoenol pyruvate (PEP)** in the presence of **PEP–carboxykinase** (*ie;, a* **cytosolic enzyme**). The enzyme **guanosine–5–triphosphate (GTP)** actually serves as a **'phosphate donor'** in the said reaction. However, **PEP–carboxykinase** represents a highly **'specific enzyme'** in the domain of *phosphoneogenic pathway.*

Thus, we may have the following expression:

$$2, \text{Oxaloacetate 2 GTP} \xrightarrow[\text{Mg}^{2+}]{\text{PEP Carboxypinase}} 2 \text{ Phosphenol pyruvate} + 2 \text{ GDP} + 2 \text{ CO}_2$$

14.6.4.2. Conversion of Phosphoenol Pyruvate to Glucose

The actual **conversion of phosphoenol pyruvate to glucose** essentially engages *three* sequential steps *viz.,* **Step-3** through **Step–5** as given under:

Step-3: It represents the *reversible steps* involved in the **glycolytic pathway** duly catalyzed by a series of *five* vital and specific enzymes, namely:

- **aldolase A,**
- **enolase,**
- **glyceraldelyde–3–phosphate dehydrogenase,**
- **phosphoglycerate mutase, and**
- **phosphoglycerate kinase,'**

in order to afford the conversion of **phosphoenol pyruvate** to the corresponding **fructose–1, 6-diphosphate.**

Thus, we may express the reaction as under:

$$2 \text{ Phosphoenol Pyruvate} + 2 \text{ ATP} + 2 \text{ NADH} + 2\text{H}^+ 2\text{H}_2\text{O}$$
$$\downarrow \text{Several Reversible Steps in Glycolysis}$$
$$\text{Fructose-1,6-biphosphate} + 2 \text{ NAD} + 2 \text{ ADP} + 2 \text{ Pi}$$

Step-4: The resulting **fructose–1, 6-diphosphate** gets duly *dephosphorylated* to the corresponding **fructose–6–phosphate,** which being an extremely critical and vital **'regulatory enzyme'** in the **gluconeogenic pathway** (see Fig: 1. 1.8).

Further, **fructose–1, 6–diplosphate** undergoes dephosphorylation to yield **fructose–6–phosphate,** which being an **irreversible reaction;** and, therefore, it essentially *bypasses* the **irreversible reaction** duly catalyzed by **phosphofructokinase (PFK 1).**

Besides, **gluconeogenesis** may also be accomplished by the specific enzyme **fructose–1, 6–diphosphatase** usually found in **kidneys** and **liver.**

$$\text{Fructose-1, 6-diphosphate} + H_2O \xrightarrow{\text{Fructose-1, 6-diphosphatase}} \text{Fructose-6-phosphate} + Pi$$

Ultimately, the **fructose-6-phosphate** as obtained above gets duly **isomerized** by the enzyme *phosphohexose is omerase* to produce the desired **glucose–6–phosphate** there by designating a *reversible step* in the **glycolytic pathway.**

Step–5: The **glucose–6–phosphate** as obtained above in *step–4* gives rise to the formation of '**glucose**' due to the *dephosphorylation* by the *enzyme* **glucose–6–phosphatase** (see Fig: 1.8).

Importantly, the dephosphorylation of **glucose–6–phosphate** to '**glucose**' designates an '**irreversible reaction**', which eventually *bypasses* the respective **irreversible reaction** duly catalyzed by the enzyme *hexokinase.*

Thus, we may have following reaction.

$$\text{Glucose-6-phosphate} + H_2O \xrightarrow{\text{Glucose-6-Diphosphatase}} \text{Glucose} + Pi$$

The Cori Cycle [or Lactic Acid Cycle]

In the '**Cori cycle**' (or **Acetic Acid Cycle**) the *lactic acid* gets actively and profusely generated in the vigorously worked out* (*ie., contracting*) **skeletal muscles** in the course of the **anaerobic oxidation** of *blood glucose* as shown in Fig. 1.9. Interestingly, the '**lactate**' produced critically may not undergo *further metabolism* in the **skeletal muscles**; and, therefore, escapes out *via.* '**diffusion**' from these muscles right into the **blood circulation.**

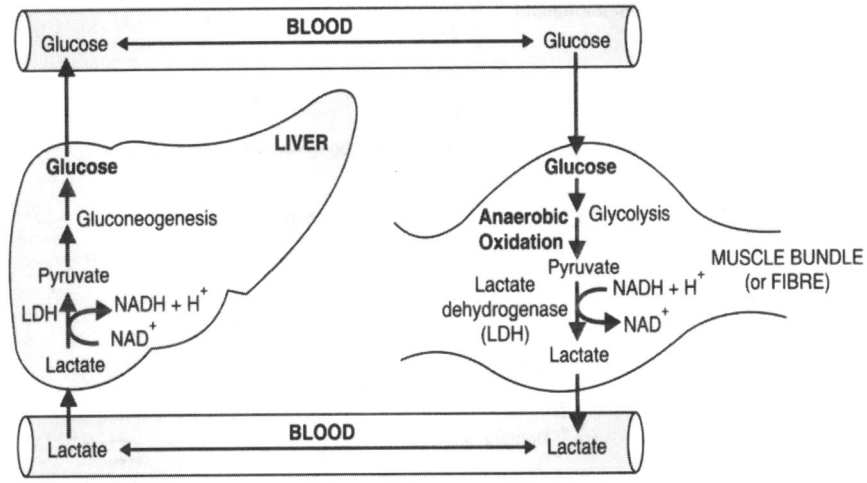

Figure: 1.9: Pathway of Cori Cycle

Consequently, from the circulating blood the specific '**lactate**' is meticulously carried to the '**liver**' wherein it gets duly and adequately converted to '*glucose*' *via* the **gluconeogenetic pathway.** In this manner, the '**glucose**' gets released right back into the ensuing **blood circulation,** and becomes readily available for its much desired circulation in the '**skeletal muscles**'.

---

* That is, excess performance of the muscles during: exercise, physical labourious work, and walking up an upward gradient.

Note: 'Cori Cycle' plays a major role in the critical removal of *'lactic acid'* right from the *circulating blood* in the typical instance of *'acute lactic acidosis'*.

14.6.5. Glycogenolysis

The 'glycogenolysis' refers to the **'cleavage of glycogen'** stored in the *liver* as well as *muscles*. Importantly, the phenomenon of **glycogenolysis** is virtually of immense importance in the critical and proper maintenance of the required **'blood glucose level'** in between two **'square meals'**. Thus, it enables predominantly the **'instant energy'** (from *glucose'* to carry out the *normal* **muscular activities** as depicted explicitly in Figs: 1.10 and 1.11,

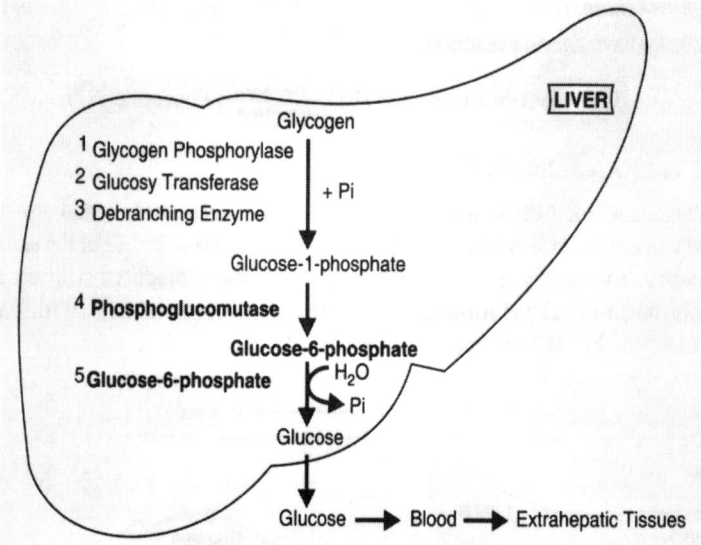

Figure: 1.10: Glycogenolysis Phenomenon in Liver

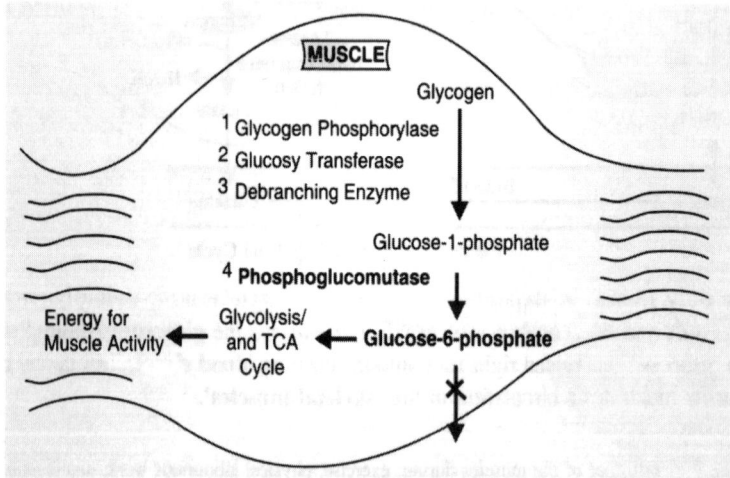

Figure: 1.11: Glycogenolysis Process in Muscles

Besides, the *two enzymes* in Figs, 1.10 and 1.11, namely:

- **Glucose–6–phosphatase** (duly **formed** from **glucose–1–phosphate**), and
- **Phosphoglucomutase** (duly **utilized** in the conversion of **glucose–1–phosphate** to glu-cose–6–phosphate,

there exists predominantly an array of **'cytosolic enzymes'** such as:

❑ **Glycogen phosphorylase,**

❑ **Glucosyl transferase,** and

❑ **Debranching enzyme,**

that participate almost exclusively in the **metabolic cleavage of glycogen** both in the **liver** (Fig. 1.10) and **muscles** (Fig. 1.11)

Importantly, **'glycogenolysis'** invariably occurs in *five* sequential steps which shall be discussed briefly as under:

Step-1: Phosphorolysis of *'glycogen'* takes place in the presence of the enzyme **glycogen phosphorylase** to produce the corresponding **glucose-1-phosphate** and there by limit dextrin. Further more, the said enzyme helps in the cleavage of the ensuing **'glucosidic linkages in the glycogen chain'** particularly from the *'non–reducing terminal'* until such time when at least 4 to 5 **'glucose units'** actually remain very much just prior a **'branching point'** *ie.,* a **(1→6) glucosidic bond** to produce **glucose–1–phosphate,** and there by **'limit dextrin',** as illustrated in Fig. 1.12.

> Note: *'Hepatic glycogen phosphorylase'* distantly differs from the *'muscle glycogen phosphorylase'*, which is based on the fact that the former happens to be *'tetramer'*, while the latter is a *'dimer'*.

Step–2: 'Limit Dextrin' branches are being removed meticulously by the help of the *enzyme* **glucosyl transferase** invariably termed as **'oligo [α (1 → 4) → α R(1 → 4)] glucan transferase'.** Interestingly, the above enzyme carefully removes an isolated part belonging to **3–4 glucose units'** that are strategically attached at a branch by affording a clearage of an **a(1 → 4) glucosidic link-age.** Thus, it ultimately transfers it to a **non-reducing terminal of another chain.**

> Note: The sequential processes involved in the *removal* and *transfer* of a *'branch'* essentially involves the critical break down of an *a(1 → 4) glucosidic bondage*, and subsequent formation of an altogether never *a(1 → 4) glucosidic bondage*.

Step-3: Debranching enzyme specifically cleaves the critical *'debranching point'* with the help of the remaining *single glucose unit.* It is also termed as **amylo–α(1 → 6) glucosidase,** which enables the *release* of **free glucose residue.** (see Fig: 1.12).

Importantly, once the **'branches'** are duly removed by the aforesaid phenomenon, the **'branches'** are duly removed by the aforesaid phenomenon, the ensuing **'glycogen chains'** do remain abundantly available for the **phosphorolysis by the** *enzyme* **'glycogen phosphorylase'** producing **glucose–1–phosphate (or glucose–1–P).**

Step-4 and 5: Glucose–1–P, thus generated by **glucogenolysis,** gets duly converted to the respective **'glucose–6–phosphate'** (see Fig. 1.12) in the presence of the enzyme **phosphoglucomutase** present both in *liver* and *muscles* (see *step–4* in Figs. 1.10 and 1.11). Furthermore, the actual fate of **glucose–6–phosphate** definitely differs in *liver* as well as *muscles.*

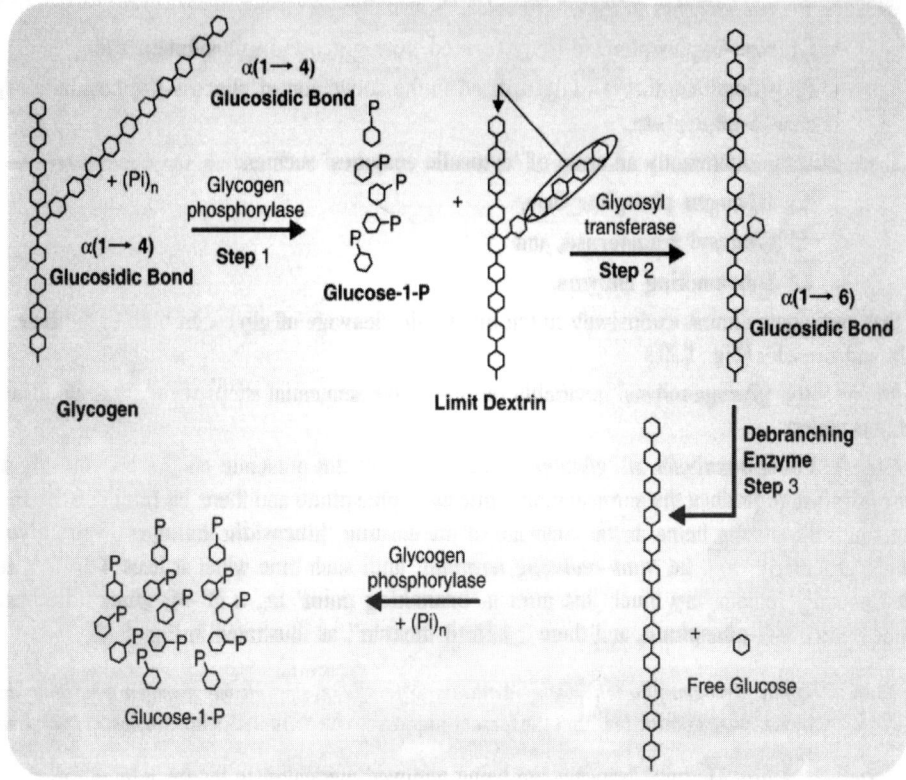

Figure:1.12: Various Steps in Glycogenolysis

Thus, we may observe critically that in:

Liver: Glucose–6–phosphate gets duly **dephosphorylated** by the *enzyme* **glucose–6–phos-phatase** exclusively available in *liver* and *kidneys*. It particularly helps to free **'glucose'**, that gets diffused from the corresponding **hepatic cells;** and hence, gains entry into the **circulating blood,** that gets consumed ultimately by the different **extrahepatic tissues** present *in vivo*.

Muscles: **Glucose–6–phosphate** gets adequately used up in the **glycolysis** process as a vital and important **'source of energy'** to carry out the much desired **muscular activity.**

Note: *Glucose–6–phosphate* **fails to get diffused outside the muscles.**

14.6.6 Glycogenesis

Glycogenesis refers to the actual **synthesis of glycogen** from **'glucose'** *in vivo*. It has been duly established that the phenomenon critically comes into play almost in *all tissues* but predominantly and genuinely in the **liver** and **skeletal muscles,** which eventually **store glycogen.** Thus, one may observe critically that:

Muscle Glycogen – serves as a *'sources of energy'* in the course of **muscle contractions.**

Glucopinase (or hexokinase), Phosphoglucomutase, UDP–glucose phrophosphorylase, glycogen synthase, and debranching enzymes collectively and exclusively participate in the critical synthesis of **'glycogen'** present both in *liver* and *mucles* (see Figs. 1.13 and 1.14)

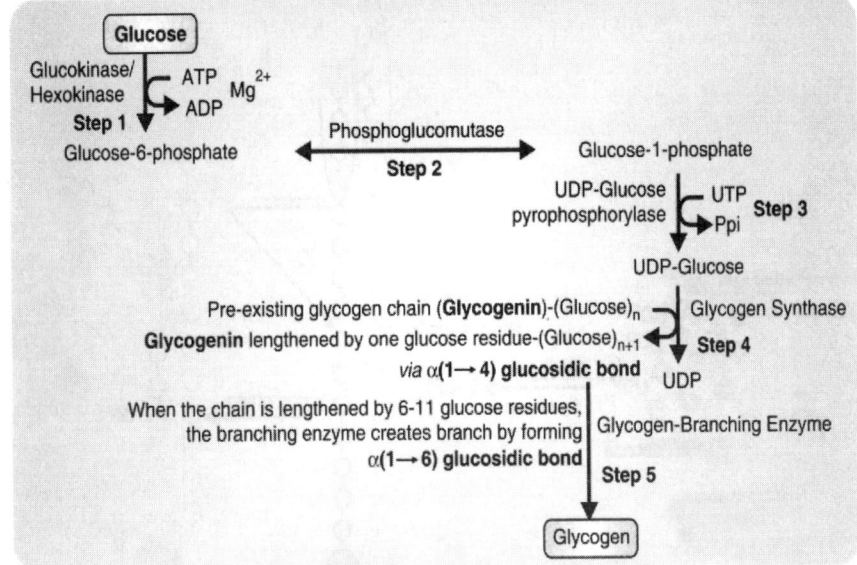

Figure: 1.13: Glycogenesis Occurring in Liver and Muscles

The various steps involved in **glycogenesis** are duly explained in the following *five* steps, namely:

Step-1: α–D–Glucose *ie.*, the precursor of *glycogen*, gets critically phosphorylated to **glucose–6–phosphate** by the *enzyme* glucokinase in **liver** and **hexokinase** in 'muscles', as depicted in Fig.1.13.

Step-2: Glucose–6–phosphate gets eventually converted to **glucose–1–phosphate** in the particular presence of *enzyme* **phosphoglucomutase** (see Fig. 1.13), which being a **reversible reaction.**

Step-3: Glucose–1–phosphate, obtained in *Step-2*, undergoes critical reaction in the presence **uridine triphosphate (UTP)** to produce the corresponding **uridine diphosphate (UDP)–glucose** along with **pyrophosphate (PP)** by the enzyme **UDP-glucose pyrophosphatase.**

In fact, **UDP-glucose** generated above essentially serves as a 'donor' of the *glucose residue* to the ever increasingly growing 'glycogen chain' in the course of 'glycogen synthesis, *in vivo*.

Step-4: Glycogen synthase predominantly aids in the crucial catalyzation process in the transfer of 'glucose' unit from **UDP–glucose** to the respective **non-reducing terminal** of the 'primer' *i.e., pre-existing glycogen chain*, thereby giving rise to the formation of α(1 → 4) *glucosidic bond* very much between the HO–C1* of 'glucose' unit (in **UDP-glucose**), and **HO-C-2**** duly attached to the 'glucose' unit of the 'primer'.

> **Note:** (1) *Glycogen synthase* fails to carry out the *initial synthesis* of 'glycogen' from a *single glucose unit;* nevertheless, it is capable of extending the 'primer' specifically (as shown in Fig. 1.14).
>
> (2) 'Glycogenic' (Fig. 1.14) critically serves as a *primer*. It is indeed a *'proline'* having a polyglucose chain of eight glucose residues joined *via* a(1→4) glucosidic bonds

---

* **HO–Cl**: *i.e.*, hydroxyl group attached to the *first C–atom.*

** **HO–C—2**: *i.e.*, hydroxyl group attached to the *second C–atom.*

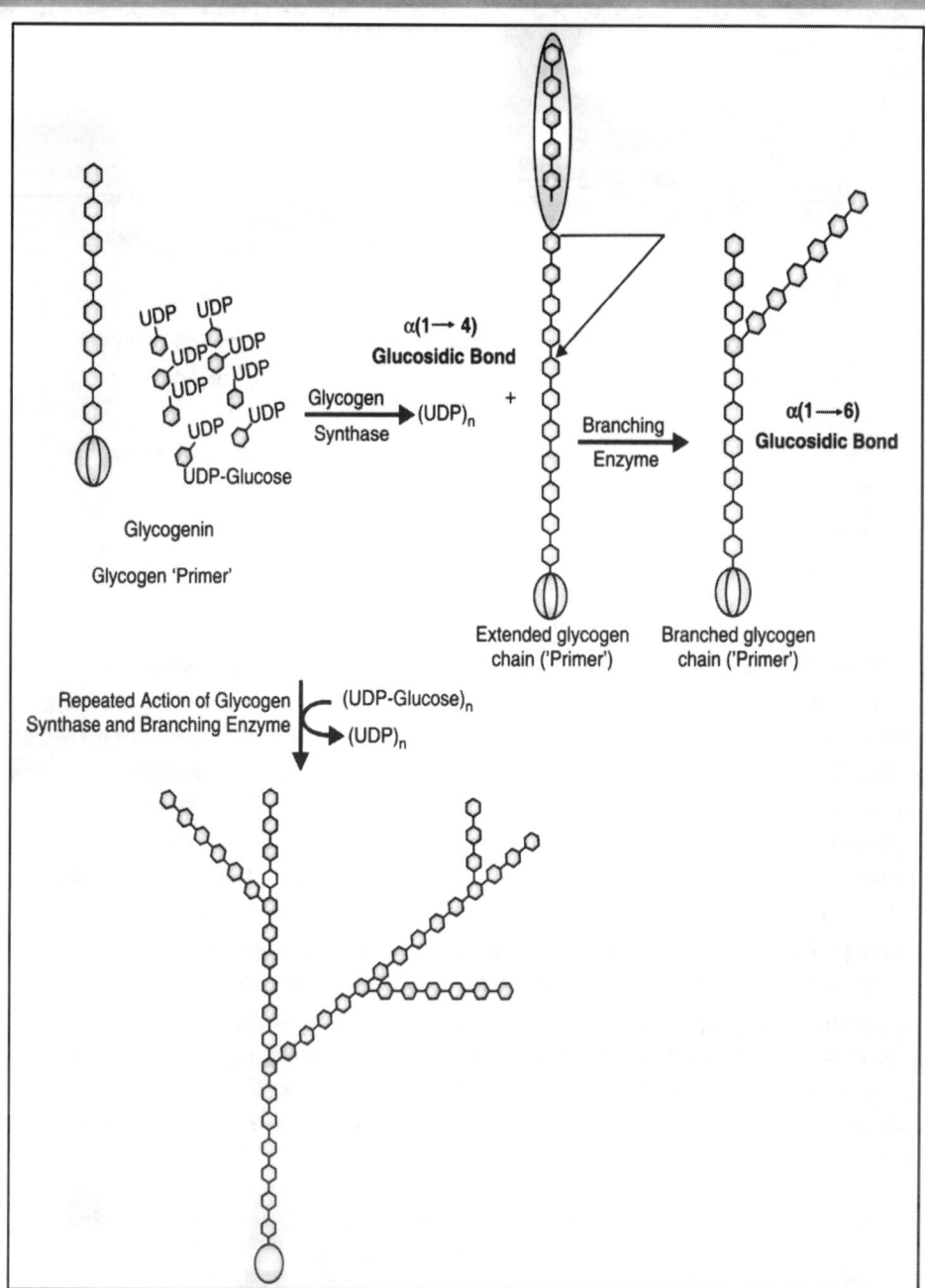

Figure: 1.14: Different Steps Involved in Glycogenesis.

Step–5: 'Glycogen Synthase' is incapable of forming a 'branch point' *viz.*, a distinet
α(1 → 6) glucosidic bond. Therefore, in a specific situation the 'primer' is duly lengthened by
glucose units varying between **6 to 11** then the ensuing 'glycogen branching enzyme' duly generates
branch by forming α(1 → 6) glycosidic bond (see Fig. 1.13).

14.6.7 Miscellaneous Pathways of Carbohydrate Metabolism

There are quite a few *miscellaneous pathways of* **'carbohydrate metabolism'** that essentially exert their own importance and significance in the biological system such as:

- Galactose metabolism,
- Fructose metabolism, and
- Sorbitol (Polyol) pathway of fructose metabolism, and
- Uronic acid pathway,

which shall be discussed briefly in the sections that follows:

14.6.7.1 Galactose Metabolism

Lactose, the *milk–sugar,* caters as the primary major source of *'galactose.* In reality, **lactose** gets duly hydrolyzed by the typical intestinal enzyme β–galactosidase (or *lactase*) to yield a mole each of **galactose** and **glucose.** However, within the cell a small quantum of **galactose** may be critically obtained by the so called **'lycosomal degradation'** of *three* **galactose** containing substances, such as:

- **mucopolysaccharides,**
- **glycolipids, and**
- **glycoproteins.**

It has been duly established that a major segment of the derived **'galactose'** is eventually converted into **'glucose'** in **liver** by means of the following *six* **steps** (as shown in Fig. 1.15):

Step–1: **Galactose** undergoes due **phosphorylation** to yield **glucose-1-phosphate** in the presence of the enzyme *galactokinase* in the **liver.**

Step–2: Resulting **galactose–1–phosphate** undergoes reaction with **UDP–glucose** to give rise to the production of **UDP–galactose** by the *enzyme* **galactose–1–phosphate uridyl transferase.**

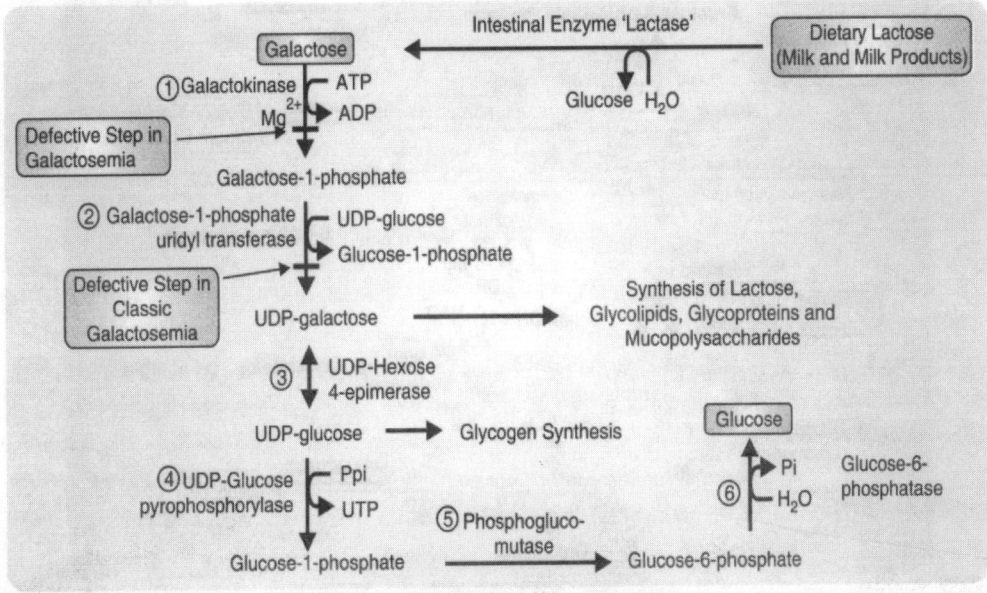

Figure: 1.15: Galactose Metabolism Pathway.

Step–3: UDP–Galactose gets duly converted to **UDP–glucose** by the *enzyme* **UDP–hexose–4–epimerase** which being an 'epimerization reaction.**

> **Note:** A certain portion of the enzyme *UDP–galactose* may also be exclusively utilized for the *'synthesis of lactose'* (during lactation), glycoproteins, glycolipids, mucoproteins, and mucopolysaccharides.

Step–3, 4 & 5: Resulting **UDP–glucose** gets adequately converted to **'glucose'** *via* a set of *three* important reaction duly catalyzed in the presence of such **enzymes as:**

- **UDP–glucose pyrophosphorylase,**
- **phosphoglucomutase,** and
- **glucose–6–phosphatase.**

Importantly, **UDP–glucose** may also yield *glycogen,* that ascertains that **galactose** is converted to **glucose** in the *liver.*

14.6.7.2 Fructose Metabolism

Sucrose, obtained from *cane sugar,* represents the main *dietary source* of **'fructose'**. In fact, **sucrose** gets duly hydrolyzed by the intestinal enzyme **'sucrase'** (or *invertase*) to produce **fructose** and *vegetables*) gets duly converted to **'glucose'** in the *liver* or undergoes typical metabolism *via* **glycolysis** as depicted in Fig. 1.16.

The actual *conversion* of **'fructose'** to **'glucose'** (in the **liver** and its subsequent *metabolism* essentially involves the following *eight* sequential steps:

Step–1: **Fructose** undergoes **phosphorylation** to the respective **fructose–1–phosphate** by the enzyme **fructokinase** in the **liver** (as shown in Fig. 1.16).

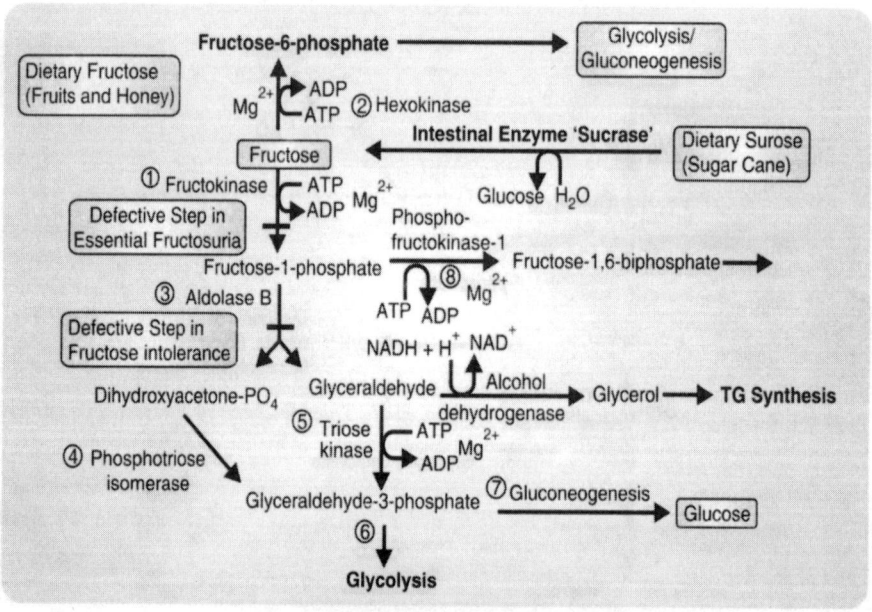

Figure: 1.16: Fructose Metabolism Pathway

Step–2: Amazingly, a major proportion of **fructose** gets also duly phosphorylated to **fructose–6–phosphate** by the *enzyme* **hexokinase.** It may be observed that the *two* aforesaid enzymes do have distinet characteristics features:

Hexokinase	:	Possesses very **low affinity** *i.e.*, high **Km for fructose,** and
Fructokinase	:	possesses extremely **high affinity** *i.e.*, low **Km for fructose.**

Hence, it evidently expatiates the fundamental reasons for the *effective phosphorylation* of **fructose** to **fructose–1–phosphate** as the most predominant **pathway (Step–1).**

Thus, the generated **fructose–6–phosphate** essentially has a tendency to enter:

- 'glycolytic pathway' or
- undergoes critical conversion to 'glucose' *via* 'gluconeogenesis'.

Step–3: The **fructose–1–phosphate** (obtained in *Step–1)* is broken into *two* different products, namely:

❑ **dihydroxyacetone phosphate,** and
❑ **glyceraldehyde,**

by the enzyme **aldolase B** (see Fig. 1.16)

Step–4: Dihydroxyacetone phosphate (from *Step–3)* gets duly isomerized to form **glyceraldelyde–3–phosphate** by the enzyme *phosphotriose isomerase.*

Step–5: Glyceraldehydes (From *Step–3)* undergoes critical phosphorylation to yield **glyceraldelyde–3–phosphate** in the presence of the enzyme *triose kinase.* Besides, **glyceraldehyde** also undergoes reduction to yield **glycerol** by the enzyme *alcohol dehydrogenase,* which gets ultimately consumed in the production of **triacylglycerol (TG) synthesis.**

Step:6 & 7: Glyceraldelyde–3–phosphate duly produced in *Steps–4 and 5,* may readily enter the 'glycolytic pathway' in *Step–6* or gets easily converted to **glucose** *via* **gluconeogensis** (in *Step–7).*

Step–8: Fructose–1–phosphate duly formed in *Step–1* can undergo **phosphorylation** to produce **fructose–1, diphosphate** by the enzyme **phosphofructokinase–1.**

> Note: The above is regarded to be a 'minor pathway' that hardly makes any viable and solid contribution to the ensuing fructose metabolism;.

14.6.7.3 Sorbitol (Polyol) Pathway of Fructose Metabolism

The **sorbitol (Polyol) pathway** of the fructose metabolism essentially involves the critical conversion of 'glucose' to 'fructose' *via* glucitol as depicted in Fig. 1.17. The reduction of **glucose** to 'sorbitol' (Step–1) specifically in the presence of the *enzyme* **aldose reductase** in several tissues in the body, such as: **lens, renal cells, retina, sperm, placenta,** and the like.

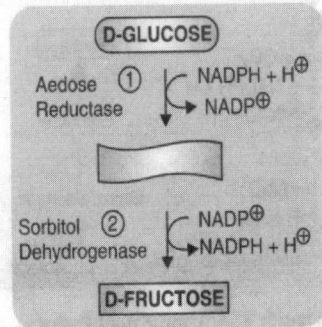

Figure: 1.17: Sorbitol Pathway of Fructose Metabolism

Ultimately, **sorbitol** gets oxidized to **fructose** by the *enzyme* sorbitol dehydrogenase (*Step–2*) duly present in **liver, sperm** and **seminal vesicle cells** in humans.

14.6.7.4 Uronic Acid Pathway [or D–Glucuronic Acid Pathway]

The **uronic acid pathway** (or D–glucuronic acid pathway) is most importantly regarded to be an **'alternative oxidative pathway'** for **'glucose'**.

It has been established beyond any reasonable doubt that in the **liver 'D–glucose'** undergoes critical oxidation to yield **D–glucuronic acid,** which may be ultimately converted to **'D–xylulose'** *via* the **uronic acid pathway** as shown in Fig. 1.18. Interestingly, **D–xylulose** strategically gains its legitimate entry to the **HMP Shunt** so as to accomplish its **further metabolism.**

There are *nine* vital and important sequential steps in the **'uronic acid pathway'** as stated under (Fig. 1.18.):

Step–1: D-Glucose–6–phosphate undergoes critical conversion to **glucose–1–phosphate** in the critical presence of the enzyme **phosphoglucomutase.**

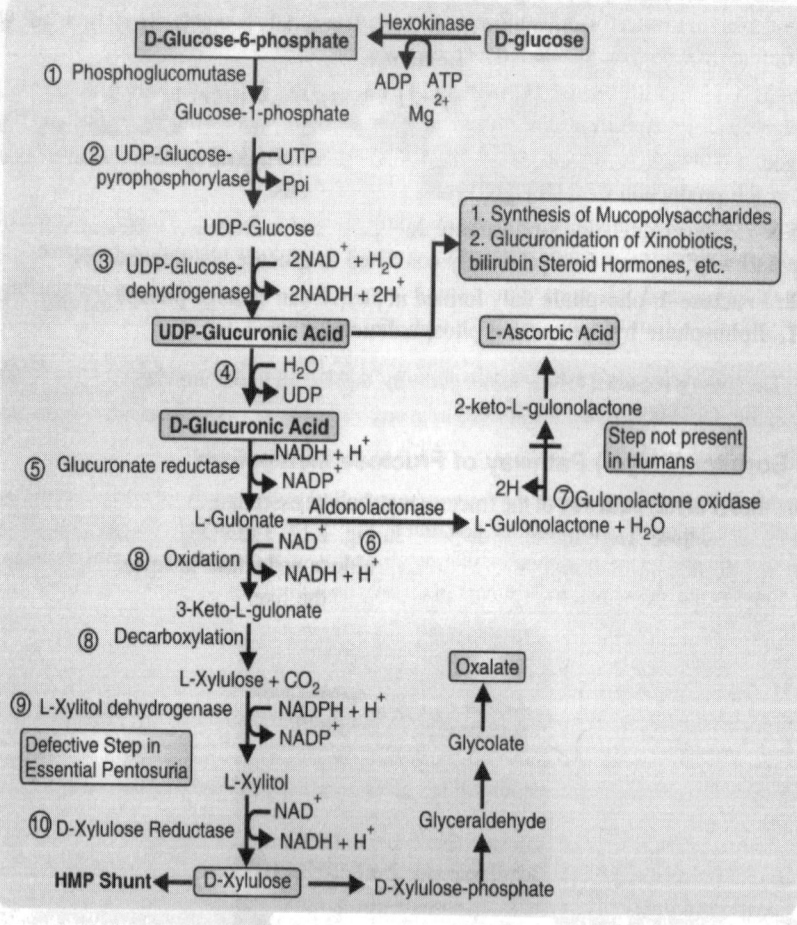

Figure: 1.18: Uronic Acid Pathway

Step–2: Resulting **glucose–1–phosphate** subsequently reacts with **UTP** to give rise to the production to *two* products, namely: **UDP–glucose,** and **pyrophosphate (PPi)** by the enzyme UDP–glucose pyrophosphorylase.

Step–3: **UDP–Glucose,** obtained in *Step–3,* gets duly oxidized to **UDP–glucuronic acid** in the crucial presence of the enzyme **UDP–glucose dehydrogenase.**

> **Notes:** (1) *UDP–Glucuronic acid* designated the metabolically active form of *glucuronic acid.*
>
> (2) *UDP–Glucuronic acid* is essentially and predominantly required for *two* cardinal functionalists, such as:
> - Synthesis of *'mucopolysaccharides',* and
> - glucuronidation of *'xenobioties' viz.,* steroid hormones, bilirubin etc,.

Step–5: **UDP–Glucuronic Acid,** obtained duly in *Step–4,* gets reduced to L–gulonate gets duly converted to **L–ascorbic acid** and **D–xylulose** respectively, which shall now be discussed separately in the sections that follows:

Formation of Vitamin C (or L–Ascorbic Acid): The formation of **Vitamin C** may be accomplished by the fallowing *two* critical steps, namely:

Step–6: **L–Gulonate,** obtained in *Step–5,* gets duly converted to its corresponding **'lactone'** known as: **Gulonolactone** in the persevere of the *enzyme* **aldonolactonase.**

Step–7: **L–Gulonolactone** gets consequently oxidized to produce **L–assorbic acid** by the *enzyme* **gulonolactone oxidase,** whereby 2–Keto–L–gulonolactone is duly obtained as an **'intermediate'**

Formation of D–Xylulose: The formation of D–Xylulose takes place by the help of the following *two* steps in a sequential manner:

Step–8: **L–Gulonate,** as obtained in *Step–5,* gets duly oxidized to the corresponding **3–Keto–L–gulonate,** that eventually undergoes decarboxylation to produce **L–xylulose.**

Step–9: **L–Xylulose** ultimately gets reduced to **L–xylitol** in the critical presence of the *enzyme* **L–Xylitol dehydrogenase.**

RECOMMENDED FURTHER READINGS

Aspinall Go (ed.) : **The Polysaccharides,** Vols. 1–2, Academic Press Inc., New York, 1982–1983.

Baird Jk : **Gums, Kirk–Othmer Encyclopedia of Chemical Technology,** Vol.12, Wiley, New York, 4th edn., pp 842–862, 1994.

Bell Wr : **Blood, Coagulants and Anticoagulants, Kirk–Othmer Encyclopedia of Chemical Technology,** Vol.4, Wiley, New York, 4th ed., pp 333–360, 1992.

Be Miller JN : **Carbohydrates, Kirk–Othmer Encyclopedia of Chemical Technology,** Vol.4, Wiley, New York, 4th ed., pp 911–948, 1992

Bugg TDH : **Bacterial Peptidoglycan Biosynthesis and Its Inhibition, In: Comprehensive Natural Products Chemistry,** Vol.3, Elsevier, Amsterdam pp 241–294, 1999.

Collins Am (ed.) : **Carbohydrate,** Chapman and Hall, London (UK), 2003.

Coppen JJW : **Gums, Resins, and Latexes of Plant Origin, FAO,** Rome, 1995.

Colwell RR *et al* : **Biotechnology Marine Polysaccharides,** Hemisphere Publishing Corpn, Washington (USA), 1985.

Dea ICM : **Industrial Polysaccharides,** *Pure Appl. Chem.,* **61** : 1315–1322, 1989.

Dewiek PM : **Medicinal Natural Products: A Biosynthetic Approach,** Wiley, London (UK), 2nd ed., 2002.

Dey PM and Harborne JB (eds.) : **Methods in Plant Biochemistry,** Vol. 2: **Carbohydrates,** Academic Press Inc., San Diego, (USA), 1990.

El–Khadem HS : **Carbohydrate Chemistry–Monosaccharides and their Oligomers,** Academic Press, New York, 1988.

Ferrier RJ and Collins PM : **Monosaccharide Chemistry,** Penguin Books, London (UK), 1972.

Hanson JR : **Natural Products: The Secondary Metabolites,** the Royal Society of Chemistry, London (UK), 2003.

Loewas FA and Tanner W. : **Encyclopedia of Plant Physiology,** Vol. 13 A, **Plant Carbohydrates I,** Springer–Verlag, Berlin (Germany), Tanner W, 1982.

Morrison WR and Karkalas J : **Starch,** In : *Methods in Plant Biochemistry,* Vol.2. **Carbohydrates,** (Day Pm ed.) Academic Press, London, P–353–352, 1980.

Percival E and Me Dowell R : **Algal Polysaccharides,** In: *'Methods in Plant Biochemistry, Vol.2 Carbohydrates',* Dey PM (Ed) : Academic Press, San Diego (USA), p ;523–547, 1980.

Pigman WW and Horton D (eds.) : **The Carbohydrates,** Vols 1 A (1972) and 1 B (1980), Academic Press, New York.

Preiss J (ed.) : **The Biochemistry of Plants,** Vol.3, **Carbohydrates: Structure and Function,** Academic Press Inc., New York, 1980.

Rochrig KL : **Carbohydrate Biochemistry and Metabolism,** The AVI Publishing Co., Inc., Westport, Connecticut (USA), 1984.

Walter RH (ed)	:	**The Chemistry and Technology of Pectin**, Academic Press Inc., San Diego, 1991.
Whistler RL and Be Miller JM	:	**Industrial Gums**, Academic Press, New York, 1993.
Whistlen RL *et al.* (eds.)	:	**Starch, Chemistry, and Technology**, Academic Press Inc., Orlando (USA), 2nd ed., 1984.
Yalpani M (ed.)	:	**Progress in Biotechnology**, Vol.3., **Industrial Polysaccharides**, Elsevier Science Publisher, B.V., Amsterdam, 1987.

REVIEW QUESTIONS

1. How would you establish the **'Cyclic Structure of Glucose'**?

2. Give a brief account of the **chemistry** and **structure** of *'glucose'* with specific reference to the **structure** and **relationship** with other carbohydrates as: **Arabinose, Fructose, Mannose,** and **Starch.**

3. Discuss the various steps involved in developing the structure of **Glucose.**

4. Explain diagrammatically the following:

    (a) **Fischer Projection of D-Glucose,**

    (b) **Haworth Projection of β-D-Glucopyranose**

5. How would you account for the '**Mutrotation of Carbohydrates**? Explain with important observations.

6. Discuss the following briefly:

    (i) **Synthesis of 'Sucrose'**

    (ii) **Inversion of 'sucrose'**

    (iii) **Constitution of 'Maltose'**

7. Write short notes on the following:

    (a) **Photosynthesis of Carbohydrates**

    (b) **Epimerization**

    (c) **Determination of 'Ring Sizes of Monosaccharide Units' present in 'Maltose'.**

8. Give a comprehensive account on the following:

    (i) **Cellobiose,**

    (ii) **Gentiobiose,**

    (iii) **Melibiose**

9. Describe the **'Polysaccharides'** with particular reference to:

    (a) **Nomenclature,**

    (b) **Structure of Polysaccharides,** and

    (c) **Isolation and Structural Analysis of Carbnohydrates**

10. Write short notes on the following:
    (i) **Dextrans**
    (ii) **Xanthan Gum**
    (iii) **Lantinan**

11. Discuss the 'Polysaccharides of Lower Plants:
    (a) **Aliginic Acid,**
    (b) **Carrageenans,** and
    (c) **Agar**

12. What are the **Honogenous Polysaccharides?** Describe any one of the following in details:
    (i) **Starch,**
    (ii) **Cellulose,** and
    (iii) **Fructans**

13. Explain the **Heterogenous Polysaccharides.** Discuss the following explicitly:
    (a) **Gum Arabic,** and
    (b) **Pectin**

14. Give a brief account on the **'Glucosides'.** Discuss the following aspects briefly:
    (i) **Glycoside Formation,** and
    (ii) **Naturally Occurring Glycosides**

15. Elaborate the following important **Glycosides**:
    (a) **β-Amygdalin,**
    (b) **Arbutin,**
    (c) **Glucovanillin,**
    (d) **Indican**
    (e) **Ruberhythric Acid,** and
    (f) **Salicin**

16. Write an essay on the **'Carbohydrate Metabolism'** with particular reference to the following:
    (i) **Tricarboxylic Acid (TCA) Cycle,** and
    (ii) **Embden-Meyerhof-Parnas (EMP) Pathway**

17. Discuss the following comprehensively:
    (a) **Gluconeogenesis,**
    (b) **The Cori Cycle,**
    (c) **Glycogenolysis,** and
    (d) **Glycogenesis**

18. Write **chemical equations** to show how **D-(+)-Glucose** may be duly converted to the following products:
    (i) **Methyl β-D-glucoside,**
    (ii) **D-Mannose,**
    (iii) **2,3,4,6-Tetramethyl-D-glucose,**
    (iv) **L-Gulose,**

(v) **D-Arabinose,** and

(vi) *meso*-**Tartaric acid**

19. Enumerate the **Miscellaneous Pathways** of the **'Carbohydrate Metabolism'** with regard to:

   (a) **Galacotose Metabolism,**

   (b) **Fructose Metabolism,**

   (c) **Sorbitol Pathway of Fructose Metabolism,** and

   (d) **Uronic Acid Pathway**

20. Give the **structures** and **names of products** given in the following typical *carbohydrate reactions*:

   (i) **α-Maltose** $\xrightarrow{\text{H}_3\text{O}^+}$

   (ii) **β-Maltose** $\xrightarrow{\text{Br}_2/\text{H}_2\text{O}}$

   (iii) **α-Cellobiose** $\xrightarrow{\text{H}_2\text{O}/\text{H}^+}$

   (iv) **β-Cellobiose** $\xrightarrow{\text{H}_2\text{O}/\text{H}^+}$

   (v) **Cellulose** $\xrightarrow{\text{NaOH},\Delta}$

Contents at a Glance

Amino Acids, Peptides and Proteins

2

1. INTRODUCTION

Amino acids may be defined as– **'the critical and typical monomer units of the peptides and proteins'.**

It is, however, pertinent to state here that in a **'peptide'** or **'protein'** molecule, the *'amino acids'* are intimately held together *via* the **'peptide linkages.'**

Amino acids are also invariably termed as the **'building blocks of proteins'.** Importantly, the **peptides** and **proteins** on being subjected to hydrolysis in:

- an **'acidic'** environment,
- an **'alkaline'** environment, and
- an **'enzymatic'** environment,

thereby giving rise to the formation of an array of **'amino acids',** which may be analyzed by the help of an **'Amino Acid Analyzer'.**

Amino acids, as the name indicates, refer to such compounds that essentially comprise both an **amino (–NH$_2$) moiety** and a **carboxylic acid (–COOH) group** as given under :

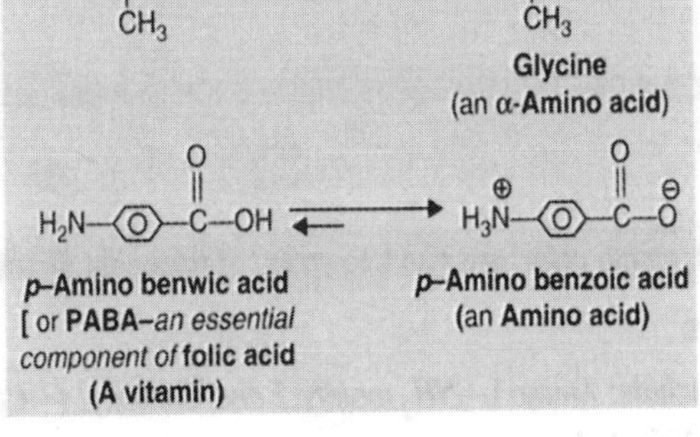

$$H_2N-CH_2-\overset{\overset{O}{\|}}{C}-OH \underset{\longleftarrow}{\longrightarrow} \overset{\oplus}{H_3N}-CH_2-\overset{\overset{O}{\|}}{C}-\overset{\ominus}{O}$$

Glycine
(an α-Amino acid)

$$H_2N-\underset{\underset{CH_3}{|}}{CH}-\overset{\overset{O}{\|}}{C}-OH \underset{\longleftarrow}{\longrightarrow} \overset{\oplus}{H_3N}-\underset{\underset{CH_3}{|}}{CH}-\overset{\overset{O}{\|}}{C}-\overset{\ominus}{O}$$

Glycine
(an α-Amino acid)

$$H_2N-\langle O \rangle-\overset{\overset{O}{\|}}{C}-OH \underset{\longleftarrow}{\longrightarrow} \overset{\oplus}{H_3N}-\langle O \rangle-\overset{\overset{O}{\|}}{C}-\overset{\ominus}{O}$$

p–Amino benwic acid
[or **PABA**–*an essential component of* **folic acid**
(A vitamin)

p–Amino benzoic acid
(an Amino acid)

Since, most of these structures reveal vividly that a **'neutral amino acid'** (*i.e.*, an amino acid having an **overall charge of** *'zero'* may predominantly possess very much **'within the same molecule' two groups of opposite charges** [*viz.*, ammonium (NH_4) +ve charge; and carboxylate (–COO) –ve charge].

Zwitterions [*German*– means a **'hybrid ion'**] :All such molecules which essentially comprise the **'oppositely charged moieties'** are termed as **zwitterions.** In other words, a **'Zwitterionic structure'** actually comes into being due to the presence of:

- **a** 'basic amino function'– that may easily **'accept a proton',** and
- **an** 'acidic carboxylic function'– which may easily **'lose a proton'.**

2. CLASSIFICATION OF AMINO ACIDS

The **amino acids** are generally classified into *three* variants, such as :

α-**Amino acid**   *eg.,* $H_3C-CH_2-\underset{\underset{NH_2}{|}}{CH}-COOH$
                                 α–**Aminobutanoic acid**

β-**Amino acid**   *eg.,* $H_3C-\underset{\underset{NH_2}{|}}{CH}-CH_2-COOH$
                                 β–**Aminobutanoic acid**

γ-**Amino acid**   *eg.,* $H_3N-CH_2-CH_2-CH_2-COOH$
                                 γ–**Aminobutanoic acid**

In a broder perspective the **amino acids** are *classified* into the following major groups, namely :

2.1. Acidic Amino Acids

These essentially comprise .

Amino (–NH₂) moiety : 1, and

Carboxyl (–COOH) moiety : 2

Examples: The typical examples are as follows:

Aspartic acid (*Asp*) : $HOOC-CH_2CH(NH_2).COOH;$

Asparagine [*Asn* or *AspNH₂*] : $H_2NCOCH_2-\underset{\underset{NH_2}{|}}{CH}-COOH$ [α–Amino succinic acid]

Glutamic Acid [*Glu*] : $HOOC\ CH_2CH_2-\underset{\underset{NH_2}{|}}{CH}-COOH$ [α–Amino glutaric acid

Glutamine [*Gln* or Glu—NH₂] $H_2NCO.CH_2CH_2.\underset{\underset{NH_2}{|}}{CH}.COOH$ [α–Amino glutaramic acid

NOTE: The *'acidic amino acids'* are also known as **'Monoamino dicarboyxylic acid'.**

2.2. Basic Amino Acids

These predominantly include: Amino [—NH₂ *moiety*:2 *and Carboxyl* (—COOH) *moiety*:1]

Examples: These mainly include:

- **Arginine** [*Arg*] : $H_2N-C-NH(CH_2)_3.CH-COOH$ [α–Amino-δ-guanidino–n–valeric acid]
  $$\underset{NH}{\overset{\parallel}{}} \qquad \underset{NH_2}{|}$$

- **Lysine** [*Lys*] : $H_2N-CH_2-CH_2-CH_2-CH_2-CH-COOH$ [α, ε–Diamino coproic acid]
  $$\underset{NH_2}{|}$$

- **Hydroxylysine** [*Hyl or Lys-OH*] : $H_2N\,CH_2\,CH\,CH_2CH_2CH_2CH-.COOH$ [α, ε–Diamino–δ–hydroxy hexanoic acid]
  $$\underset{OH}{|} \qquad\qquad \underset{NH_2}{|}$$

2.3. Neutral Amino Acids

These prominently consist of:

Amino (–NH$_2$) moiety : 1, and *Carboxyl (–COOH) moiety* : 1

Examples: The Typical examples of **neutral amino acids** are:

- **Serine** [*Ser*] : $HO.CH_2CH-COOH$ [α–Amino-β-hydroxy–propionic acid]
  $$\underset{NH_2}{|}$$

- **Threonine** [*Thr*] : $H_3C.CH(OH)-CH_2.COOH]_2$ [α–Amino-β-hydroxy-butyric acid]
  $$\underset{NH_2}{|}$$

- **Valine** [*Val*] : $(H_3C)_2CH.CH-COOH$ [α–Amin–isovaleric acid]
  $$\underset{NH_2}{|}$$

- **Leucine** [*Leu*] : $(H3C)_2CH.CH_2\,CH-COOH$ [α–Amino–isocaproic acid]
  $$\underset{NH_2}{|}$$

- **Isoleucine** [*Ile*] : $H_3C.CH_2\,CH(CH_3)-CH-COOH$ [α–Amino-γ–methylthio–n–valeric acid]
  $$\underset{NH_2}{|}$$

- **Glycine** [*Gly*] : $H_2N-CH_2\,COOH$ [α–Amino acetic acid]

- **Alanine** [*Ala*] : $H_3C.CH-COOH$ [α–Amino propionic acid]
  $$\underset{NH_2}{|}$$

Note: *'Neutral amino acids'* are also known as *monoamino monocarboxylic acid.*

2.4. Sulphur [S] Containing Amino Acids

Following are a few typical examples of the **'sulphur' containing amino acids**, namely:

• **Cysteine** [*Cys*] : HS.CH$_2$CH—COOH [α–Amino–β-mercapto-propionic acid]
  |
  NH$_2$

• **Cystine** [*Cys Cys*] : [—S.CH$_2$—CH—COOH]$_2$ [Bis-(α–Amino-propionic acid)-β-disulphide]
  |
  NH$_2$

• **Methionine** [*Met*] : H$_3$C.S.CH$_2$CH$_2$—CH—COOH [α–Amino-γ–methylthio-n-butyric acid]
  |
  NH$_2$

• **Tryptophan** [*Trp*] : CH$_2$—CH—COOH [α–Amino-β–indole propionic acid]
  |
  NH$_2$

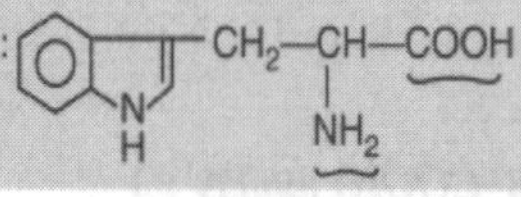

Note: (1) Most *'S-containing amino acids'* do possess either *one moiety each of amino (–NH₂)* and *carboxylic acid (–COOH)* function or a pair in them.

(2) They must essentially contain either a sulphydryl (-SH) function or a sulphide (-S-) function in them.

2.5 Aromatic Amino Acids

A few typical examples of the **'aromatic amino acids'** are as enumerated under:

• **Phenylalanine** [Phe] : —CH$_2$—CH—COOH [α–Amino-β-phenyl-propionic acid]
  |
  NH$_2$

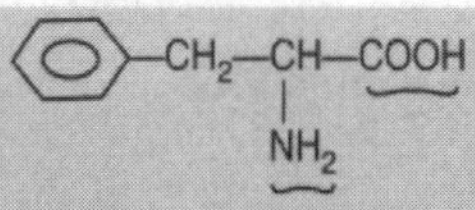

• **Tyrosine** [Tyr] : HO— —CH$_2$.CH—COOH [α–Amino-β-(*p*-hydroxy-phenyl) propionic acid]
  |
  NH$_2$

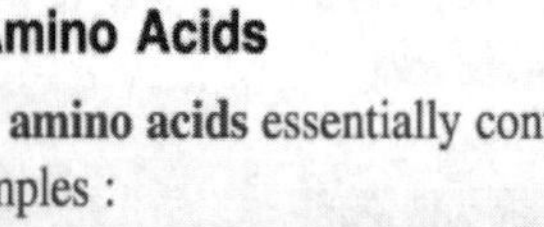

• **Thyroxine** [Tyr] : HO— —O— —CH$_2$.CH—COOH [β-3,5–Diiodo-4-(3;5'-diiodo-4'-hudnxy)
  | -α-aminopropionic acid]
  NH$_2$

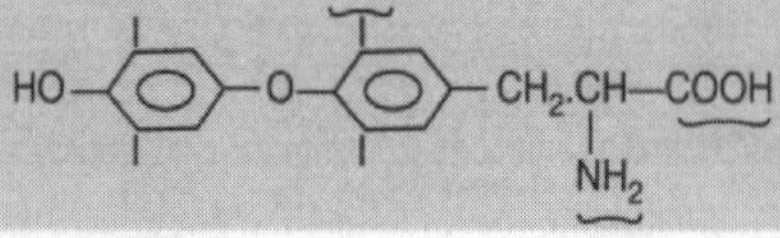

2.6. Heterocyclic Amino Acids

The **heterocyclic amino acids** essentially contain a **'heterocyclic'** entity in them, as shown in the following typical examples :

• **Proline** [*Pro*] : 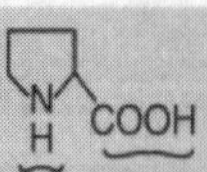   [Pyrrolidine-α–Carboxylic acid]
  N
  H COOH

• **Hydroxyproline** [*Hyp*] : 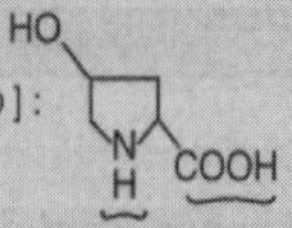   [γ-Hydroxypyrrolidine-α–Carboxylic acid]
  N
  H COOH

• **Histidine** [*His*] : CH$_2$—CH—COOH [α-Amino-β–imidazo-propionic acid]
                              |
  HN   N   NH$_2$

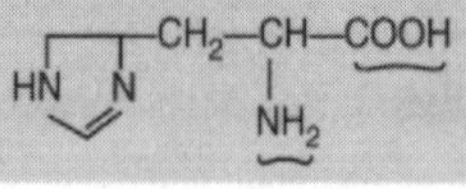

Note: *"Monosodium Glutamate (MSG)'–a* flavour enhancer, commonly used in 'Chinese Fast Food Products', and duly obtained by the careful hydrolysis of *'glutin'* or *'soyabean cake'.* MSG has been proved to be producing a number of undesirable side effects and carcinogenic as well, has been banned in Food Items. perhaps, due to the 'Chinese Restaurant Syndrome'.

Important Highlights: Based on the facts and the examples of various cadre of **amino acids** exemplified in sections from 2.1 through 2.6 (as above) one may take into account the following **important highlights** squarely:

(1) The critical occurrence of β**-hydroxyglutamic acid** and **norleucine** in naturally occurring protein, which still remains a mystery.

(2) **Hydrolysine** occurs exclusively in **'collagen'** and **'gelatin'**.

(3) Presence of **'ornithine'** in proteins is still not yet been fully ascertained; however, it may be obtained by the crucial hydrolysis of **'arginine'**.

(4) **Thyroglobulin** (a *'protein'* obtained from **'thyroid gland'**) solely comprises **thyroxine**.

(5) *Thyroid globulin* (a **protein**) essentially contains **3,4–diiodotyrosine** which prevents *hypothyroidism* in patients.

(6) The α**–amino function** in an **'amino acid'** is always located strategically on the *left-hand side* of a compound *eg.,* in **N-acetyl glycine (Ac-Gly),** the entity–**Gly** is usually designated by the entity –HN–CH$_2$–COOH.

(7) Likewise, the term **'gly-'** is invariably employed for the radical (H$_2$N–CH$_2$–CO–), for instance : 'Methyl glycinate' *ie.,* **Gly-OMe.** In usual practice, the substituents located in the *side chains* are duly designated as stated above or the **amino acid symbol** written at the *terminal end* of a **'vertical line drawn from the first letter'.**

Example: N-Acetyl lysine may be represented as: **'Ac–Lys'.**

(8) The **'natural L–forms'** are normally expressed with the *first letter* written as **'capital'** *viz.,* **Ala : L-Alanine.**

(9) The **'antipodal D–forms'** of *optically active amino acids* are expressed usually with a **'small first letter'** and written as: **ala : D–alanine.**

(10) The actual and proper use of the various recognized symbols [as in (8) and (9) above] may be further expatiated by the structure of **'glutathion (Y–L–glutamyl–L–cysteinyl–glycine)** that is expressed as under:

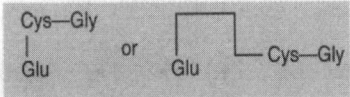

3. CHEMICAL AND ENZYMATIC HYDROLYSIS OF PROTEINS TO PEPTIDES

First and foremost one must have a clear basic concept with regard to the so called **'Derived Proteins'.** the **'derived proteins'** may be defined as–'**products that are formed critically by the action of physical, chemical and enzymatic agents upon the naturally occurring proteins'.**

Another school of thought vehemently postulates that the **'derived proteins'** may be regarded as the meticulously procured *'intermediate hydrolysis products',* which are further divided into several **different products** exclusively based upon their *progressive cleavage* as stated under:

    ❑ denatured proteins,

    ❑ metaproteins, (*viz.*, peptones, pri– and sec–proteoses),

    ❑ polypeptides, and

    ❑ simple peptides.

    (1) **Denatured Proteins:** The **'proteins'** that are critically formed by the *action of heat* on them are termed as **denatured proteins.**

    Explanation– In reality, almost all **'proteins'** essentially have unique *3D structure* (*viz.*, either *tertiary* or *quaternary structure*) that are invariably observed to be **'active biologically'** (or **functional effectively**) very much in their *'original native conformation'*. In other words, the **'Protein 'denaturation' actually refers to the specific and critical loss of 3D structure;** thereby, rendering complete disorganization of the inherent *native conformation* of the so called **protein molecule'** due to the *uncoiling* (or *unfolding*) of the *secondary* as well as *'tertiary'* structures, as depicted in Fig.2.1.

    Besides, the acquired **'protein denaturation'** necessarily and particularly fail to involve the actual **'loss of primary structure'** ie., the phenomenon of **'denaturation'** is not always accompanied by **cleavage of the** *peptide bonds* (see fig: 2.1).

    Examples: *Two* typical examples are as stated under:

    (*a*) An **'oligometric protein'** may take place even by the **denaturation** of its embodied **'subunits',** that may ultimately give rise to the formation of the **quaternary structure;** and hence, the ensuing **'biological activity'** of the *protein.*

    (*b*) A possible **'disruption'** of the *major non-covalent bondings* duly present in the protein may cause **denaturation** which are responsible for the ulitmate *stabilization* of the **structural organization of the protein moliecule.**

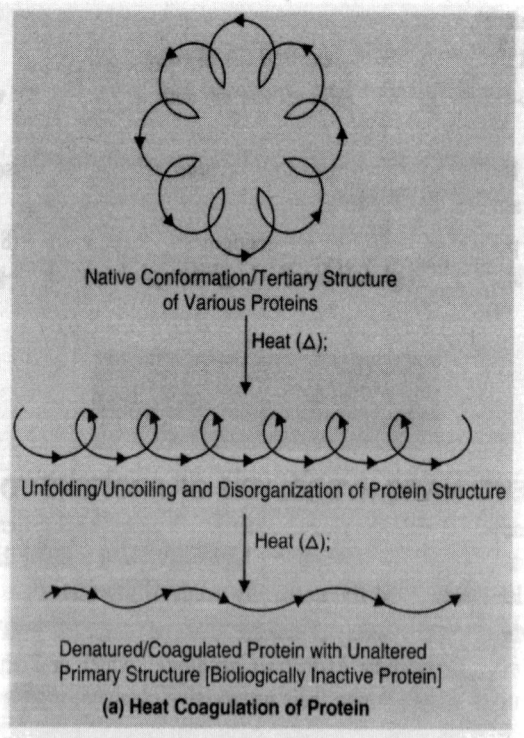

Native Conformation/Tertiary Structure
of Various Proteins

Heat (Δ);

Unfolding/Uncoiling and Disorganization of Protein Structure

Heat (Δ);

Denatured/Coagulated Protein with Unaltered
Primary Structure [Bioliogically Inactive Protein]

(a) Heat Coagulation of Protein

Native conformation/tertiary structure of Protein duly stablized by Non-Covalent Bonds

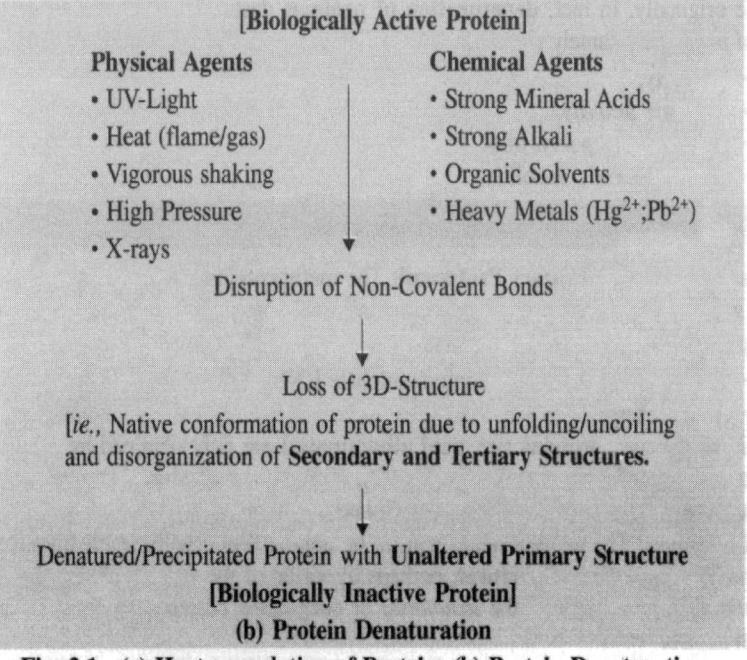

[Biologically Active Protein]

Physical Agents	**Chemical Agents**
• UV-Light	• Strong Mineral Acids
• Heat (flame/gas)	• Strong Alkali
• Vigorous shaking	• Organic Solvents
• High Pressure	• Heavy Metals (Hg^{2+}; Pb^{2+})
• X-rays	

Disruption of Non-Covalent Bonds

Loss of 3D-Structure

[*ie.*, Native conformation of protein due to unfolding/uncoiling and disorganization of **Secondary and Tertiary Structures.**

Denatured/Precipitated Protein with **Unaltered Primary Structure**

[Biologically Inactive Protein]

(b) Protein Denaturation

Fig: 2.1 : (a) Heat coagulation of Protein; (b) Protein Denaturation

It has been duly established that the **'protein denaturation'** is usually brought into effect by a good number of **physical and chemical means and ways:**

- **Physical Agents**– which predominantly includes:

UV–Light; X–Rays; Heat (flame/Gas); Vigorous shaking; High Pressure';

- **Chemical Agents**– they mostly consists of:

Strong mineral acids; strong alkalies; Souluble salts of heavy metals (Hg^{2+}, Cd^{2+}, Pb^{2+}, Cu^{2+}, etc.) ; Organic Solvents (Ethanol, Isopropanol, Acetone etc.,); Urea; Salicylic Acid; Sulphosalicylic acid; Guanidine Hydrochloride; Picric Acid. Phosphomolybdic Acid; Detergents; Phosphotungstic acid; Trichloroacetic acid, and the like.

It is, however, pertinent to state here that the **'denaturation of protein'** practically leads to the following *two* glaring situations:

(*a*) **An Irreversible Reaction Process:** Majority of *proteins* once being denatured cannot be transformed again to their **original conformation** (*i.e.*, **3D-structure**).

Note: *'Renaturation'*[*] refers to the phenomenon whereby certain specific 'globular proteins' *viz.*, *ribonuclease, Hb* etc., after due denaturation caused by *urea, salicylic acid,* and *extreme ph levels*– may ultimately help in the *'refold'* of the said proteins right into their 'original native conformation'; and hence, regain virtually their desired *'biological functions'*.

(*b*) **Results Apparent Loss of Solubility and Bioloical Activity:**

---

[*] **Renaturation Phenomenon:** It was first and foremost reported by **Anson and Mirsky (1931)** for *'haemoglobin'* and by **Anfensen (1950s)** for *'ribunuclease'*.

The **'denatured proteins'** invariably found to be absolutely **insoluble** in the medium in which it was soluble originally. In fact, **denaturation of proteins** destinctly attributes several *physical* and *biochemical properties,* namely :

- **viscosity,**
- **catalytic activity,**
- **antigenic property, and**
- **electrophoretic mobility.**

Note: Denatured proteins are *'not crystallisable'* at all.

(2) **Metaproteins [or Primary Proteoses]:** The **metaproteins** are found to be *insoluble in water,* and *dilute salt solutions.* However, these are usually soluble in **acids** or **alkalis.** Besides, these are invariably precipitated in *two* ways, such as:

- half saturation using ammonium sulphate $[(NH_4)_2SO_4]$, and
- mild heating.

In addition, the **metaproteins** (*viz,* **acid albuminates**) are duly obtained by the action of **dilute mineral acid (H_2SO_4), heavy metal salts ($HgSO_4$, $CuSo_4$),** and **alkaloidal reagents (trichloroacetic acid, sulphosalicylic acid)** upon the natural proteins.

(3) **Polypeptides:** The **protein denaturation** or **coagulation** usually result from the unfolding of protein structure *eg.,* **tertiary structures,** perhaps by virtue of the ensuing **disruption of weak non-covalent bonding.** Nevertheless, the *denatured* or coagulated protein critically retain their normal **primary structure.**

(4) **Simple Peptides:** These are mostly obtained from the careful hydrolysis of the *conjugated proteins* to such *proteins* that are free from other **structural components** *viz.,* **simple proteins** (or **simple peptides).** Thus, based on their **'solubility profile'** the **proteins** may be categorized into *two* groups, for instance:

(*a*) **Soluble Proteins:** such as–**albumins,** and **globulins** (available only in the *'animal kingdom'*); whereas, **glutelins,** and **prolamines** (available only in the *'plant kingdom'*), and

(*b*) **Insoluble Proteins:** such as–**seleroproteins** that are usually referred to as the **'fibrous proteins'** due to their specific *shape* and **insolubility in common solvents** *eg.,* **keratins** (in *human hair/ skin);* and **collagens** (in *elastins of skin/connective tissues).*

In general, the **'peptides'** are duly obtained from the further *hydrolysis of peptones*, i.e.,* they are made up of relatively *fewer amino acids strategically joined together via* the **peptide bonds.**

Nomencluature of peptides: The **'peptides'** are usually named according to the number of *amino acids* present, such as:

- **Dipeptides   :**   comprising *two* 'amino acids'
- **Tripeptides :**   comprising *three* 'amino acids';
- **Tetrapeptides:**   comprising *four* 'amino acids'.

Note: Peptidases help in the production of *amino acids* from 'peptides'.

3.1 Progressive Chemical (Acid Hydrolysis) or Enzymatic Hydrolysis of Proteins to Peptides and Amino Acids:

It is, however, pertinetn to state here that the critical, progressive, and meticulous **chemical/enzymatic hydrolysis** of the so called *'natural proteins'* shall result in the production of an array of **hydrolytic products,** namely :

- metaprotein,
- proteose,
- peptone,
- peptides, and
- amino acids,

as depicted explicity in Fig.2.2.

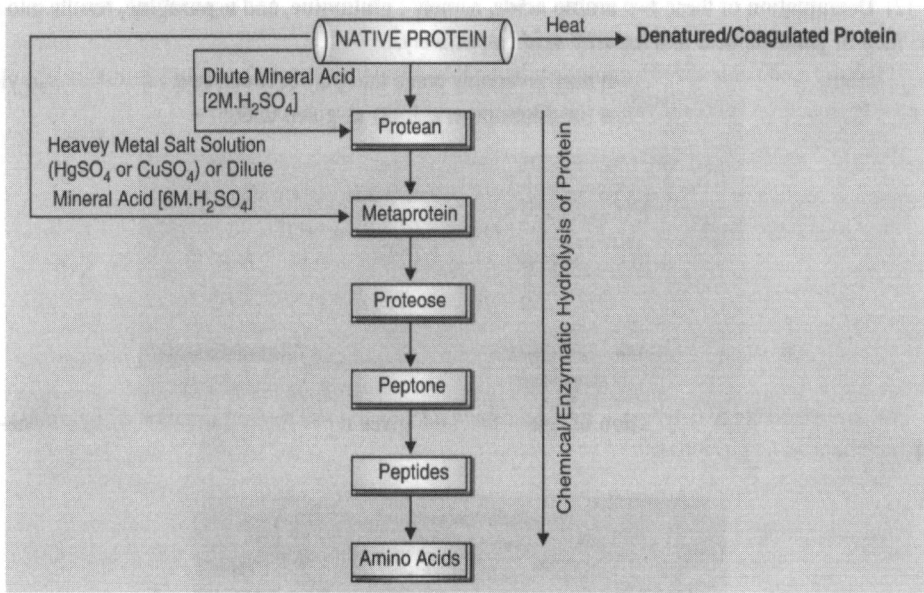

**Fig: 2.2. Progressive Chemical or Enzymatic Hydrolysis of Proteins: Production of Denatured/
Coagulated Protein, Protean, Metaprotein, Proteose, Peptone, Peptides, and Amino Acids.**

Importantly, the various derivatized hydrolytic products *viz.*, protein, metaprotein, proteose, peptone etc, are obtained duly by the progressive cleavage of the *'peptide bonds in proteins'*. Furthermore, these *three* **cleaved products,** namely : **proteoses, peptone,** and **peptides** actually exhibit their clear-cut differences and variations solely based upon the following cardinal factors:

- **molecular size,**
- **molecular weights,** and
- **amino acid composition.**

3.2. Acidic Hydrolysis:

The hydrolysis of proteins can be accomplished by means of **6M HCl** with constant gentle boiling. In a strict scientific manner the said **'acid hydrolysis'** is carried out in an evacuated thermatically sealed tube duly maintained at a temperature of **110°C** for a duration ranging between **12 to 96 hours.** Importantly,

---

* **Peptones:** They are usually derived from further hydrolysis of **'proteoses'** (*ie.*, the *immediate hydrolytic product of the native proteins.*).

under these experimental parameters the **'peptide linkages'** present are almost hydrolyzed quantitatively to produce the respective **'amino acid hydrochlorides'**.

Nevertheless, there exist *four* noticeable **adverse side-effects** of the **'acidic hydrolysis'**, namely:

(1) Gradual destruction of certain **amino acids,** such as: **cystine, cysteine, tryptophan, serine,** and **threonine.** Besides, the critical formation of **humin** (*ie., a residual cleaved product of tryptophan*) appearing mostly as a **black polymer.**

(2) **Deamination** of these *two* **amino acids,** namely : **glutamine,** and **asparagine,** results into the formation of *glutamic acid* and *aspartic acid* respectively.

(3) **Intermolecular dehydration** may invariably come into play between **two amino acids** to yield the corresponding **cyclic anhydrides** (or **diketopiperazines**) as given under :

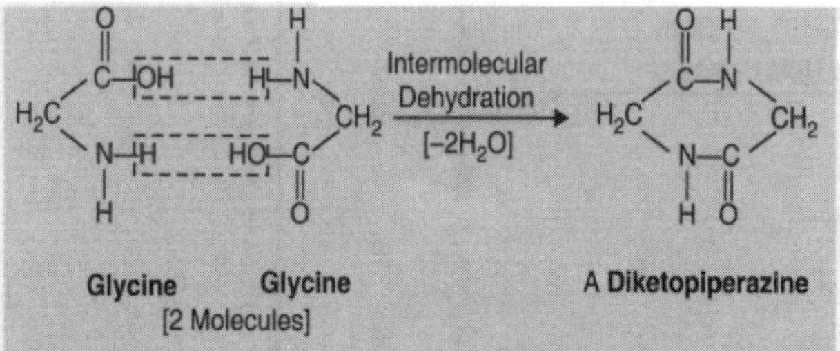

Glycine   Glycine   A Diketopiperazine
        [2 Molecules]

(4) **Intermolecular dehydration** of *'glutamic acid'* gives rise to the formation of **pyrrolidine–5–carboxylic acid** as given under:

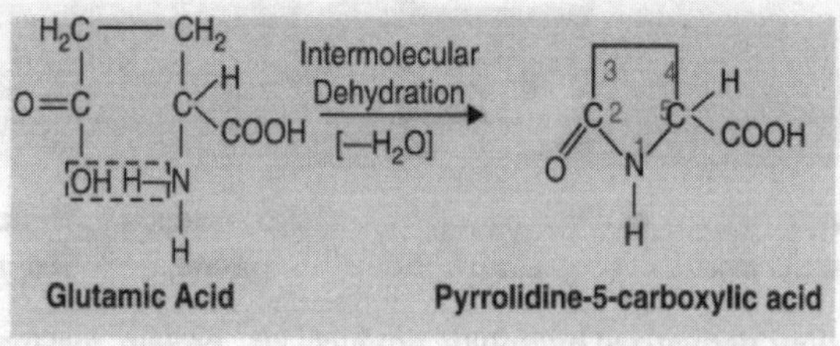

Glutamic Acid   Pyrrolidine-5-carboxylic acid

3.3. Alkaline Hydrolysis:

It is possible and feasible to complete the hydrolysis of protein in an alkaline medium as stated under:

- 2M sodium hydroxide solution,
- 4M sodium hydroxide solution, or
- 5M barium hydroxide solution,

at 100°C for a total duration ranging between **4 to 8 hours.**

In a broader perspective, this type of hydrolysis is of **limited applicability** by virtue of the following reasons, namely:

❑ **arginine, eystine, cysteine, serine** and **threonine** get virtually destroyed by **alkaline hydrolysis.**

❑ **Racemisation** phenomenon is quite prevalent in amino acids.

❑ **Deamination** may cause the partial destruction of certain amino acids.

3.4 Enzymatic Hydrolysis

In true sense, the **enzymatic hydrolysis** seems to be a rather slow and tedious process due to the fundamental fact that most enzymes do attack exclusively certain particular **types of linkages** quite rapidly. It is an usual practice to carry out the desired degradation of a **'protein'** into the **smallest possible fragments** by engaging several enzymes sequentially so as to obtain the ultimate **amino acid residues satisfactorily.**

Typical Examples: These essentially include:

- **Pronase and Subtilisin:** These **microbial peptidases** do catalyze the hydrolysis of practically most **peptide bonds;** however, the overall reaction is rather slow and sluggish in nature *vis-a-vis* the **acid hydrolysis.**

- **Chymotrypsin and Trypsin:** These **proteolytic enzymes** critically carry out the hydrolysis of certain *peptide linkages* in a relatively shorter duration only.

Note: In a broader sense, the *'enzymatic hydrolysis,* of proteins seem to be relatively more complicated due to certain unavoidable contamination caused by the *proteolysis of the enzymes themselves.*

4. PROTEIN STRUCTURE

Proteins are known to be a conglomerate of a large number of *amino acids* that have been joined to each other by **'peptide bonds'.** In order to establish the structure of a **'protein molecule'** one may have to take the cognizance of the following cardinal aspects, namely:

(a) Exact **'nature'** of *amino acid* entities duly present,

(b) Precise **'number'** of each specific amino acid duly present in *'one mole of protein'*,

(c) **'Sequencing pattern of different amino acids'** that are arranged meticulously in a **'protein molecule'.**

(d) **'ultimate shape of the *peptide chain**,**

(e) **'Various Forces'** that are solely responsible for *holding the peptide chains,*

(f) Mode of critical arrangement of **'individual peptide chains'** in a definite manner into either a **'folded'** or **'refolded'** typical individual shape of a **macromolecular protein,** and

(g) Total number of the **'peptide chains'** as well as their *crucial arrangement* with respect to the **'natural protein.**

Based on the fundamental factual reasons that each and every **'protein molecule'** possesses an unique **3D-structure** that ultimately gives rise to *four levels of the* **structural organization** of the **polypeptide chain(s),** namely:

- **Primary Structure** [*i.e.,* refers to (a), (b), and (c)],
- **Secondary Structure** [*i.e.,* refers to (d)],
- **Tertiary Structure** [*i.e.,* refers to (e) and (f)],
- **Quaternary Structure** [*i.e,* refers to (g)].

---

* In other words, whether the **'peptide chain'** present in a protein molecule is **linear, cyclic, branched,** or a **helical structure.**

The aforesaid *four* different structural variants in the **protein molecule** shall be treated separately in the sections that follows:

4.1. Primary Structure

It is worthwhile to state here that each **'protein'** essentially has a *typical* and *specific* **'linear sequence of amino acids'** that may be estimated precisely by the actual **sequence of nuclecatide bases** (or **codons**) duly present in the *gene* **encoding that protein**. In other words, the **'primary structure'** of a protein invariably signifies the so called **linear sequence of amino acids** joined together *via the* **peptide bonds** strategically located in a particular **polypeptide chain** as illustrated in Fig. 2.3.

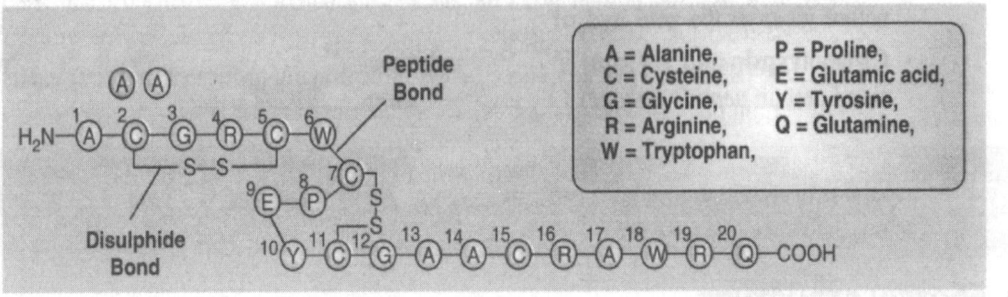

A = Alanine,	P = Proline,
C = Cysteine,	E = Glutamic acid,
G = Glycine,	Y = Tyrosine,
R = Arginine,	Q = Glutamine,
W = Tryptophan,	

Fig: 2.3. Linear Sequence of Amino Acids Duly Illustrated in the Primary Structure of a Protein.

In Fig: 2.3, the various **amino acids** have been numbered duly in an **ascending order** from 1 to 20; whereas, the presence fo the **disulphide bonds (–S–S–)** shown explicitly between *two* **cysteine residues.'**

Note: (1) Primary structure of *'several proteins'* are observed to be changed significaulty in a host of 'inherited disorders', such as : *sickle-cell anacmia, HbC disease,* and *Hb SC diseases.*
(2) Interestingly, the *conagulated and denatured proteins* do invariably retain theri *'normal primary structure'.*

Following are some of the cardinal aspects pertaining to the **'primary structure of a protein,** namely:

- Purification,
- Amino acid composition,
- End-group determination,
- Disulphide (–S–S–) bonds:
  - ☐ Locating their position,
  - ☐ cleavage of –S–S– bonds.

4.1.1. Purification

In general, there are *two* types of **proteins** which are available in a **primary structure,** such as:

(a)**Soluble Proteins:** In usual practice, the **soluble proteins** usually pose a serious problem due to their inherent relatively **weak binding forces** that both sustain and maintain the *real integrity* of the **3D-structure.** Besides, the *biological activity* of these **soluble proteins** get imbalanced quite easily and abruptly due to the exposure to **heat, organic solvents,** and **extreme limits of pH.**

Nevertheless, the crude protein sample is carefully precipitated very close to the *'isoeletric point'*. The resulting separated product on being subjected to *dialysis* removes most of the salts present. The *dialysed product* may be fractionated by known **chromatographic techniques.** Purified sample of **'soluble protein'** is isolated by **'freeze drying'**, and its purity determined by any one of the following techniques:

- **Capillary Electrophoresis (CE),**
- **Ultracentrifugation**, and
- **Counter–current Distribution.**

(b) **Insoluble Proteins:** The *non-protein contaminants* are duly discarded by the careful extraction. Obviously, any attempt to *solubilize the protein* may cause the rupture of the **'covalent bonds'** thereby distorting the **original primary structure.** However, the **'native insoluble protein'** may be established structurally by the help of the **X-Ray Diffraction Analysis.**

4.1.2. Amino Acid Compostion

It may be accomplished by the following *two* common methods, namely:

(a) **Acid/Enzymatic Hydrolysis:** Protcin on being subjected to critical hydrolysis by **acid** (6M Hydrochloric Acid) or **enzyme** (Peptidase) results into an array of **amino acids** (except the **'Tryptophan'**, which gets destroyed completely by this procedure).

(b) **Alkaline Hydrolysis:** Protein upon hydrolysis with an **alkali** (2M NaOH) gives rise to 'Tryptophan'; whereas, a host of other **amino acids gets** *destroyed promptly e.g.,* **arginine, cystine, serine,** and **threonine.**

It is, however, pertinent to state here that there are quite a few *known methodologies* that may indeed help not only in the **critical separation** but also in the **exact determination** of a host of **'amino acids'** obtained in (a) and (b) above.

4.1.2.1. Ion-Exchange Chromatography

In this particular instance, the **'chromatographic columns'** are duly packed with an inteligently designed and meticulously developed **resin material** usually known as the **'Sulphonated Polystyrene Divinyl Resin',** that essentially possesses several *highly polarized sulphonic acid [–SO₃H]* moieties. The *'mixture of amino acids'* intended to be separated are treated in the following steps in a chromatographic column packed duly with the aforesaid resin material:

(1) The solution of **'amino acid mixture'** is carefully introduced on the top end of the column [see Fig : 2.4. (a)].

(2) Amino acid undergoes *critical conversion* into the respective **ammonium (NH_4^+) ion,** and the *resin* into the corresponding **sulphonate (SO_3^-) ion.** In this manner, the *ammonium ions* emanated from different amino acids are adequately adsorbed on the surface of the **sulphonic acid resin anion** having *different acidic strength* of the respective hydrolysed **amino acids** [see Fig : 2.4.(b)]

(3) Subsequent **'elution'** of the column with an appropriate **'buffer solution'** (particular pH range) affords the desired separation of **amino acids** that may be collected separately and weighed [see Fig. 2.4 (c)].

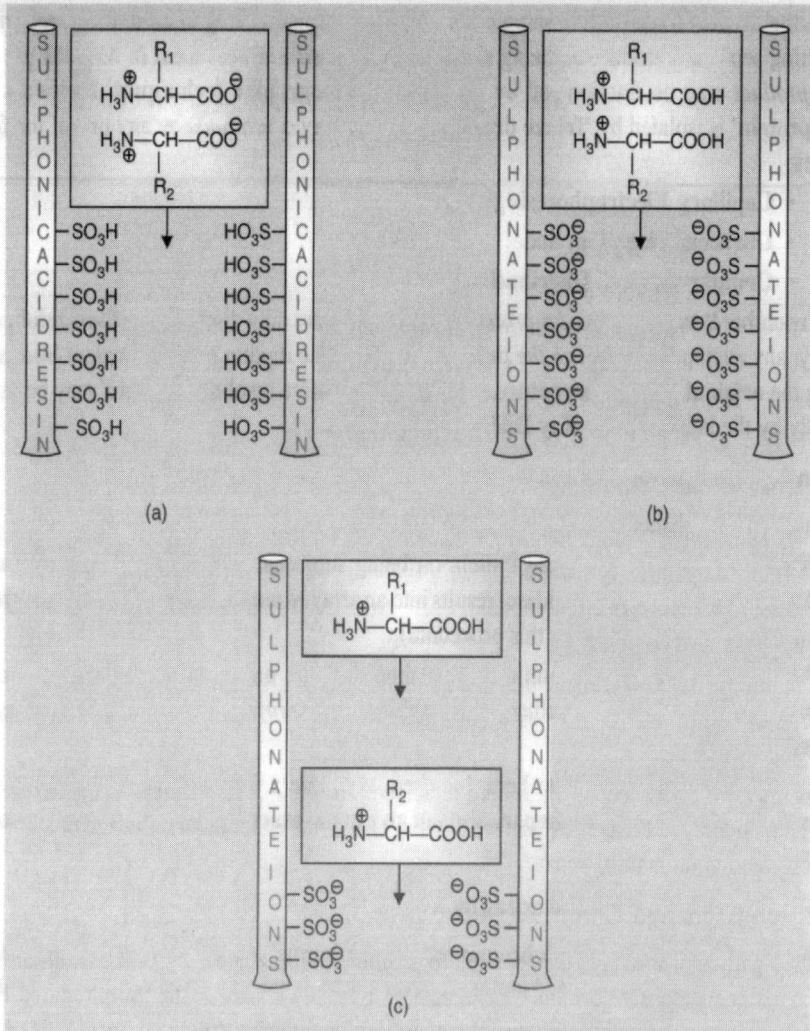

Fig: 2.4 (a), (b), and (c) Showing the Stepwise Separation of Two Amino Acid Mixture on a Sulphonic Acid Resin Column (*i.e.*, Ion-Exchange Chromatography).

Note: The *'eluted amino acids'* from the column may be treated with a *'Ninhydrin Solution'* and estimated colorimetrically.

4.1.2.2. Capillary Electrophoresis [EC]:

The exact quantum of the carefully hydrolyzed protein sample consisting of a mixture of **amino acids** may be determined by the **'isotopic dilution method'** that essentially includes the skilful incorporation of an **aliquot** (*i.e.*, a **known amount**) of a [14]C–**labelled species of an amino acid.** The resulting mixture of amino acids, thus obtained, is duly assayed by making use of the radioactivity (of amino acids) by **Geiger-Molar Counter** or **Scintillation Counter.** Obviously, the chemical characteristic features of the *naturally occurring amino acid** and the *^{14}C–labelled amino acid species*** isolated almost simultaneously from which the **naturally occuring amino acids** may be quantified precisely by estimating its newly acquired *level of radioactivity*. However, the exact amount of each **amino acid** present may be assayed conveniently by **ninhydrin based colorimetric assays.**

Example: The protein hydrolysate of one of the **insulin chains (B)** critically give rise to the following **chain of amino acids:**

(Gly)$_3$–(Ala)$_2$–(Val)$_3$–(Leu)$_4$–(Pro)–(Phe)$_3$–(CySH)$_3$–(Arg)–(His)$_2$–(Lys)–(Asp)–(Glu–NH$_2$)– (Glu)$_2$–(Ser)–(Thr)–(Tyr)$_2$.

4.1.2.3. pH Endorsed Precipitation

Based on the observation that the **amino acids** are *almost insoluble* in an aqueous medium at their respective **'isoelectric points'** indicating thereby that they essentially possess specifically different solubilities; and, therefore, may be duly separated from each other. However, in actual practice one may critically take cognizance of the underlying fact that the **ensuing pH** of either *protein* or *peptide hydrolysate* gets altered rather slowly that ultimately gives rise to the *successive precipitation of the amino acids* (of course, at the corresponding **isoelectric point**).

4.1.2.4. End-Group Estimation

In fact, the **'end-group estimation'** invariably refers to the exact and precise corresponding *N-terminal* and *C–terminal* **'amino acids'** of a given *polypoptide chain* by either of the following *two* methods:

- Amino-End Degradation in Amino Acids, and
- Carboxyl-End degradation in Amino Acids.

4.1.2.4.1. Amino-End Degradation in Amino Acids

There are in all *five* **known techniques whereby it is possible to determine the** *amino-end degradation in 'amino acids',* **namely:**

- Sanger's Method,
- Dansyl Chloride Method,
- Enman Degrdation Method,
- Enzymatic Method, and
- Infrared Spectrosocopic Method.

4.1.2.4.1.1. Sanger's Method

An epoch making revolution in **protein chemistry** was the determination of the entire **amino acid sequence** duly present in the **'insulin molecule'** be **Frederick Sanger** at the renowmned Cambridge University team, who was honoured with the Nobel Prize in 1958. Since, then the *exact number* as well as the ensuing *complexity of completely* **'mapped proteins'** has grown into a rapid expansion:

Example: The typical example being the four distinet chains of **'haemoglobin'** *e.g;*, each chain comprising **140-odd amino acid residues; 'chymotrypsinogen'** having a single chain **246 units** long; and **'immunoglobulin'** (γ–globulin) with *two* chain of **446 units** each *two chains of* 214 units each *i.e.,* a sum total of **1320 odd amino acid residues.**

---

* Obtained duly from the actual hydrolysis of proteins.

** The **radio-labelled amino acid(s)** added from outside in *known quantities.*

Note: However, the *'ultimate confirmation'* of the structure duly assigned to a *'peptide'* rests upon its actual *'total synthesis'* by a method that should critically and unambiguously yield a compound of the *'assigned structure'*.

In actual practice, one may make use of the **Sanger's Reagent** *i.e.*, **1–fluoro–2,4–dinitrobenzene (FDNB)** so as to establish and determine that which particular amino acid precisely be the part of **'amino-end of the polypeptide'**.

The various steps involved in the **Sanger's Method** are as enumerated under:

(1) 'Polypeptide' under investigation is treated with **FDNB** in the presence of **sodium bicarbonate** [NaHCO$_3$] solution at $20 \pm 2°$ C (*i.e.*, room temperature) to give rise to the production of the corresponding **2,4–dinitrophenyl (DNP) derirative** of the respective **polypeptide.**

(2) **DNP-derivative of polypeptide,** obtained in (1), is carefully subjected to hydrolysis by the help of a dilute mineral acid (that essentially affords the critical cleavage of the **'peptide linkage'** prevailing between the **N-terminal amino acid** and the **remaining polypeptide molecule** or the **corresponding amino acid residues.**

The various **chemical reaction** involved may be sequentially shown as under:

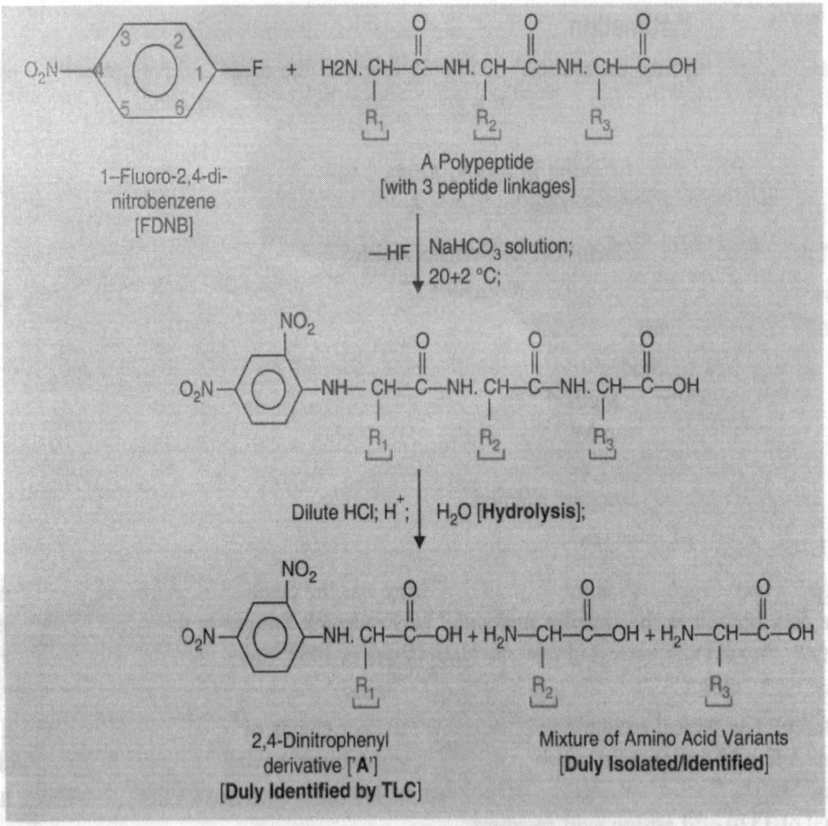

1–Fluoro-2,4-di-nitrobenzene [FDNB]

A Polypeptide [with 3 peptide linkages]

–HF   NaHCO$_3$ solution; 20+2 °C;

Dilute HCl; H$^+$;   H$_2$O [Hydrolysis];

2,4-Dinitrophenyl derivative ['A'] [Duly Identified by TLC]

Mixture of Amino Acid Variants [Duly Isolated/Identified]

Special Points: These *special points* may be considered; such as:

(1) Importantly, the resulting **dinitrophenyl (DNP)** derivative ('A'), thus obtained by **Sanger's Method,** does not comprise any *'free amino moiety'* as shown above.

Therefore, **'lysine'**, which is a **diamino monocarboxylic acid,** shall

$$H_2N-(CH_2)_4-CH(NH_2)-COOH$$

Lysine

produce a corresponding **DNP–derivative,** even though it necessarily possesses the **amino (–NH₂)** function *not* located strategically upon the **terminal group.**

(2) Likewise, the following **amino acids,** for instance:

- **Tyrosine** HO—⟨O⟩—CH₂—CH—(NH₂).COOH *ie.,* **hydroxyl (OH) moiety;**

- **Cysteine** HO—S—CH₂—CH—(NH₂).COOH *ie.,* **thiol (OH) moiety;**

- **Histidine** ⌐CH₂ CH (NH₂).COOH *ie.,* **imidazole moiety;**
  HN N

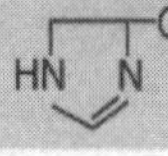

with their indicated **functional moieties** may also under go typical reaction with **1–fluoro–2,4–dinitrobenzene (FDNB);** however, at an unusual slow pace. Therefore, the **Sanger's Method** may eventivially land up with a good number of **DNP–derivatives** that can be isolated convneiently, and identified subsequently be **thin layer chromatographic (TLC) techniques.**

> Note: *Sanger's Method* **may not be utilized successively (or repititively), because in this particular instance the emanated DNP-derivative gets hydrolyzed almost completely (*i.e.,* a marked difference from *Edman's method*). Thus, the former method specifically determines the N–terminal amino acid present duly either in a *'protein'* or a *'polypeptide'* entitiy.**

4.1.2.4.1.2. Dansyl Chloride Method

Dansyl chloride is 5-dimethylamino-1-naphthalene sulphonyl chloride (**DNS-Cl**) which reacts with **amino moieties** to:

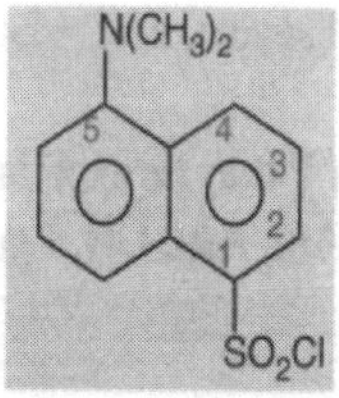

give a distinct **fluorescent derivative.**

In this particular instance, a **polypeptide** is reacted with **DNS-Cl** to form an adduct with the elimination of a mole of HCl. The *intermediate product* on being subjected to hydrolysis in an **acidic**

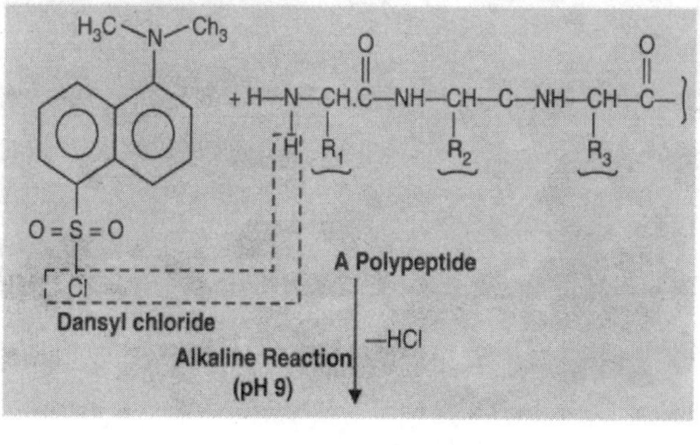

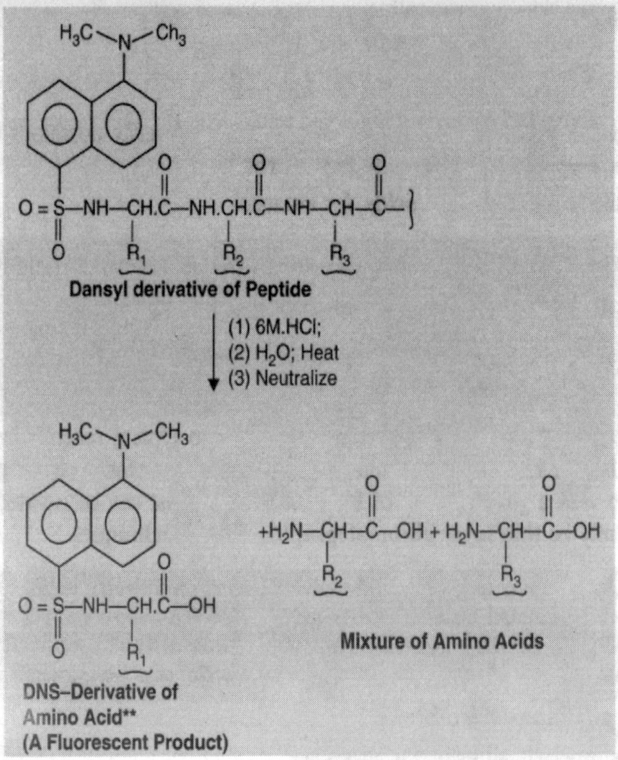

Dansyl derivative of Peptide

(1) 6M.HCl;
(2) H$_2$O; Heat
(3) Neutralize

DNS–Derivative of Amino Acid** (A Fluorescent Product)

Mixture of Amino Acids

medium (6M. HCl) at a temperature of **100°C** yields the resulting **dansyl derivative of the amino acid** (*i.e.*, **DNS–derivative**)* plus an admixture of several **amino acids.** the various reactions that usually come into play are deseribed as under:

Furthermore, the **dansyl chloride method'** may be expatiated by carrying out the reaction particularly with a **'peptide P'** having the composition [Arg, Asp, Gly, Leu, Thu, Val) which reacts critically with

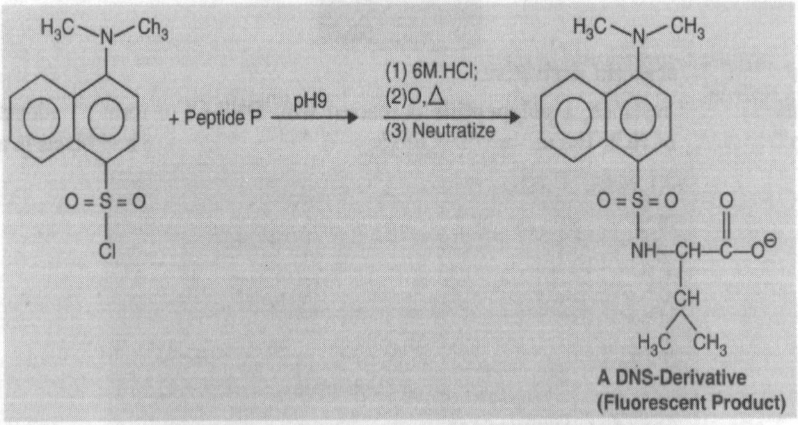

+ Peptide P → pH9

(1) 6M.HCl;
(2)O,Δ
(3) Neutralize

A DNS-Derivative (Fluorescent Product)

---

* **DNS–Derivative** may be further identified by means of known **TLC Procedures.**

dansyl chloride at pH 9, and is duly hydrolyzed in 6M aqueous HCl. In fact, the resulting DNS–derivative, as given in the following reaction is detected by its **fluorescence** property; and is isolated carefully after neutralization, along with the generated 'free amino acids' viz., Arg, Asp, Gly, Leu, and Thr.

The above reaction explicitly show an **anino-end degradation** of the *amin acids*.

> Note: The *'dansyl method'* is employed abundantly in combination with the *'Edman degradation method'* (see section 4.12, 4.13) As we remove each successive residue, an *aliquot* may be duly treated by the *'dansyl method'* to establish precisely the exact nature of the *next residue*, invariably making use of *thin-layer chromatography* (TLC).

4.1.2.4.1.3. Edman Degradation Method

The standard technique frequently employed for the critical implementation of the so called 'sequential degradation of peptides' is termed as the **Edman degradation method***. Thus, in an **Edman degradation** the 'peptide' under study is duly treated with a specific reagent, *phenyl iso thiocyanate [Ph–N=C=S]* invariably known as the **Edman Reagent**. Interestingly, the 'peptide' crucially reacts with the **Edman Reagent** at its *amino moiety* to yield the respective **thiourea derivatives.**

> Note: Importantly, the *Edman regent* reacts most critically at the respective *side-chain amino functions* of the *'lysine units'*; however, the reaction exclusively taking place at the *'terminal amino group'* remains absolutely to the degradation.**

Thus, we may have the following reaction:

A Thiourea Derivative

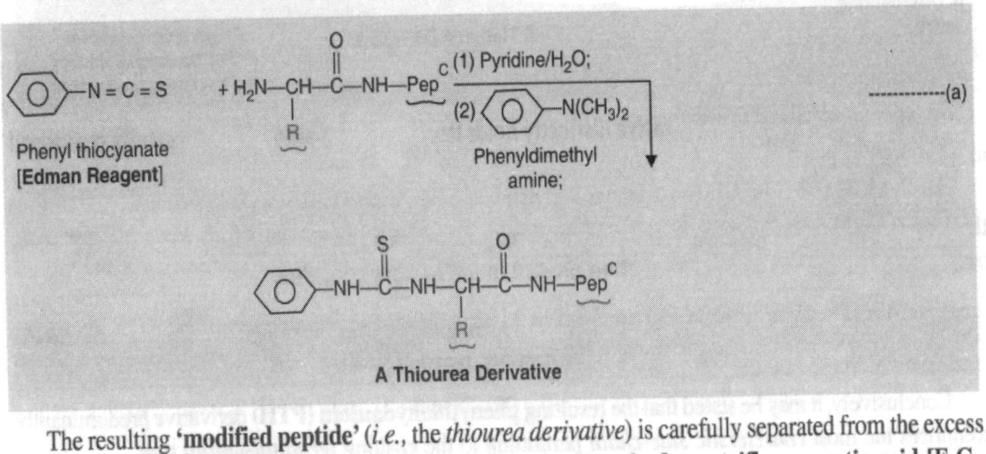

The resulting 'modified peptide' (*i.e.*, the *thiourea derivative*) is carefully separated from the excess of unreacted **Edman Reagent,** and subsequently treated with **anhydrous trifluoro acetic acid [F$_3$C–COOH].**

---

* After **Pehr Victor Edman** (1916–1977)–a Swedish *biochemist,* who meticulously devised the technique in the year 1952.

** Both in this as well as subsequent equations, the abbreviation **Pep^N** is usually employed for the amino-terminal portion of a **peptide**; whereas, the abbreviation **Pep for the carboxy terminal portion**

In this way, the S-atom of the *thiourea derivative*, which being **nucleophilic** in nature, helps sepcifically on the **adjacent residue** to give rise to the production of a **5–membered heterocycle** known as the **thiazolinone**; whereas, the second product of the said reaction being essentially a 'peptide which is actually one residue shorter in size.

The aforesaid reaction may be depicted as under:

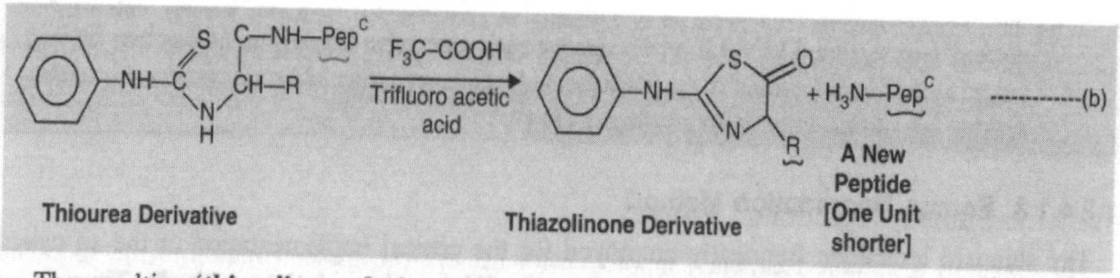

| Thiourea Derivative | Thiazolinone Derivative | A New Peptide [One Unit shorter] |

The resulting 'thiazolinone derivative' on being treated with an *aqueous acid*, results into the formation of an *'isomer'* known as **phenylthiohydantion** as given below:

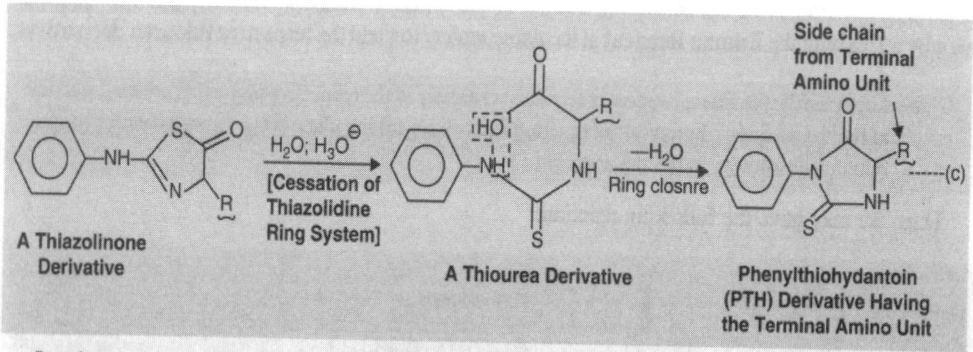

A Thiazolinone Derivative — A Thiourea Derivative — Phenylthiohydantoin (PTH) Derivative Having the Terminal Amino Unit

In other words, the **PTH–dervative** distinctly holds the **terminal amino unit** strategically positioned in the *side-chain*.

Mechanism: The underlying probable mechanism may be duly expatiated by critical **ripening of the thiazolinone ring** to the corresponding *thiourea form*, which is eventually followed up by the *'ring fromation'* critically involving the **N-atom** present in the **thiourea residue**.

Note: However, one may observe particularly the magnificent *intramolecular formation* of the five-membered rings in both this and the previous reaction stated above.

Conclusively, it may be stated that the resulting **phenythiohydantoin (PTH)** derivative predominantly comprises the most *charcterstic side-chain* pertaining to the ensuing *terminal-amino unit.*

Nevertheless, the careful **identification of PTH*** confirms the particular **amino acid unit** which was removed ultimately.

Obviously, the *peptide* liberated in **Eqn. (b)** may be duly subjected to the aforesaid **Edman degradation** once again to produce another **PTH dervative** of the *next amino acid;* and, therefore, to yield and altogether *'new peptide'* which certainly is shorter by yet another **amino acid residue**.

---

* **PTH Derivatives** may be duly identified by several known chromatographic methods.

Limitations of Edman Degradation Method: Theoretically, one may continue with the stepwise **Edman degradation** indefinitely up to as many amino acid residues as normally and effectively achievable so as to define completely the ensuing **sequence of the peptide chain.**

It is, however, pertinent to state here that in actual practice, one may **not** accomplish a *perfect quantitative yield* at each and every **Edman degradation step.** Therefore, a definitely more complicated and complex mixture of the **peptides** is generated with each successive step in the cleavage, which renders the results more or less absolutely **'ambiguous'** evirdently after a series of such steps.

Thus, one may come to an established fact and conclusion that the actual, feasible, and possible number of residues present in **sequence of the peptide chain** that may be determined by the **Edman degradation method** is *limited.*

Note: With the advent of more sophisticated instruments the *Edman chemistry* has perecptively become a lot easier and precise for the exact *determination of peptides* with highly *standardized, automated,* and *reproducible* manner. In fact, such instruments do help in:
  • sequential degradation of almost 20 residues, and
  • degradation of amino acid residues ranging between 60 to 70.

4.1.2.4.1.4. Enzymatic Method

It has been duly proved and established that the *enzyme* **'leucine aminopeptidase'** critically affords a sudden onset of action upon the **peptides** or **proteins** or the **ends comprising the 'free amino moiety'.** Therefore, once the **terminal amino acid** gets liberated completely, there prevails an ample apportunity for the **'new emanated terminal with free amino moiety'** being attacked both *subsequently* and *successively* by **'leucine aminopeptidase'.**

4.1.2.4.1.5. Infrared Spectroscopic Method

The **infrared spectroscopic** studies have essentially revealed the typical characterstic band of the **peptides** present duly in the crystalline portion of the **'silk-fibroin'** showing the strategic presence of **glycine** and **alanine** residues critically arranged at alternate position in the said **peptide chain.**

4.1.2.4.2. Carboxyl-End Degradation in Amino Acids

The carboxyl (–COOH)–end [or C–terminal] degradation in **amino acids** may be accomplished by any one of the following *five* methodologies, namely:

  • Hydrazinolysis,
  • Reduction Technique,
  • Schlack and Kumpf Method,
  • Racemization Method, and
  • Carboxypeptidase method.

The aforesaid methodologies shall now be treated individually in the sections that follows:

4.1.2.4.2.1. Hydrazinolysis

The *hydrazinolysis* is regarded to be the most abundantly employed technique put forward by Akabori *et al.* (1956), whereby the **peptide** (or **protein**) is carefully heated along with *anhydrous hydrazine* at 100°C when almost the entire **amino acids** except the *carboxyl–end one* are duly converted to the corresponding **amino acid hydrazine** as given under:

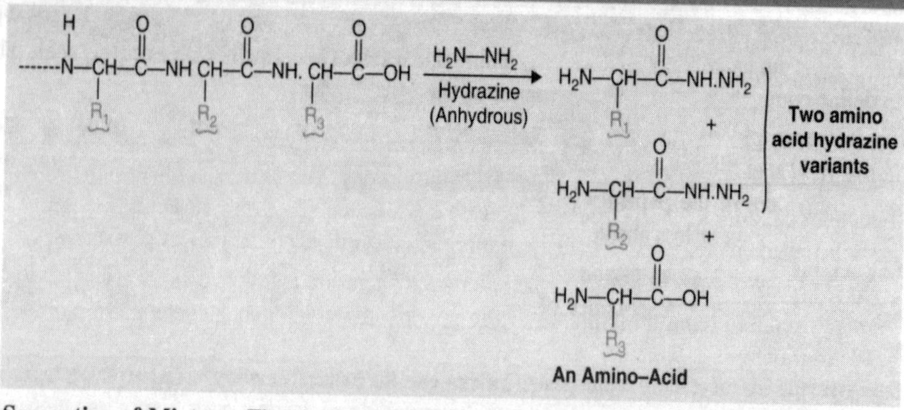

Separation of Mixture: The resulting mixture of reaction products (containing *two amino acid hydrazine variants* plus an *amino acid*) is separated meticulously by **ion-exchange chromatography** using a packed column of a **'strong cation-exchange resin'.** Upon subsequent elution with an appropriate solvent (as carrier), the remarkably strong **basic hydrazides** are adequately retained by the said *cation-exchange resin;* whereas, the **'free amino acid'** produced simultaneously gets eluted and which may be *identified ultimately*.

Demerit: Hydrazinolysis has a major *demerit* based on the fact that:

- **Cystine and tryptophan undergo complete destruction,**
- **Arginine gets converted to ornithine,** and
- **Methionine undergoes oxidation to its corresponding sulphoxide.**

4.1.2.4.2.2. Reduction Techniques

The *reduction technique* essentially involves the critical reduction of the **peptide** or **protein** with *two* commonly used **reducing agents,** namely:

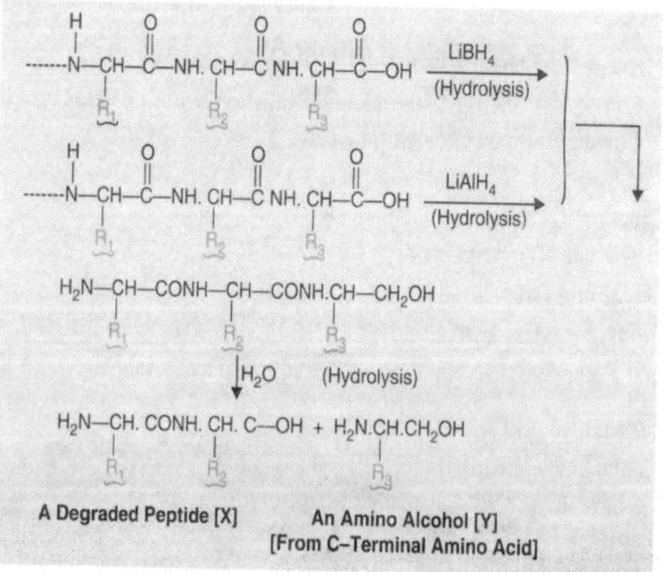

A Degraded Peptide [X]   An Amino Alcohol [Y]
[From C-Terminal Amino Acid]

Lilthium borohydride : LiBH$_4$, and

Lithium-aluminium hydride: Li Al H$_4$,

whereby the **free terminal carboxyl-end (in peptide** or **protein** to a *primary alcoholic function after careful hydrolysis* as indicated above.

Ultimately, the generated **amino alcohol [Y]** is separated duly, and identified critically by **thin-layer chromatography** or **paper chromatography.** Furthermore, subsequent hydrolysis of the **degraded peptide [X]** to obtain the sucessive **amino alcohols** having **R$_1$** and **R$_2$** moieties.

4.1.2.4.2.3. Schlack and Kumpf Method

The **Schlack and Kumpf method wherein** the peptide or protein is first and foremost subjected to:

- **benzoylation** (with C$_6$H$_5$COCl),
- **heating with ammonium thiocyanate (NH$_4$CNS),**
- **hydrolysis in an alkaline medium (NaOH), and**
- **treatment with barium hydroxide [Ba (OH)$_2$].**

Step-1: *Benzoylation–* causes protection of the **'free amino moiety'** present in *peptide* or protein (I).

Step-2: *Ammonium thiocyanate (NH$_4$CNS) reaction–* of the product (I) obtained from *Step-1* with heat and acetic anhydride [(CH$_3$CO)$_2$O] to form the converted compound **thiohydantoin (II).**

Step-3: *Alkaline hydrolysis–* of the resulting product (II) with NaOH solution gives rise to the production of the corresponding **peptide** or protein having a *lesser amino acid residue* together with the desired **'thiohydantion'** *molecule* **essentially bearing the C–termial amino acid unit (III), and**

Step-4: *Treatment with Ba(OH)$_2$–* yields the specific **'amino acid unit' IV** due to the cleavage of the *pentavalent heteroeyclic ring system* of the **'thiohydantion'** nucleus.

The aforesaid *four steps* are explicitly elaborated as under:

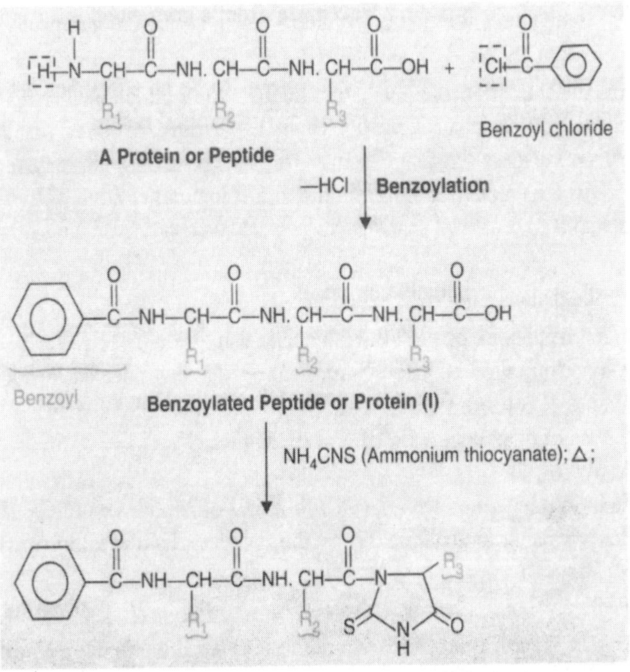

A Protein or Peptide

Benzoyl chloride

—HCl | Benzoylation

Benzoyl   Benzoylated Peptide or Protein (I)

NH$_4$CNS (Ammonium thiocyanate); Δ;

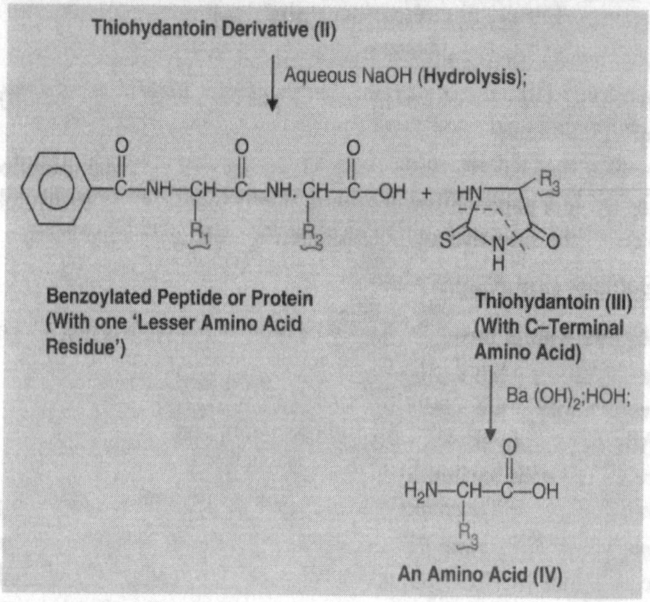

Note: The *'amino acid (IV)'* having R_3 moiety is duly purified and identified by known *'standard procedures'* or by the *'amino acid analyzer'*. Likewise, the benzoylated peptide or protein having one *'lesser amino acid residue'* on successive treatment yields *amino acids* with R_2 and R_1 moieties, and idientified duly.

4.1.2.4.2.4. Racemization Method

The critical phenomenon of forming a **'racemate'** from a *pure enantiomer* is invariably termed as *racemization*.

Importantly, the **racemization method** is considered to be an altogether *newer technique* for the particular identification *vis-a-vis* characterization of the **C–terminal residues,** first and foremost reported be a team of renowned Japanese Organic Chemists. In fact, the underlying methodology essentially and predominantly involves the crucial racemization of the **C–terminal residue** of a **peptide** or **protein** and subsequently the *simultaneous* as well as *concurrent* **selective labelling of the said residue with tritium** (^{3}H).

The various cardinal steps essentially comprise:

- treatment of peptide or protein with acetic anhydride [$(CH_3CO)_2O$],
- oxazolone formation between C-terminal residue and adjacent residue,
- removal of H^+ion from the α–C–atom of C–terminal amino acid,
- hydrolysis of oxazolone with tritium oxide [$3H_2O$], and
- acidic hydrolysis.

Step–1 and Step–2: *Acetylation with acctic anhydride*–of peptide or protein after careful reflux for a stipulated duration yields an **'oxazolone'** (**I**) by the fusion of **C–terminal residue** and the **adjacent residue** from the *amino acid*.

Step–3: *Removal of H^+ion from the α–C–atom of the oxazolone (I)*–involves the **C–terminal amino acid** to undergo a **'keto'–'enol' tautomerism,** and produce an **intermediate compound (II).**

Step–4: *Intermediate (II) upon hydrolysis in the presence of* **tritium oxide [^{3}H$_2$O]**–cleaves the oxazolone residue into an open-chain entity which specifically introduces the **'tritium label'** upon the α–C–atom strategically (**III**).

Step–5: *Acidic hydrolysis–* of the resulting **'tritium labelled'** *product* (**III**) critically yields *three* different types of **amino acid fragments** *viz.*, **first:** with '**R**' attached to it (**IV**) **secondly:** with '**Rn–1**' attached to it (**V**) and **thirdly:** with both '**Rn**' and **'tritium label'** attached to it (**VI**).

All the aforesaid *five steps* elaborated are explicityly stated as under.

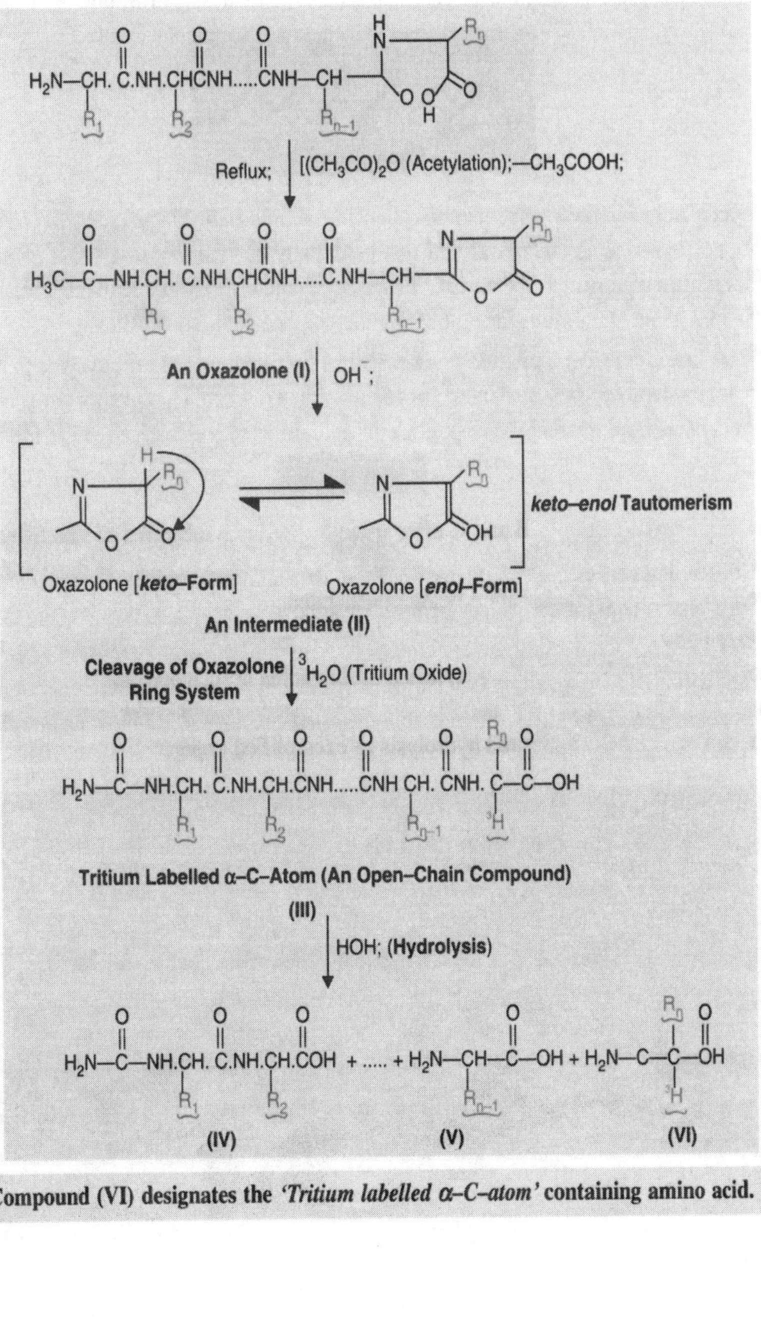

Note: Compound (VI) designates the *'Tritium labelled α–C–atom'* **containing amino acid.**

4.1.2.4.2.5. Carboxypeptidase Method

Carboxypeptidase belongs specifically to the well recognized class of *'exopeptidases'* that predominantly designate the enzymes which catalyze the critical **hydrolysis of peptide bonds** at the end of the chain (*viz.*, in this instance : the **carboxylend degradation in amino acids**).

Importantly, the so called *pancreatic* **carboxypeptidases** do hydrolyze the *'peptide linkage'* in which the **C–terminal amino acid** is involved exclusively. Grassmann *et al.* (1930) was pioneer in the determination of the precise and exact structure of the *amino acid* **'Glutathione'** by using this method.

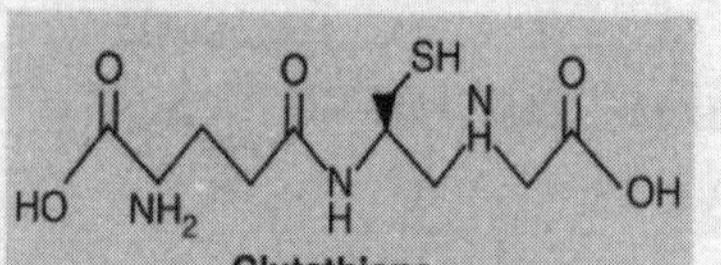

Glutathione

It has been duly observed that **'carboxypeptidase'** invariably attacks the *peptides* or *proteins* strategically positioned at the **terminal-end** that essentially has the *free* α–COOH group. Obviously, the **terminal amino acid unit** on being duly liberated, the newly exposed and available **'free carboxyl moiety'** shall be liable to an immediate attack by the enzyme **carboxypeptidase.**

Example: Let us consider a **peptide** or a **protein** having a number of **amino acids** in it *e.g.*, **A, B, C.** Now, on treatment with the *enzyme* **carboxypeptidases,** one may obtain a plethora of successive **amino acids** duly liberated from the respective **peptide** or **protein** in quatities given in the following order:

$$C > B > A$$

The aforesaid amino acids A, B and C after being isolated are subjected to **identifiecation, assay** (*ie.*, **quantitative determination**) which may *evidently* and *definitely* suggest the most probable *'amino acid sequence'* in the **peptide** or **protein** being investigated

Boyer (1972) proposed a modified **'carboxypeptidase method'** for the most reliable, accurate, and precise identification of the prevailing **C-terminal amino acids** in such *proteins* that critically comprise **more than one polypetide chain** in them. In this specific instance, water with a **radiolabelled O-atom** $[H_2{}^{18}O]$ is used to carry out the desired hydrolysis as exemplified under:

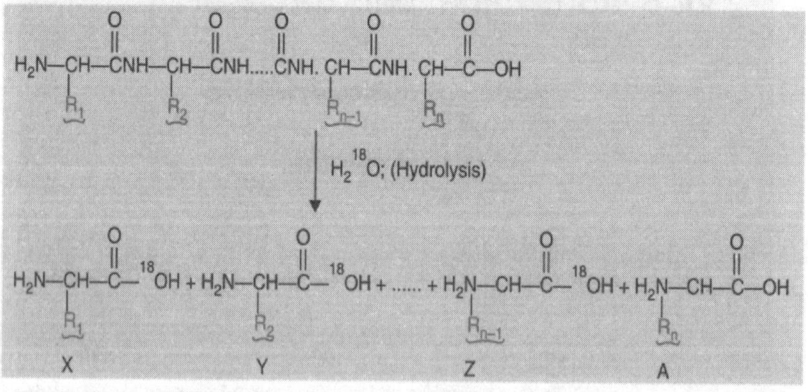

Therefore, with the help of a meticulous 'radio-isotopic critical analysis' one may definitely and easily arrive at the ultimate idification of the C–terminal amino acid duly present in a peptide or protein.

4.1.2.5 Characteristic Features of Amino Acids

In a broader perspective, the characteristic features of amino acids may be studied both intensively and extensively under the following *two* heads, namely:

- physicochemical properties of 'amino acids', and
- chemical properties of 'amino acids, and
  - ❑ l due to 'carboxyl moiety (–COOH), and
  - ❑ l due to 'amino moiety (–NH$_2$).

These aspects shall now be described briefly in the sections that follows:

4.1.2.5.1. Physicochemical Properties of Amino Acids

In general, based upon the various array of functional moieties (–R) duly present in the *twenty* standard amino acids, one may observe that they do exhibit an apparent perceptive difference in their physicochemical properties, namely:

- solubility profile,
- inherent electric charge,
- specific 'isoclectric pH'
- molecular weight,
- observed pH,
- actual particulate size,
- hydrophilic nature,
- hydrophobic profile,
- electrophoretic mobility, and
- melting points (mp).

Keeping in view the interesting revelations mentioned above it would be worthwhile to examine in detail the following *four* aspects, such as:

(a) Iomization *vis-a-vis* isoelectric pH,

(b) Amphoteric Nature,

(c) Isomerism,

(d) Solubility and Melting point, and

(e) Titration Profile of Amino acid Variants.

4.1.2.5.1.1 Ionization *vis-a-vis Isoelectric pH*

The 'amino acids' invariably comprise at least *'two ionizable moieties'* embedded in it, namely:

- an amino (–NH$_2$) moiety, and
- a carboxyl (–COOH) moiety.

Thus the *amphoteric* amino acids *i.e.,* the Zwitter Ion form of α-amino acid can undergo *two* intramolecular transformations due to:

❑ release of **'one proton'** from the **carboxyl moiety** of an **amino acid** forming an **'anion'**, and

❑ adaptation of **'one proton'** by the **amino moiety** of an **amino acid** forming a **'cation'**, as given under explicitly:

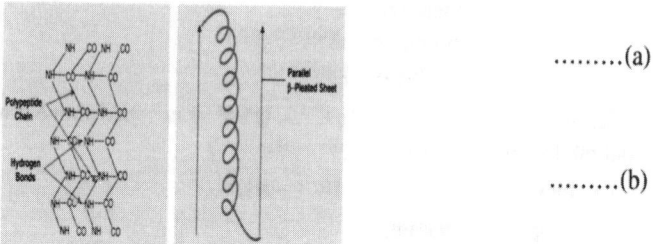

Anionic Form of α-Amino acid	**Zwitter Ion Form** of α-Amino acid	**Cationic Form** of α-Amino acid

Importantly, at the critical *'isoelectric pH (designated as pI)'*, an **amino acid** largely prevails as a **'Zwitterion'** or also known as the **'Dipolar Ion'**.

Isoelectric pH (*i.e.*, *pI*) of an **amino acid** may be defined as– **'the specific pH at which the amino acids essentially possess equal quantum of +ve and –ve charges, and are found to be absolutely neutral electrically.'**

> **Note:** Isoelectric pH (pI) attributes *three* cardinal characteristic features to the *amino acids*, such as:
> - **They fail to migrate in an electric field at its pI,**
> - **They tend to be least soluble at its pI, and**
> - **Each amino acid does *exhibit a particular pI i.e.*, isoelectric pH e.g.,**
>   **Aspartic acid : pI 2.9, and**
>   Lysine : pI 9.7

Obviously, the following *two* **'dissociation reactions'**, namely:

.........(a)

.........(b)

These actually correspond to an **'equilibria'** duly monitored by the well-known **'Law of Mass Action'**. Eventually, it would render the proportions of **ionized and unionized amino acids** to maintain a balance based upon the ensuing concentration of H^+ ions. In this manner the *two* **dissociation constants**, *viz:* **Ka** (for *acidic portion*–COOH); and **Kb** (for *basic portion* –NH$_2$), which correspond to the *two* *equilibria*, as shown in Eqns. (a) and (b) above, may be expressed as under:

$$Ka = \frac{[R-COO^{\ominus}][H^{\ominus}]}{[R-COOH]} \quad \text{and} \quad Kb = \frac{[R-NH_2][H+]}{[R-NH_3^{\ominus}]}$$

Nevertheless, the exact values of **'Ka'** and **'Kb'** do vary for a wide range of substances; but it may be generalized that:

Ka for a **carboxylic acid moiety** range between 10^{-4} to 10^{-6} *, and

Kb for an **amine moiety** ranges between 10^{-8} to 10^{-10} .

If one knows the values of **Ka** and **Kb**, one may easily calculate for **each H⁺ion concentration,** the ensuing perceantage of the *'ionized molecules'*.

Example: At a specific **pH 7**, we may have the respective **carboxyl (–COOH) moiety** (considering **Ka** having a **mean value of 1 × 10⁻⁵**):

$$1\times10^{-5} = \frac{[R\text{–}COO^-]1\times10^{-7}}{[R\text{–}COOH]}$$

$$\text{or} \quad = \frac{[R\text{–}COO^-]}{[R\text{–}COOH]} = 10^{-2}$$

From the above derivations one may certainly infer that at the **specific pH 7**, there would be present **only one non-ionized molecule** for every *'100 anion molecules'*. Thus, alternatively, the actual percentage of the *ionization of the respective carboxyl (–COOH) moiety* **will be 99%**.

Likewise, the corresponding **amino (–NH₂) moiety** at the same pH 7, we may have the values as (taking for **Kb** a mean value of 1×10^{-9}):

$$1 \times 10^{-9} = \frac{[R\text{–}NH_2]1\times10^{-7}}{[R\text{–}NH_3^{\oplus}]}$$

$$\text{or} \quad = \frac{[R\text{–}NH_2]}{[R\text{–}NH_3^{+}]} = 10^{-2}$$

Hence, at pH 7, the percentage of ioization of **amino moiety** is **99%**.

In a situation, when the **H⁺ion concentration** (*i.e.,* pH) is almost equal to 'Ka', we have the following expression:

$$[R\text{-}COOH] = 1$$

which overwhelmingly suggests that the **number of anion** duly present is almost equivalent to the number of non-ionized **molecules**. Thus, if in the aforesaid instance $Ka = 10^{-5}$, we may also have [H⁺] $= 10^{-5}$. Eventually, the said situation actually comes into being at **pH = 5** *i.e.,* when the pH attained is equal to **pk (–log K)** of the **ionizable moiety.**

In true sense, the **'pK corresponds to the pH of half–dissociation'**; thereby indicating the great ease with which the **dissociation commences,** and predominantly *show up the most realistic measure of the strength of the acid.*

In fact, the same very *ideology, concept,* and *explanation* holds good when the prevailing **H⁺ion concentration** is very much equal to '**Kb.'** Thus, we may have the following expression:

---

* The **carboxyl (–COOH) moiety** of an amino acid is found to be **100 times** *more strongly acidic vis-a-vis* of **acetic acid** on account of the ensuing **proximity of the α–amino moiety;** and, therefore, **Ka** is normally of the order of 10^{-2} to 10^{-3}.

$$\frac{[R\text{-}NH_2]}{[R\text{-}NH_2]} = 1$$

In other words, the **cation and non-ionized molecules** are in *equal number;* and when $K_a = 10^{-9}$ (as in the above cited example), one may eventually attain this situation at $pH = 9$ Hence, the **pH of the half-dissociation** is known as **pK**, and evidently *shows up the measure of actual strength of the base.*

In another exemplary situation, when the **pH of a solution of an amino acid** is raised gradually from a **low value** to a **high value,** we may crtically observe the **transformations** stated as under:

CATION [Acidic pH]	Zwitterion [Dipolar Form]	ANION [Alkaline pH]
Net Overall Charge = +1	Net Overall Charge = 0	Net Overall Charge = –1

Importantly, one may critically take cognizance of the fact that at a specific pH, the **amino acid molecules** are invariably found to be in the **Zwilter ion form** (or the **dipolar form);** whereby the net overall change on the said molecule is **zero** (*i.e.,* **nil).** In fact, it designates the *'isoelectric'* or *'isoionic'** *point* of the **amino acid.**

However, at this **particular pH,** the following aspects of **amino acids** do become glaringly note worthy, such as:

- **solubility** remains at a low ebb,
- **migration** remains almost absent on being placed in an **'electric field'.**

4.1.2.5.1.2. Amphoteric Nature

Interestingly, based upon the characteristic inherrent **'isoclectric pH'** the respective *amino acids* may either **donate** or **accept protons** (H^+) by virtue of their **ionizable functional moities** *viz.,* **amino** (**–NH₂**) and **carboxyl (–COOH).**

The following **universal fact** may be noted emphatically: **"an *'amino acid'* accepts proton when pH of a solution is found to be *'less than pI,* and donates proton when pH of solution is *'more than pI'.*"**

Based on the above statement of facts one may observe that:

- an **'amino acid'** behaves as a *base* particularly on the **acidic segment of its pI;** and, therefore, *accepts proton* to render into a **positively charged cation, and**
- an **'amino acid'** caters as an *acid* specifically on the **basic segment of its pI;** and, therefore, *donates proton* to become **negatively charged anion.**

---

* Isoionic Point. represents the crical **pH** at which the net overall charge reamins almost **'nil'** When the *analyte* is dissolved in water, and essentially bears the possibility of **fixing protons** almost exclusively. However, in most instances there exist other **ions in solution,** that may also be fixed by the *analyte.* Therefore, it renders the **molecule** still very much in a **'charged state'** at the pH corresponding to the **'isoionic point'.** Hence, the terminology **'isoclectric point'** is used frequently to represent the particular **pH** at which the charge of a molecule hairing **'fixed foreign ions'** is 'nil'.

In conclusion, it may be added that– 'an *amino acid* serves as both *'acid'* and *'base'* and is termed as an **'ampholyte', which essentially renders it (amino acid) to exhibit a typical amphoteric nature and feature.'**

4.1.2.5.1.3. Isomerism

It is observed critically that the *'α–C–atom'* usually present in a host of standard **'amino acids'** (with **'glycine'** as an exception) is indeed a **chiral carbon** *i.e.,* an *asymmetric carbon* to which *four* different moieties are attached with *'covalent linkages',* as given under:

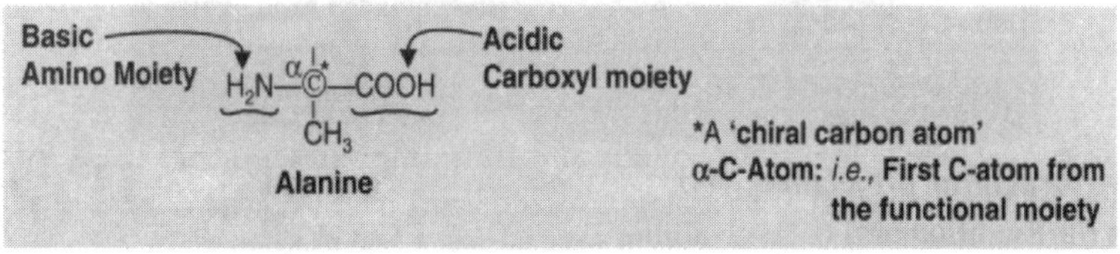

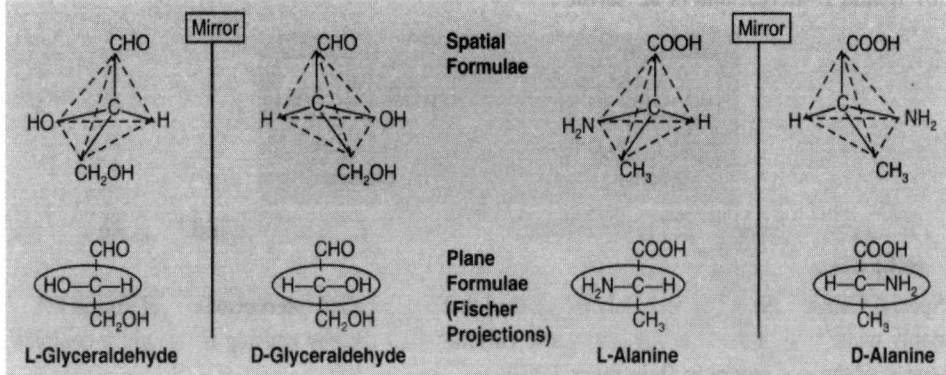

Importantly, the critical *spatial arrangement (i.e., with respect to space)* of these *four* entirely different functional moieties present duly on the asymmetric C–atom allows specifically the formation of **'isomers'**, such as:

- **optical isomers,** and
- **stereoisomers.**

(*a*) **Optical Isomers:** In general, all such **amino acids** having **'chiral carbon atoms'** usually exhibit *'optical activity'* (*i.e.,* they rotate the **plane of polarized light** either to the *'left'* or *'right'*), and exist as :

- **dextrorotatory–** designated as **'*d*',** and
- **levorotatory–**designated as **'*l*'** isomers.

The structure of the **'optical isomers'** of *glyceraldehyde* and *alanine* are as illustrated below:

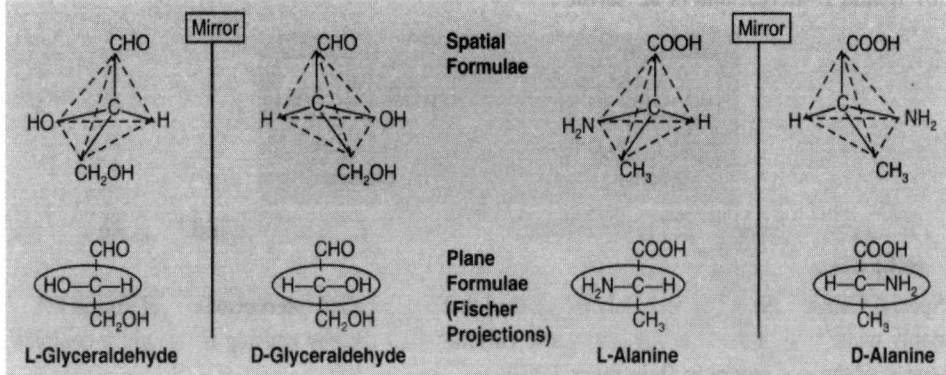

Thus, in **'alanine'–** the spatial formula of *two optical isomers* or **enantiomers** (that form a **'racemate'** when mixed) are *non-superimposable* since one is a **plane mirror image** of the other, and whose *physical* and *chemical* properties are identical (eccept the *rotaray power*). Therefore, one may easily distinguish **each amino acid,** an **isomer D,** and an **isomer L,** depending solely on whether the spatial structure is found to be absolutely identical to that of **D–glyceraldehyde** or **L–glyceraldehyde** (whose *absolute configuration (is known).*

Interestingly, very much akin to 'saccharides', the ensuing *d*–amino acids, and *l*–amino acids predominantly rotate the **plane of polarized light** to **right** or **left** respectively, as given under:

> **Note:** As 'glycine' the simplest amino acid [H_2N-CH_2-COOH] fails to show optical activity due to the critical absence of the *'chiral carbon'*.

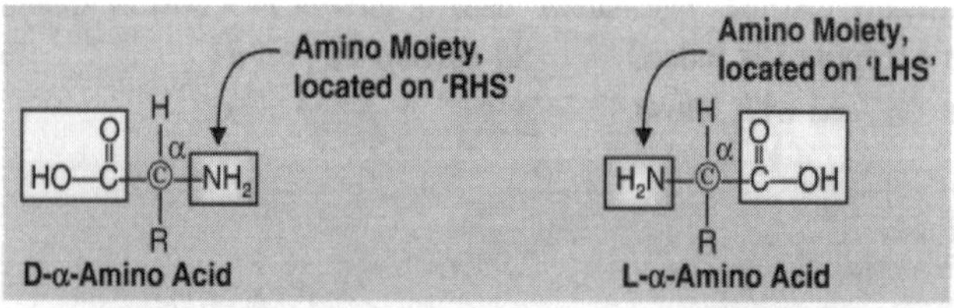

D-α-Amino Acid   L-α-Amino Acid

The above *structural projections* vividly illustrates the **'general structure'** of **D–** and **L–stereoisomers** (or **Enantiomers**) of an α–amino acid having a **chiral centre.**

Examples: Following are *two* typical examples of **amino acids** showing their **specific steric relationship:**

(*a*) **D–and L–stereoisomers of 'alanine':**

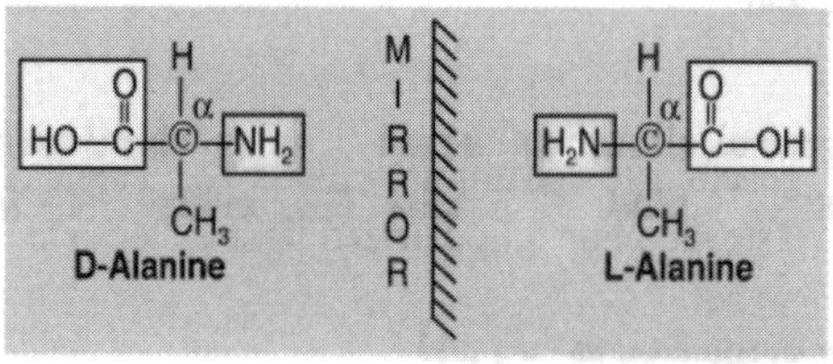

D-Alanine   L-Alanine

In **D–alanine** the *amino moiety* is on the **RHS** of the **chiral C–atom;** whereas, in **L–alanine** It is on the **LHS.**

(*b*) **D–and L–stereoisomers of 'serine':**

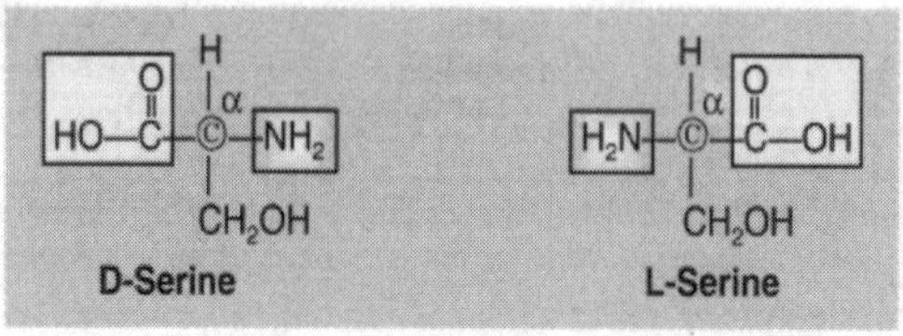

D-Serine   L-Serine

In **D–serine** the *amino moiety* is strateqically located on the **RHS** of the **chiral C–atom;** whereas, in **L–serine** it is on the **LHS.**

Special Comments: It has been duly established that the so called **'stereoisomers of amino acids'** invariably undergo **spontaneous nonenzymatic racemization** thereby causing ultimately a complete conversion of the **L–isomers** to **D–isomers,** and *vice-versa.*

Example: *L–Aspartic Acid* pertaining to the *'dentinne-protein'* undergoes racemization at a critical rate of 0.10% each year, which fact has been duly used to estimate precisely the age of living humans, animals, and the fossil bones as well.

(*b*) **Stereoisomers:** Based upon the exact and precise orientation of the **hydroxyl (–OH) moiety** strategically positioned upon the **β–carbon atom,** there exists *two* distinct stereoisomers (*viz.,* **L–, and**

D–isomers) of each and every β–amino acid. Having conceptualized the aforesaid line of thoughts one may look at an **amino acid** in its *L–isomer* and *D–isomer* form only if the **hydroxyl (–OH) moiety** present duly on the **β–carbon atom** is on ther **RHS** or **LHS** repectively, as shown earlier for **alanine** and **serine.**

In addition, the *two* **stereoisomers** *viz.*, the **L–** and **D–isomers,** of an **amino acid** are observed to be existing as the **'non-superimposable mirror images'** of each other as depicted above and are normally termed as **'enantiomers'.**

In a broader perspective, the **L–α–amino acids** are present exclusively in the **'mammalian proteins'**; whereas, the **D–α–amino acids** are exclusively noticeable in :

- **non-mammalian peptides,**
- **microbial cell walls, and**
- **peptide antibiotics** *e.g.,* **Actinomycin D,** and **Gramicidin S**

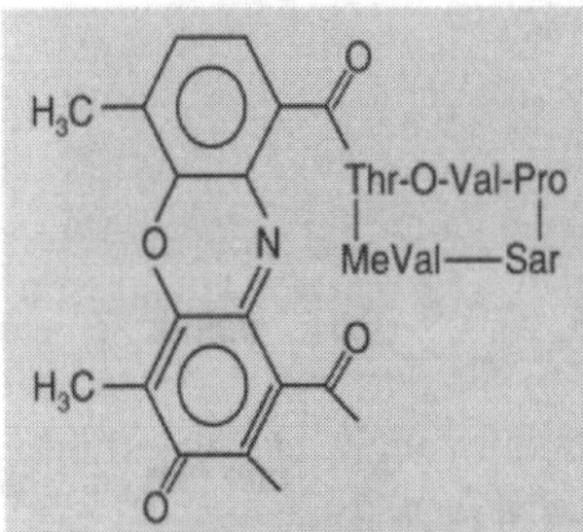

Actinomycin D
[or Dactinomycin]

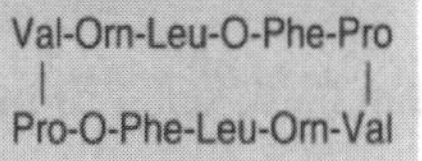

Gramicidin S

Note: Brain tissues are likely to comprise *D–glutamic acid* and *D–serine*.

4.1.2.5.1.4. Solubility and Melting Point

The **amino acids** are mostly water-soluble; however, they are found to be insoluble in **organic solvents,** such as: **benzene** (a nonpolar solvent), and **ether** (a *slightly polar solvent.*)

In general, the *melting points* of **amino acids** do fall in the range **beyond 200°C.**

4.1.2.5.1.5. Titration Profile of Amino Acid Variants

In actual practice, the dissociation pattern of an array of **polar functional moieties** of an **amino acid** may be determined by the addition of **HCl** or **NaOH** to the respective **'analyte solution;,** and measuring carefully the pH after each addition.

In this manner, it is quite convenient to plot the **'titration curves'** with different *shapes and pattern* depending exclusively whether the **amino acid** under investigation is:

- acidic,
- basic, or
- neutral

---

* That is, such **peptide bonds** wherin certain **'amino acids'** are duly engaged.

** Kar, A : **Pharmaceutical Analysis–Vol.II,** CBS–Publishers and Distributors, New Delhi, 2009.

(*a*) **Titration of an Acidic Amino Acid** (*e.g.*, **Aspartic Acid**): The **aspartic acid,** a *dicarboxylic acid*, shows a typical *'titration curve'* as given in Fig.2.5.

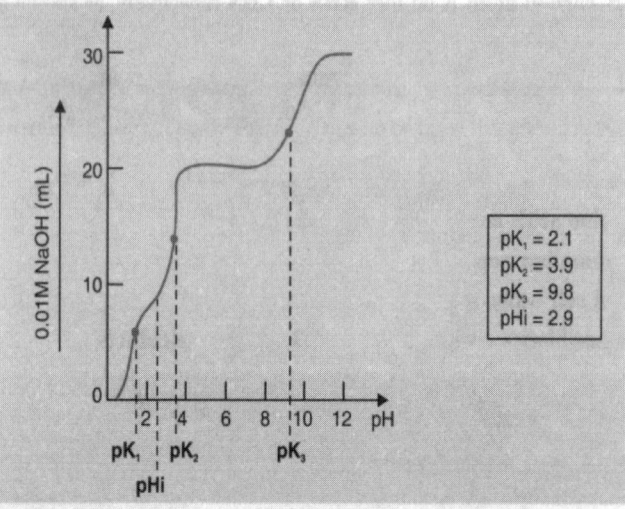

Fig: 2.5. Titration Curve obtained by Aspartic Acid [10m L of a 0.01M solution] *Vs* 0.01 M NaOH Solution.

In a particular situation when the solution remains **acidic strongly,** one may critically observe that the **aspartic acid** shall exist almost entirely in its *'protonated form'*.

Now, on the addition of **0.01M NaOH** gradually into the **aspartic acid solution,** the former will react prominently with all the **available protons** being provided by the respective **aminoacid*.** Thus, in the course of the aforesaid **titration by NaOH** we may come across **various distinct forms of aspartic acid** as enumerated under:

COOH
|
CH₂
|
H₂N—CH—COOH

Aspartic Acid
[Net Change = +1]

First Equivalent of NaOH
pK₁#2.1

COOH
|
CH₂
|
H₃N⁺—CH—COO⁻

Aspartic Acid
[Net Change = 0]

Second Equivalent of NaOH
pK₂#3.9

COOH⁻
|
CH₂
|
H₃N⁺—CH—COO⁻

Aspartic Acid
[Net Change = –1]

COOH⁻
|
CH₂
|
H₂N—CH—COO⁻

Aspartic Acid
[Net Change = –2]

Third Equivalent of NaOH
pK₃#9.8

---

* That is, the carboxyl moieties plus the protonated amino function (–NH₃⁺); are all **'acids'** strictly as per the **Bronsted's definition,** since they may all yield a proton (H⁺).

Important Observations: These essentially comprise:

(1) **pK Values** corresponding to the respective 'half dissociations' of the *three* protons (at the *first*, *second*, and *third equivalents*) are found to be **2.1, 3.9,** and **9.8** respectively.

(2) The most prevalent acidic functional moiety (*i.e.*,–**COOH group**) being the one attached duly to the **α–carbon atom,** since it is strategically located in close proximity to the –NH_3 moiety that by causing repulsion to the +ve charges, helps to promote the **critical dissociation** of the α–carboxylic **moiety.**

(3) However, at **pH 2.1** (*i.e.*, at the very 'first equivalent of NaOH*) we may specifically have, in **equal quantities,** the *form* having a net change of +1, and the other having a net change O.

(4) Likewise, at the critical **pH 3.9,** we may have in **equal quantities** the *two* forms bearing a charge O and –1.

Thus, it is pertinent to state here that the observed 'net charge O' shall be apparently **most** predominant at pH 2.9, and this represents the **isoelectric pH of aspartic acid.**

(*b*) **Titration of a Basic Amino Acid:** Though the 'titration curve' of **basic amino acids** *viz.,* *Lysine* and *Arginine* has not been duly shown under, but one may have a good idea of it provided one knows the respective **pK values** of the *said two amino acids.* Thus, we may have the **pK_1, pK_2, pK_3** of **Lysine** and **Arginine** along with their corresponding **pHi** values as given below :

Lysine : **pK_1 [COOH] = 2.2 ; pK_2 [α–NH_3^⊕] = 9 ;**

   pK_3 [ε–NH_3^⊕] = 10.5 ; pHi = 9.7 ;

   Arginine : pK_1 [COOH] = 2.1 ; **pK_2 [α–NH_3] = 9 ;**

   pK_3 [H_2N–C = NH_2^⊕] = 12.5 ; pHi = 10.8 ;

At this point in time, let us assume that we have a *mixture of amino acids* in a solution at pH6, for instance:

- *Basic amino acids viz.,* **Lysine** [$H_2N–(CH_2)_4–CH–COOH$] Shall be available in the form of **2+A–(charge = +1),**
- *Acidic amino acids viz.,* **Aspartic acid** [$HOOC–CH_2CH(NH_2)–COOH$] Shall occur in the form **+A^{2-} (charge +–1),**
- *Neutral amino acids viz.,* **Glycine** [Glycine [$H_2N–CH_2–COOH$] shall be available in the form of *zwitterion* **+A^- (charge = O).**

Bearing in mind the above facts one may muster a logical explanation with regard to the '**fractional separation of amino acids exclusively based upon their inherent charge differences.** *eg.,* by *capillary electrophoresis** and **ion–exchange chromatography.***

(*c*) **Titration of Neutral Amino Acid:** When the *basic component* [$R–COO^-$] of an *amino acid e.g.,* **Glycine** is carefully titrated** with an **acid (HCl),** we may have the following equation :

$$H_3\overset{\oplus}{N}–CH_2–CO\overset{\ominus}{O} + \overset{\oplus}{H}\overset{\ominus}{Cl} \longrightarrow \overset{\ominus}{Cl}\ H_3\overset{\oplus}{N}–CH_2–COOH$$

Protonated Glycine   **Chloride Salt of Glycine**

---

* Kar, Ashutosh : **'Pharmaceutical Analysis'** – Vol II, CBS–Publishers and Distributors New Delhi, 2009

** That is, according to **Bronsted's** definition–a '*base*' refers to a compound which has the **ability to capture one proton.**

When the *weakly acidic component* [R–NH$_3$+] of an *amino acid e.g.,* **Glycine** is titrated***
meticulously with an **alkali (NaOH),** we may have the following expression :

$$\overset{\oplus}{H_3}N-CH_2-\overset{\ominus}{COO} + Na\ \overset{\oplus}{OH}\overset{\ominus} \longrightarrow H_2N-CH_2-\overset{\ominus}{COO}\ \overset{\oplus}{Na} + H_2O$$

Protonated Glycine   **Sodium Salt of Glycine**

In Fig.2.6. there are *two* distinctly different **zones** *viz.,* **zone '1'** and **zone '2';** where in, the careful
addition of **HCl** or **NaOH** critically generates only a **very small variation in observed pH;**

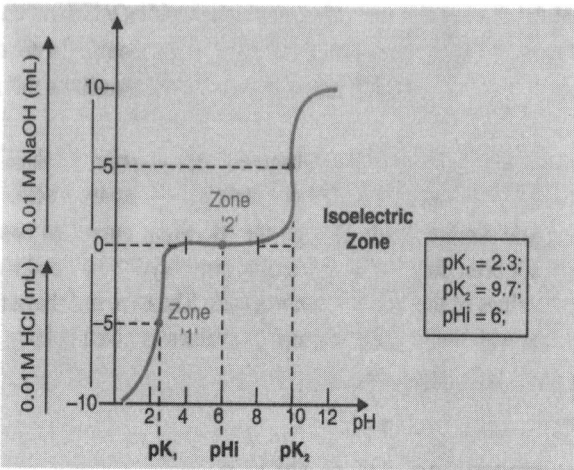

Fig: 2.6: Titration Curve Obtained by Glycine [10mL of a 0.01M Solution of HCL and NaOH.]

and, therefore, renders a sort of **'Buffer Effect'.** Furthermore, **Glycine** is actually employed to give rise
to *two* **'different types of buffer',** namely:

* *Glycine–HCl Buffer–* wherein **'glycine'** very much exists in the form of **hydrochloride salt;**
* *Glycine–NaOH Buffer–wher in* **'glycine'** usually occurs in the form of **sodium salt.**

Interestingly, just at the **centre of buffer 'zone1'** there exist a critical **pH value** corresponding to
the **pK values** of the *two* **different functional moieties** (*e.g.,* **H$_3$N–** and **– COO**) thereby giving rise to:

pK$_1$ [*i.e.,* the **'half-dissociation'** of the *carboxyl (—OOOH)moiety*] = ~**2.3,** and

pK$_2$ [*i.e.,* the **'half–dissociation'** the *amine (–NH$_2$moiety]* = **9.7**

Note: For the sake of just comparison, the pK of acetic acid stands at 4.8; and, therefore, it evidently
suggests that the *carboxyl (–COOH) moiety* present in 'glycine' is an *acid* which turns out to be
almost *100 folds stronger vis-a-vis* that of *acetic acid (H$_3$C–COOH).* This vast difference in the
pK values of glycine is on account of the crucial presence of the α–amino functional moiety in
'glycine'.

Therefore, the **Isoelectric pH or pHi of Glycine** $= \dfrac{Pk_1+Pk_2}{2}$

$$= \dfrac{2.3 + 9.7}{2}$$

$$= 6$$

*** Against an acid, because it has the **ability of producing one proton.**

Fig.2.6. clearly depicts that **this pH** (*i.e.,* **6**) lies very much at the centre of an *'isoelectric zone'*.

At a **pH** = pK_1 + 2 (we have here: 2.3 + 2 = **4.3** *i.e.,* the **acidic** (–COOH) moiety gets duly ionized even up to **99%**, and

At a **pH** = pK_2 – 2 (we have here: 9.7 – 2 = **7.7** *i.e.,* the amino (–NH_3) moiety gets adequately ionized to almost **99%**,

from which one may rightly infer that **specific pH zone** prevailing between **4.3** and **7.7** (*see above*) for **'glycine'** predominantly and totally occurs as the **Zwitterion state.**

4.1.2.5.2. Chemical Properties of Amino Acids

In usual practice, the ensuing **chemical properties of amino acids** are solely attributed by the following *two* different functional groups, namely:

- ❑ **Amino** (–NH_2) **group,**
- ❑ **Carboxyl** (–COOH) **group,**
- ❑ **Side–chain R,** and
- ❑ **Reactions involving both amino** (–NH_2) **and carboxyl** (–COOH) **moieties.**

which shall now be discussed briefly as stated under :

4.1.2.5.2.1. Chemical Properties Due to Amino Group

The **chemical characteristic features** due to the *amino group* in **amino acids may be ascertained** by *two* well–recognized **reactions,** such as:

(*a*) **Ninhydrin Reaction:** The **amino acids** comprising α–amino functional moiety specifically reacts with **Ninhydrin Reagent** to produce a distinct and prominent **blue–violet colour** or **Ruhemann's purple** (λ–max 570nm) In fact, this reaction forms the fundamental basis of the so called **'Ninhydrin Test',**

Ninhydrin

Ruhemann's Purple
[λmax = 570 nm]

The intensity of the resulting **purple colouration** is directly proportional to the **exact quantum** of the **amino acid** present.

Example: Let us consider an obsolutely imaginative hypothetical **peptide 'P',** which has been duly **hydrolyzed, tagged with AQe** [*i.e.,* **1–[[6–quinolylamino) carbonyl]oxy]–2,5–pyrrolidibnedione],** and subsequently subjected to C18–HPLC, and the results are as stated under:

'P' : (Asp or Asn), Gly_2, His, NH_3, Arg, Ala_3, Pro, Tyr, Val, Met, Lys, Ile, Leu, Phe, Trp

Observations: These essentially include:

(1) **Peptide 'P'** comprises *three folds* as much **Ala,** and *twice* as much **Gly** as **Arg, His, Lys,** or the *other amino acids* present.

(2) Obviously, the absolute number of each **amino acid residue** is not fully known unless and until one determines the exact *'molecular mass'* of the said **peptide 'P'.**

(3) It may also be noticed that the relative **order of the amino acid residues** very much within the **peptide 'P'** is **not known.**

> **Note:** Thus, in this particular context, *amino acid analysis* duly accomplished actually refers to the *amino acid composition of the peptide 'P',* which seems to be fairly comparable to the *elemental analysis vis-a-vis* the *molecular formula* of an *organic compound.*

(*b*) **AQC–NHS*–Amino Acid Method:** The most commonly used technique as to date is the **AQC–NHS amino acid method,** which essentially involves the critical conversion of the **mixture of amino acids**** duly generated by the hydrolysis of a **peptide** or **protein** into the respective **derivatives** which are easily and precisely detected by various **spectroscopic methods.**

Example: The reaction of the mixture of amino acids with **AQC–NHS** [*i.e.,* a compound whose name in routine-common usage has been shortened drastically to the corresponding acronym *'AQC–NHS'*].

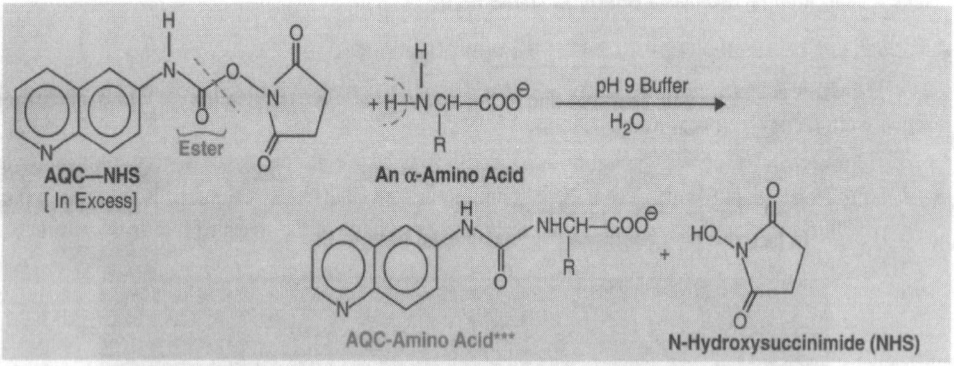

It may be observed that **AQC–NHS** is really an **'ester'**; and, therefore, the reaction represents and **'ester-aminolysis reaction'.** Besides the **esters of N–hydroxysuccinimide (NHS)** are found to be exceptionally reactive particularly in the desired **aminolysis reactions** as shown with the *dotted–lines* in the desired **aminolysis reaction,** as shown with the *dotted–lines* in the above reaction. In this manner, the aforesaid reaction gives rise to the phenomenon of **'tagging'** of each **amino acid** produced in a **'hydrolysis mixture'** with the specific **AQC moiety,** that eventually affords a relatively **strong absorption** at **254nm** in **UV spectrophotometric determinations.**

Interestingly, the resulting **AQC–moiety** also possesses remarkable **'fluorescent properties'.** Hence, it evidently suggests that the **AQC–moiety** when critically absorbs **UV-light** at a *specific wavelength,* it invariably *emits light* at a comparatively *longer wave length,* which stands at **397nm,** in the *'blue region'* of the visible spectrum.

Thus, after the eventful separation of the **AQC–amino acids,** these individual products can be measured quantitatively by using:

---

* **AQC** = 1-[[6–Quinolylamino) carbonyl] oxy]–2,5–pyrrolidinedione;

  NHS = N–Hydroxysuccinimide;

** Obtained as the hydrolysate from **peptides** and **proteins.**

*** Fluorescent at **395 nm** with excitation at **254nm.**

- **UV-absorption at 254 nm,** or
- **Flurescence at 395 nm.***

> **Note:** Importantly, before the relative quantum of *AQC–amino acids* produced in a hydrolysis mixture may be assayed. These essentially need to be separated by sophicticated analytical techniques e.g., High Performance Liquid Chromatography (HPLC) using *C–18 HPLC column.*

(*c*) **Transamination Reactions:** The **transmination reactions** emphatically make use of the **α–amino moiety** of an **α–amino acid₁** which being transferred meticulously to an **α–ketoacid₂,** thereby ultimately producing an altogether new **α–amino acid₂** together with another **α–ketoacid¹.** ** It may be observed that the **'transamination reactions'** are usually catalyzed by the enzyme **transaminases,** and prove to be absolutely **'reversible',** as given under:

(*d*) **Alkylation of Amino Acids:** The *Alkaline solution* of **amino acids** interact with **'alkyl halides'** to result into the formation of **N–alkyl substitued amino acids** as given under:

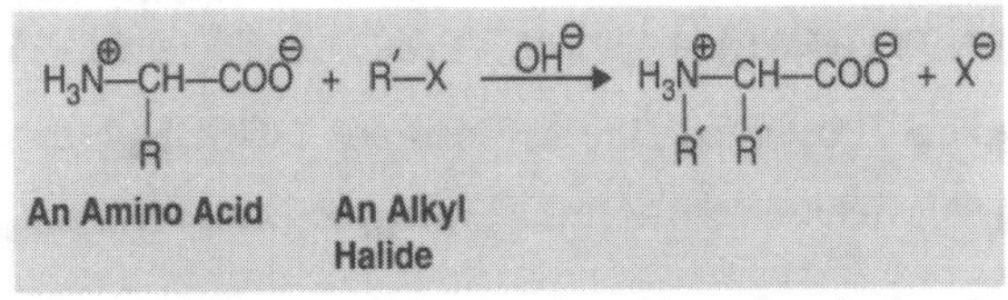

$$H_3\overset{\oplus}{N}-CH-\overset{\ominus}{COO} + R'-X \xrightarrow{O\overset{\ominus}{H}} H_3\overset{\oplus}{N}-CH-\overset{\ominus}{COO} + X^{\ominus}$$

$$\underset{R}{\mid} \qquad\qquad \underset{R}{\mid}\underset{R'}{\mid}$$

**An Amino Acid   An Alkyl
Halide**

Large excess of **'alkyl halides'** give rise the formation of typical and characteristic **'internal quaternary alkylammonium salts',** which essentially exhibit the **Zwitterion character,** and are usually termed as *'betaines':*

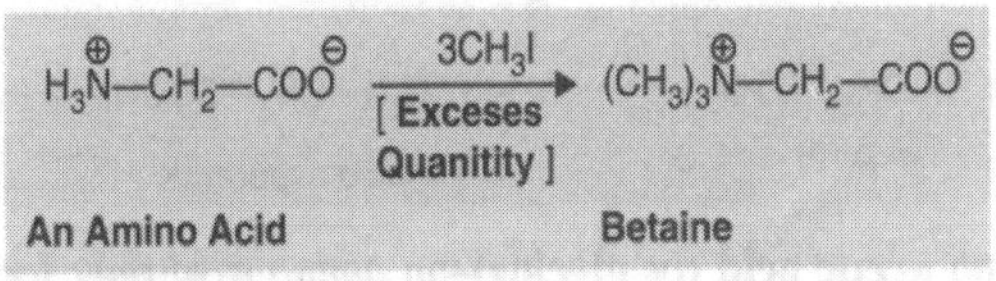

$$H_3\overset{\oplus}{N}-CH_2-\overset{\ominus}{COO} \xrightarrow[\text{[Excess]}{3CH_3I}]{} (CH_3)_3\overset{\oplus}{N}-CH_2-\overset{\ominus}{COO}$$

An Amino Acid   Betaine

(*e*) **Acylation of Amino Acids:** The **amino moiety** in an **amino acid** may be suitably acylated using **acetic anhydride** or **acid chloride:**

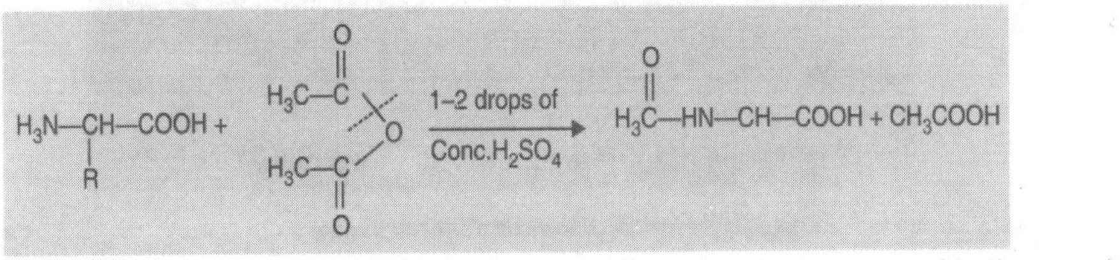

$$H_3N-CH-COOH + \underset{H_3C-C}{\overset{H_3C-C}{\big\rangle}}O \xrightarrow[\text{Conc.H}_2SO_4]{\text{1–2 drops of}} H_3C-HN-CH-COOH + CH_3COOH$$

(*f*) **Reaction with Nitrous Acid [HONO] :** The **α–amino acids** critically react with **nitrous acid** (produced in the reaction mixture by reacting together sodium nitrite and dilute HCl) to yield **α–hydroxy acids** and **N₂** is liberated as a **'gas',** as given under:

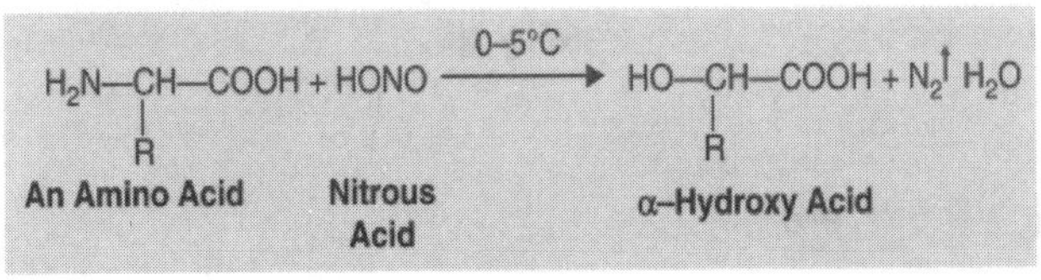

$$H_2N-CH-COOH + HONO \xrightarrow{0-5°C} HO-CH-COOH + N_2{\uparrow} H_2O$$

$$\underset{R}{\mid} \qquad\qquad\qquad \underset{R}{\mid}$$

**An Amino Acid   Nitrous
Acid   α–Hydroxy Acid**

---

* Since, both these techniques depend exclusively upon the concentration of the **'absorbing'** or **'fluoresuing'** *species.*

** Originally derived from **α–amino acid¹.**

Note: The above reaction forms the basis for the *assay of the amino acids* after due collection and measurement of the liberated N_2-gas (50% of which is contributed by amino acid). It is usually known as '*van Slyke amino–nitrogen estimation*'.

$$\begin{array}{c}
\text{L–Alanine} \\
\text{[}\alpha\text{–Amino Acid}_1\text{]}
\end{array} + \begin{array}{c}
\alpha\text{–Ketoglutarate} \\
\text{[}\alpha\text{–Keto Acid}_2\text{]}
\end{array} \xrightleftharpoons[\text{Pyridoxal Phosphate}]{\begin{array}{c}\text{Glutamate–Pyruvate} \\ \text{Iransaminase}\end{array}}$$

$$\begin{array}{c}
\text{L–Glutamate} \\
\text{[}\alpha\text{–Amino Acid}_2\text{]}
\end{array} + \begin{array}{c}
\text{Πψρυϖατε} \\
\text{[}\alpha\text{–Keto Acid}_1\text{]}
\end{array}$$

(*g*) **Formation of Salts:** The basic amino acids do react specifically with the **mineral acids** (*viz.*, HCl) in *stoichiometric proportions* to yield their **respective salts.**

(*h*) **Formation of Amino Alcohols:** The **amino acids** are adequately subjected to reduction with **lithium-aluminium hydride [LiAlH₄]** to produce the corresponding **amino alcohols.**

4.1.2.5.2.2. Chemical Properties Due to Carboxyl (–COOH) Group

The various **chemical characteristic features due to carboxyl group** in *amino acids* are as dissussed under:

(*a*) **Amino Acids Produces Salts with Alkalies:** The **amino acid** on being treated with an **alkali (NaOH)** forms its respective **sodium salt'** plus a mole of water, as given below:

$$\overset{\oplus}{H_3N}-CH_2-\overset{\ominus}{COO} \xrightarrow{\overset{\oplus\ \ominus}{NaOH}} H_2N-CH_2-COONa + H_2O$$

An Amino Acid   **Sodium glucinate**

Interestingly, α–amino acetic acid (or glycine)–an *amino acid* duly forms '**chelate compounds**' with **bivalent metals** *eg.*, Ca^{2+}, Zn^{2+}, Cu^{2+}, Fe^{2+}, Mg^{2+} etc., as depicted under:

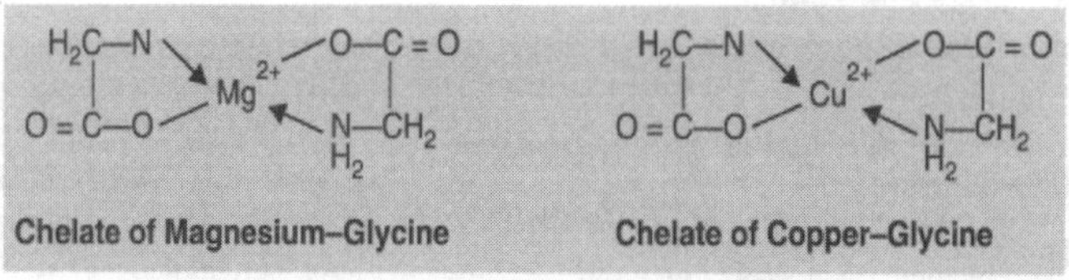

Chelate of Magnesium–Glycine   **Chelate of Copper–Glycine**

Explanation: The N–atom in the *amino function* is pentavalent (*i.e.*, N has *five valency*) of which a **single pair of electrons** forms a '**coordinate bond**' with the respective **metal bivalent ion** *viz.*, Mg^{+2} and Cu^{2+} respectively as shown above

(*b*) **Decarboxylation of Amino Acids:** The **decarboxylation** [*i.e.*, removal of the carboxyl (–COOH) moiety from an *amino acid* may be accomplished by several methods, namely:

- acids,
- bases,
- specific enzymes, and
- heat,

to the corresponding '**primary amines**'.

Exmples: A few typical examples are as stated under:

(1) Decarboxylation of Alanine to Ethyl amine: In the presence of

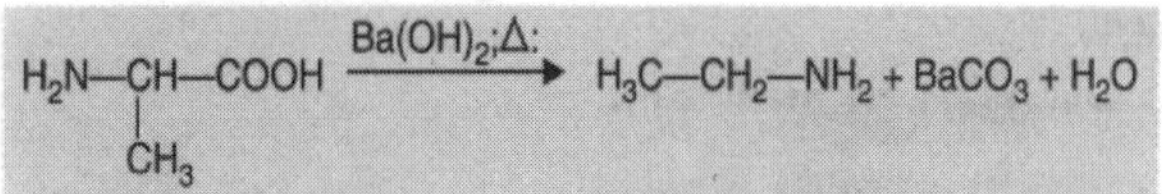

$$H_2N-CH-COOH \xrightarrow{Ba(OH)_2;\Delta:} H_3C-CH_2-NH_2 + BaCO_3 + H_2O$$
$$\underset{\displaystyle CH_3}{|}$$

Alanine   Ethyl anine

barium hydroxide [Ba(OH)$_2$] and heat, **alanine** gets decarboxylated to the respective **aliphatic primary amine** *viz.*, **ethyl amine.**

(2) Decarboxylation of Cysteine to Mercaptoethylamine

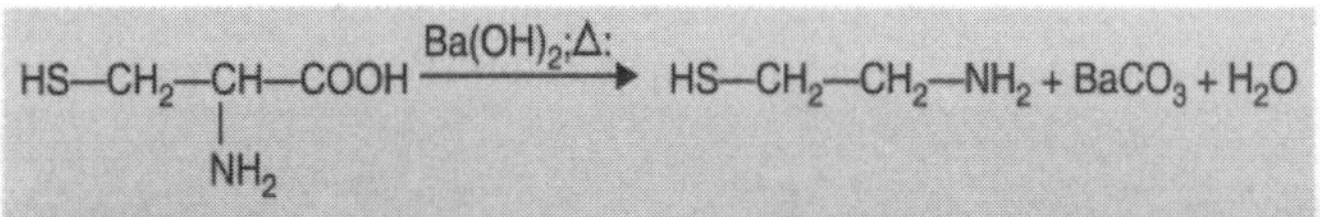

$$HS-CH_2-CH-COOH \xrightarrow{Ba(OH)_2;\Delta:} HS-CH_2-CH_2-NH_2 + BaCO_3 + H_2O$$
$$\underset{\displaystyle NH_2}{|}$$

Cysteine   Ethyl anine–2–lhiol
                            [or Mercaptoclhylamine]

The decarboxylation of **cysteine** with barium hydroxide and heat gives **mercaptoethylamine.**

(3) Importantly, some of these **'amines'** (as *by products*) do play a vital and important **physiological** or **pharmacological** role; and, therefore, are invariably termed as *'biogenic amines'*.

Example: Decarboxylation of **histidine** to **histamine** (a **biogenic amine**) which causes *hypotensive effect* and *allergic reactions* (*viz.*, severe itching all over the body, swelling of lips, and rashes on the body). However, in extreme cases one may even get a **'histamine shock'**. The **antihistaminies** *e.g.*, **chlorpheniramine maleate, pyrilamine maleate, meclizine hydrochloride, thenalidine tatrate** etc.

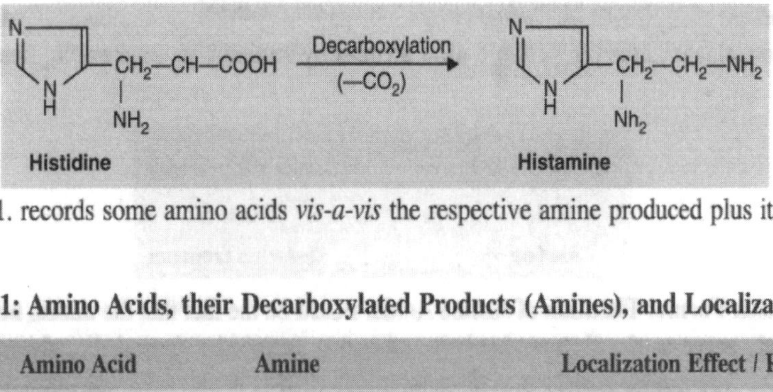

Histidine   Histamine

Table: 2.1. records some amino acids *vis-a-vis* the respective amine produced plus its localization effect or role

Table: 2.1: Amino Acids, their Decarboxylated Products (Amines), and Localization Effect

S.No.	Amino Acid	Amine	Localization Effect / Role
1	Aspartic acid	α–Alanine and β–Alanine	In proteins and CoA.
2	Cysteine	Mercaptoethylamine	In CoA.
3	Glutamic acid	GABA (α–Amino buty-ric acid)	Mediator of CNS.
4	Histidine	Histamine	Hypotensive effect, allergic conditions
5	DOPA (3,4–Dihydroxy phenyl alanine)	Dopamine	Precursor of 'adrenaline'.
6	Serine	Ethanolamine In 'phophatides'.	
7	5–HT(5–Hydroxy tryptophan)	Seratonine	A vasoconstrictive tissue hormone.

(c) **Esterification:** The **amino acid** on being dissolved in **absolute ethanol** (*'AnalaR'–Grade*) and subjected to treatment:

- **saturation with HCl gas,** and
- **careful heating,**

causes **smooth and critical esterification.** Nevertheless, the above observations suggest predominantly that **'ethanol'** fails to react with the **Zwitterion form** of the *amino acid the amino acid"* to give rise to the formation of the desired **'ester salt of the amino acid'.** Thus, one may segregate the *'free ester'* by subsequent treatment with **aqueous Na$_2$CO$_3$** or **aqueous Ag$_2$O** upon the corresponding **ester salt,** as given under:

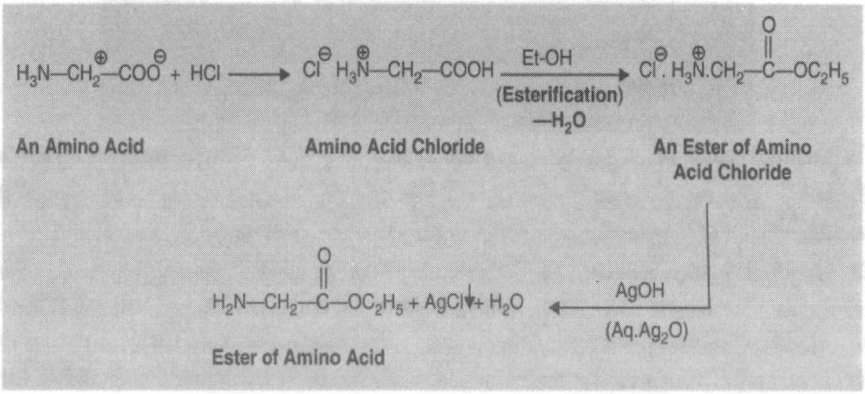

An Amino Acid   Amino Acid Chloride   An Ester of Amino
                                                     Acid Chloride

Ester of Amino Acid

(d) **Reduction of Amino Acid:** *Reduction of amino acid* with **lithium-aluminium hydride [LiAlH$_4$]** helps in its conversion to the respective **amino alcohol** by retaining its inherent **'optical activity',** as given under:

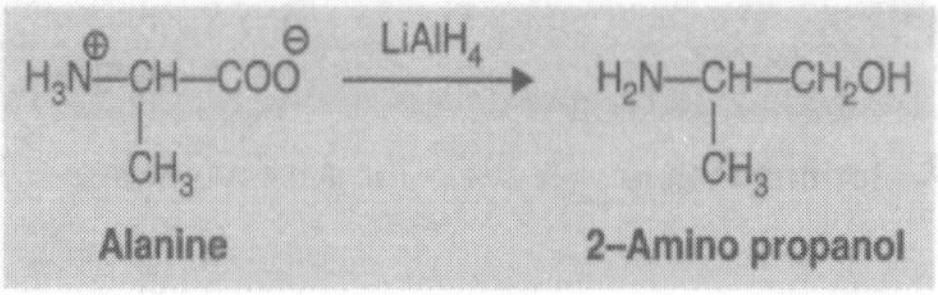

Alanine   2–Amino propanol

(e) **Sorensen Formol Titration of Amino Acids:** Based on the fact that the **amino acids** invariably and extensively serve as the **'internal salts',** and hence, cannot be titrated directly with the **bases.** However, if the **free amino moiety** present in an *amino acid* is **protected meticulously** by the aid of **formaldehyde (HCHO)** thereby forming the respective **secondary amine** *e.g.,*

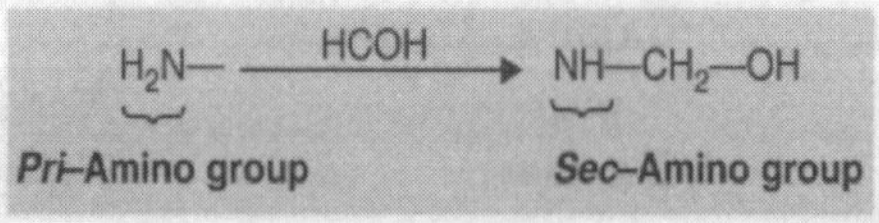

Pri–Amino group   *Sec*–Amino group

Thus, the **carboxyl (–COOH) moiety** present in the **amino acid** is rendered *'normal'* (*i.e.,* not influenced by the **strong *pri*–** amino functional moiety anymore; and, therefore, the resulting **'acid'** may be literated conveniently using an appropriate *indicator* (*viz.,* **methyl orange, methyl red** etc.) Thus, we have the following expression:

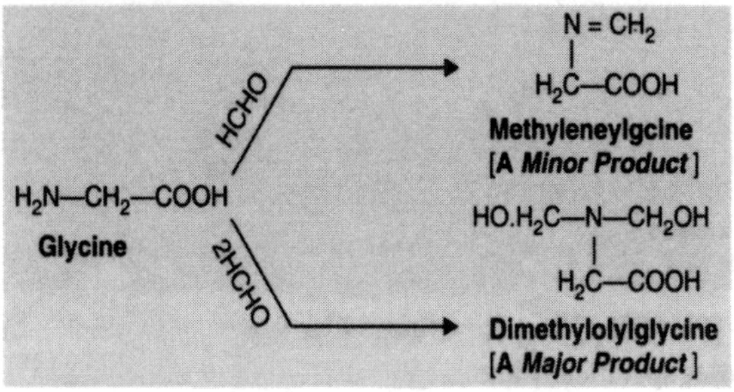

Glycine on treatment with **one mole of formaldehyde (HCHO)** yields **methyleneglycin** as a *'minor product';* and with **two moles of HCHO** produces **dimethylolylglycine** as a *'major product'*.

(*f*) **Dakin-West Reaction*:** It refers to the typical reaction of α–amino acids with *acietic anhydride* in the presence of a **'base'** (pyridine) to produce α–**acetamido ketones.** However, the reaction usually comes into play *viz* the *intermediate* **azlactone** as stated under:

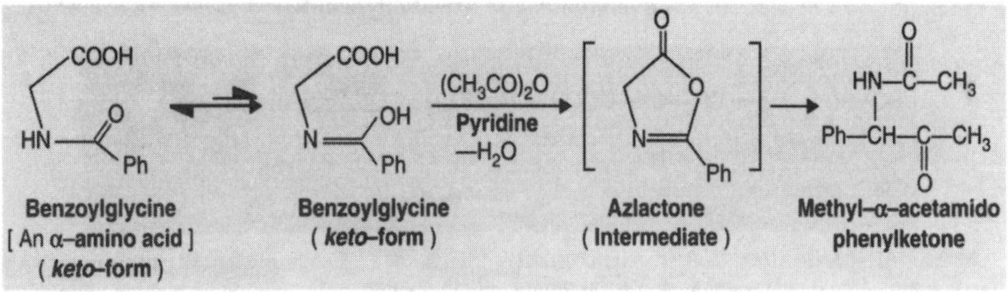

ultimately, the **azlactone** gives rise to the formation of methyl α–**acetamidophenylketone.**

4.1.2.5.2.3. Chemical Properties Due to Side Chain R

The very presence of a variety of **specific side–chains** (R) in different **amino acids,** they do react with an array of reagents to produce highly critical **colour reactions.** Following are a few vital and important **colour reactions:**

(*a*) **Chelating Feature of Amino Acids:** It has been duly observed that most *amino acids* do chelate certain bivalent **methyl ions** *viz,* Ca^{2+}, Mg^{2+}, Cu^{2+}, Mn^{2+}, Fe^{2+} etc, to give rise to the formation of some **chelated coordination complexes,** that are supposedly **non-ionic in nature.** Interestingly, the inherent charagteristic chelating characteristic feature of an amino acid is critically observed to be due to the presence of both **amino (–NH₂)** and **carboxyl (–COOH)** moieties as illustrated below:

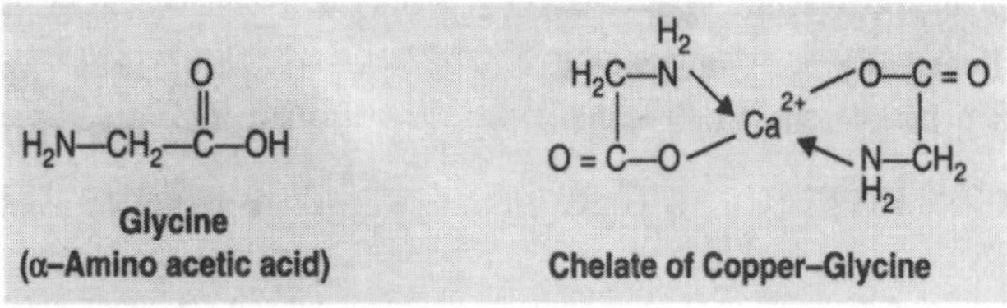

(*b*) **Hopkins–Cole Reaction:** It is solely meant for **tryptophan** that critically contains the **heterocyclic ring, indole,** in its cyclic side chain. In fact, **tryptophan** gets specifically condensed with the **aldehydes** in the presence of *sulphuric acid* together with traces of certain *oxidizing agents* to produce a distinct **purple coloured complex.**

---

* Dakin HD and West R : *J Biol. Chem,* **78** : (91) 745, 757, 1928.

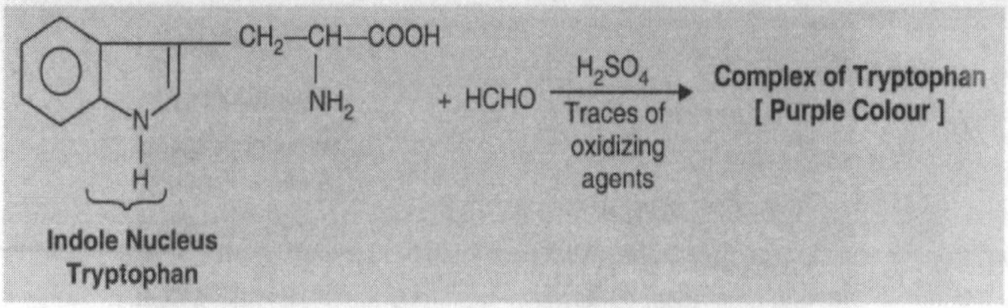

(c) **Lead Acetate [Pb (CH₃CO)₂] Reaction:** The S–containing amino acids, such as : **Cystine, cysteine,** and **methinonine** specifically interact with a solution of **lead acetate** in an *alkaline environment (NaOH)* and ultimately gives rise to the formation of a **brown** or **black** insoluble residue of '**lead sulphide**'.

(d) **Millon's Reaction:** It is particularly meant for the presence of '**phenyhydroxy moiety**' in *tyrosine* located in its **side chain.** tyrosine interacts with **Millon's Reagent*** to produce the respective **hydroxyphenyl mercuric complex.** The resulting **Hg-complex** on treating with **sodium nitrite** gives rise to the production of a distinct **red colour** due to **nitrohydroxyphenyl mercuric complex.**

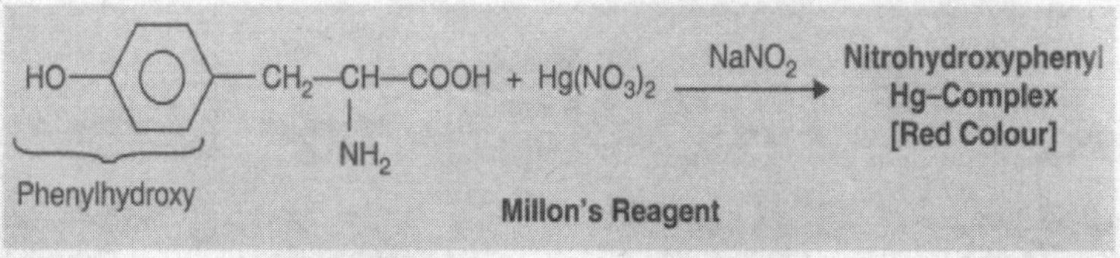

(e) **Nitroprusside Reaction:** It is specifically meant for **cysteine** essentially containing a '**free–SH moiety**' duly present in its **side chain. Sodium nitroprusside** reacts with the inherent **–SH moiety** to result into the formation of corresponding **sodium thionitroprusside** which gives an intense **red** or **magenta** colour.

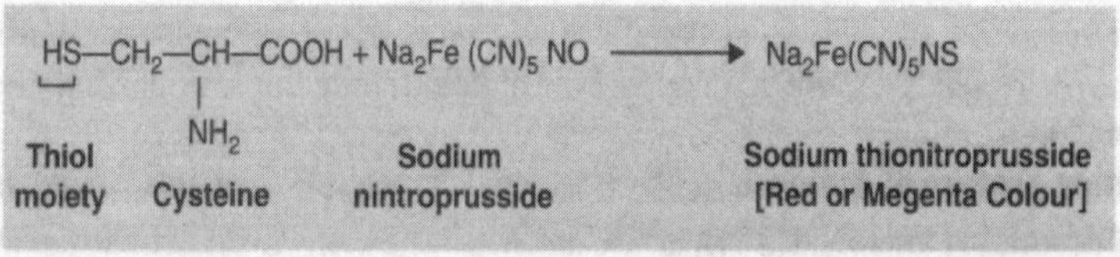

(f) **Xanthoproteic Reaction:** In fact, the **xanthoproteic reaction** is found to be highly specific for the **aromatic amino acids** *e.g.,* **phenylalnine, tryptophan,** and **tyrosine** which essentially comprises an *aromatic nucleus* in their **side chains.** Thus, the **aromatic amino acid** reacts with *fuming* **nitric acid (HNO₃)** to give a distinctly **yellow colouration** that almost develops due to the formation of certain **nitro–derivatives.** Eventually, when a **strong alkali (NaOH)** is slowly added to the resulting solution, the *intital red/magenta colour* turns into '**orange**' due to the formation of the respective '**nitro-derivatives**.

4.1.2.5.2.4. Chemical Properties Involving Both Amino (–NH₂) and Carboxyl (–COOH) Moieties

Following are some of the typical chemical properties involving both amino- and carboxyl-moieties present in an **amino acid,** such as:

---

* **Millon's Reagent (or Mercuric Nitrate Solution) :** Dissolve 3 mL of pure Hg in 27mL of **cold fuming HNO₃,** and dilute the solution with an equal volume of DW.

- **action of heat,**
- **hydantoin formation,** and
- **Ninhydrin reaction.**

(a) **Action of Heat:** The **action of heat** on **'amino acids'** is critically and gainfully exploited as an extremely advantageous characteristic feature for distinguishing remarkably the various **amino acid types** *viz., α, β, γ,* and δ, since the production of divergent products upon heating.

Example: A few typical examples of **amino acid variants** are as enumerated under:

(*i*) **α–Amino Acids:** It has been duly observed that *two* molecules of **α–amino acids** (*eg., alanine*) usually react in such a manner that the *amino function* of one critically forms *amide function* with the corresponding *carboxyl function* of the other, thereby giving rise to the product **2, 5–diketopiperazine –3,6–dimethyl** (*i.e., 'cyclic diamide'*) as shown below:

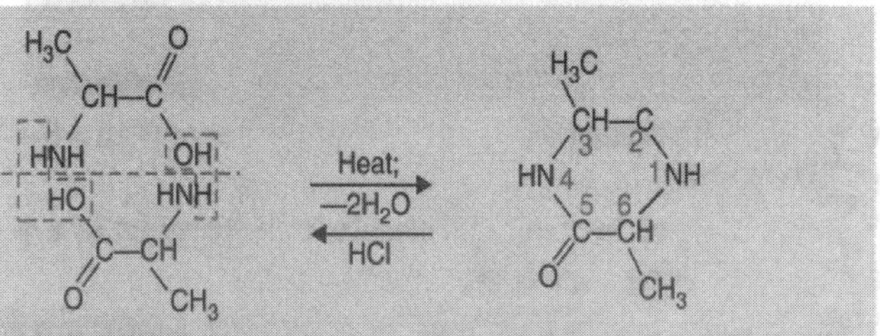

Alanine [2 Moles]   2,5–Diketo piperazine–3,6–dimethyl
(An α–amino acid)   [A 'Cyclic Diamide']

(*ii*) **β–Amino Acids:** The **β-amino acids** (*eg., β–aminobutanoic acid*) upon heating crtically undergoes the **β–elimination of the amino moiety** thereby forming an **α, β–unsaturated acid** as given under:

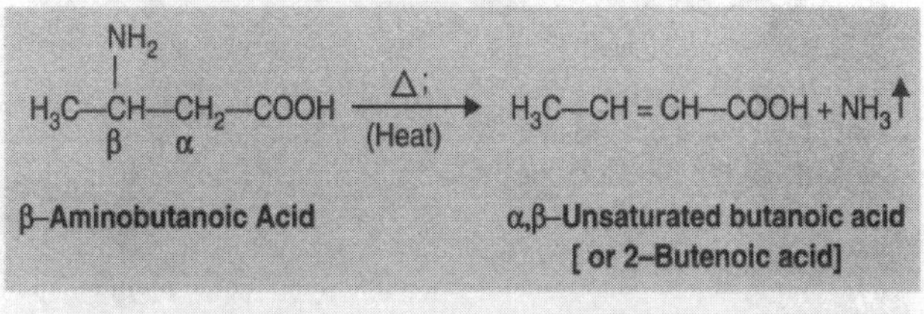

β–Aminobutanoic Acid   α,β–Unsaturated butanoic acid
[or 2–Butenoic acid]

Note: The above typical reaction reveals the exact *mechanism of the β–elmination of amino (–NH₂) moiety.*

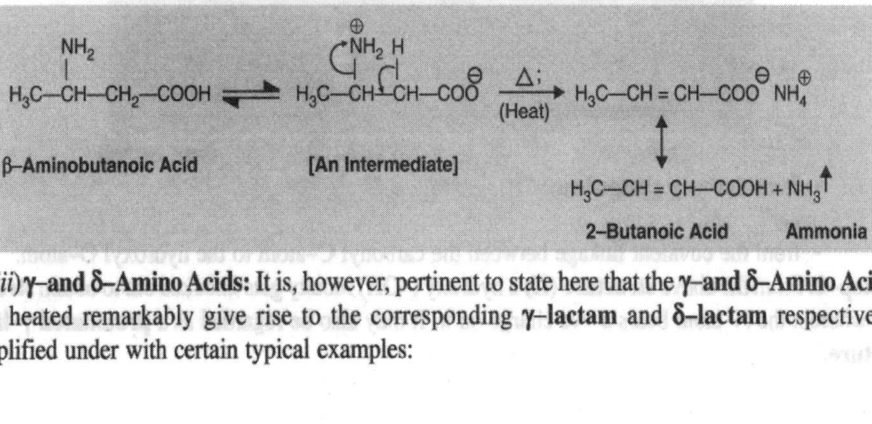

β–Aminobutanoic Acid   [An Intermediate]

H₃C—CH = CH—COOH + NH₃↑

2–Butanoic Acid   Ammonia

(*iii*) **γ–and δ–Amino Acids:** It is, however, pertinent to state here that the **γ–and δ–Amino Acids** on being heated remarkably give rise to the corresponding **γ–lactam** and **δ–lactam** respectively as exemplified under with certain typical examples:

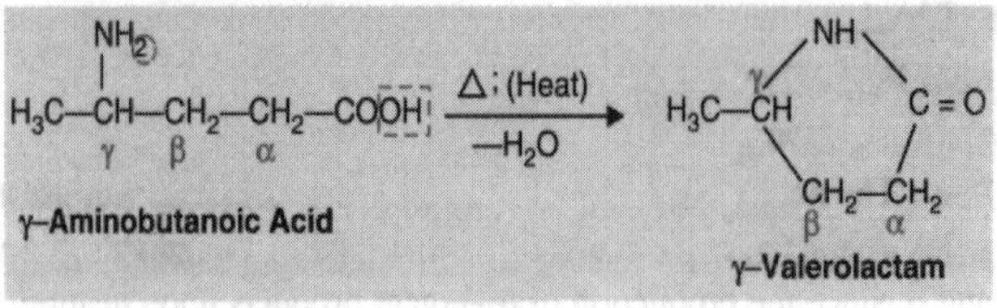

γ-Aminobutanoic Acid → γ-Valerolactam

i.e., **γ-aminovaleric acid** on heating loses a molecule of water to form a **γ-valerolactam.**

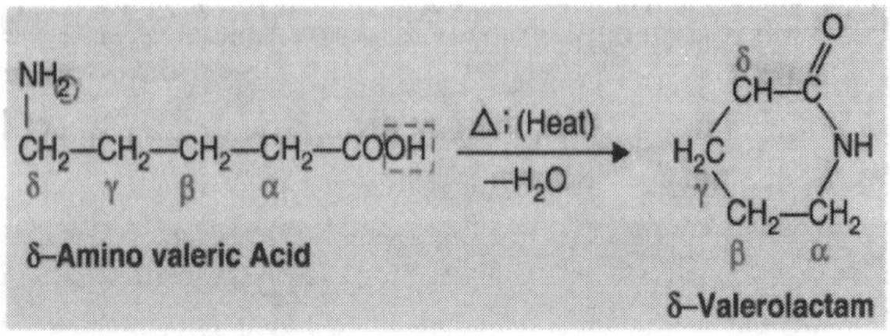

δ-Amino valeric Acid → δ-Valerolactam

i.e., **S-aminovaleric** acid on heating loses a molecule of water ot form a **S-valerolactam.**

How are the 'Lactams' formed?

The **'lactams'** are duly formed by the intramolecular transformation of a **γ-amino acid** *e.g.,* **γ-amino valeric acid** *via* the following *three* steps, namely:

Step-1: The N-atom bearing a **'lone pair of electron'** in the **γ-aminovaleric acid (I)** gets shifted to the carbonyl. C-atom, thereby establishing a covalent linkage between the carbonyl C-atom and the N-atom to form a **cyclic structure (II).**

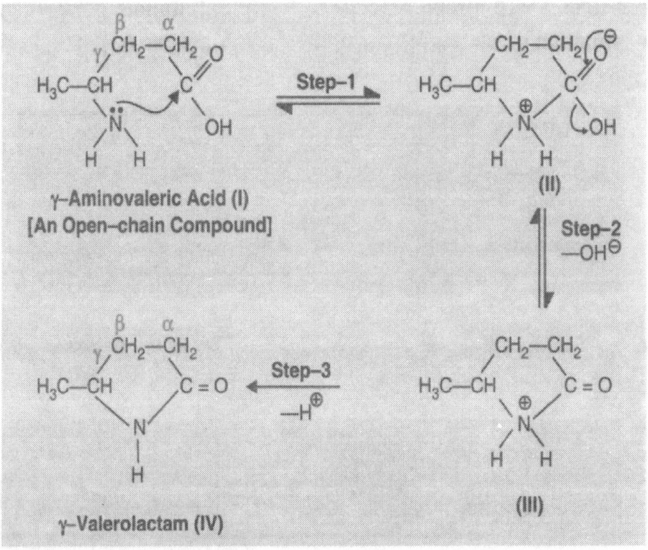

Thus, the resulting **cyclic structure (II)** shows *two electronic migrations,* namely:

- from the-vely charged carbonyl O-atom to the covalent linkage between the C- and O-atoms, and
- from the covalent linkage between the carbonyl C-atom to the hydroxyl O-atom.

Step-2: From the above **structure (II)** a hydroxy (-OH) moiety gets knocked out to obtain **structure (III),** wherein the N-atom bears a +ve charge on it. It may also be regarded as a **protonated γ-lactam structure.**

Step–3: Ultimately, the resulting **structure (III) loses a proton (H⁺)** to yield the desired **γ–valerolactam (IV).**

Lactam–Lactim Tautomerism: Interestingly, the ensuing **'lactams'** distinctly exhibits an altogether special kind of *tautomerism,* invariably termed as the **'lactam-lactim tautomerism;,** as depicted under:

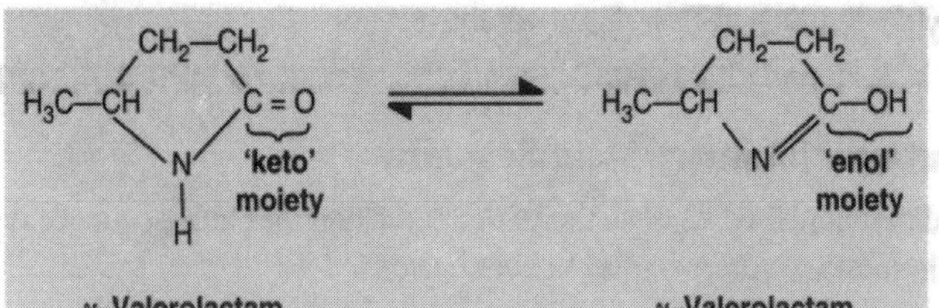

γ–Valerolactam
[*Lactam*–form]

γ–Valerolactam
[*Lactim*–form]

(b) **Hydantoin Formation: Hydantoin** is a *diketo–derivative of tetrahydroimidazole* **(imidazolidine).** In fact, the **hydantoin ring** numbering (with the N–atoms, of course, at 1 and 3) follows the direction which gives the lowest available number to the **carbonyl (>C=O) moieties,** as given under:

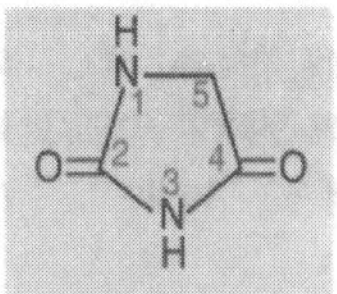

Hydanatoin

The *α–amino acids* (*e.g.,* **alanine**) react with isocyanates to produce the respective **'carbamides'** that eventually form the desired **'hydantoins'** on warming gently with dilute HCl. The resulting **'hydantoins'** may then be utilized for the subsequent synthesis of **'other amino acids'.**

We may have the following reactions to expatiate the above pronouncements:

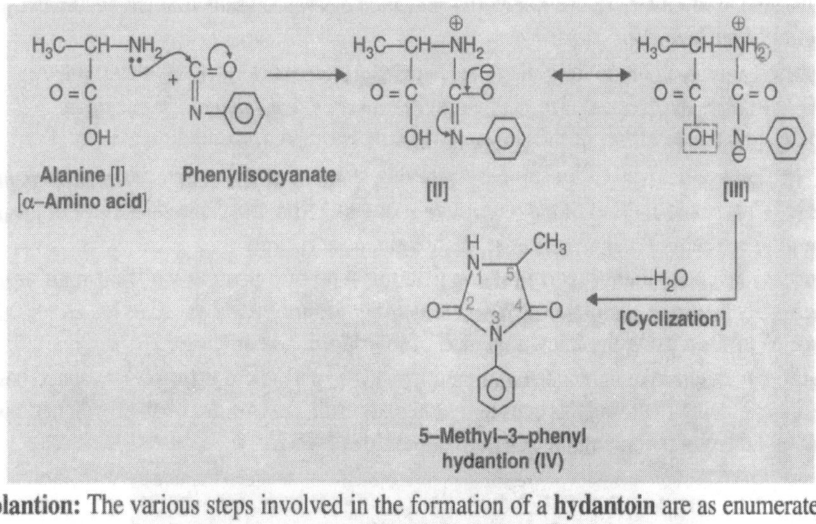

Alanine [I]
[α–Amino acid]

Phenylisocyanate

[II]

[III]

—H₂O
[Cyclization]

5–Methyl–3–phenyl
hydantion (IV)

Explantion: The various steps involved in the formation of a **hydantoin** are as enumerated below:

(1) **Alanine (I)** an *α–amino acid,* reacts with phenylisocyanate to form an **intermediate phenyl isocyanate salt (II)** bearing a **+ve change** on the *amino function,* and a **–ve change** on the *carbonyl O–atom* belonging to *phanylisocyanate.*

(2) The resulting **intermediate (II)** undergoes **intramolecular transformation** whereby the –ve **charge** shifts from the **carbonyl O–atom** to the **N–atom** present in the *isocyanate entity* to form **another intermediate (III).**

(3) The resulting **product (III)** undergoes *cyclization* by the loss of a molecule of water to yield 5–methyl–3–phenyl hydantion **(V)** as shown above.

(c) **Ninhydrin Reaction:** It has already been discussed under **section 4.12.5.2.1 (a).**

4.1.2.6. Determination of Amino Acid Sequence

In a generalized manner, let us consider to determine the probable sequence of a protein comprising a 'single peptide chain' whose *two* vital criteria, such as:

- molecular weight, and
- amino acid composition,

are known supposedly; and which has the following **sequence:**

$$H_2N—A\,.B.\,C.\,D.\,E.\,F.\,G.................S.\,T.\,U.\,V.\,W.\,X.\,Y.\,Z.—COOH$$

Eventually, based on the meticulous application of the so called **end-group methods** (see section 4.1.2.4.1. and 4.1.2.4.2), the **amino–** and **carboxyl–** end residues, 'A' and 'Z' respectively, have been duly identified with reasonably good precesion and accuracy. It is, however, pertinent to state here that by the help of:

- Edman stepwise procedure (section 4.12. 4.1.3–the **end amino-residual sequence** *viz.,* A. B. C. D. has been duly ascertained and established, and   •
- Carboxy peptidase stepwise procedure (section 4.1.2. 4.1.5)–the **end carboxy–residual** sequence *viz.,* W. X. Y. Z. has been adequately confirmed and established.

Having intelligently accomplished the above cited two important criteria with great satisfaction, it is now equally vital to **elucidate** the rest of the sequence of **amino acids** that are strategically present in the 'protein' (*under investigation*).

Therefore, one may adopt the following 'general approaches' in a sequential manner:

(1) **Partial hydrolysis** of the protein either by *chemical reagents* or enzymes.

(2) **Critical isolation** of the resulting **peptide residues (or segments).**

(3) **Precise determination** of the **prevailing sequences of the relatively** *smaller peptides.*

(4) **Careful deduction** of the **'complete sequence'** from the *derived peptides* of the consequent 'overlapping sequences' using utmost skill, thought, and wisdom.

In usual practice, one must carry out the determination of a **protein** *via* 'thorough and complete sequencing' by at least *two altogether different types of partial hydrolysis so as to deduce* precisely the **structure of protein** using the known method of **overlapping sequences.**

Method 1: Let us take into consideration a typical **'hypothetical example'**, as stated below, where in the critical points of hydronlysis caused by one particualr enzyme are explicitly demarked by means of arrow (↑), thereby producing the respective **peptides I→VI.**

$$H_2N—A\,.B.\,C.\,D.\,E.\,F.\,G.................S.\,T.\,U.\,V.\,W.\,X.\,Y.\,Z.—COOH$$

```
          ↑        ↑  ↑  ↑   ↑
          1        2  3  4   5        6
```

Observations: These essentially include:

(1) **Peptides 1** and **6** may be assigned tentatively the respective strategic positions provided we know the **exact amino (–NH$_2$) and carboxyl (–COOH) end sequences**.

(2) So far, there exist no means and ways of positioning aptly the remaining **peptides II, III, IV, and V.**

Method 2 : Further hydrolysis of the protein may be duly accomplished by *'another enzyme'* to obtain definitely a **'different series of peptides'** by virtue of the fact that the enusuing action of this **specific enzyme** is crucially directed at a *different set of the peptide bonds* as given in Table .2.1.

Table: 2.1 Specificity of Certain Typical Proteolytic Enzymes.

S.No.	Proteolytic Enzyme	Source of Enzyme	Major sites of Action +	Minor Sites of Action
1	Carboxypeptidase–A	Pancreas	C–terminal of Tyr, Trp, Phe.	No action at Arg, Lys, Pro.
2	Carboxypeptidase–B	Pancreas	C–terminal Arg, Lys.	None
3	Carboxypeptidase Y	Yeast	C–terminal	Most terminals
4	Chymotrypsin	Pancreas	Trp, Phe, Tyr	Asn, His, Leu, Met
5	Elastase	Pancreas	Neutral aliphatic terminal	None
6	Leucine aminopeptidase	Kidney, intestinal mucosa etc.	NH$_2$–terminal bond of different residues.	No action at **X–Pro bond^{++}**.
7	Pepsin	Gastric mucosa	Trp, Phe, Tyr, Met, Leu	Different Acidic residues.
8	Papain	Papaya (Fruit)	Arg, Gly, Lys etc.	Broad specificity. No action at acidic residues
9	Subtilisin	*Bacillus subtilis*	Aliphatic and Aromatic residues	Various residues.
10	*Staphylocoeus protease*	Staphylococus	Glu.	Certain specific Asp bonds.

+ Except for the **carboxypeptidases,** the sites of action actually refer to the residues having the **carboxyl (=C=O) moiety** pertaining to the peptide bond *viz.,* trypsin catalyses the hydrolysis of **arginyl and lysyl linkages.**

++ × = Any other **'amino-acid residue'.**

Assuming that a particular **'peptide'** having the sequence B.C. D. E. F . . .., is duly obtained, it would certainly overlap the corresponding sequences of the ensuing **peptide I** and **peptide II,** thereby establishing this **sequence.** Exactly in a similar manner, one may appropriately deduce the entire **'amino acid sequence'.**

Example: Typical example of a hexapeptide [Gross formula: AB$_2$C$_2$D]: Most importantly the **Sanger's Method** (section 4.12.4.1.1) of **2, 3–dinitrophenyl (DNP)–amino acids** clearly indicated that W is the *N–terminal amino acid.* Thus, by the subsequent action of **carboxypeptidase, X,** and then Y are duly detached successively and satisfactorily. Nevertheless, it enables to write down the **probable formula of the hexapeptide** in the following fashion, depicting **within parentheses the amino acids** whose *sequence* still remains to be established:

W [XYZ] YX

Several *partial hydrolyses* are carried out skillfully and meticulously as given under:

    ❏ **enzymatically** *i.e.,* making use of certain **proteolytic enzymes** *viz.,* chymotrypsine, pepsine, trypsine etc., (see Table : 2.1.), that eventually hydrolyze in a preferential manner some **peptide linkage*,** and

❑ **chemically** *i.e.,* by decreasing the prevailing **acid strength, temperature** or **duration of heating.**

Thus, one may categorically lay hands on to some **dipeptides** and **tripeptides** that could be fractionated effectively by a plethora of well-known **analytical techniques***, namely.

- **Capillary Electrophoresis (CE),**
- **Counter-Current Distribution (CCD), and**
- **Ion–Exchange Chromatography (IEC)**

Nevertheless, the deisired **sequence of amino acids** may be determined quite easily and conveniently from the following description from these ensuing **dipeptides** and **tripeptides** one may judiciously afford a **logical deduction by comparison** of the ultimate '**sequence of the hexapeptide':**

$$
\begin{array}{ll}
\textbf{Dipeptides Derived} & \left\{
\begin{array}{l}
\text{W—Y} \\
\text{Y—Z} \\
\text{Z—X} \\
\text{X—Y} \\
\text{Y—X}
\end{array}
\right. \\[2em]
\textbf{Tripeptides Derived} & \left\{
\begin{array}{l}
\text{Y—Z—X} \\
\text{Z—X—Y}
\end{array}
\right. \\[1em]
\text{"Sequence Hexapeptide"} & \text{W—Y—Z—X—Y—X}
\end{array}
$$

Fig: 2.7 illutrates the complete 'amino acid sequence' and the strategic location of the specific 'disulphide linkages' present in 'porcine' *i.e.,* the *pancreatic secretory trypsin inhibitor 1.*

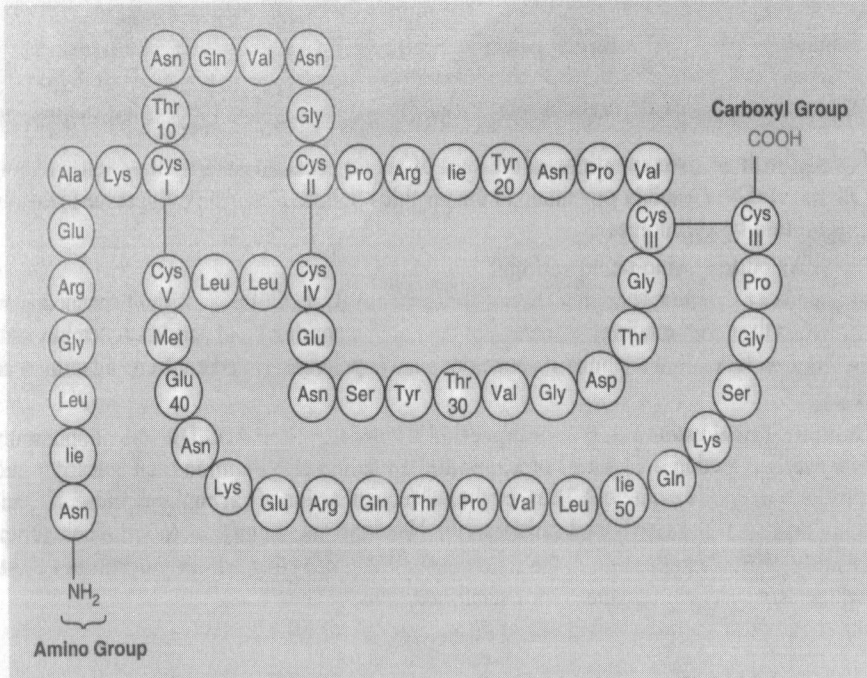

Fig: 2.7: Complete Amino-Acid Sequence of Disulphide Linkages in Porcine
[Adapted from : Guy O *et al. J.Biol Chem,* 246: 7740, 1971]

* Kar, A; **Pharmaceutical Analysis,** Vol II, CBS Publishers & Distributers Pvt. Ltd., New Delhi, 2009.

4.1.2.7. Synthesis of α-Amino Acids

The **Synthesis of α-amino acids** can be accomplished by almost *nineteen* different highly specific procedures as described under in a sequential manner. An attempt has been duly made to provide the typical synthesis of certain **amino acids** and also their **'mechanisms'** wherever necessary and applicable:

4.1.2.7.1. Acylamidomalonic Ester Synthesis

The **acylamidomalonic ester synthesis** involves the use of an **acyl–or aroyl-amido** structural analogue that may be prepared easily from the **malonic ester** in the following manner:

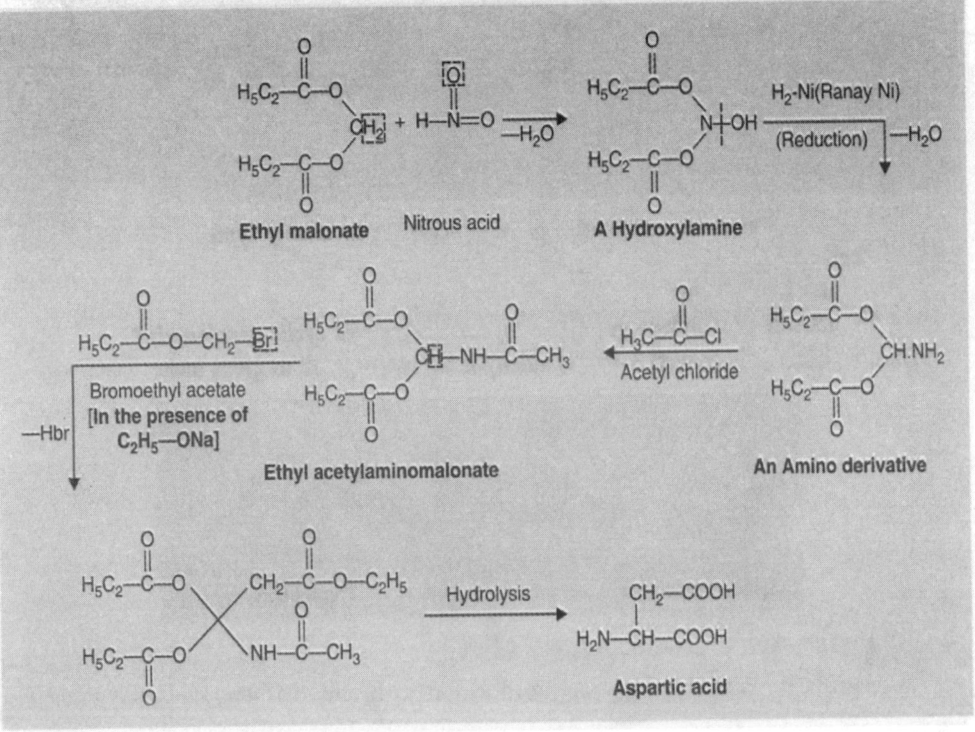

Ethyl malonate + Nitrous acid $\xrightarrow{-H_2O}$ A Hydroxylamine $\xrightarrow[\text{(Reduction)}]{H_2\text{-Ni(Ranay Ni)}}\Big\downarrow -H_2O$

An Amino derivative $\xleftarrow{\substack{H_3C-C-Cl \\ \text{Acetyl chloride}}}$ Ethyl acetylaminomalonate

Bromoethyl acetate [In the presence of C_2H_5—ONa] $-HBr$

Ethyl acetylaminomalonate $\xrightarrow{\text{Hydrolysis}}$ Aspartic acid

$$CH_2-COOH$$
$$H_2N-CH-COOH$$

Nevertheless, the skillful extension of the aforesaid **acylamidomalonic ester synthesis** could lead to the formation of a host of useful α-*amino acids*, namely: **glutamic acid, leucine, lysine, methionine, ornithine,** and **serine** effectively.

Greenstein and Wintz (1961)* synthesized **tryptophan** by reacting **gramine methosulphate** and **benzamidomalonic ester** in the presence of **sodium ethoxide** (freshly prepared) to obtain the corresponding **benzamido malonic diethyl ester of gramine (I).**

The resulting product (I) on being subjected to hydrolysis first with NaOH and then with HCl yields the disired **tryptophan.**

Thus, we may have the following reactions:

* Greenstein JP and Wintz M : **Chemistry of the Amino Acids,** Vol. 1–3, John Wiley & Sons, New York, pp, 2316–2347, 1961.

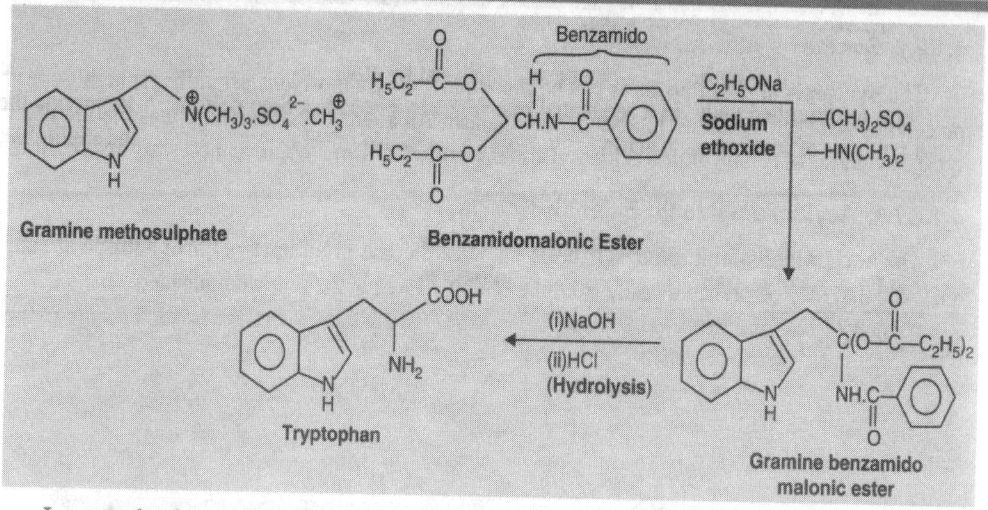

Importantly, the aromatic containing **α-amino acids** are invariably synthesized by **Perkin–type reactions** between:

- **aromatic aldehydes**, and
- **activated methylene groups** of *'hydantoin'/related cyclic compounds.***

Thus, we may easily convert **3–indole aldehyde** to **'tryptophan'** as given under:

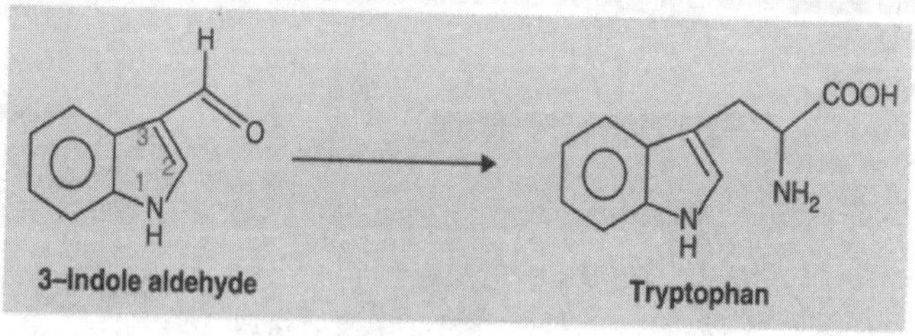

4.1.2.7.2. Amination of α-Halogenated Acids

In this particular instance, an **α-halogenated acid** *viz.*, **α-Bromo, α-Chloro acids,** duly accomplished by:

- **direct halogenation of carboxylic acids,** or
- **bromination of corresponding malonic acids,**

may be aminated easily and conveniently by alcoholic, aqueous, or concentrated ammonia (NH_4OH). It has been observed largely that the **'bromo acids'** do exhibit a market and pronounced **'reactivety'** *vis-a-vis* other halogenated acids.

Thus, we may have:

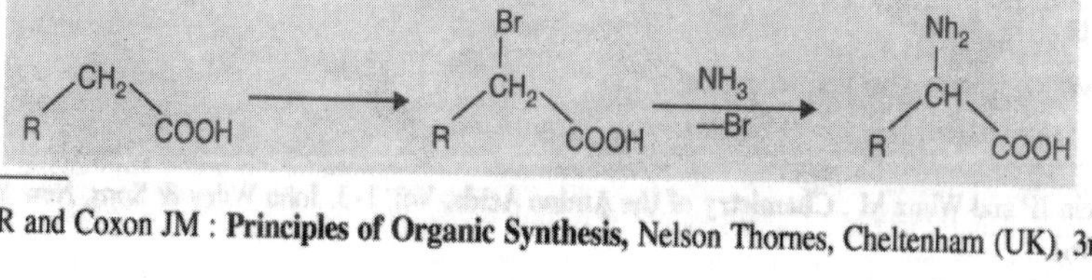

---

* Norman R and Coxon JM : **Principles of Organic Synthesis,** Nelson Thornes, Cheltenham (UK), 3rd ed,. p. 330, 2005.

Note: As the C–atom is chiral in nature, it produces an equal mixture of L– and D– isomers or a 'racemate'.

In a broader perspective the *Hell–Volhard–Zelinsky method* is invariably used only for first step (or the *'initial step'*).

The various *procedural steps* essentially include:

(1) Treatment of the acid with Br_2 in the presence of a small amount of **phosphorus (P)** to produce **phosphorus tribromide (PBr$_3$).**

(2) Resulting **(PBr$_3$)** helps to convert the acid into its corresponding **acid bromide,** which in turn undergoes **(electrophlic)** bromination at the specific α–position *via* its *'enol–tautomer'*

(3) Interestingly, the α–bromo acid bromide, thus generated, does exchange with more of the acid to yield α–bromo acid plus more of *acid bromide* to cause further **successive bromination** effectively.

Thus, we may have the following reaction:

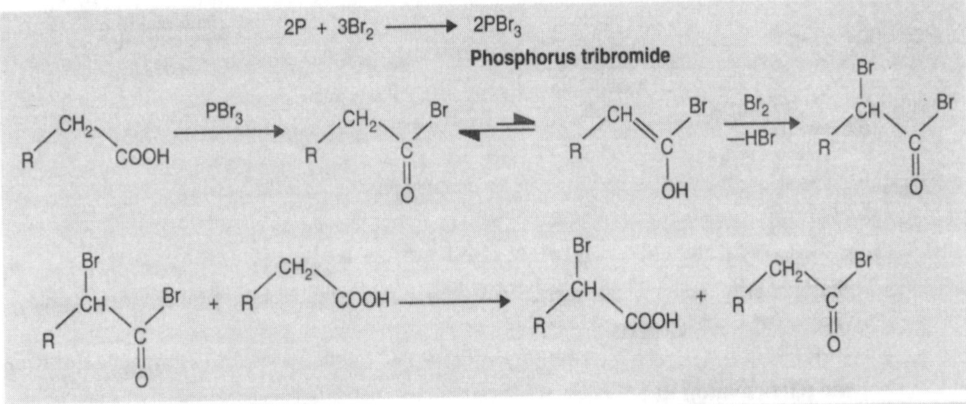

Note:However, the conversion of the respective α–bromo acid into the corresponding *α–amino acid* may be obtained by making use of an *'excess of ammonia'*; but, one may still accomplish *better yields* and *purer products* by the help of *Gabriel's phthalidomide procedure* (see section 4.1.2.7.8)

4.1.2.7.3. Albertson Method

In the **Albertson method** the *acylamido structural analogues of malonic ester* have been employed to carry out the synthesis of **amino acids.** The most commonly used **acylamido malonicester** derivative being the **'acetamidomalonate'** duly prepared from the respective **dielthyl malonate.**

Procedure: The various cardival steps involved are as stated under:

(1) The starting material, **diethyl malonate,** is skilfully **'nitrostated'** by reacting with HNO_2 *i.e.,* nitrous acid.

(2) Indeed the aforesaid *mechanism of action* may be expatiated by the fact that generally most commonly enolizable **β–dicarbonyl compounds** invariably occurs due to the critical reaction of an adequately *'resonance–stabilized carbanion'* and the *'conjugated acid obtained from nitrous acid'.*

It is, however, pertinent to state here that the *'nitroso compound'* although duly formed as the first and foremost step undergoes critical *'tautomerisom'* with particular reference to the respective *'more stable oxime'* thereby yielding the product known as: *diethyl oximinomalonate.*

(3) The resulting product obtained in (2) above is carefully hydrogenated over a catalyst to give rise to the formation of the desired 'amino compound' that is subsequently subjected to **acetylation** by means of *acetic anhydride* to obtain **the acetylated amino malonic compound (I)**

The various reactions are described as under:

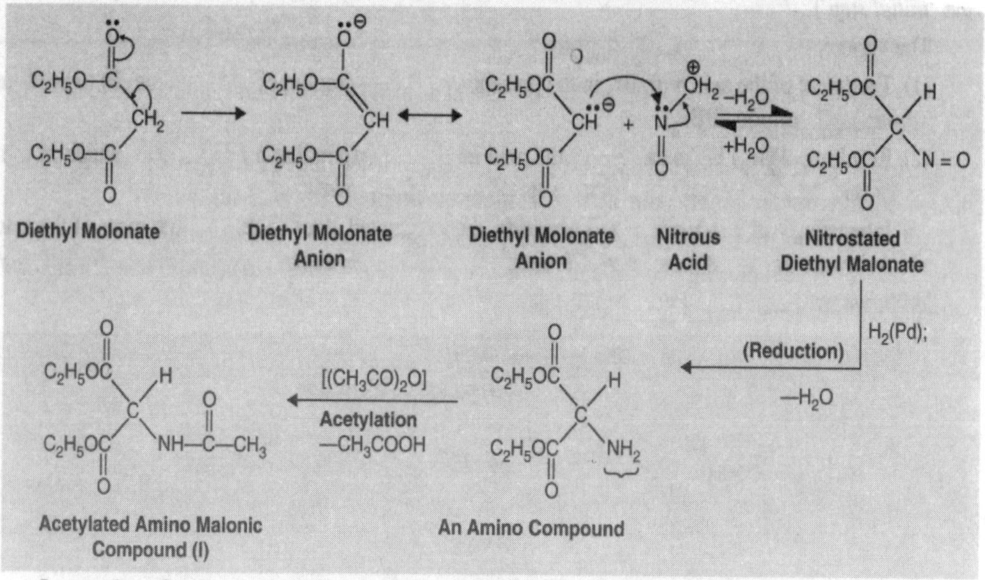

Diethyl Molonate   **Diethyl Molonate Anion**   **Diethyl Molonate Anion**   **Nitrous Acid**   **Nitrostated Diethyl Malonate**

$H_2(Pd);$
(Reduction)
$-H_2O$

$[(CH_3CO)_2O]$
Acetylation
$-CH_3COOH$

Acetylated Amino Malonic Compound (I)   **An Amino Compound**

It can be observed that the **acetylated ethyl amino malonic compound (I)** *i.e.,* the acetylaminomalonic ethyl ester retains meticulously these *two* vital characteristic features, namely:

- one 'acidic α-hydrogen atom', and
- ability to form a typical 'resonance–stablized carbanion', which will render an **immediate participation in :**
  - ❏ nucleophilic displacement reaction, or
  - ❏ addition reaction.

Thus, we may have the following reaction:

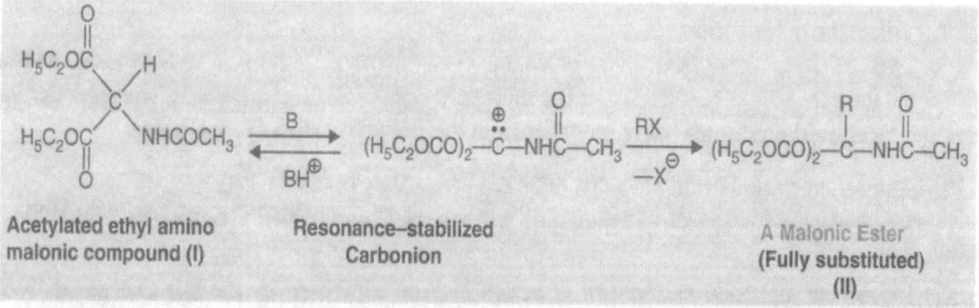

Acetylated ethyl amino malonic compound (I)   **Resonance–stabilized Carbonion**   **A Malonic Ester (Fully substituted) (II)**

Therefore, the resulting **fully substituted malonic ester (II)**, obtained above as the resulting product, gets readily converted to the respective *α–amino acid* by means of *hydrolysis* in an **acidic medium.**

Mechanism: The α–amino acid is duly obtained by the careful 'decarboxylation' of one of the *two* carboxyl (–COOH) groups present in the **fully substituted malonic ester (II)** in an *acidic environment, after its* **acidic hydrolysis.** The resulting **decarboxylated product (III)** upon *further* **acidic hydrolysis** yields the desired α–amino acid as given below:

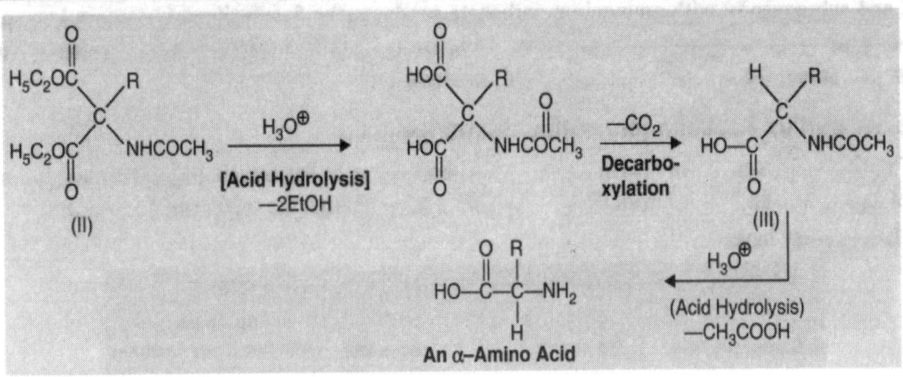

An α–Amino Acid

4.1.2.7.4. Bucherer Hydantoin Synthesis

Bucherer (1934) modified the previously discussed method under section 4.12.7.1, whereby an 'oxo compound' is duly converted to an 'amino acid' *via* the following cardinal sequential steps, namely:

- treatment with **sodium cyanide (NaCN)/HCN** to obtain the respective cyano derivative (I),
- the resulting **compound (I)** is reacted with **ammonium carbonate $[(NH_4)_2CO_3]$** to produce the desired **5–substituted hydantoin (II)**, and
- **hydrolysis** of compound (II) with *dilute mineral acid* (**HCl**) yields the α–amino acid, **alanine (III)**.

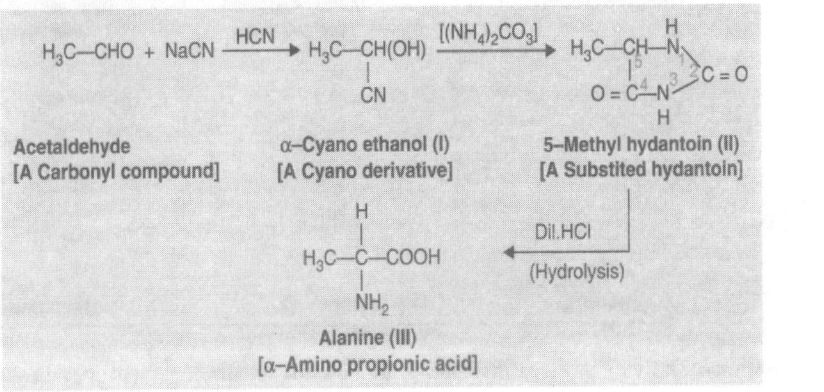

Methionine (a **S-containing** α-amino acid) may also be synthesized in the same manner:

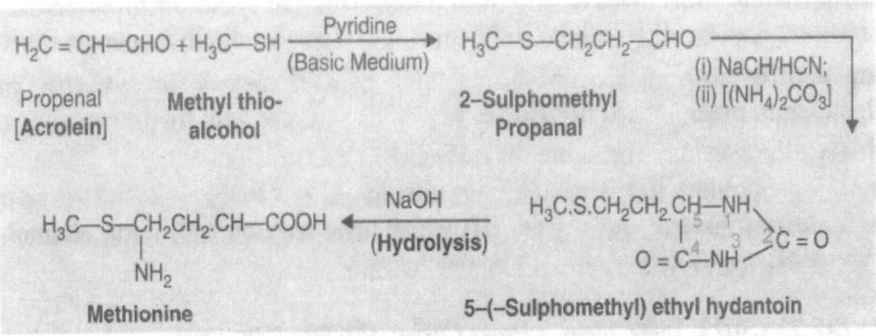

Acrolein (or propenal), an unsaturated aliphatic aldehyde is reacted with methylthio alcohol in the presence of pyridine (a base) to yield 2–sulphomethyl propanal, which on being treated with NaCN/

HCN and subsequently with ammonium carbonate produces the 5–substitued hydantoin known as **5–(2–sulphomethyl) ethyl hydantoin.** The resulting compound on hydrolysis in an alkaline medium yields the desired **S–containing α–amino acid, methionine.**

4.1.2.7.5. Curtius Reaction [or Curtius Rearrangement]*

The **Curtius reaction** (or **Curtius rearrangement**) refers to the actual stepwise conversion of a **carboxylic acid** to an 'amine' with *one lower carbon unit,* through the **azide** and **isocyanate.**

Thus, we may have:

$$R.COOH \longrightarrow R.C-NH_2 \longrightarrow R.N = C = O \longrightarrow R.NH_2$$

| A Carboxylic acid | An Amide | An Isocyanate | An Amine (with one lower C–atom) |

In actual practice, the **Curtius reaction** may produce an array of important **amino acids** *viz.,* **alanine, glycine, phenylalanine,** and **valine.**

Alternatively, the respective **K–salt** fo the specific **malonic acid monoazide** prepared meticulously as per the following sequential reaction:

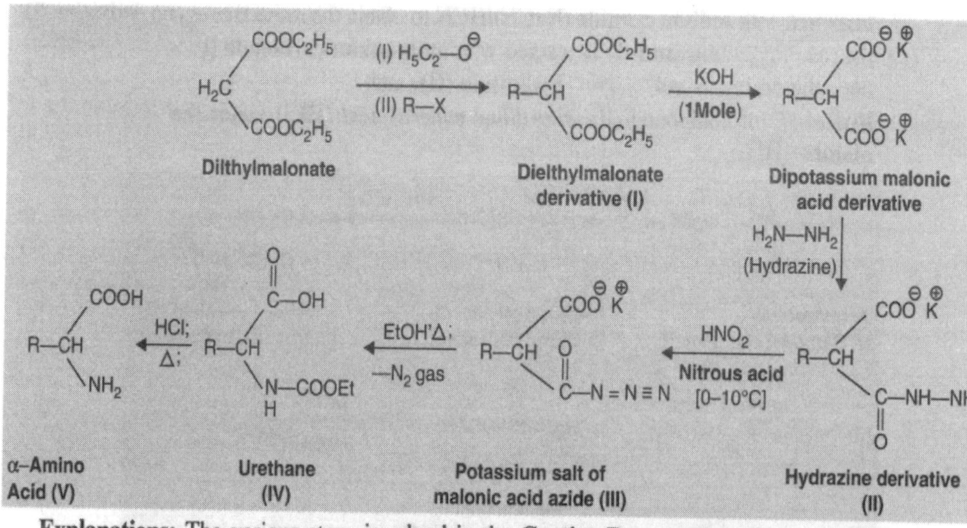

Explanations: The various steps involved in the **Curtius Rearrangement** may be explained as under:

(1) **Diethylmalonate** when made to react with *sodium ethoxide* yields *diethylmalonate,* which on subsequent treatment with an **alkyl halide (R–X)** gives the respective **diethylmalonate derivative (I).**

(2) **Compound (I)** upon interaction with **one mole of KOH** (*i.e.,* in *stoichiometric proportion*) yields the **dipotassium malonic acid derivative,** which on treatment with **hydrazine** gives rise to the formation of the corresponding **hydrazine derivative (II).**

(3) The resulting product **(II)** is treated with **nitrous acid (HNO$_2$)**** at **0–10°C** to obtain the **monopotassium salt of malonic acid azide (III)** which upon reaction with warm **ethanol** produces **urethane (IV)** and **N$_2$–gas** is duly released from the reaction mixture.

---

* Curtius T : *Ber.* **23** : 3023, 1890; *idem, J Prakt Chem;,* [2], **50** : 275, 1894.

** Nitrous acid (HNO$_2$) being **highly volatile** and hence 'unstable' is usually prepared in the reaction mixture itself by dilute HCl and NaNO$_2$.

(4) Ultimately, the resulting **product (IV)** undergoes hydrolysis with slight heating to produce the desired **α–amino acid (V)**.

Tyrosine: It can be prepared by the aforesaid **Curtius Reaction :**

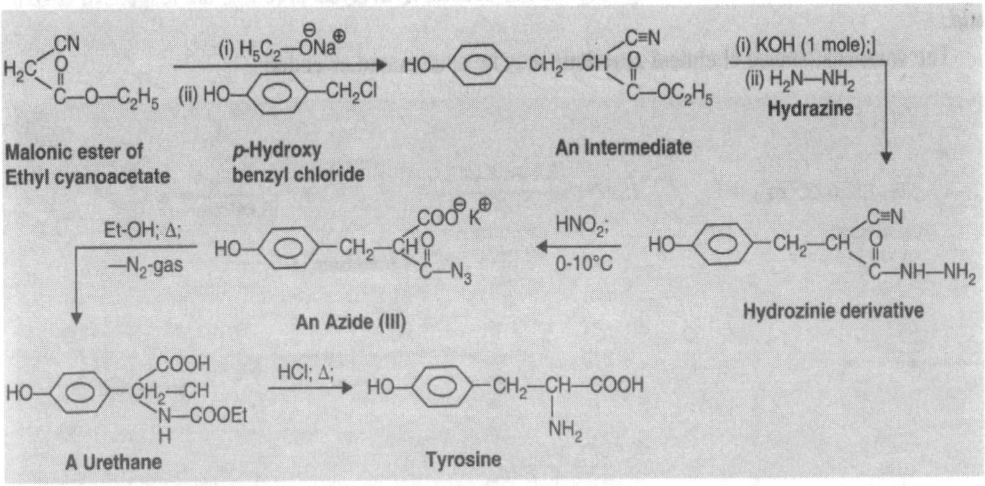

**Malonic ester of
Ethyl cyanoacetate**

**p-Hydroxy
benzyl chloride**

An intermediate

Hydrazine

An Azide (III)

Hydrozinie derivative

A Urethane

Tyrosine

Note: *Phenylalanine* can also be prepared by using this method.

4.1.2.7.6. Darapsky Synthesis

Darapsky synthesis essentially comprises the following steps sequentially to obtain **α–amino acids:**

- Condensation of an **aldehyde** and **malonic ester of ethyl cyanoacetate**
- simultaneous reduction with **H$_2$–Ni** to obtain a **malonic ester of ethyl cyanocetate having one more C-atom,**
- treatment with hydrazine (H$_2$N–NH$_2$),
- reaction with nitrous acid (HNO$_2$),
- **Curtius reaction,** and
- acidic hydrolysis (HCl) with heating.

Thus we may have :

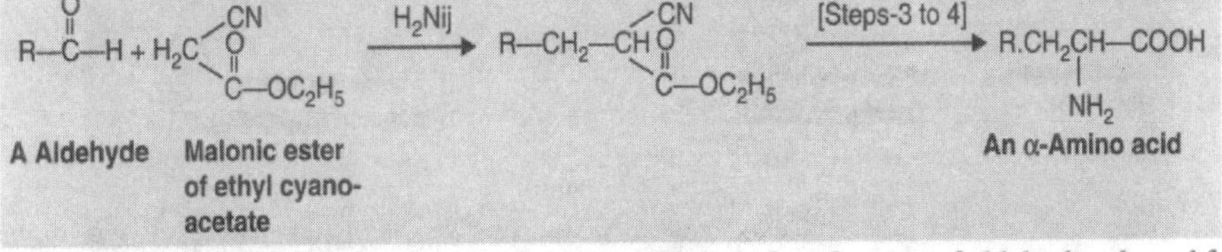

A Aldehyde **Malonic ester
of ethyl cyano-
acetate**

[Steps-3 to 4]

An α-Amino acid

In fact, we may synthesize a variety of **α–amino acids** based on the type of aldehyde selected for the **Darapsky synthesis.**

4.1.2.7.7. Erlenmeyer–Plöchl Azlactone and Amino Acid Synthesis*

The unique formation of **'azlactones'** by the specific **intramolecular condensation** of *acylglycines* in the critical presence of *acetic anhydride* broadly refers to the **Erlenmeyer–Plöchl** azlactone

---

* Erlenmeyer E: *Ann*, **275** : 1, 1893; Plöchl J, *Ber*, **17** : 1616, 1884.

vis-a-via **amino acid synthesis.** Summararily, the ensuing reaction of **azlactones** with the **carbonyl compounds** immediately followed by hydrolysis to the corresponding unsaturated **α–acylamino acid,** and the subsequent **reduction** gives rise to the formation of an **α–amino acid.** However, the resulting product (*i.e.*, **α–amino acid**) on being subjected to **drastic hydrolysis** produces the respective **α–oxo acid.**

The various aforesaid **chemical reactions** may be enumerated as under:

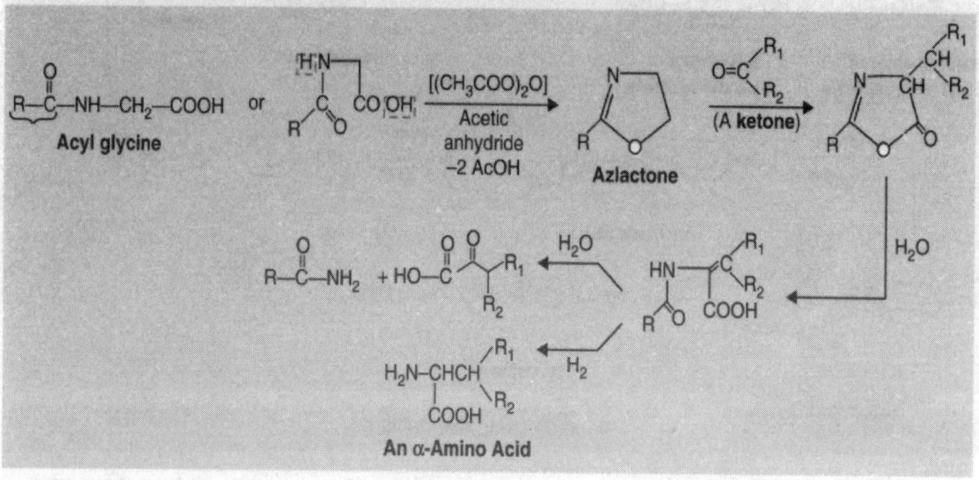

Nevertheless, it is pertinent to state here that the **Erlenmeyer-Plöchl azlactone synthesis** may be duly affected by boiling the mixture consisting of :

- an **aromatic aldehyde** *eg.,* benzaldehyde, and
- a **benzoylglycine** (or **hippuric acid**),

dissolved in *acetic anhydride* along with an effective **condensing agent,** such as : **potassium carbonate, sodium acetate.**

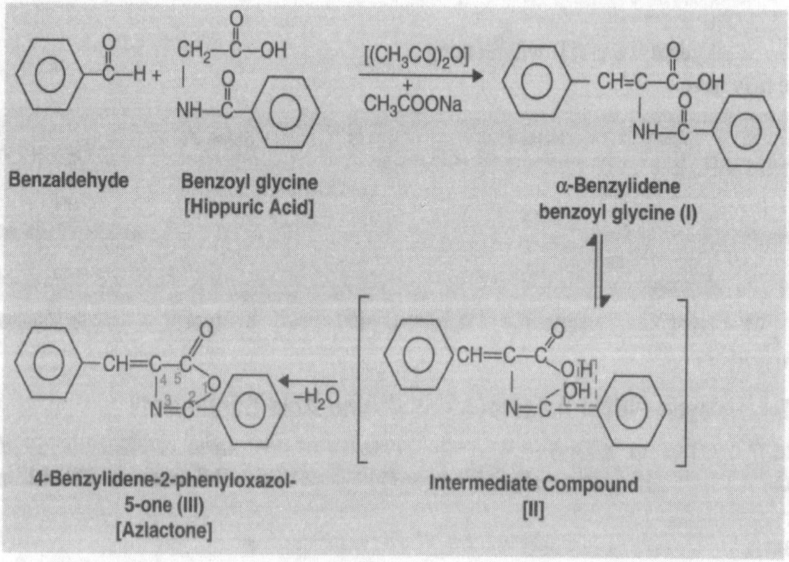

Benzaldehyde   Benzoyl glycine [Hippuric Acid]   α-Benzylidene benzoyl glycine (I)

4-Benzylidene-2-phenyloxazol-5-one (III) [Azlactone]   Intermediate Compound [II]

Example: (+) and (–) Tyrosine: Both **D–and L–Tyrosine** may be prepared by the interaction of benzaldehyde and benzoyl glycine in acetic anhydride and sodium acetate to yield **α–benzylidene benzoyl glycine (I).** Subsequently, (I) undergoes transformation to form and **intermediate compound (II),** which eventually loses a molecule of water to produce a cyclic compound known as **4–benzylidence–2–phenyl–oxacol–5–one (III.)**

Importantly, one may take cognizance of the underlying fact that the careful **partial hydrolysis** of azlactones *e.g.,* (III), with hot NaOH [1% (w/v)] by means of:

- **Catalytic hydrogenation,** or
- **Soduim–amalgam (Na–Hg),**

invariably, gives rise to the formation of the **'free amino acid'** after due **hydrolysis** either in an *acidic* or *alkaline medium.*

Thus, we may have:

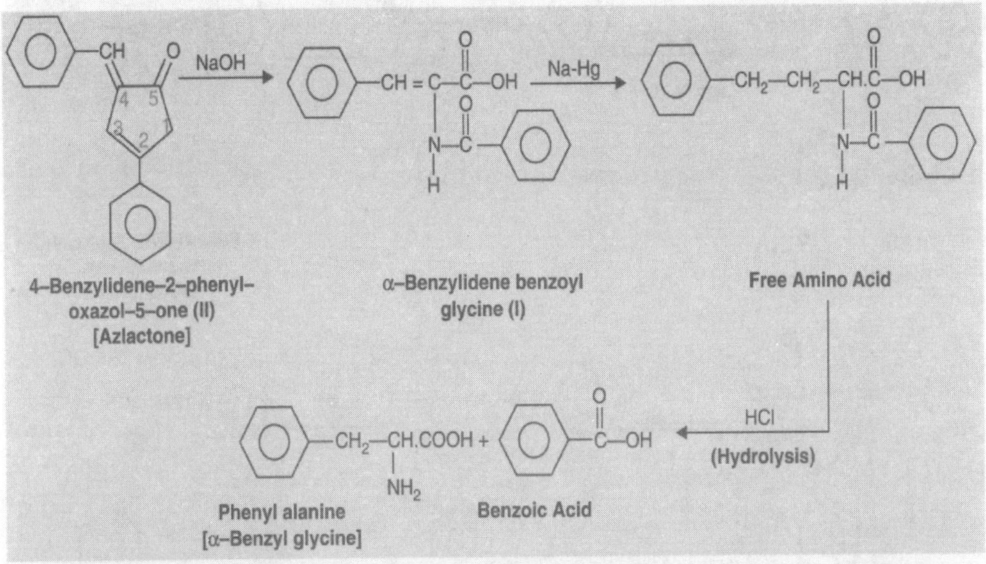

4–Benzylidene–2–phenyl–
oxazol–5–one (II)
[Azlactone]

α–Benzylidene benzoyl
glycine (I)

Free Amino Acid

Phenyl alanine
[α–Benzyl glycine]

Benzoic Acid

HCl
(Hydrolysis)

Note: The *Erlenmeyer–Plöchl azlactone synthesis* provides an easy and convenient means of preparing a number of *amino acids,* such as : phemylalanine, thyroxime, tryptophan, and tyrosine.

Mechanism : Based on the various evidences cited duly in the literatures one may have a closer look at the *proposed mechanism* for the **'critcal formation'** and the **'ring–opening phenomenon in azlactone'** as illustrated explicitly below:

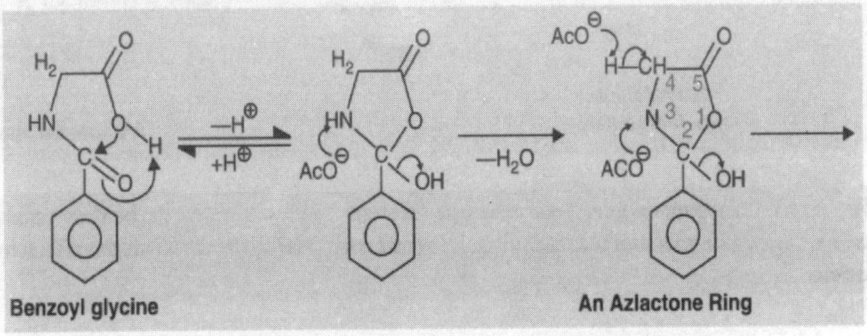

Benzoyl glycine

An Azlactone Ring

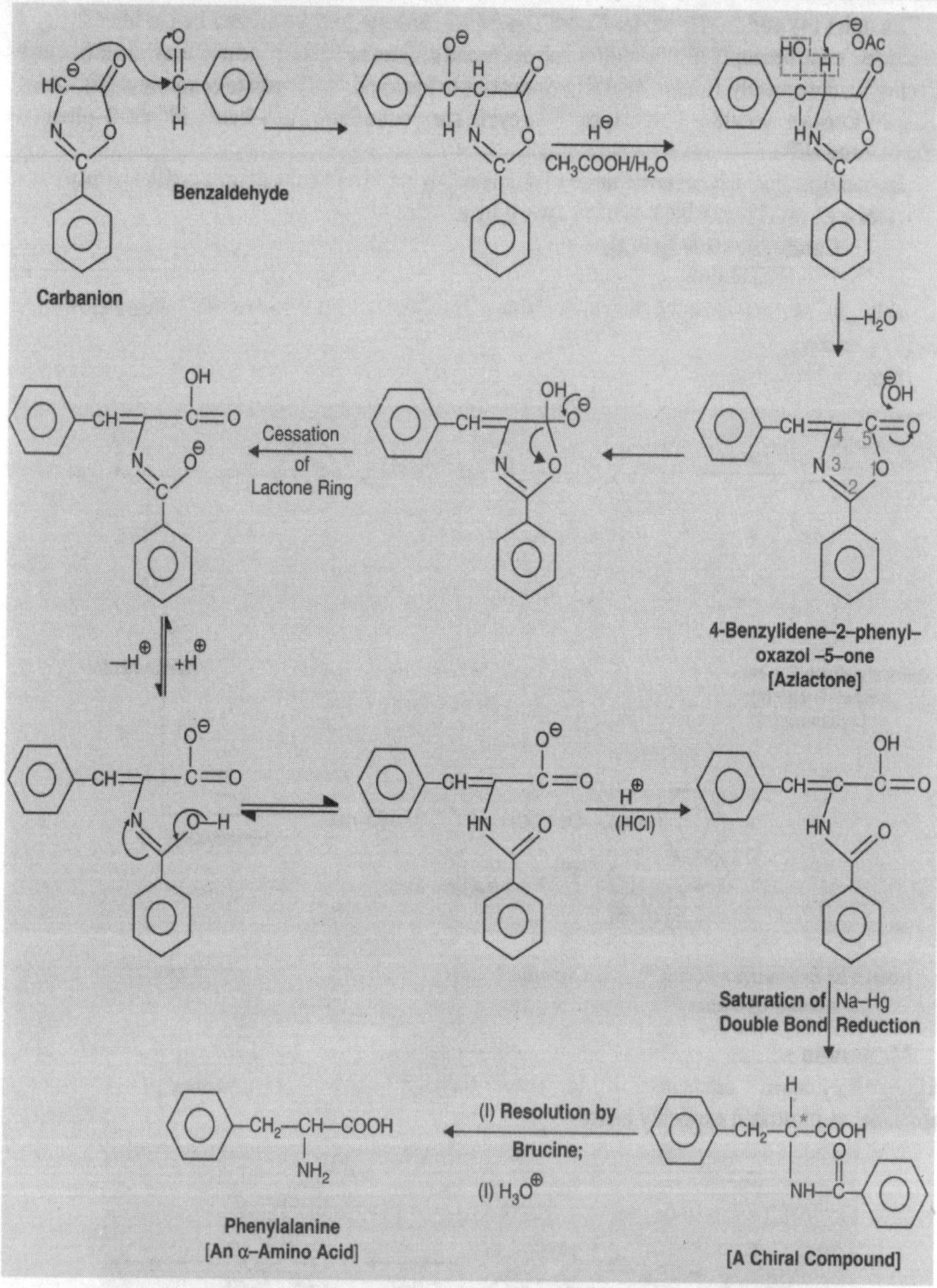
Benzaldehyde
Carbanion
$H^{\ominus}$
CH_3COOH/H_2O
$-H_2O$
Cessation of Lactone Ring
4-Benzylidene-2-phenyl-oxazol -5-one [Azlactone]
$-H^{\oplus}$
$+H^{\oplus}$
$H^{\oplus}$ (HCl)
Saturaticn of Double Bond
Na-Hg Reduction
(I) Resolution by Brucine;
(I) $H_3O^{\oplus}$
Phenylalanine [An α-Amino Acid]
[A Chiral Compound]

Note: Being a *'chiral compound'*, it can form two different isomers which may be duly resolved by using *brucine* (an *alkaloid*) that may be recovered completely after causing the effective separation.

4.1.2.7.8. Gabriel Phthalimide Synthesis

The commonly encountered problem that arises critically in usual attempts to prepare 'primary amines' by the known *monoalkylation of ammonia** is duly circumvented (overcome) by a newly devised procedure by Gabriel (1887)**.

In fact, the **Gabriel phthalimide synthesis** is solely based upon the fact that the 'phthalimide entity', essentially possessing an 'acidic N–H group' specifically reacts with a *base* to produce a N–containing anion which, being a rather **strong nucleotide,** critically displaces upon the available *alkyl halides.* Thus, the **hydrolysis** of the resulting compound in the presence of **alkali** yields the desired **primary amine.**

Thus, we may have the following reaction:

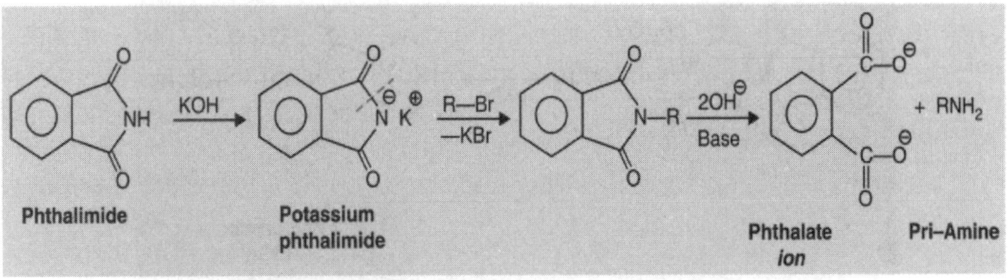

Phthalimide   Potassium phthalimide   Phthalate ion   Pri–Amine

Interestingly, the **Gabriel Phthalimide synthesis** affords definitely a better yield. In other words, an α–halogenated acid ester is carefully treated with **potassium phthalimide (I)** to give rise to the formation of a **substituted phthalimide ester (II).** The resulting ester (II) critically yields the desired 'amino acid[(III) by one of these methods, namely:

- **drastic hydrolysis,**
- subjecting to *two* sequential reactions stepwise:
  - ❑ treatment with **hydrazine** or **4M NaOH,**
  - ❑ hydrolysis in an **acidic environment.**

Thus, we may have:

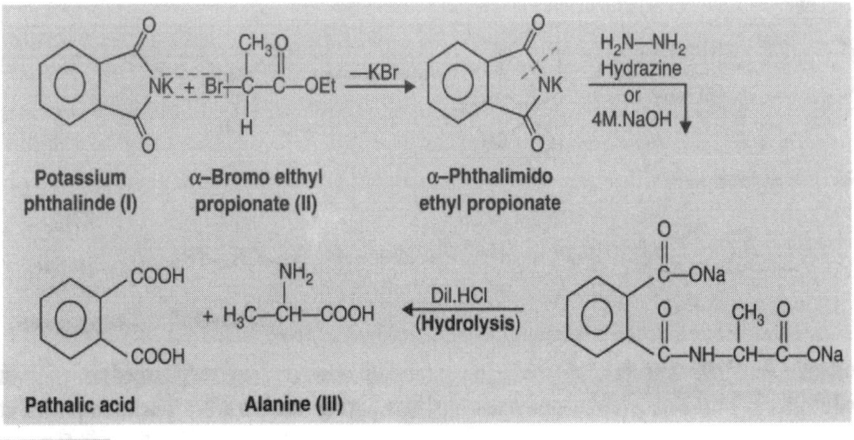

Potassium phthalinde (I)   α–Bromo elthyl propionate (II)   α–Phthalimido ethyl propionate

Pathalic acid   Alanine (III)

---

* Because, further *'alkylation'* in certain degree is quite unavoidable.

** Gabriel S : *Ber.,* **20** : 2224, 1887.

Mechanism: The proposed mechanism of the **Gabriel phthalimide synthesis** may be explained as under:

- a lone pair of electrons residing on the **phltalimide N-atom** attacks the α–C–atom of the respective bromo ethyl propionate (*i.e.,* the **ester**) thereby eliminating a mole of KBr and forming the corresponding α–**phthalimido ethyl ester,**
- the resulting ester undergoes hydrolysis in an acidic enviornment to produce the desired **amino acid** (*alanine*) and **phthalimic acid** (a *dicarboxylic aromatic acid*) plus a mole of **ethanol eliminated** *i.e.,* in all **3 moles of H_2O** are innolved duly.

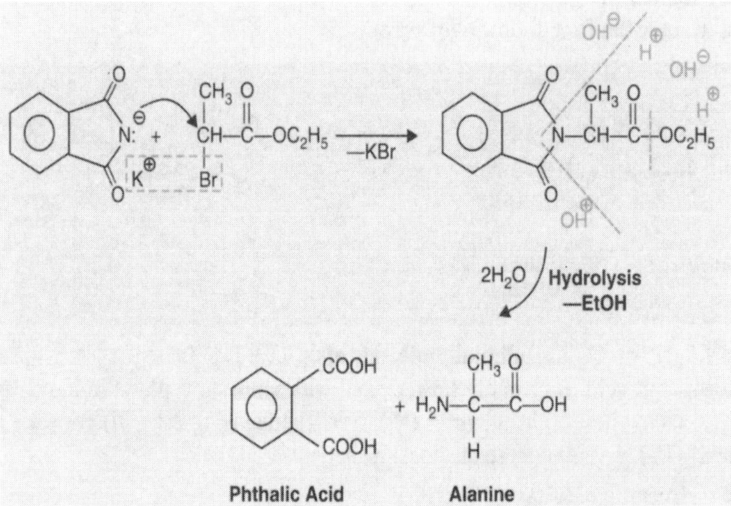

Phthalic Acid   **Alanine**

4.1.2.7.9. Hofmann Degradation Method [or Exhaustive Methylation Method]

Hofmann (1881)* suggested the critical formation of an **'olefin'** and a **'tertiary amine'** by the *pyrolysis* of a **quaternary ammonium hydroxide** as given under:

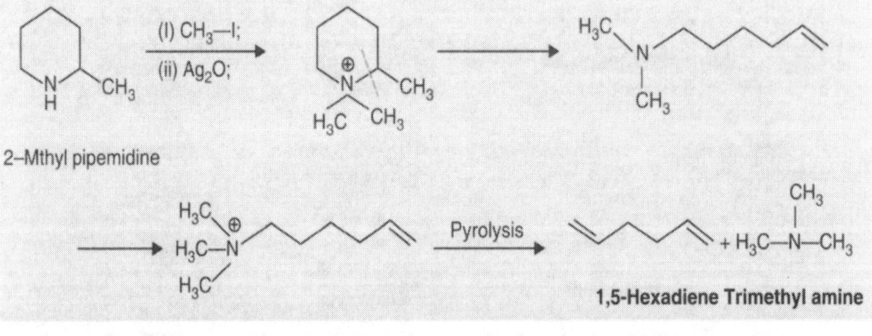

2–Mthyl pipemidine

1,5-Hexadiene Trimethyl amine

Importantly, the **Hoffmann degradation** of the α–**methyl malonic ethyl ester amide** also essentially designated one of the commonly employed methods of preparing **'amino acids',** as exemplified under:

---

* Hoffmann AW : *Ber,* **14** : 659 (1881)

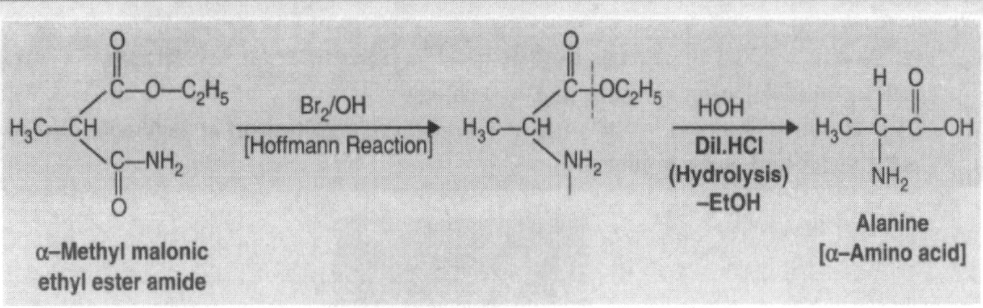

α–Methyl malonic
ethyl ester amide

Alanine
[α–Amino acid]

4.1.2.7.10. Hydantoin Synthesis

It is worthwhile to state here that the **'hydantoin synthesis'** is found to be almost analogous to the **Erlenmeyer Plöchl Azlactone Synthesis** (disucussed under section 4.1.2.7.7.). Thus, in the **'hydantoin synthesis'** the so called **aromatic aldehydes** are invariably condensed with **hydantoin.** The resulting product so obtained is caused to reduction with **sodium amalgam (Na–Hg)** *or* **ammonium hydrogen sulphide** (NH₄HS) followed by hydrolysis gives rise to the formation of an α–amino acid.

We may obtain **indole–3–aldehyde** from **indole** as stated under:

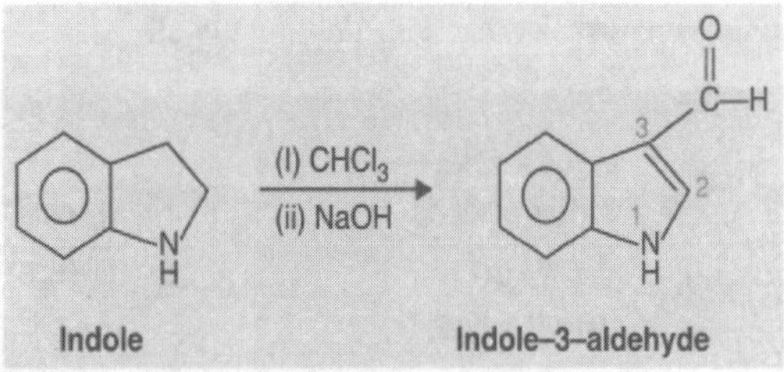

Indole   Indole–3–aldehyde

Tryptophan, an *aromatic containing α–amino acid,* is usually prepared by the **Perkin–type reactions** taking place tetween an *aromatic aldehyde* and the *'activated methylene moiety'* present in the *hydantoin residue.* The reduction with sodium–amalgam (Na–Hg) saturates the methylene linkage, and subsequent hydrolysis with dilute mineral acid (HCl) yields **tryptophan** and a molecule each of NH₃ ↑ and CO₂↑,

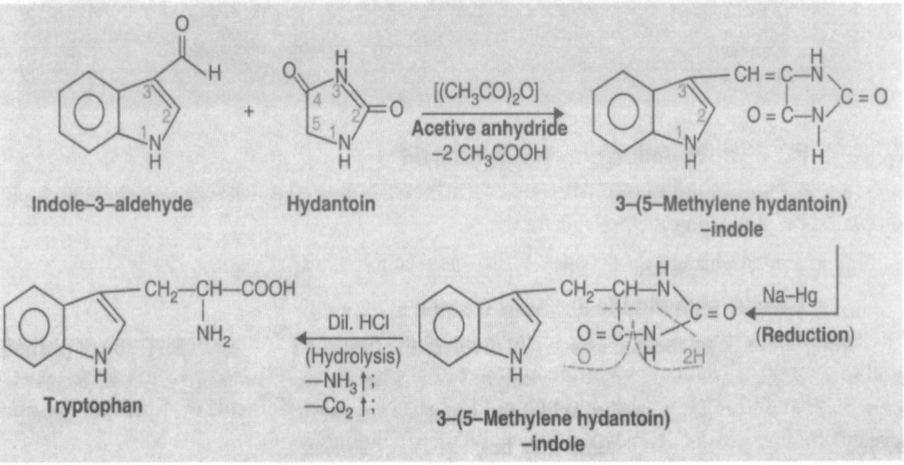

Indole–3–aldehyde   Hydantoin   3–(5–Methylene hydantoin)
                                                                      –indole

Tryptophan   Dil. HCl
                        (Hydrolysis)
                        —NH₃↑;
                        —Co₂↑;   3–(5–Methylene hydantoin)
                                            –indole

                                                           Na–Hg
                                                           (Reduction)

which escape as gases, with a startegical cleavage of the **hydantoin residue** (as shown by the *'dotted line'*). Thus, we may have the above sequential steps engaged in the synthesis of *'tryptophan'* starting from **indole–3–aldehyde:**

> Note: (1) *'Hydantoin synthesis'* is effectively used for the preparation of an array of amino acids, such as: phenylalanine, tyrosine, and methionine.
>
> (2) It has been observed that the use of *1–acetylhydantoin* instead of *hydantoin* usually gives better yields and purer products.

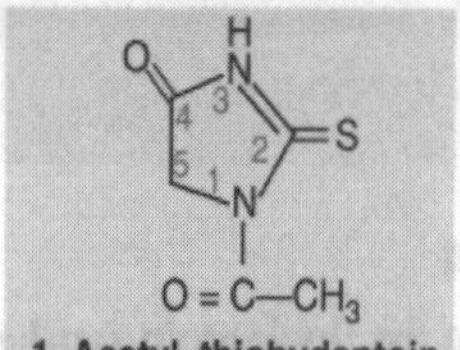

1–Acetyl–thiohydantoin

4.1.2.7.11. Malonic Ester Synthesis

In fact, the **malonic ester synthesis** it is both feasible and possible to prepare an **'α–halogeno acid'**, which may be judiciously reacted with **ammonia** to yield a host of common α–amino acids for instance : **proline, phenylalanine, leueine, isoleucin,** and **methionine**.

We may have the following explict reactions duly involved:

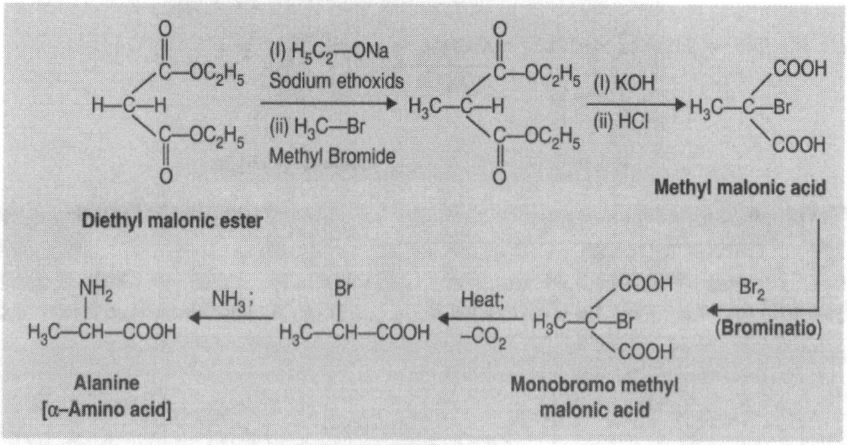

4.1.2.7.12. Phthalimidomalonic Ester Synthesis

In fact, the **'phthalimidomalonic ester synthesis'** is indeed a marvellous technique of the already discussed *two* following methods, namely:

- **acylamidomalonic ester synthesis** (section 4.1.2.7.1), and
- **Gabriel phthalimide synthesis** (section 4.1.2.7.8.)

However, the present method essentially involves the first and foremost preparation of **phthalimido malonic ester,** which being carefully hydrolyzed to produe the respective **amino acid.** Interestingly, a host of vital and important **amion acids,** for instance : **aspartic acid, cystine, lysine, methionine, phenylalanine, proline,** and **tyrosine** may be prepared effectively.

Following are the typical synthesis of *three* amino acids namely, **Cystine, Phenylalanine,** and **Proline** along with their brief course of reaction being undertaken :

[A]. Cystine: The various steps involved in the synthesis of **cystine** by this method are as follows:

(1) **Diethyl malonic acid ester phthalimide (I)** on being treated with sodium ethoxide (freshly prepared) and benzylethiomethyl chloride yields **phthalimido diethyl malonic acid ester benzylthiomethyl (II)** with the elimination of a mole of HCl.

(2) The resulting product **(II)** upon treatment with NaOH and HCl sequentially produces α–thiobenzyl alanine **(III)**, which upon reduction with liquid *ammonia and Na* yields the racemic mixture of **cysteine (IV).**

(3) The **product (IV)** under oxidation with **air** yields the desired (±)– **cystine (IV).**

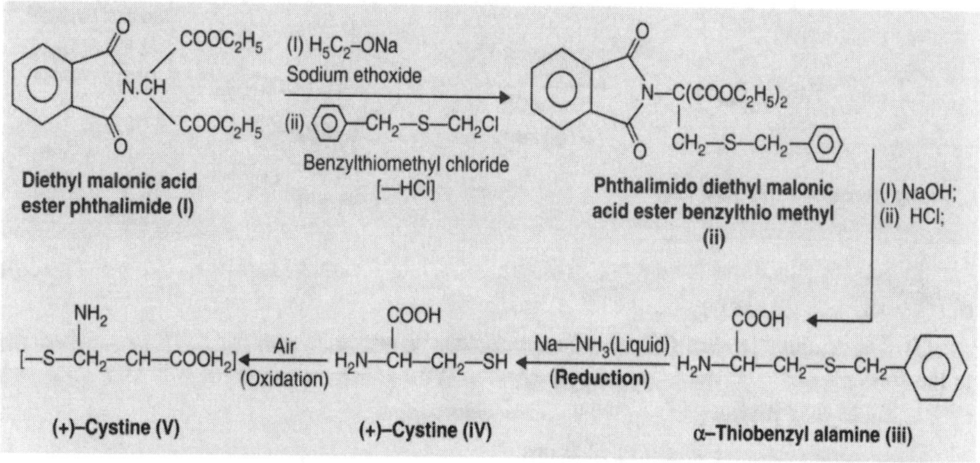

Diethyl malonic acid ester phthalimide (I)   Benzylthiomethyl chloride [—HCl]   **Phthalimido diethyl malonic acid ester benzylthio methyl (ii)**

(+)–Cystine (V)   **(+)–Cystine (iV)**   α–Thiobenzyl alamine (iii)

[B] Phenylalanine: Following are the different steps engaged in the synthesis of **phenylalarine** by the **phthalimido ester method:**

(1) The interaction between potassium phthalimide and bromo–diethyl malonate yields **diethyl phlthalimido malonic ester (I)** with the release of a mole of **KBr.**

(2) The resulting product **(I)** on being subjected to treatment with freshly prepared **sodium ethoxide** followed by **benzyl chloride** gives rise to the formation of **benzyl diethyl phthalimido malonic ester (II).**

(3) The product (II), when treated with KOH/HCl/heat in a sequential manner releases one mole of **phthalic acid** and the desired **phenylalnine** *i.e.,* an α–amino propionic acid.

[C] Proline: The cardinal steps that are usually followed stepwise in the synthesis of **proline** are as enumerated under:

(1) The treatment of **diethyl phthalimide malonate (I)** with **sodium ethoxide** and 1, 3–dibromo propane yields **3–bromo–diethyl phthalimide malonate (II)** with the release of one mole of **HBr** as shown below.

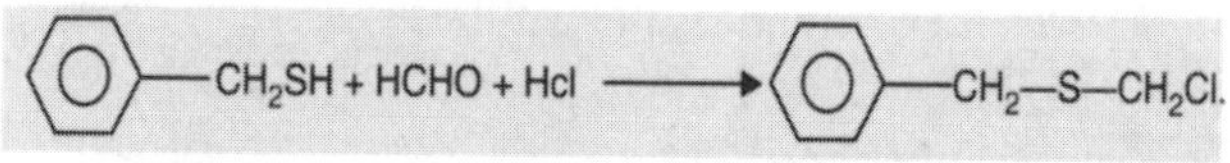

---

* **Benzylthiomethyl Chloride** is usually prepared by the interaction of **benzylthiol** with *formalin* and *HCl:*

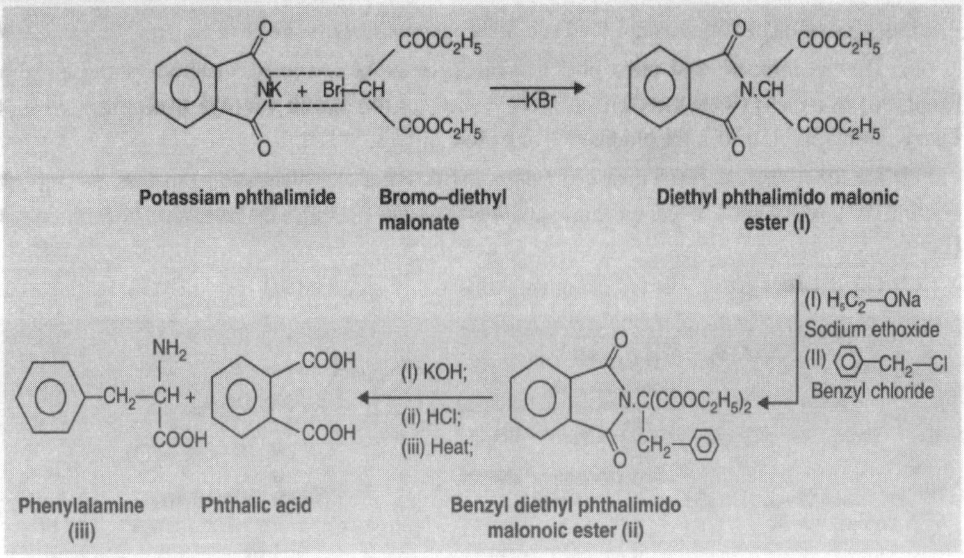
Potassiam phthalimide
Bromo–diethyl malonate
–KBr
Diethyl phthalimido malonic ester (I)
(I) H5C2—ONa
Sodium ethoxide
(II) Benzyl chloride
(I) KOH;
(ii) HCl;
(iii) Heat;
Phenylalamine (iii)
Phthalic acid
Benzyl diethyl phthalimido malonoic ester (ii)

(2) Esterification of resulting prduct **(II)** with **potassium acetate** yields the **acetyloxy derivatiqve of II** as **(III)** with the elimination of a mole of KBr.

(3) The resulting product **(III)** on subsequent stepwise treatment with NaOH/HCl/Heat gives rise to the formation of **α–(3–hydroxypropyl)–glycine (IV),** which on treatment with HCl undergoes **cyclization** to yield **proline [V]** *i.e.*, the targetted **amino acid.**

Thus, we may have the following reactions:

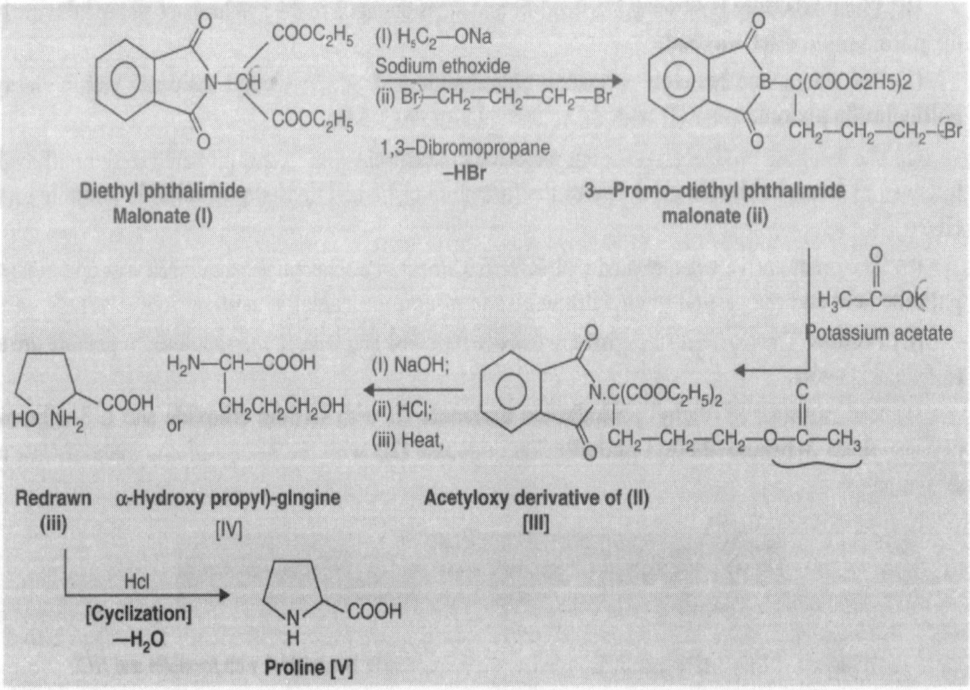
(I) H5C2—ONa
Sodium ethoxide
(ii) Br—CH2—CH2—CH2—Br
1,3–Dibromopropane
–HBr
Diethyl phthalimide Malonate (I)
3—Promo–diethyl phthalimide malonate (ii)
Potassium acetate
(I) NaOH;
(ii) HCl;
(iii) Heat,
Redrawn (iii)
α-Hydroxy propyl)-glngine [IV]
Acetyloxy derivative of (II) [III]
Hcl
[Cyclization]
–H2O
Proline [V]

4.1.2.7.13. Modified Phthalimidomalonic Ester Synthesis

The precise and articulated mechanism for the specific **phthalimidomalonic ester synthesis** may be explicitly described by the following sequence of adequately modified reactions:

(1) Interaction between **potassium phthalimide** and **monobromo diethyl malonic ester** invariably gives rise to the formation of **diethyl phthalimidomalonic ester (I)** that essentially provides the requisite group necessarily required for the 'amino acid' molecule.

(2) The resulting product (I) is duly converted to the respective **carbanion (II)** by careful treatment with freshly prepared **sodium ethoxide [H$_5$C$_2$–ONa]**.

(3) The **carbanion (II)**, thus accomplished is subsequently subjected to reaction with a requisite **alkyl/aryl halide**, such as: **benzyl chloride** $\left[\bigcirc\!\!-CH_2-Cl \right]$ to yield the corresponding **diethyl phthalimido benzyl malonic ester (III)**.

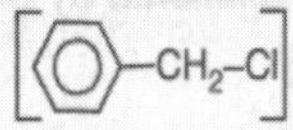

(4) Meticulous **hydrolysis** of the **substituted product (III)**, followed by **partial decarboxylation** produces N–(α–benzyl) phthalimido acetic acid (IV).

(5) The resulting **product (IV)** on treatment with **hydrazine [H$_2$N–NH$_2$]** gives rise to the production of the desired *amino acid,* **phenylalanine (V)** plus a highly basic **cyclic diaza product**.

The various reactions described from (1) through (5) are given under:

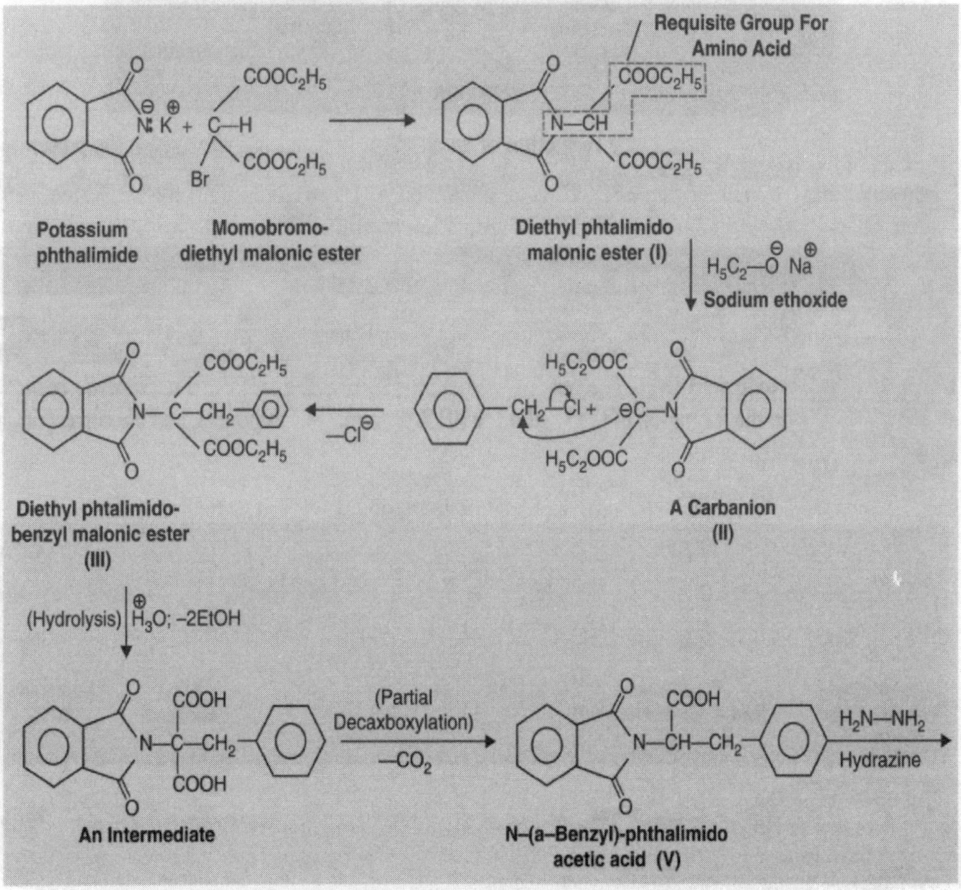

Potassium phthalimide   Momobromo-diethyl malonic ester   Diethyl phtalimido malonic ester (I)

Diethyl phtalimido-benzyl malonic ester (III)   A Carbanion (II)

An Intermediate   N–(a–Benzyl)-phthalimido acetic acid (V)

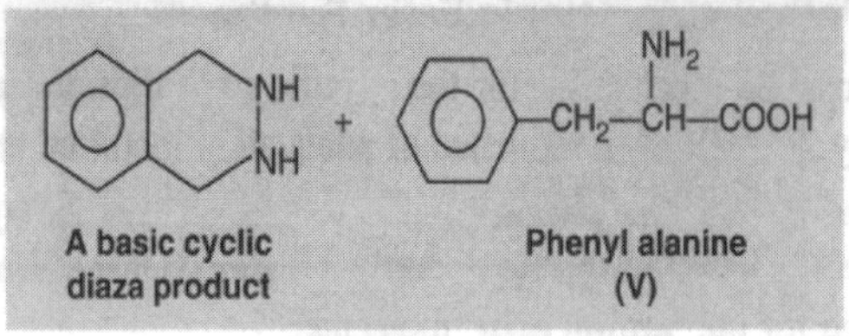

A basic cyclic diaza product	Phenyl alanine (V)

4.1.2.7.14. Reducation of α–Ketonic Acids

It is, however, feasible and possible to prepare various *amino acids* by causing the critical reduction of α–ketonic acids in the presence of *ammonia* by *two* well recognized and practised methods, namely:

- **catalytic reduction,** and
- **sodium ethoxide (freshly prepared) treatment.**

Though the exact mechanism of the aforesaid *chemical reaction* has not yet been fully ascertained and established; however, it most probably takes place *via* the formation of an *'amino acid'* as shown under:

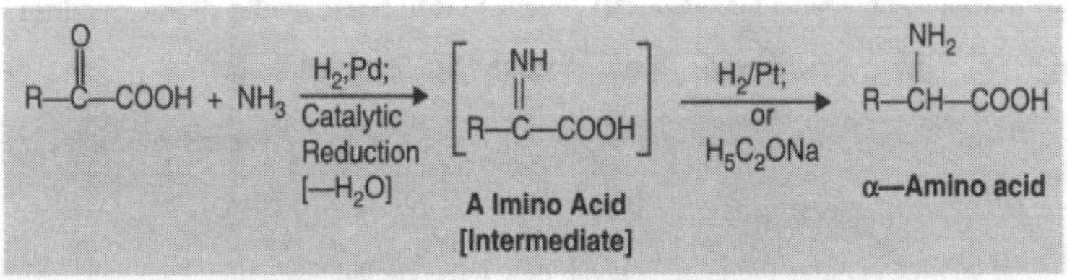

The *first step* being the **catalytic reduction** of an *alkyl carbonyl carboxylic acid* in the presence of *ammonia (gas)* to yield an *'imino acid'* as an intermediate. The *second step* being the further catalytic reduction with **H₂/Pt** to give rise to the formation of an **α–amino acid.**

> **Note: This method may be used gainfully for the synthesis of alanine and glutamic acid.**

L–Alanine: The synthesis of **L–alanine** may be accomplished *via* an **asymmetric synthesis,** wherein an *'optically active amine'* eg., **L–α–methyl benzyl amine (I),** instead of ammonia, yields a **Schiff's base (III)** with **pyruvic acid (II)*.** The resulting product (III) on being subjected to *catalytic reduction (H₂/Pd)* gives rise to the formation of the **L–alanine (91%)** as a major component, and the corresponding **D–alanine (9%)** only.

The various steps involved in the synthesis of **L–alanine** are given as under:

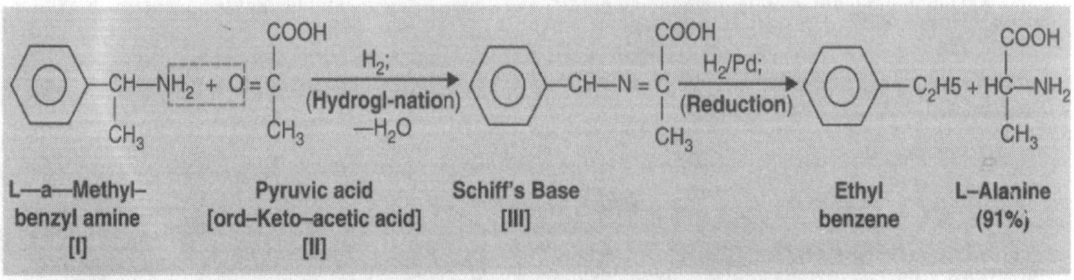

| L—a—Methyl– benzyl amine [I] | Pyruvic acid [ord–Keto–acetic acid] [II] | Schiff's Base [III] | Ethyl benzene | L–Alanine (91%) |

---

* **Pyruvic Acid :** That is, α–keto acetic acid undergoes reduction at the **carbonyl–CO–) moiety** to form an adduct, as an **intermediate.**

** Amorosa R *et al* ∴ *J Org Chem,* **57** : 1082, 1992.

D(–)–Alanine: Amorosa *et al.* (1992) put forward the synthesis of **D–(–)–alanine,** whereby the optically active amino acid is prepared by means of the following sequential steps:

(1) Interaction between **pyruvic acid (I)** and **D (+) or L–(–)α–methyl benzylamine (II)** to produce the **benzylimine salt (III)** *via azomethine* as an *intermediate.*

(2) The resulting **product (III)** upon **hydrogenolysis** in the presence **protonated α–methyl benzylamine [C$_6$H$_5$–CH (CH$_3$).NH$_3$$^\oplus$;designated as BH$^\oplus$]** into a corresponding **benzylamine salt (IV)].**

(3) The resulting **product (IV)** on being subjected to *reduction* with **palladous hydroxide [Pd(OH)$_2$]** produces the respective **ammonium salt (V)** plus the release of *two* moles of ethyl benzene.

(4) Ultimately, compound (V) loses, a mole of **ammonia** (as a gas) and forms **D–(–)alanine (VI)** upto **78%.**

The various reactions are stated as under:

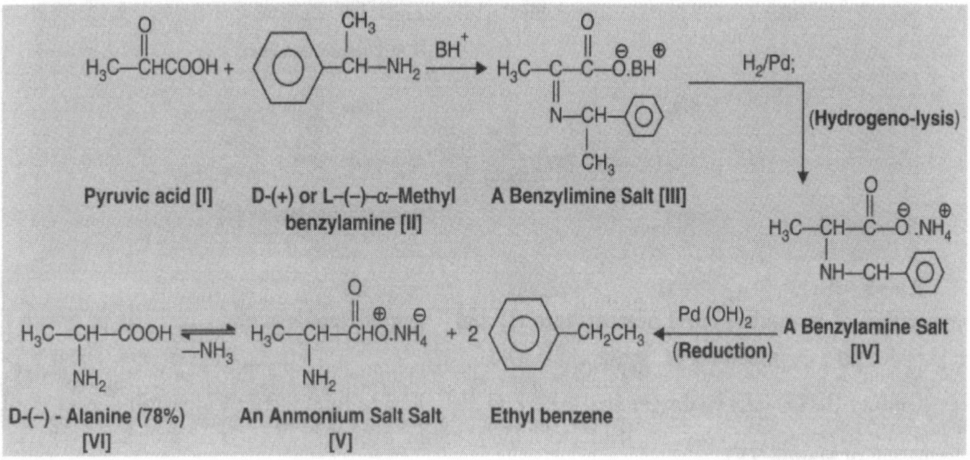

4.1.2.7.15. Reduction of α–Imino Acids

In fact, there prevails another most convenient and easier mode of synthesis for the **α–amino acids** that critically entails the *chemical reduction* or the *catalytic reduction* of an array of such vital and important organic entities as:

- **Oximes,** or
- **Phenylhydrazones of α-keto acids.**

The various steps involved essentially comprise:

(1) Interaction of an **α–keto alkyl carboxylic acid** *e.g.,* **pyruvic acid (I)** with liquid ammonia to form an **imine (II)** by the elimination of a mole of water.

(2) The resulting **imine (II)** on hydrogenation with Pd yields the respective **amine (III)** *i.e.,* the α–amino acid (*alanine*).

Thus we may have the following reactions:

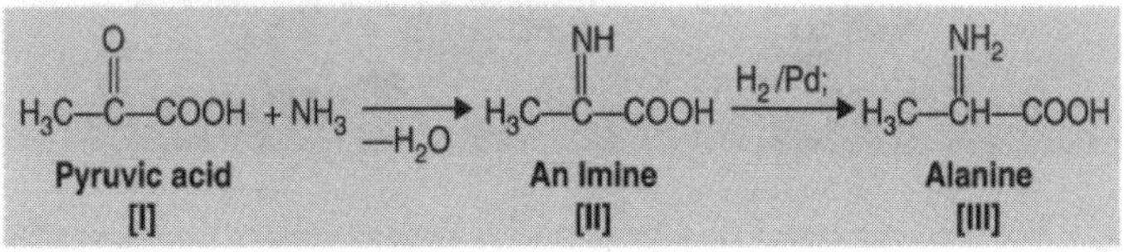

---

* From α–Keto Acids, Oximes, and **Hydrazones of α–Keto acids.**

Alanine from Ethyl Nitrite:

Importantly, one may also obtain **'alanine'** by duly treating **α–methyl ethyl aceto acetate (I)** with **ethyl nitrite (II)** in the presence of an acid to produce a **hydroxyl imine (III)** plus a mole each of *acetic acid* and an *aliphatic alcohol*. The resulting **product (III)** on being subjected to **hydrogenation** (H_2/Pd) yields the corresponding **ethyl ester of alanine (IV),** which upon hydrolysis produces ultimately the desired **alanine** plus a mole of ethanol, as shown below:

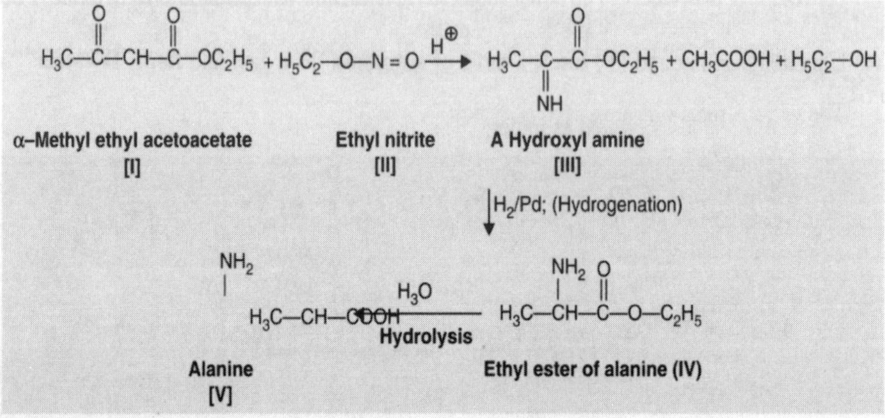

$$H_3C-\overset{\overset{O}{\|}}{C}-CH-\overset{\overset{O}{\|}}{C}-OC_2H_5 + H_5C_2-O-N=O \xrightarrow{\overset{\oplus}{H}} H_3C-\overset{\overset{O}{\|}}{C}-\underset{\underset{NH}{\|}}{C}-OC_2H_5 + CH_3COOH + H_5C_2-OH$$

α–Methyl ethyl acetoacetate	Ethyl nitrite	A Hydroxyl amine
[I]	**[II]**	**[III]**

$\Big\downarrow H_2/Pd;$ (Hydrogenation)

$$\underset{\text{Alanine}}{\underset{[V]}{H_3C-\underset{\underset{NH_2}{\|}}{CH}-COOH}} \xleftarrow[\text{Hydrolysis}]{H_3O} \underset{\text{Ethyl ester of alanine (IV)}}{H_3C-\underset{\underset{NH_2}{\|}}{CH}-\overset{\overset{O}{\|}}{C}-O-C_2H_5}$$

Alanine from Phenyldiazonium Chloride: Interestingly, **alanine (V)** may also be prepared by the interaction of **α–methyl ethyl acetoacetate (I)** and **phemyldiazoniem chloride* (II)** to obtain the respective **phenyldiazonium derivative (III)** with the elimination of a mole of *acetic acid.* The resulting **product (III)** is duly **hydrogenated** with Zn/CH_3COOH to form the respective **ethyl ester of alanine (IV)** plus one mole of *aniline*. Ultimately, the resulting **product (IV)** upon hydrolysis gives rise to the formation of **alanine (V).**

The various reaction may be summarized as given below:

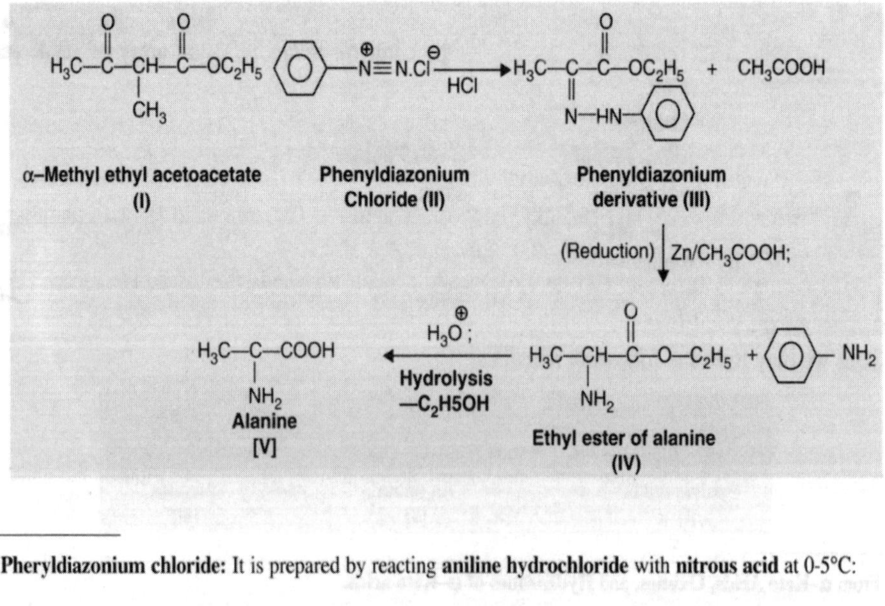

$$H_3C-\overset{\overset{O}{\|}}{C}-\underset{\underset{CH_3}{\|}}{CH}-\overset{\overset{O}{\|}}{C}-OC_2H_5 \quad \underset{HCl}{\bigcirc -\overset{\oplus}{N}\equiv N.\overset{\ominus}{Cl}} \longrightarrow H_3C-\overset{\overset{O}{\|}}{C}-\underset{\underset{N-HN-\bigcirc}{\|}}{C}-OC_2H_5 + CH_3COOH$$

α–Methyl ethyl acetoacetate	Phenyldiazonium	Phenyldiazonium
(I)	Chloride (II)	derivative (III)

(Reduction) $\Big\downarrow Zn/CH_3COOH;$

$$\underset{\underset{\text{Alanine}}{\underset{[V]}{\text{Alanine}}}}{H_3C-\underset{\underset{NH_2}{\|}}{C}-COOH} \xleftarrow[\substack{\text{Hydrolysis} \\ -C_2H5OH}]{\overset{\oplus}{H_3O};} \underset{\text{Ethyl ester of alanine}}{\underset{(IV)}{H_3C-\underset{\underset{NH_2}{\|}}{CH}-\overset{\overset{O}{\|}}{C}-O-C_2H_5}} + \bigcirc - NH_2$$

---

* **Pheryldiazonium chloride:** It is prepared by reacting **aniline hydrochloride** with **nitrous acid** at 0-5°C:

4.1.2.7.16. Strecker Synthesis

It has been demonstrated beyoned any reasonable doubt that an important procedure largely adopted to synthesize the **'carboxylic acids'** is caused by the careful *hydrolysis of nitriles.* In the same vein, the α–amino nitriles may be duly hydrolyzed to produce **α–amino acids.**

The **'Strecker Synthesis'** puts forward the preparation of α–amino acids by the help of the following steps sequentially:

(1) **Acetaldehyde (I)** when treated with **ammonia (or NH$_4$Cl)** in the presence of **cyanide ion** (*i.e.,* Na$^+$CN$^-$) yields an **α–amino nitrile** termed as **2–amino propane nitrile (II)** plus a mole each of NaCl and water.

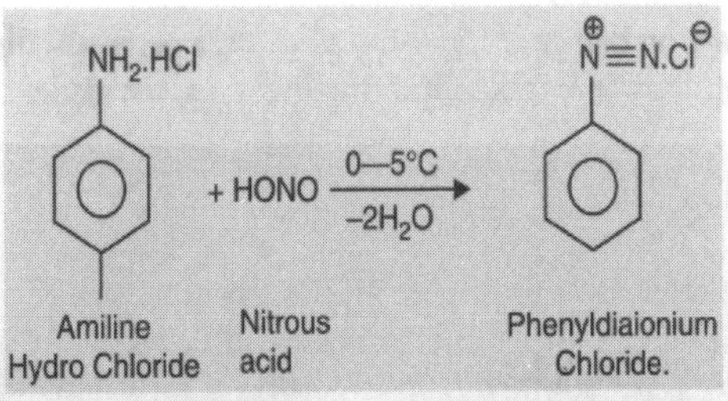

	Amiline Hydro Chloride	Nitrous acid		Phenyldiaionium Chloride.

(2) The resulting **compound (II)** on heating with HCl and hydrolysis in a slightly alkaline environment gives rise to the formation of **alanine, (III)** upto **52–60% yield** with the release of ammonia (NH$_3$↑) gas.

These reactions may be explained as stated under:

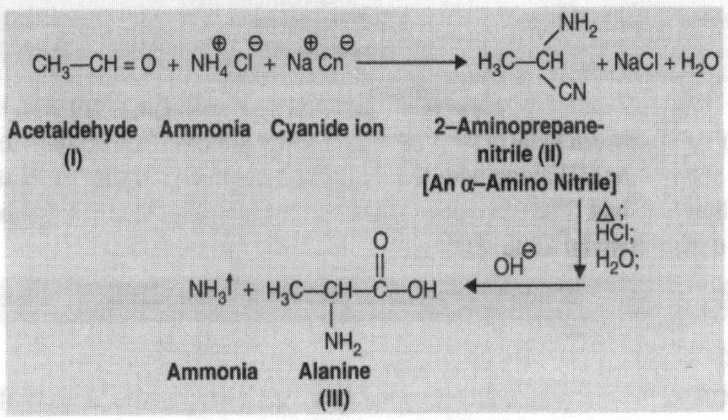

Mechanism: The probable mechanism of the **α–amino nitrile formation** critically involves the formation of an **'imine intermediate' (A)** as given under:

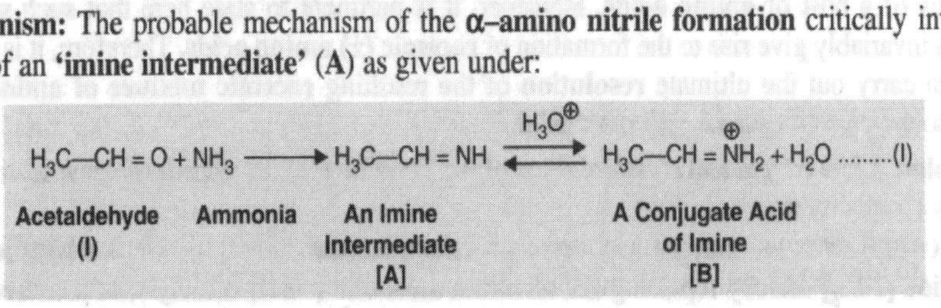

The resulting **product (A)** in the presence of a **hydronium ion [H$_3$O$^\oplus$]** produces a **conjugate acid of imine (B),** which eventually reacts with a **cyanide ion (:CN)** under the specific parameters of the reaction to produce the desired **α–amino nitrile (C)** as stated below:

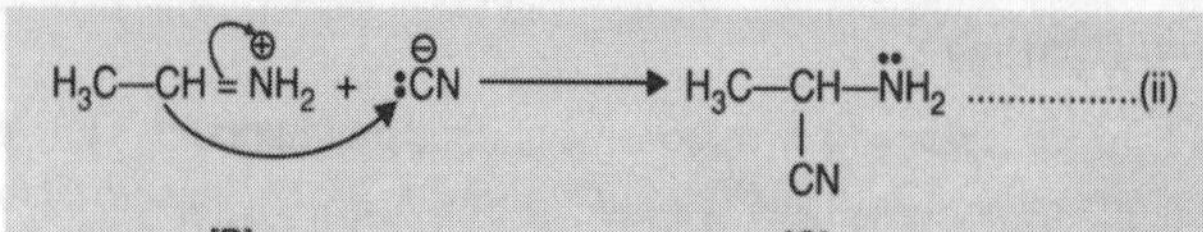

$$H_3C-CH=\overset{\oplus}{N}H_2 + :\overset{\ominus}{C}N \longrightarrow H_3C-CH-\overset{..}{N}H_2 \quad\ldots\ldots\ldots(ii)$$

$$\underset{[B]}{} \qquad \underset{[C]}{\overset{|}{C}N}$$

Notes: (1) In the Streeker Synthesis, the nucleophile happens to be the respective cyanide ion [:CN]

(2) Except for '*alanine*', one may accomplish rather poor yields by this well-known method.

(3) It helps in the large scale preparation of several amino acids *e.g.*, alanine, glycine, glutamic acid, leucine, isoleucine, serine, valine, methinonine, and phenylalanine.

4.1.2.7.17. Synthesis *via* Diketopiperazine

Interestingly, **aspartame***, on being stored for an **extended period of time** in an aqueous medium, gives rise to the formation of a *cyclic* 'diketopiperazine' as given under:

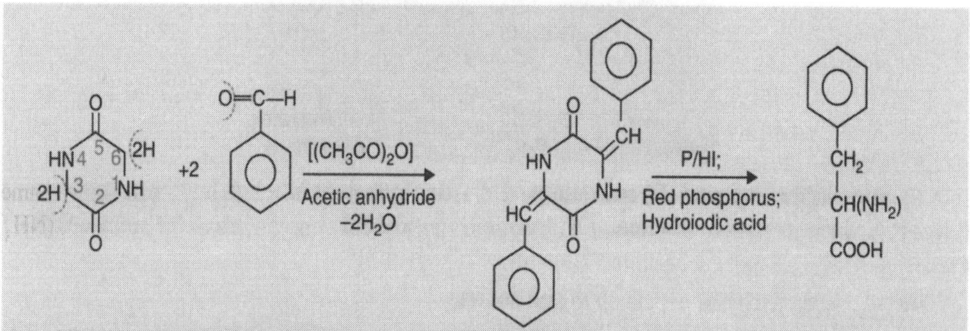

2,5—Diketo piperazine [I]   **Benzaldehyde**   **2,5—Benzylidene piperazine [II]**   **Phenyladanine (An α–Amino acid) [III]**

Explanation: The deterioration of 'aspartame' gives the **2, 5–diketopiperazine (I)** which eventually reacts with *two* moles of **benzaldehyde** in the presence of **acetic anhydride** to produce the corresponding **2, 5–benzylidene piperazine (II)** along with the loss of *two* moles of water as indicated in the above reaction. The resulting **product (II)** on being treated with red phosphorus and hydroiodic acid yields the desired *α–amino acid*, **phenylalanine (III)**.

Note: By using the above synthesis it is possible to prepare '*tyrosine*' and '*methionine*'.

4.1.2.7.18. Specialized Procedures

In actual practice, there are quite a few remarkably **specialized procedures** that ultimately lead to the synthesis of a host of **amino acids.** However, it is pertinent to state here that such **specialized procedures** invariably give rise to the formation of **racemic (±) amino acids**. Therefore, it is absolutely necessary to carry out the ultimate **resolution** of the resulting **racemic mixture of amino acids** to obtain the so called '**natural L–amino acids**' duly.

Examples: A few *typical* and *classical* **examples** pertaining to the synthesis of some selected **amino acids** are as stated under:

[A] L–(+)–Cysteine: It may be synthesized by the treatment of **L–(–)–serine** with **phosphorous pentachloride [PCl₅]** thereby replacing the **terminal alcoholic (–OH) moiety** with a **chloro group**

---

* **Aspartame:** It is an **artificial sweetner,** which has been withdrawn from the market by **US–FDA** because of its **deterioration upon storage** in an aqueous medium for a long duration.

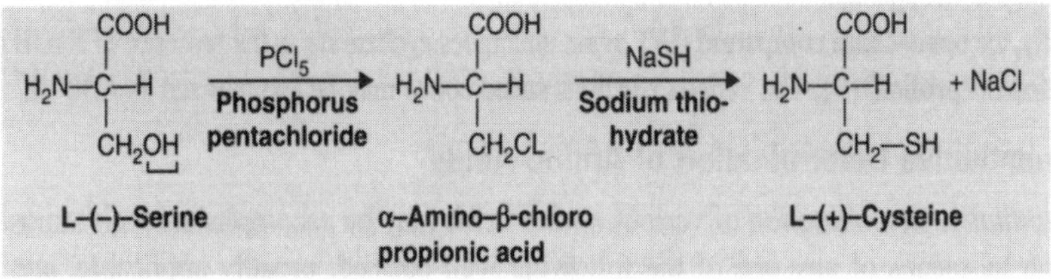

L-(-)-Serine   α-Amino-β-chloro propionic acid   L-(+)-Cysteine

to yield **α–amino–β–chloro–propionic acid.** The resulting product on reacting with **sodium thiohydrate (NaSH)** form the desired **L–(+)–Cysteine** plus a mole of **sodium chloride.**

[B] **Tyrosine: Tyrosine** may be obtained directly from the *amino acid* (**phenylalanine**) on being

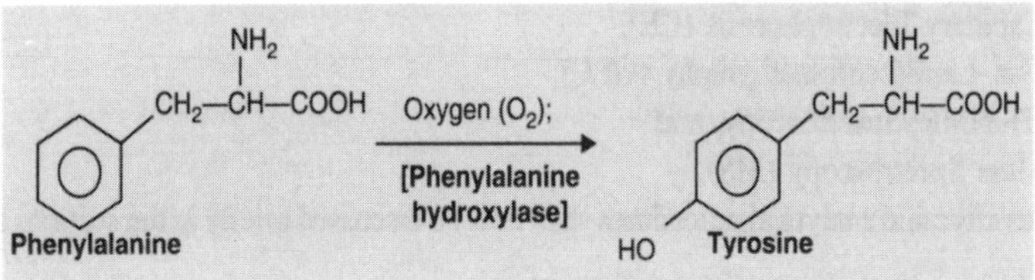

subjected to treatment with **oxygen (O₂)** in the presence of an *enzyme*, **phenylalanine hydroxylase.**

[C] **Proline : Proline** is synthesized by carrying out the following sequential steps:

(1) The reaction between **ethylene cyanide (or ethylene nitrile)** and **deithyl malonic ester** in the presence of freshly prepared **sodium ethoxide (H₅C₂–ONa),** usually known as **Michael Condensation,** yields the corresponding **diethyl (ethyl nitrile) malonic ester (I).** Obviously, the **active H–atom** abstracted duly from the diethyl malonic ester saturates the *ethylenic double bond* to produce the **condensed product (I).**

(2) The **resulting product (I)** undergoes reduction with **Raney–Ni** to yield a *cyclized product* usually termed as **2–keto–3–ethyl carboxylate peperidine (II).**

(3) The **cyclized product (II)** upon treatment with **thionyl chloride [SO₂Cl₂]** produces the corresponding *chloro derivative* known as **2–keto–3–chloro–3–ethyl carboxylate piperidine (III).**

(4) The respective **chloro derivative (III)** with acidic hydrolysis loses a molecule each of *EtOH* and *carbon dioxide* to obtain an **open–chain structural analougue 1–methyl–4–amino–pentanoic acid hydrochloride (IV).**

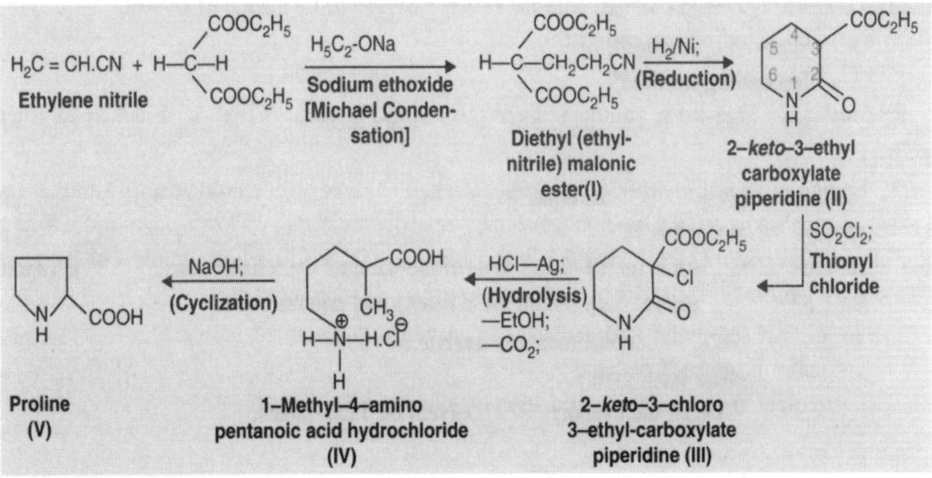

(5) Finally, the **open–chain compound (IV)** again undergoes **cyclization** in the presence of NaOH to give rise to the formation of '**proline' (V)**. The various reactions stated above may be summarized as ABOVE.

4.1.2.8. Quantitative Determination of Amino Acids

The *quantitative determination* of various *amino acids* may be accomplished with utmost accuracy and precision by means of any one of the following well-defined, broadly applicable, and profusely recognized **analytical techniques**, such as:

- Thin layer chromatography (TLC),
- Ion–exchange chromatography (IEC),
- Capillary Electrophoresis (CE),
- Gas–Liquid chromatography (GLC),
- ^{1}H–NMR–Spectroscopy, and
- Mass Spectroscopy (MS).

The various aforesaid analytical procedures shall now be discussed briefly in the sections that follows:

4.1.2.8.1. Thin Layer Chromatography (TLC)

Thin Layer Chromatography (TLC) is usually employed to perform the *quantitative determination* of **amino acids.** In order to accomplish maximu *uniformity,* as a rule, plates of length × breadth (20 × 20 cm) are used invariably with a thin layer of 0.5 mm thickness. However, one may critically observe the following *three* cardinal requirements, namely:

- Component mixtures (*i.e.,* '*analyte'*) be applied as one spot only with its known concentration using a micro-syringe,
- separation is normally accomplished by multiple development, and
- localization of segreagated components be carred out by visual inspection under UV-light (**preferably in a UV-chamber**).

TLC–Variant: There are several **TLC–variants** which shall be discussed briefly as under:

(a) **One–Dimesnsional TLC:** The spotted TLC plates, after evaporation of the sample solvent, is duly placed in a closed chamber saturated with vapours of the devoloping solvents (s). Now, one end of TLC–plate is carefully dipped (wetted) with the selected developer by the aid of either:

- ❑ an '**ascending method',** or
- ❑ a '**descending method',**

Precautionary Measures: Following *three* precautionary measures need to be adhered to rigidly, such as:

(1) *Equilibration of Chamber:* The **equilibration of chambers** (or **chamber saturation**) is a *vital factor* to accomplish reproducible **Rf values.** It may be easily achieved by allowing the solvent to remain in the closed '**chromatank'** for at least **1–2 hrs** so as to pre–saturate the '**chromatank'** with the vapours of the solvent uniformly. Thus, one may obtain these *three* vivid criteria:

- distinct separation of constituents in analyte,
- uniform solvent front, and
- prevent and check the evaporation of solvent on TLC–plates.

as illustrated explicitly in Figs.2.7. (a) and (b) respectively.

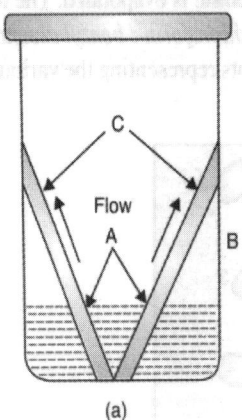

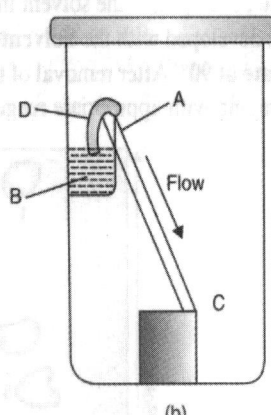

| (a) | (b) |

Fig: 2.7 (a) Ascending Flow   **A = Starting position of sample(s) ; B = Solvent System;**
                                    C = Chromatographic Surface ;

     (b) Descending Flow A   **= Starting position of sample(s) ; B = Developer ;**
                                    C = Chromatographic surface ; D = Cotton wick ;

(2) *Prevention of Oxidation:* It has been duly observed that both **temperature** and **light** afford distinct augmentation of oxidation; and, therefore, the underlying **experimental conditionalties** should be observed to accomplish the desired maximum development of **thin-layers,** for instance:

- **Temperature**   :   18–23°C, and
- **Light**   :   Diffused daylight *'both natural and artificial'.*

> **Note: Direct sunlight (*i.e.,* UV–light) or drought may result in the *'oblique formation'* particularly of the solvent front.**

(3) *Visulization of Developed Spots:* The critical and exhanstive research has revealed that a host of **amino acids** have been duly identified by means of certain highly specific *organic and inorganic substances* which demonstrate positively an **'improved visulization'** of the separated compounds. In fact, such substances are usually known as the **'fluorescent indicators'.**

Examples: A few typical **fluorescent indicators** are:

- ❏ Barium diphenylamine sulphonate,
- ❏ 2, 7–Dichloro fluorescein,
- ❏ Fluorescein (0.2% *w/v* in Ethanol),
- ❏ Morin (0.1% *w/v* in Ethanol),
- ❏ Sodium fluorescinate (0.4% *w/v* in water),
- ❏ Rhodamine B,
- ❏ Zinc silicate, and
- ❏ 7–Methylumbelliferone.

(b) **Two Dimensional TLC (or 2D–TLC) : 2D–TLC** is also commonly known as **2D–planar chromatography.** In this particular instance, the **'analyte sample'*** is carefully spotted in one corner of a **20 cm × 20 cm TLC plate** as depicted in Fig. 2.8.

---

* **Analyte Sample :** It consists of mixture of **ten** amino acids.

The **TLC–plate** is first developed using the **Solvent–1** comprised of *toluene : 2–chloroethanol : pyridine*. After due development the solvent in the first mobile phase is evaporated. The resulting TLC–plate is now again developed with the **Solvent–2** consisting of *chloroform : benzyl alcohol : acetic acid* by **turning the plate at 90°**. After removal of the solvent the spots representing the various **amino acids** are located by spraying with appropriate reagents.

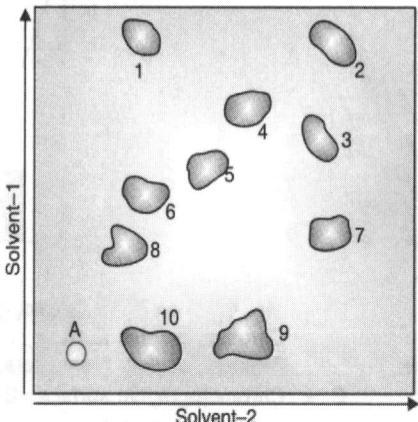

Fig: 2.8 : Two-Dinensional Thin-Layer chromatogram of some Amino-Acids : A = Sample spot *Slovent-1* : Toluene : 2-Chloroethanol : Pyridine; *Solvent -2* Chloroform : Benzyl alcohol : Acetic Acid

In fact, the mixture of *'amino acids'* duly obtained from the *protein hydrolysate* are separated effectively by **2D–TLC** method*; and the resulting spots are located correctly by using **ninhydrin reagent** which essentially gives rise to a *pink to purple* spot with the **amino acids.**

(c) **Wedged–Tip TLC Method:** Both, Reindel *et al.* (1953)** and Mathias (1954)*** reported the wonderful **wedged–tip TLC method** that clearly and emphatically show the following *two* **advantageous aspects**, namely:

- **improved and distinct separation,** and
  - **constituents forced to assume a typical band-like path.**

Fig : 2.9 vividly show the **TLC-plate** having **wedged-tip divisions.** However, the following sequential steps must be adopted, such as:

(a) Draw dividing lines 0.5–1 mm broad upon the surface of the coated layer using a sharp-narrow-metal spatula,

(b) Pentagons are duly facilitated by the aid of a *'stencil'* made up of a **transparent plastic material,** and

(c) Sample mixture are carefully applied to the narrow portion of the wedge to accomplish the best possible results.

---

* Von Arx E and Nehr R A : *J Chromatog*, **12** : 329. 1963.

** Reindel *et al. Naturwissenschaften,* **53** : 454, 1953

*** Mathias W, *Naturwissenschaften,* **41** : 18, 1954.

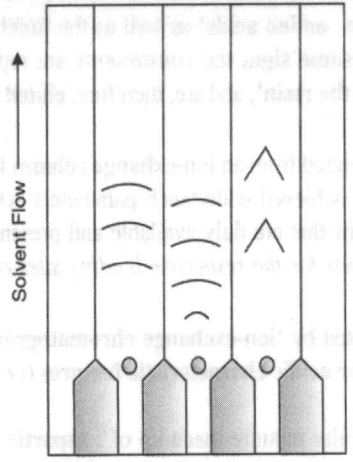

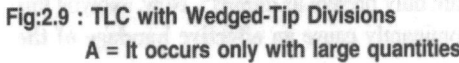

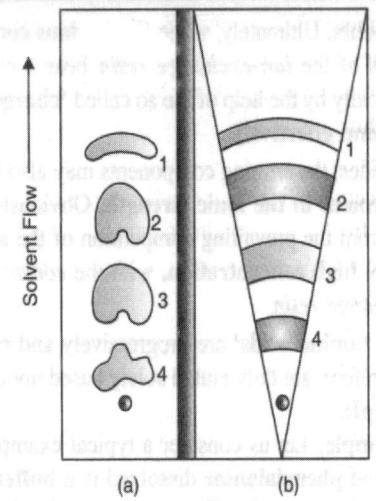

(a)　　　　　(b)

Fig:2.9 : TLC with Wedged-Tip Divisions　　**Fig: 2.10 (a) TLC–Normal technique**
　　A = It occurs only with large quantities　　　2.10 (b) TLC–Wedged-Tip Technique
　　　　　　　　　　　　　　　　　　　　　　1 = Tyrosine; 2 = Phenylalanine;
　　　　　　　　　　　　　　　　　　　　　　3 = Sterine; 4 = Alanine;

In addition, Fig. 2.10 (a) depicts TLC of a protein hydrolysate by means of the *normal* **TLC–technique** *vis-a-vis* the **wedged-tip technique** [Fig. 2.10. (b)]. From these critical observations one may, vividly visualize the **'remarkably separated beautiful bands'** in the latter in comparison to the **several odd and irregular shaped spots'** in the former.

Importantly, not only the *'clarity of separation'*, but also the *'reproducibility of results'* are observed to be absolutely *predominant, clear,* and *distinct* in the latter technique.

Fig: 2.10 (a) and (b) duly represent the typical analysis of a mixture of **amino acids** comprising *tyrosine, phenylalnine, serine* and *alanine.*

4.1.2.8.2. Ion–Exchange Chromatography (IEC)

In **ion–exchange chromatography (IEC),** the various **amino acids** present in an **'analyte'** are duly separated on the basis of their *net charge.* The **stationary–phase** being an *ion–exchange resin* (*i.e.,* a **cross–linked polymer having several charged functional moieties**); whereas, the respective **mobile–phase** being an **appropriate aqueous solution.** Importantly, **IEC** is invariably carried out on a **'column'**. The various components are duly retarded in their critical movment *via* the **column** solely depending upon the **'sign'** as well as **'magnitude'** of their charge.

Nevertheless, one may observe the *'electrostatic binding'* of the separated entites to the *'oppositely charged'* moieties located strategically upon the **ion-exchange resin.** Therefore, the actual observed **disruption of these ensuing bonds,** does come into being almost repeatedly as the various components move down the column length.

Methodology: Ideally, the *'analyte sample'* is carefully applied to the **column** at a *specific pH* at which the *molecules (components)* possess and **'opposite charge'** to that of the various moieties present on the *resin,* thereby the **components get bound to the column temporarily.** Now, the **pH** of the **'eluting buffer'** is altered subsequently, so as to **change the net overall charge existing on the**

components. Ultimately, when **the various components** (*i.e.,* **amino acids**) as well as the **functional moieties** of the *ion-exchange resin* bear a charge of the **same sign,** the *constituents* are repelled categorically by the help of the so called **'charged groups of the resin',** and are, therefore, **eluted from the column effectively.**

Besides, the ensuing components may also be gainfully eluted from an **ion-exhange column by an enhancement in the ionic strength.** Obviously, the elution achieved under such parameters actually results from the prevailing competition of the **added salt ions** that are duly available and present at a relatively **high concentration,** with the *constituents of interest for the respective binding sites on the ion-exchange resin.*

The **'amino acids'** are progressively and rapidly separated by **'ion-exchange chromatography';** and, therefore, are duly eluted solely based upon their **basic or acidic characteristic features** (*i.e.,* **low or high pI).**

Example: Let us consider a typical example of a particular mixture made up of : **aspartic acid, lysine,** and **phenylalanine** dissolved in a **buffer of high pH** (*viz.,* **phosphate buffer pH 9).]**

Amazingly, at this **pH,** the respective **'amino acids'** are duly present as *anions*.* Now, applyng this mixture on to an **'anion–exchange column'** shall predominantly cause an **effective bondage of the amino acids to the column.**

Consequently, when the column is subjected to **'elution'** with buffers having **'decreasing pH',** the *three* aforesaid **amino acids** shall emerge out of the column as given under:

Lysine : It will get **'protonated'** first and foremost, which is due to the fact that the **'ε-NH$_2$ moiety'** of **lysine** possesses the **maximum pKa value** of all the functional moieties that are present in these *three* **amino acids.** Consequently, **'lysine'** will accomplish a **net zero charge; and hence, a 'net +ve charge'.*** * Thus, in a situation when **lysine** has a **net zero charge** it never gets retained intimately to the column; and when it possesses a **net + ve charge,** it gets repelled squarely by the respective resin moieties and is eluted ultimately.

Aspartic Acid : It is observed to be the **most acidic amino acid** (*i.e.,* having the **lowest PI);** and hance, may pass through all these stages at the end.

Phenylalanine : It is found to pass through all these stages just in between *lysine* and *aspartic acid.*

> **Note: The** ultimate *'order of elution'* **under the aforesaid experimental paramenters shall turn out to be:**
> ■ **First :** *'Lysine';* ■ **Second :** *'phenylalanine';* ■ **Third :** *Aspartic Acid.*

4.1.2.8.3. Capillary Electrophoresis [CE]

The *electrophoretic mobility* designated by **'u'** is the velocity per unit of the electric field. It may be expressed by the following equation:

$$\frac{v}{E} = \frac{q}{f}$$

where, v = Velocity of the component (cm. sec^{-1}),

E = Electric field (volts. cm^{-1}),

---

* The **amino acids** are present on the **'alkaline side'** of their **isoelectric points;** they do possess a **net negative charge.**

** **Lysine** has the **maximum pI value** of all the *three* **amino acids.**

q = Net overall charge of the particle (electrostatic units), and

f = Frictional coefficient (*i.e.*, a funtion of the size and shape of the particle).

In actual practice, however, one may come across *two* **major electrophoresis variants** as illustrated in Fig : 2.11 (a) and (b) showing:

- Free Electrophoresis [or Solution Electrophoresis; or Moving Boundary Electrophoresis], and
- Zone Electrophoresis [or Zonal Electrophoresis].

(a) **Free Electrophoresis :** This refers to Fig : 2.11 (a) wherein the various constituents (*viz.,* **amino acids**) are initially present throughout the total volume of the solution. The subsequent application of the electric field ranging between **6–30 KV** ultimately leads to the critical establishment of a **'boundary'.*** In fact, the movement of this boundary is duly monitored by the careful passage of light *via* the solution and the resultant pattern being photographed precisely.

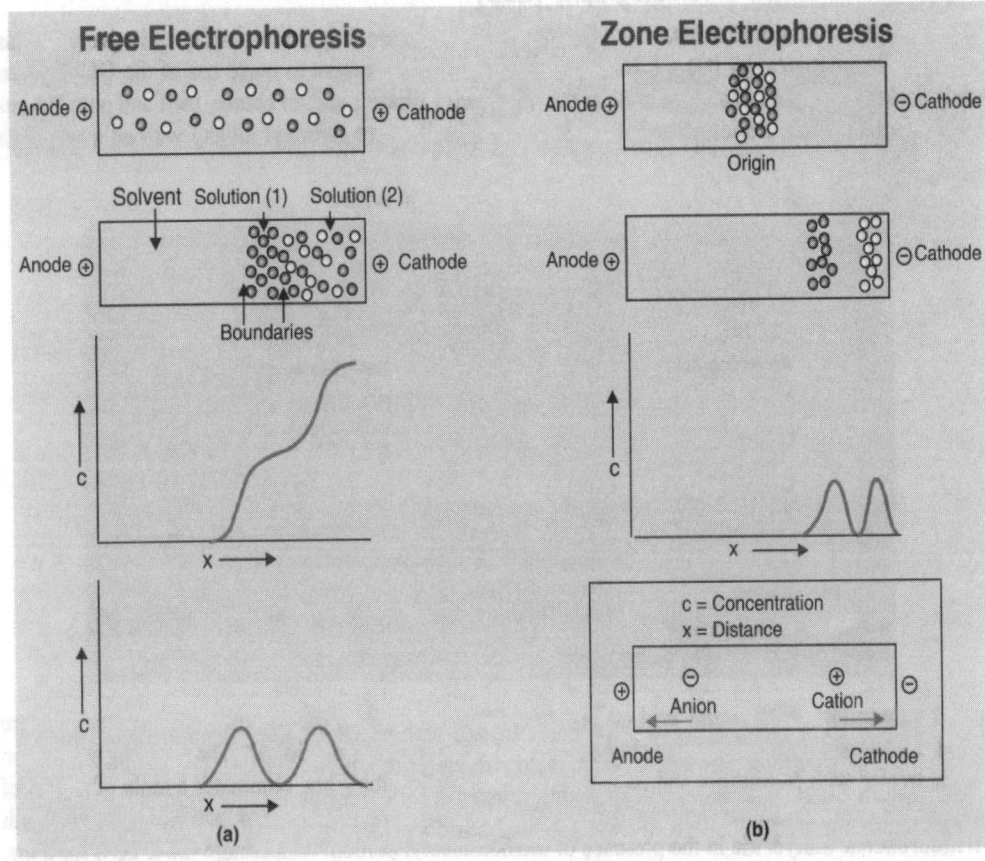

Note: The *'Free Electrophoresis'* raises a good number of *theoretical complications* that ultimately render the actual interpretation of the ensuing patterns rather difficult and cumbersome. Hence, this method has been largely replaced by the second type of electrophoresis termed as *'zone electrophoresis'*.

(b) **Zone Electrophoresis:** The **zone electrophoresis** is normally carried out by placing a small aliquot of analyte solution in contact with certain **'support medium'**, for instance : **gel, paper,** or **cellulose acetate.** The eventual application of an **electric field (between 6–30 kv)** would render the *analyte components* to move upon the said *support medium* as **zones** or **spots.**

Interestingly, there do exist a plethora of highly special **'zone electrophoresis'** variants, namely:

- **Disc gel electrophoresis,**
- **Sodium dodecyl sulphate polyacrylamide gel electrophoresis (SDS–PAGE),** and
- **Isoelectric focusing.**

4.1.2.8.4. Gas Liquid Chromatography [GLC]

The **esters* of amino acids,** which being *thermo-labile,* are duly determined quantitatively by **Gas Liquid Chromatography [GLC].** In actual practice, it is convenient to make use of the **GC–column** duly packed with **diethylene succinate** as the **'stationary phase';** and, of course, inert and pure N_2–gas as the mobile phase. Besides, one may also esterify the inherent carboxyl moiety with an appropriate alcohol.

Following are the usual common reactions of the **amino acids:**

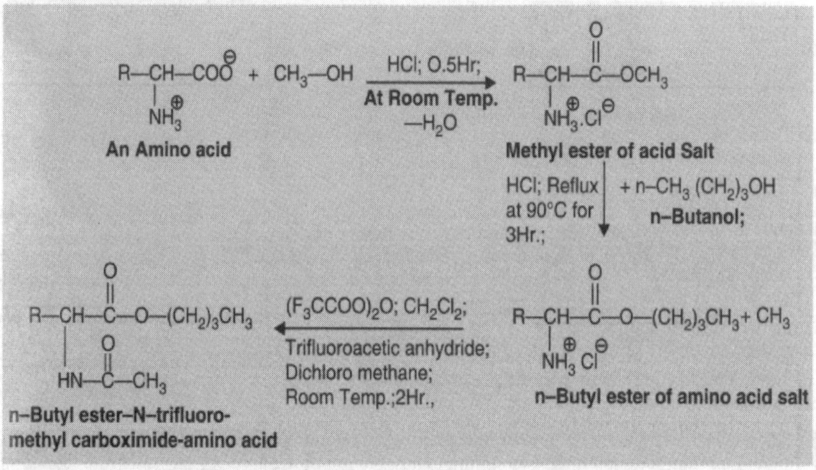

Explanation : The amino acid on being treated with HCl for half an hour at room temperature $(20 \pm 2°C)$ loses a mole of water to obtain the **methyl ester of amino acid salt.** The resulting product on being treated with **N–butanol** and HCl, and refluxed at 90°C for 3 hrs. eliminates a mole of *methanol* and yields the corresponding **methyl ester of amino acid salt.** This product when reacted with **trifluoroacetic anhydride** in the presence of dichloromethyl at room temperature $(20 \pm 2C°)$ for 2 hrs. produce the desired *thermolabile amino acid derivative i.e.,* **n–butyl ester–N–trifluoromethyl carboxymide– amino acid.**

---

* **Esters :** The *acetyl esters* of the respective **amino acids** may be obtained by treating the latter with **acetic anhydride** whereby the **free amino moiety** gets duly acelylated, which being thermo–labile in nature.

Fig: 2.12 depicts clearly the **gas liquid chromatographic** quantitative analysis of the respective **'bovine serum protein hydrolysate'**.

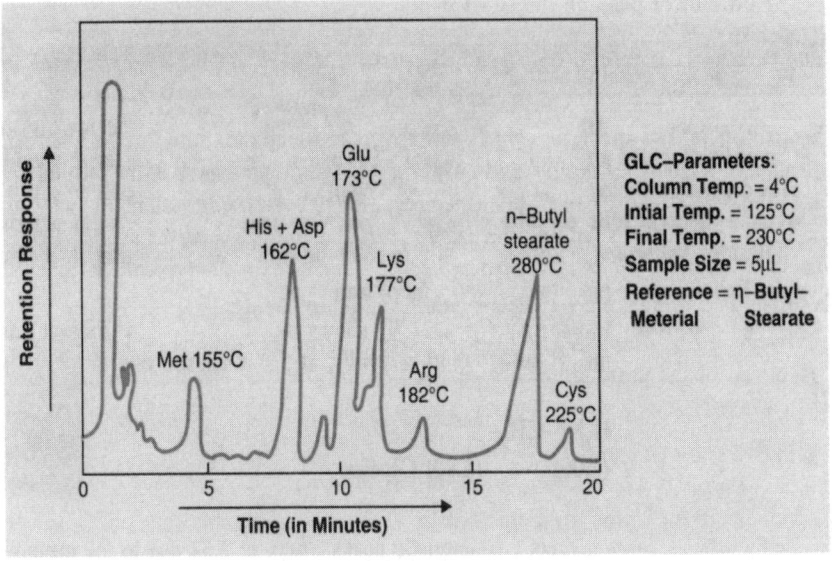

Fig: 2.12 : Gas chromatogram of Bovine Serum Albumin Hydrolysate of its N–Butyl–N–trifluoroacetyl Derivatives.

4.1.2.8.5. ¹H–NMR– Spectroscopy

The general procedure adopted invariably for carrying out the exact and precise measurement of a **nuclear magnetic resonance (NMR)** spectrum is as stated under:

- ❑ pure **amino acid** to be studied needs to be dissolved in an equally pure **deuteriated organic solvent** (*i.e.*, an *organic* solvent that *ideally* and *absolutely* is free from any **proton** *viz.*, **CDCl₃, deuteriated chloroform**).

- ❑ dissolved **'amino acid'** is carefully held in position in between the poles of the **electric magnet**.

- ❑ **trimethylsilane (TMS)** is incorporated directly into the above solution in a quantum that gives a **marked and pronounced signal** to ascertain the **'base-peak'**.

- ❑ most **NMR-instrument** are usually designed in such a fashion so as to make use of a **'standard chart paper'** duly callibrated over a range approximately, **700 cycles. sec⁻¹ (Hz);** whereas, the corresponding *'TMS–signal'* is invariably adjusted so as to fall at the position of the **'chart'** marked O Hz.

- ❑ **'variable field strength'** pertaining to the *main static magnetic field* being changed continuously in a **'sweep'** that eventually covers the entire **800–Hz range of the callibrated chart'.***

---

* Kar A : **Pharmaceutical Analysis** Vol. II, CBS Publishers & Distributors Pvt. Ltd., New Delhi, 2009.

❑ occurance of *'resonance'* affords a distinct and sharp change in the **NMR–Signal** as detected by the strategically positioned **'radio-frequency receiver'**, which being duly recorded as a **prominent peak on the chart paper.**

> **Note:** The actual difference in frequency between the *'TMS peak'* and the *'absorption peak'* is normally termed as the **'chemical shift'** of that specific proton.

The **proton NMR–spectroscopy** is employed most abundantly for the identification and quantification of fractionated **amino acids**. The *Tau* **(T) values** pertaining to the **'amino acid protons'** are found to be dependent exclusively upon the **prevailing pH of the solution.** Importantly, in a particular instance having a **'neutral solution'**, the ensuing **'dipolar-ion peak'** serves as a characteristic of the **amino acid.** Besides, the *protons attached to the side-chain* do exhibit a **typical characteristic pattern.**

Example: Following are *two* typical **examples** of amino acids:

(a) **Glycine (or α–Amino acetic acid):** It vividly shows *two* distinct peaks: *first*, at **2.06τ** due to **three H–atoms** of the **ammonium anion;** and *secondly*, at 5.80τ caused due to carboxyl H–atom.

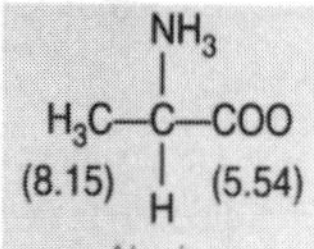

$$H_3N^{\oplus}-CH_2-COO$$
$$(2.06) \qquad (5.80) \text{ Glycine}$$

$$H_3C-\underset{\underset{H}{|}}{\overset{\overset{NH_3}{|}}{C}}-COO$$
$$(8.15) \qquad (5.54)$$
Alanine

(b) **Alanine (or α–Amino propionic acid) :**

It explicitly exhibits *three* distinct characteristic peaks : *first*, at **2.52** due to the **ammonium ion;** *secondly*, at **5.54** due to the **carboxyl H–atom;** and *thirdly*, at **8.15** due to the **three methyl protons.**

4.1.2.8.6. Mass Spectrometry (MS)

Mass spectrometry (or Mass Spectroscopy) determine the **molecular ion peak** that refers to the **molecular weight** of **amino acid** together with the **most preferred** *'zone of fragmentation'* of the molecule.

Based on the fact that the actual *'volatility of amino acids'* happen to be very low; and, therefore, it becomes almost an absolute necessity to have them converted duly into their **respective volatile derivatives** *i.e.,* **methyl and ethyl esters).** However, one may explicitly indicate the *'most preferred zone of fragmentation'* are duly represented by the **bonds** (shown with *dotted line*) at 'x' and 'y' located strategically on either sides of the **carbonyl (>C=O) moities.**

Example: Valine (or α–Amino isovaleric acid):]

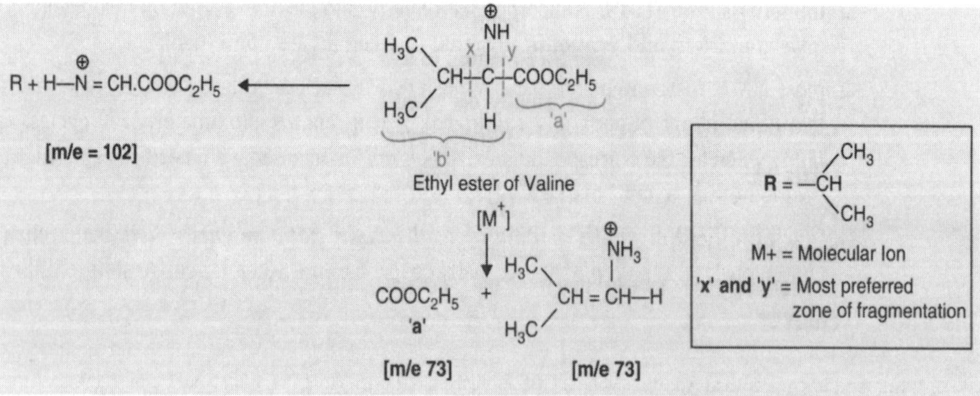

$$R + H-\overset{\oplus}{N} = CH.COOC_2H_5 \longleftarrow$$

[m/e = 102]

Ethyl ester of Valine
$$[M^+]$$

$$COOC_2H_5 \ + \ \overset{H_3C}{\underset{H_3C}{>}}CH = CH-H$$
'a'

[m/e 73] \qquad [m/e 73]

$$R = -CH \overset{CH_3}{\underset{CH_3}{<}}$$

M+ = Molecular Ion
'x' and 'y' = Most preferred zone of fragmentation

Observations: Following cardinal **observations** may be noted:

(1) **M$^+$ peak** *i.e.*, the *molecular ion peak*, is observed distinctly; however, it exhibits a **'weak'** intensity in the **mass spectrum.**

(2) Intensity of the **fragmented ion** [H$_2$N = CH COO C$_2$H$_5$] which stands at **m/e 102** appears to be fairly **strong or medium.**

(3) Intensity of the resulting **amine ion** [(CH$_3$)$_2$ CH = $\overset{+}{N}$H$_2$] (m/e 73) is found to be usually either **medium** or **strong.** Obviously, it may now be quite feasible and possible to **'deduce'** the exact and precise *magnitude of the side chain i.e.,* to **substract 29 mass units,** caused due to *CHNH$_2$* from M–102 to obtain m/e or M 73 (*i.e.,* 102–29 = **73**).

In the light of the aforesaid *three* critical observations **(1) through (3)** one may identify the *unknown amino acid easily.* Thus, the prominent **'mass peak'** at **m/e 102** is invariably accompanied by a critical **'mass peak'** observed at **m/e 73**, which is caused due to the **McLafferty Rearrangement*** as vividly explained under :

| Mass Peak at m/e 102 | [m/e = 28]   [m/e = 74] |

> **Note:** MS is used frequently in establishing the structure elucidation of rather *'small peptides'*. Besides, the derivatives of relatively *'large peptides'* exhibit not only *low volatility*, but also afford *narow operational spectrum in mass spectrometry (MS).*

4.1.2.9. Estimation of Net Charge : Amino Acids

It has been proved and established that the *'net charge of a molecule'*, as a specific function of pH, may be determined quantitatively, by taking into consideration the **'critical charges actually borne'** by various *functional moieties* strategically located in the **molecule.** A brief and comprehensive discussion shall be undertaken with particular reference to **amino acids, peptides,** and **proteins,** in a sequential manner, in the section that follows:

There are, in actual practice, *two* different methods which would be treated individually as under:

Method: 1: It is essentially refers to a relatively easy and simple technique that is employed to determine precisely the **'net charge of an amino acid'** at a *critical and specific pH.* Method 1 comprises *three* cardinal steps, namely :

- First and foremost the particular structure of the **amino acid,** is drawn under investigation, in a **'schematic form'** thereby explicitly exhibiting the following *two* aspects:
  - ❑ **ionizable functional moieties,** and
  - ❑ **their respective pre-determined pKa values.**
- Duly assign specific charges of -1, $-\frac{1}{2}$, O, $+\frac{1}{2}$, $+1$ to each and every **ionizable functional moiety** by carrying out a critical comparison pertaining to the **pK of each moiety** *vis-a-vis* the **given pH.**

---

* Kar A : **Pharmaceutical Drug Analysis**, New Age International, New Delhi, 3rd. ed, 2010.

- Finally, determine the **'net charge of the molecule'** with respect to the nearest '$\frac{1}{2}$ unit' by meticulously summing up all the respective **positive and negative charges** *algebraically*.

Importantly, one may have critically observed in the **'Henderson-Hasselbach Equation'** that :

- **at 1 pH unit above the pK:** nearly **90%** of a *functional moiety* is usually present in the **'deprotonated state'**, and

- **at 2 pH units above the pK:** approximately **99%** of a *functional moiety* is normally present in the **'deprotonated form.**

In the same vein, at **less than 1 and 2 pH units** the pK, nearly **90%** and **99%** respectively, of the *functional moiety* is duly available in the **'protonated state'** exclusively. However, based upon these **'calculations'**, it is quite possible to make an **important and vital assumption** that a *functional moiety* is present solely either in the **deprotonated or protonated** state in a situation when the **ensuing pH is one or more pH units either** *above* (or *below*) the observed **pK value.** Besides, the said *functional moiety*, **if charged,** does essentially contribute **one full charge unit** to the amino acid molecule. Hence, the *functional moiety is categorically assigned either a charge of* **–1** or **+1**, depending on the specific instance, provided that :

- pH is either **one unit above, or below the pK of the group,**

- It remains **more pH units above or below the pK of the moiety.**

Assuming a specific condition, when the pH of the *'analyte sample'* **is** *exactly equivalent to the* **pK value, we may have:**

- $\frac{1}{2}$ **of the 'functional moiety'** present duly in the respective *'deprotonated state'*, and

- $\frac{1}{2}$ **of the 'functional moiety'** adequately available in the **'protonated form'.**

Thus, based on the aforesaid *two* assumptions, we may legitimately assign one **'half of a full charge unit'** (viz;, +½ or –½) to the **specific functional group,** when the pH falls very much within the range of pK ± 1.

In order to expatiate generously the actual application of this method, let us estimate the **'net charge of cysteine' at a prevailing pH6.** There are in all *three* cardinal **steps** that need to be followed rigidly in a sequential manner :

Step–1: Structure of Cysteine in Schematic form:

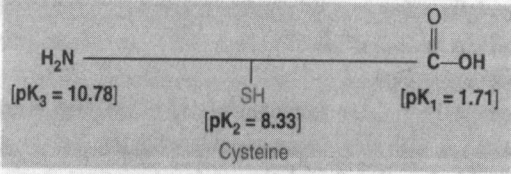

In the above **structure of cysteine in schematic form** the *functional groups* [*viz.*, **amino (–NH$_2$)**; **sulphhydryl (–OH)**; and **carboxylic (–COOH)**] are duly written with their respective *'uncharged forms'*; whereas, the *numbers* in parentheses are the corresponding **pKa values.**

Step–2: Assigning charges on Functional Groups: It may be noted evidently that a **pH of 6** stands at nearly **4 pH units** over and above **pK$_1$**; and, therefore, the respective **carboxy (–COOH)** moiety may be regarded to be present almost *exclusively* and *entirely* in its *'charged state'* (*i.e.*, –COO), and consequently may be assigned a **charge of –1.**

Nevertheless, in an absolutely **contrast situation** the **pH 6** stands at **2pH units** very much below pK_2 in order that the respective **sulphhydryl moiety (–SH)** may be duly regarded to be present in totality in its **uncharged state (SH)**; and, therefore, may be assigned a **charge of zero.**

Finally, **pH 6** obviously stands at nearly **5 pH units** *well below* pK_3, in order that the **amino (–NH$_2$) functional moiety** may be regarded to be present solely in its respective **charged form (+NH$_3$);** and, hence, may be legitimately assigned a **charge of +1.**

Thus, making use of the **schematic representation of cysteine,** one may conveniently and logically indicate the *'charges'* as stated under :

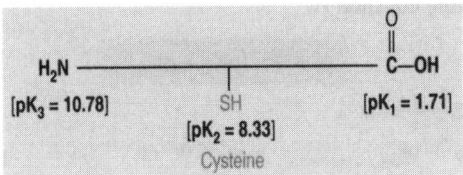

$[pK_3 = 10.78]$ SH $[pK_1 = 1.71]$

$[pK_2 = 8.33]$

Cysteine

Step–3: In fact, on the basis of **Step-3** it may be possible to conclude that at **pH6,** the *amino acid* **cysteine** critically possessess a **'net charge of zero'.**

Further, proceeding in the same manner, one may logically and meaningfully assign the following charges to **'cysteine'** at several altogether different **pH values** as stated below:

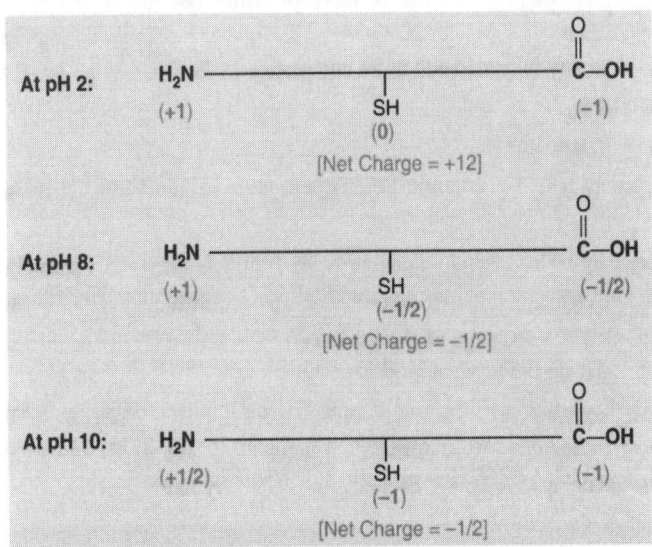

Method: 2 : In a broader perspective, **Method–2,** in *principle* is very much akin to **Method–1;** however, the former being more *elaborate* and *practicable.* Besides, it definitely affords a distinct **charge–pH scale** in order that the **net overall charge of the** *'amino acid'* may be determined almost directly for any desired **pH value.** Essentially, **Method–2** involves the following steps sequentially.

---

* The **'Charge-scale'** usually extends from +2 to –2, because that designates the maximum charge of a *basic* and an *acidic* **amino acid respectively.**

Step: 1– First and foremost construct a **'charge–scale'** extending between **+2** to **–2**, having a **'midpoint at zero'.** Now, divide the entire charge scale into **eight (8) equal segments,** employing **each increment of ½ charge units.*** Besides, a **charge–scale unit of ½** *charge unit* is so selected since ½ of a charge unit is duly assigned to a functional moiety at a **pH** almost equivalent to the **pK** (as we have seen in **Method–1').**

Step:2– Secondly, carefully calculate the **isoclectric point** of the *'amino acid'* under investigation, and duly assign this particular **pH** to the **respective uncharged species on the scale.** Subsequently, the **isoclectric point** is calculated meticulously right from the equation for a **neutral, acidic,** or **basic** *amino acid* as per the following equation :

$$pI = 1/2\ (pK_m + pK_{m+1})$$

where,   pI = Isoelectric point,

m = Maximum number of +ve charges of the compound in a **strong acid solution.***

Step–3: Now, assign **pH values** intelligently to all the different available *'charged species'* of the amino acids :

(a) **pH values** may be assigned that critically correspond to **pk values** *vis-a-vis* the suitably **charged species.** Because, practically in all instances, the actual difference between **pk₁** and **pk₂**, or the prevailing difference between **pk₂** and **pk₂** is either one or more **ph units,** one may conveniently **maintain the same assumption** as that employed in *'Method–1'.* It implies obviously that at a **pH equal to the pK,** the ensuing *functional moiety* is considered to be completeal in the following *two states:*

- **protonated form,** and
- **deprotonated form.**

(b) **pH value** may be assigned almost equivalent to ½ of the **'flanking pk values',** to the **fully charged species.**

(c) **pH values** may be assigned to the ends of the charge scale either by **enhancing** or **decreasing** the last **pK** by **1 unit,** thereby *eliminating non–applicable* **segments of the charge scale.**

Step–4 : Ultimately, the **net charge of the amino acid** is duly estimated using the **'charge scale'** for any **pH value** *vis-a-vis* the **nearest ½ charge unit.**

Example: We may expatiate the various steps that are involved critically in **'Method–2'** by citing the example of the *same amino acid* **'cysteine'** (as shown in **'Method–1';** and really commence by constructing, **'charge scale'** strictly as per **Step–1.** Thus, we may have:

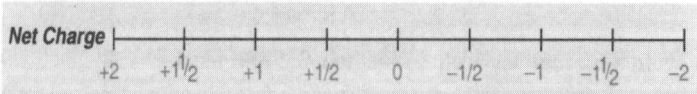

In **Step–2,** the **isoelectric point (pI)** of *cysteine,* as calculated duly from the already stated **'equation;,** stands at **5.02.** Now, this specific **pH value** is precisely inserted in the **'charge scale'** at the strategic point where the **net charge** is almost equal to **zero.**

---

* Always remember that both **cysteine** and **tyrosine** are treated as **acidic amino acids.**

Thus, we may have:

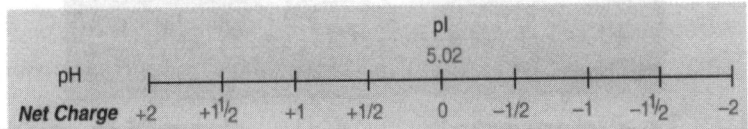

The **Step-3** may now be adopted to assign specific **pH values** to the different 'charged species'. It is invariably illustrated explicitly in the following sections:

(a) **At pH 1.71 [pK$_1$] : Cysteine** possesses the following **charges:**
- For **carboxyl (–COOH) group : –½ ;**
- For **sulphydryl (–SH) group : O ;**
- For **amino (–NH$_2$) group : +1 ;**

in order that the overall **net charge** boils down to +½ **only.**

At pH 8.33 [pK$_2$] : Cysteine possesses the following **charges:**
- For **carboxyl (–COOH) group : –1 ;**
- For **sulphhydryl (–SH) group : ½ ;**
- For **amino (–NH$_2$) group : + 1,**

so that the **overall net charge** boilds down to –½ **only.**

At pH 10.78 [pK$_3$] : Finally, **cylteine** bears the following charges:
- For **carboxyl (–COOH) group : –1 ;**
- For **sulphydryl (–SH) group : –1 ;**
- For **amino (–NH$_2$) group : +½ ;**

in order that the **overall net charge** remains at –1½ **only.**

Based on the above acutal findings, we may, therefore, rightly inscribe the **pH values of 1.71, 8.33, and 10.78** respectively at the corresponding points **+½, –½,** and **–1½ on the folloiwng** charge scale :

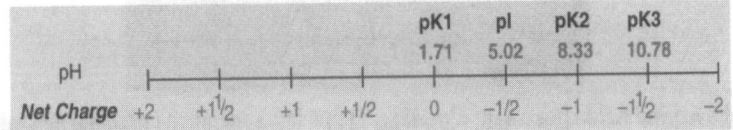

(b) **At pH 9.56** [*i.e.,* ½ (8.33 + 10.78)], may be assigned a value to the –1 charged **species** (of *amino acids*) which being critically flanked by **pK$_2$** and **pK$_3$,** as given under :

(c) **At pH 11.78** [*i.e.,* 10.78 + 1], may be assigned a value to the **+ve end,** and a **pH value** of (1.71–1 = 0.71) to the respective –ve end of the **charge scale.** Importantly, at these **pH values,** the molecule (**amino acid**) bears a *net charge* of **+1** and **–2** respectively, that eventually designates the 'maximum **+ve and –ve charges**' prevailing upon **cysteine.**

Example: At rather **'Low pH'** (*i.e., strongly acidic solution*), **cysteine** possesses a definite net charge of **+1**; whereas, at a critical *strongly basic solution* (*i.e.,* **higher pH values**), **cysteine** shows a net charge of **–2.** From the aforesaid statement of facts, one may safely infer that **'cysteine'** the *charge scale ranging between +1 to +2* is *not* applicable at all; and hence, may be eliminated altogether. Thus, we may finally do have the following **pH/charge scale for cysteine:**

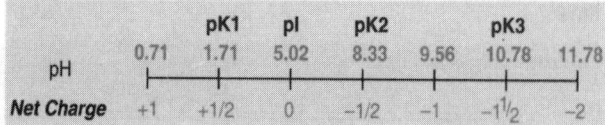

		pK1	pI	pK2		pK3	
	0.71	1.71	5.02	8.33	9.56	10.78	11.78
pH	⊢——+——+——+——+——+——+——⊣
Net Charge	+1	+1/2	0	−1/2	−1	−1½	−2

According to **Step–4**, one may now make use of the above **'charge scale'** to determine exactly the **overall net charge** of **'cysteine'** to the nearest ½ **charge unit**. Following *data* shows an evidence of the same net charges as *'computed earlier'* for the respective **pH values** employed for *'Method–1'*:

pH	Net Charge
2	+1/2
6	0
8	−1/2
10	−1½

Note: Based on the above derived scale for *'cysteine'*, one may promptly and directly estimate the *net charge at any pH* without any further consideration of the ionication involved.

Application of Method–2 in pH/charge scale of Neutral, Acidic, and Basic 'Amino Acids': The fundamental basics as applicable to the various types of **amino acids** by *Method–2* are illustuated along with one specific example drawn in each instance as disussed under :

[A] Neutral Amino Acids [*viz.,* Ala, Gly, Ile, Met, Phe, Pro, Ser, Thr, Trp, Val]: The generalized **pH/charge scale** of the *neutral amino acids* are as given under :

	(pK₁−1)	pK₁	pI	pK₂	(pK₁+1)
pH	⊢——+——+——+——+——⊣
Net Charge	+1	+1/2	0	−1/2	−1

Example: Alanine [H₃C–CH (NH₂)–COOH] *i.e.,* α–amino propionic acid.

	pK₁	pI	pK₂		
	1.34	2.34	6.02	9.69	10.69
pH	⊢——+——+——+——+——⊣
Net Charge	+1	+1/2	0	−1/2	−1

[B] Acidic Amino Acids [*viz.,* Asp, Cys, Glu, Tyr] : The generalized **pH/charge scale** of the *acidic amino acids* are as stated under :

	(pK₁−1)	pK₁	pI	pK₂		pK₃	(pK₃+1)
pH	⊢——+——+——+——+——+——+——⊣
Net Charge	+1	+1/2	0	−1/2	−1	−1½	−2

Example: Glutanic Acid [HOOC–CH₂–CH₂CH (NH₂)–COOH] *i.e.,* α–amino glutaramic acid,

	pK₁	pI	pK₂		pK₃	
	1.19	2.19	3.22	4.25	6.96	9.67	10.67
pH	⊢——+——+——+——+——+——+——⊣
Net Charge	+1	+1/2	0	−1/2	−1	−1½	−2

[C] Basic Amino Acids [*viz.,* **Arg, His, Lys**]: The generalized **pH/charge scale** of the *basic amino acids* are as given below:

Example: Histidine CH$_3$–CH (NH$_2$)–COOH] *i.e.,* α–amino–β–imidazo–propionic acid.

		pK$_1$	pI	pK$_2$		pK$_3$	
	0.82	0.82	3.91	6.0	7.59	9.17	10.17
pH	├────	──┼──	──┼──	──┼──	──┼──	──┼──	──┤
Net Charge	+1	+1/2	0	–1/2	–1	–1½	–2

Interestingly, based upon these aforementioned *pH/charge scales,* we may now exactly and precisely estimate the **overall net charge** of these **amino acids** at *pH 5.0* to be approximately :

Amino Acid		Overall Net Charge
Alanine	:	0
Glutanic Acid	:	–½
Histidine	:	+½

4.1.2.10. Estimation of Net Charge : Peptides

It has been duly demonstrated that both of the methods (*viz.,* **Method–1,** and **Method–2**), as discussed above for the **'amino acids',** may also be employed to estimate the **'overall net charge'** of a *peptide.*

Necessarily, we may have to make the following *three* **cardinal assumptions,** namely :

❑ **Observed pk values** are absolutely identical to those present in the *'free amino acids,*

❑ **Observed pk values** are *not* affected by the strategical location of various *'functional moieties'* very much present within the **'peptide residue',** and

❑ **Observed pk values** may *not* be grossly *influenced* and *affected* by the **state of ionization of the neighbouring moieties.**

Incidentally, for both methods it is always gainful to draw the correct *structure of the peptide* in the usual **conventional manner** by making use of:

• *Standard abbreviations* for the ensuing **amino acids** and

• placing the respective **'N–terminal'** upon the **'left',** while the **'C–terminal'** on the **'right'** hand side.

In addition, the different **ionizable functional moieties** are duly indicated schematically, and the corresponding **pk values** are written just next to each moiety.

Example: Peptapeptide is a typical example, and the **two methods** shall be employed to determine the overall **net charge** at **pH 7.0** as given below:

Cys—His—Asp—Tyr—Ser

Method–1 : It has *three* essential steps, such as:

H$_2$N—Cys—His—Asp—Tyr—Ser—COOH
(10.78) | | | | (2.21)
Step–1 :   SH IM COOH OH
(8.33) (6.0) (3.86) (10.07)

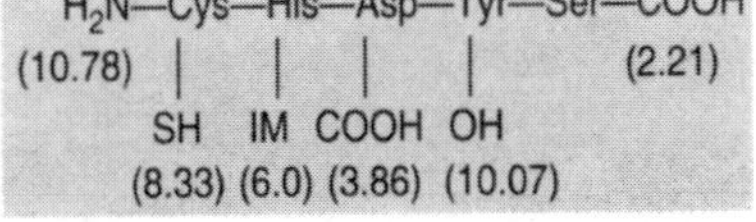

where, **Im–** represent (COOH) *Asp* and (OH) respectively; besides;

- **imidazole moiety (*His*),**
- **side–chain (β–COOH moiety), and**
- **phenolic hydroxyl (OH) group.**

Step–2:

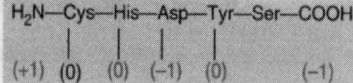

Step–3: The overall net charge is given by (+1–2) = **–1.**

Method–2: It essentially comprises *four* **cardinal steps :**

Step–1 : To enable the proper use of **Method–2** for *peptides,* the appropriate **'charge scale'** should be carefully adjusted so as to correctly reflect the **'range of changes'** which may occur in the specific **'peptide'** under investigation. In the present example of a **'pentapeptide'**, one may critically *construct* the follooing **'charge–scale':**

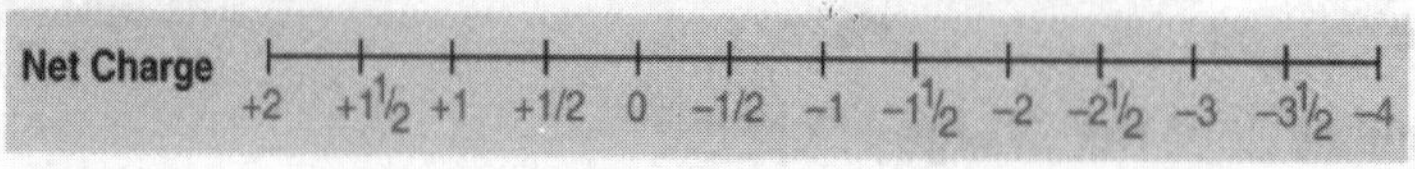

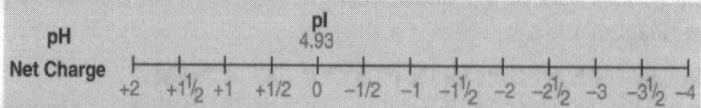

Step–2: The **Isoclectric point (pI)** is duly calculated based on the earlier mentioned **'Equation;** *viz.,*

$$pI = \frac{1}{2}(pK_m + pK_{m+1})$$

In this instance, we have:

$$pI = \frac{1}{2}(pK_2 + pK_3)$$
$$= \frac{1}{2}(3.86 + 6.0)$$
$$= \mathbf{4.93}$$

Therefore, the following **'charge scale'** may be constructed aptly:

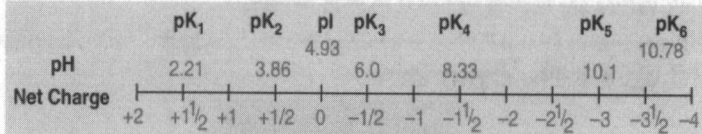

Step–3: There are in all *three* segments in **Step–3,** which shall be discussed briefly as under:

Segment (a) :

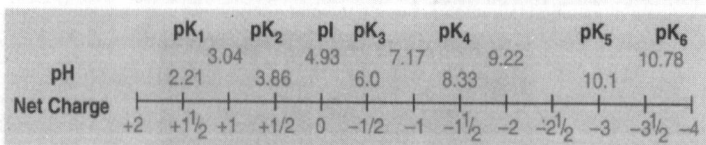

Segment (b) :

Segment (c) :

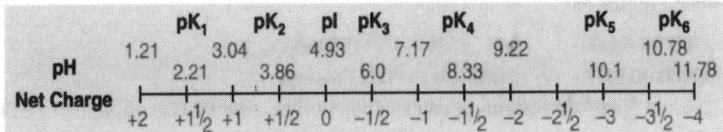

Step–4: Based on the above *pH/charge scale* one may arrive at:

- pH of the 'peptide' to be 7.0, and
- overall net charge of the 'peptide' to be –1.

4.1.2.11. Estimation of Net Charge : Proteins

The overall **net charge** of a **'protein'** may be exclusively determeined by an appropriate extension of *Method–1,* as employed above for the **amino acids** (*section 4.1.2.9*) and the **peptides** (*section 4.1.2.10*). Nevertheless, the exact and precise application of the ensuing **method for proteins** predominantly involves the identical and similar *three* cardinal assumptions already made in the case of **'peptides'**. The approach is vividly depicted for the case of **'peptides'**. The approach is vividly depicted for the typical enzyme *'lysozyme'* by means of the *'computational method'* as given in Table: 2.2.

Table: 2.2: Charge Characteristics of Lysozyme at Different pH Values

Functional Moiety			Charge Characteristics			
Number	Variants	pK*	pH 7		pH 9	
			Group	Total	Group	Total
1	α–Carboxyl [Leu]	2.36	–1	–1	–1	–1
8	β–Carboxyl [Asp]	3.86	–1	–8	–1	–8
2	γ–Carboxyl [Glu]	4.25	–1	–2	–1	–2
1	Imidazole [His]	6.0	0	0	0	0
8	Sulphhydryl [Cys]	8.33	0	0	–1	–8
3	Phenolic [Tyr]	10.07	0	0	0	0
1	α–Amino [Lys]	8.95	+1	+1	+½	+½
6	ε–Amino [Lys]	10.53	+1	+6	+1	+6
11	Guanido [Arg]	12.48	+1	+11	+1	+11
Net Charge				+7		–1½

4.2. Secondary Structure

It has been established beyond any reasonable doubt that the ensuing **polypeptide chain (s)** critically undergo **'regular folding or coiling'** to give rise to the formatin of a **secondary structure of protein.** In reality, such coiling of the typical **'polypeptide chain'** is invariably generated as well as stabilized by *two* characteristic type of **bonds,** such as:

- **weak hydrogen bonds,** and
- **low–energy non–covalent hydrogen bonds.**

---

* Pertaining to the **'free amino acid'**.

Generally, the **'hydrogen bonds'** are duly formed by sharing mutually a **single hydrogen atom** between *two* such entities as:

- **electronegative 'O'–atom (of = CO) moiety,** and
- **electronegatve 'N' atome (of = NH) moiety,**

belonging to the **'peptide bonds'** present either in the *same* or *two* altogether **different adjacent polypeptide chains.** Interestingly **congregation of H–bonds** may serve as a reasonably **strong entity** so as to stablize the so called **'secondary structure'.** One may usually come across *two* most predominant and prominent types of the **'secondary structure of proteins',** namely :

- ❏ **α–Helix structure,**
- ❏ **β–pleated sheet structure, and**
- ❏ **Triple Helix structure**

which shall be discussed in the sections that follows:

4.2.1. α–Helix Structure

Pauling and Corey (1951)* was pioneer in putting forward the **α–helix structure** whereby the *'polypeptide chain'* gets meticulously **coiled helically** to produce a *regular helix structure* as shown in Fig: 2.13, and is termed broadly as the **'α–helix'.**

In true sense, the so called **'right–handed α–helix'** is obviously found to be the most abundantly available *secondary structure in proteins.* In addition, it is regarded to be the one which is permitted to occur very much within the ensuing restrictions duly imposed by the **geometry of the peptide bone,** plus the **allowed rotation** about the φ and ψ angles.

> **Note: The above finding was suggested evidently to exist in proteins but only estabished as taking place in a** *'globular protein'* **when the structure of** *myoglobin* **was determined duly.**

Fig: 2.14. shows explicitly the diagramatic representation of the folding of a polypeptide chain of carp muscle calcium–binding protein. Besides, the **amino (NH₂) terminus** as well as the **carboxyl**

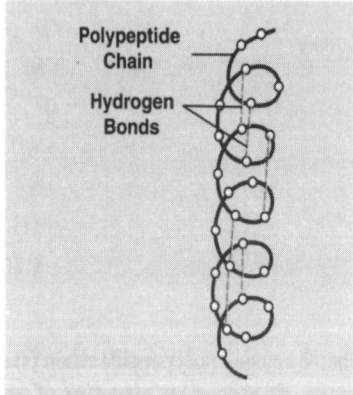

Fig: 2.13: Structure of α–Helix

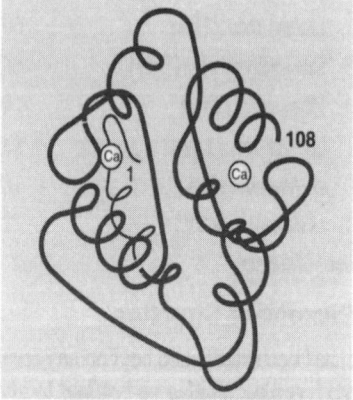

Fig: 2.14: Schematic Diagram Showing Folding of a Polypeptide Chain of Carp Muscle Calcium-Binding Protein. [Adapted From : White A et al.: Principles of Biochemistry, Tata Mcgraw Hil, New Delhi, 6th ed, 2004.]

* Pauling L and Corey RB : *Proc. Natl Acad Sci.,* **37** : 729-740, 1951.

(COOH) terminus are clearly indicated by the help of the numerals **1 and 108.** In all there are *six helical segments* are duly present in the said molecule.

Note: Here, the *R–group* are not shown, including those which essentially bind the *two Ca²⁺ atoms*.

Fig: 2.15. illustrates explicitly the diagramatic folding of the ensuing **polypeptide chain** pertaining to the **Ca²⁺–binding protein** from the *carp muscle,* comprising the *six α–helical regions,* **duly formed by residues 40 to 51 in the respective molecule.** In this Fig. the ribbon (*marked with colour*) precisely follows the **inherent polypeptide back bone** for one turn of the α–helix. However, the respective *bond angles* and *bond distances* are adequately demarcated by the help of 'sticks' joining **C, N, or O atoms,** that are duly represented by the 'balls'. Besides, the **H–bonds** are shown prominently by means of the *'dotted lines',* and are duly formed between two **O atoms** indicated duly.

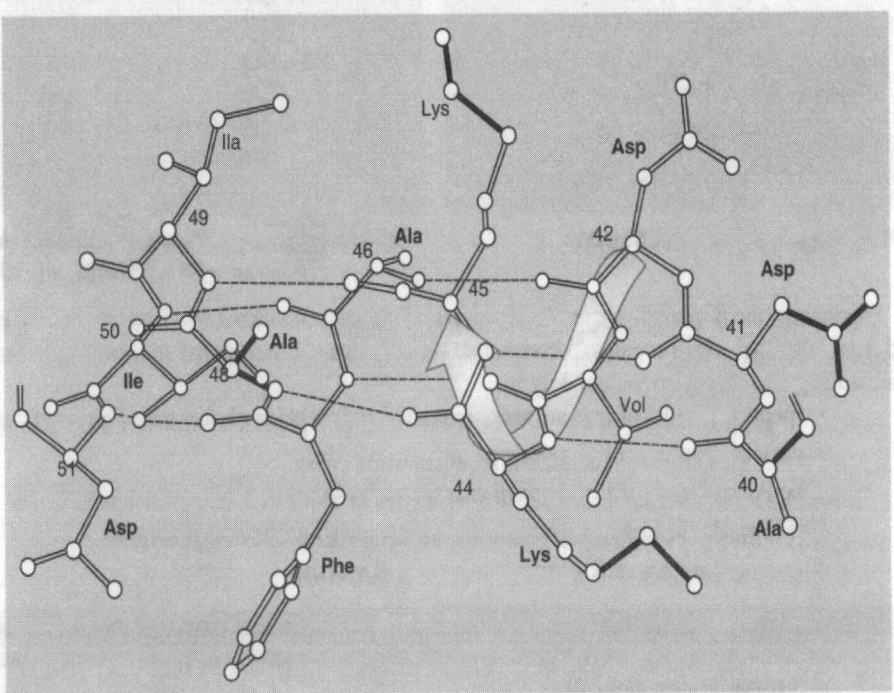

Fig: 2.15 : Representation of a Model Showing the Helical Segments (Residues 40 to 51) Present in Carp Muscle Ca²⁺ Binding Protein.

Here, the H–atoms are not shown. The side of the **helix,** which being nearest to the reader is **hydrophilic in character;** whereas, the one located in the **opposite side** (*i.e.,* away from the reader) is **hydrophobic in nature,** and hence comprise **side–chains** that solely contribte to the **hydrophobic region.**

4.2.2. β–Pleated Sheet Structure

The terminology 'β' designates that it was indeed the **'second structure'** duly proposed by **Pauling and Corey (1951)** soonafter the **'α–Helix Structure'** (section 4.2.1.). In reality, the β–pleated **sheet structure** is essentially composed of **two or more** altogether *separate polypeptide chains* which are arranged meticulously side–by–side (unlike the *α–helix structure*) thereby resembling a series of *'pleats'* usually known as **'β–plcated sheat',** as depicted in Fig: 2.16.

Besides, the so called β–**pleated sheet** gets duly formed between various segments of the same prevailing *'polypeptide chain'*, as shown in Fig. 2.17. Obviously, the β–**pleated sheat** invariably shows the ensuing **polypeptide chain** being critically extended into a **zig-zag manner** rather than a **helical structure** (Fig: 2.16).

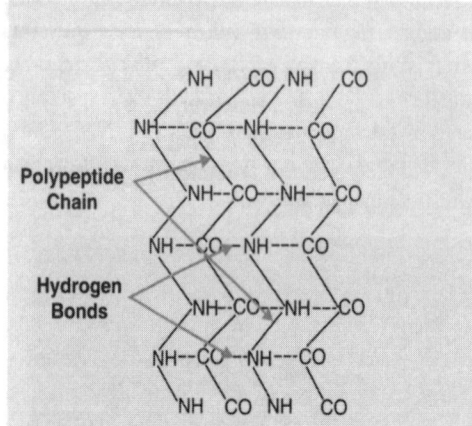

Fig. 2.16: β-Pleated Sheet Structure

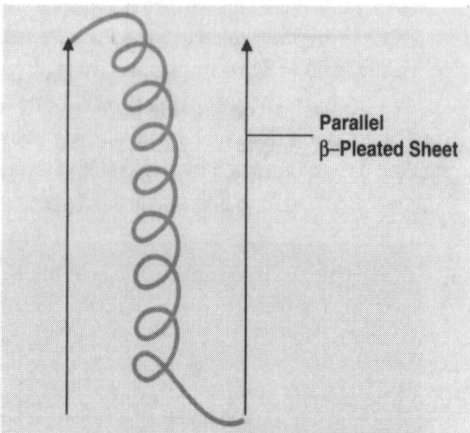

Fig. 2.17: Mixed α-Helix and β-Pleated Sheet Present in Same Polypeptide Chain

Importantly, the β–**pleated sheet structure** gets adequately *stabilized* by virtue of the '**interchain hydrogen bonding**' which eventually comes into being between the *carbonyl oxygen* (= CO), and the **amide nitrogens** (–CONH$_2$) present duly in:

- '**peptide bonds' of two adjacent extended polypeptide chains** (see **Fig: 2.16**) , and/or
- **various segments of the same polypeptide chain.**

Examples: Following are the *two typical* examples of β–**pleated** sheet structures, such as :

❑ '**Fibroin**' of spider-web comprises **anti-parallel β–plcated sheets,** and

❑ '**β–Keratins**' *e.g., silk fibroin* present in **silk-worms.**

> **Note:** *'Globular Proteins' viz.,* **Fibrinogen (in blood), Immunoglobulins (Igs) and Haemoglobin (Hb)** essentially do possess both the '*α–helix'* as well as the '*β–pleated sheet structure'* duly occurring in the same protein molecule.

Pauling and Corey (1951) made on exceptionally clear diagramatic sentation of the β–**pleated sheet structures** exhibiting both :

- **antiparallel–chain structure–as shown on the 'left', and**
- **parallel–chain structure–as shown on the 'right',**

in the following Fig: 2.18.

Antiparallel pleated sheet structure (or Autiparallel chain structure)–is duly formed by the aid of the **extended polypeptide chains** whose strategically positioned *'sequences'*, with regard to the actual direction of the chain i.e., **amino (–NH$_2$)** to **carboxy (–COOH)** *terminal residues,* usually run in the **'opposite directions'.**

Parallel pleated sheet structure (or Parallel chain structure)–is duly formed by chains running in the **'same direction'.**

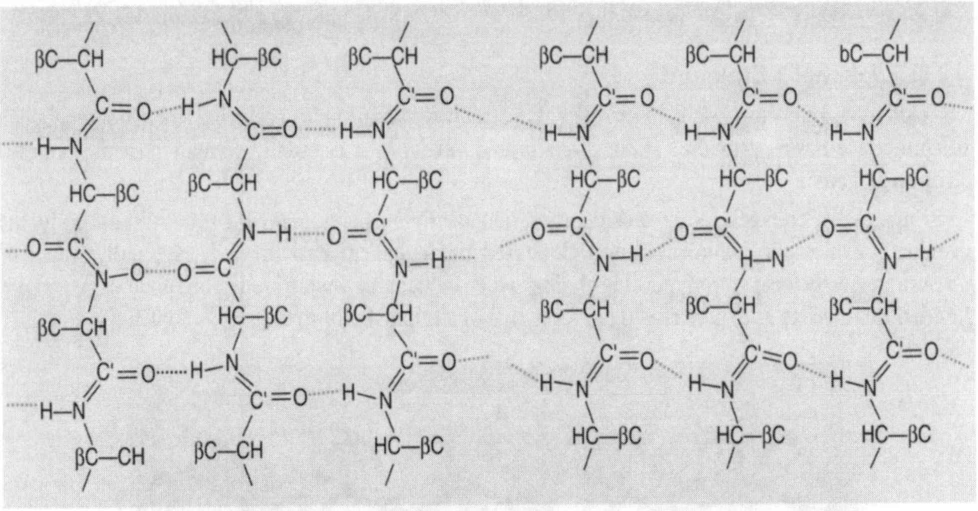

Fig: 2.18: Pauling and Corey's Diagramatic Representations of the β–Pleated Sheet Structures.

Amazingly, these *two* aforesaid structures are fabulously **stablilized** by dint of the exceptional formation of the–'**extensive network of H–bonds caused due to the atoms present in the** *peptide bonds* **of adjacent chains'.**

Interestingly, Fig. 2.19 illustrates diagramatically the folding of the *polypeptide chains* that are critically present in the '**flavodotoxin**' (*Clostridium MP*), that categorically comprise an appreciable

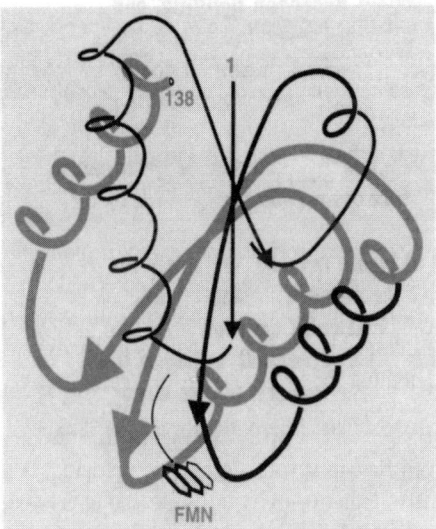

Fig: 2.19: Flavodotoxin: Showing Schematic Diagram of its Polypeptide Chain.

quantum of β–**pleated structure.** The '**flavodotoxin**' essentially contains in all **138 residues** together with a molecule of *noncovalently–bound* **flavin monoucleotide (FMN).** In fact, the arrow do point out the actual direction of the **parallel pleated sheet strands** [*e.g.,* NH$_2$ to COOH terminus].

Note: It has been amply observed that the *'parallel pleated sheets'* invariably fail to form *'cylinders'*.

4.2.3. Triple Helix Structure

Collagen, refers to a family of *extracellular, closely related proteins* occurring as a major component of **connective tissue,** providing it due *strength* and *flexibility.* it is usually composed of molecules of **'tropocollagens'.**

Importantly, **collagen,** designates the most abundantly available **protein in the human body,** that eventually exhibits an **'elaborated and elongated triple–helical structure'.** It essentially comprises *three* strategically located **'polypeptide chains'** which meticulously wrap around one another in a **'right–handed manner'** so as to give rise to the formation of a **triple–helix structure,** as depicted in Fig. 2.20.

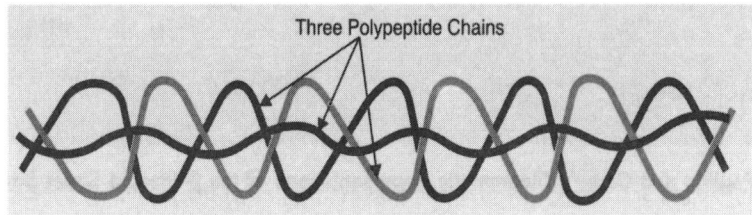

Three Polypeptide Chains

Fig: 2.20: Triple–Helix Structure of Collagen.

4.2.3.1. Stabilization of Triple–Helical Structure of Collagen

It may be accomplished adequately by means of the following *two* factors:

* **non-covalent interchain hydrogen bonding,** and
* **covalent bonds existing between the *three* polypeptide chains.**

–How does 'collagen' form triple–helix structure? It does so *via* the following means, namely:

* **Collagen** being rich in such *amino acids* as : proline, **hydroxyproline**, **lysine**, and **hydroxy–lysine**.
* **hydroxy–proline residues** do play a pivotal role in causing the stabilization of the ensuing **triple–helical structure** of the *collagen molecule* by making it possible to form **additional H–bonding.**
* **hydroxy-lysine and lysine residues** are solely responsible for establishing the **covalent cross-linkages** with the **adjacent collagen molecules.**
* **neighbouring individual collagen molecules** do associated critically *via* the ensuing **covalent cross-linkages** to generated **'collagen fibrils',** which proves to be an absolute necessity for attributing the **'superb tensile strength of collagens'.**

4.2.3.2. Secondary Structures Can Form Supersecondary Motifs:

It has been duly observed that in several **'globular proteins'** (*e.g., albumins, globulines*), the **secondary structural motifis** of either α–helix or β–pleated sheet structure usually give rise to the recognizable **'super secondary motifs',** as shown in Fig. 2.21. This clearly illustrates the presence of several **supersecondary motifs,** such as :

* **β–α–β** *i.e.,* two trands of β–**pleated sheets** connected by an α–helix,
* **β–hairpin–** which is duly composed of **antiparalled β–pleated sheets** by short regions of loop, and
* **'Greek key' motif*.**

Amazingly, the obvious repititions of these **'super secondary motifs'** may then generate such structures that appears to be the **regular repitive β–α–β units** of an *entire protein* as depicted in in Fig. 2.22.

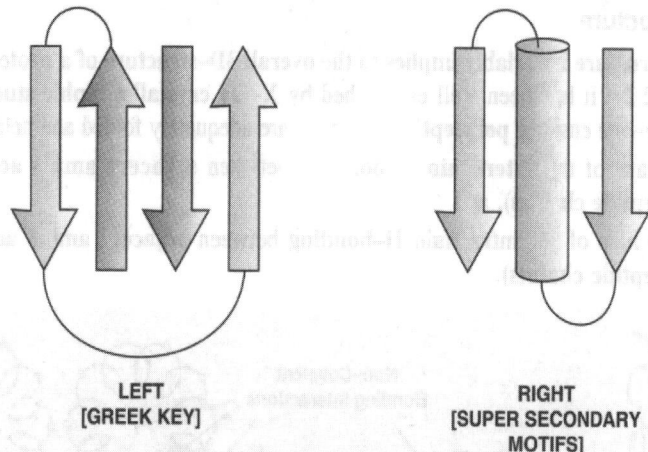

LEFT
[GREEK KEY]

RIGHT
[SUPER SECONDARY
MOTIFS]

Fig: 2.21 : Supersecondary Motifs. [The α–helices and β–pleated sheets of several 'Globular Proteins' have been duly arranged in *repeatng units viz.,* (Left-hand side–*Greek key*), and (Right -hand side-*supersecondary motifs*).

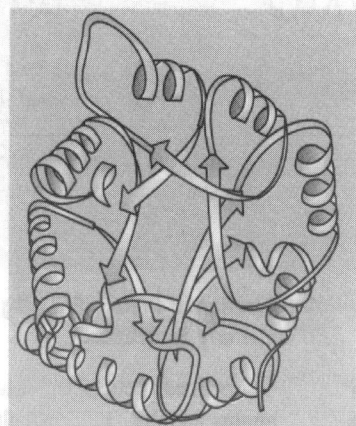

Fig: 2.22: Tertiary Structure. [*Triose phosphate isomerase,* depicated duly as an *'end-on view'*– is made from *four* β–α–β units which are found to be consectire in both *'primary'* and 'tertiary' structural sense.

In other words, the terminology **'tertiary structure'** usually refers to the'**spatial relationships'** taking place inbetween the so called *secondary structural structures* of relatively larger proteins invariably are organized as the articulated *'domains' i.e.,* such compact units which are duly connected by the so called *'polypeptide backbone'*. In fact, the **tertiary structure** predominantly elaborates the **prevailing relationships of these domains** *i.e.,* the means by which the phenomenon of *protein folding* may bring together various **amino acids :**

* Actually named for its resemblance to a decorative motif on the ancient **Greek Flower Vases.**

- poles apart in a 'primary structural sense', and
- bonds which critically stabilized these conformations.

4.3. Tertiary Structure

The **tertiary structure invariably implies to the overall 3D–structure of a protein molecule**, as illustrated in Fig. 2.23. It has been well established by **X-ray crystallographic studies** that in the *secondary structure–* the ensuing **polypeptide chains(s)** are adequately folded and held together :

- by means of the **interchain H–bonding** between **adjacent amino acid residues of polypeptide chain(s)**, and
- by the help of the **intrachain H–bonding** between **adjacent amino acid residues of polypeptide chain(s)**.

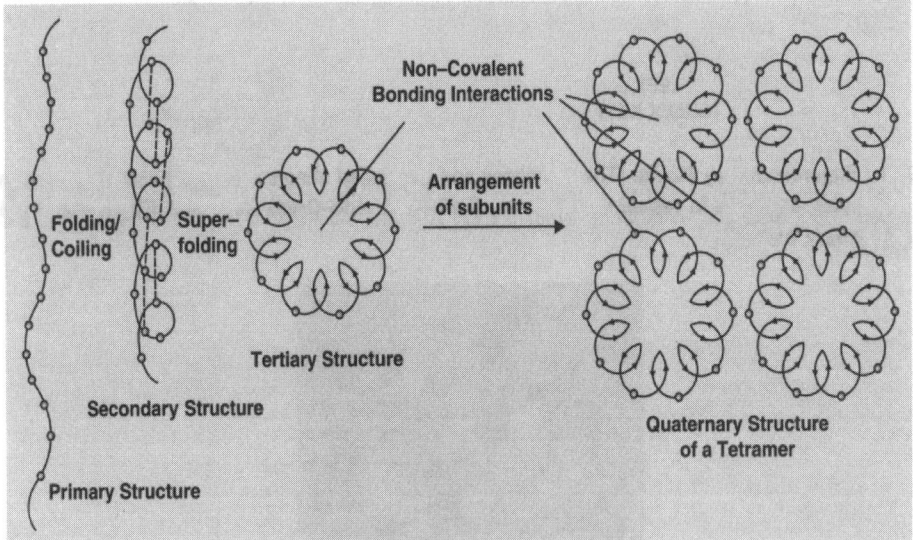

Fig: 2.23 : The Diagramic Structural Organization of Protein Showing Primary–Secondary–Tertiary– Quaternary Structures.

Nevertheless, the **tertiary structure** is obtained meticulously due to the intricate **folding** and **superfolding** of the ensuing **polypeptide chains** thereby exhibiting ultiately the so called **secondary structure.**

4.3.1. Stabilization of Tertiary Structure

It ay be accomplished by several procedures, such as :

- **making use of different weak non-covalent bonding interactions**, and
- **due to the presence of covalent disulphide (–S–S–) bonds,**

that are invariably produced between different moieties of the **amino acid residues,** which incidentally are located strategically *far away* in **polypeptide sequence** and those present quite adjacent very much within the completely **folded structure of a protein.**

The different cardinal **non-covalent bonding interactions** that prominently maintain and stabilize the **tertiary structure** are, namely :

- hydrophobic interactions,
- ionic interactions,
- von der Waal's forces, and
- hydrogen bonds,

which shall be discussed briefly in the sections that follows:

4.3.1.1. Hydrophobic Interactions

These specific reactions are duly caused due to the presence of the **non-polar side chains** of **amino acids** *e.g.,* **alanine, isoleucine, leucine, methionine, phenylalanine** and **valine.** In a broader perspective, these **amino acids** crtically promote the specific **hydrophobic** interactions in the *protein molecule;* and, therefore, do play a pivotal role in causing **stabilization** of the *'tertiary structure.* However, one may obviously observe the predominant presenc of :

- the *non-polar amino acids cluster'*– found in the **interior surface** of the folded protein molecule, and
- the *'polar amino acids cluster'*– found on the **exterior surface** of the folded protein molecule.

4.3.1.2. Ionic Interactions (or Electrostatic Bonds)

In actual practice, these highly specialized and typical bonds are adequately established between the **positively charged amino** $(-N^{\oplus}H_3)$ **moieties** of *basic amino acids* (*e.g.,* **Arg; Lys; Lys-OH**), and the **negatively charged carboxylic** $(-COO^{\ominus})$ **moieties** of *acidic amino acids* (*e.g.,* **Asp ; Glu ; Asn ; Gln;**).

4.3.1.3. von der Waal's Forces

They do represent relatively very weak forces which usually develop between the non-polar side chains of the **neutral amino acids.**

4.3.1.4. Hydrogen Bonds

They designate the **non-covalent bonds** that are invariably formed by the mutual sharing of a **single H–atom** by another **electronegative O–atom** (*e.g.,* of = CO moiety) as well as **electronegative N–atom** (*e.g.,* of = NH group) present commonly in the *peptide linkages.* However, it is pertinent to state here that the 'hydrogen bods' do play a very important and vital role in the overall stabilizaton of the so called *'structural organization of protein'.*

4.4. Quaternary Structure

Based on scientific evidences it has been proved that **'proteins'** possessing *two* or *more* **polypeptide chains,** namely :

- **Lactate dehydrogenase (LDH),**
- **Creatine phosphokinase (CPK),**
- **Glycogen phosphorylase (GPP),**
- **Asparate transcarbamoylase (ATCM),**
- **Aldolase, and**
- **Haemoglobin.**

---

* However, the **peptide** and **disulphide bonds** are not ruptured.

invariably exhibit a prominent **4th level of structural organization** usually termed as the **'quaternary structure'**. Besides, such highly specialized **proteins** are frequently known as:

- ❏ **oligomeric proteins** (or **'oligomers'**),
- ❏ **multimeric proteins** (or **'multimers**).

Besides, a **separate/individual polypeptide chain** is invariably termed as a *'subunit'* or a *'monomer'*.

There are several cardinal aspect pertaining to the **'quaternary structure'** which may be considered briefly as stated under :

4.4.1. Oligomeric Proteins : Multiple Polypeptide Chains

In this specific instance the **electrostatic bonds** and the **hydrogen bonds** critically formed in between the strategic surface residues of adjacent subunits actually cause the **stabilization** of the association of **monomers** (or **subunits).**

There are several variant in the **'oligomeric proteins',** such as:

❏ **Dimeric or Tetrameric Proteins**	:	Composed of 2 or 4 **subunits** respectively,
❏ **Homodimers, Homotetramers**	:	Comprise several **identical subunits,** and
❏ **Heterooligomers**	:	**Consist** of **dissimilar subunits.**

In a broader perspective, one may discretely understand that the various subunits of *'heterooligomeric proteins'* may perform **typically distinct function.**

It may be observed that while **one subunit** or **a set of identical subunits** may particularly carry out a *catalytic function;* at the same time another **set of a subunit** may categorically perform:

- • **a regulatory role,** or
- • **a ligand recognition role.**

Importantly, a host of exceptional roles taking place in the *'intracellular regulation'* are remarkably governed by the altogether different **'spatial orientation'** of the ensuing **subunits** thereby attributing specifically changed characteristic features upon the **'oligomer'** and **'multimeric proteins'.**

4.4.2. Disruption in Secondary, Tertiary, and Quaternary Structure Due to Protein Denaturants:

There are quite a few highly specific regents (*protein denaturants*), for instance : **sodium dodecyl sulphate (SDS), urea, mile H$^+$, and OH$^-$** which categorically help in the complete/partial rupture of such bonds as:

- • **hydrogen bonds,**
- • **electrostatic bonds,** and
- • **hydrophobic bonds*.**

In this manner, the aforesaid **protein denaturants** (or **reagents**) usually bring forth complete disruption in *all orders of protien structure* thereby destroying **'bioligical activity'**, with the *primary structure* being an exception as shown below in Fig : 2.24.

Amazingly, the extremely **'thermostable enzymes'** duly obtained from the so called *'extreme theremophilies'* which usually inhabit in **hot–spring, hot–sulphur springs,** and **hot–deep ocean vents** quite intimately show a resemblance to the corresponding proteins duly obtained from the **'mesophilic organisms'.**

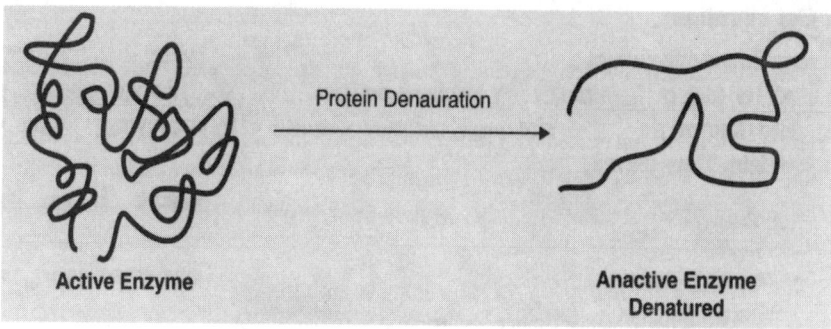

Fig: 2.24 : Diagramatic Representation of Denaturation of a Protein.

In fact, there have been no clear cut generalized *rules, norms,* and *guidelines* to accomplish theremostability caused by such methodologies as :

- **shorter surface–loops,**
- **increased packing of hydrophobic residues,**
- **optimized α–helix composition,**
- **optimized hydrogen bonds,** and
- **optimized ion-pairs.**

4.4.3. Revelation of Molecular Weight and Quaternary Structure: Due to Physical Methodologies:

The fundamental basis for the exact and precise estimation of the **'quaternary structure'** pertaining to the *oligomeric proteins* essentially involves the following cardinal steps :

- **determining the number and kind of promoters present articulately,**
- **mutual orientations found amongst them,** and
- **critical interaction which unite them intimately.**

Assuming that the **oligomers** fail to undergo the phenomenon of *denaturation* in the course of the definite procedure being adopted to determine **molecular weight,** several techniques may provide molecular weight data.

> **Note:** It is possible to make use of these very methodical procedures so as to determine the *'protomer molecular weight'*, if the *'oligomer'* is first determined accurately.

Following are *four* well-known and recognized *physical methodologies* used for the actual determination of *molecular weight* as well as *quaternary structure,* namely:

4.4.3.1. Electron Microscopy:

It vividly visulizes the **macromolecular complexes.** Interestingly, a versatile **'electron microscope'** may distinctly **enlarge** (or **magnify**) up to 10^5 **diameters**; and thus visualized a plethora of important objects, such as :

- **enzyme complexes,**
- **virus particles,**
- **oligomeric proteins,** and
- **high-molecular-weight oligomers.**

4.4.3.2. Gel Filtration:

The versatile and effective technique of **'gel filtration;** may be adopted generously by using columns of **Sephadex**[(R)] or such other commercially available **matrices** which essentially do possess **'pores'** of pre-determined (**known**) *particulate size range,* and duly **'calibrated'** by employing **proteins of known molecular weight.** Thus, we may have:

- molecular weight of an *'unknown protein'* duly calculated based critically on its *'elution status' vis-a-vis* to these standard, and
- necessary precautionary measures may be taken to avoid *'huge errors'* in such instances wherein the **protein** (under investigation) remain to be :
  - ❑ **highly asymmetric in nature,** and/or
  - ❑ **interacts overwhelmingly with the material*.**

4.4.3.3. Polyacrylamide Gel Electrophoresis [PAGE]:

In actual practice, both **standard** and **unknown proteins** may be separated effectively by **polyacrylamide gel-electrophoresis [PAGE].** It may be carried out in a **5-15% cross-linked polyacrylamide gels** having *'varying porosity',* that are subsequently treated with :

- **Coomasic Blue Stain–** For protien, or **Ag⁺ ions.**

SDS–PAGE is invariably employed to estimate the exact and precise **subunit size of oligomeric proteins.**

Procedrue: The various steps adopted are as follows:

(1) First of all **denature** the **proteins** by carefully boiling it with:

- **β–mercaptoethanol [HS CH₂CH₂OH],** and
- **sodium dodecyl sulphate [SDS]** *i.e.,* **a –vely charged ionic detergent.**

(2) The **denatured proteins**** thus obtained are duly separated on the basis of their actual respective **dimension** (*not their charge*) by **SDS–PAGE** upon gels which contain **SDS.**

4.4.3.4. Sucrose Density Gradient Centrifugation [SDGC]:

In reality, this related methodology is best applicable to the **'globular proteins** *e.g.,* **α–globulin, γ–globulin.**

SDGC may be accompished by using **analytes** (*unknowns*) and **protein standards** that are layered with extreme care and skill upon a gradient of **buffered sucrose (5-20%),** usually prepared by **repeated freeze-thawing of 20%),** surcose. *Centrifugation* is done at $10^5 \times$g after a duration of **16-20 hrs** where upon the clear contents are removed slowly (dropwise), and subjected to **protein analysis.**

Calculation of Molecular Weight of Analyte: The mobility of the **'analyte'** (*i.e.,* **unknown protein**) is meticulously expressed *vis-a-vis* the **protein standards** of *pre-determined molecular weight* using a *computer-data base,* and its **molecular weight** is calculated to its **nearest correct value.**

---

* That is, from which the **'molecular sieve'** has been duly manufactured.

** They are all duly coated with **SDS**; and, therefore, are **charged negatively.**

5. AMINO ACID (PROTEIN) METABOLISM

How does Protein (Amino Acid) Metabolism take place in humans?

Protein is an important integral component of diet, which being provided by a host of natural resources *e.g.,* milk, milk products, pulses, Soyabean, **Spurulina**[(R)*], meat, egg, fish, crabs, and poultry products. In general, a normal human being receives approximtely **50-70g** of *dietary protein* each day, and most of which get duly digested in the **gastrointestinal tract (GIT)** by the critial presence of the so called **'proteolytic enzymes',** such as : **pepsin, chymotrypsin, trypsin** and the like. These proteins are eventually converted into their respective constituent *L–amino acids* that usually get obsorbed into the **portal blood system** *via* the long ailimentary canal (*i.e.,* **intestine**), and finally get transported to liver.

> Note: All *'dietary proteins'* do serve as an exogenous source of most amino acids.

Salient Features of Proteins: Following are the **salient features of proteins** which are of prime importance and significance:

(1) **Normal Protein Turnover:** A larger segment of the **'body proteins'** including the *extracellular* and *intracellular proteins,* invariably undergo regular and constant **synthesis** and **cleavage** almost simultaneously thereby attributing the **normal protein turnover.** In actual practice, a normal human body is capable of causing simultaneous degradation and synthesis of nearly **300-400g body proteins each day.**

(2) Biological half-lives of various **body proteins** usually vary from **½ hr to several years.**

(3) **PGST–Sequence:** The **proteins** which are found to be rich in the **PGST-sequence** (*i.e.,* **Proline-Glutamine-Serine-Threonine Sequence**) get degraded rather promptly.

(4) **Ubiquitin:** It designates a **'small polypeptide' (Mol. Wt. 8.5kd)** that is critically present in most **eukaryotic cells** and solely responsible for **protein turn over.** Besides, **ubiquitin gets** intimately associated with a host of **intracellular proteins;** and, therefore, does facilitate their ultimate degradation *in vivo.*

(5) **Endogenous Source of Amino Acids:** These **amino acids** are, in fact, exclusively liberated by the breakdown of the **inherent body proteins.** In addition, they are also synthesized from the so called **non-essential amino acids**** in the body.

Availability of Amino Acids in Humans: It may be largely observed that a good number of **amino acids** are derived meticulously from such means as:

- **Degradation of** *'body proteins',*
- **Hydrolysis of** *'dietary proteins',* and
- **Endogenous synthesis critically and collectively give rise to the** *'amino acids pool'* in humans.

Ultimately, these available amino acids are duly metabolized both in the *liver* and the *extrahepatic tissues;* however, a majority of them undergo metabolism exclusively in the forms, as illustrated in Fig: 2.25. Importantly, as the protein are obviously *hydrolyzed* to the respective **amino acids** just prior to their *actual metabolism,* the **'amino acid metabolism'** is overwhelmingly recognized as the **'protein metabolism'.**

---

* **Spurulina**[(R)] : Prepared from the dry sea-weeds which contain up to 28% protein on dry basis, and hence serves as a protien supplement for elderly patients.

** **Non-essential Amino Acids :** They refer to the *eleven* α–amino acids that can be synthesized by humans, and are not specifically required in the diet.

Explanations: The 'metabolic fate' pertaining to array of **exogenous** and **endogenous amino acids** may be explained vividly as stated under *three* major categories, namely:

(a) **Biosynthesis of Proteins and Peptides:** It essentially designates an elaborated process that predominantly comprises **synthesis of most of the 'body proteins'**, such as:

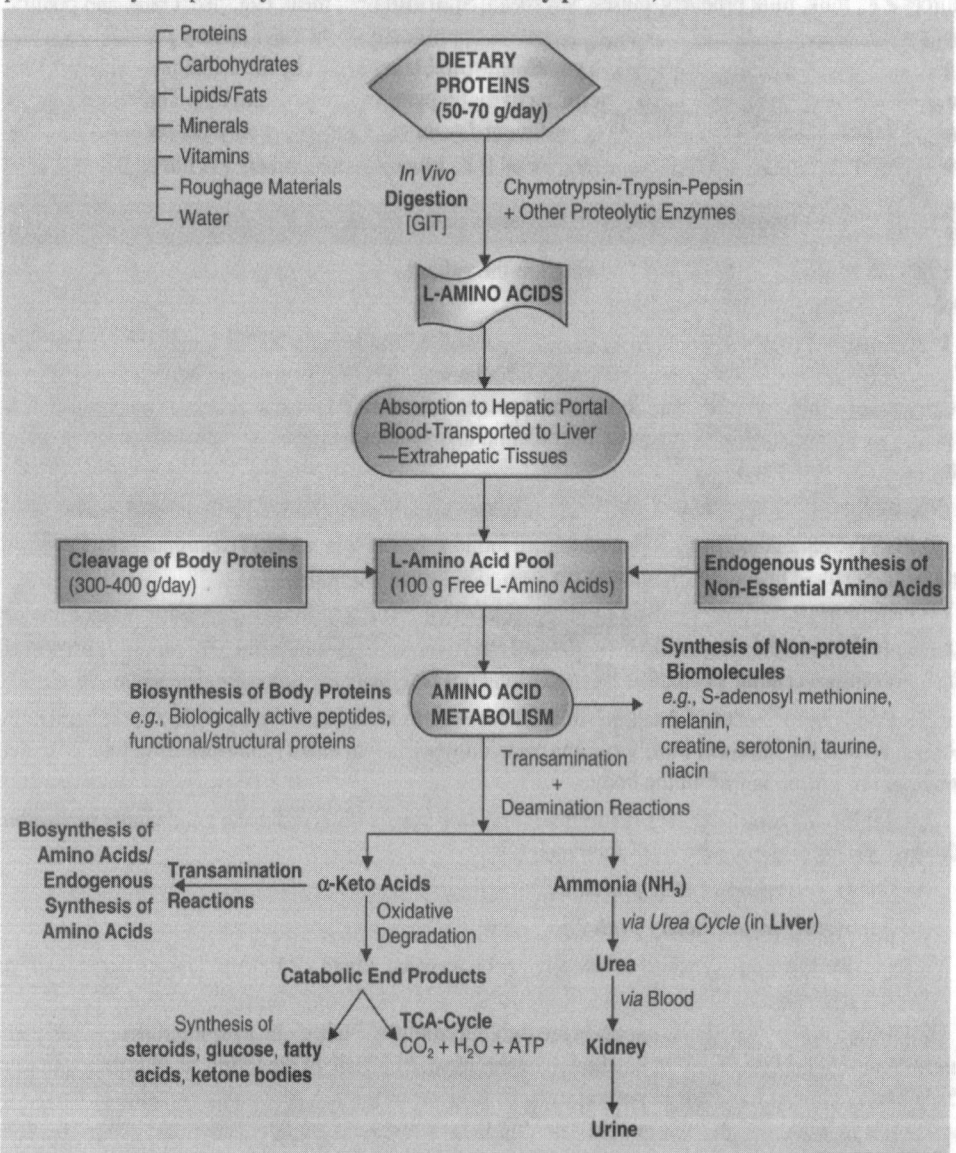

Fig: 2.25 Exogenous and Endogenous Amino Acid Spectrum vis-a-vis their metabolic fate.

- *structural proteins viz.,* **collagen, elastin, membrane proteins, keratins** etc.,
- *functional proteins viz.,* **contractile proteins, plasma proteins, hormones, enzymes** etc., and

- *biologically active peptides viz.,* **hormonal peptides, angiotensin, glutathione, bradykinin,** and **atrial natriuretic factor (ANF).**

Notes: (1) Nearly 75% of all available amino acids that are eventually released from the breakdown of body proteins get remarkably *'reutilized'* solely for the *protein biosynthesis'.*

(2) Interestingly, the *'ribosomes'* are recognized as the wonderful *'work benches'* for the ensuing *'protein synthesis'.*

(b) **Synthesis of Non-Protein Biomolecules:** It has been duly established that a certain portion of the available *amino acids* are gainfully consumed for the syntheis of **N-containing non-protein biomolecules** (invariably referred to as : *'specialized products'*) *e.g.,* **carnitine, creatine, choline, heme, histamine, melanin, melatonin, niacin, serotonine,** and **taurine.**

Note: The spectrum of amino acids that are critically and essentially needed for the synthesis of *'body proteins* as well as as *'non-protein biomolecules'* are mostly derived from the breakdown of either:
- **dietary proteins • body proteins**

(c) **Amino Acid Catabolism:** Importantly, the *'amino acids'* that are normally present as *over and excess* of the actual **biosynthetic requirements** in the human body are duly subjected to **catabolism** based on the fact the *'surplus amino acids'* can never be **stored in the body at any cost.**

In fact, the phenomenon of **catabolism** prominently engages the simultaneous critical removal of the α–amino moiety ($-NH_2$) by means of *two* marked and pronounced **reactions,** namely:

- **transamination reactions,** and
- **deamination reactions,**

thereby giving rise to the formation of *two* vital product as shown in Fig. 2.25, for instance:

ammonia ($-NH_3$), and

❑ α–keto acids.

How does 'ammonia' eliminated from human body? It is an universal fact that the *'ammonia'* duly released from the α-amino acids turns out to be exhorbitantly **toxic in nature;** and, therefore, needs to be discarded from the body by such *God-gifted* natural processes, such as:

- In 'liver' – *ammonia* gets converted into relatively less toxic *'urea',*
- In 'urine' – *urea* is duly excrated *via* **kidney,** and
- Free 'Amminia' – a smally quantum of *ammonia* gets excreted as such in the **urine** (*i.e.,* a **biological fluid**).

How do the 'α–keto acids' get eliminated from human body? It has been duly established that the 'α–keto acdis' (*i.e.,* the *'C–skeletons of amino acids'*) get subjected to critical **oxidative degradation,** and ultimately give rise to the formation of:

- **pyruvate due to catabolism,**
- **tricarboxylic acid (TCA) cycle intermediates,**
- **enzyme : acetoacetyl CoA,**
- **enzyme : acetyl CoA,** and
- **Acetoacetic acid (see Fig : 2.25)**

Glucogenic 'amino acids'– are mostly *catabolized to* **pyruvate,** plus **one of the intermediates** pertaining to the **TCA–cycle.**

Ketogenic 'amino acdis'– are abundantly *catabolized* to such specific *enzymes* as : **acetoacetyl CoA, acetyl CoA,** and **acetoacetic acid.**

Importantly, these *catabolized end-products* have a tendency to undergo the following *two* different and proved pathways, namely:

- ❑ 'oxidized' *via* TCA cycle to carbondioxide (CO_2), water (H_2O), and energy, or
- ❑ 'converted' to glucose, ketone bodies, fatty acids, and steroids.

Transamination Reaction(s): As per the *essential* and *critical* need of the situation *in vivo* the so called 'α-keto acids' do cater as the most direly required *'precursors'* for carrying out the 'resynthesis of non-essential amino acids' *via* the 'transamination reactions', which categorically monitors the actual transfer of a **pre-existing amino (–NH₂) moiety** rightly derived from an 'amino acid' to a corresponding α–keto acid so as to produce an altogether 'new amino acid'.

5.1. Variants in Amino Acids Metabolic Pathways

A survey of literature reveals the presence of several **variants in amino acids metabolic pathways;** however, the following *four* cardinal aspects shall be discussed briefly in the sections that follows:

(a) Transamination Reactions,

(b) Oxidative Deamination Reactions,

(c) Oxidation-Reduction Reactions, and

(d) Metabolism of Nitrogenous Compounds (or Dicarboxylic Amino Acids and their Respective Amides).

5.1.1. Transamination Reactions:

Transamination is usually regarded to be the very *first step* in the specific **'catabolism of free amino acids',** but **lysine, threonine** and **proline** are an *exception.*

Transamination reaction may be defined as–**reversible reaction in which the α–amino moiety (–NH₂) gets transferred enzymatically from an *amino acids–1* to an *α–keto acid-1,* thereby resulting in the formation of an altogether *new amino acid–2* plus another *α–keto acid–2'.***

Fig. 2.26. Illustrates the **'transamination reaction'** as stated under:

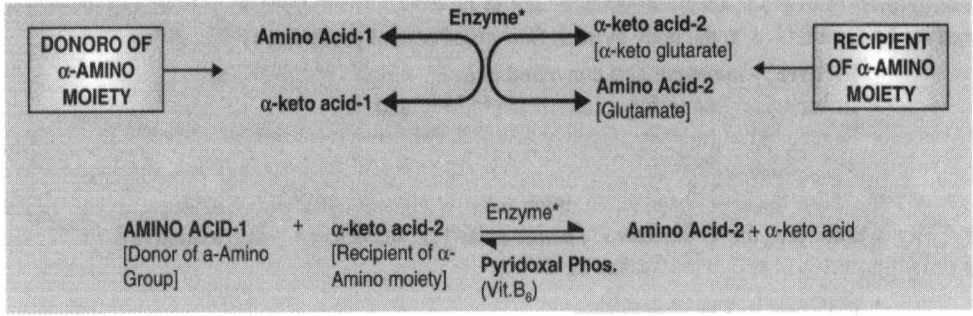

* Enzyme: Transaminase **(or Aminotransferase)**

Fig. 2.26: Diagramatic Representation of Transamination Reaction

Mechanism: The mechanism of **transamination reaction** essentially involves the transfer of **α-amino moiety exclusively,** but not the total removal of **α-amino moiety,** as a mole of **'ammonia',** from the amino acid. It is duly catalyzed by the specific enzymes usually termed as **'transaminases'** (or **'aminotransferases'**). The *'coenzyme' i.e.,* **pyridoxal phosphate (PLP)** is invariably derived from the water-soluble **vitamin-B₆.** Principally this reaction comes into play in **liver, brain,** and **muscles.**

Various cardinal steps essentially comprise:

- All **transaminases** (or **aminotransferases**) critically need *pyridoxal phosphate* (**PLP**) as the requisite coenzyme,
- **PLP** gets bonded covalently to **ε-amino moiety** of the respective **'lysine residue'** *via* the **Schiff's base linkage** (*i.e.*, the **'imine bond'**) strategically located at the **'active site of the enzyme'**.
- **PLP** predominantly caters as an **'intermediate carrier of the respective amino moiety'** located at the specific site of **transaminase**,
- In the course of the **transaminase reaction**, **PLP** critically accepts an **α-amino moiety** (obtained from **'amino acid'**); and gets duly aminated to the corresponding **pyridoxamine phosphate**,
- **Resulting product PLP** subsequently donates its inherent **amino, (-NH₂) group** to an *α-keto acid-2* to produce a *'new amino acid-2'* and therapy the **coenzyme PLP** gets duly regenerated.

Note: Importantly, the 'transamination reaction' designates an 'intracellular reaction, which particularly exemplifies a sort of typical 'Ping-Pong Reaction'.

The entire **mechanism of the 'transamination reaction'** may be illustrated judiciously in Fig: 2.27.

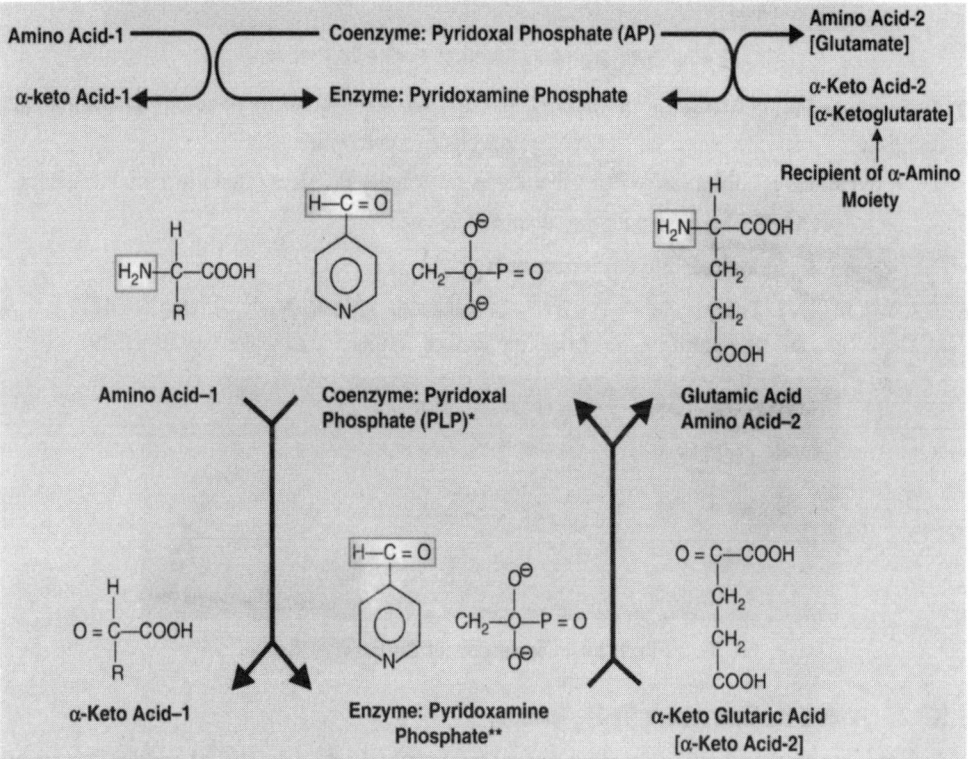

* Amino moiety *'acceptor'* form of *'coenzyme'*.
** Amino moiety *'donor'* form of *'coenzyme'*.

Fig. 2.27: Illustration of the 'Mechanism of Transamination Reaction'

Examples: There are *two* typical examples of the **transamination reaction'**, namely:

(1) **Metabolism of Serine:** Under the critical action of **serine dehydratase** enzyme, the amino acid '*serine*' gets duly **deaminated** (or **transaminated**) to pyruvate as shown in Fig. 2.28.

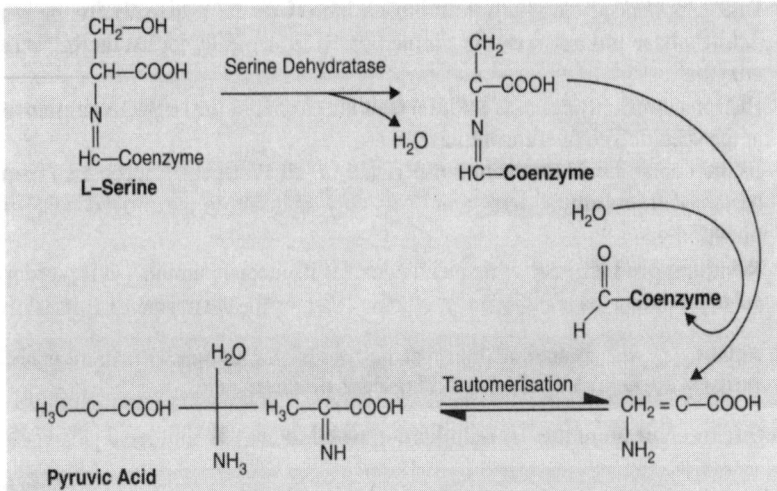

Fig. 2.28: Transamination of Serine to Pyruvic Acid.

(2) **Interconversion of Serine to Glycine:** In fact, the interconversion of **serine** to **pyruvic acid** is catalyzed by a specific enzyme, and critically needs *two* **coenzymes:**

- **pyridoxal phosphate (PLP)** that forms intermediately gives rise to *two* products, namely:
  - **Schiff's base** with the '*amino acid*', and
  - **Tetrahydrofolate derivative (FH$_4$),**

as shown in Fig. 2.29. It may be observed obviously that the specific N_5-N_{10} **methylene (FH$_4$)**, which is duly produced by the critical removal of a molecule of water (see Fig. 2.29).

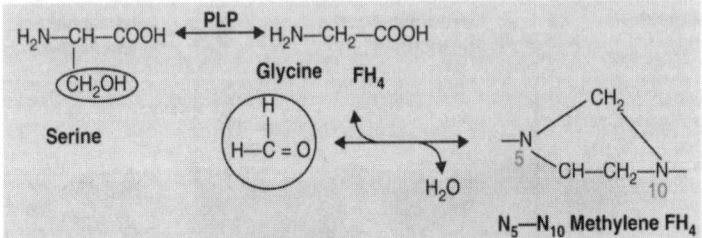

Fig. 2.29: Interconversion of Serine-Glycine.

5.1.2. Oxidative Deamination Reactions

In contrast to the **transamination-reaction** that predominantly involves the critical transfer of **α-amino moiety** right from an *amino acid* to **α-keto acid**, the ensuing **oxidative deamination reaction** specifically causes the removal of, **α-amino moieties** from the respective *amino acids* as '*free ammonia*'.

In reality, the **deamination reactions** are broadly categorized into *two* distinct variants:

- **oxidative deamination,** and
- **non-oxidative deamination.**

As illustrated in Fig. 2.30, the **oxidative deamination** entails essentially an *irreversible phenomenon* that eventually occurs in *two* glaring steps, such as:

Step-1: The enzyme catalyzes a **dehydrogenation** of the *amino acid* into an **imino acid.** Besides, the prevailing enzyme is a *'flavoprotein'*, and the **2H-atoms** duly removed from the substrate cause reduction of:

- **FAD to FADH$_2$** or
- **FMN to FMNH$_2$**

Oxidative deamination is, therefore, regarded to be a **non-enzymatic consequence of the enzymatic process of dehydrogenation.** Ultimately the resulting amino acid is duly hydrolyzed into the following **two products:**

- **α-keto acid,** and
- **ammonia**

Step-2: The **'flavin coenzyme'** usually gets reoxidised by the aid of **molecular oxygen (O$_2$)** which leads to the formation of **hydrogen peroxide (H$_2$O$_2$)** that may be subsequently decomposed by a **catalase;** however, other electron acceptors can also cause the **reoxidation of FADH$_2$.**

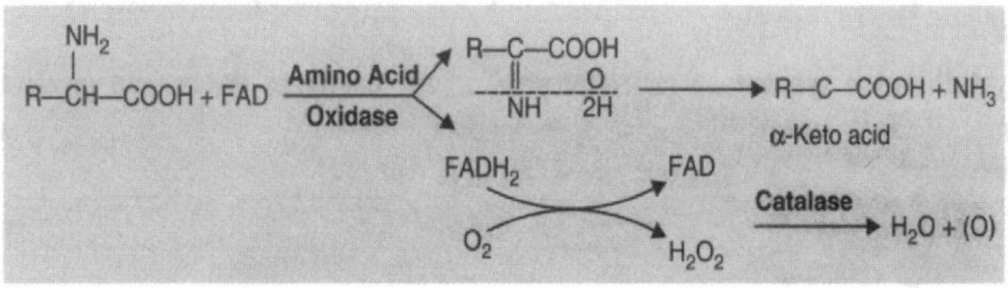

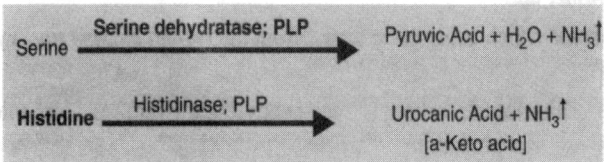

Fig. 2.30: Oxidative Deamination of Amino Acids.

Non-Oxidative Deamination: In this particular instance, **the deamination of** *amino acid* **is never accompanied by its oxidation.** Amazingly, a good number of *amino acids,* such as: serine, histidine, cysteine, and threonine are **non-oxidatively deaminated** by certain highly **specic enzymes** into the respective α-keto acids and ammonia, as depicted in Fig. 2.31:

Fig. 2.31: Non-Oxidative Deamination of Serine and Histidine.

5.1.3. Oxidation-Reduction Reaction

This **oxidation-reduction fraction** categorically serves as the *metabolism* of **S-containing amino acids** *viz.,* **cysteine, methionine,** and **cystine.**

Importantly one may take cognizance of the following cardinal aspects, namely:

- Both **cysteine** and **cystine** are found to be easily and conveniently **interconvertible** by means of the **oxidation-reduction reaction;** hence, they may be duly regarded as **one single entity.**

- **Methionine** designates an **'essential amino acid'** for the *humans*; whereas, **cysteine** is *not so.*

- Nevertheless, when particularly **'cysteine'** is duly provided in the *human system,* the requirements for **methionine** gets depleted simultaneously. This critical and significant observation suggests predominantly that:

  □ a certain portion of **'methionine'** gets aptly used up in the formation of **cysteine,** and

  □ The aforesaid critical observation was duly ascertained by performing exhaustive and in-depth, studies by the aid of the **'radioactive isotopes'**

Examples: There are *three* most typical examples to expatiate the **oxidation-reduction reaction,** such as:

- Formation of cysteine from methionine,
- Metabolism of methionine, and
- Metabolism of cysteine,

which shall be discussed individually in the sections that follows:

(1) **Formation of Cysteine from Methionine:** Importantly, in humans, this specific mechanism is termed as **'transulphuration',** which essentially implies *two* such enzymes that comprises the **pyridoxal phosphate (PLP),** namely:

- **Cystathionine synthetase (E_1)**-that prominently catalyzes the critical production of an **'intermediate compound'** known as **'crystathionine'** duly obtained from **serine** and **homocysteine***. Thus, E_1, exerts its action in *two* ways:

  - catalyzes *dehydration* of **serine** (*i.e.,* the *first substrate*) as per the mechanism described in Fig: 2.32, and

  - acts on the double bond thus formed between the α-and β-C-atoms, thereby the generation of the respective **homocysteine molecule** (*i.e.* the **second substrate**).

- **Cystathionase (E_2)**-which brings about the cleavage of **crystathionine** thereby giving rise to *two* distinct **amino acids** as stated under:

  - **cysteine**-produced from the **C-and N-atoms** derived from **serine,** and the **S-atom** from **methionine,** and

  - **homoserine**-which being not released from the *second enzyme* [*i.e.,* **cystathionase (E_2)**], and subsequently undergoes a *deamination reaction***.

(2) **Metabolism of Methionine:** In a broader perspective, the overall critical role of **'methionine'** *constitutive of proteins,* and its predominant function in initiating the phenomenon of **protein biosynthesis,** it largely acts as a big source for the supply of **methyl (-CH_3) moieties.**

---

* **Homocysteine:** It refers to a *higher homologue* of **'cysteine'** invariably formed by the **demethylation** of **methionine.**

** That is, the ensuing **α-γ elimination reaction,** whose probable **mechanism** matches quite similar to the one already discussed under **serine** (Fig. 2.32).

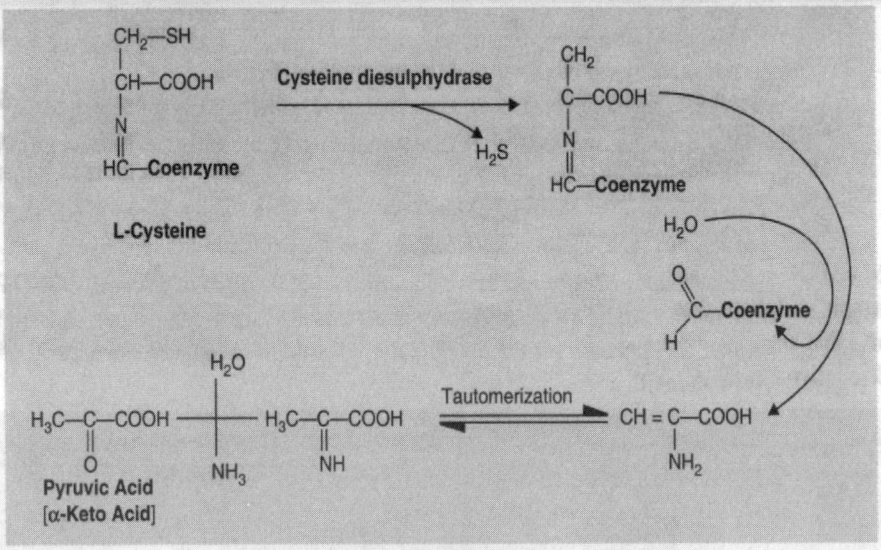

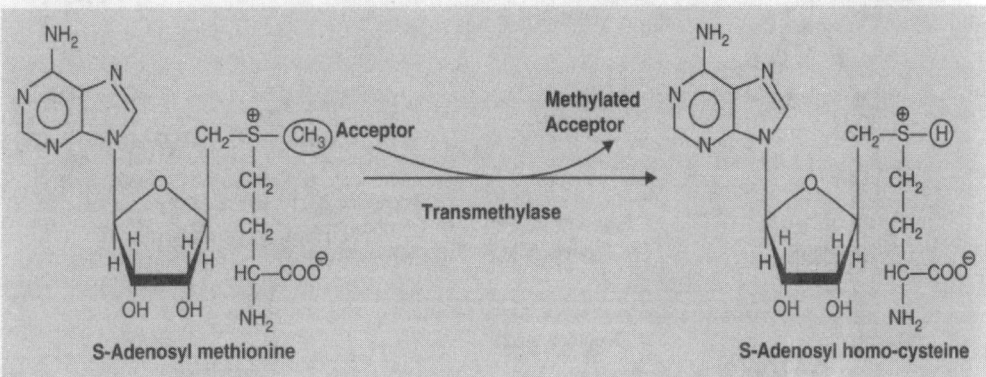

Fig. 2.32: Oxidative Deamination of L-Cysteine

In general, **methionine** should be adequately activated by **ATP** to produce the corresponding **S-adenosyl methionine***. Thus, it is capable of producing its **methyl (-CH₃) moiety** to some extremely 'diverse natured compounds'; and ultimately gets converted into **S-adenosyl-homocysteine** as shown in Fig. 2.33.

Fig. 2.33: Mechanism of Transmethylations Obtained from S-Adenosyl Methionine

(3) **Metabolism of Cysteine:** The metabolism of **'cysteine'** may be explained vividly by the help of the following *two* pathways, namely:

- Transformation of cystine into pyruvic acid, and
- Formation of 'taurine',

which shall be discussed briefly in the section that follows:

(a) **Transformation of Cysteine into Pyruvic Acid:** Interestingly, the **transformation of cysteine into pyruvic acid** (an α-keto acid) may be duly accomplished by *two* pathways, namely:

---

* **S-Adenosyl Methionine:** It represents the real agent that essentially takes part in the phenomena of **'transmethylation'**.

- *First*—which has already been examined adequately (see Fig. 2.32), comprising a **deamination process** duly catalyzed by the enzyme **cysteine desulphydrase.**
- *Secondly*—it essentially consists of an oxidation of **cysteine** to **cysteine-sulphinic acid** that eventually loses its **amino (-NH$_2$) functional moiety** by means of **transamination**, and ultimately yields **sulphinyl-pyruvic acid** as the *penultimate product.* Thus, the resulting product yields **pyruvic acid** by the critical liberation of a **S-atom** which is observed usually in the form of *sulphite*, and finally as *sulphate*, as depicted explicitly in Fig. 2.34.

Nevertheless, the resulting *'sulphate'* gets duly activated in the form of **adenosine-3`-phosphate-5`-phospho-sulphate,** that would eventually react with several diversified compounds, such as: **phenols, steroids** that get finally eliminated *via* **urine** in the form of **sulpho-conjugated structural analogues (or derivatives)** by **detoxification.**

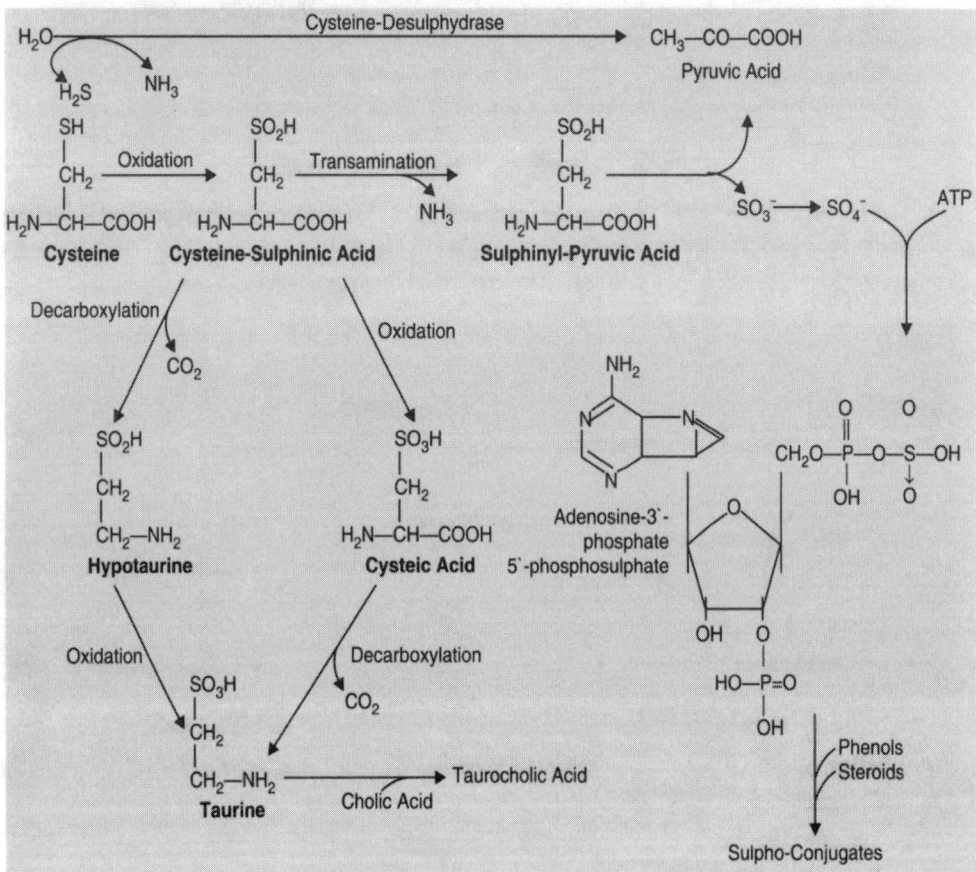

Fig. 2.34: Metabolic Pathway of Cysteine.

(b) Formation of Taurine: Fig. 2.34 evidenlty reveals the **cysteine-sulphinic acid** whose critical formation by the articulated oxidation of **cysteine,** as stated above, may instead of undergoing the phenomenon of **transamination,** be decarboxylated to the respective **hypotaurine,** which eventually gets:

- oxidized to **'taurine'***
- **'taurine'** may be conjugated the **'bile acids'** *in vivo.*

5.1.4. Metabolism of Nitrogenous Compounds [or Dicarboxylic Amino Acids and Their Respective Amides]

It has been proved beyond any reasonable doubt that **aspartic acid [HOOC-CH$_2$-CH(NH$_2$)-COOH]**, **glutamic acid [HOOC.CH(NH$_2$)CH$_2$CH$_2$.COOH]**, and their respective **amides** do occupy an important and pivotal status with regard to the ensuing **metabolism of the 'nitrogenous compounds'**.

Importantly, these two **amino acids** *viz.*, **aspartic** and **glutamic acids** are also recognized to be of prime importance based upon the following *two* cardinal reasons, namely:

- due to their ensuing **'intermediate metabolism'** in a rather broader perspective, and
- **'Krebs Cycle'*** do have the corresponding **α-keto acids** as **marked intermediates**;

and, therefore, predominantly constitute various strategic contact points between the **'carbohydrate metabolism'** and the **'protein metabolism'**.

Example: The **metabolism of 'nitrogenous compounds'** may be duly exemplified by the following typical example:

Metabolic Inter relationships Amongst Glutamic Acid-Ornithine-Proline as illustrated in Fig. 2.35.

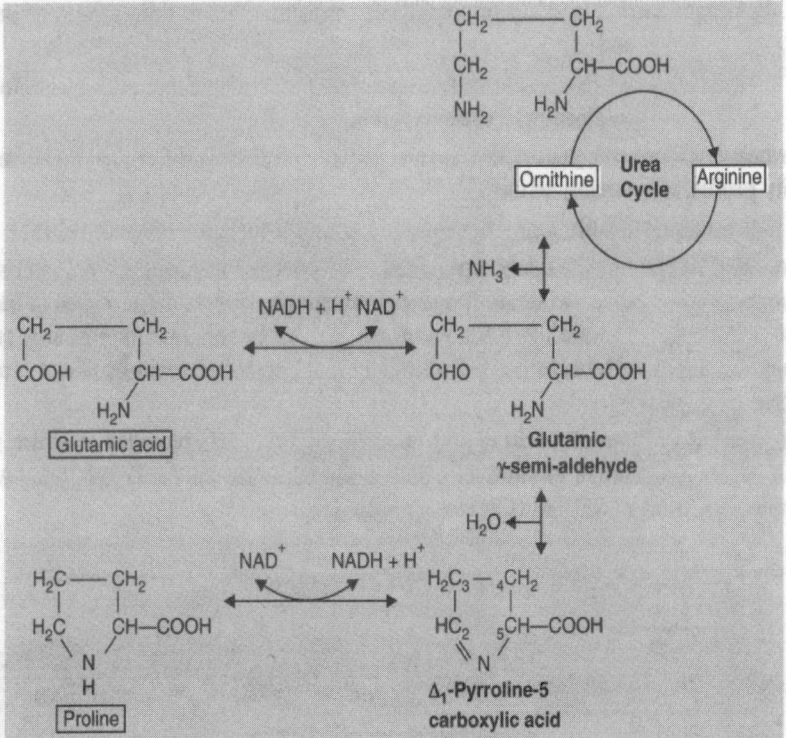

Fig. 2.35: Metabolic Interconversions Amongst Glutamic Acid, Ornithine, and Proline.

---

* **Krebs Cycle [or Krebs Urea Cycle]:** It refers to the **'cyclic pathway'** that specifically converts waste ammonia molecules along with CO_2 and **aspartate** into **urea**; named after its discoverer **Hans Krebs**.

In fact, **glutamic γ-semialdehyde,** which eventually comes into being from **glutamic acid** by reduction of its **γ-carboxyl moiety,** may lead to:

- **inherent cyclization and yield 'proline', or**
- **undergo transamination reaction thereby causing its conversion into ornithine.**

6. CHEMISTRY OF OXYTOCIN

Oxytocin designates a **'hormone secreted by the posterior pituitary gland'.** Amazingly, **'oxytocin'** happens to be a **nonapeptide,** which categorically **stimulates the critical contraction of the uterine smooth muscles,** as depicted in Fig. 2.36.

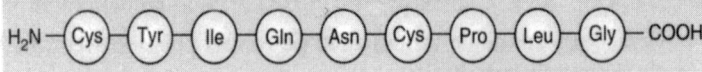

Fig. 2.36: Oxytocin-a 'Nonapeptide' i.e.; a peptide with Nine-Amino Acids.

The Neurohypophysis: It has been amply demonstrated that the **mammalian neurohypophysis** are, namely:

- **Vasopressin**—that is solely responsible for causing either **antidiuretic** or **pressor** effects, and
- **Oxytocin**—which is exclusively responsible for causing either **milk-ejection (lactation)** or smooth-muscle **contraction.**

Nevertheless, the *two* **polypeptides,** each essentially possessing the critical **melanocyte-stimulating activity profile,** is invariably termed as:

- **β-melanocyte stimulating hormones** or **α-and β-melanotropins** or **β-MSH,**

and *coherin* (MW4,000) has been duly reported to be effective in causing appropriate *stimulation of the coordinate contractions* in the **small intestine.** Interestingly, *coherin* is proved to be distinct from both **'oxytocin'** and **'vasopressin'.** Besides, **vasotocin** (another hormone) is meticulously produced by most **non-mammalian vertebrates,** and distinctly exhibits both **'oxytocin-like'** and **'vasopressin-like'** pharmacological activities.

The structures of human **oxytocin** and **vasopressin** are evidently provided as under, which also includes the critical structure of **vasotocin,** duly produced by several *lower vertebrates e.g.,* **amphibians** (*frogs, toads*), and **bony fishes** (*Cirrhina mrigala, pomphrct*).

```
 1                          9
Cy — Tyr —Ile — Gln — Asn — Cy — Pro — Leu — Gly — NH₂
 └──────── S—S ─────────┘


       Cy — Tyr — Phe — Gln — Asn — Cy — Pro — Arg — Gly —⁹ NH₂

               └──────── S—S ────────┘
                       Vasopressin
 1                                      9
Cy — Tyr — Ile — Gln — Asn — Cy — Pro — Arg — Gly — NH₂
 └──────── S—S ─────────┘
```

Vasotocin

Importantly, the so called **melanocyte-stimulating activity (MSA)** is appreciably detectable in the **adenohypophysial extracts as well. In actual practice, the melanocyte-stimulating hormone (MSH)** is secreted specifically by the **pars intermedia**, except in such species that critically devoid of this structure *e.g.,* **whale, chicken,** and **porpoise**, wherein the activity is solely confined to the **adenohypophysial extracts**.

> **Note:** 'Oxytocin' distinctly differs from he *'vasopressin'* exclusively in residues at positions 3 and 8. Besides, each of the above structures essentially posses 'H_2N-terminal half-cystine and glycinamide located at the 'HOOC-terminal position'.

Salient Features: A plethora of the structural analogues of **oxytocin, vasopressin,** and **melanocyte-stimulating hormone (MSH)** have been **synthesized meticulously** and **studied exhaustively** for their *biological activity profile*. Following are the **salient features** observed, namely:

(1) The following structural requirements are an absolute necessity for attributing the *'biological activity'* of **oxytocin** and **vasopressin,** such as:

- **cyclic structure,**
- **chain-length of 9-residues,**
- presence of *'proline residue'* at **position** 7, and
- presence of *'amide moieties'* at **positions 4 and 5**.

(2) The existence of the *'aromatic hydroxyl (-OH) group'* at position 2 may not prove to be essential for showing the **full activity;** nevertheless, the **phenyl moiety** is critically needed for showing the **'hormonal action'**.

(3) Such **amino acids** that essentially that are having *'branched aliphatic side-chains'* at position 3 prove to be of prime importance in exhibiting the **oxytocic activity.**

(4) The presence of a **'phenylalanine residue'** at position 3 increases prominently the **vasopressor activity profile**. Incidentally, it shows a clear cut contrast to the fact that the so called **aliphatic amino acid residues** prove definitely to be more *significance* and *importance* in showing the **'oxytocic activity'** *vis-a-vis* the **'phenylalanine residue'** for the **activity of vasopressin.**

(5) In order to have the optimized **oxytocic activity**, it is absolutely necessary to have the *amino acids* having *'aliphatic side-chains* at position 8; whereas, the presence of the **basic amino acids** (*viz*, **Arg; Lys; Lys-OH**) at the same position are essentially vital and important for the **vasopressor activity**.

(6) The removal of the **H_2N-terminal half-cysteine** by the help of *aminopeptidases* results in the total loss of both **oxytocin activity** and **vasopressin activity.**

(7) Amazingly, such structural analogues that essentially have additional residues at the **amino (-NH_2) terminal end** do have a longer duration of activity.

(8) A remarkable and significant separation of the specific **antidiuretic and pressor activity** of *vasopressin* has been duly accomplished by the meticulous substitution of the *amino acid threonine* for *phenylalanine* at position 3 of the said molecule having substantial overall enhancement in the particular ensuing **ratio of antidiuretic to pressor potency profile.**

(9) Ultimately, the judicious incorporation of a **'tyrosine residue'** to the **carboxyl (-COOH) terminal end** of *'oxytocin'* gives rise to a molecule which eventually turns out to be an **'inhibitor'** of this *hormone*.

6.1. Structure and Synthesis of Oxytocin

A number of **researchers*** between *1953 to 1965* have put forward the **'structure and synthesis of oxytocin'**.

Based on the fact that **oxytocin** is produced without affecting any sort of *chain-fission* suggests and affirms that the critical presence of the distinct **disulphide bond [-S-S-]** in the former chemical entity. Now, assuming that the specific **α-carboxy moieties of Glu and Asp** are directly involved in the so called *'peptide bondages'*, the probable structure of **'oxytocin** may be written as given under in Fig. 2.37:

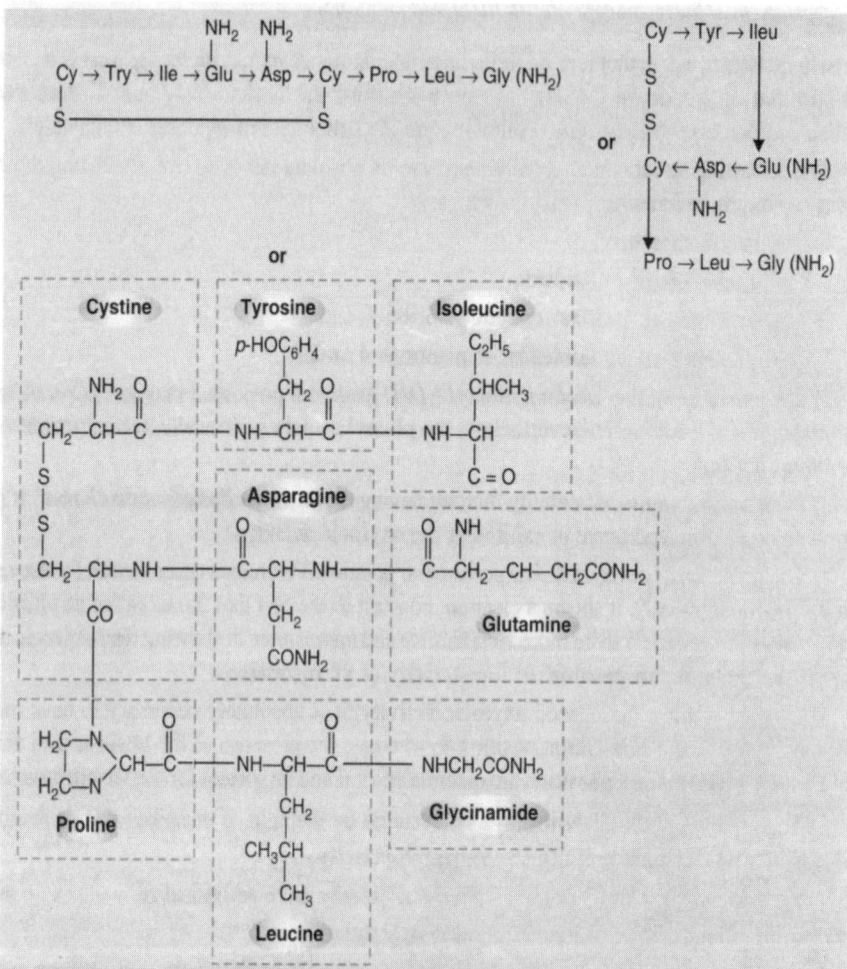

Fig. 2.37: Probable Abbreviated and Comprehensive Structure of Oxytocine

* Tuppy and Michl: *Monatsh*, **84**: 1011, 1953.

Tuppy, *Biochem, Biophys. Acta*, **11**: 449, 1953.

du Vigneaud *et al., J. Am. Chem. Soc.,* **75**: 4879, 1953; **76**: 3115, 1954.

Bodanszky and du Vigneaud, *ibid,* **81**: 2504, 1959.

Cash *et al. J Med Pharm Chem,* **5**: 413, 1962.

Sakakibara *et al. Bull Chem* **50c** (Japan), **38**: 1965.

Note: Dotted lines explicitly show the various critical points of cleavage of the *'oxydized oxytocin'* by the *microbial proteinase* to the corresponding tetrapeptides

The du Vigneaud's Procedure: Importantly, the **du Vigneaud's procedure** is, in fact, found to be absolutely different from the **Tuppy's method**. However, **du Vigneaud's procedure** essentially comprise the following cardinal steps as given under:

(1) The structure of **oxytocin** is invariably established by the critical and close exhaustive study of different associated *'fragments'* that is duly accomplished due to the **'partial hydrolysis of oxytocin'** to yield:

- **the performic acid [] oxidation product,**
- **Br$_2$-water oxidation product,** and
- **desulphurized** *'oxytocin'*

(2) Treatment with the appropriately selected *'ion-exchange resins* are gainfully employed so as to segregate and isolate the so called **'residual peptides'** into the *acidic* as well as *neutral* components. Ultimately, these *resulting peptides* are meticulously separated by the well-known **Paper Chromatographic Methods.***

(3) The **DNP (dinitrophenyl) method** revealed that **'oxidized oxytocin'** predominantly possessed only one *terminal moiety,* which was found to be that of **cysteine**.

(4) Subsequent, treatment of the **'oxidized oxytocin'** with **Br$_2$-water** *two* products came into existence, namely:

- a **'dibromopeptide'**, and
- a **'heptapeptide'**.

Dibromopeptide (DBP)—actually determined to be:

$$Cy\text{-}SO_3H \rightarrow Tyr\text{-}Br_2 \ i.e. \ \text{a } \textbf{3,5-Dibromo derivative}$$

Its structural status has been duly ascertained by **hydrolysis,** followed by **end-group analysis**.

Heptapeptide (HP)—on being subjected to hydrolysis [see (ix)—(xii)], it resulted into the formation of:

$$Cy\text{-}SO_3H, \ Leu, \ Ileu, \ Pro, \ Glu, \ Asp, \ Gly + Ammonia$$

Its **end-group analysis** should that *'isoleucine'* was duly present as the **N-terminal residue**. Nevertheless, the **'oxytocin'** only possesses **one terminal functional amino (–NH$_2$) moiety;** and, therefore, the *amino moiety* strategically located in *isoleucine* should inevitably result into the formation of the actual **peptide linkage** with **tyrosine** critically. Thus, it overwhelmingly confirms the prevailing *sequence of three residual entities* as given under:

$$\boxed{Cy - So_3H \rightarrow Tyr \rightarrow ILeu}$$

Furthermore, when the ensuing **'heptapeptide'** is carefully hydrolyzed under strictly controlled experimental parameters, it essentially produced *four articulated fragmented entities* [from (*xiii*) through (**xvi**) as given below:

---

* Kar Ashutosh: **Pharmaceutical Analysis, Vol. II,** CBS-Publishers and Distributors, New Delhi, 2009.

[xiii] Asp–Cy–SO$_3$H; [xiv] Cy–SO$_3$H, Pro; [xv] Cy–SO$_3$H, Pro, Leu, Gly; and [xvi] Cy–SO$_3$H, Asp, Glu.

Likewise, when the *desulphurized oxytocin* is subjected to hydrolysis, it essentially yielded *four distinct fragments* [from (xvii) through (xx) as stated under :

[xvii] Tyr, Cy–S–S–Cy, Asp, Glu, Leu, Ileu; [xviii] Ala, Asp; [xix] Glu, Leu; [xx] Ala, Asp, Glu, Leu, Ileu.

> **Note:** In fact, no explicit differentiation could be made between *Leu* and *Ileu* in the *two peptides* [xvii] and [xx] respectively based on the fact that both of them critically comprised *only one of these* acids. It was rather difficult to determine exactly and precisely which one could be present as both these 'amino acids' emphatically appeared together on the 'chromatogram'.

(5) In a situation, when **DNP (dinitrophenyl) method** was duly applied to [ix], the critically observed sequence was Asp → Cy – SO$_3$H. Thus, by carefully examining the *amino acids* duly present in [ix] through [xiii], it may be observed that the prevailing sequence of *five peptide residues* present categorically in the 'oxidized oxytocin' might be almost similar to the one present in [xiii].

Summararily, '**oxytocin**' prominently contains the following **sequence of amino acids,** provided:

- **Tyr is particularly joined to ½ of cysteine residue,** and

- **Asp is joined to other ½ of cystines.**

Thus, we may have :

> [Cys—Try—Ile—Gln—Asn—Cys—Pro—Leu—Gly—NH$_2$]

In other words, the above probable structure of '**oxytocin**' evidently accounts for the presence of **eight amino acids.**

Evidence for Cyclic Structure of Oxytocin: It may be explained by the fact that *cysteine* contains the **only free amino (–NH$_2$) functional moiety,** and further supported by the fact that **oxidation fails to cause fission at all;** which may be fully justified by strategically joining together **Tyr** and Asp.

Besides, the prevailing '**Gly–terminal**' (see above *probable structure*) proves to be quite unsuitable due to the fact that this specific residue is normally present as a '**carbonamide**'. Importantly, this particular analogy has been duly substantiated and confirmed by means of the **Edman's Method** pertaining to the *end-terminal analysis* applied to the '*oxidized oxytocin*', wherein the first *four amino acids,* present in the following observed sequence, have been removed completely :

> Cy–SO$_3$H, Tyr, Ileu, Glu

6.1.1. Synthesis of Oxytocin

The remarkable concerted efforts of du Vigneaud *et al.* (1954)* put forward the **total synthesis** of 'oxytocin' thereby confirming its structure satisfactorily.

---

* du Vigneaud *et al.: J Am chem Soc,* **76** : 3115, 1954.

Following are the *four* most vital and important aspects in the 'total synthesis' of *oxytocin,* namely :

(1) The presence of all **free basic amino (–NH₂) moiety** is duly guarded and protected by the critical **benzyloxy carbonyl [or carbobenzoxy ((⊙)–CH₂. O. C–) or designated as 'Z' group],** which is introduced carefully by the treatment of the candidate *amino acid* with **carbobenzoxy chloride**

[(⊙)–CH₂. O. C–Cl,]

Thus, one may prepare **carbobenzoxy chloride** by the interaction of **benzyl alcohol** and **phosgene** as indicated below:

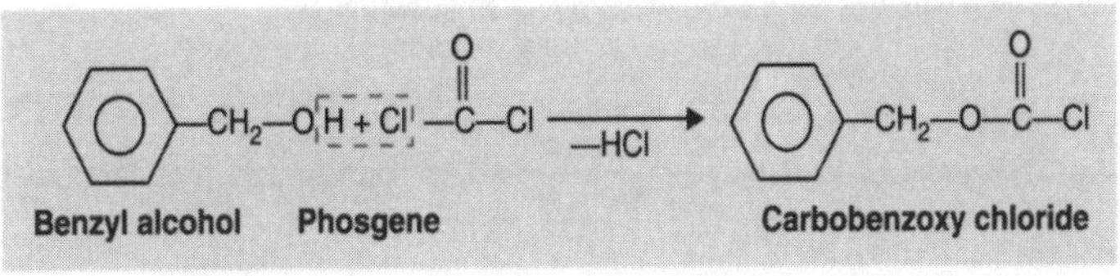

Benzyl alcohol   Phosgene   Carbobenzoxy chloride

(2) In this manner, the *amino acid* with an adequately protected **amino (–NH₂) functional** moiety by the respective (**Z**) is carefully condensed with certain **ester (s) of other amino acid molecules** as:

- **OET** : Ettyl ester ;
- **Bzl** : Benzyl ;
- **NP** : *para*-Nitro phenyl ;
- **Z** : Benzyloxycarbonyl [or Carbobenzoxy] ;

(3) Subsequently, '**Z**' *i.e.,* the protected moiety being removed with utlmost care by the help of a mixture comprising **HBr–CH₃COOH** (1.1); and the resulting product is duly treated with the corresponding **ester of other amino acids** having a genuinely **protected amino functional moiety** by '**Z**'.

(4) Ultimate removal of the so called '**protective Z moiety**' in the form of a mole each of **toluene, carbon dioxide** by adopting any one of the following *two* methods:

- **freshly cut Na–metal in liquid ammonia,** or
- **catalytic reduction,**

as may be shown under :

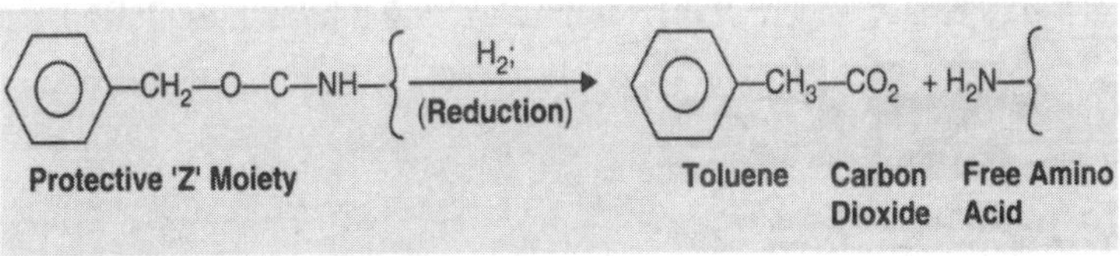

Protective 'Z' Moiety   Toluene   Carbon   Free Amino
                                                              Dioxide   Acid

Thus, the *free amino acid* is available once again in the reaction mixture to take part in the '**total synthesis of oxytocin**'.

Total Synthesis of 'Oxytocin': The various symbols that are used in detailing the **total synthesis of 'oxytocin'** have been provided under section 6.1.1 (2) earlier.

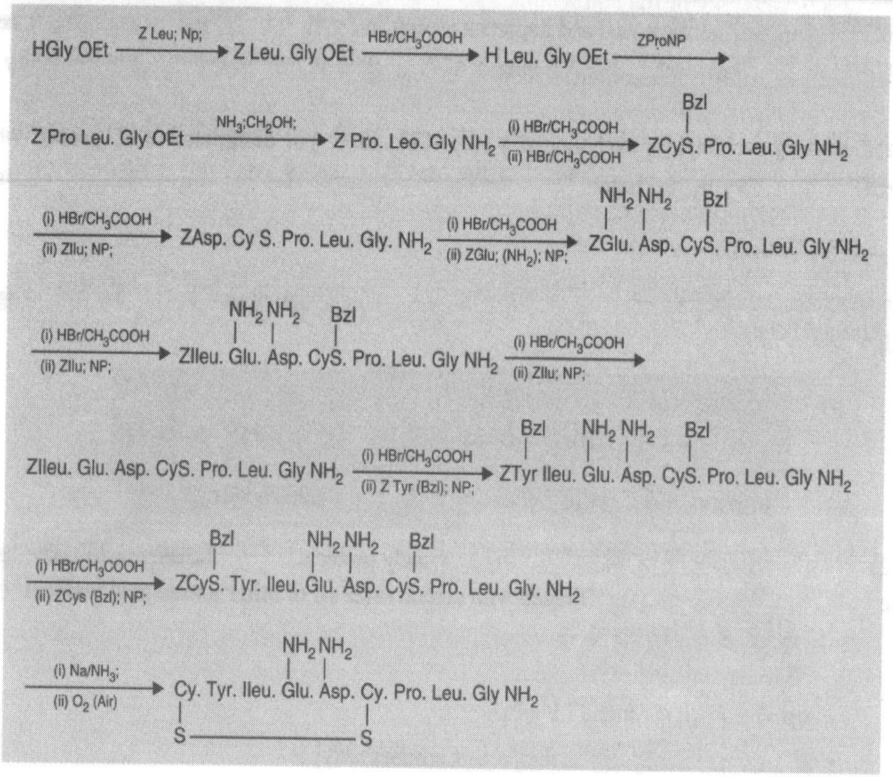

HGly OEt $\xrightarrow{\text{Z Leu; Np;}}$ Z Leu. Gly OEt $\xrightarrow{\text{HBr/CH}_3\text{COOH}}$ H Leu. Gly OEt $\xrightarrow{\text{ZProNP}}$

Z Pro Leu. Gly OEt $\xrightarrow{\text{NH}_3;\text{CH}_2\text{OH};}$ Z Pro. Leo. Gly NH$_2$ $\xrightarrow[\text{(ii) HBr/CH}_3\text{COOH}]{\text{(i) HBr/CH}_3\text{COOH}}$ ZCyS. Pro. Leu. Gly NH$_2$
$\overset{\text{Bzl}}{\overset{|}{}}$

$\xrightarrow[\text{(ii) ZIIu; NP;}]{\text{(i) HBr/CH}_3\text{COOH}}$ ZAsp. Cy S. Pro. Leu. Gly. NH$_2$ $\xrightarrow[\text{(ii) ZGlu; (NH}_2);\text{ NP;}]{\text{(i) HBr/CH}_3\text{COOH}}$ ZGlu. Asp. CyS. Pro. Leu. Gly NH$_2$
$\overset{\text{NH}_2\,\text{NH}_2\;\;\text{Bzl}}{\overset{|\;\;\;\;|\;\;\;\;\;\;|}{}}$

$\xrightarrow[\text{(ii) ZIIu; NP;}]{\text{(i) HBr/CH}_3\text{COOH}}$ ZIleu. Glu. Asp. CyS. Pro. Leu. Gly NH$_2$ $\xrightarrow[\text{(ii) ZIIu; NP;}]{\text{(i) HBr/CH}_3\text{COOH}}$
$\overset{\text{NH}_2\,\text{NH}_2\;\;\;\text{Bzl}}{\overset{|\;\;\;\;|\;\;\;\;\;\;|}{}}$

ZIleu. Glu. Asp. CyS. Pro. Leu. Gly NH$_2$ $\xrightarrow[\text{(ii) Z Tyr (Bzl); NP;}]{\text{(i) HBr/CH}_3\text{COOH}}$ ZTyr Ileu. Glu. Asp. CyS. Pro. Leu. Gly NH$_2$
$\overset{\text{Bzl}\;\;\;\;\text{NH}_2\,\text{NH}_2\;\;\;\text{Bzl}}{\overset{|\;\;\;\;\;\;|\;\;\;\;|\;\;\;\;\;\;|}{}}$

$\xrightarrow[\text{(ii) ZCys (Bzl); NP;}]{\text{(i) HBr/CH}_3\text{COOH}}$ ZCyS. Tyr. Ileu. Glu. Asp. CyS. Pro. Leu. Gly. NH$_2$
$\overset{\text{Bzl}\;\;\;\;\text{NH}_2\,\text{NH}_2\;\;\;\text{Bzl}}{\overset{|\;\;\;\;\;\;|\;\;\;\;|\;\;\;\;\;\;|}{}}$

$\xrightarrow[\text{(ii) O}_2\text{ (Air)}]{\text{(i) Na/NH}_3;}$ Cy. Tyr. Ileu. Glu. Asp. Cy. Pro. Leu. Gly NH$_2$
$\overset{\text{NH}_2\,\text{NH}_2}{\overset{|\;\;\;\;\;|}{}}$
$\underset{\text{S}\text{————————}\text{S}}{}$

Oxytocin

Note: There are in all *twelve steps* that need to be followed in a sequential manner with extreme precautionary measures under recommended experimental conditions meticulously.

7. THYROTROPHIN–RELEASING HORMONES (TRHs) [or THYROTROPIC REGULATORY HORMONES (TRHs)]

The **thyrotrophin-releasing hormone (TRH)** refers to the particular **thyroid-stimulating hormone (TSH)**. *i.e.,* a hormone pertaining to the **anterior pituitary gland** *having an affinity for,* and *specifically stimulating the thyroid gland.*

There are *two* extremely important **thyrotrophin-releasing hormones (TRHs),** namely :

- **The Thyrotropic hormone,** and
- **The Gonadotropic Hormone,**

which shall be discussed briefly in the sections that follows:

7.1. The Thyrotropic Hormone:

It has been proved evidently that **hypophysectomy*** particularly in mammals critically gives rise to the **'involution of the respective thyroid gland',** thereby distinctly exhibiting these *two genuine observations:*

---

* **Hypophysectomy :** That is, removal of the **pituitary gland** (in **experimental animals**).

- **flattening of the epithelium,** and
- **syndrome of hypothyroidism.**

Consequently, the careful administration of the **hypophysial extracts** remarkably caused an induction of *reparative effects* in the ***thyroid gland*** in such animals, with a gradual return to the **normal condition.**

7.1.1. Thyrotrophin (or Thyrotropin) Secretions in Humans:

In fact, the **thyrotropic-regulator hormone (TRH),** that has been isolated meticulously from the specific **hypothalamic tissue of** *cattle* and *hogs* is found to be a *'tripeptide amide',* **pyroglut amylhistidyl proline'** having the following **chemical structure:**

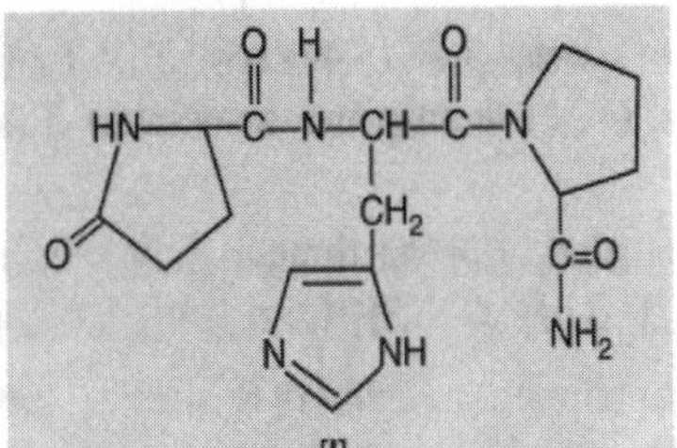

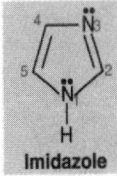

[I]

Important Feature of [I] : These essentially comprise:

(1) **Excellent Potency:** A few **mcg quantum** on being administered intravenously (IV) in humans cause immediate stimulation in this critical secretion of **TSH.**

(2) Unexpectedly, [I] is found to be **'active'** when administered orally, though a relatively greater amounts are needed *via* this route.

(3) Incidentally, (I) is also found to be **'active'** upon the so called **'hypophysial slices'** *in vitro.*

(4) **Thyroxine** and **Triiodothyronine** are observed to *inhibit* both

- **release profile of TRH by the hypothalsmus,** and
- **TRH's actions at the level of the adenohypophysis.**

(5) TRH also regulates the secretion of **somatotropin,** and in turn, **somatostatin** (*i.e.,* an **inhibitor of somatotropin release**) besides, it also causes the inhibition of the *'TRH–induced TSH–release'.*

(6) **Synthetic TRH–Analogue Variants:** A plethora of intelligently designed **synthetic TRH– analogue variants** have been made that eventually produced compounds having:

- **methyl (–CH₃) moiety) at C–1 of the imidazole nucleus of the histidine residue–thereby reducing the activity almost to 1/5000 the** *vis-a-vis the naturally occurring TRH.*

Imidazole

- judicious replacement of the *end–prolinamide* by means of the **proline** affords significant reduction in the ensuing activity almost to a **similar extent.**

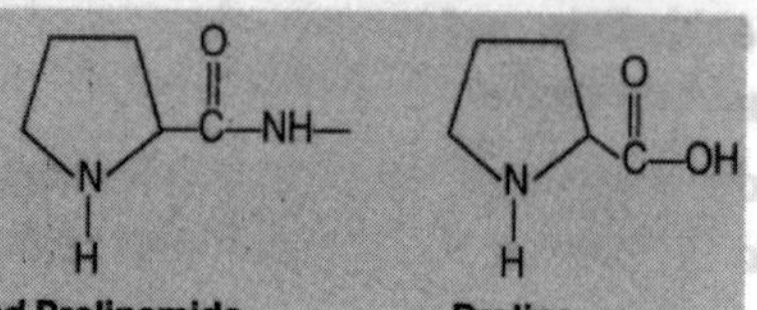

end-Prolinamide   Proline

- Contrarily, the thoughtful substitution of **methyl (–CH₃)** moiety at C–3 of the imidazole nucleus distinctly exhibits an activity which stands at **several thousand times superior per mg.** *vis-a-vis* the '**unsubstantiated tripeptide**'.

(7) **Characteristics of TRH:** These essentially comprise:

- enhances the rates of **synthesis** and **release profile of TSH by the adenohypophysis** caused due to the stimulation of the **membrane–associated adenylate cyclase** pertaining to the '**target cells**',

- both **CAMP** or **epinephrine,** even though duly present in *higher concentrations* in comparison to **TRH,** also release TSH from the **adenohypophysial tissue** *in vitro,*

- **TRH** with regard to '**clinical investigative studies**' has been reported to produce a **quick antidepressant effect** in such patients who already show a history of having certain **types of depression,**

- importantly, the so called behavioural types of actions of **TRH** have also been duly reported both in:

  ❑ **normal mice,** and

  ❑ **hypophysectomized mice*,**

  and are, therefore, predominantly prove to be quite **independent** pertaining to the **release profile of TSH** from the **hypophysis,** and

- **TRH's** action pattern upon the ***nervous system*** may appear to be of immense interst due to the fact that it is also specifically present in certain **segments of the brain** (other than the '**hypothalamus**') and hence, has been shown to be present throughout the :

  ❑ **spinal cord,** and

  ❑ **human cerebrospinal fluid (CSF).**

Note: The '*tripeptide amide*' seems to exert an additional '*neurotransmitter role*' in modulating the TSH secretion.

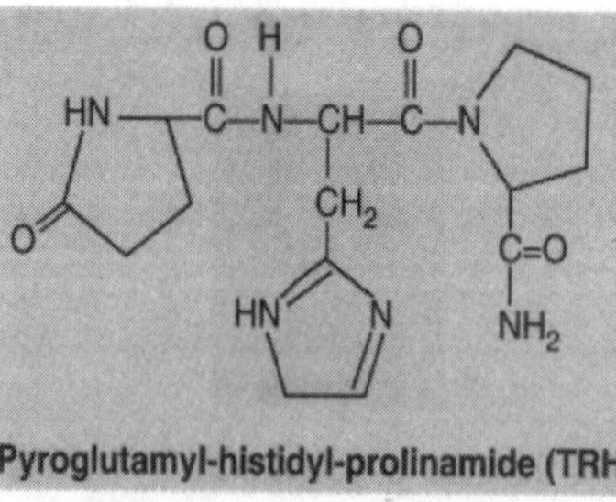

Pyroglutamyl-histidyl-prolinamide (TRH)

---

* **Hypophysectomized Mice :** A mice having an **excision** of the **pituitary gland** (*i.e.,* hypophysis).

Fig: 2.38 interestingly depicts yet another variant of **TRH,** where in the **amio (–NH₂) end** glutamate is duly cyclized to the corresponding **proglutamic acid,** and the specific **carboxyl (–COOH) end** propyl carboxyl get **amidated squarely.**

Besides, a *mammalian* **polypeptide** may essentially comprise even more than one *physiologically potent polypeptide.*

β–Lipoprotein– Very much within its **primary structure resides a hypophysial hormone** which critically stimulates the release of **'fatty acids'** derived from the *adipose tissue–* do designate **sequences of 'amino acids'** that are found to be common to many other **polypeptide hormones** possessing prominently altogether **diversified physiologic activity profiles.**

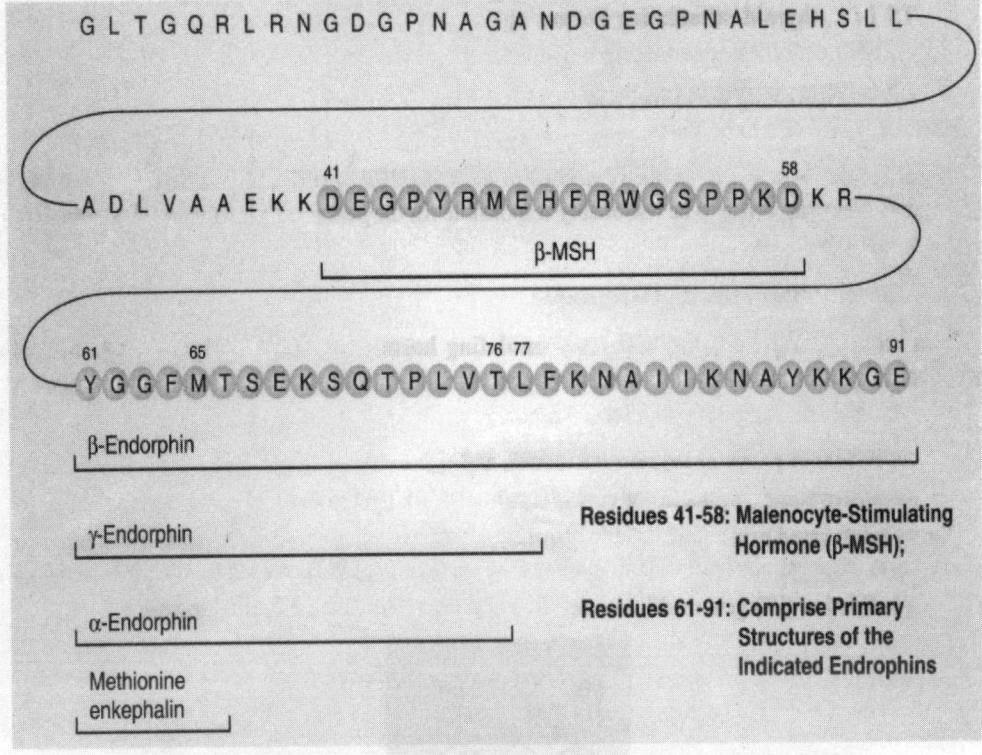

Fig: 2.38 : Diagramatic Representation of Primary Structure of β–Lipoprotein.

7.1.2. Chemistry of Thyrotropin (TSH):

In a broader perspective the **'chemistry of TSH'** may be adequately expatiated by carrying out first and foremost its meticulons *'isolation'* from the specific **bovine hypophysial tissue;** and ultimately establishing its crucial **'amino acid sequence'.**

Interestingly, the **'glycoprotein'** (with **Molecular Weight = 28,300**) essentially comprise *two* **noncovalently associated subunits** invariably known as :

TSH–α   :   **Molecular Weight = 13,600** and

TSH–β   :   **Molecular Weight = 14,700.**

Interestingly, the *hormone* (*i.e.,* thyroid–stimulating hormone) **is found to be among the** proteins richest **in S–content specifically.** It actually comprises **eleven disulphide [–S–S–] linkages,** but practically devoid of any **sulphydryl moieties;** however, the latter are present mostly as the **'interchain–bridges'** as given under:

α–subunit hairing **5–interchain–bridges,** and

β–subunit hairing **6–interchain–bridges.**

Nevertheless, one may critically observe that the **'same species'** essentially do have the following characteristic features, such as:

- α–subunits– are found to be more or less the *same* as in :

  TSH *i.e.,* **thyroid–stimulating hormone;**

  FSH *i.e.,* **follicle–stimulating hormone;**

  LH *i.e.,* **leuteizing hormone;** and

  HCG *i.e.,* **human chorionic gonadotropin.**

- β–subunits– do differ distinctly in each of these **hormones** *viz.,* **TSH, FSH, LH,** and **HCG;** and, therefore, prove to be reasonably responsible for the *perceptible* **hormonal characteristic features** of each of them.

7.1.3. Effects of Thyrotropic Hormone:

In a broader perspective, the **thyroid–stimulating hormone (TSH)** largely exerts its critical influencel(s) upon the rates of the following *three* **process** taking place in the **'thyroid gland'** for instance:

- critical removal of $I^\ominus$ **(iodide) from blood,**
- conversion of iodide to **thyroid hormones,** and
- modulated release of **hormone (TSH)** from the **'thyroid gland'.**

It may be observed explictly that the **hypophysectomized mice** shows a *remarkable* and *significant* lowering in the prevailing **'rate of iodide uptake'** by the *thyroid gland,* however, following its **immediate entry** the **iodide ion** ($I^\ominus$) gets readily converted to the corresponding **3,5-diiodotyrosine.***

3, 5-Diiodotyrosine

Importantly, the actual rate of conversion of 3,5–diiodotyrosine into the corresponding *thyroxine* **is mostly and invariably depressed in the critical and absolute absence of the hypophysis, as depicted clearly in fig: 2.39.**

It has been duly observed that the **lowered hormonal I_2**** in the plasma does not cause a direct stimulus to the **'thyroid'** so as to release its **hormones;** and, therefore, is solely dependent upon **TSH.**

---

* Also Found in the *skeletal proteins* of **corals, sponges,** and other marine organisms.

** Due to **hypophysectomy** in rats.

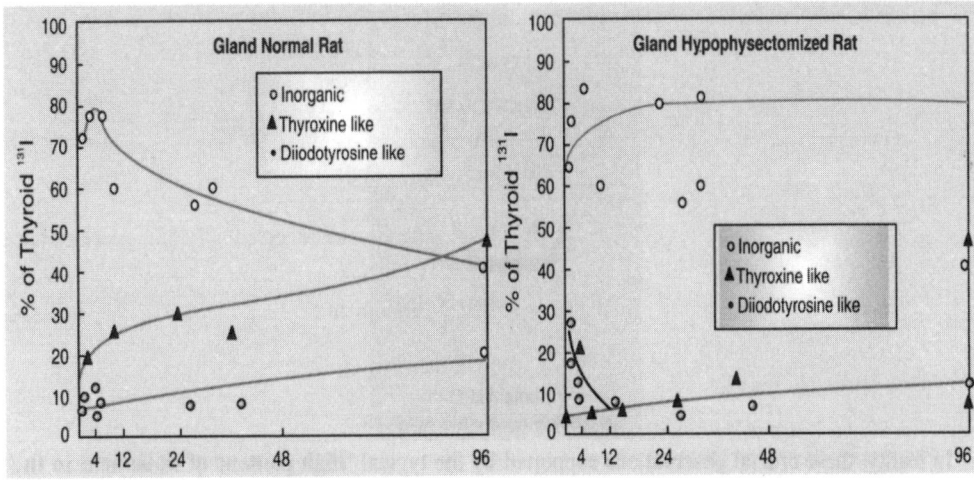

Fig: 2.39 : Diagramatic Distribution of Radioactive Iodine [^{131}I] Present in the Thyroid
Glands of 'Normal; [LHS– 'a'], and of Hypophysectomized [RHS–'b'] Rats.

7.1.4. Long–Acting Thyroid Stimulator [LATS]:

First and foremost an *'abnormal thyroid-stimulating substance'* was duly described in the blood of
such patients having a history of **hypothyroidism.** Later on this particular substance was critically
distinguished from **TSH** by its apparently *longer duration of action,* and duly termed as the **long-acting
thyroid stimulator [LATS].**

Salient Feature: Following are some of the **salient features** of **LATS:**

- **indistinguishable from immunoglobulin [IgG],**
- **serves as an 'antisera to normal human IgG' (but not the 'anti–TSH serum'), and**
- **neutralized exclusively the critical hormonal activity of LATS.**

7.1.5. Mechanism of Action of TSH:

In fact, the exact and precise action of **TSH** upon both:

- **thyroid gland,** and
- **target cells,**

is particularly initiated by binding of the **'hormone'** to the specific **TSH–receptors.** Thus, the receptor
ensuing from the corresponding **bovine-thyroid plasma membranes** designated a **'glycoprotein'** essentially consisting of **sialic acid*,** that is absolutely essential to the on-going **receptor function.** However,
the subsequent treatment with the so called *neuraminidase* (an **enzyme**) completely eliminates the ability
of the receptor to bind **TSH.**

---

* **Sialic Acid :** Acid generally occurs in throughout the **'animal kingdom',** and appear to be as *regular components* of
   all types of **neuroproteins, mucopolysaccharides,** and some **mucolipids.** It invariably happens to exist as:
   - Bovine and ovine submaxillary gland–as **diacetyl esters of sialic acid,** and
   - Pig submaxillary Mucin–as **N–glycolyl moiety in sialic acid.**

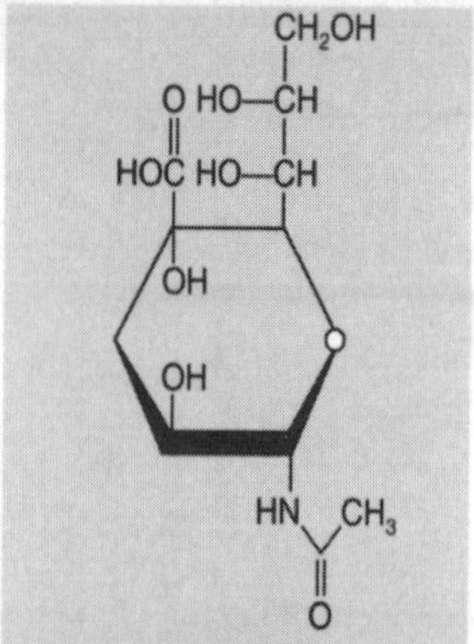

Sialic Acid
[N-Acetylneuraminic Acid]

In reality, these crucial observations supported by the typical 'high content of sialic acid in the 'gangliosides' remarkably and appreciably inhibits the *binding profile of TSH to the respective target tissues,* such as :

- **thyroid glands,** and
- **adipocytes** (*i.e.,* **fat-cells**).

Nevertheless, and apparent **conformational alteration in TSH profile** may come into play due to the resulting overall inhibition of the interaction between **TSH** and **ganglioside.** In addition, the extent of progression of the so called **'conformational change'** almost equalled the ensuing observed order of **'ganglioside effectiveness'** particularly for causing the inhibition in the binding of **TSH to the receptors**

7.2. The Gonadotropic Hormones:

These hormones cause specific stimulation of the **'gonads'** *viz.,* as applicable to the respective hormones produced by the **anterior pituitary lobe.**

Importantly, one may obviously take cognizance of the following aspects immediately after **'hypophysectomy':**

- **In Adults :** critical atrophy of the *gonads* and other *related organs,* and
- **In Younger Individuals :** delayed **maturation phenomenon.**

Note: *Adenohypophysis* in the humans exemplary cause such distinct disorders, such as : *impotence, amenorrhea,* and *atrophy of the gonads.*

In all, there are *four* recognized **'gonadotropic hormones'** as stated under :

- **FSH–follicle stimulating hormone,**
- **LH or ICSH–luteinizing hormone or interstitial cell-stimulating hormone,**
- **Prolactin – an altogether separate adenohypophysial principle**

These are usually secreted by the adenohypophysis

- **HCG–human chorionic gonadotropin**

It is produced by human placenta.

_____

* **galactose oxidase** or **incubation with neuraminidase** as detailed under :

> **Note:** *HCS i.e., chorionic somatomammotropin* essentially possesses and *'amino acid sequence'* very much akin to that of the *human growth hormone.*

7.2.1. Gonadotropin Secretion *in vivo:*

It has been amply proved and demonstrated that the highly specific secretion of the **hypophysial gonadotropins** gets articulately modulated thereby showing a broad-spectrum of the *neurogenic activities* upon the sexual profile of a subject both:

- **experimentally,** and
- **clinically.**

Isolation of a **'decapeptide'** either from **porcine** or **ovine** *hypothalamic tissue* predominantly gives rise to both **LH** and **FSH–releasing activity profile** on being administered to appropriate **laboratory test animals,** and the **female subjects.** Therefore, it has been logically and legitimately designated as **LRH–FRH*** as given under :

pGlu–His–Trp–Ser–Tyr–Gly–Leu—Arg–Pro–Gly–NH$_2$–LRH–FRH

Evidently, the actual secretion of **'prolactin'** (secreted by the **adenohypophysis),** very much similar to that of the **human growth hormone,** as well as **MSH** may be appropriately modulated by both *stimulating and inhibiting substances.* In this manner, the ensuing **hypothalamic stimulatory substance** is invariably known as **prolactin–releasing factor (PRH)** ; whereas, the **hypothalamic inhibitory substance** is commonly termed as **prolactin-inhibitory factor (PRIH).**

7.2.2. Chemistry of the Gonadotropins:

Based on the credences from the literature one may broadly observed that the so called *'hypophysial gonadotropins'* **LH and FSH,** and also the **placental HCG** are *glycoproteins* comprising both **'α–subunits'** and **'β–subunits';** nevertheless, the **former** being common to each hormone emphatically.

Importantly, the relationship (s) being found in the *structural analogues* of the following *two* **thyrotropins :**

- **TSH–α,** and [The **noncovalently bound 'subunits']**
- **TSH–β,**

explicitly suggests these *two underlying aspects:*

- critically observed **'homologies'** in the **amino-acid sequence and**
- presence of **biological and immunological specificity** of the *hormones* in the β –subunit.

Table: 2.3. Records certain chemical characteristic features of the **gonadotropic hormones** along with their **molecular weight** and the **carbohydrate content.**

In fact, the **gonadotropins** do contain certain array of **carbohydrate,** for instance : **fucose, mannose, galactose, N–acetylglucosamine, N–acetylgalactosamine,** and **sialic acid.** Furthermore, typical tailor-made alterations in the specific *carbohydrate segment* of the **glycoprotein hormones** may be accomplished by either

---

* **LRH–FRH :** It refers to **leutinizing hormone regulatory hormone–follicl–stimulating hormone regulatory hormone.**

Table : 2.3 :Certain Chemical Characteristic Features of Gonadotropic Hormones[†]

S.No.	Hormone Variant	Molecular Weight	Carbohydrate Content
1	**LH or ICSH** • Human	28,500	15.5
2	**FSH** • Human	34,000	16.0
3	**Prolactin** • Sheep	23,500	0
4	**HCG**	40,000	30-33
5	Pregnant Mare's Serum Gonado Tropin	23,000	45.0

- enhanced elimination rate of the *changed hormones* right from the blood by the **liver after due IV administration,**
- an apparent lowering in **general behavioural pattern** *vis-avis* directly proportional to the total loss of carbohydrate, and
- an observed variable extent in the **'loss of specificity'** with regard to the **immunologic behavior for each of the hormones.**

7.2.3. Mechanism of Action of the Gonadotropins:

In a broader perspective the known **gonadotropins** *viz.,* **LH, FSH,** and **HCG** get bound to the *specific receptors* rather intimately to the membranes of their respective **'target cells'** *i.e.,* the **ovaries** and **testes.**

There are *two* marked and pronounced mechanisms of the **gonadotropins,** which shall be discussed briefly as under:

First, the crucial binding of the **gonadotropic hormones** to the ensuing **'target cells'** is duly followed by on enhanced observed activity of the **plasma membrane-bound adenylate cyclase.** Besides, the exhaustive *in vitro* investigative studies with specific reference to **HCG** evidently bring forth the following *two* suggestions, namely :

- absolute devoid of *strict correlation* in the **'intracellular [cAMP]',** and
- prevailing **'steroidogenesis'.**

Secondly, LH besides exhibiting its **stimulatory activity** *in vitro* upon the **adenylate cyclase** present in the *corpora lutea,* also enhance the specific synthesis of one of the **prostaglandins** belonging to **Group–E.** Interestingly, the ensuing **LH–action** upon the **luteal tissue** does stimulate the overall activity of *two* vital and important **enzymes,** such as:

- **Cholesterol esterase,** and
- **Cholesterol acyl transferase,**

thereby attributing the presence of **'arachidonic acid'***, which could be the ultimate logical basis for the enhancement in the yield of the *'intracellular prostaglandins'.*

---

* **Arachidonic Acid :** It is regarded to be **precursor** of the **prostaglandins.**

+ **[Adopted From :** White *et al.* : **Principles of Biochemistry,** Tata McGraw Hill Publishing Co., LTD., 6th ed., 2004]

RECOMMENDED FURTHER READINGS

Bhown AS	:	**Protein-Peptide Sequence Analysis : Current Methodologies,** CRC–Press, New York, 1988.
Biemannn K	:	**Mass Spectrometry of Peptides and Proteins,** *Ann Rev Biochem,* **61** : 977, 1992.
Bollag DM and Edelstein SJ	:	**Protein Methods,** Weley–Less, New York, 1990.
Branden C and Tooze J	:	**Introduction to Protein Structure,** Garland, New York, 1991.
Copeland RA	:	**Methods of Protein Analysis : A Practical Guide to Laboratory Protocols,** Chapman & Hall, Philedelphia (USA), 1993.
Creighton TE	:	**Protein Structure : A Practical Approach,** Oxford Press, London (UK), 1990
Crieghton TE	:	**Protein Structure,** IRL Press, London, (UK), 1993.
Dunn MJ	:	**Gel Electrophoresis of Proteins,** Wright, London (Uk), 1986.
Hames BD and Rickwood D	:	**Gel Electrophoresis of Proteins : A Practical Approach,** IRL Press, London, (UK), 1990.
Hunkapiller MW, Strickler JE, and Wilson KJ	:	**Contemporary Methodology for Protein Structure Determination,** *Science,* **226** : 304, 1984.
Landry SJ and Giersach LM	:	**Polypeptide Interactions with Molecular Chaperones and their Relation ship to** *in vivo* **Protein Folding,** *Annu Rev Biophys Biomol Struct,* **32** : 645,1994
Richardson JS	:	**Principles and Patterns of Protein Conformation. In :** *Prediction of* and Richardson DC *Protein Structure and the Principles of Protein Conformation,* Fasman GD (ed.), Plenum Press, London (Uk), 1989.
Rose GD *et al.*	:	**Turns in Proteins and Peptides,** *adv. Protein Chem.,* **37:** 1, 1985.
Scholz JM and Baldwin RL	:	**The Mechanism of α–Helix Formation by Peptides,** *Annu Rev Biophys Biomol Struct,* **21** : 95, 1992.
Scopes RK	:	**Protein Purification : Principles and Practices,** Springer, New York, 1993.
Stein RL	:	**Mechanism of Enzymatic and Nonenzymatic Prolyl** *cis-trans* **Isomerization,** *Adv Protein Chem.,* **44** : 1, 1993.
Stoschek CM	:	**Quantitation of Proteins,** *Methods Enzymol,* **182** : 50, 1990.
Wagner G *et al*	:	**NMR Structure Determination and Solution : A Critique and Comparison with X-Ray Crystallography,** *Anny Rev Biophys Biomol Struct;,* **21** : 167, 1992.
Wilson KJ	:	**Micro-level Protein and Peptide Separations,** *Trends. Biol. Sci.,* **14** : 252. 1989.

REVIEW QUESTIONS

1. Discuss the following types of 'Amino Acids' by giving a few typical examples from each class:
   - (a) Acidic Amino Acids,
   - (b) Basic Amino Acids,
   - (c) Neutral Amino Acids,
   - (d) Sulphur Containing Amino Acids,
   - (e) Aromatic Amino Acids, and
   - (f) Heterocyclic Amino Acids.

2. Give an account of the **Chemical and Enzymatic Hydrolysis of Proteins to Peptides** with specific reference to:
   - (a) **Heat Coagulation of Protein,** and
   - (b) **Protein Denaturation**

3. Discuss the 'Protein Structure' as a 3D-Model.

4. How would you determine the **Amino Acid Composition'** in the **protein hydrolysate** by:
   - (i) **Ion-Exchange Chromatography (IEC),**
   - (ii) **Capillary Electrophoresis (CE),** and
   - (iii) **pH-Endorsed Precipitation,**

5. Describe the following with **suitable examples:**
   - (a) **Amino-End Degradation in Amino Acids,** and
   - (b) **Carboxyl-End Degradation in Amino Acids.**

6. Discuss the 'Physicochemical Properties of Amino Acids' with reference to:
   - (i) **Ionization** *vis-a-vis* **Isoelectric pH**
   - (ii) **Amphoteric Nature,**
   - (iii) **Isomerism,**
   - (iv) **Solubility and Melting Point,** and
   - (v) **Titration Profile of Amino Acid Variants**

7. Give a comprehensive account of the **Chemical Properties** of the **Amino Acids** support your answer with appropriate examples.

8. Describe **Sanger's Method** for the determination of **amino acid sequence** in a *protein* or *polypeptide chain.* How the structure of 'Insulin' was determined by this method? Explain.

9. Discuss the **Synthesis** of the following **Amino Acids:**
   - (i) **Alanine from Ethyl Acetate,**
   - (ii) **Leucine from Ethyl-3-methylbutanoate,** and
   - (iii) **Threonine from Crotonic Acid**

10. Enumerate the **synthesis of C-Amino Acids.** How 'aspartic acid' may be synthesized from **Ethyl Malonate?** Explain.

11. Write short notes on the following:
   (a) **Amination of α-Halogenated Acids**
   (b) **Albertson Method**
   (c) **Bucherer Hydantoin Synthesis**
   (d) **Curtius Reaction**
   (e) **Darapsky Synthesis**
   (f) **Edenmeyer-Plochi Azlactone and Amino Acid Synthesis**
   (g) **Gabriel Phthalimide Synthesis**
   (h) **Hoffmann Degradation Method**
   (i) **Hydantoin Synthesis**
   (j) **Malonic Ester Synthesis**
   (k) **Phthalimidomalonic Ester Synthesis**
   (l) **Modified Phthalmidomalonic Ester Synthesis**
   (m) **Strecker Synthesis**

12. How would you carry out the **Quantitative Determination of Amino Acids** by:
   (i) **Thin-Layer Chromatography (TLC)**
   (ii) **Ion-Exchange Chromatography (IEC),**
   (iii) **Capillary Electrophoresis (EC),**
   (iv) **Gas-Liquid Chromatography (GLC),**
   (v) **^{1}H-NMR Spectroscopy,** and
   (vi) **Mass Spectroscopy (MS)**

13. How would you estimate the **'Net Charge in Amino Acids'?**

14. Give the two Methods recommended for the estimation of **'Net Charge in Peptides'**.

15. Discuss briefly the Method for the Determination of **'Net Charge ion Proteins'**.

16. Describe the **'secondary Structures'** of **Protein** with particular reference to:
   (a) **α-Helix Structure,**
   (b) **β-Pleated Sheet Structure,** and
   (c) **Triple Helix Structure**

17. Discuss the Stabilization of **'Tertiary Structure of Protein'** and **'Quaternary Structure of Protein'**.

18. Write a descriptive essay on the **Amino Acid (Protein) Metabolism**.

19. Give a comprehensive account on the **'Transamination Reaction'.** Explain the **Mechanism of Transamination Reaction.**

20. Discuss the **'Chemistry of Oxytocin'** and outline its **Synthesis** briefly.

21. Write short notes on the following:
   (a) **Thyrotrophin-Releasing Hormones (TRH),**
   (b) **Thyrotropic Regulatory Hormones (TRHs).**

Contents at a Glance

1. Introduction

2. General Methods of Structure Elucidation of Alkaloids.
3. Classification of Alkaloids.

   3.1 Pyrrolidine Alkaloids.

   3.2 Pyridine Alkaloids [or Piperidine Alkaloids]

   3.3 Pyridine-Pyrrolidine Alkaloids

   3.4 Tropane Alkaloids.

   3.5 Quinoline Alkaloids.

   3.5 Isoquinoline Alkaloids.

Alkaloids

1. INTRODUCTION

Richter (1993)*, a well-known German phytochemist, was pioneer in the legitimate proclamation that the vast number of products which have already been duly described in the literature(s), their structural diversity, and the possible scope of their inherent pharmacological activity profiles do render the *alkaloids* as one of the most cardinal categories of naturally occurring substances of remedial therapeutic interest, whose actual number **'is almost unfathomable'.**

It is, however, pertinent to state here that the above powerful statement of facts and ideology does stand like a rock in view of the exceptional characteristic features with regard to the **structural** as well as **biosynthetic pathways** of the **'alkaloids'.**

1.1. Definitions

Meisner (1819) first and foremost introduced the common and well-known terminology **'alkaloid'** to designate critically and specifically all such *natural substances* almost interacting like:

- **bases,** or
- **alkalis.**

In fact, the term **'alkali'** has been duly derived from the *Arabic* word *al kaly* = soda; and from the *Greek* word *eidos* = appearance. Nevertheless, **Meisner's definition** shows clear cut ambiguity between the following *three* **chemical entities,** namely:

- **alkaloids,**
- **protoalkaloids,** and
- **pseudoalkaloids**

Intellectually sound but logically controversial definition of 'alkaloid': Meisner's definition also fails to explain justifiably such anomalies as given under:

- **Colchicine (an alkaloid):** has its **N–atom** strategically located in the **amide functional moiety,** but *not* within the so called **heterocyclic ring;**

_____

* Richter G: **Stoffwechselphysiologic der pflanzen,** George Thieme Verlag, Sttutgart (Germany), 1993

- **Caffeine and Theophylline:** Can these be also called as **alkaloids,** but they belong to the xanthine class;
- **Chaconines:** *i.e., an amine-containing glycosides* be termed as **alkaloids** or rather merely as the 'N–containing saponins'.*

Colchicine   Caffeine   α–Chaconine

R = α–L–Rha (1⟶4)–β–d–Glc(1⟶)
↑ (1⟶2)
α–L–Rha

Hence, one may visualize the peculiar and difficult status of '**alkaloids**' *vis-a-vis* to assign a logical and befitting and definitive homogenous category of compounds belonging to **biochemical** or **chemical** or **physiologic** point of view.

As a result, it is advisable and preferable to induct cautious reservation pertaining to any **generalized definition** or to refer '**alkaloids**' as representing *all organic complex nitrogenous chemical entities* having a **restricted distribution in nature.**

Comprehensive Definition of Alkaloid(s): A rather comprehensive definition of alkaloid(s) was duly put forward as– '**basic nitrogenous naturally occurring plant products, mostly optically active, and essentially possessing nitrogen heterocycles as their structural units, with a marked and pronounced physiological activity profile.**'

Importantly, the aforesaid comprehensive definition of alkaloid still has several serious '**lacunae**'; and, therefore, need to be interpreted with great caution and restrain, since there exists quite a few such **chemical entities (compounds)** that are **not actually alkaloids,** however, do confine to the said *definition.* It may further be expatiated with the help of certain more specific, typical, and critical examples, such as:

(1) **Thiamine (Vitamin B$_1$)** does possess a **heterocyclic nitrogenous base** but never classified as '**alkaloids**' perhaps because it is distributed universally in most of the '**living matter**'.

(2) **Betaines, choline, ephedrine, hordenine, muscarine, stachydrine,** and **tryptamine** may **not** have their **N–atom** very much confined to a **heterocyclic system,** but are duly classified as '**alkaloids**' or '**protoalkaloids**'.

(3) **Amino acids, cholines** and **phenylethylamines** invariably occur in nature as typical **open-chain basic compounds** and do exhibit marked and pronounced **physiologic activity profile,** but are **not classified amongst the alkaloids.**

(4) **Piperine** (obtained from *black pepper*) amazingly neither possesses a **basic characteristic profile** nor exhibits any **physiologic activity;** however, is genuinely included in the list of **alkaloids.**

---

* **Saponin:** It refers to a group of **amorphous colloidal glycosides** that essentially form *soap-like aqueous solutions.*

At the end, it may however, be inferred that it is still difficult and an exceptionally delicate task to determine whether a compound is an **alkaloid** or not.

1.1.1. Protoalkaloids

They designate *simple amines* wherein the N–atom is **not part of a heterocyclic ring structure**. Nevertheless, the **protoalkaloids** are:

- **fundamentally basic in character,** and
- elaborated adequately *in vivo* from the 'amino acids'.

Importantly, a plethora of chemical entities do comply with this **definition justifiably,** such as:

❑ **Simple amines** *e.g.,* serotonin and **mescaline** from *peyote;* **cathinone** from **Abyssinian tea;** and **betaine*** from plants and animals.

> **Note:** Some researchers usually include *'betalains'* in this particular class of *'protoalkaloids'* *e.g.,* **betanin.**

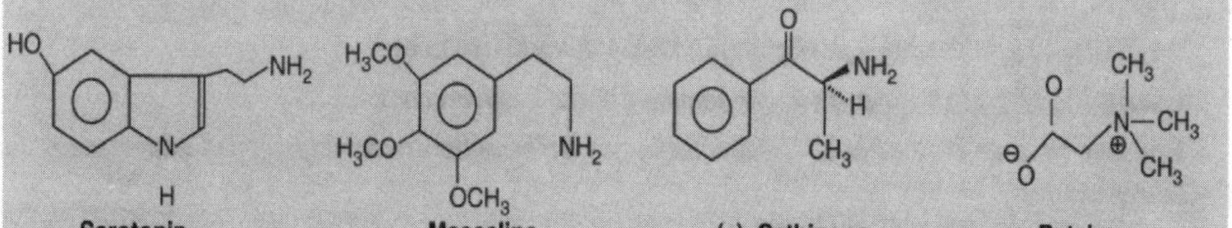

Serotonin   Mescaline   (–)-Cathinone   Betaine

1.1.2. Pseudoalkaloids

They certainly do meet all the properties of the **'true alkaloids',** but are **not** derived from the *'amino acids'.*

Examples: These essentially comprise:

- ❑ **Monoterpenoid Alkaloids** *e.g.,* β–skythantine,
- ❑ **Steroidal Alkaloids** *e.g.,* paravallarine; and
- ❑ **Heterocyclic N–containing substances derived from 'acetate metabolism'** *e.g.* (+)-connine.

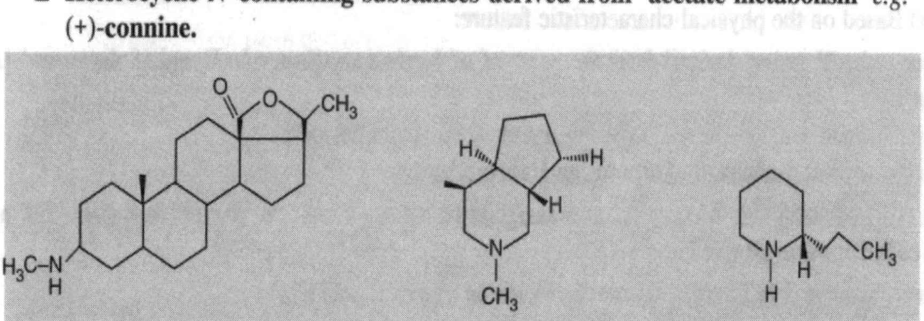

Paravallarine   β–Skythantine   (+)–Connine

---

* Resulting from the **quaternization of the N–atom** of **amino acids.**

Note: Following compounds are not recognized as 'alkaloids', such as: amino sugars, porphyrins, alkylamines, and arylalkylamines.

1.2. Nomenclature of Alkaloids

In a broader perspective, the cardinal characteristic features related to the **nomenclature of alkaloids** is the absolute lack of any commonly agreed systematic guidelines. Hence, based on a **general agreement*** the *chemical rules* invariably represent the names of most alkaloids should essentially end with the *suffix* (*–ine*). Thus, the names of the **'alkaloids'** are normally assigned in several means and ways, such as:

(*i*) Based on the **generic name** of the plant from which they are actually derived:

Examples: Hydrastine from *Hydrac Stis canadenesis* L. *(Ranunculaceae);* and **quinine** from *Cinchona officinalis* L. (*Rubiaceae*).

(*ii*) Derivation from specific name of the plant producing them:

Examples: Cocaine from *Erythroxylum coca Lam.* (*Erythroxylaceae*); **Berberine** from *Hydrastis canadensis* L. (*Hydrastidaceae*); and **Belladonine** from *Atropa belladona* L. (*Solanaceae*).

(*iii*) Based on their highly **specific physiological activity profile:**

Examples: Morphine from *Papaver somniferum L.* (*Papveracac*);

Emetine from *Hedera helix* L. (*Araliaceac*); and **Trigonelline** from *Trigonella foenumraecum* L. (*Trigoniaceae*).

(*iv*) Derived from the common name of the **'drug'** producing them:

Examples: Ergotamine from *Claviceps purpurea* (Er.) Tul. (*Hypocreals*): **'ergot'** being the common name of the **herbal drug;** and **Ginseng** from *Panax ginseng* CA May. (*Araliaceae*): **'ginseng'** being the common name of the *'herbal drug'*.

(*v*) From the name of the discoverer of the plant species:

Example: Pelletierine from the barks of *Punea granatum L.* (*Punicaceae*)**:   used as an **anthelmintic (Cestodes).**

Note: In the early literature the names *'pelletierine'* and *'isopelletierine'* were used interchangeably.

(*vi*) Based on the physical characteristic feature:

Example: Hygrine derived from the roots of *withania somniferum* L. Dunal (*Solanaceae*) [*Hygro* = moist].

(*vii*) From related **'bases'** invariably named by **'transpositions':**
Examples: Narcotine, Cotarnine, and **Tarconine.**

(*viii*) Minor alkaloids have been duly assigned by attaching one *'prefix'* or *'suffix'* to the name of the so called **'major alkaloids':**

Examples: The famous **cinchona alkaloids** as given under:

- **Quinine;**   • **Quinidine ;**
- **Cinchonine;**   • **Cinchonidine;**

---

* Kar A: **Pharmacognosy and Pharmacobiotechnology,** New Age International, New Delhi., 2nd ed., 2007.

** Gilman and Marion: *Bull Soc. Chim. France,* 1931, 1993.

1.3. Physiological Actions

Alkaloids are specifically potential naturally occurring substances due to their **multifarious physiological activity profile.** An array of such distinct, remarkable, and recognized alkaloids along with their typical **physiological actions** shall be enumerated as under:

- **CNS Activities:** These are of *two* important types:
  - **CNS–stimulants** *e.g.,* strychnine, caffeine.
  - **CNS–Depressants** *e.g.,* morphine, scopolamine.
- **Antimalarial Activity:** These essentially comprise:
  - **Cinchona alkaloids** *e.g.,* quinine, quinidine, cinchonine, cinchonidine.
  - **Unexpanded flower heads of *Artemisia annua L.,*** *e.g.,* **Artemisinin** (or **Qinghaosu**) (a *Chinese* discovered **Herbal Drug**)
- **ANS Activities*:** These are of *five* cardinal types, namely:
  - **Sympathomimetics** *e.g.,* **ephedrine,**
  - **Sympatholytics** *e.g.,* **yohimbine, ergot alkaloids.**
  - **Para sympathomimeties** *e.g.,* **eserine, pilocarpine,**
  - **Anticholinergics** *e.g.,* atropine, hyoscyamine, and
  - **Ganglioplegics** *e.g.,* **sparteine, nicotine.**
- **Local Anaesthetics:** *e.g.,* **cocaine** (a *banned narcotic agent*).
- **Antineoplastics:** *e.g.,* **vinblastine, ellipticine.**
- **Antibacterials:** *e.g.,* **berberine.**
- **Amebicides:** *e.g.,* **emetine.**
- **Antidote in Strychnine Poisoning, Diagnostic aid in Myasthenia Gravis, Adjunct in Electroshock Treatment,** and **Control convulsions in neuropsychiatry,** } *e.g.,* **tubocurarine HCl**
- **Antipsychotics** *e.g.,* **reserpine.**

Note: The overall functionality of alkaloids may be regarded to be the *'end-products of detoxication mechanisms',* rest their accumulation in various plants would be responsible solely for causing undesired damage in them substantially.

1.4. Occurrence of Alkaloids

Mckee (1962)** made a critical survey of literature and reported that approximately **1000 alkaloids,** that are known and recognized, do belong to almost **100 families, 500** genera, and stretched over 1200 species. Nevertheless, the **alkaloids** are not only distributed squarely in the *plant kingdom,* but also observed to be present in **fungal alkaloids** *e.g., ergot alkaloids* (with an exception).

It is worthwhile to mention here that the alkaloids are distributed abundantly in *'higher plants',* **dicotyledons** in particular, such as:

---

* Mehta SC and Kar Ashutosh: **Pharmaceutical Pharmacology,** New Age International, New Delhi, 2009.

** Mckcc HS: *Nitrogen Metamobolism in Plants,* 1962

- Apocynaceae   • Papilonaceae   • Papaveraceae   • Ranunculaceae
- Rubiaceae   • Rutaceae   • Solanaceae

and markedly less frequent in *'lower plants' viz.,* **fungi.**

(1) **Various Parts of Plants:** It has been duly observed that the **alkaloids** are strategically located in various parts of plant, namely:

❑ **Nicotine** – produced in the **roots** of **'tobacco plant'** (*Nicotiana tabacm*) which eventually gets *translocated* to the **leaves** where it *acculmulate richly*. However, the seeds of **tobacco plant** are completely devoid of **nicotine.**

❑ **Morphine** – obtained from the **latex** of **'opium poppy'** (*Papaver sominferum*) duly collected from fresh fruit; whereas, the **poppy seeds** are almost *devoid'* of alkaloids (one of the principal drugs used for pain relief).

❑ **Quinine** – produced from the **bark** of **cinchona tree** (*Cinchona officinalis*) exclusively up to **5%.**

❑ **Colchicine** – obtained from the **seed** and **corm** of **colchicum** (*Colchicum autumnale*).

(2) **Structurally Related Alkaloids:** The structurally related alkaloids are invariably found in the form of **'salts'** pertaining to the **same plants acids**, such as:

- **aconite alkaloids** – with *aconitic acid,*
- **cinchona alkaloids** – with *quininic acid,* and
- **opium alkaloids** – with **neconic acid.**

(3) **Fluctuations in Alkaloids Content with Seasonal Variance:** It has been duly observed that in certain typical instances one may actually obtain appreciable and significant fluctuations in the alkaloidal content in different parts of the plant during the **various stages of its** *growth, seasonal variance,* and *even between day or night.*

Note: In certain *perennial,* both *accumulation* as well as *localization* of the inherent *'alkaloids'* in one or two specific organs, remarkably and distinctly appear to be more critical' with the normal advancement of the *actual age of the plant itself.*

(4) **Confinement of 'Alkaloids' to a Specific Plant Family:** In a broader perspective, the specific **alkaloids** having *complex chemical structures* are invariably confined to **specific plant families,** for instance:

- **hyosciamine**   :   in *Solanaceae ;*
- **colchicine**   :   in *Liliaceae ;*
- **morphine**   :   in *Papaveaceae ;*
- **pellctrierine**   :   in *Punicaceae* etc.,

(5) **Concentration of Alkaloids in Plants:** It differs overwhelmingly in various types of **plant species** and solely depends upon the following *three* cardinal aspects, such as:

- **season of the year,**
- **geographical location,** and
- **approximate 'age' of the plant.**

Examples: There are *three* glaring examples, such as:

(a) **Hyoscyamine** – is found in seven different genera of the family, *Solanaceae;*

(b) **Twenty 'Opium Alkaloids'** – having a close similarity are found in the **same plant species.**

(c) A rather **complex alkaloid** is found in **one particular genus of a family** (*i.e.,* in **one species only**); whereas, a **simple alkaloid** is mostly found in **several different plants species.**

1.5 Isolation of Alkaloids

The **isolation of alkaloids** from various segment of a plant source *viz.,* **leaves , bark, fruits, seeds, latex** etc., may not be always a simple and easy task perhaps due to their **complexity of individual structure** or their **presence along with other congeners** (or **structural analogues**). Besides, the coexistence of other plant constituents *e.g., glycosides, tannins,* and *plants acids* (viz., **quinolinic acid, quinic acid, caffeic acid, balsamic acid, apiolic acid, ascorbic acid, benzoaric acid, brahmic acid** etc.,) in plant materials do render the process of isolation into a rather complicated one.

General Methods: The general methods of isolation of an **'alkaloid'** essentially comprise the following steps in a sequential manner:

1.5.1. Presence of Alkaloids with Specific Alkaloidal Precipitating Reagents:

First, it has to be ascertained that the particular part of the plant under investigation does contain an **alkaloid** by the help of the following **alkaloidal precipitating reagents***, namely:

- **Mayer's Reagent,**
- **Wagner's Reagent,**
- **Dragendorff's Reagent,**
- **Kraut's Reagent** (*i.e,* **Modified Dragendorff's Reagent**),
- **Marme's Reagent,**
- **Scheibler's Reagent,**
- **Hager's Reagent,**
- **Sonnenschein's Reagent,**
- **Bertrand's Regent,**
- **Marquis' Reagent,**
- **Erdmann's Reagent,** and
- **Frohde's Reagent.**

Besides, there are some typical **inorganic acids** which give rise to useful characteristic colour reactions with **'alkaloids'** even in the presence of *very small quantities,* such as:

❑ Chloroauric acid [HAu Cl$_4$],

❑ Chloroplatinic acid [H$_2$ Pt Cl$_6$],

❑ Phosphomolybdic acid [24 MoO$_3$. P$_2$O$_5$. x H$_2$O], and

❑ Phosphotungstic acid [24 WO$_3$, 2H$_3$PO$_4$. 8H$_2$O].

1.5.2. Extraction of Alkaloids with Solvents

Secondly, the critical *separation of* **alkaloids** essentially involves the separation of relatively *small percentage* (on **strictly dry weight basis**) of **'alkaloids'** from a *large quantum* of *extraneous plant material.*

---

* Kar A: **Pharmacognosy and Pharmabiotechnology,** New Age International, New Delhi., 2nd ed., 2007.

Examples: Opium: consists of *10% morphine* ;

Cinchona: consists of *5–8% quinine* ;

Atropa Belladona: consists of *0.2% hyoscyamine* ;

Rauwolfia Serpentina: consists of *0.1–0.2% reserpine* ;

Procedure: The out-line of the procedure involving the systematic **extraction of alkaloids with solvents** are as stated under:

(1) The **dried*** and pulverized *plant portions* is first extracted with **solvent either** (*i.e.,* diethyl ether)** ; and subsequently, filtered for the complete removal of soluble fat components.

(2) The resulting **'defatted residual mass'**, known as *'marc'*, is extracted with a polar solvent **methanol** [CH₃-OH] in order to get rid of *two* undesired materials:

   • **cellulosic components**, and

   • **insoluble components**.

   The resulting filtrate is evaporated completely and carefully under reduced pressure (vacuum).

(3) The **evaporated residual mass**, obtained duly from step (2), is made to dissolve in DM water, acidified to pH 2, and ultimately subjected to **stem-distillation** to remove **methanol.**

Fig: 3. Illustrates the **'schematic flow chart for alkaloid extraction'** from a plant source:

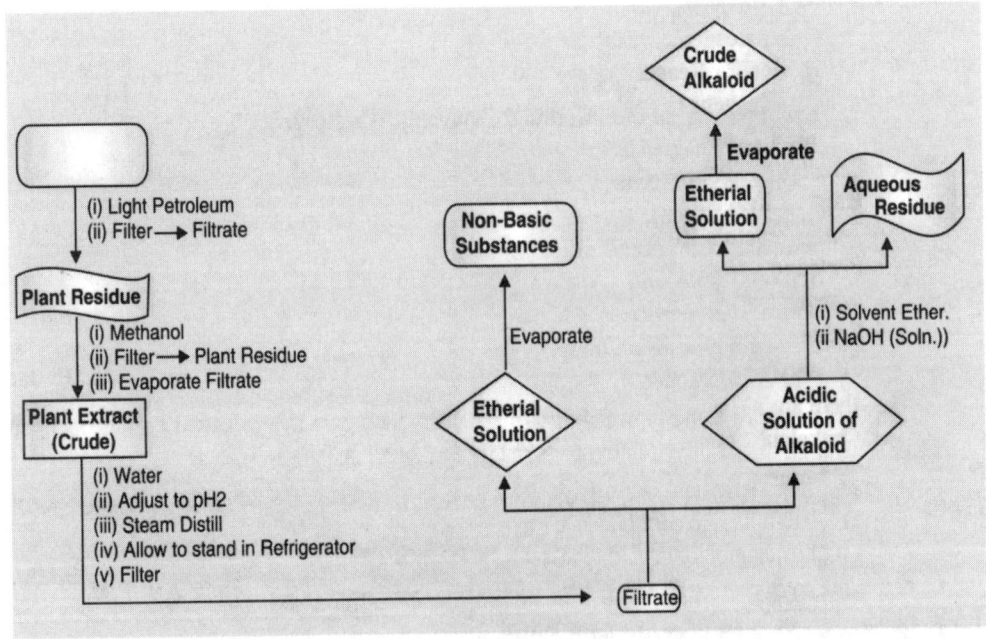

Fig. 3.1: Schematic Flow Chart for the Extraction of Alkaloids.

---

* The drying of medicinal plants for the extraction of **alkaloids** must be carried out under an **airy shaded place** and not directly in the **sun rays.**

** **Defatting** of plant material is an **important step** prior to the extraction of **alkaloids.**

(4) **Removal of Suspended Impurities:** The evaporated residual solution (usually *'dark'* in colour) is treated generally in *two* ways:

- allowed to stand undisturbed from **8–10 days** in a refrigerated condition **(5–10C°),** or

- warmed gently with *molten paraffin,* so as to remove the **discardable suspended impurities** by careful filtration.

(5) The resulting almost transparent/clear **'filtrate'** is extracted duly with **solvent ether** or **chloroform** to get rid of most of the **soluble nonbasic organic materials.**

(6) The remaining liquid mixture is subjected to **'steam distillation',** whereupon the **steam-volatile alkaloids** get separated in the **'distillate'.**

(7) The **'rest of the alkaloidal salts'** in the resulting solution is made **alkaline** and extraction protocol repeated with **solvent ether or chloroform, solvent evaporated** to obtain a **crude mixture of alkaloids.**

2. GENERAL METHODS OF STRUCTURE ELUCIDATION OF ALKALOIDS

Introduction: The *first-half of the 19th century* could not accomplish an appreciable progress in the elucidation of the complex structure of the **'alkaloids'.** However, the second-half *of the 19th century,* with the advent of a tremendous progress in the remarkable development in the field of **physico-chemical methods of analysis*,** these complex structures were assigned a definite structure (which were duly ascertained by **'total synthesis').**

Based on the fundamental principle that the exact and precise structure of most of the **'alkaloids'** could now be established by such means and ways as:

- selective typical reaction, and

- critical analytical techniques to locate and determine even quantitatively the various substituents and functional moieties strategically present in an unknown **'alkaloid'.**

In a broader perspective, the *methodical elucidation of the structure* of **'alkaloids'** can be determined by the following methods:

- ❏ Determination of Molecular Formula,
- ❏ Functional Group Analysis,
- ❏ Functional Nature of O–atom,
- ❏ Functional Nature of N–atom,
- ❏ Estimation of C–Methyl Moieties, and
- ❏ Degradation of Alkaloids.

The aforesaid aspects shall now be treated individually in the sections that follows:

2.1. Determination of Molecular Formula

After carrying out the isolation of the alkaloid, it is duly subjected to fractionation by several recognized methods, namely:

- Preparative TLC,
- Preparative HPTLC,

---

* That is, C, H, N - analysis, FTIR, UV, ^{1}H–NMR, MS,ORD, X–Ray Diffraction, Circular Dichroism, and the like.

- GLC–Method (for **thermolabile alkaloids**),
- HPLC–Methods, and
- Column Chromatography Procedures.

The pure **alkaloid,** thus obtained, is subjected to **elemental analysis** by using **automated CHN–analyzer** to assign the **'empirical formula'**. Now, its **'molecular weight'** is duly determined by:

- **Mass spectrometer (MS) analyzer, or**
- **Rast method** (due to the observed depression of the *'freezing point'*).

Important points: The following *two* **important points** need to be appreciated and used intelligently:

- Based on the observed **molecular formula of the alkaloid** one may precisely calculate the exact number of **double bonds** equivalent to the former, and
- Predominantly the very **presence of a double bond** or **critical cyclization of the chain** affords an apparent *reduction of the molecular formula* by at least **2H–atoms** *vis-a-vis* the corresponding **'saturated aliphatic hydrocarbon'**.

Double Bond Equivalency: One may critically visualize and examine the prevailing differences amongst the following *three* compounds showing explicitly the so called **double-bond-equivalency**

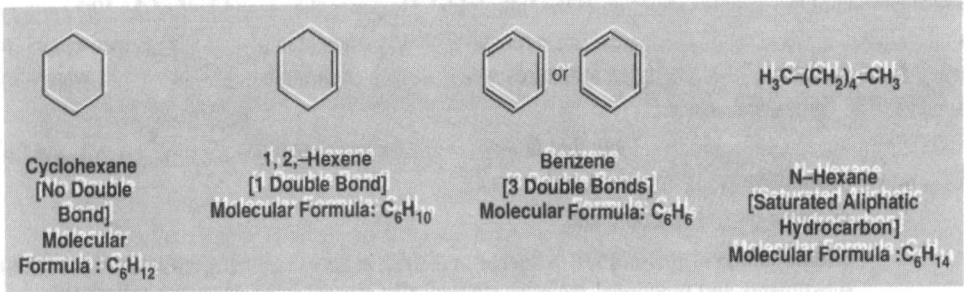

| Cyclohexane [No Double Bond] Molecular Formula : C_6H_{12} | 1, 2,–Hexene [1 Double Bond] Molecular Formula: C_6H_{10} | Benzene [3 Double Bonds] Molecular Formula: C_6H_6 | N–Hexane [Saturated Aliphatic Hydrocarbon] Molecular Formula : C_6H_{14} |

From the above structures, we may conclude that:

- difference between **1, 2-Hexcne** and **Cyclohexane** is of 2H–atoms only, which stands for only *one* **double bond equivalent,** and
- difference between **Benzene** and **N–Hexane** is of **8 H–atoms,** that actually corresponds to 8/2 or **4 double bond equivalents,** and this is acccomodated by the presence of **3–double bonds plus one cyclic ring** in Benzene.

Method for Complex Molecular Formulae: In a typical situation, when the **complex alkaloidal formulae** bears more than *three* elements *viz,.* C, H, N, O, the prevalent methodology pertaining to the generalized formula *Ca Hb Nc Od* *vis-a-vis* the number of **double bond equivalents** may be invariably expressed as given under:

$$a - 1/2b + 1/2 c + 1 \qquad \text{...(i)}$$

The above **expression (***i***)** is vehemently used for the needful precise calculation for the presence of the actual **'number of cyclic rings present in a given alkaloidal structure'.**

Examples: A few typical examples based on the above analogy are as given under:

(1) **Atropine: Mol. Formula [C$_{17}$H$_{23}$NO$_3$]**
According to **Eqn. (i),** we have: 17–23/2+1/2+1 = **7**

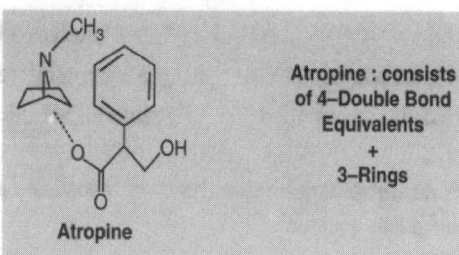

Atropine : consists
of 4–Double Bond
Equivalents
+
3–Rings

Atropine

(2) **Hygrine: Mol. Formula [C$_8$H$_{15}$NO]:**
According to **Eqn. (i),** we have: 8–15/2+1/2+1 = **2**

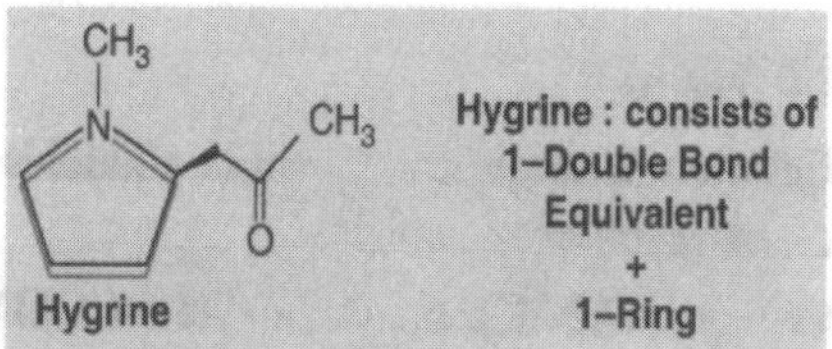

Hygrine : consists of
1–Double Bond
Equivalent
+
1–Ring

Hygrine

(3) **Mescaline: Mol. Formula [C$_{11}$H$_{17}$NO$_3$]**
According to **Eqn. (i),** we have: 11–17/2+1/2+1 = **4**

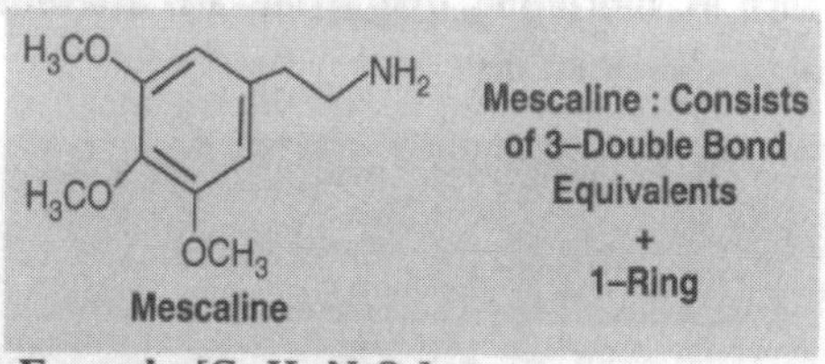

Mescaline : Consists
of 3–Double Bond
Equivalents
+
1–Ring

Mescaline

(4) **Neostigmine: Mol. Formula [C$_{12}$H$_{19}$N$_2$O$_2$]**
According to **Eqn. (i),** we have: 12–19/2+1+1 = **5**

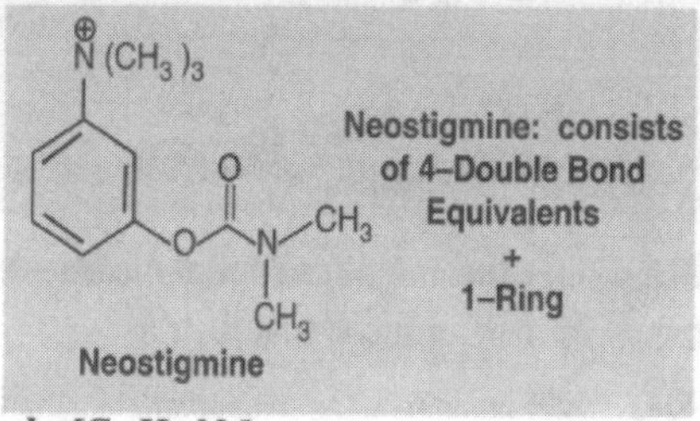

Neostigmine: consists
of 4–Double Bond
Equivalents
+
1–Ring

Neostigmine

(5) **Nicotine: Mol. Formula [C$_{10}$H$_{14}$N$_2$]**
According to **Eqn. (i),** we have: 10–14/2+1+1 = **5**

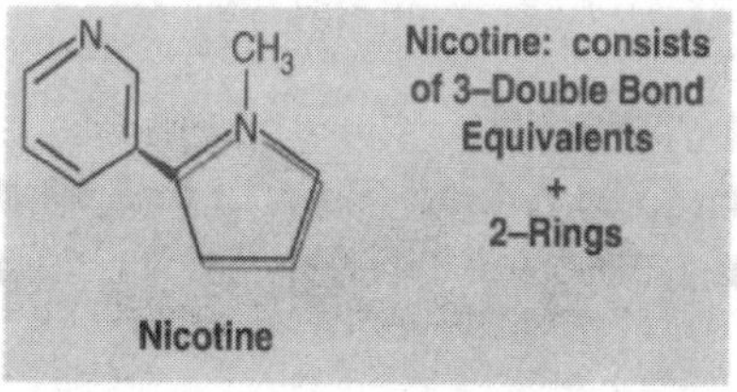

Nicotine: consists
of 3–Double Bond
Equivalents
+
2–Rings

Nicotine

2.2. Functional Group Analysis

The presence of various functional group duly present in a **simple or complex alkaloid** *viz.,* **amino**

(-NH$_2$), **imino** (=NH), **amide** (-CONH$_2$), **carboxyl** (-COOH), **ester** (-COOR), **acid chloride** (-$\overset{O}{\overset{\|}{C}}$-Cl), **alcoholic** (-OH), **phenolic** (OH), **carbonyl** (-CO) etc., may be confirmed by these *two* methods:

- **FI–IR spectral data:** provided the pure isolated alkaloid is only available in small quantum, and
- **classical techniques of 'organic analysis':** provided the pure isolated alkaloid is duly available in sufficient quantum.

Besides, the **functional group analysis** also helps in revealing as well as ascertaining the aliphatic/ aromatic nature of the **alkaloid,** presence of **unsaturation** in an **alkaloid,** and also quantify the concentration of an **'alkaloid'** in an unknown sample.

2.3. Functional Nature of O–Atom

The **functional nature of O-atom** strategically present in **alkaloids** due to the various variants as stated in section 2.2.; besides, other moieties, namely:

methoxy (-OCH$_3$) ; **acetoxy** (-OCOCH$_3$) ; **carboxylate** (-COONa); **benzoxy** (-OCOC$_6$H$_5$) ; **phenoxy** (-OC$_6$H$_5$), and **methylenedioxyl** (-OCH$_2$O-)

Importantly, one may critically come across the specific presence of the *'lactone ring systems'* in certain *alkaloidal entities,* such as: **morphine, hydrastine,** and **artemisinin,** as given under, which have been duly ascertained by **FT–IR studies.**

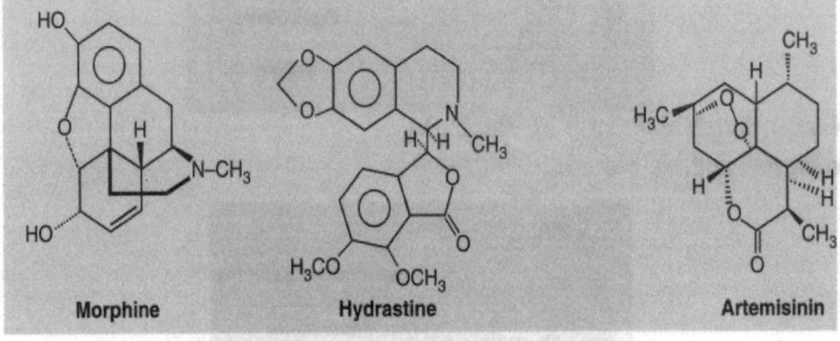

Morphine	**Hydrastine**	**Artemisinin**

Importantly, the array of **O–atom containing functional moieties** may be characterized meticulously with the help of the following specific determinations, such as:

- Alcoholic hydroxy group,
- Phenolic hydroxyl group,
- Carboxylic group,
- Methoxy group,
- Methoxylenedioxyl group, and
- Miscellaneous group (*viz,*. Amide, Ester, Lactone, Lactam groups),

That shall be treated individually with specific examples in the sections that follows:

2.3.1. Alcoholic Hydroxyl Group

The critical formation of an **'acetate'** by the interaction of an **'alkaloid'** with *acetic anhydride or acetyl chloride* (an *acetylating agent*) confirms the presence of an **alcoholic hydroxyl group**. However, it may also be ascertained by reacting with **benzoyl chloride** (a *benzolyating agent*) in the presence of NaOH. The various reactions may be expressed as under:

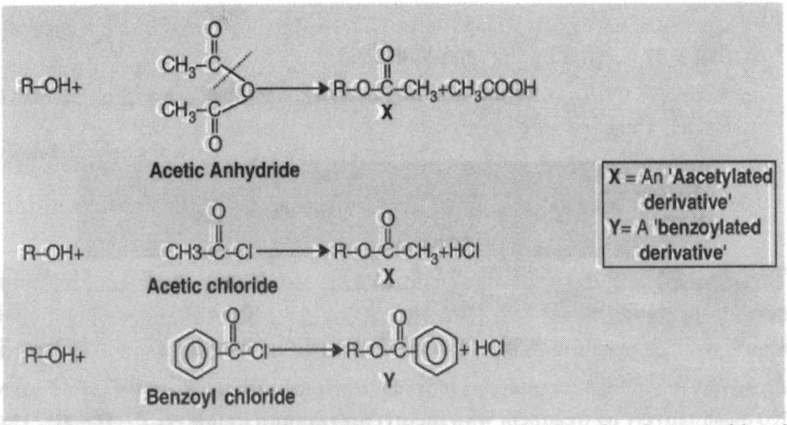

Acetic Anhydride

Acetic chloride

Benzoyl chloride

X = An 'Aacetylated derivative'
Y= A 'benzoylated derivative'

Important Points: The following **important points** may also be taken into consideration duly:

(a) **Primary Alcoholic Function [–CH₂OH]:** The **pri–alcoholic function** can be confirmed by carrying out its oxidation sequentially to obtain,

- *first* – aldehyde (**–CHO**), and
- *secondly* – a **carboxyl (–COOH)**,

both essentially having the **'same number of C–atoms'** *vis–a–vis* the **'parent alcohol'** as given under:

$$-CH_2OH \xrightarrow{Oxd.} -CHO \xrightarrow{Oxd.} COOH$$

Alcohol   Aldehyde   Carboxylic Acid

(b) **Secondary Alcoholic Function [=CH. OH]** : The **sec–alcoholic function** may be ascertained in *two* **critical steps**, namely:

- *First*–**oxidation** to yield corresponding **'ketone'** with **same number of C–atoms;** and
- *secondly*– when the *sec*–**alcoholic function** is duly constituted by the **inherent cyclic C–atoms**, the said compound get **oxidized** adequately to give rise to the formation of an **open-chain dicarboxylic acid,** of course having the **same number of C–atoms.**

Example: The above analogy may be expatiated with the help of the following typical examples:

Isopropanol (an aliphatic secondary alcohol): It undergoes oxidation to yield **acetone** (a *ketone*) as the **first step,** which upon further oxidation produces a **carboxylic acid** with **fewer number of C–atoms.**

Isopropanol   Acetone   Acetic Acid
[With fewer number of C–atoms]

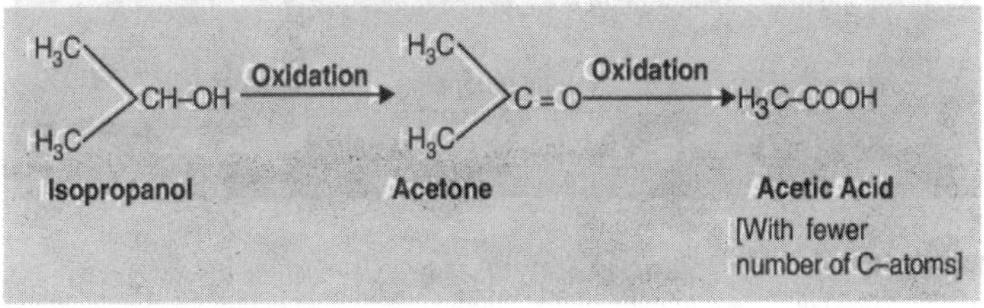

Caution: It is, however, pertinent to state here that the aforesaid test for the presence of **O–atom** must be carried out with extreme care and caution based on the fact that the **'primary imines'**, if present in an *'alkaloid'* shall also give rise to the production of **acetyl and benzoyl structural analogues.** Therefore, in such critical and typical instances the hydroxyl (–OH) moieties is judiciously determined by **'Acelylation Method'** or **'Zerewitnoff's Method'***. The actual number of the **hydroxyl groups** is estimated:

- *first* – by *acetylating* the **'alkaloid'**; and
- *secondly*– by *hydrolyzing* the resulting acetylated product with a known volume of 1M. NaOH. Thus, we may have:

$$R\text{–}OH \xrightarrow[\text{Acetylation}]{CH_3COCl} ROCOCH_3 \xrightarrow[\text{Hydrolysis}]{NaOH} R\text{–}OH + CH_3COONa$$

Alcohol   Acetylated   Alcohol
                           Product

Back Titration: The excess of **alkali added (1M. NaOH)** is duly determined by **back titration** with a **previously standardized 1M. HCl.** Hence, the exact number of **acetyl or hydroxyl moieties** may be calculated precisely from the volume of alkali used up for carrying out the **hydrolysis.**

Alternatively, the **hydroxyl moieties** plus other **primary imino moieties** (>N–H) may be detected and estimated quantitatively by treatment with **methyl magnesium iodide [H_3C–Mg–I]***. The reactions take place as given under:

$$-O\text{–}H + H_3C\text{–}Mg\text{–}I \longrightarrow -O\text{–}Mg\text{–}I + CH_4\uparrow$$
An Alcohol

or

$$=N\text{–}H + H_3C\text{–}Mg\text{–}I \longrightarrow =N\text{–}Mg\text{–}I + CH_4\uparrow$$
An Imine

Thus, **methane (CH_4)** gets eliminated as a *gas,* and escape from the *reaction mixture,* which may eventually be estimated quantitatively by the **volumetric method** thereby confirming the actual number of **hydroxyl (–OH)** or **imino (=NH)** moieties present in the **alkaloid.**

Thus, at standard temperature and pressure (STP) we have:

1 Hydroxyl (–OH) Group = 1 Im no (=NH) Group = 22.4 L of N_2

2.3.2. Phenolic Hydroxyl Group

It essentially undergoes **oxidation** to produce a *ketone* (**cyclohexanone**) which upon further **oxidation** yields an **open–chain dicarboxylic acid** with the **same number of C–atoms** as given under:

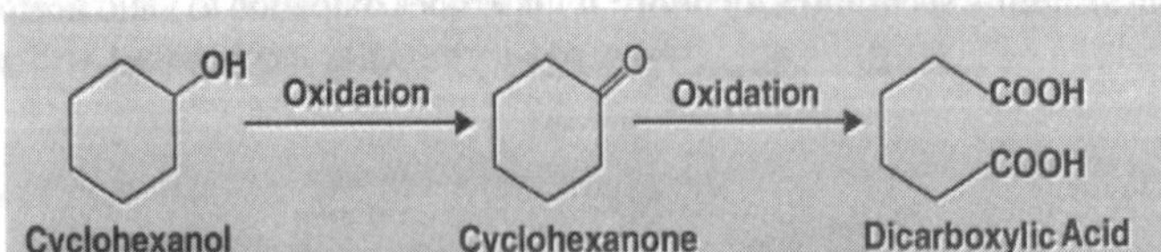

Cyclohexanol   Cyclohexanone   Dicarboxylic Acid
[Secondary   [Armatic Ketone]   [With same number of
Aromatic Alcohol]   C–atoms]

---

* It is also commonly termed as *'Zerewitnoff Active Hydrogen Determination'.*

Tertiary Butanol [an **aliphatic tertiary alcohol**]: The **tertiary alcohol** *e.g., tert*–butanol on oxidation give rise to the formation of a **ketone** and a **carboxylic acid** having *lesser number of* C–atoms.

2.3.3. Carboxylic Group

The presence of the **carboxylic (–COOH) group** in an 'alkaloid' may be indicated by its solubility in an aqueous **sodium carbonate (Na_2CO_3)** or **ammonia (NH_4OH)**. Besides, the quantitative formation of an equivalent quantum of an 'ester' obtained after careful treatment with an **alcohol** also reveals the presence of a **carboxyl moiety**.

> Note: The exact number of carboxyl moieties in an *'alkaloid'* may be estimated volumetrically by titration with a previously standardized barium hydroxide [$Ba(OH)_2$] solution employing *'phenolphthalein'* as an indicator or even by *'gravimetric determinations'* using 'Silver Salt Method'.

2.3.4. Methoxy Group

In general, the **alkoxy group** *viz.,* **methoxy (–OCH$_3$) group**, ethoxy (–OC$_2$H$_5$) group do come into being in an **alkaloid** overwhelmingly. **Zeisel Method*** helps both in the **detection** and *estimation* of an **alkaloid** *via* the following steps:

- boiling of the **alkaloid** with *concentrated hydroiodic acid (HI)* at its bp 126°C thereby enabling the complete conversion of the inherent **alkoxy moieties** to the corresponding **alkyl halides**, and
- subsequent **estimation** of the resulting *alkyl halides* as **silver iodide (AgI)** by reacting carefully with **ethanolic freshly prepared silver nitrate (AgNO$_3$) solution.**

Thus, we may summarize the reactions as stated under:

$$R-OCH_3 + HI \xrightarrow{\sim126°C} \underset{\substack{\text{Methyl} \\ \text{Iodide}}}{R-OH+CH_3I} \xrightarrow[\substack{AgNO_3 \\ \text{Soln.}}]{\text{Ethanolic}} \underset{\substack{\text{Silver} \\ \text{Iodide}}}{AgI \downarrow}$$

An Alkaloid

The exact and precise number of **silver iodide** moles thus generated is an indicative of the actual number of **methoxy (–OCH$_3$) moieties** duly present in an **alkaloid**. Analysis, and halogen determination.

2.3.5. Methylenedioxyl Group [–OCH$_2$O–]

Importantly, the presence of the **methylenedioxyl group** in an *alkaloid* is determined conveniently by the following steps:

- **alkaloid** on being heated with **H$_2$SO$_4$** or **HCl** yields **fromaldehyde (HCHO)**, and
- resulting **formaldehyde** is quantitatively converted to the **dimedone derivative**,

and thus the said reaction affords a useful means of a 'Gravimetric Method' for the *quantitative determination of formaldehyde.*

---

* **Zeisel Method:** It enables a degree of accuracy for the alkoxyl determination in an **alkaloid** which being fairly comparable to other such analytical methods *e.g.,* **C, H, N analysis,** and **halogen determination.**

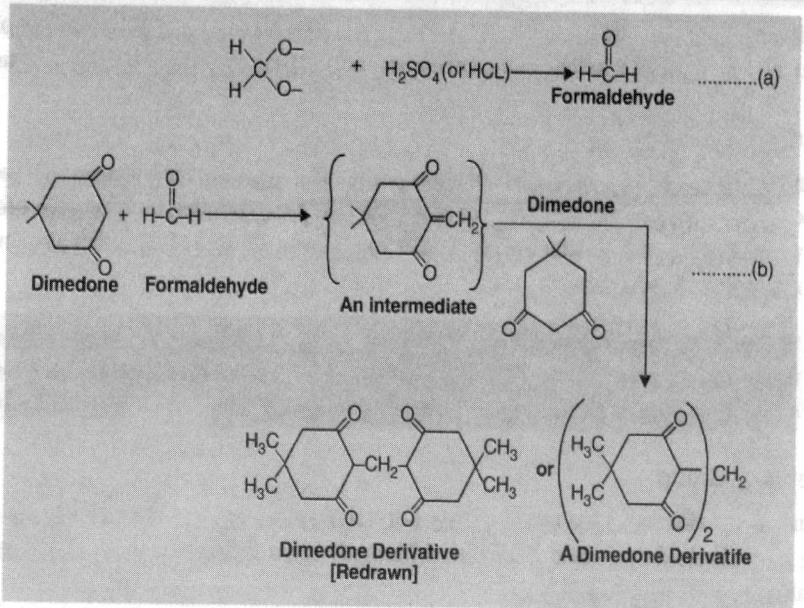

The aforesaid reaction **Eqn. (b)** probably comes into play in *two* **different stages** as illustrated above:

  □ '**Aldol Condensation**' of the aldehyde with the '*active methylene moiety*' present duly in '**dimedone**' to form an **intermediate,** and

  □ '**Michal Addition**' of a '**second molecule of dimedone**' to yield a '**dimedone**' deriva-tive' that may be estimated by '**gravimetric method**' of analysis.

2.3.6. Miscellaneous Groups

It essentially comprises such moieties as: **amide, ester, lactone,** and **lactam** groups, which shall now be treated separately in the sections that follows:

2.3.6.1. Amide Group

It can be detected and estimated by observing the product of its **alkaline hydrolysis** with gentle heating:

$$\text{R–CONH}_2 + \text{NaOH} \xrightarrow{\Delta;} \underset{\text{Sodium Salt}}{\text{R–COONa}} + \text{NH}_3\uparrow$$

$$\underset{\text{An Amide}}{}$$

2.3.6.2. Ester Group

It may be detected and estimated by observing the products of its **alkaline hydrolysis** with slow heating:

$$\underset{\text{An Ester}}{\text{R–COOR}} + \text{NaOH} \xrightarrow{\Delta;} \underset{\text{Sodium Salt}}{\text{R–COONa}} + \underset{\text{Alochol}}{\text{R–OH}}$$

2.3.6.3. Lactone Ring

The '**lactone ring**'* is detected and estimated by observing the **cleaved product** with an **alkaline hydrolysis:**

---

* Both '**Lactone Ring**' and '**Lactam Ring**' undergoes *cleavage* to form **open-chain aliphatic substitued compounds.**

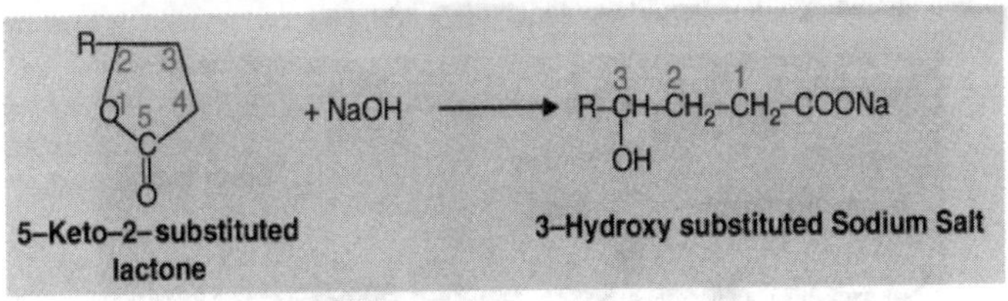

5-Keto-2-substituted lactone +NaOH ⟶ **3-Hydroxy substituted Sodium Salt**

$R-\overset{3}{C}H-\overset{2}{C}H_2-\overset{1}{C}H_2-COONa$
  |
  OH

2.3.6.4. Lactam Ring

Just like the **'lactone ring'** (section 2.3.6.3) the **'lactam ring'*** also can be detected and estimated by observing the **cleaved product** with an **alkaline hydrolysis:**

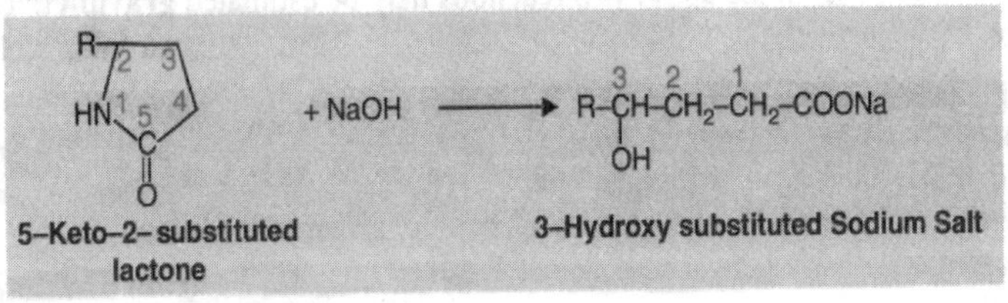

5-Keto-2-substituted lactone +NaOH ⟶ **3-Hydroxy substituted Sodium Salt**

$R-\overset{3}{C}H-\overset{2}{C}H_2-\overset{1}{C}H_2-COONa$
  |
  OH

2.4. Functional Nature of N-atom

It has been observed that all **alkaloids** of the naturally occurring plant origin essentially contain **N-atom** either embedded in the:

- **heterocyclic ring,** or
- **present in the side-chain.**

Nevertheless, a large segment of the **alkaloids** the *nature of N-atom* could vary from:

❑ a **secondary N-atom** *e.g.,* = NH, or

❑ a **tertiary N-atom** *e.g.,* = N–CH$_3$ (N–methyl (group), or = N-.

Interestingly, there are some **phenylalkyl type of alkaloids,** such as: **adrenaline** and **ephedrine,** that do **not** contain the **N-atom** as a part of the **integral heterocyclic ring,** but present in the form of a **primary amino (–NH$_2$) moiety** as shown under:

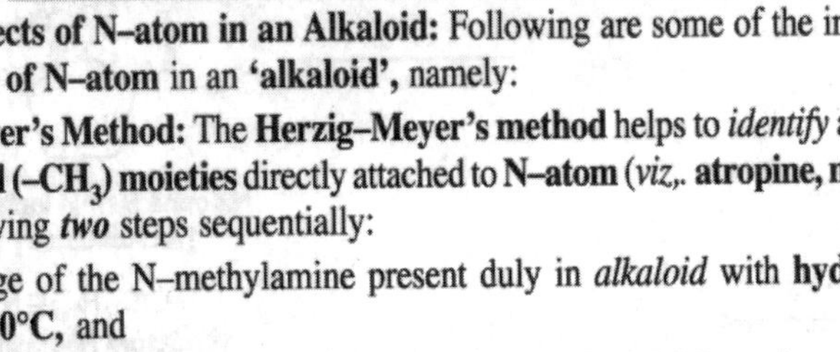

Adrenaline [Epinephrine] **Ephedrine**

Important Aspects of N-atom in an Alkaloid: Following are some of the important aspects with regard to the **nature of N-atom** in an 'alkaloid', namely:

(a) **Herzig–Meyer's Method:** The **Herzig–Meyer's method** helps to *identify* and determine exactly the number of **methyl (–CH$_3$) moieties** directly attached to N–atom (*viz,.* **atropine, nicotine**). It essentially comprises the following *two* steps sequentially:

(i) Cleavage of the N–methylamine present duly in *alkaloid* with **hydroiodic acid (HI)** at **150–300°C,** and

(ii) Determining the **exact quantum** of **methyl iodide (CH$_3$I)** so formed by critical conversion to **silver iodide (AgI)** on treatment with **ethanolic AgNO$_3$ solution:**

Thus, we may express the above *two* reactions as given under:

>N – ĊH₃ —HI→ 150–300°C >NH + CH₃I —AgNO₃→ [EtOH] Ag I↓

N–Methyl Group   **Methyl**   **Silver Iodide**
[In an 'Alkaloid']   **Iodide**

>N–CH₂ ₅ —HI→ 150–300°C >NH + CH₂ ₅I —AgNO₃→ [EtOH] Ag I↓

N–Methyl Group   **Ethyl**   **Silver Iodide**
[In an 'Alkaloid']   **Iodide**

The **Ag I** thus obtained from the above *two* reactions may be estimated **gravimetrically** to have an accurate access to the number of **N–methyl** or **N–ethyl** moieties present in an 'alkaloid'.

Note: (1) The corresponding methoxyl (–OCH₃) moiety present in an *'alkaloid'* is duly estimated by the Herzig–Mayer's method; however, the temperature should be maintained at *126°C*.
   (2) *NMR–Spectroscopy* may also be applied for the rapid and precise detection of the N–methyl moiety and *N–ethyl* physico–chemical method of analysis).

(b) Distillation of 'alkaloid' is invariably detected by carrying out its **steam distillation** with **soda–lime (CaO)** to obtain the following products in the volatile fraction.

 • **methylamine [H₃C–NH₂]:** indicates **1–alkyl group** attached to N–atom;

 • **dimethylamine** $\begin{bmatrix} HC_3 \\ HC_3 \end{bmatrix} > NH$: indicates **2–alkyl groups** attached to N–atom; and

 • **triethylamine [CH₃C)₃N]:** indicates **3–alkyl groups** attached to N–atom.

(c) Detection of Tertiary N–atom: The presence of the **tertiary N–atom** in an 'alkaloid' may be detected explicitly by *two* **methods:**

 ❏ By treatment with **hydrogen peroxide [H₂O₂; 30%],** when the respective **tertiary N–atom** gets oxidized to **amine oxide** as shown under:

≡N   + H₂O₂ ⟶   ≡N→Ö + H₂O
Tertiary   Hydrogen   **Amine**
N–atom   Peroxide   **Oxide**

 ❏ By **additive reaction** of an 'alkaloid' with *one mole* of **methyliodide (CH₃I)** to yield the respective **'quaternary crystalline salt'** largely indicated that the N–atom in the alkaloid is certainly a **tertiary,** as given below:

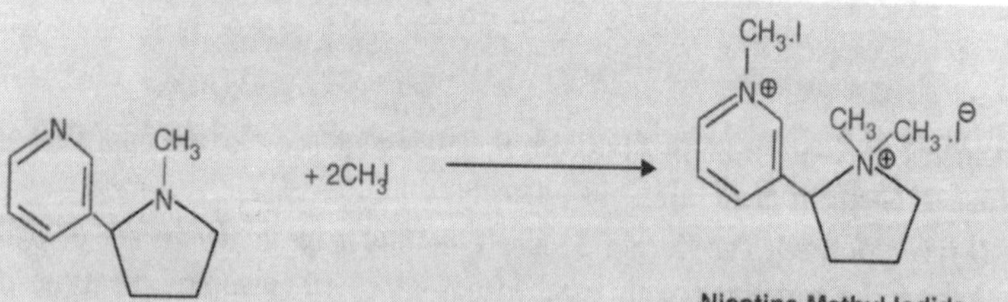

Nicotine
or
[N≡(C₁₀H₁₄)≡N]
(Structure Redrawn)

Nicotine Methyl Iodide
[An Adductor]
[⊖I. H₃C.N⊕≡(C₁₀H₁₄)≡N.CH₃.I⊖]
(Structure Redrawn)

*****Amine Oxide [NO]:** In this N–atom is **'pentavalent'** and the linkage between N and O atom is *via* a **'coordinate bond'.**

2.5. Estimation of C–Methyl Moieties

One of the most important, versatile, and useful *analytical devices* frequently employed by an 'organic chemist' is the critical determination of the C–linked methyl moieties duly present in the alkaloids. In true sense, the method solely depends upon the fact that when an 'alkaloid' predominantly containing C–methyl moieties is carefully *oxidized* with chromic acid under severe and vigorous experimental parameters, the inherent methyl groups fail to undergo oxidation and remain very much in the resulting solution as acetic acid (CH_3COOH); whereas, the *rest of the molecule* is mostly converted into carbon dioxide (CO_2). By the method of 'steam distillation', the generated acetic acid may be isolated/separated and quantitatively titrated for the presence of *'acetic acid'*.

In short, the above comprehensively expatiated method may be served as a means of estimating precisely the number of C–methyl moieties (or *C–linked methyl groups*) duly present in an 'alkaloid' (or an organic compound), and is invariably referred to as the Kuhn–Roth determination* (or C–methyl determination).

2.6. Degradation of Alkaloids

There are indeed a vast number of analytical sequential steps, as discussed in sections 2.1 through 2.5, that eventually help in establishing such vital information(s) as:

- nature of *N–atoms* present in an alkaloid,
- nature of *O–atoms* present in an alkaloid,
- number of *C–methyl moieties* present in an alkaloid, and
- presence of various functional groups *viz.,* aldehyde, ketone, amide, ester, carboxylic, methoxyl, lactone, lactim, phenolic, alcoholic, methoxylendioxyl etc.

In spite of these enormous, valuable, and vital informations with regard to the structure of an *unknown alkaloid*, it still remains to assign and discover the most preferred structural system that *emphatically* and *essentially* incorporates these substituent moieties. Therefore, ultimately the organic chemists are left with no other choice than to perform:

- ❑ degradation of *unknown alkaloid* molecule in a systematic manner, and
- ❑ critical identification of the *fragmented entities* duly obtained from each degradation step performed meticulously.

The array of such extremely crucial and vehemently vital *reactions* employed in the so called 'degradation of alkaloids' are as stated under:

- Hoffmann's exhaustive degradation method,
- Emde's degradation method,
- von Braun's method for tertiary cyclic amines,
- Reductive degradation,
- Alkali Fusion,
- Oxidation,
- Dehydrogenation, and
- Zinc (dust)/ alkali distillation.

The aforesaid *eight* reactions shall now be discussed individually in the sections that follows:

---

* Norman R and Coxon JM: **Principles of Organic Synthesis,** Nelson Thornes, Cheltenham (Uk), 3rd. ed., 2001,

2.6.1. Hoffmann's Exhaustive Degradation Method:

Hoffmann's exhaustive degradation method is also known as **Hoffmann's exhaustive methylation method,** which designates an extremely cardinal step in the **'alkaloidal chemistry'**. It essentially enables *two* potential **clues and information,** namely:

❑ *elimination of N–atom* due to the **cessation of the heterocyclic ring system** present mostly in an **alkaloid,** and

❑ **residual C–skeleton** obtained from above helps in decephering the actual prevalent **nature of the original heterocyclic ring** in the alkaloid.

Principle: The underlying *principle* of **Hoffmann's exhaustive degradation method** being that such organic compounds having the structural unit as given under:

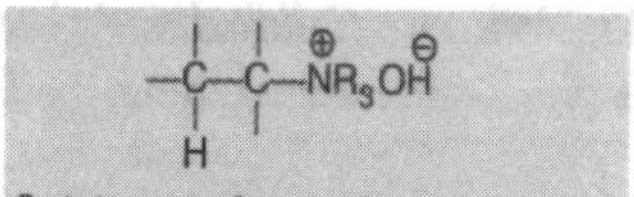

Qutaternary Ammonium Hydroxide

duly eliminates a **trialkylamine** [*viz.,* trimethylamine $\{N(CH_3)_3\}$] upon **pyrolysis*** at $200\pm5°C$ to give rise to the formation of an *'olefin'* (*e.g.,* CH_3 CH = CH-propylene).

The reactions may be expressed as below:

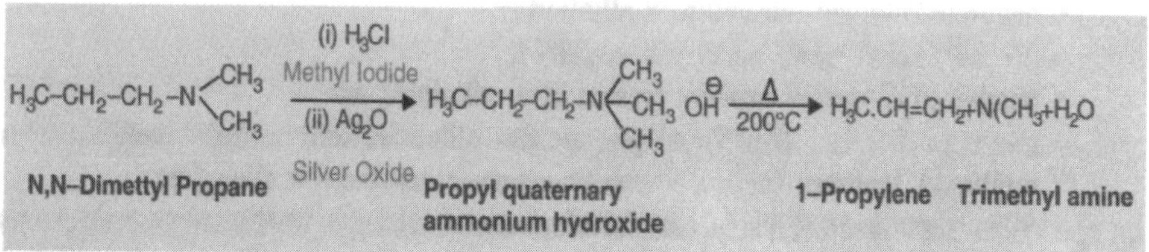

Mechanism: The aforesaid sequel of reactions obviously come into play by an E_2 **mechanism** (or **bimolecular elimination)*,** whereby the **requisite β–hydrogen** and **quaternary N–atom** (or group) are strategically located in the *trans–antiparallel* **configuration** as depicted under:

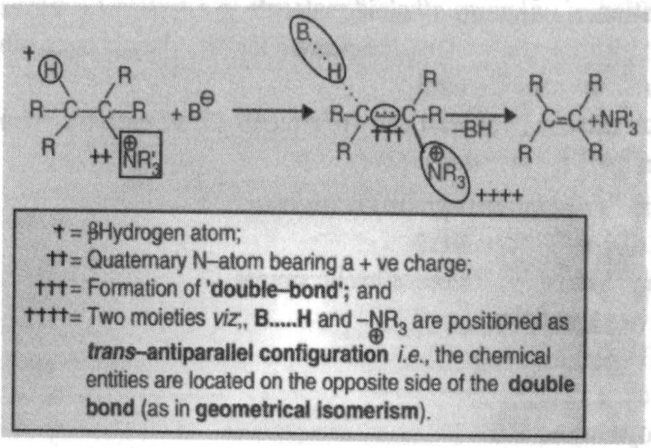

    † = βHydrogen atom;

    †† = Quaternary N–atom bearing a + ve charge;

    ††† = Formation of **'double–bond';** and

    †††† = Two moieties *viz,,* B.....H and $-\overset{\oplus}{N}R_3$ are positioned as

         trans–antiparallel **configuration** *i.e.,* the chemical entities are located on the opposite side of the **double** bond (as in **geometrical isomerism).**

---

* E_2**–Mechanism:** Hughes and Ingold proposed this mechanism specifically for such reaction that proceeds by **second–order kinetics.** Reaction involves a **single step:** base pulls a proton away from carbon; simultaneously, a halide ion departs and the double–bond forms. Halogen takes the electron pair with it; H leaves its electron pair behind, to form the **double bond.**

[Morrison RT and Boyd RN:**Organic Chemistry,** Prentice–Hall of India, New Delhi, 6th ed, 1997) .

In actual practice, the **'alkaloidal amine'** undergoes the following reaction in a sequential manner:

(1) Hydrogenation of **alkaloidal amine** is first carouid out

(2) Treatment with an excess of **methyl iodide (CH₃I)** leads to the **conversion of the hydrogenated alkaloidal amine** into the corresponding **quaternary iodide.**

(3) The resulting **'salt'** is subsequently converted into the *more basic hydroxide* by interaction with **silver oxide (Ag₂O).**

(4) **Basic hydroxide** (obtained in **step–3**) is gradually heated up to **200°C** to obtain an **'olefin'** by the elimination of a **tertiary amine.**

(5) Ultimately, the proper careful identification of the **'olefin'** by subjecting to *further degradation* invariably gives rise to the exact location of the N–atom present duly in the **original alkaloid.**

Importantly, in a situation when the **N–atom** forms an **integral part** of a *cyclic structure* (usually a **heterocyclic nucleus),** it is always advisable to perform at least 2 or 3 **repeated cylces** in order to accomplish distinctly.

- **liberation of the N–atom,** and
- **exposure of the inherent C–skeleton in the nucleus.**

Note: Interestingly, the *Hoffmann's exhaustive degradation method* is exclusively applicable to the *'reduced ring systems' viz.,* piperidine. Nevertheless, it does not hold good to the specific analogous unsaturated compounds *viz.,* pyridine; and, therefore, the *latter* has got to be converted to the *former* before proceeding to the Hoffmann's degradation method.

Degradation of Pyridine: The actual degradation of *pyridine* to *piprylene via piperidine* may be duly accomplished by certain important steps as described under:

(1) **Pyridine (I)** first and foremost gets reduced to **piperidine** by means of **catalytic dehydrogenation.**

(2) **Piperidine (II)** on being treated with an excess of **methyl iodide (CH₃I)** yields the quaternary ammonium iodide *i.e.,* **dimethyl piperidinium iodide (III).**

(3) The **resulting product (III)** is carefully treated with:
- **moist silver oxide (Ag₂O), or**
- **aqueous KOH,**

to yield **quaternary ammonium hydroxide** *i.e.,* **dimethyl piperidinium hydroxide (IV).**

(4) **Product (IV)** on gentle heating affords *two* **changes perceptively:**
- critical loss of a mole of water, and
- cleavage of the C–N linkage at the right hand side from which the strategically positoned **β–H atom** being eliminated as a mole of water, to produce **Δ⁴**–pentenyl dimethylamine **(V).**

(5) **Product (V)** again subjected to **methylation**–followed by treatment with **moist Ag₂O** to yield the corresponding **quaternary ammonium hydroxide** (an **open–chain compound)** *i.e.,* **Δ⁴**–pentenyl **quaternary trimethyl ammonium hydroxide (VI).**

(6) Resulting **product (VI)** upon heating gives rise to the formation of a mole of **trimeltyl amine** **[(CH₃)₃N]** plus **1, 4–pentadiene (VII).**

(7) Ultimately, **1, 5–pentadiene (VIII)** undergoes *isomerization* (i.e., shifting of the double bond from C–4 & C–5 to C–3 & C–4 position) to give **piperylene (VIII)** as the end product. The various reactions from **step–1 through step–7** may be summarized as given under:

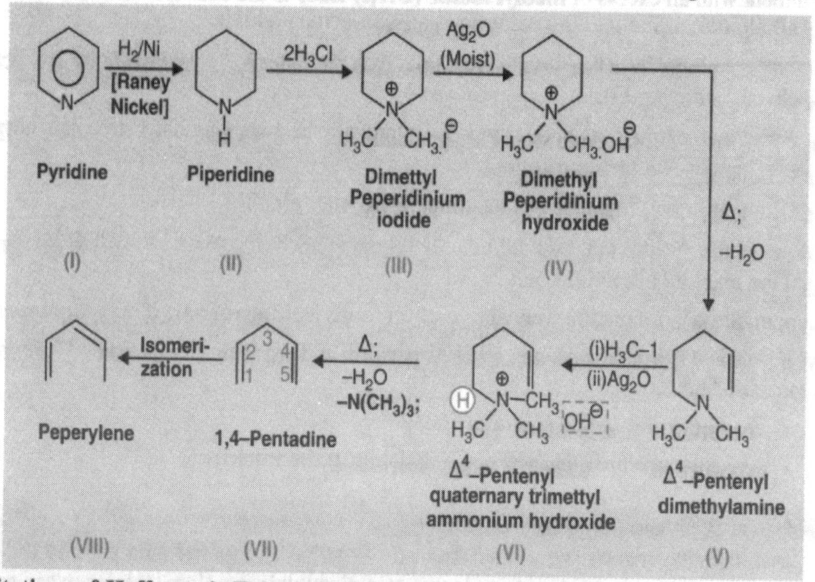

Limitations of Hoffmann's Exhaustive Degradation Method:

In **step–6** from the 'degradation of pyridine' it may be observed that the critical elimination of water is usually feasible and possible only upon the availability of the **H–atom** strategically positioned at the β–position in the **quaternary ammonium hydroxide (VI)**. In case, there exists **no β–Hydro-gen atom**, the **Hoffmann's exhaustive degradation reaction** *fails miserably*.

Example: The above vital and important critical findings may be further expatiated from the typical instance of the 'degradation of isoquinoline' as depicted under:

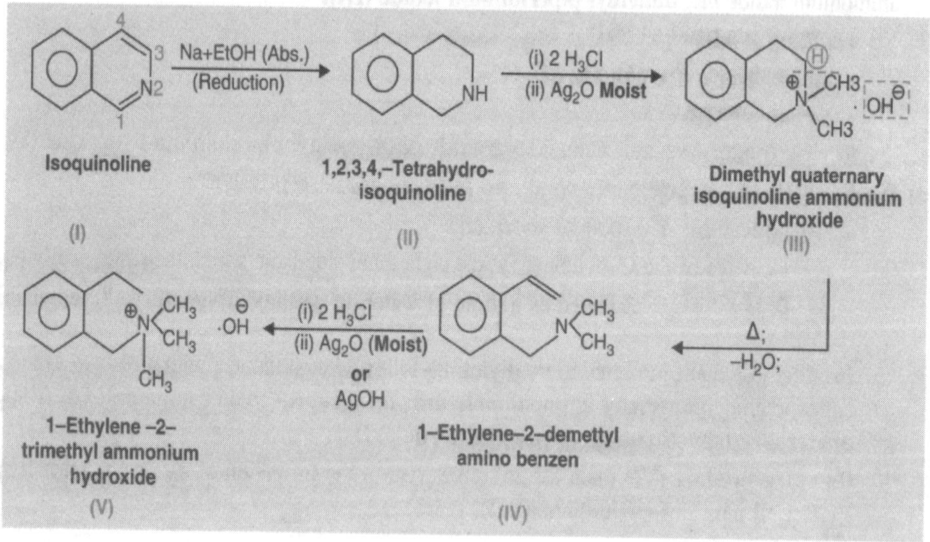

Explanation: The various steps involved in the above sequence of reactions pertaining to the 'degradation of isoquinoline' are given under:

(1) **Isoquinoline (I)** on reduction with sodium metal and absolute ethanol yields **1, 2, 3, 4-tetrahydro-isoquinoline (II)** due to the more convenient hydrogenation of the *basic ring* than the *phenyl ring*.

(2) **Product (II)** on treatment with **2 moles of methyl iodide** and a subsequent treatment with moist **Ag$_2$O (or silver hydroxide)** yields **dimethyl quaternary isoquinoline ammonium hydroxide (III).**

(3) The resulting **product (III)** on heating eliminates a mole of H$_2$O by causing a cleavage between **N–atom and C–3** in the **isoquinoline nucleus** thereby producing **1–ethylene–2–dimethyl amino benzene (IV).**

(4) Finally, the **product (IV)** with another cycle of treatment with **CH$_3$I** and **AgOH** gives rise to the production of **1–ethylene–2–trimethyl ammonium hydroxide (V).**

> **Note:** Due to the *non–availability* of any further *β–H–atom* the *product (V) i.e.,* 1–ethylene 2–trimethyl ammonium hydroxide *fails to undergo any further degradation.*

Typical Example of an Alkaloid *vis-a-vis* **Hoffmann's Exhanstive Degradation Method:**

Hordenine is a *germination inhibitory alkaloid* duly obtained from **'barley'** (a *cereal*), *Hordeum vulgare,* belonging to the natural order *Graminae/Poaceae.*

Importantly, the **Hofmann's degradation method** may be judiciously applied to **'hordenine methyl either'** that eventually produces *para–*methoxy styrene as given below:

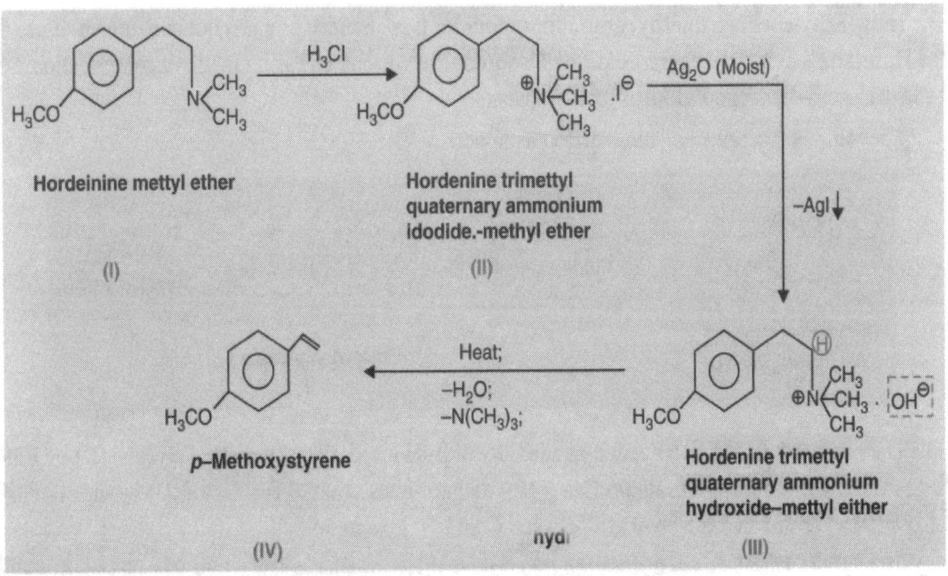

Hordeinine mettyl ether (I)

Hordenine trimettyl quaternary ammonium idodide.-methyl ether (II)

Hordenine trimettyl quaternary ammonium hydroxide–mettyl either (III)

p–Methoxystyrene (IV)

Explanation: The above reactions commencing from **'hordenine methyl ether'** to *p–*methoxystyrene may be explained as under:

(1) **Hodenine methyl ether (I)** on treatment with CH$_3$I results into the formation of **hordenine trimethyl quaternary ammonium iodidemethyl ether (II).**

(2) The resulting **product (II)** on further reaction with **moist Ag$_2$O** [*i.e.,* **AgOH**] rapidly forms **hordenine trimethyl quaternary ammonium hydroxide methyl ether (III)** and a mole of **AgI.**

(3) The **product (III)** on being subjected to heating critically loses a mole each of **water** (as indicated by *dotted lines*) and **trimethyl amine** with the production of *p*–**methylstyrene (VI)**.

2.6.2. Emde's Degradation Method

The **Emde's degradation method** holds good only in such typical naturally occurring '**alkaloids**' that are completely devoid of a β–**hydrogen atom** *i.e.*, when the **Hofmann's exhaustive methylation method** fails. It is, however, pertinent to state here that in the **Emde's degradation method** the '**final step**' essentially and critically involves the reductive cleavage of the '**quaternary ammonium salts**' by *three* known methods:

- **Na–Hg** *i.e.*, **sodium–amalgam**, or
- **Na in liquid ammonia**, and
- **Catalytic hydrogenation.**

Thus, we may have the following expression:

$$\overset{\oplus}{R—CH_2—N.R_3}.\overset{\ominus}{X} \longrightarrow R—CH_3 + NR_3 + HX$$

Examples: There are two specific examples that vividly explains the **Emde's degradation method**, namely:

(1) **1–Ethylene–2–trimethyl ammonium iodide:** It is obtained as a last but one **intermediate** from the **Hofmann's exhaustive degradation of isoquinoline** (discussed earlier in this chapter: section 2.6.1.) in which the β–**hydrogen atom** is totally absent.

The various reactions are summarized as under:

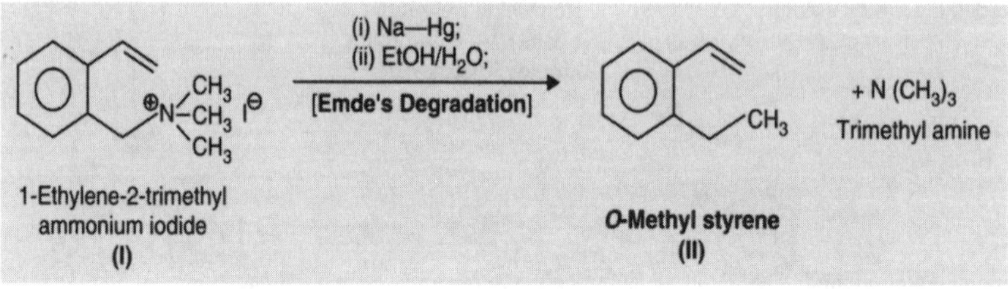

1-Ethylene-2-trimethyl
ammonium iodide
(I)

(i) Na—Hg;
(ii) EtOH/H$_2$O;
[Emde's Degradation]

O-Methyl styrene
(II)

+ N (CH$_3$)$_3$
Trimethyl amine

The **compound (I)** due to the non-availability of β–**H atom** gives *O*–**methyl styrene (II)** by Emde's degradation with **sodium–amalgam (Na – Hg)** and **aqueous ethanol (Et OH/H$_2$O)** by the elimination of **trimethyl amine** and the cleavage between C and N linkage.

(b) **Dimethyl quaternary quinolinium iodide:** In this particular instance it may be observed evidently the presence of β–**hydrogen atom**, which also undergoes **Emde's degradation** by **Na–Hg** followed by **aqueous ethanol** to produce the following *two* products:

- λ–**Dimettylamino propyl benzene (A)**, and
- *O*–**Propyl dimethyl aniline (B)**,

as given under:

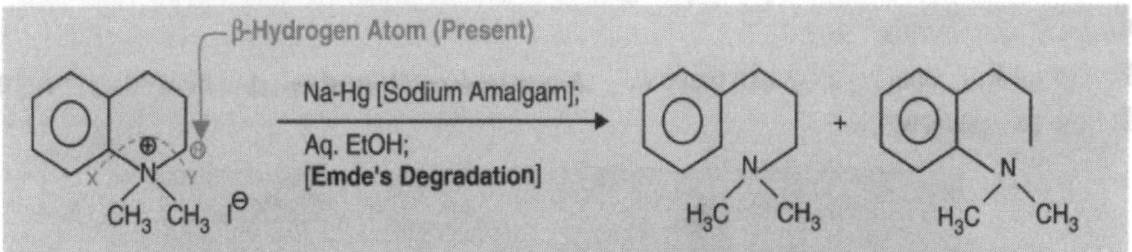
β-Hydrogen Atom (Present)
Na-Hg [Sodium Amalgam];
Aq. EtOH;
[Emde's Degradation]
X
Y
CH_3 CH_3 $I^{\ominus}$
H_3C CH_3
H_3C CH_3

Dimethyl quateunary quinolinium iodide **α-Dimethylamino-propylbenzene (A)** **O-Propyl dimethylamine (B)**

Note: (1) ***Compound (A)* is formed due to the cleavage (X) shown with a dotted line in the above starting material.**
(2) ***Compound (B)* is formed due to the cleavage (Y) shown with a dotted line in the above starting material.**

2.6.3. von Braun's Method [or Tertiary Cyclic Amines]

The **von Braun reaction** specifically makes use of **cyanogen bromide (CNBr)** which critically brings about the **dealkylation reaction** to a ***tertiary amine*** according to the following reaction:

$$\underset{\text{A Tertiary amine}}{R_3N} + \underset{\text{Cyanobromide}}{BrCN} \longrightarrow \underset{\substack{\text{Tertiary amine cyano bromide}\\ \text{(A salt)}}}{(R_3\overset{\oplus}{N}\text{—}CN)\,Br^{\ominus}} \quad \text{--------(a)}$$

Eqn. (a) is immediately followed by:

$$Br^{\ominus} + R\text{—}\overset{\oplus}{N}(R)_2\text{—}CN \longrightarrow R\text{—}Br + \underset{\text{N-Cyanoamine}}{R_2N\text{—}CN} \quad \text{--------(b)}$$

The **N–cyanoamine** obtained in **Eqn (b)** undergoes hydrolysis to give rise to the formation of a **'carbamic acid structural analogue'**, that rapidly knocks out a mole of **CO_2** to form the respective ***'secondary amine'*** as shown under:

$$\underset{\text{N–Cyano amine}}{R_2N-CN} \xrightarrow[H+]{H_2O\text{ (Hydrolysis)}} \underset{\substack{\text{Carbamic Acid derivative}\\ \text{[An Intermediate]}}}{[R_2N-COOH]} \xrightarrow{-CO_2} \underset{\text{A Secondary Amine}}{R_2\text{—}NH} \quad \text{--------(c)}$$

It is, however, pertinent to mention at this point in time that the particular **'alkyl moiety'** which is being removed from a **tertiary amine (R_3N),** [see **Eqn. (a)],** in the prominent **cyanogen bromide reaction** is the one most susceptible to the so called **'nucleophilic attack'** by the **bromide ion (Br^-)** [see **Eqn. (b)].**

Conclusively, one may draw a concrete inference that the **von Braun's method** is extremely useful in the removal of **methyl, allylic,** and **primary alkyl moieties** most rapidly.

Applicability in Alkaloids: In true sense, the **von Braun's method** has been gainfully, judiciously, and extensively used in carrying out the studies of **naturally occurring alkaloids.** Interestingly, a plethora of these substances (*i.e.*, **alkaloids**) are **tertiary amines,** as a step towards their **degradation** to rather *simpler compounds.*

Example: Ring opening can take place in **octahydro–2H–quinolizine** by means of **cyanobromide (CNBr)** as given below:

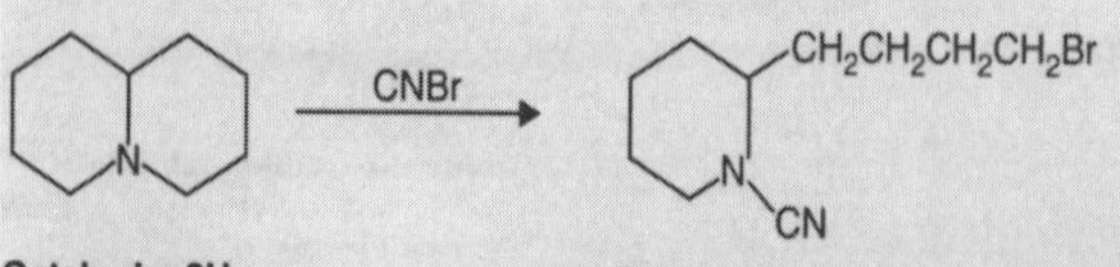

Octahydro–2H–
quinolizine

N–Cyano–2 (butyl bromide) peperidine

In the above reaction, **cessation of quinozoline ring** takes place as indicated by the *dotted line,* thereby giving rise to the formation of **N–cyano–2 (butyl bromide) piperidines**.

Many simpler **'quinolizidine alkaloids'**, such as:

'lupinine', the naturally occurring *l–form* isolated from the seeds of *lupinus luteins* L., and other **L., species** (*Leguminous*)

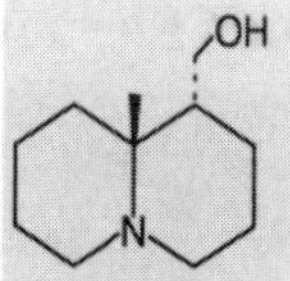

Lupinine

2.6.4. Reductive Degradation

It has been duly observed that in certain typical instances whereby the *'ring'* may be opened by heating with **hydrologic acid (HI)** at nearly 300°C.

Examples: There are *two* typical examples cited below:

(*a*) **Reductive Degradation of Pyridine and Piperidine:**

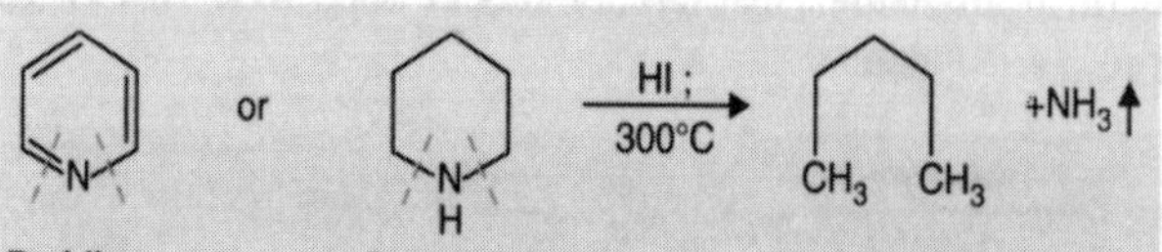

Pyridine

Peperidine

N–Pentane   Ammonia

The **heterocyclic rings** get ruptured as indicated by the *dotted lines* due to the reductive degradation with HI at~300°C to yield *n*–pentane and **ammonia gas** is liberated.

(*b*) **Reductive Degradation of Coniine:** Coniine is an alkaloid usually referred to as the **toxic** *principle alkaloid* of the **poisonous hemlock,** *Connium maculatum* L., (*Umbelliferae*). It gets degraded accordingly, after the cessation of the hetero-cyclic ring, to yield *n*–octane and **ammonia**.

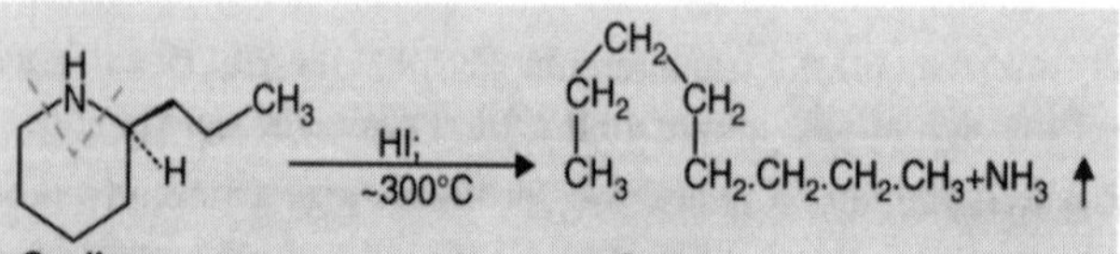

Coniine

n–Octane

Ammonia

Zinc-Dust Distillation:

Importantly, the **reductive degradation** of an *alkaloid* may also be accomplished by **zinc-dust distillation** alternatively, which eventually gives rise to the formation of rather small and simple **chemical entities** (or *fragments*) from which one may logically draw the valuable inference with respect to the **C-skeleton of the unknown alkaloid**. Besides, the critical **Zinc-dust distillation** usually brings forth the following *two aspects,* such as:

- **dehydrogenation,** or
- **removal of O-atom present.**

The above reaction may be expressed as under:

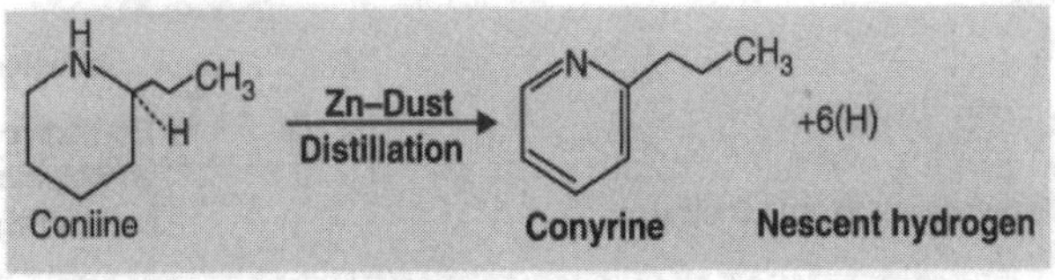

Coniine   Zn–Dust Distillation   Conyrine   Nescent hydrogen   +6(H)

> **Note:** Great care and caution must be exercised in the interpretation of duly obtained data based on such drastic/critical experiments, since these may invariably lead to possible and subtle skeletal changes in the parent alkaloid molecule.

2.6.5. Alkali Fusion

The **'alkali fusion'** is regarded to be an extremely **severe and drastic procedure** that is used most frequently in causing the *cleavage* of the **complex structure of an alkaloid** into several *simpler fragments*. Further investigative studies upon the generated *simpler fragments* do reveal valuable information(s) pertaining to the exact and precise nature of the *nuclei* duly present in the **alkaloid molecule.**

Examples: Adrenaline* is a *catechol* **structural analogue** which on fusion with solid KOH gives rise to the formation of **protocatechuic acid** as shown under:

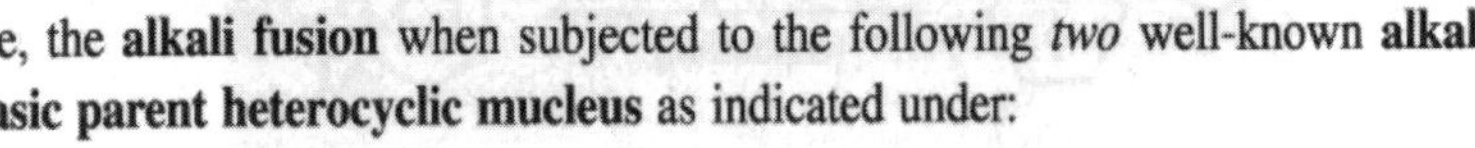

Adrenaline [or Epinephrine]   KOH–Fusion +H_2O Protocatechuic Acid   Methyl Amine   +CH_3–NH_3

Likewise, the **alkali fusion** when subjected to the following *two* well-known **alkaloids** usually yields the **basic parent heterocyclic mucleus** as indicated under:

---

* **Adrenaline :** A **catechol amine** compound (a *neurohormone*) duly secreted by the **adrenal medulla.**

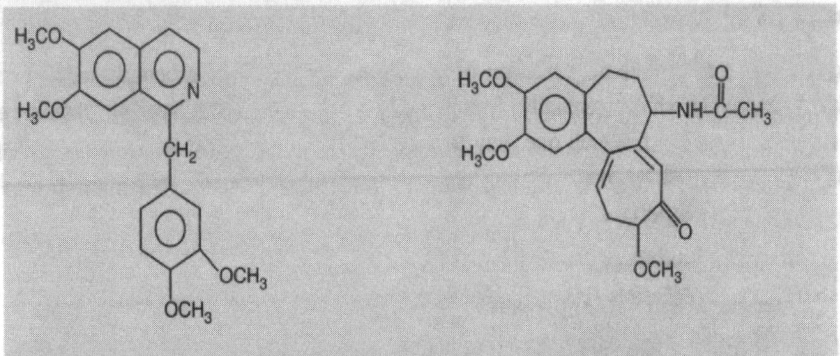

Papaverine
↓ Alkali-fusion

Isoquinoline Derivative
It indicates that **papaverine** essentially comprises an isoquinoline nucleus.

Colchicine
↓ Alkali-fusion

Quinoline Derivative
It reveals that **colchicine** consists of a **quinoline nucleus.**

2.6.6. Oxidation

The critical **'oxidation'** of an *alkaloid* provides valuable and informative knowledge with regard to:

- **basic structure** of an **alkaloid,**
- nature and exact position of the **side-chain,** and
- nature and position of the **functional moieties** (*viz.,* –CHOH; –C=C–; –CH$_2$OH etc.)

(a) **Vigorous Oxidation:** It is usually performed with such extremely potent vigorous oxidizing reagents, such as:

- K$_2$Cr$_2$O$_7$–H$_2$SO$_4$* mixture,
- Cr$_2$O$_3$–H$_2$SO$_4$* mixture,
- MnO$_2$–H$_2$SO$_4$* mixture, and
- Fuming concentrated HNO$_3$.

Example: Nicotine undergoes oxidation with a mixture of K$_2$Cr$_2$O$_7$ H$_2$SO$_4$ to yield nicotinic acid as shown below:

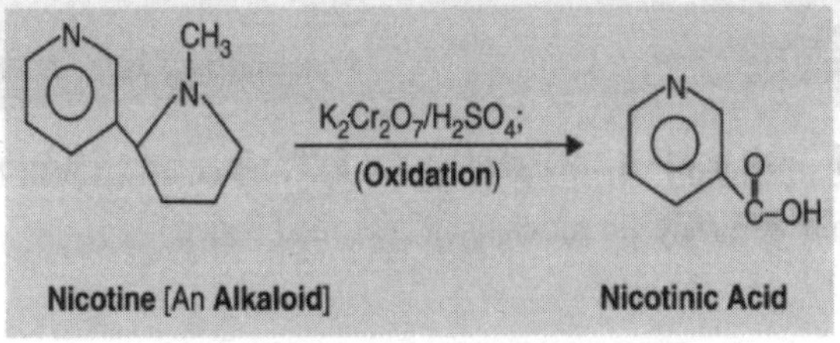

Nicotine [An Alkaloid]

$$\xrightarrow[\text{(Oxidation)}]{\text{K}_2\text{Cr}_2\text{O}_7/\text{H}_2\text{SO}_4;}$$

Nicotinic Acid

---

* H$_2$SO$_4$: It is usually concentrated with **36M (sp. gr. 1.88).**

From the aforesaid reaction it may be concluded safely that the **'nicotine'** comprises a basic **'pyridine nucleus'** with a side-chain at β–position.

(b) Moderate Oxidation: It is normally carried out with the help of such reagents as:

- Cr_2O_3–CH_3COOH,
- $KMnO_4$–NaOH, and
- $KMnO_4$–H_2SO_4.

Example: Morphine upon oxidation with Cr_2O_3/CH_3COOH yields **codeinone** whereby the **secondary alcoholic function** gets converted into a respective **'ketonic function'**, as stated under:

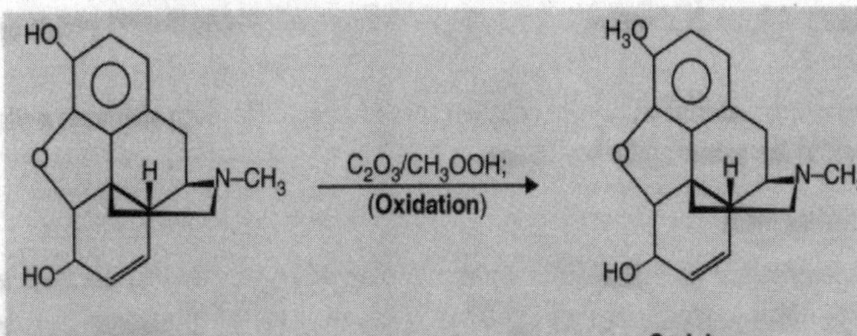

Morphine
[An Opium Alkaloid]

C_2O_3/CH_3OOH;
(Oxidation)

Codeinone

(c) Mild Oxidation: It is invariably performed by relatively **mild oxidizing reagents,** for instance: hydrogen peroxide (H_2O_2); ozone (O_3); $K_4Fe(CN)_6$–NaOH (alkaline potassium ferrocyanide), and iodine (I_2) in aqueous ethanol.

Example: Coniine (an *alkaloid*) on being subjected to mild oxidation with I_2/aqueous ethanol yields picolinic acid (*i.e.*, a **'plant acid'**).

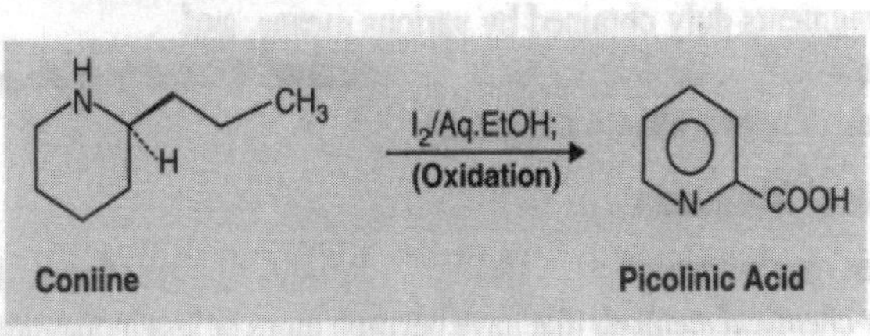

Coniine

I_2/Aq.EtOH;
(Oxidation)

Picolinic Acid

The oxidation of **coniine** converts the **piperidine nucleus** to a **pyridine nucleus** and also reduces the side chain at C–2 to a **carboxyl moiety.** In fact, such valuable informations do help in the **elucidation of complex structure of alkaloids.**

Nevertheless, in certain typical instances the following *two steps* need to be performed in a sequential manner for an **alkaloid,** namely:

- *first,* dehydrating the *alkaloidal compound* to the **unsaturated form,** and
- *secondly,* to carry out the **oxidation** at the *double bond* so created.

The above reaction may be expressed as given below:

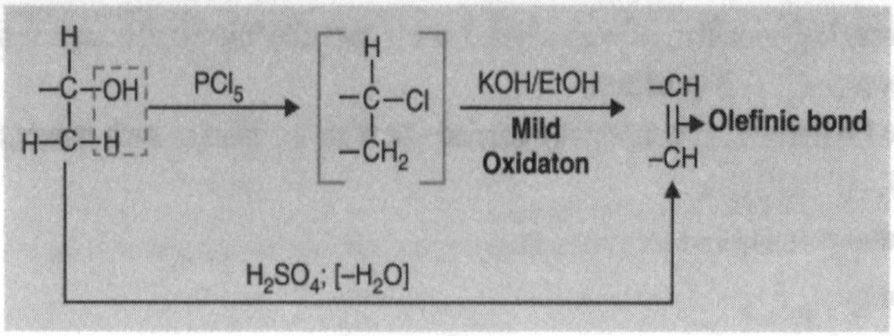

Explanation: (1) PCL_5 replaces the free alcoholic (–OH) moiety to the respective **chloro (Cl) group.**

(2) The resulting **chloro derivative** upon mild oxidation with **KOH/EtOH** yields an **unsaturated aliphatic compound.**

(3) Even the starting material with concentrated sulphuric acid (H_2SO_4) abstracts a mole of water, as shown above, to yield the same **olefinic linkage.**

2.6.7. Dehydrogenation

The **dehydrogenation** of an **'alkaloid'** comes into play when it is usually distilled with a **suitable catalyst,** such as: **sulphur, selenium,** or **palladium.** However, the phenomenon of **dehydrogenation** occurs to yield/afford a host of such recognizable chemical end-products as:

- **gross-skeleton of an 'alkaloid',**
- specific **elimination of peripherally attached functional moieties** e.g., **hydroxyl (–OH)** and **C–methyl (C–CH_3).**

In a broader perspective, the aforesaid **'degradation of an alkaloid'** one may certainly establish a plethora of important informations, namely:

- ❑ exact nature of the nucleus,
- ❑ different fragments duly obtained by various means, and
- ❑ certain vital and critical linkages in an **'alkaloid',** thereby enabling the most preferred probable structure of an **unknown alkaloid** logically.

2.7. Physico-Chemical Methods

The last *four* decades have witnessed a sea-change in the development of several *reliable, dependable,* and *trustworthy physico-chemical methods* that have become more or less indispensable means to identify and assign the **correct molecules structures** to an **'alkaloid'.**

Following are a few recognized **physico-chemical methods** that are used invariably in the elucidation of the **molecular structure,** namely:

- Ultraviolet spectrophotometry,
- Infrared spectrophotometry,
- Nuclear magnetic resonance spectroscopy,
- Mass spectrometry,
- Circular dichroism and optical rotary dispersion,
- X-ray diffraction analysis,
- Emission spectroscopy, and
- Conformational analysis.

These *eight* **physico-chemical methods of analysis** shall be treated briefly and individually in the sections that follows:

2.7.1. Ultraviolet (UV) Spectrophotometry

The **UV-spectrophotometry** measures the *electromagnetic radiation* passed through *sample* (an alkaloid), and the resulting absorption of the radiation (on account of the interactions with the **analyte sample**). In fact, such determinations are duly conducted either:

- at a **fixed wavelength,** or
- at a **varying wavelength,**

almost over a **specified region.** Besides, the **UV-spectrum** of an **alkaloid** does not reflect upon the characteristic features of the molecule as a whole, but solely confined to the particular **chromophoric system(s) present.**

In a broader perspective, the **UV–spectral data** of a good number of **different types of alkaloids** are **recorded, analyzed,** and **categorized** meticulously with regard to the ensuing **structural correlation** *vis-a-vis* the **probable chromophoric systems(s)** present in the *unknown alkaloid.*

Based on the retrived valuable information(s) pertaining to the **alkaloid,** it becomes really a lot easier and convenient to assign the exact and precise **'aromatic'** or **'heterocyclic system'** in the forms, such forms as given under:

- benzene system,
- pyridine system,
- pyrimidine system,
- quinoline system,
- iso-quinoline system
- indole system, and
- pyrrolidine system.

2.7.2. Infrared Spectrophotometry

The **infrared spectrophotometry** (or **IR spectrophotometry**) measure the absorbance of varying frequencies of *infrared light* due to the presence of a host of **functional moieties** duly present in an **alkaloid,** such as: carbonyl, ester, carboxyl, hydrodxyl, amino, imino, imido, amido, nitro, methoxy, ethoxy, ketone, aldehyde phenol, lactone etc. In other words, **IR-spectrophotometry** ascertains the absence/presence of the aforesaid critical functional groups in an **alkaloid** presence of the aforesaid critical functional groups in an **alkaloid.**

Besides, it also helps in the characterization of the **methyl ($-CH_3$) group** either attached to a **N– or O-atom,** the aromaticity of the sample; nevertheless, the most reliable and dependable spectral analysis pertaining to the **quantitative determinations of such groups** is best achieved by **Nuclear Magnetic Resonance (NMR)** spectroscopy:

2.7.3. Nuclear Magnetic Resonance (NMR) Spectroscopy

The **NMR–spectroscopy** is exclusively dependent upon the respective **'vibrational frequencies'** emanated by a *radio-frequency signal* strategically present in the **'nuclear protons of ^1H, ^{13}C, ^{19}F'**

among others, which are observed to be processing critically in an **applied magnetic field.** It is also sometimes referred to as the **Proton Magnetic Resonance (PMR) spectroscopy.**

Importantly, the **NMR-spectroscopy** is employed most abundant by for the critical and specific **identification** and even **quantitative** estimation of such vital *functional moieties* usually found in an **'unknown alkaloid'** as:

- O-methyl [O-CH₃],
- N-methyl [N-CH₃],
- Dioxymethylene [-O-CH₂-O-],
- C-methyl [C-CH₃], and
- HO- phenolic [HO-Ar].

Table: 3.1. records the important **'chemical shifts'** and **'multiplicity'** of the above cited functional moieties invariably come across in an **alkaloid:**

Table: 3.1: Chemical Shifts and Multiplicity of Some Important Functional Moieties

S.No.	Functional Moieties	Chemical Shift (r)	Multiplicity
1	O–CH₃ [O-Methyl]	6.2-6.5	Singlet
2	N-CH₃ [N-Methyl]	7.0-7.9	Singlet
3	-O-CH₂-O-[Dioxymethylene]	3.8-4.2	Singlet
4	-C-CH₃ [C-Methyl] aliphatic	8.9-9.1	Singlet/Doublet/or Triplet*
5	Ar-OH [Phenolic OH]	-2 to +5	Singlet
6	-C-CH₃ [C-Methyl] aromatic	7.5-7.7	Singlet

* Exclusively depends upon the number of exact α–H-atom available.

Notes: (1) *NMR-spectroscopy* critically determines:
- a precise estimation of the number of *aromatic and heteroaromatic protons,* and
- the exact *substitution profile(s)* of the *ring system(s).*

(2) *'Chemical Shift'* of *phenolic protons* changes so drastically perhaps solely due to:
- solution concentration,
- extent of intermolecular association, and
- degree of intramolecular association.

(3) *Presence of Phenolic (–OH) Moiety* present duly in alkaloid may be ascertained by the simple incorporation of 'deuterium oxide' (D₂O) to the solution, and re-examining its NMR-spectrum. An obvious rapid exchange of H-atom (in phenolic OH) by the deuterium (-OD) gives rise to the *absence of 'signal'* previously *obtained due to the phenolic (–OH).*

2.7.4. Mass-Spectrometry (MS)

Massspectrometry designates a destructive method of analyzing a molecular structure ,whereby the molecules **(alkaloids)** are duly subjected to high energy-electrons (or protons), breaking them into **'charged fragments'** whose spectra are analyzed meticulously by their respective differences in mass (m/e).

In *alkaloidal structure elucidation* **MS** plays a vital and important role in establishing the type of **nucleus** (*viz.*, **aromatic** or **heterocyclic**). Besides, it also helps in determining the actual **'size'** as well as 'structure' of the *side-chains* critically present in the various *categories of alkaloids* or alternatively in the corresponding **Zn-dust-distillation products** (*i.e.*, the respective degraded products of the **'alkaloid'**).

The magnificent and spectacular success of **MS** is solely due to the following *two* critical instances to decepher the complex structure of the **unknown alkaloids** as given under:

(a) **Polycyclic Indole Alkaloids:** It is known that the **'indole nucleus'** provides evidently and **abundant stable molecular ion** that in turn gets adequately decomposed by extremely critical as well as specific **'bond fission'** essentially cause to participate the *'alicyclic segment'* of the molecule comprising other N-atom(s), and

(b) **Total Structure of an Alkaloid:** Furthermore, it is quite feasible and possible to determine the **total structure of an alkaloid** by critically performing other equally important analysis *viz.*, MS, UV, FT-IR, NMR etc., so as to establish with ample evidential support *vis-a-vis* the nature and position of the various inherent functional moieties attached to the **'parent alkaloid'**.

2.7.5. Circular Dichroism (CD) /Optical Rotary Dispersion (ORD)

Circular dichroism (CD) refers to the specific type of instrumentation wherein the **molar elipticity of an optically active substance** (*viz.*, an **alkaloid**) *is determined precisely in solutionis varying wavelengths.*

Optical rotary dispersion (ORD) usually refers to the results of a **measurement of the angle of rotation of polarized light at different wavelengths** as it passes *via* an **alkaloid** or a **substance in solution.** Nevertheless, its overall applicability is grossly restricted to only such compounds that are found to be **optically active** *i.e.*, compounds wherein the ensuing *rotation-reflection symmetry axis remains absent virtually.*

Example: Based on the above fundamental facts the application of **ORD** is confined to the elaborated studies of such alkaloids as: **Apomorphine, Benzylisoquinoline, Morphine,** and **Yohimbine**

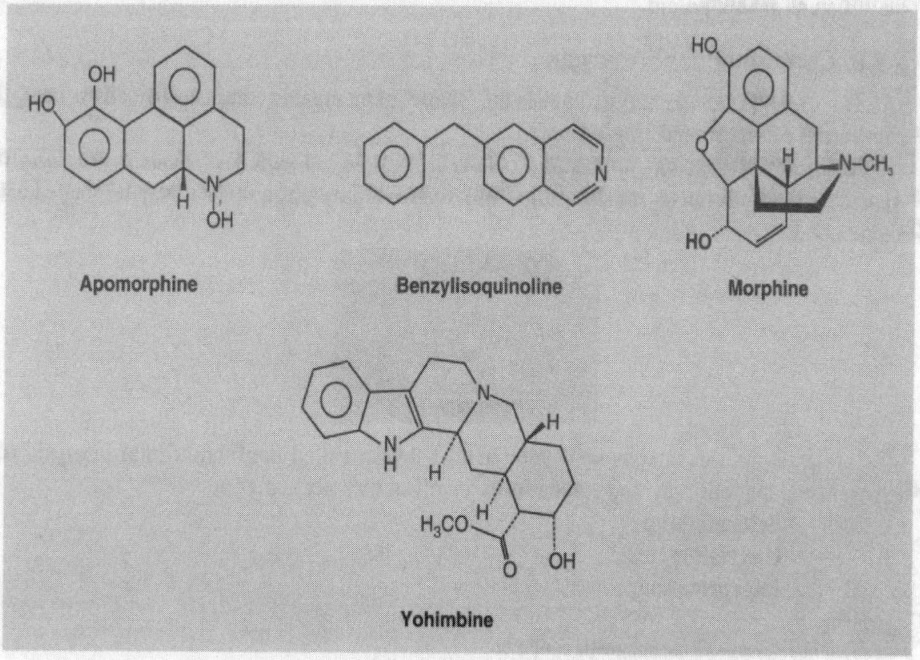

Apomorphine

Benzylisoquinoline

Morphine

Yohimbine

2.7.6. X-Ray Diffraction Analysis [X-Ray Crystallography]

The **X-ray diffraction analysis** of crystalline **alkaloids** by the critical observation of the **diffraction patterns** which take place when a **beam of X-rays** is made to pass *via* a **crystal.** It comes into play principally as a result of the application of **X-ray diffraction analysis** that the structure of certain **alkaloids** have been analyzed duly.

Besides, it provides specific vital informations with respects to the structure of an **alkaloid,** such as:

- **bond angles and bone lengths,**
- **relative stereochemistry,**
- **overcrowding of electrons due to the presence of 'twisted bonds',** and
- **'absolute configuration'** of the *alkaloidal molecule.*

Obviously, the advent of a tremendous quantum jump in the remarkable development in **'Computer Technology',** it has now become a lot easier and reasonably simpler task to rapidly perform the intricate calculations from the retrieved data from **X-ray diffraction analysis.** Thus, it enables one to assign the **complete stereochemical structure** to an **unknown alkaloid** from a *single pure crystal.*

Example: A classical example is of an **alkaloid,** *thelepogine,* $(C_{20}H_{31}NO)$-the complete structure of which has been duly established by **X-ray diffraction analysis** exclusively without carrying out any sort of **chemical analysis,** whatsoever.

2.7.7. Emission Specotroscopy

The *emission spectroscopy* refers to the critical study of the **composition of materials** and identification of elements by the observation *vis-a-vis* measurement of wavelengths of radiation they emit* invariably when they actually get back to the **'normal state'** after *suitable excitation* by an **external energy source.**

However, it may be added that **emission spectroscopical data** does contribute profusely in assigning the structure of an **alkaloid.**

2.7.8. Conformational Analysis

The *conformational analysis* implies the **'shape'** of an organic compound usually accomplished by **rotation of atoms** around *single bonds.*

Example: Muscarine– an *alkaloid* obtained from the red variety of *Amanita muscaria* (L) Pers. *Agaricaceae (a* **poisonous mushroom). The trimethylammonium moiety** may be rotated around the single bond.

Muscarine

Interestingly, the fundamental approach of the so called **conformational analysis** is chiefly experimental that critically and emphatically involves such acclivities as:

- **determination,**
- **correlation,** and
- **interpretation,**

---

* That is, release of **electromagnetic radiation.**

of both *product ratios* and *kinetics* from the ensuing **'chemical transformation'**, for instance:

- ❑ **reduction:** double bonds, carbonyl moieties,
- ❑ **oxidation:** alcohols,
- ❑ **hydrolysis:** esters,
- ❑ **esterification:** alcohols,
- ❑ **quaternization**: amines, and
- ❑ **epimerization:** *i.e., reversible interconversion of epimers.*

3. CLASSIFICATION OF ALKALOIDS

An attempt has been made to classify the **'alkaloids'** solely based upon the **N-heterocyclic basic rings,** namely:

(a) Pyrrolidine alkaloids,

(b) Pyridine alkaloids (or Piperidine alkaloids),

(c) Pyridine-Pyrrolidine alkaloids,

(d) Tropane alkaloids,

(e) Quinoline alkaloids,

(f) Isoquinoline alkaloids,

(g) Indole alkaloids, and

(h) Imidazole alkaloids.

These aforesaid *eight* typical **N-heterocyclic basic ring** catergories of **'alkaloids'** shall now be treated individually in the sections that follows and duly exemplified with appropriate example(s) from each **class of alkaloids:**

3.1. Pyrrolidine Alkaloids

There are *three* most prominent examples of the **pyrrolidine alkaloids,** such as: **hygrine, cuscohygrine,** and **stachydrine.**

3.1.1. Hygrine

It occurs abundantly in the *leaves* of *Erythroxylon coca* Lam. (*Erythroxylaceae*) (Coca); and the *roots* of *Withania somniferum* (L.) Dunal (*Solanaceae*) (**Ashwagandha**).

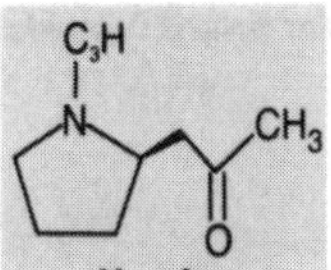

Characteristic Features

- Liquid alkaloid;
- bp_{11} 76.5, bp_{14} 81°C;
- n_D^{20} 1.4555
- Soluble in : mineral acids, ethanol chloroform, and sparingly in water.

3.1.1.1. Constitution of Hygrine

The various vital and important aspects of the **constitution of 'hygrine'** are as described under:

(1) **Molecular Formula:** Its molecular formula is $C_8H_{15}NO$.

(2) **Presence of Carbonyl (C=O) Moiety:** It reacts with **hydroxylamine** to yield an **oxime [C=N–OH],** which shows evidently that it has a *ketonic functional moiety.* Nevertheless, its critical presence is revealed further based on the fact that **hygrine** on being subjected to oxidation with **chromic**

acid [CrO_3] to form **hygrinic (or hygric) acid** [$C_6H_{11}NO_2$] that apparently bears lesser number of C–atoms *vis-a-vis* **hygrine** (*i.e.*, a difference of 2 C-atoms); and confirms the presence of **keto moiety** in the *side-chain* of **hygrine**.

> Note: In case, the *keto moiety* been duly located in the *heterocyclic ring system of hygrine*, one would have expected a *'carboxylic acid'* with the *same number of C-atoms* (*i.e.*, eight C-atoms)

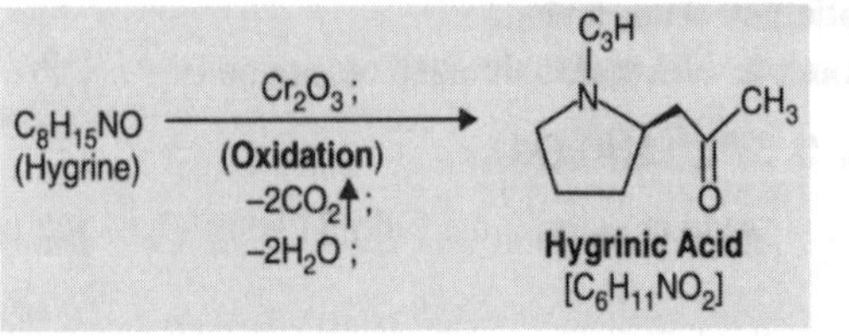

$$C_8H_{15}NO \text{ (Hygrine)} \xrightarrow[\substack{\text{(Oxidation)} \\ -2CO_2\uparrow; \\ -2H_2O;}]{Cr_2O_3;} \text{Hygrinic Acid } [C_6H_{11}NO_2]$$

(3) **Presence of *tert*- N-atom:** The **hygrinic acid** [step (2)] upon heating gives rise to the formation of one mole each of **N-methyl pyrrolidine** and **carbon dioxide** which escapes from the product as shown under:

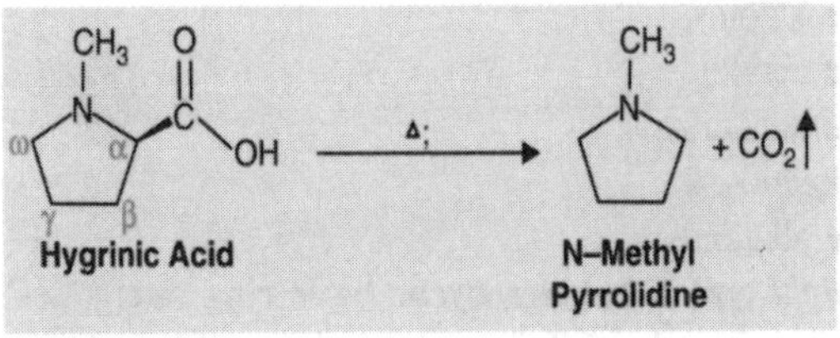

Hygrinic Acid $\xrightarrow{\Delta;}$ N–Methyl Pyrrolidine $+ CO_2\uparrow$

Remarks: The ease and convenience by which the CO_2 molecule gets knocked out from **hygrinic acid** to form **N-methyl pyrrolidine** strongly suggests that just like the *α–amino acid* the **hygrinic acid** *i.e.*, **N-methyl pyrrolidine-α-carboxylic acid,** which may be further adequately substantiated by its **total synthesis** (Willstater, 1900)*.

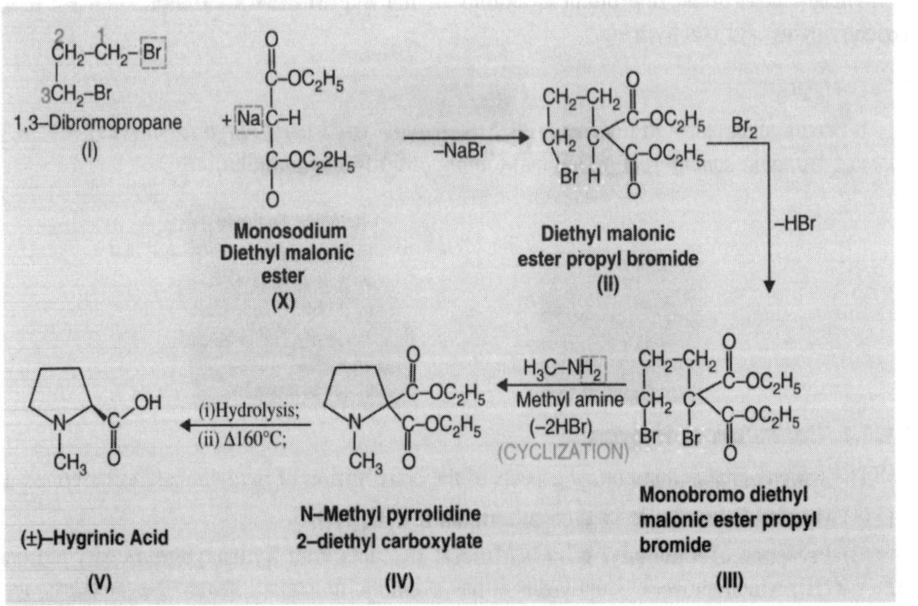

1,3–Dibromopropane (I) + Monosodium Diethyl malonic ester (X) $\xrightarrow{-NaBr}$ Diethyl malonic ester propyl bromide (II) $\xrightarrow{-HBr}$ Monobromo diethyl malonic ester propyl bromide (III) $\xrightarrow[\text{(CYCLIZATION)}]{\substack{H_3C-NH_2 \\ \text{Methyl amine} \\ (-2HBr)}}$ N–Methyl pyrrolidine 2–diethyl carboxylate (IV) $\xrightarrow[\text{(ii) }\Delta160°C;]{\text{(i)Hydrolysis;}}$ (±)–Hygrinic Acid (V)

* Willstater: *Ber.*, **24:** 513, 1900.

Explanation: The various steps include:

(*a*) **1,3-Dibromopropane (I)** on reaction with **(X)** yields **diethyl malonic ester propyl bromide (II).**

(*b*) **Product (II)** on bromination yields **(III)** with the loss of one mole of HBr.

(*c*) **Product (III)** undergoes **cyclization** with methylamine and loses two moles of HBr as indicated with dotted lines to yields **(IV).**

(*d*) The resulting **product (IV)** on being subjected to hydrolysis followed by heating gives rise to the formation of a racemic mixture of **hygrinic acid (V).**

(4) **Assigned Structure of Hygrine**

Based on the aforesaid evidential results one may possibly assign the following *two* probable sturctures for **'hygrine'** *viz.,* [A] and [B] as shown under:

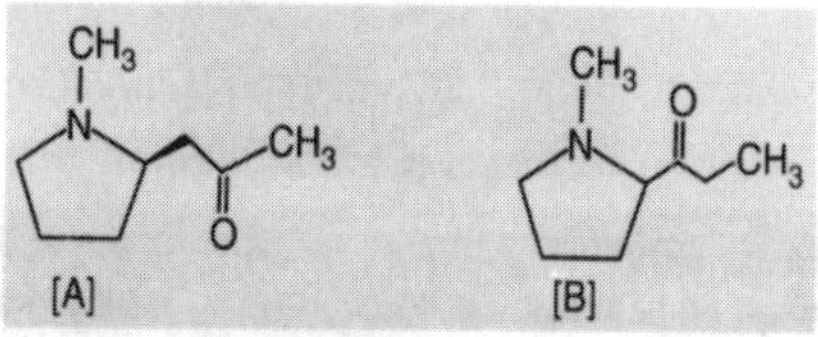

[A] [B]

Of the *two* assigned structures, the **structure [A]** was found to be the correct structure of **'hygrine'**, which was further proved by its **total synthesis.**

(5) **Synthesis:** The total **synthesis** of (+) **hygrine** may be accomplished by *four* different methods using altogether specific **starting materials,** namely:

(a) **Hess's Method (1913)** – from **pyrrole,**

(b) **Sorm's Method (1947)** – from **N-methyl pyrrole,**

(c) **Anet's Method (1949)** – from **N-butaldehyde methyl amine,** and

(d) **Mannich Condensation Method.**

The above synthesis shall now be treated individually in the sections that follows:

A. Hess's Method

Hess (1913) carried out the synthesis of **hygrine** from **pyrrole (I)** and reacted with **Grignard's reagent** *methyl magnesium bromide* to obtain **pyrroyl magnesium bromide (II).** The resulting product (II) on being treated with **propylene oxide** yields **2-(2-hydroxypropane)- pyrrole (III),** which on reduction with **H₂-Pt** saturates the pyrrole ring completely to produce **2-2(hydroxy propane) pyrrolidine (IV).** The **product (IV)** when treated with **formaldehyde** gives rise to the formation of the desired racemic (±)-hygrine (V). The various reactions are as shown under:

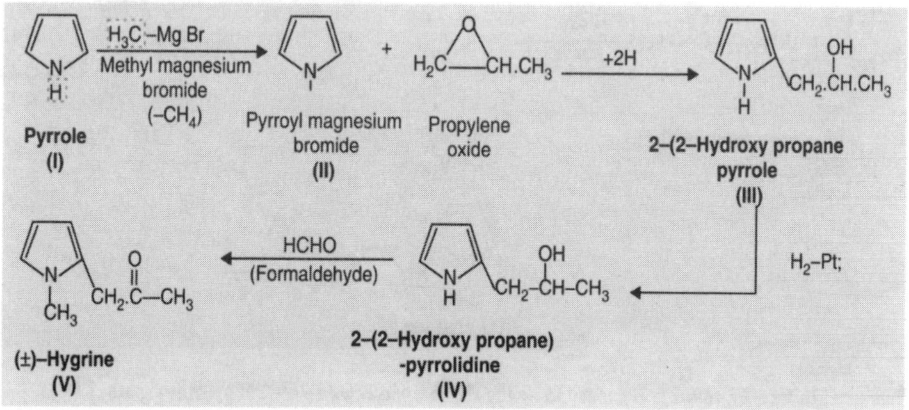

Notes: (1) The last step in *Hess's Method* for the synthesis of (±)-hygrine essentially involves the *Eschiveiler–Clarke methylation*, wherein the simultaneous oxidation of the ensing *hydroxy methylene group* (*–CHOH–*) takes place predominantly *i.e.*, in this particular method the pri– or *secondary* amines are *methylated* reductively with *formaldehyde* and *formic acid*.

(2) Lukes *et al.* (1959) simply repeated Hess's Method, and observed that the product duly obtained in the aforesaid synthesis is *NOT hygrine*, but instead an altogether different chemical entity known as *tetrahydro-oxazine (VI)*. They also advocated strongly that the interpretation of the *'last step in Hess's method'* has been interpreted wrongly.

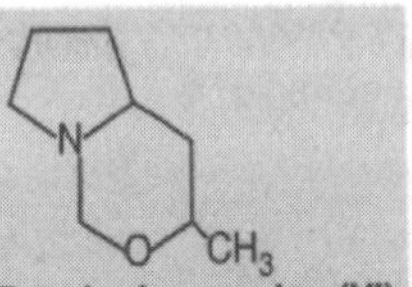

Tetrahydro oxazine (VI)

(3) *Eschiveiler-Clarke methylation* essentially requires the careful heating of a specific pri- or *secondary amine* with an excess of *formaldehyde (HCHO)*, whereby the Hydrogen required absolutely is adequately provided by the critical *oxidation* of the excess of *HCHO* to CO_2 or *HCOOH* (formic acid). Thus, we may have:

$$-NH_2 + 2\,HCHO \longrightarrow -N-CH_3 + H-\overset{O}{\underset{}{C}}-OH$$

A *Pri*–amine   Formalin   **Methyl   Formicacid**
                                                amine

B. Sorm's Method (1947)

Sorm adopted a *three* step synthesis of (±)–hygrine as stated under:

Step-1: The interaction of **N-methylpyrrole** and **acetodiazomethane** in the presence of **Cu** (as a *catalyst*) yields **N-methyl pyrrole-2-acetone (I)**, with the elimination of a mole of N_2-gas.

Step-2: The resulting **product (I)** upon reduction with H_2-Pt *i.e.*, **catalytic hydrogenation** gives rise to the formation of **hygroline (II)**.

Step-3: The **product (II)** upon oxidation produces ultimately the desired product (±)-**hygrine (III)**.

The various reactions are as given under:

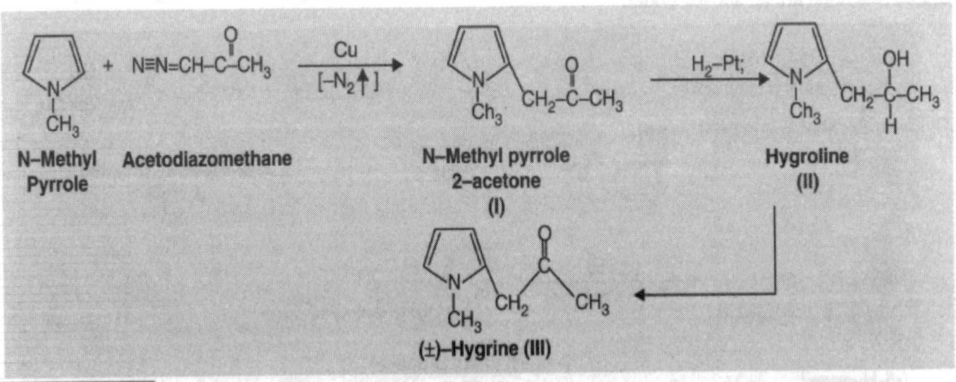

N–Methyl   Acetodiazomethane
Pyrrole

N–Methyl pyrrole
2–acetone
(I)

Hygroline
(II)

(±)–Hygrine (III)

---

*Note   : In fact, *hygroline* is regarded to be the *secondary alcohol* related to hygrine (Spath, 1943).

C. Anet's Method

Anet *et al.* (1949) proposed a condensation reaction involving only one step between **3-methyl amino butyral-dehyde** and **ethyl acetoacetate** in a *buffered medium at pH 7*. The reaction takes place as given below which eventually also ascertains the structure (I) for (±)-**hygrine**:

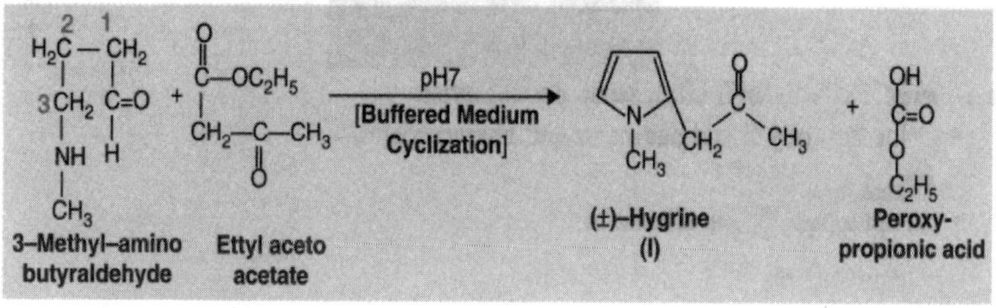

| 3-Methyl-amino butyraldehyde | Ettyl aceto acetate | (±)-Hygrine (I) | Peroxy-propionic acid |

D. Mannich Condensation Method

Mannich (1912)* proposed a condensation method for the synthesis of **hygrine** by the interaction of **acetoacetic acid** and **methyl pyrrolidinium (I)** (an *ammonium salt*). The latter may be obtained by:

- **dehydrogenation of N-methyl pyrrolidine**, or
- **partial reduction of N-methyl pyrrolidone**, as given under:

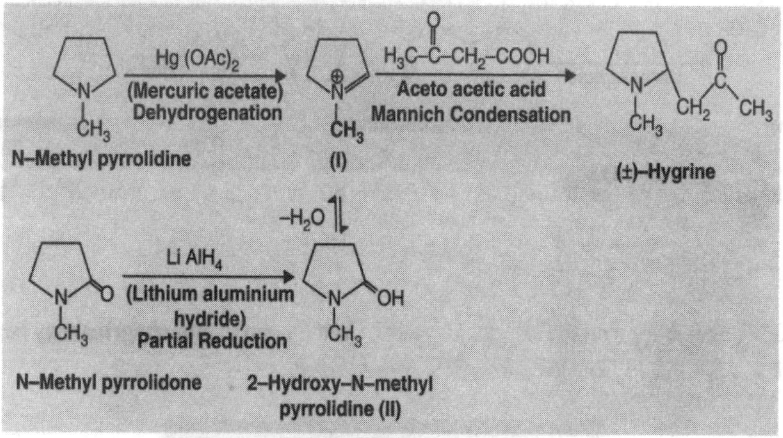

N–Methyl pyrrolidine   (I)   (±)-Hygrine

N–Methyl pyrrolidone   2–Hydroxy–N–methyl pyrrolidine (II)

Note: The respective products of reaction (I) obtained by the *dehydrogenation* and (II) by the *partial hydrolysis* are found to be reversible (or interchangeable) with the loss of one mole of H_2O from (II) to (I) as indicated above.

3.1.2. Cuscohygrine [or Cuskhygrine]

Cuscohyrine is found in the **cusko** and **coca leaves** of various origin and also in '**crude hygrine**'. It may be readily converted to **hygrine** by *acids* and *bases*.

---

* Mannich C and Krosche W: *Arch. Pharm,* **250:** 647, 1942).

** For a review of **ammonium formate in organic synthesis**, see Ram; Ehrenkaufer, *Synthesis, pp 91-95, 1988.*

The structure of **cuscohygrine** has been suggested by several researchers* as given below:

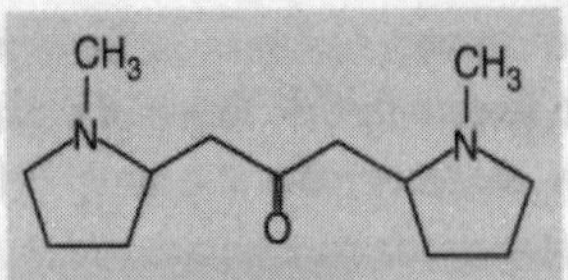

Cuscohygrine

Characteristic Features: These essentially comprise:

- colourless oily liquid having a faint characteristic,
- bp_{32mm} 185°C, and
- an optically inactive dibasic base.

3.1.2.1. Constitution of Cuscohoygrine

The different aspects of the constitution of **cuscohygrine** are as enumerated below:

(1) **Molecular Formula:** Its molecular formula is $C_{13}H_{24}NO_2$.

(2) **Oxidation:** Interestingly, **cuscohygrine** undergoes *oxidation* with.

    (*a*) **Chromic acid [CrO₃]**– to produce **N-methyl proline (I)**, and

    (*b*) **Nitric oxide [N₂O₂]**–to yield **homohygric acid (II)**,

as given under:

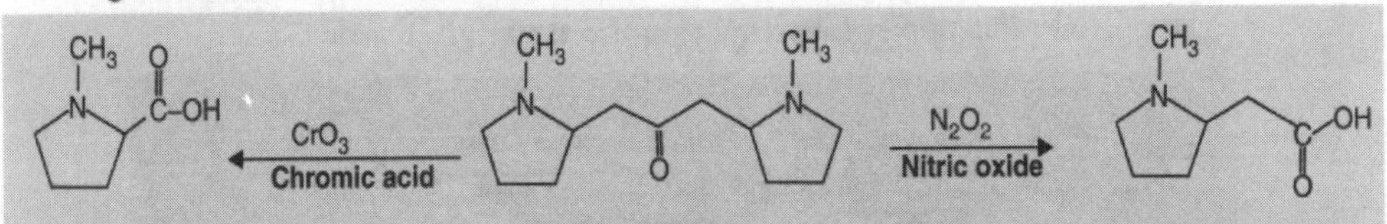

| **Methyl proline** (I) | **Cuscohygrine** | **Homohygric acid** (II) |

Conclusion: The above *two* critical oxidation reaction reveals the presence of **keto-dimethylene**

$(-CH_2-\overset{O}{\underset{\|}{C}}-CH_2-)$ moiety in **cuscohygrine.**

(3) **Keto [$-\overset{O}{\underset{\|}{C}}-$] Group:** The critical presence of a **keto group** in **cuscohygrine** may be established by reacting with hydroxylamine (**NH₂OH**) to form an **'oxime'.**

Thus, we may have:

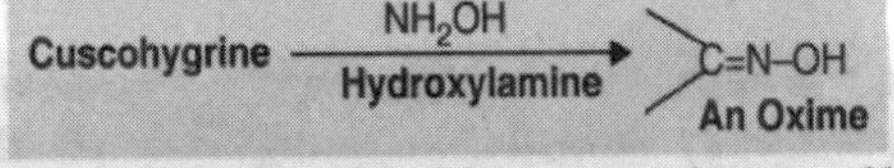

(4) **Tertiary N-atom:** The presence of **tertiary N-atom** may be proved by the fact that **cuscohygrine** usually adds on to **two moles** of **methyl iodide [H₃CI]** to give rise to the production of a **diquaternary ammonium salt,** as shown under:

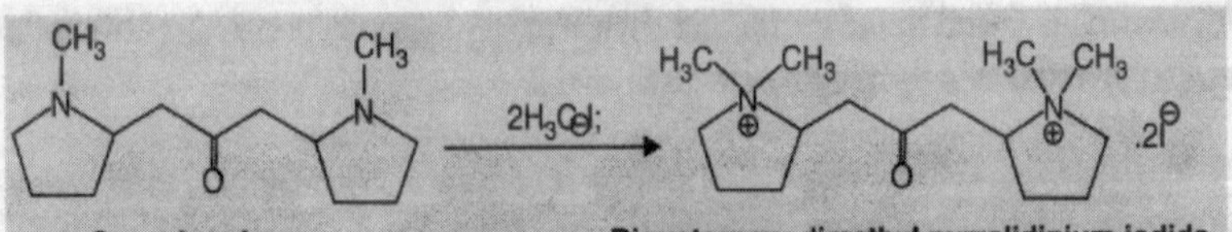

| **Cuscohygrine** | **Diquaternary dimethyl pyrrolidinium iodide** [A Diquaternary Ammoniun Salt] |

*** Hess, Fink: *Ber.* **53**: 794, 1920; Sohl, Shriner: *J Am Chem. Soc.,* **55**: 3829, 1933; Rapoport, Jorgensen: *J Org Chem:* **14**: 664, 1949.

Conclusion: Cuscohygrine definitely possesses a 'ditertiary base'.

Note: It has also been duly substantiated that the *cusochygrine* does contain *two* N-methyl (–N–CH₃) moieties by the help of *Herzig-Meyer's method*.

(5) Exposure to KOH: An exposure to *solid KOH* or *solution of KOH* aids in the **partial conversion of** cuscohygrine **into** (±)–hygrine.

$$\text{Cuscohygrine} \xrightarrow{\text{KOH}} (\pm)\text{–Hygrine}$$

(6) Liebermann's* Proposed Structure of Cuscohygrine: The proposed structure of cuscohygrine **(I) is as given under:**

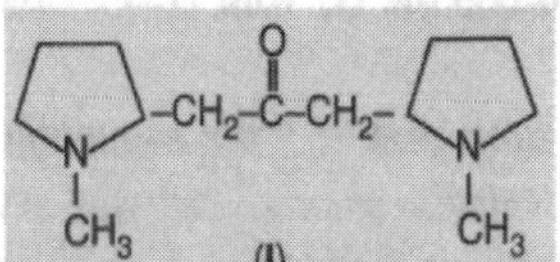

(I)

Hess and Fink (1920)** first and foremost confirmed the structure (I) for **cuscohygrine** solely based upon the remarkable outcome of their meticulous synthesis performed duly by '**Hofmann's exhaustive methylation**' of (I) to **n-decan-6-ol** ultimately.

The various steps involved in the '**Hoffmann's exhaustive methylation**' are as stated under:

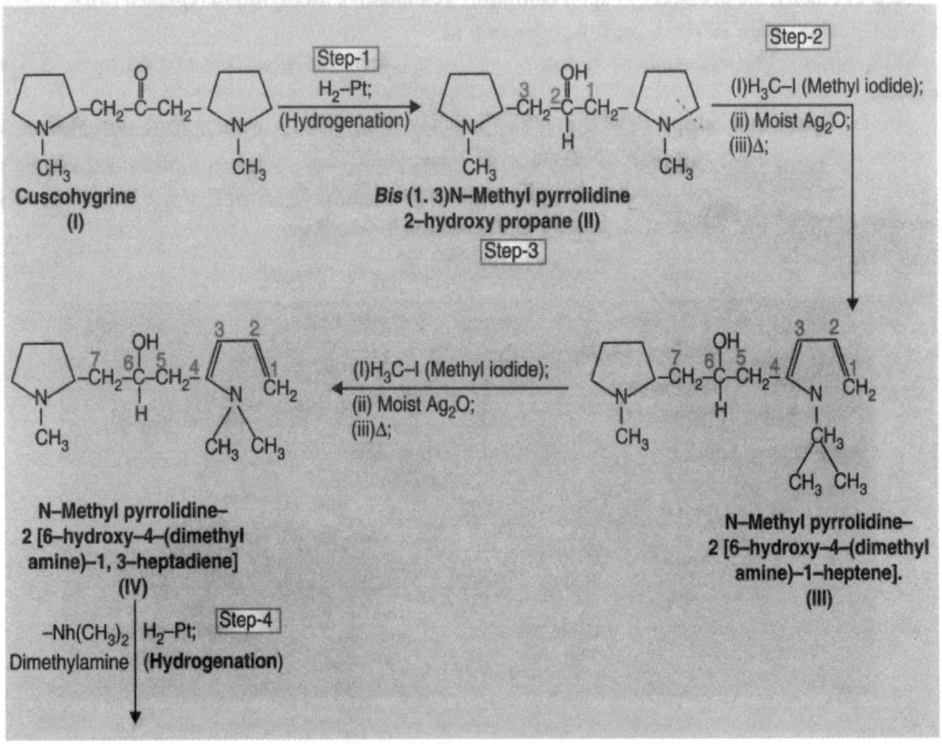

* Libermann : *Ber.* **22** : 679, 1898; Liebermann, Cybulski : *ibid,* **28** : 578, 1895.

** Hess and Fink ; *Ber;,* **53** : 794, 1920.

Step-5

(i)Hofmann' Exhaustive Methylation (2 Times)
(ii) H₂–Pt (Hydrogenation)

N–Methyl pyrrolidine–2–(6–hydroxy) pentane (V)

N–Undecan–6–ol
VI

Explanation: The various **steps** involves are duly explaned below :

Step-1 : Hydrogenation of **cuscohygrine (I)** with H_2-Pt yields **bis (1, 3)–N–methyl pyrrolidine–2–hydroxy propane (II).**

Step-2: Product **(II)** on being subjected to *three* sequential treatments, namely: methyl iodide, moist Ag_2O (or Ag OH), and heating causes specific *cleavage* of the **pyrrolidine nucleus,** as shown by dotted line, to yield **N-methyl pyrrolidine-2 [6-hydroxy-4 (dimethyl amine)–1–heptene] (III).**

Step-3: The resulting product **(III)** is a repeat of *three* sequential treatments (as in **step-2**) to obtain **N-methyl pyrrolidine-2 [6-hydroxy-4-(dimethyl amine)-1,3-heptadiene] (IV).**

Step-4: Now, the **product (IV)** on being subjected to hydrogenation with H_2-Pt loses a mole of **dimethylamine** and gives rise to the formation of **N-methyl-pyrrolidine-2-(6-hydroxy) pentane)V).**

Step-5: Finally, the **product (V)** upon **Hofmann's exhaustive methylation** *repeated twice,* followed by hydrogenation with H_2-Pt, yields **n-undecan-6-ol.**

(7) **Synthesis:** The total synthesis of 'cuscohygrine' has been carried out successfully by *two* different methods:

- **Spath and Tuppy (1948)*;** and Galinovsky *et al.* (1951)–from **methyl pyrrolidine;** and
- **Tuppy and Faltaous (1960)****–from *two* moles of **γ–methylamino-butaraldehyde** and one mole of **acetone dicarboxylic ester** *enzymatically followed by decarboxylation,* which shall be discussed individually in the section that follows:

7.1. First Method [Spath and Tuppy (1948); Galinovsky *et al.* (1951)]:

+N≡N=CH–C–OC₂H₅
Diazomethane propioester

N–Methyl pyrrole

Cu;
[–N₂↑]

Hygroline
(I)

H₂O;
(Hydrolysis) –EtOH

N–Methyl pyrrole–2–acetate
(II)

(I)Dry distillation with Pb–salt;
(ii) H₂–Pt (Reduction)

Cuscohygrine
(III)

Explanation: The various steps involved in the above synthesis may be explained as under:

(1) Interaction of **N-methyl pyrrole** and **diazomethane propioester** presence of *Cu* as a catalyst

* Spath and Tuppy : *Monatsh,* **79** : 119, 1948.

** Tuppy and Faltaus : *Monatsh,* **91** : 167, 1960.

yields **N-methyl pyrole-2-methane propioester** (I) upon hydrolysis yields **N-methyl pyrrole-2-methane propio ester** (I) with the loss of a mole of N_2.

(2) The resulting **product (I)** upon hydrolysis yields **N-methyl pyrrole-2-acetate (II)** with the elimination of a mole of EtOH.

(3) The **product (II)** on being subjected to '**dry distillation**' with *Pb-salt*, followed by reduction with H_2 Pt gives rise to the formation of **cuscohygrine (III)**.

7.2. Second Method [Tuppy and Faltaus (1960)]:

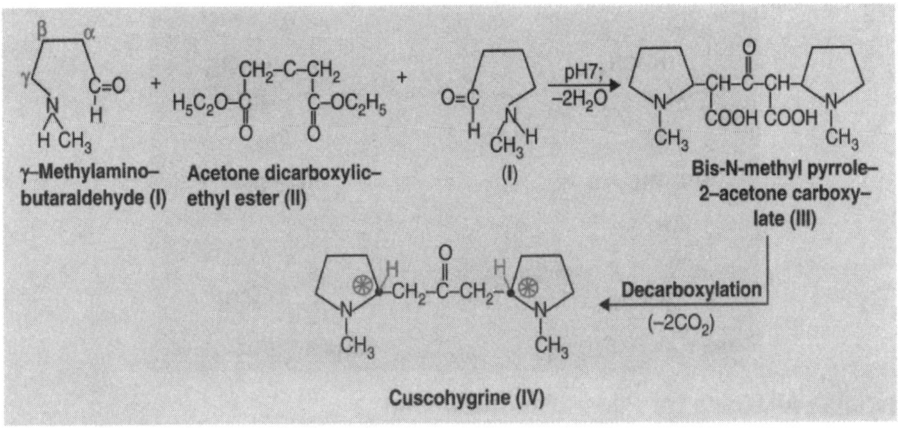

γ-Methylamino–butaraldehyde (I)   Acetone dicarboxylic–ethyl ester (II)   (I)   Bis-N-methyl pyrrole–2-acetone carboxy–late (III)

Cuscohygrine (IV)

They proposed a *two* step synthesis:

- Interaction with two moles of **λ-methyl amino butaraldehyde (I)** with **acetone dicarboxylic ethyl ester (II)** at *pH7* yields **Bis-N-methyl pyrrole-2-acetone carboxylate (III)** with the loss of two moles of water, and

- *Decarboxylation* of the resulting **product (III)** gives the desired alkaloid **cuscohygrine (IV)** with the elimination of **2 moles of CO_2**.

(8) **Stereochemistry:** Evidently, the **cuscohygrine (IV)** structure has *two* **identical chiral centres***, *marked with encircled coloured asterik marks* (see section 7.2 above). Therefore, it may form [$2^2 = 4$] **four isomers** or **an enantiomeric pair plus a** *meso* **form.** Since, the **naturally occurring cuscohygrine** is observed to be *inactive* optically, it may be inferred that **cuscohygrine** occurs as:

- a *meso* **form,** or
- a **racemic mixture** (*i.e.,* an **equal proportion of +ve and –ve isomers or dl-mixture**).

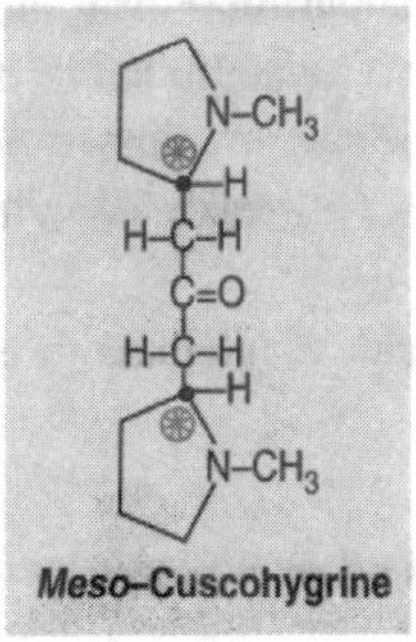

***Meso*-Cuscohygrine**

---

* Just like '**tartaric acid**'.

Furthermore, the aforesaid *meso*-form of **cuscohygrine** may be substantiated based upon the following *two* cardinal reasons, namely:

❑ **Resolution** of **'natural cuscohygrine'** into its respective **'optically active forms'** is **not** possible at all, and

❑ **Reduction** with **Na-metal/absolute ethanol** yields a mixture of *two* **epimeric alcohols** *viz.*, α–and β–hydroxy cuscohygrine* (the **carbonyl moiety** produces a **mono-hydroxy product only**).

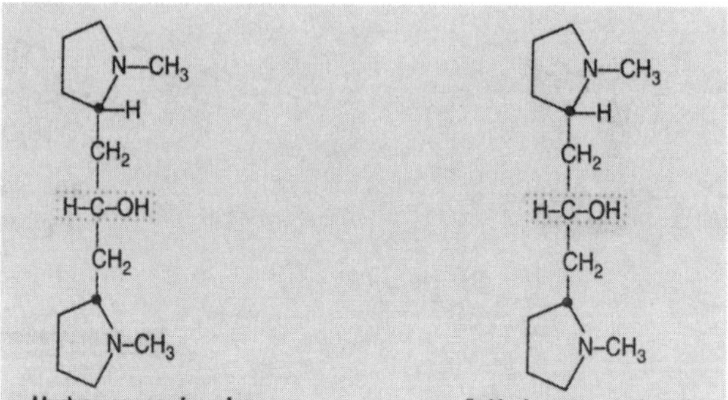

α–Hydroxycuscohygrine   β–Hydroxycuscohygrine

3.2. Pyridine Alkaloids [or Peperidine Alkaloids]

The **pyridine (or piperidine) alkaloids** was discovered in an array of important plant products, namely:

- **Ricinine**,
- **Coniine**,
- **Piperine**,
- **Pelletierine**,
- **Isopelletierine**, and
- **Pseudopelletierine**.

3.2.1. Ricinine

Böttcher (1918)**7 was pioneer in the extraction and isolation of **ricinine** from the *seeds* and *leaves* of the **castor plant**, *Ricinus communis* L., *Euphorbiaceae*, which being **toxic in nature**. It is a **2–pyridone structure** and essentially comprises-a **nitrile (–CN) grouping**, probably formed by **dehydration of a nicotinamide derivative.** posseses an intensely bitter acrid taste. It is found to be **inactive optically. Ricinine** usually separates from mixture of ethanol/water as almost colourless crystals having mp 201.5°. **Interestingly,** the **alkaloids** is so weakly basic incharacter that it rarely forms any salts with acids.

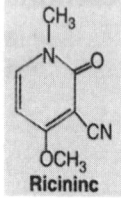

Ricininc

Ricinine is obtained from **nicotinic acid** *via* **nictinamide** and ultimately to the desired compound:

---

* This is due to the fact that the critical **'reduction of racemic (or dl–mixture** yields explicitly a **single racemic alcohol.**

** Böttcher : *Ber,* **51** : 673, 1918.

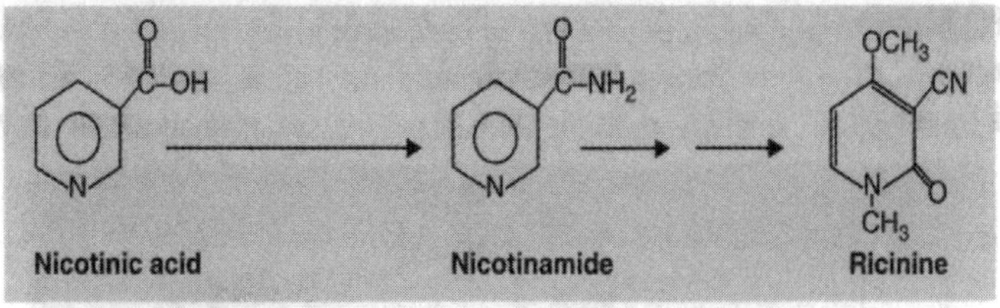

Nicotinic acid → Nicotinamide → → Ricinine

3.2.1.1. Constitution of Ricinine

The elucidation of structure of **ricinine** is mainly due to the excellent researches by Schroeter *et al.*(1932)* starting with the polymerization of **cyanoacety chloride.**

(1) Its molecular formula is $C_8H_8N_2O_2$.

(2) **Zn-Dust Distillation: Ricinine** when subjected to **Zn-dust distillation** yields pyridine, which glaringly reveals that the **alkaloids** had essentially a **'pyridine nucleus'.**

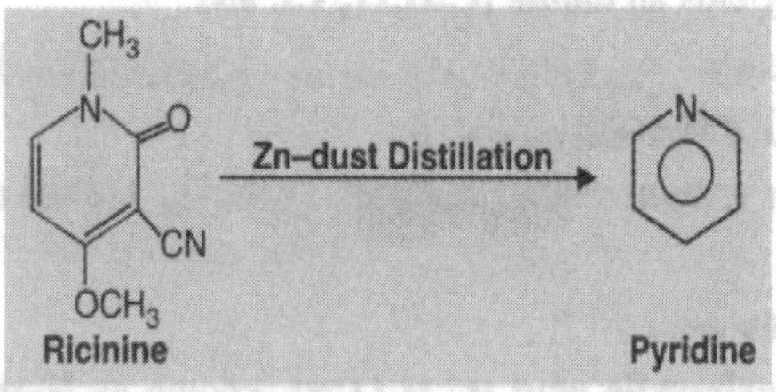

Ricinine — Zn–dust Distillation → Pyridine

> **Note:** Besides, *ricinine,* on being subjected to *catalytic reduction* produces corresponding *tetrahydro structural analogue* that takes up *four H-atoms;* and, therefore, shows evidently the presence of a *dehydrogenated pyridine nucleus system in ricinine.*

(3) **Presence of Methoxy (–OCH₃) Group:** The presence of the **methoxy (–OCH₃) group** in *ricinine* is duly indicated by the formation of *'methanol'* and **ricinine acid**** by carrying out the **alkaline hydrolysis,** as given under:

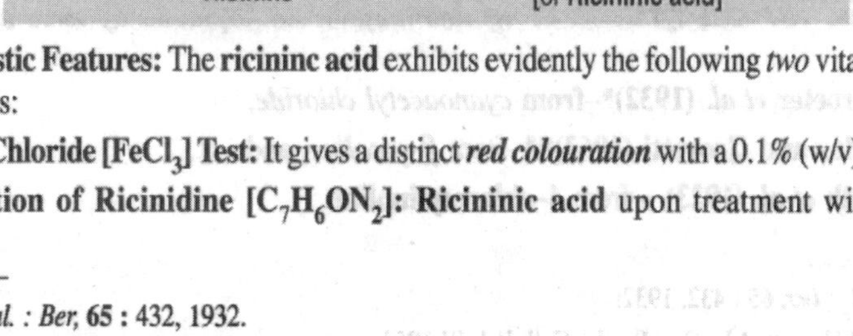

Methoxy Group → Ricinine — NaOH; (Hydrolysis) → Ricinine acid [or Ricininic acid] + H_3C–OH Methanol

Characteristic Features: The **ricininc acid** exhibits evidently the following *two* vital **characteristic features,** such as:

(*a*) **Ferric Chloride [FeCl₃] Test:** It gives a distinct *red colouration* with a 0.1 % (w/v) Fe Cl₃ solution.

(*b*) **Formation of Ricinidine [$C_7H_6ON_2$]: Ricininic acid** upon treatment with **phosphorus**

---

* Schrocher *et al.* : *Ber,* 65 : 432, 1932.

** It is also known as **'Ricininic Acid'.**

oxychloride [POCl₃]–followed immediately by *reduction* gives rise to the formation of **ricinidine** (a *cyano derivative*). The above **structural transformation(s)** clearly suggests that the *hydroxyl (–OH) moiety* in **ricininic acid** is solely responsible for causing the said changes, as given below:

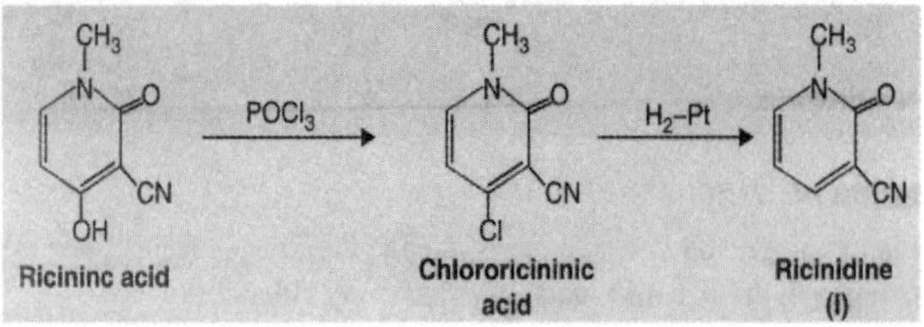

Ricininc acid   Chlororicininic   Ricinidine
                            acid   (I)

(4) **Probable Structure of Ricinine:** It is worth while to state at this point in time that the structure of *ricinine* is fundamentally based upon the structure of **ricinidine (I)**. The **compound (I)** may be subjected to **hydrolysis** at *two* distinct stages, namely:

- Stage-I: which yields an *amide* [$C_7H_8O_2N_2$], and
- Stage-II: further hydrolysis of the *amide* gives rise to an **acid** [$C_7H_7O_3N$], and a **mole** of ammonia (NH_3) gets eliminated.

Importantly, the above reactions/results indicate explicitly the following critical and cardinal observation, such as:

- presence of **cyano (–CN) moiety**,
- presence of a **carboxylic acid** [$C_7H_7O_3N$], known as **N–methyl–2–one–pyridine–3–carboxylic acid (A)**, which has also been confirmed by **synthesis. Hence, ricinidine (B)** may be assigned the **structure,**
- oxidation of ricinine with *chromic acid* (*vigorous oxidation*)-produces **hydrogen cyanide (HCN)** that vividly shows the **ricininic acid (C)**; and; therefore, **ricinine** should be(**D**).

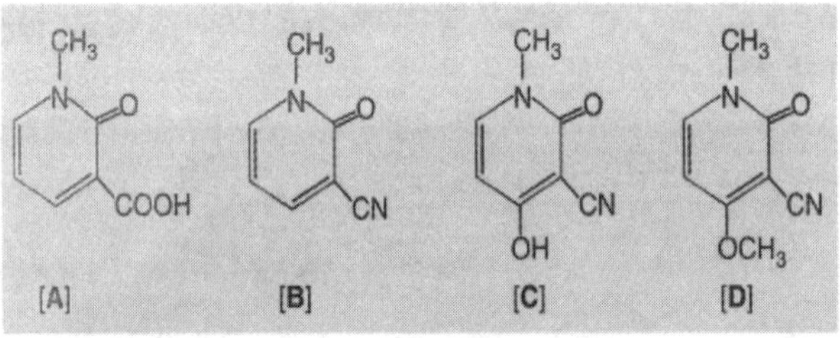

[A]   [B]   [C]   [D]

From the aforesaid statement of facts and evidential support one may conclude that 'ricinic acid' and 'ricininc' shall have the structures [C] and [D] respectively.

(5) **Synthesis:** The **total synthesis** of *ricinine* may be accomplished by *three* known methods, namely:

- **Schroeter *et al.* (1932)*–from** *cyanoacetyl chloride,*
- **Taylor and Crovetti (1963)****–from *β–picoline,* and
- **Spath *et al.* (1923)** – from **4–chloroquinoline.**

---

* Schroster *et al.* : *Ber*, **65** : 432, 1932.

** Taylor EC and Crovetti AJ : *Org. Synth., Coll. Vol. IV*, 1963

5.1. Schroeter's Synthesis

The starting material **cyanoacetyl chloride, I,** (an *acid chloride*) undergoes **dimerization*** upon prolonged standing to **2,4–dihydroxy–6–chloronicotino nitrle (II),** which on being subjected to 'methylation', with *dimethyl sulphate* $[(CH_3)_2SO_4]$ gives rise to the formation of a corresponding **dimethyl derivative (III).** The resulting **compound (III)** is subjected to **reduction** with ZnB CH_3COOH to yield **ricinine (IV)** ultimately.

The various reaction involved are expressed as under:

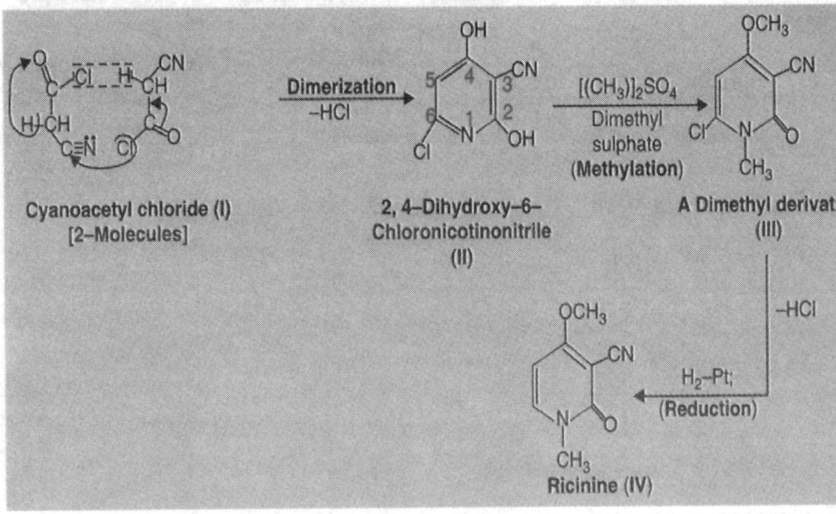

Cyanoacetyl chloride (I)
[2–Molecules]

2, 4–Dihydroxy–6–
Chloronicotinonitrile
(II)

A Dimethyl derivative
(III)

Ricinine (IV)

Since, **compound (II)** contains exactly *twice* the number of **carbon** and **hydrogen atoms** as the original **compound (I)** *i.e.*, C_6H_{14} in **(II)**, and C_3H_2 in **(I)** respectively, they are commonly termed as 'dimers', and such a reaction is known as **dimerization.**

> Note: Radio tracer studies have established that specific experiments carried out on the *castor oil plants* with *radio-labelled $H^{14}CN$* takes up the *cyano (–CN) moiety of ricinine.*

5.2. Taylor and Crovetti Synthesis

The critical and specific reactivity of **pyridine-1-oxide** to the respective **electrophilic substitution at C-4** has been duly exploited quite in contract to the **pyridine nucleus itself.** Eventually, the latter undergoes *strong deactivation* in favour of:

- **electrophilic substitution,** and
- **specific reactivity at C-3.**

In addition, the **nitro (–NO) moiety at C-4 in pyridine** gets easily and conveniently displaced by the aid of **'nucleophilic substitution'.**

The various steps in the **Taylor and Crovetti synthesis** are as given under:

---

* **Dimerization :** Means, **dimers** [*di* = two; *mer* = part] of **cyanoacetyl chloride** gives one mole of **ricinine,** and the reaction is termed as **dimerization.**

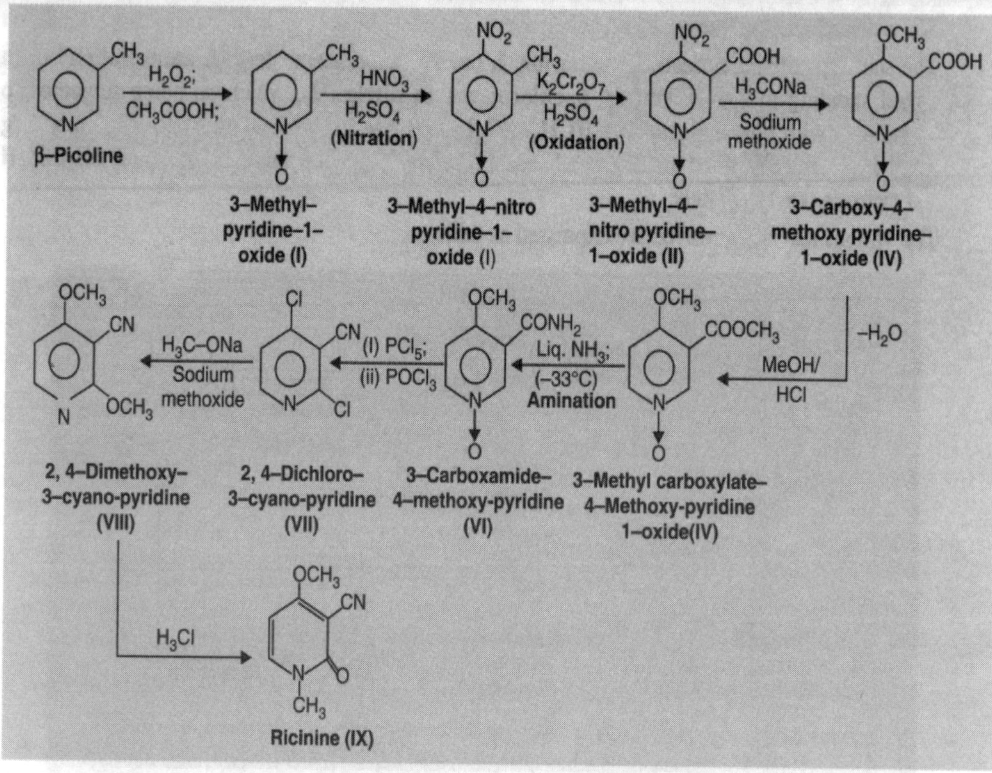

β–Picoline → (H₂O₂; CH₃COOH) → 3–Methyl–pyridine–1–oxide (I) → (HNO₃ H₂SO₄ Nitration) → 3–Methyl–4–nitro pyridine–1–oxide (I) → (K₂Cr₂O₇ H₂SO₄ Oxidation) → 3–Methyl–4–nitro pyridine–1–oxide (II) → (H₃CONa Sodium methoxide) → 3–Carboxy–4–methoxy pyridine–1–oxide (IV)

2, 4–Dimethoxy–3–cyano-pyridine (VIII) ← (H₃C–ONa Sodium methoxide) ← 2, 4–Dichloro–3–cyano-pyridine (VII) ← ((i) PCl₅; (ii) POCl₃) ← 3–Carboxamide–4–methoxy-pyridine (VI) ← (Liq. NH₃; (–33°C) Amination) ← 3–Methyl carboxylate–4–Methoxy-pyridine 1–oxide (IV) ← (–H₂O, MeOH/ HCl)

Ricinine (IX)

Explanation: The various aforesaid steps may be expatiated as under:

(1) The starting material, a **pyridine–1–oxide derivative (I)** *ie.,* **3–methyl pyridine–1–oxide,** is duly obtained by treating **β-picoline** with **hydrogen peroxide (H₂O₂)** followed by *acetic acid.*

(2) **Product (I)** is nitrated with HNO₃/H₂SO₄ (*via* **nitronium ion** to yield **3–methyl–4–nitro pyridine–1–oxide (II).**

(3) The resulting product **(II)** upon oxidation with K₂Cr₂O₇/H₂SO₄ produces **3–carboxy–4–nitro pyridine–1–oxide (III)** *ie.,* the *methyl group* gets oxidized to a *carboxyl (–COOH)* group

(4) The **product (III)** undergoes methoxylation with freshly prepared **sodium methoxide** to yield **3–carboxy–4–methoxy pyridine–1–oxide (IV).**

(5) The resulting **product (IV)** on treatment with acidified methanol loses a mole of water to produce **3–methyl carboxylate–4–methoxy pyridine–1–oxide (V).**

(6) The **product (V)** undergoes amination with **liquid ammonia at–33°C to yield 3–carboxamide–4–methoxy pyridine–1–oxide (VI).**

(7) The resulting **product (VI)** on being treated with **PCl₅** followed by **phosphorus oxytrichloride** yields **2, 4–dichloro–3–cyano pyridine (VII).**

(8) The **product (VII)** when treated with freshly prepared **sodium methoxide** gives rise to the production of **2, 4–dimethoxy–3–cyanopyridine (VIII).**

(9) Finally, **product (VIII)** when treated with *methyl iodide (CH₃I)* yields the desired alkaloid *ricinine (IX)*.

5.3. Spath's Syntnhesis

Spath *et al.* (1993) put forward a multistep synthesis of *ricinine* starting from **4–chloroquinoline** as stated under:

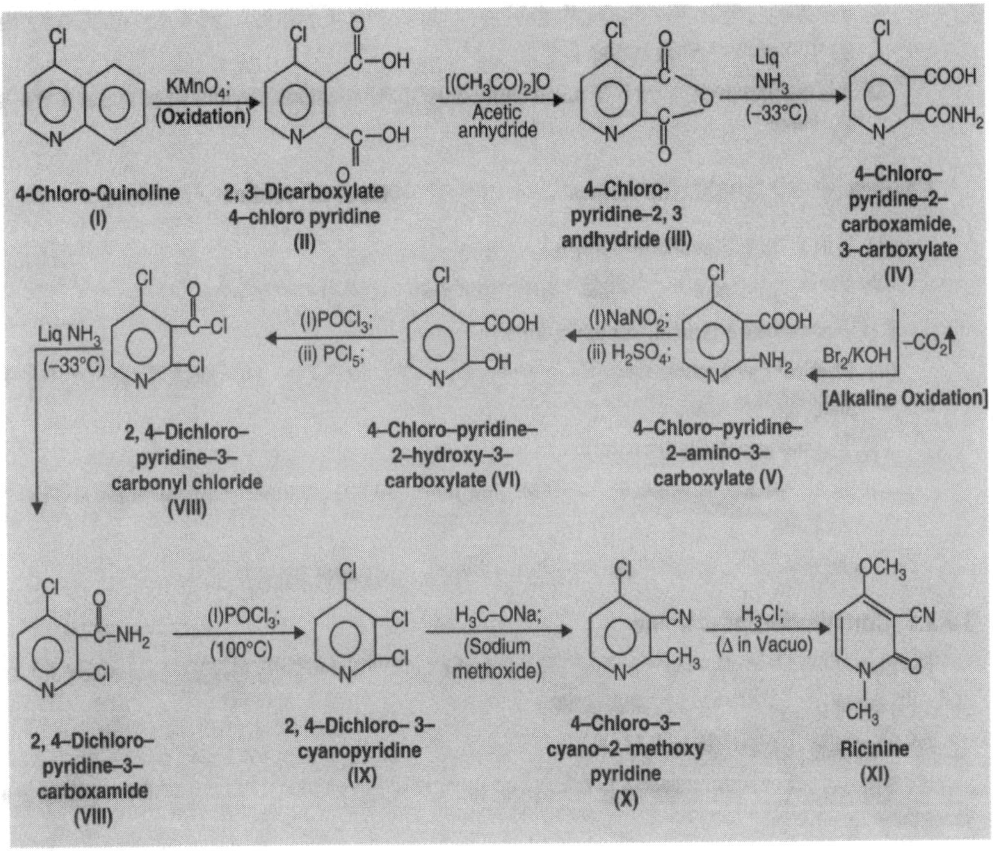

3.2.2. Coniine

Preamble: **Coniine** is the '**toxic principle**' of the *poison Hemlock, (which Socrates was made to drink), Conium maculatum* L., *Umbelliferae.*It occurs naturally as the **S-(+)–isomer**. In fact, the **Hemlock plant** comprises a host of highly **potent alkaloids**, namely:

γ–coniceine conhydrine, **N-methyl coniine**, and **pseudoaconhydrine.**

There are *two* other plant sources that contains **coniine:**

- **Fool's Parsley** *i.e., Aethusa cynapium* L.,*(Apiaceae)*; and
- **Water Hemlock** *i.e., Cicuta maculata* L., *(Apiaceae).*

Coniine enjoys having the widest recognition across the globe for *two* glaring reasons, namely:

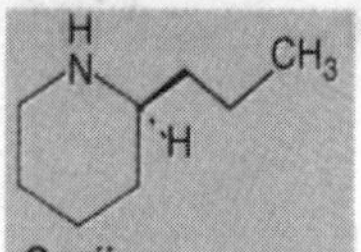

Coniine

□ designates and forms the *chief constituent* of the **Hemlock alkaloids,** causing *critical paralysis* of the **Hemlock alkaloids,** causing *critical paralysis* of the **central nervous system (CNS) in humans,** and

□ it is the *first* ever natural chemical entity duly and meticulously synthesized by **Ladenburg in 1986*.**

Characteristic Features: The various **characteristic features** of coniine are as detailed under:

(1) It is a colourless alkaline **liquid alkaloid.**

(2) An exposure to air and light turns it **'dark'** due to *polymerization.*

(3) It possesses a typical **'mousy odour'.**

(4) **Physical parameters** essentially include: mp = –2°C; bp 760 = 166 – 166.5 °C; D_4^{20} = 0.844; and $[\alpha]_D^{25}$ = + 8.4°.

(5) It is steam-volatile in nature.

(6) Pooly soluble in water (1 in 90 ml), but fairly soluble in an array of **'organic solvents',** such as:

acetone, amyl alcohol, chloroform, ethanol, ether, and benzene.

3.2.2.1. Constitution of Coniine

The pure alkaloid, **coniine,** was isolated by Landenberg (1986) and Koller (1926) (see *section 3.2.2*), and subsequently its structure was duly established.

(1) Its molecular formula is $C_8H_{17}N$.

(2) **Test for a Secondary Amine [>NH]:** The specific tests conducted to confirm that **coniine** is a secondary amine:

• with **nitrous acid (HNO_2)** it gives a **nitrous (–N = O) derivative,** and

• with *two* moles of **methyl iodide ($H_3C–I$)** it forms a **quaternary ammonium iodide.**

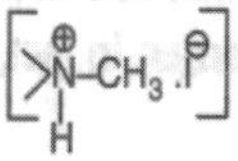

$$\left[\overset{\oplus}{\underset{H}{>N}}-CH_3 \right] . \overset{\ominus}{I}$$

(3) **Zn–Dust Distillation:** **Coniine** on being subjected to **Zn-dust distillation** causes **dehydrogenation** to form **conyrine ($C_8H_{11}N$)** which upon further oxidation with **$KMnO_4$** gives rise to the formation of α–picolinic acid (or **pyridine–2–carboxylate),** as given under:

---

* Ladenberg, *Ber,* **19** : 439, 1886; Koller, *Montash,* **47** : 393, 1926.

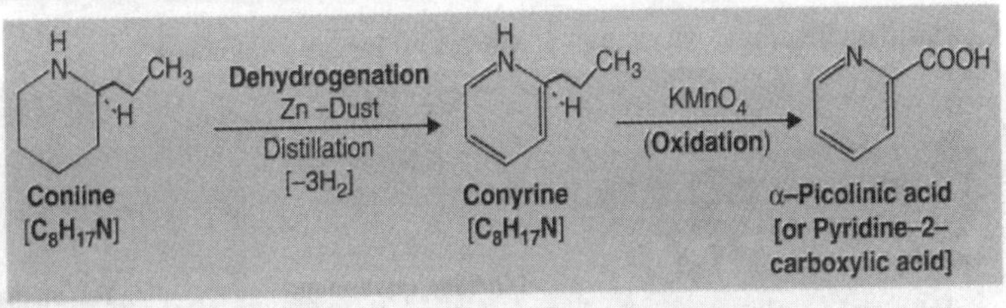

Coniine [C$_8$H$_{17}$N] →(Dehydrogenation Zn –Dust Distillation [–3H$_2$])→ Conyrine [C$_8$H$_{17}$N] →(KMnO$_4$ Oxidation)→ α–Picolinic acid [or Pyridine–2–carboxylic acid]

Inference Drawn: From the above set of reactions it may be inferred safely that generation of l–picolinic acid clearly suggests **conyrine** to be a **pyridine derivative positively** having a **side chain** (–C$_3$H$_7$) at C–2 position. Nevertheless, coniine, being a *'secondary amine'* loses **six H–atoms** to yield **conyrine**, which indicated overwhelmingly that the former is a **'piperidine derivative'**.

(4) **Confirmation of Side-Chain as *'n-Propyl'* or *'Isopropyl'*:** At this critical point in time it has become almost necessary to ascertain whether the side chain in **coniine** is:

N-propyl *i.e.,* –CH$_2$–CH$_2$–CH$_3$, or

isopropyl *i.e,* –CH$\begin{smallmatrix}CH_3\\CH_3\end{smallmatrix}$

Based on the above pronouncement one may have the following *two* probable structures of **coniine** [A] or [B]:

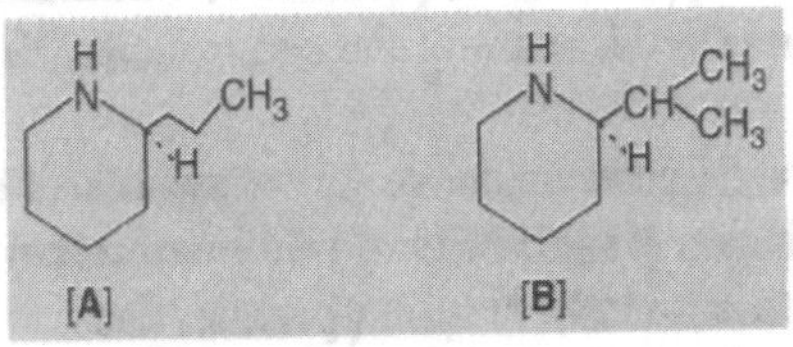

[A]   [B]

Of the *two* probable assigned structures for **coniine**, the **structures [A]** remains the **acceptable form,** which is exclusively based upon the following evidential and supportive facts , such as:

(a) **2-n-Propylpiperidine:** Coniine on being heated with **HI at 300°C** under pressure produces only **n-octane** and **not iso-octane** (that could have been possible if [B] would be the assigned structure of coniine). Thus, one may conclude that **coniine** is **2-n-propyl piperidine.**

(b) **Hofmann's Exhaustive methylation:** The Hofmman's *exhaustive methylation* of **coniine** followed immediately by reduction produces **n-octane.**

(c) **Von Braun Degradation:** It cleaves the **peperidine ring** to give an **open-chain compound** *viz.,* **1, 5-dichloro octane.**

All the *three* reactions **from (a) through (c)** may be expressed as given under:

For (a): Coniine on distillation with HI at 300°C gives **n-octane** with the loss of a mole of **ammonia.**

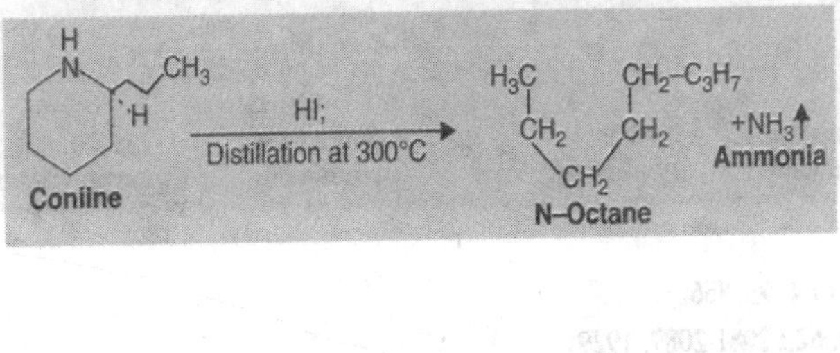

Coniine →(HI; Distillation at 300°C)→ N–Octane + NH$_3$↑ Ammonia

For (b): Coniine on treatment with methyl iodide followed by moist Ag_2O

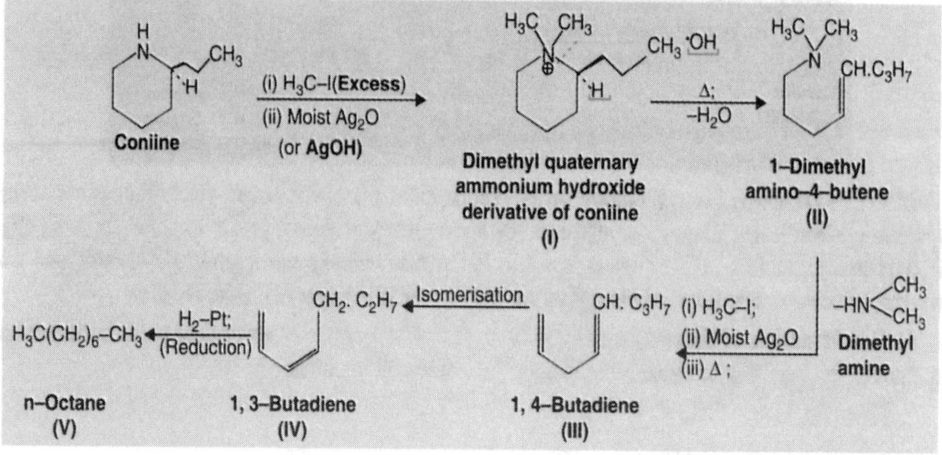

gives rise to the formation of **compound (I)**. The resulting **product (I)** upon heating loses a mole of water (as indicated) with the cessation of **piperidine ring** (shown by *coloured dotted line*) to yield **product (II)**. Subsequently, the **product (II)** on sequential treatment with **methyl iodide, moist Ag_2O,** and **heating** loses a mole of **dimethyl amine** and produces **1, 4-butadiene (III)**, which eventually get *isomerized* to yield **1, butadiene (IV)**, Ultimately, the **end-product (IV)** undergoes reduction with H_2/Pt to yield **N-octane (V)**.

5. **Synthesis:** There are several known methods for the synthesis of **coniine**, such as:
- **Ladenberg's synthesis (1886)*:** from **pyridinium methyl iodide;**
- **Diels Alder's synthesis (1929)**:** from **pyridine;** and
- **Bergmann's synthesis (1932):** from **2-methyl pyridine.**

5.1. Ladenberg's Synthesis (1886)*

Ladenberg suggested a *three step* synthesis for **coniine:**

Step-I: Heating **pyridinium methyl iodide** at 300°C yields **2-methyl pyridine (I),**

Step-II: Treatment of the resulting **product (I)** with 3 moles of acetaldehyde in the presence of zinc chloride at **250°C** gives rise to the production of **propylene pyridine (II).** and

Step-III: Product (II) on treatment with freshly cut sodium metal in absolute ethanol cause **reduction** to yield **(±)-coniine.**

These reactions may be expressed as under:

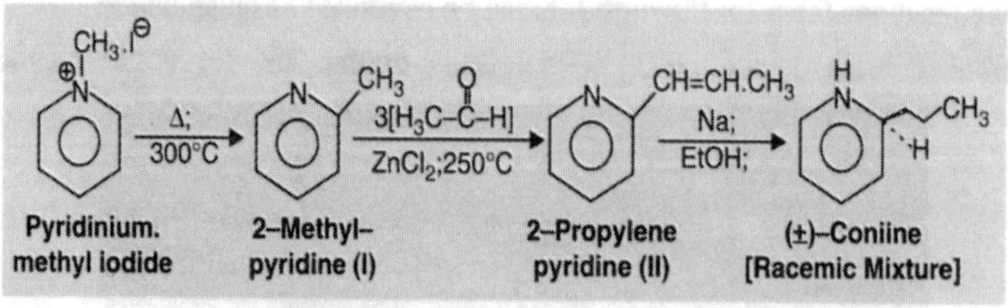

---

* Landenberg, *Ber,* **19** : 439, 1886;

** Diels K Alder : *Ber,* **62** : 2081-2087, 1929.

Separation of Racemic Coniine:

The end-product in **Landenberg's synthesis** is a *racemic mixture* of **coniine** that may be resolved effectively by the help of (±)–**tartaric acid**. The (±)–**coniine tartrate** crystals will get separated at the very first instance due to its poor solubility, which is removed consequently, and decomposed with dilute alkali carefully to obtain (±)–**coniine**. The product may be recrystallized, and found to be absolutely similar to the **natural alkaloid**.

5.2. Diels Alder's Synthesis (1929)*

The various steps involved are:

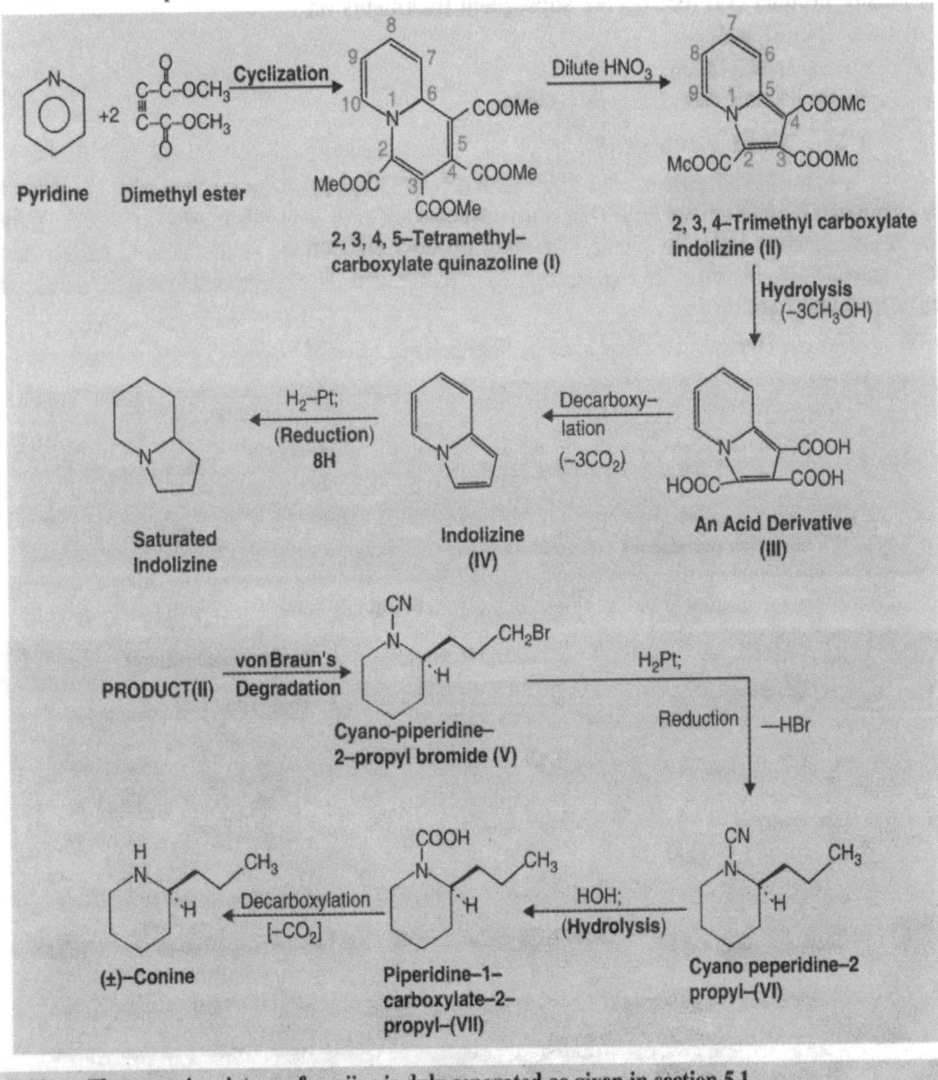

Note: The racemic mixture of *coniine* is duly separated as given in section 5.1.

* Diels K Alder : *Ber,* **62** : 2081-2087, 1929.

Explanation: These essentially comprise:

(1) Interaction of **pyridine and 2 moles of a dimethyl ester** yields **product (I)** due to cyclization *i.e.*, formation **quinazoline ring.**

(2) Treatment with dilute HNO_3 shrinks one of the **6–member rings** to a **5–member ring** to yield **indolizine ring** *i.e.*, **product (II).**

(3) **Product (II)** upon hydrolysis yields a corresponding *acid derivative (III)*, which on **decarboxylation** gives **indolizine (IV);** and this on reduction with H_2–Pt produces the **saturated indolizine nucleus.**

(4) Interestingly, the **product (II)** on sequential treatment with **von Brauns' degradation** yields **product (V)**, which on reduction loses a mole of **HBr** to produce **product (VI).**

(5) Finally, **product (VI)** receives *two* **subsequent treatments** viz.,

  • **Hydrolysis,** and
  • **Decarboxylation,**

to yields the *racemic mixture* of (±)–coniine.

5.3. Bergmann's Synthesis (1932)

Bergmann's synthesis involves the interaction of **2–methyl pyridine (I)** with *phenyl lithium* to yields **2–methyl lithium pyridine (II)**, which on treatment with *ethyl bromide* gives rise to the formation of **2–propyl pyridine (III).** Finally, the resulting **product (III)** on treatment with freshly prepared *sodium ethoxide (i.e., Na–C_2H_5OH)* produces a **racemic mixture of coniine** that may be duly separated as elaborated under section 5.1.

The various reactions taking place may be expreseed as under:

| 2–Methyl pyridine (I) | 2–Methyl lithium pyridine (II) | 2–Propyl–bromide (III) | (±)–Coniine (IV) |

Following are the *four* important **hemlock alkaloids** along with their *characteristic features:*

	Name of Alkaloid	Chemical Structure	Characteristic features
1	α–Coniceine	$[C_8H_{15}N]$	Alkaline liquid; Mousy odour; bp 171°C ; bp$_{15}$ 63°; Steam volatile; D_4^{15} 0.8753; N^{16}$_D$ 1.4661; freely soluble in ethanol, ether, and chloroform.
2	Conhydrine		Crystals from ether; mp 121°C; bp 226°C [α]$_D$+ 10°; slightly water soluble; freely soluble in ethanol, ether, and chloroform.
3	Pseudoconhydrine		Hygroscopic needle from. absolute ether; mp 106°C; bp 236° /cl [α]$_5^{20}$ + 11° (C+=10 in alcohol); pk (18°) : 3.70;
4	N–Methylconiine		Water in water plus most organic solvents. dl-form bp10.5 56.6°; d-form bp 173–174; [α]$_D^{24}$ + 81°; l- form [a]$_5^{20}$ –84°;

Stereoselective Synthesis of (S)–(+) Form and (R)–(–)–Form of Coniine: These *two* aforesaid forms of highly stereoselective **coniine** have been duly accomplished as stated under:

(S)–(+) form of coniine: Apeta K *et al.: Chem. Pharm. Bull,* **24**: 621, 1976.

(R)–(–) form of coniine: Lathbury D and Gallagher T: *Chem. Commun,.* 114, 1986.

3.2.3. Piperine

Piperine occurs in unripe fruits of **black pepper** (*Piper nigrum L.,*) and also in the **root bark** of *Piper geniculatum* Sw., belonging to the natural order *Piperaceae.*

However, the exact content of **peperine** in the **black pepper** ranges between **6 to 11%**.Besides, **peperine** is also found to be present in certain other **Piper species,** namely:

Piper longum: ~ 5 %; and *Piper lowong: ~ 1.5 %.* Piperine has a sharp buring taste.

Piperine

3.2.3.1. Discovery

Piperine was extracted for the first time by Cazeneuve and Calliot (1877)*, who obtained the **alkaloid** as *monoclinic* crystals with a strong inherent *flavour* as well as *taste* of **black pepper.**

3.2.2.2. Characteristic Features

These essentially comprise:

(1) It is duly obtained as **monoclinic prisms** from ethanol.

(2) Its mp is 130°C.

(3) It is *tasteless* at first instance, but gives a typical **burning after taste.**

(4) It is almost neutral to litmus.

(5) Its pK value (18°C) stands at 12.22.

(6) It is sparingly soluble in water *viz.,* 40 mg. L^{-1} at 18°C, and in petroleum ether.

(7) It is fairly soluble in organic solvents *viz.,* **ethanol (1 in 15 mL)**, **chloroform (1 in 1.7 mL)**, solvent ether (1 in 36 mL).

3.2.3.3. Extraction Procedure

Piperine may be extracted by *two* common methods, namely:

Method-1: It is usually obtained by mixing together the powdered **black pepper** with *'milk of lime'* and heating gently for 60-90 minutes. Evaporate the water, preferably under vacuum, to almost dryness. The *alkaloid* is extracted finally from the residual mass with solvent ether (or **diethyl ether**).

Method-2: It makes use of the *extraction procedure* with the powdered **black pepper** with **ethanol [95% (v/v]** in a continuous **Soxhlet Apparatus****, and the extract is subsequently evaporated to **almost dryness.** The resulting residue is carefully extracted with **ether,** and the ethereal extract thus obtained is

---

* Cazeneuve and Calliot : *Bull. Soc. Chim,.* [2], **27** : 291, 1877.

** Kar A : **Pharmacognosy and Pharmacobiotechnology,** New Age International, New Delhi, 2nd. ed., 2007.

washed successively first with **dilute NaOH solution,** and then using **DM water.** The ethereal solution may be subjected to evaporation *under vaccum,* and the residue is taken up duly in *ethanol* to obtain *pure crystals of* **piperine.**

3.2.3.4. Constitution of Piperine

Piperine after its *extraction, purification,* and *crystallization* is duly subjected to the determination of its *probable structure,* followed by confirmation of the structure by **'total synthesis'.**

(1) **Molecular Formula:** Based on the *elemental analysis* and *molecular weight determination,* the *molecular formula* of *piperine* is found to be $C_{17}H_{19}NO_3$.

(2) **Hydrolysis Reaction:** *Hydrolysis* of *piperine* accomplished by heating with KOH yields **piperic acid** and **peperidine,** as shown under:

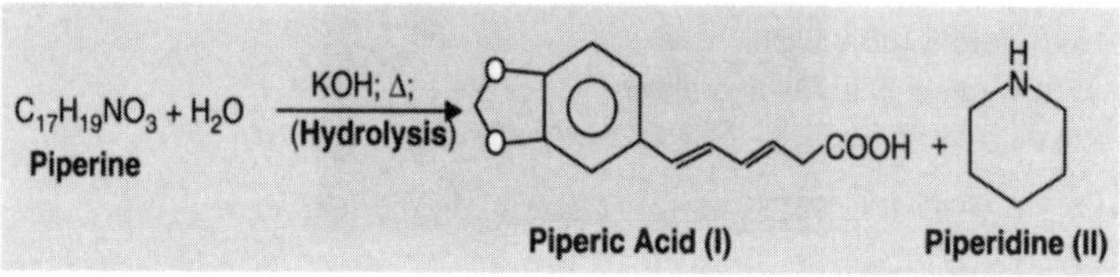

Piperic Acid (I)   **Piperidine (II)**

Inference: It may be inferred from the *products of hydrolysis* **(I)** and **(II)** that both **piperic acid** and

peperidine are duly joined by an **amide** $(-\overset{O}{\overset{\|}{C}}-NH-)$ moiety. Nevertheless, it has been adequately confirmed by virtue of the fact that the **acid chloride of piperic acid*** when heated with **peperidine** **(II)** in *benzene* produces the alkaloid **'piperine',** as shown under:

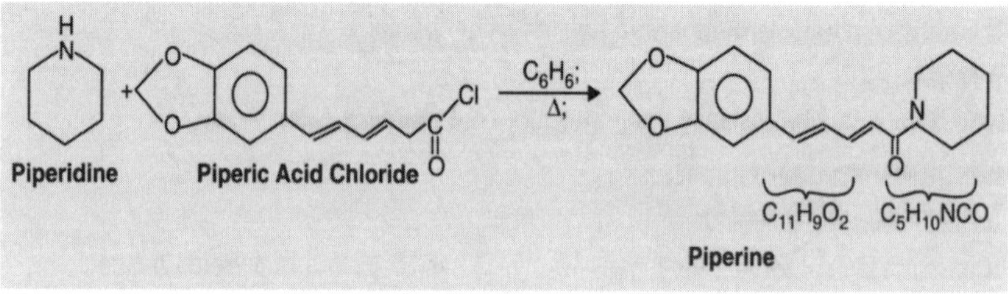

Piperidine   **Piperic Acid Chloride**   **Piperine**

It is, however, pertinent to state at this point in time that the structure of:

· **piperidine** is well established, and
· **piperic acid** remains to be *proved logically* and *established duly,*

in order to assign a proposed structures of **'piperinc'.**

(3) **Establishing Structure of Piperic Acid:** The structure of **piperic acid** may be established *via* the following cardinal steps adopted in a sequential manner:

(*i*) **Molecular formula:** Its molecular formula is found to be $C_{12}H_{10}O_4$.

(*ii*) **Catalytic Reduction:** The *catalytic reduction of* **piperic acid** with either **Pd/C** or **Raney Ni** yields **tetra hydropiperic acid,** as given under:

---

* It is prepared by the action of PCl_5 (phosphorus pentachloride) on **peperic acid.**

$$C_{12}H_{10}O_4 \xrightarrow{\text{Pd/C or Raney Ni;}} C_{12}H_{14}O_4$$

Piperic Acid   **Tetrahydropiperic Acid**

It shows that **piperic acid** contains **two double bonds.**

(*iii*) **Presence of Two Double Bonds:** It may be duly established alternatively by carrying out the **bromination** in *carbon tetrachloride* (Cl$_4$) to obtain **tetrabromopiperic acid,** as shown below:

$$C_{12}H_{10}O_4 + 2Br_2 \xrightarrow{\text{CCl}_4} C_{12}H_{10}O_4Br_4$$

Piperic Acid   **Tetrabromopiperic Acid**

(*iv*) **Presence of One Carboxyl) COOH) Moiety:** The presence of the **carboxyl moiety** in **piperic acid** may be ascertained by the interaction with **sodium bicarbonate (NaHCO$_3$)** to produce **sodium salt of piperic acid** and a mole each of **water** and **CO$_2$.** Thus, it duly accounts for *two* of the *four* O-atoms present in **piperic acid,** as expressed under:

$$C_{11}H_9O_2(COOH) \xrightarrow[\text{Sodium bicarboate}]{\text{NaHCO}_3} C_{11}H_9O_2(COONa) + H_2O + CO_2\uparrow$$

(*v*) **Oxidation Reaction:** **Piperic acid** on being oxidized with **alkaline KMnO$_4$** (potassium permanganate) gives rise to *two* products, namely:

- **piperonal** [*ie.,* **3,4–(methylendioxybenzaldehyde)**], and
- **piperonylic acid** [*i.e,.* **1,3–Benzodixole–5–carboxylic acid**],

together with small quantum of *oxalic acid* and *tartaric acid,* as given under:

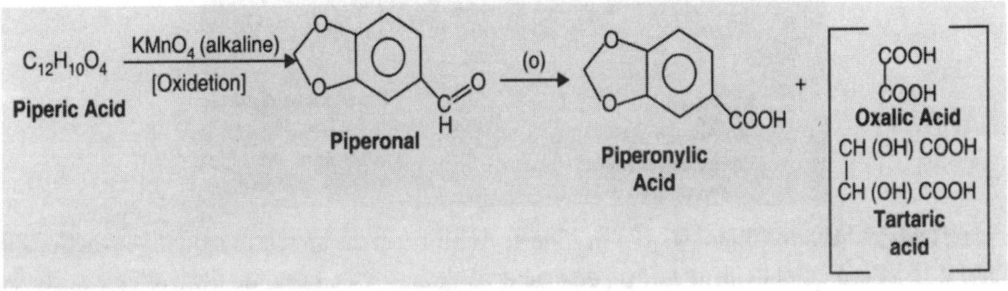

It is, however, worthwhile to mention here that the structure of **piperonylic acid** may be duly deduced from the fact that on being subjected to **heating with HCl at ~ 200°C** *under positive pressure* yields **formaldehyde** and **protocatechuic acid*** as shown below:

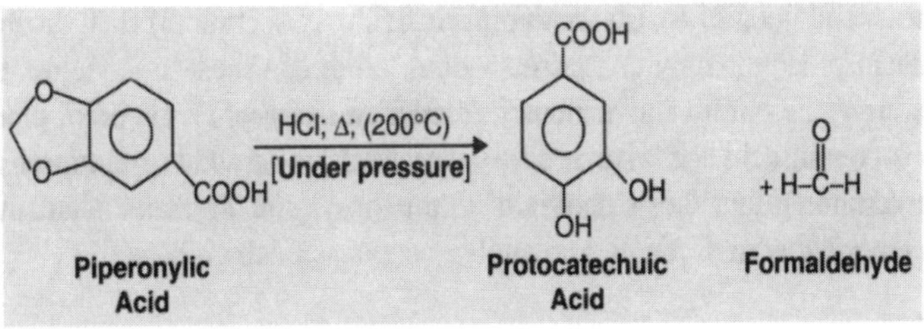

---

* That is, 3,4–dihydroxy benzoic acid.

Conclusions: These essentially comprise *two* vital aspects, namely:

(a) **Presence of Methylenedioxy [–CH$_2$–O–] Moiety in Piperonylic Acid:**

Since **piperonylic acid** is completely *devoid* of any **hydroxyl (–OH) moiety**, it suggests obviously that the actual **genesis** (*origin*) of the **two hydroxyl (–OH) moieties** in protocatechuic acid plus the formation of **formaldehyde (HCHO)** as a *by–product* may only be expatiated both logically and predictively, in case, one assumes that the former possesses a **methylenedioxy moiety (–O–CH$_2$–O).**

Furthermore, the above analogy can be substantiated by the formation of **piperonylic acid** by treating **3,4–dihydroxy benzoic acid** with **diiodomethylene [CH$_2$I$_2$]** in the presence of an alkali and heating as expressed under:

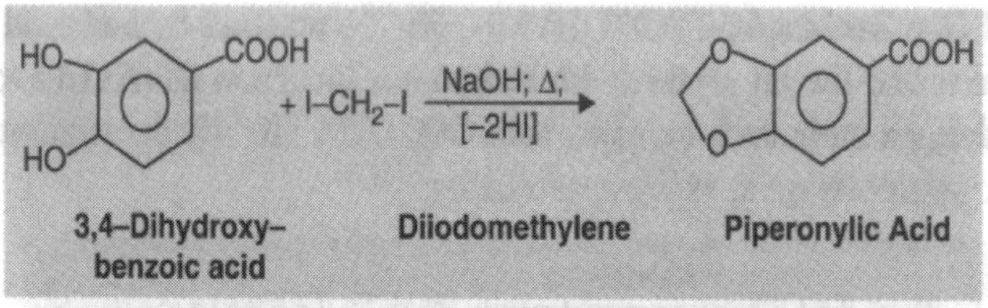

3,4–Dihydroxy–   Diiodomethylene   Piperonylic Acid
benzoic acid

(b) **Affirmation of Piperonal (an *'aldehyde'*) as 3,4–Methylenedioxybenzaldeyde:** It can be adequately explained based on the fact that **piperonal** on being subjected to *mild oxidation* produces **piperonylic** acid thereby ascertaining the former to be **3,4–methylenedioxybenzaldehyde,** as given under:

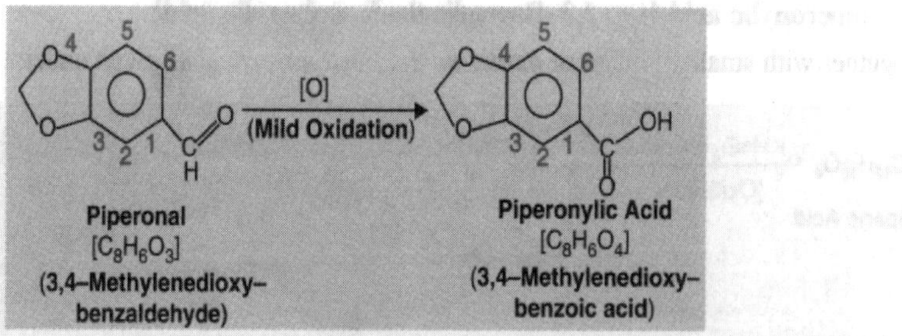

Piperonal   Piperonylic Acid
[C$_8$H$_6$O$_3$]   [C$_8$H$_6$O$_4$]
(3,4–Methylenedioxy–   (3,4–Methylenedioxy–
benzaldehyde)   benzoic acid)

(*vi*) **'Straight Side-Chain' (C$_4$H$_4$) in Piperic Acid:** From the aforesaid **oxidative reactions** in (iv) and (v), it is indeed quite evident that **piperic acid** designates a **benzene derivative** essentially that **piperie acid** designates a **benzene derivative** essentially contains:

• **methylenedioxy moiety,** and
• **one side-chain (with an acidic moiety) exclusively.***

Besides, instant addition of **Br-atoms** to **piperic acid** [see section (3) (iii)] distinctly reveals that its inherent side-chain prdeominantly comprises **2 double-bonds** (because it yields **'tetrabromopiperic acidic).** Interestingly, the careful and meticulous *oxidation* [see section (3) (v) of **piperic acid** gives rise to the formation of **oxalic acid** and **tartaric acid** (as *by products*) in addition to *piperonal* and *piperonylic acid* thereby ascertaining that the **'side-chain'** in *piperonylic acid*. This eventually suggests that the **'side-chain in *piperic acid* is C$_4$H$_4$** and invariably occurs as a **'straight-chain'.**

---

* Perhaps this is caused due to the development of only one aldehyde (–CHO) or **carboxyl (–COOH)** moiety in the course of **oxidation rection.**

Thus, we may have the following reactions:

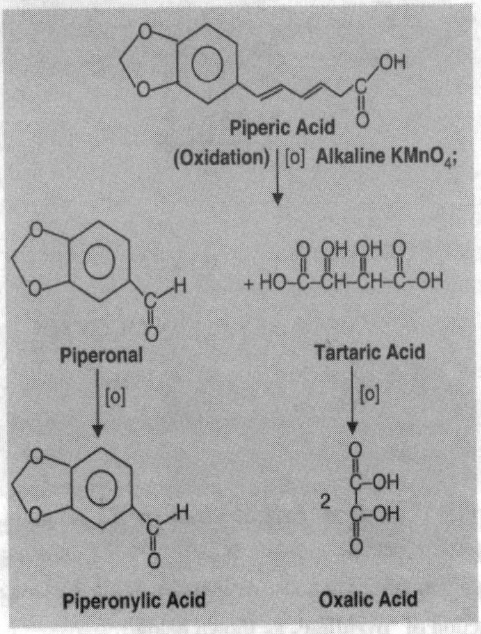

Piperic Acid

(Oxidation) | [o] Alkaline KMnO₄;

Piperonal   Tartaric Acid

Piperonylic Acid   Oxalic Acid

(vii) **Synthesis of Piperic Acid:** Ladenburg and Sholtz (1894)*, ultimately confirmed the structure of **piperic acid** [*E, Z–form*] by its *total synthesis.* In fact, they used the **Reimer–Tiemann Reaction*** *i.e.,* formation of **phenolic aldehydes** from *phenols, chloroform* and *alkali.*

Thus, **piperic acid,** is synthesized duly from **catechol** by the help of *three* sequential well-known *organic reactions,* namely:

- **Reimer-Tiemann Reaction (1876),**
- **Claisen–Schmidt Reaction (1881)**,** and
- **Perkin Reaction (1868)**.**

Explanation: The above course of reactions commencing from **catechol** to **piperic acid** may be explained as under:

(1) Interaction between **catechol (I)** and **chloroform** in the presence of NaOH undergoes **Reimer-Tiemann reaction [A]** to yield **3, 4–dihydroxy benzaldehyde (II).**

(2) The resulting **product (II)** upon treatment with **diiodomethylene** and **NaOH** produces **piperonylic acid (III)** due to the conversion of **aldehydic group** to the **carboxylic group.**

(3) The **product (III)** upon treatment with *acetaldehyde/NaOH* undergoes **Claisen-Schmidt condensation reaction [B]** to yield as condensed **product (IV)** known as **piperonyl acrolein** (with the loss of a mole each of H₂O and O₂)

*   Ladenburg A and Sholtz M : *Ber,* **27** : 2598, 1894.

**   Reimer T and Tiemann F : *Ber.,* **9** : 824, 1876.

***   Claisen A *et al.* : *Ber,* **14** : 2460, 1881; Schmidt JG : *Ber,* **14** : 1459, 1881

**** Perkin WH : *J Chem Soc,* **21** : 53, 1868; **31** : 388, 1877.

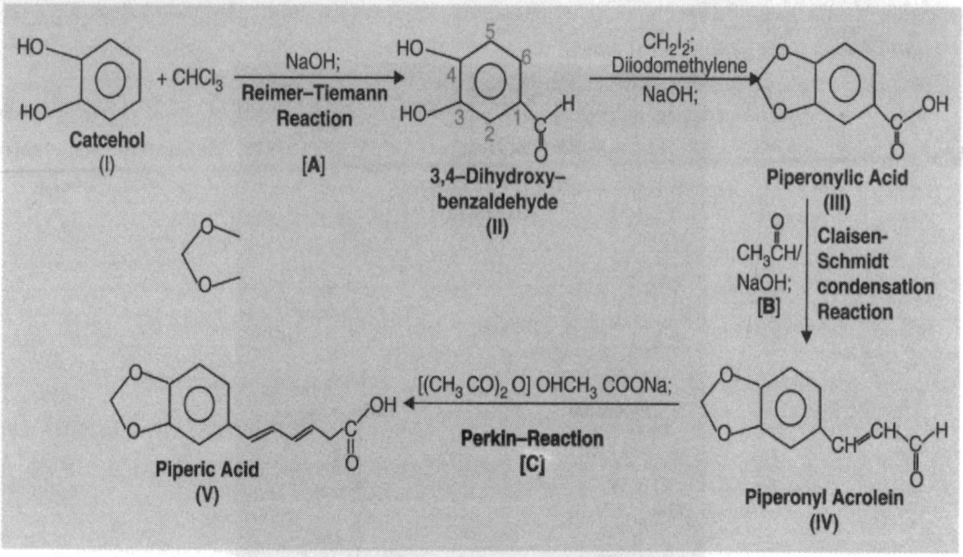

(4) Finally, **product (IV)** undergoes **Perkin-Reaction [C]** in the presence of a mixture of **acetic anhydride** and **sodium acetate** to give rise to the desired product **'piperic acid' (V)**.

(*viii*) **Synthesis of Piperdine:** Obviously, **piperidine** being a saturated **hexahydro pyridine** duly accomplished by the reduction of **'pyridine'**, as shown below:

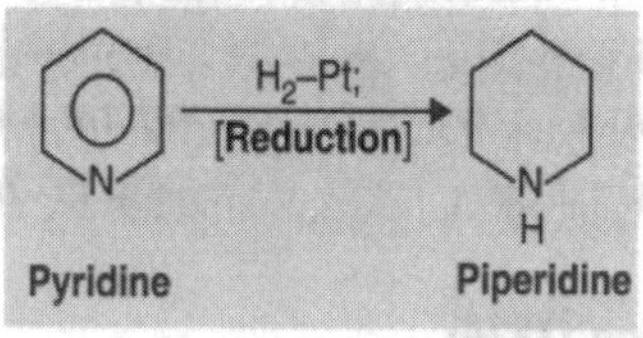

However, the actual synthesis of **piperidine** has been duly confirmed by heating carefully **pentamethylene diamine hydrochloride** with the loss of a mole of **ammonium chloride [NH$_4$Cl]**, as given under:

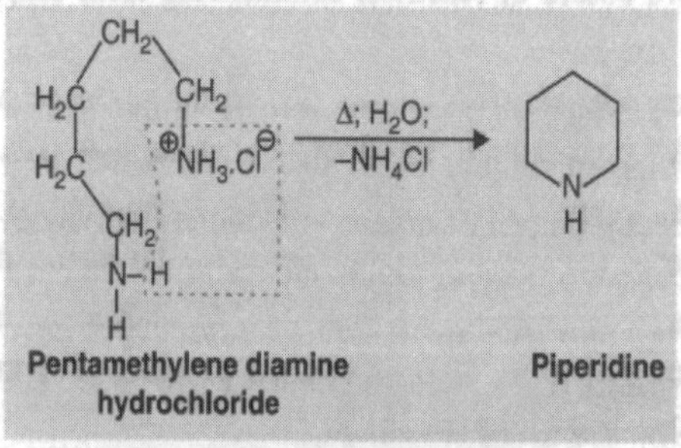

Note: Pentamethylene diamine being insoluble in water is first dissolved in dilute HCl in stoichiometric proportion so as to convert one of the *two* free amino groups into the respective *quaternary ammonium chloride* (as shown above), which upon heating loses a mole of NH$_4$Cl (indicated by a dotted line) to yield *'piperidine,*.

(*ix*) **Structure of Piperine:** An intelligent derivation based on the aforesaid evidential facts and statements it may be concluded almost certainly that 'piperine' should be the *'condensation product'* duly obtained from **'piperic acid'** and **'piperidine',** which eventually explains its formation. Indeed, an additional evidence can also be mustered by carrying out the synthesis by *heating* **acid chloride of piperic acid*** with **piperidine** in a *non-polar organic solvent* **benzene****. Importantly, this synthesis categorically indicates that the *alkaloid* **'piperine'** is nothing but a **'piperidine amide of piperic acid'** as expatiated under:

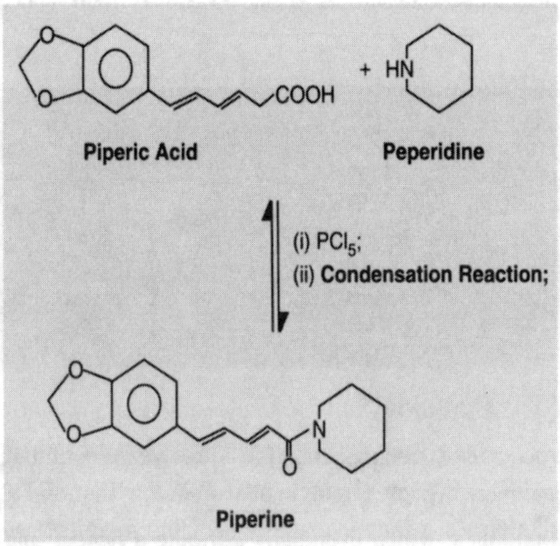

Piperic Acid   **Peperidine**

(i) PCl₅;
(ii) **Condensation Reaction;**

Piperine

(*x*) **Synthesis of Piperine:** The aforesaid structure of **'piperine'** in (*ix*) may be critically proved by synthesiszing it from:

- **acid chloride of 'piperic acid'*,** and
- **condensation with piperidine.**

(*xi*) **Interrelationship with 'charieine'** It is, however, pertinent to state here that the above structure for **'piperine'** and **'chavicine'** (*i.e.,* the **two isomeric alkaloids**) differ only in the **stereochemical arrangement** of the *function moieties.* Evidently, the *double bonds* strategically located in the **'specific region of the acidic group'** duly formed in the course of the **'hydrolysis of alkaloids'** in an *alkaline environment,* for instance:

- **Piperic Acid** from **Piperine,** and
- **Cavicinic Acid** from **Chavicine.**

Piperic Acid = (E,E)–Form ; *Chavicinic Acid* = (Z, Z)-form;

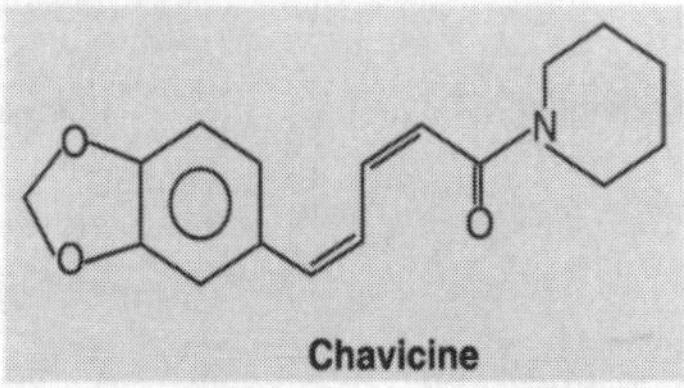

Chavicine

---

* It is prepared by reacting **piperic acid** with **phosphorus perntachloride (PCl₅).**

** **Benzene :** It is **carcinogenic** in nature; and hence, must be used with **'CAUTION'.**

[One of the most *active constituents* of 'black pepper'].

It has been observed that *'piperic acid'* duly exhibits:

❑ *trans-trans:* arrangement of functional groups; whereas, *'chavicinic acid'* (from **chavicine**) duly exhibits:

❑ *cis-cis:* arrangement of functional groups, and

❑ *cis-trans:* arrangement of functional groups.

All these aforesaid of *geometrical isomeric forms* of **'piperic acid'** and **'chavicinic acid'** are shown as under:

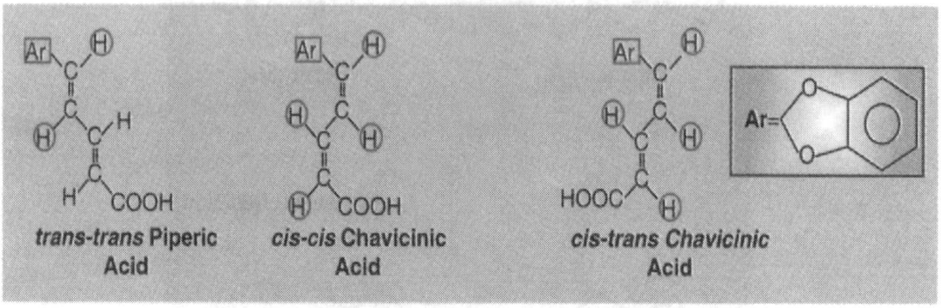

| *trans-trans* Piperic Acid | *cis-cis* Chavicinic Acid | *cis-trans* Chavicinic Acid |

3.2.4. Pelletierine [*Syn: Punicine*]

Preamble: *Pelletierine* is the alkaloid obtained from the *rootbark* of the **pomegranate tree**, *Punica granatum* L., *Punicaceae.* The **French phytochemist-***Pelletier* first and foremost discovered the so called **'pomegranate alkaloids's**, which eventually attribute a *typical and characteristic vermicide properties* to the **pomegranate bark** sine a long duration. In fact, there are *four* known **pomegranate alkaloids**, namely:

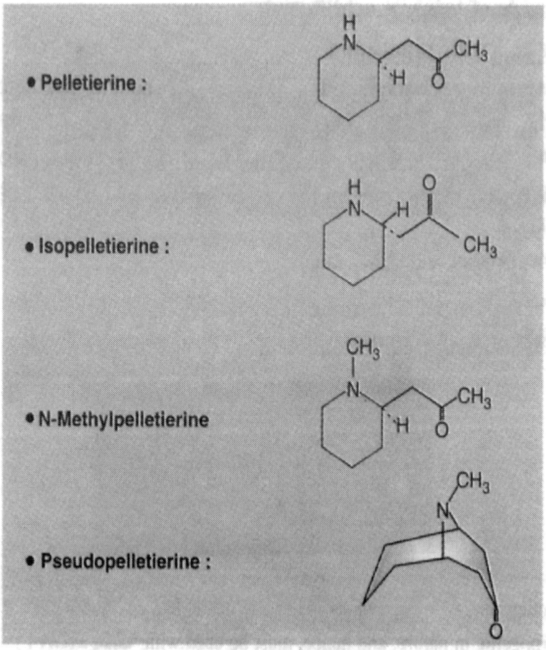

● Pelletierine :

● Isopelletierine :

● N-Methylpelletierine

● Pseudopelletierine :

Interestingly, out of these *four* **pomegranate alkaloids,** the very first one, **pelletierine,** is found to be the most important.

3.2.4.1. Characteristic Features

These essentially comprise:

(1) It is a colourless oily liquid.

(2) It has bp_{21} 106°C.

(3) It usually gets darkened on exposure to air.

(4) It invariably occurs in the nature as the racemic mixture *i.e.* (±)-form; and, therefore, shows **no optical activity.**

> **Notes: (1)** The racemic form of *pelletierine* may be resolved duly into (+) and (–) form.
> **(2)** In the early literature the *two* names: pelletierine, and isopelletierine were used *interchangeably.*

(5) **Solubility Profile:** *Pelletierine* is freely soluble in **water, ethanol, solvent ether,** and **chloroform.**

3.2.4.2. Constitution of Pelletierine

Pelletierine is duly subjected to an array of critical determinations in order to assign it a most *befitting and probable structure*, and finally its **'total synthesis'** so as to confirm the structure.

(1) **Molecular Formula:** It has the molecular formula $C_8H_{15}NO$.

(2) **Presence of a Secondary Amino Function (>NH): Pelletirine** is found to react with *two* **moles of methyl iodide** to yield a *quaternary ammonium salt* that eventually suggests it to have a **secondary amino functions moiety,** as shown under:

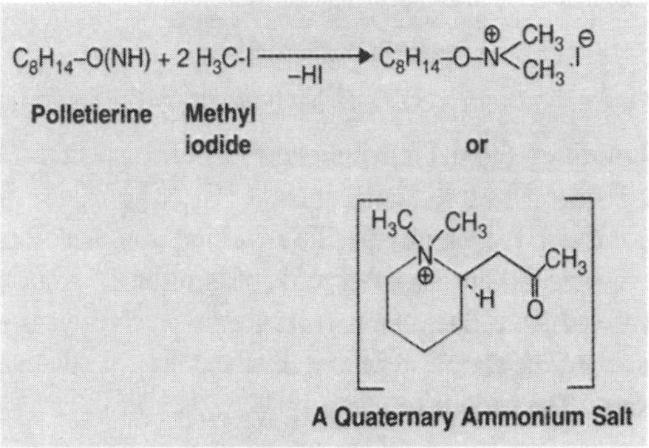

A Quaternary Ammonium Salt

Thus, one mole of **hydroiodic acid [HI]** gets eliminated by abstracting the *N-hydrogen atom* the **piperidine nucleus.**

(3) **Wolf-Kishner Reduction (1911, 1912)*:** It completely causes reduction of the *carbonyl group* in **Pelletierine** to *methyl/methylene* groups upon heating with **hydrazine hydrate [$H_2N\ NH_2$]** and a **base (KOH),** as stated under:

---

* Wolff L : *Ann.*, **394** : 86, 1912; Kishner N : *J Russ Phys Chem Soc*, **43** : 582, 1911.

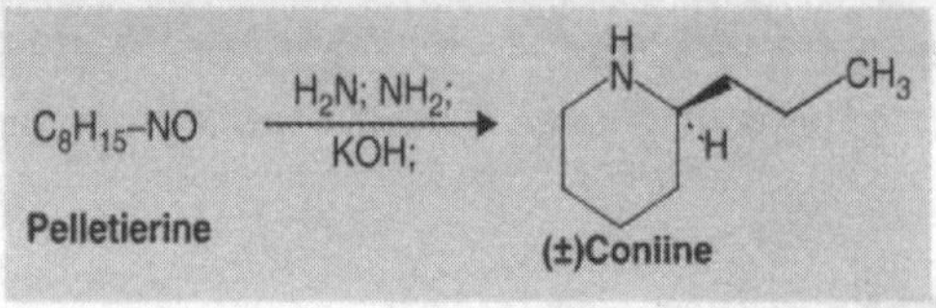

C_8H_{15}–NO $\xrightarrow[\text{KOH;}]{\text{H}_2\text{N; NH}_2;}$

Pelletierine

(±)Coniine

Conclusions: (*i*) The aforesaid reaction suggests that **pelletierine** is structurally related to **coniine** (see *section 2.6.4.*).

(*ii*Pelletierine has essentially a **'piperidine nucleus'**

(*iii*) It has only one **straight side-chain** consisting of **3–C–atoms.**

(4) **Modern Conceptualized Views on Pelletierine:** It has been amply proved and shown that **pelletierine** never existed as such in nature. Perhaps it is the (*–*)–*isopelletierine* that was wrongly termed as **'pelletierine'.** The above statement of facts have been adequately substantiated by the following glaring evidential supports and explanations.

(a) **Similarity in Physical Constants:** There exist an apparent close similarity in **pelletierine** (plus its *derivatives) vis-a-vis* **isopelletierine** (plus its *derivatives*)

Examples: There is a typical classical example whereby the chemical entity **1–benzoylpelletierine** duly:

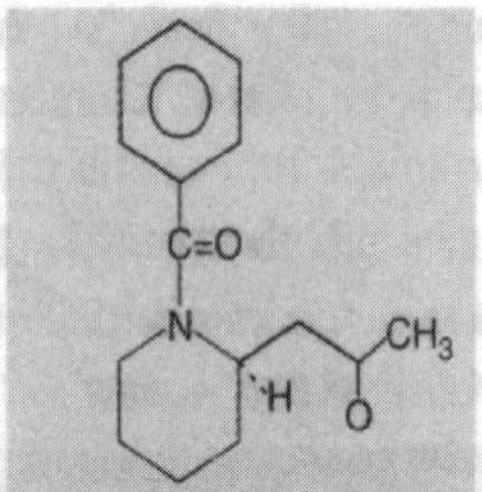

1–Benzoyl pelletierine

- obtained from either **pure natural pelletierine,** or
- obtained by **pure synthetic route,** do have precisely the **same mp.**

(b) **Formation of Amidine* from Isopelletierine:** The IR–spectrum of the *hypothetical* (or *presumed*) **nitrile** [$C_{17}H_{14}N$. C = N; Molecular formula: $C_8H_{14}N_2$] obtained due to the *dehydration of pelletierine oxime,* which failed to exhibit the specific ***IR-absorption peak*** for **nitrile (–CN)** moiety at ~1685–1640, but instead showed **IR–absorption profile** of **'amidine'.** Interestingly, **amidine** was duly formed by means of the well-known **Beckmann rearrangement**** causing effective *dehydration* to yield an **isopelletierine oxime.** The various reactions involved may be summarized as under:

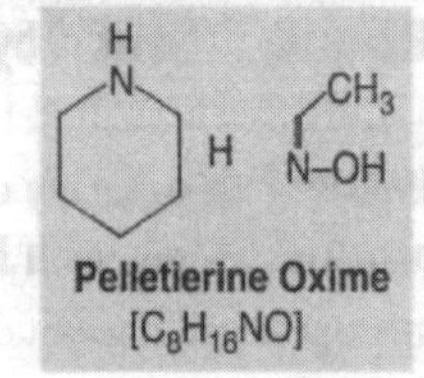

Pelletierine Oxime
[$C_8H_{16}NO$]

---

* It is an **isopelletierine oxime.**

** Beckmann E : *Ber,* **19** : 98, 1986.

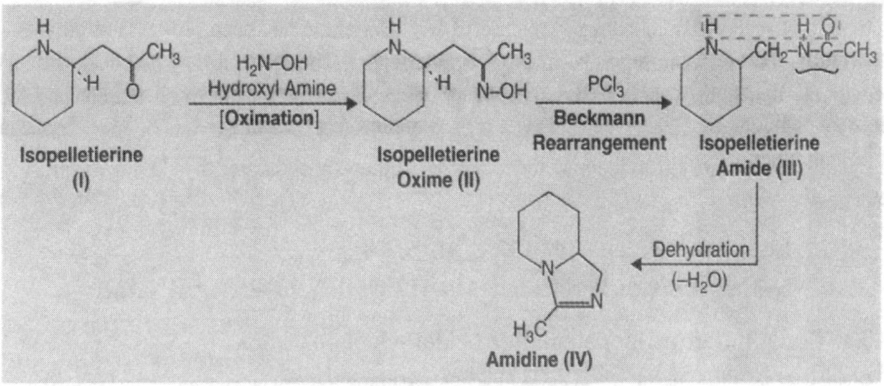

Isopelletierine (I) → [Oximation] with H₂N–OH Hydroxyl Amine → Isopelletierine Oxime (II) → Beckmann Rearrangement with PCl₃ → Isopelletierine Amide (III) → Dehydration (–H₂O) → Amidine (IV)

Explanation: Formation of **amidine (IV)** from **isopelletierine (I)** may be explained explicitly as under:

(*i*) Oximation of **isopelletierine (I)** with **hydroxyamine** yields the respective **isopelletierine oxime (II)**, which upon **Beckmann rearrangment** (in the presence of **PCL₅**) into an **acid-mediated isomerization** of the *oximes* to *amides i.e.,* **isopelletierine amide (III).**

(*ii*) The resulting **compound (III)** undergoes dehydration with a loss of one mole of water (H_2O) as indicated by dotted line to form a cyclized compound known as **amidine (IV)**

(*c*) **Similarity between Pelletierine and Isopelletierine:** Galinorsky *et al.* (1954) put forward an exemplary evidential proof based on the meticulous **partition chromatographic studies** that both **pelletierine** and **isopelletierine** (characterized as its corresponding **picrate salt**) are virtually the **similar compounds.**

(*d*) **Optical Activity Features:** Bayermann *et al.* (1965) examined the **optical activity features** of **pelletierine** and **isopelletierine** and observed duly that the former happens to be the **(–)–form of the latter** only. Besides, the latter seems to undergo a *shift* and *rapid racemization** via the **open = chain structure analogue,** as given under:

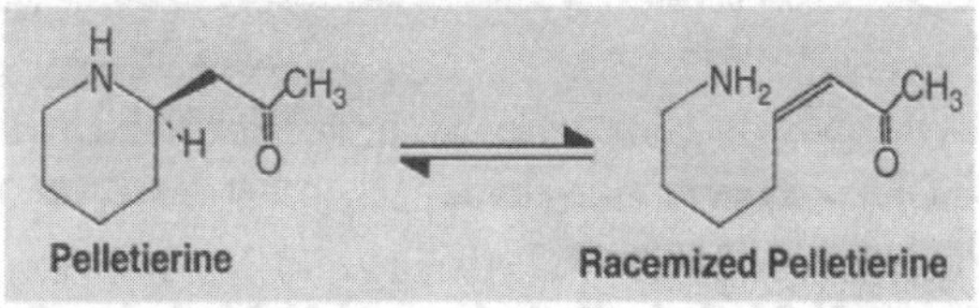

Pelletierine ⇌ Racemized Pelletierine

(*e*) **Absolute Configuration:** The absolute configuration of **(–)–pelletierine** picrate, a *fairly stable adduct* (1:1), revealed explicitly that it was a **'D–isomer'**, which is just the *opposite configuration vis-a-vis* the **amino acids.**

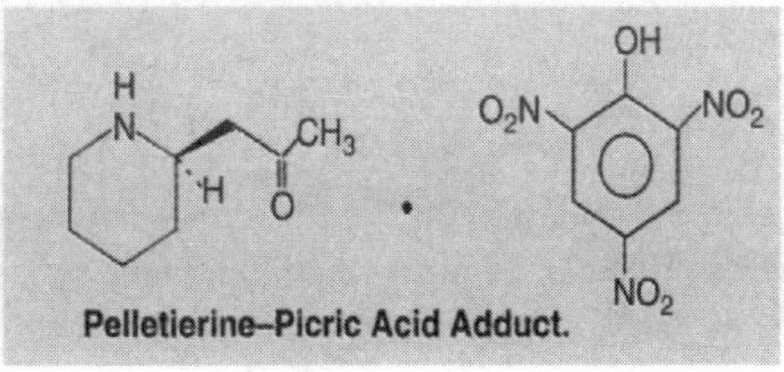

Pelletierine–Picric Acid Adduct.

---

* **Racemization:** It refers to the **'transformation'** of 1/2 of the molecules of an optically active compound into molecules that are mirror image configuration of each other the resultant **optical rotation** becomes **'zero'**.

3.2.5. Isopelletierine and N–Methylisopelletierine

Importantly, the basic chemical structure of **isopelletierine** has been proved and duly established *via* its **close structural analogue N–methylisopelletierine.** Therefore, it may be concluded safely and affirmatively that both **isopelletirine** and **N–methylisopelletirine*** seem to be closely and intimately related to each other.

Besides, the aforesaid evidential fact may be further substantiated by their very close physical properties:

Isopelletierine: bp$_{11}$ 102-107°C; D$_4^{20}$ 0.988;

N–Methyl isopelletierine: bp$_{26}$ 114–117°C; D$^{20}_4$ 0.948; N$_D{}^{20}$ 1.46737.

3.2.5.1. Constitution of Isopelletierine/N-Methylpelletierine:

In order to establish the chemical structures of **isopelletierine** and **N–methylpelletierine** the following vital and important aspects need to be considered, namely:

(1) **Molecular formula:** The **molecular formula** of these *two* alkaloids have been determined precisely as:

Isopelletierine: C$_8$H$_{15}$NO;

N–Methyli isopelletierine: C$_9$ H$_{17}$ NO;

(2) **Keto (–$\overset{\text{O}}{\underset{}{\text{C}}}$–) Functional Moiety:** It has been duly observed that **N–methylisopelletierine** gives rise to the formation of a **hydrazone derivative** when the former is treated with **hydrazine (H$_2$N–NH$_2$)**– as given under:

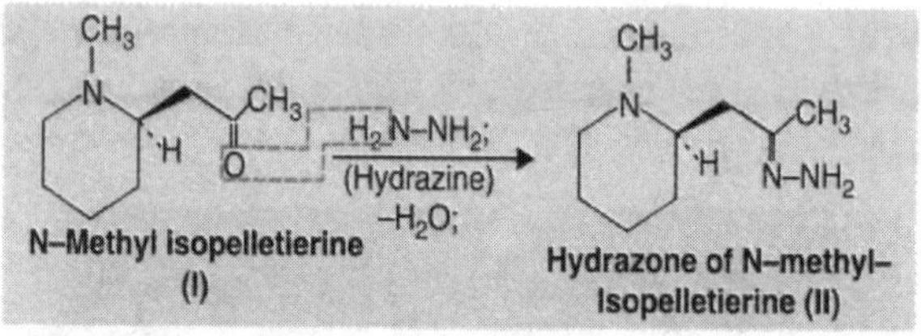

N–Methyl isopelletierine
(I)

Hydrazone of N–methyl–
Isopelletierine (II)

Inference: Based on the above reaction it may be inferred that **N–methylisopelletierine (I)** contains essentially a keto (>C = O) moiety in its **side-chain.**

(3) **Relationship between N–Methylpelletierine and N–Methylconiine:** The **hydrozone** of N–methylisopelletierine **(A)** on being subjected to **reduction** with *Na/Et OH* yields the respective N–methyl-coniine **(B)** as given under:

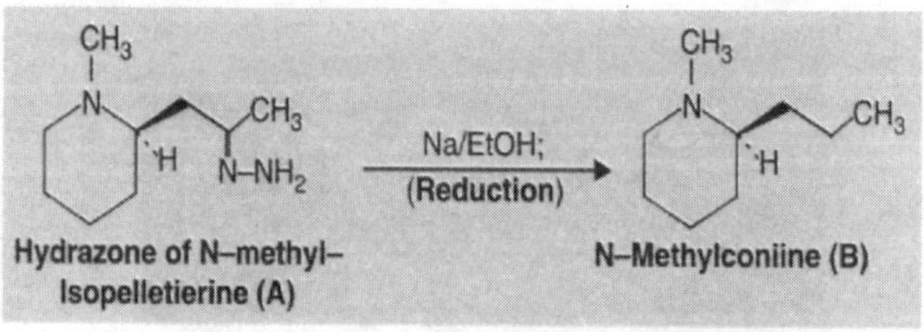

Hydrazone of N–methyl–
Isopelletierine (A)

N–Methylconiine (B)

_____

* **N-Methylisopelletierine:** It is prepared by treating **pelletierine** with **dimethyl sulphate [(CH$_3$)$_2$ SO$_4$]** to afford methylation.

Inference: From the above reaction it may be inferred that both compounds (**A**) and (**B**) are related to each other **structurally.**

(4) **Oxidation Reaction:** N–Methylisopelletierine on being oxidized with **chromium–6–oxide [CrO₃)** gives rise to the formation of **N-methyl piperidine–2–carboxylic acid** [or **N–methylpipecolic acid],** as given under:

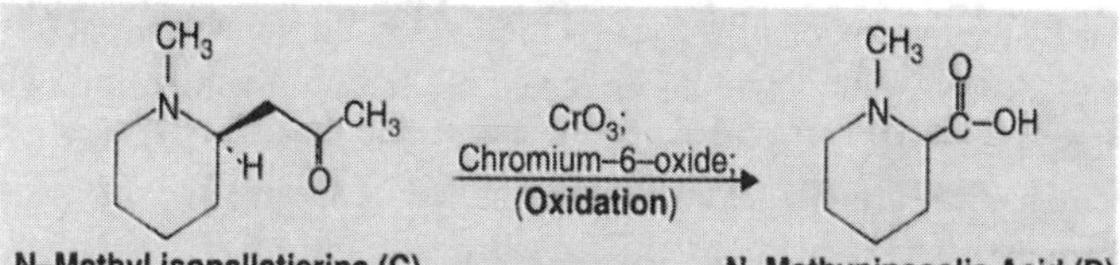

N–Methyl isopelletierine (C)   N–Methypipecolic Acid (D)

Inference: Based on the aforesaid reaction it may be concluded that the structure of **[C]** to be **1–methylpiperidine** having a specific **side-chain at C-2.**

(5) **Exact location of *'keto (>C=O) Moiety'* in the side-chain:** At this critical point in time it is, however, absolutely necessary and important to locate exactly the position of the **carbonyl [>C=O] function** in the *side-chain* of **N-methyl isopelletierine.** Thus. One may have the following *two* options, namely:

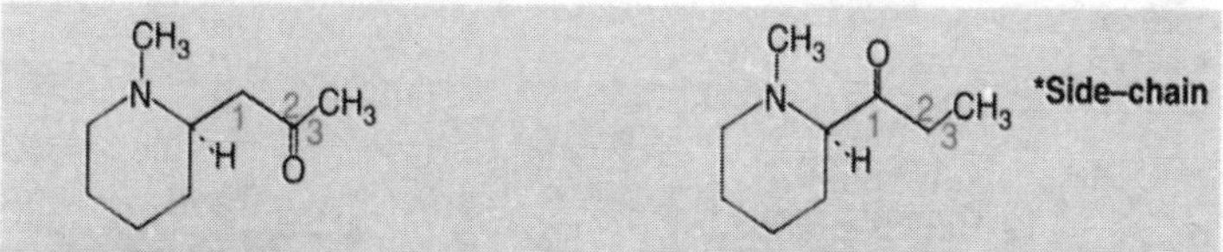

'*Keto*'–Group at C–2*   '*Keto*'–Group at C–1*
[X]   [Y]

(6) **Synthesis:** The total synthesis of **N-methylisopelletierine** was accomplished in *two* different ways, namely:

- Meisenheimer *et al.* (1928): from **1–(2–pyridyl)–propane–2–ol;** and
- Wilbaut (1944): from **2–methyl lithium pyridine.**

6.1. N–Methylisopelletierine From 1–(2–pyridyl) propane–2–ol:

Meisehheimer *et al* first of all proposed the synthesis of **N-methyl isopelletierine (I)** by carrying out the **reduction (H₂/Pt)** of **1–(2–pyridyl) propane–2–ol** to obtain **piperidyl–2–hydroxy propane.** The resulting product on being subjected to treatment at *two* different stages, namely:

- **oxidation** with *chromium–6–oxide [CrO₃)and acetic acid [CH₃ COOH),* and
- **methylation** with *dimethyl sulphate [(CH₃)₂SO₄],*

yields the desired **product (I)** as shown below:

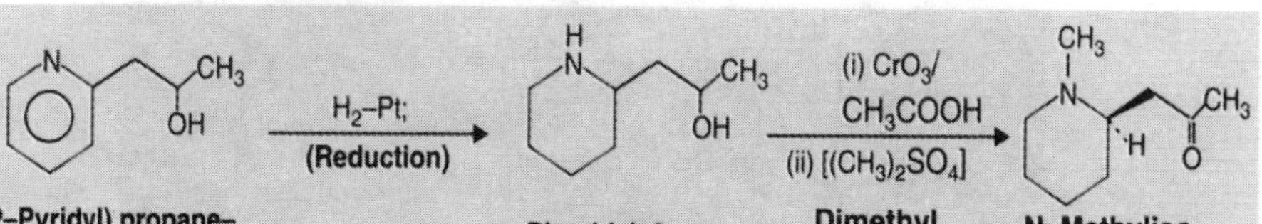

1–(2–Pyridyl) propane–   Piperidyl–2–   Dimethyl   N–Methyliso-
2–ol   hydroxy propane   sulphate   pelletierine (I)

Note: The end-product (I) *i.e.*, *N–methylisopelletieretine* on being subjected to treatment with a *base* gives rise to the formation of *isopelletieretine*.

6.2.N–Methylisopelletierine from 2–Pyridyl methyl lithium *via Isopelletierine*:

Wilbaut (1944) synthesized the aforesaid *two* **alkaloids** *viz.*, **N–methylisopelletierine** and **isopelletierine** by the interaction of **2–pyridyl methyl lithium (I)** with acetic anhydride **[(CH₃O)₂O]** in the presence of **Raney-Ni** to obtain an **intermediate addition compound**, which eventually loses a mole of **lithium acetate (CH₃COOLi)** to yield **pyridyl–2–acetone (II)**. The resulting **product (II)** on reduction with **H₂–Pt (catalytic reduction)** forms one of the desired alkaloids, **isopelletierine (III)**, which subsequently upon **methylation with dimethyl sulphate [(CH₃)₂ SO₄)]** gives rise to the formation of **N-methylisopelletierine (IV)**.

The various steps of reactions are as stated under:

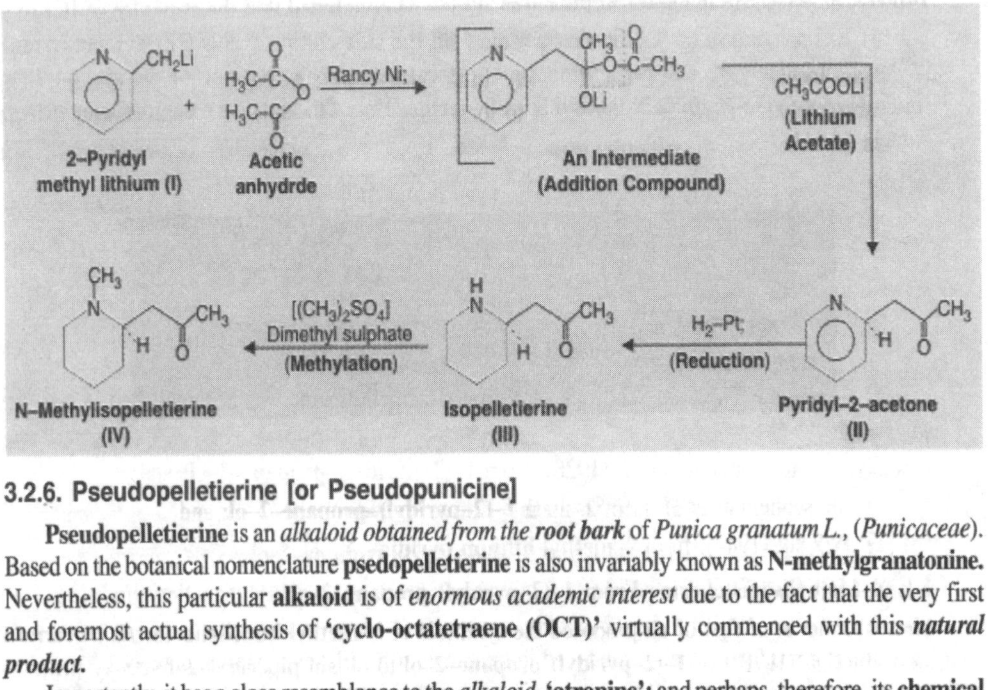

3.2.6. Pseudopelletierine [or Pseudopunicine]

Pseudopelletierine is an *alkaloid obtained from the **root bark** of Punica granatum L., (Punicaceae)*. Based on the botanical nomenclature **psedopelletierine** is also invariably known as **N-methylgranatonine**. Nevertheless, this particular **alkaloid** is of *enormous academic interest* due to the fact that the very first and foremost actual synthesis of '**cyclo-octatetraene (OCT)**' virtually commenced with this *natural product*.

Importantly, it has a close resemblance to the *alkaloid*, '**atropine**'; and perhaps, therefore, its **chemical structure** was meticulously elucidated quite on the same lines as that of '**atropine**', such as:

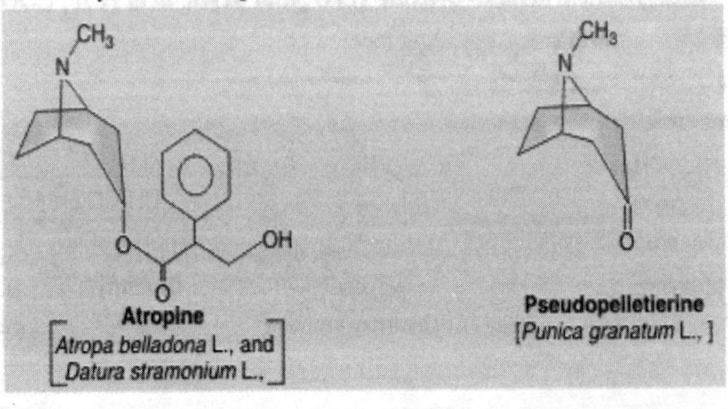

- **quaternization with CH₃I,**
- **oxidation,**
- **Hofmann's exhaustive methylation,** and
- **catalytic reduction** (to yield 'suberic acid' [HOOC (CH₂) COOH]

3.2.6.1. Constitution of Pseudopelletierine

The various vital and important points that must be taken into consideration with regard to the elucidation of the structure of **pseudopelletierine** are as stated under:

(1) **Molecular Formula:** It is found to be $C_9H_{15}NO$.

(2) **Nature of N-Atom:** The exact and precise **nature of N-atom** may be ascertained by the fact that **pseudopelletierine (I):** up one mole of **methyl iodide [H₃C–I]** to give rise to the production of **N-methyl quaternary ammonium pseudopelletierine iodide (II).** Containing (=N–CH₃) moiety.

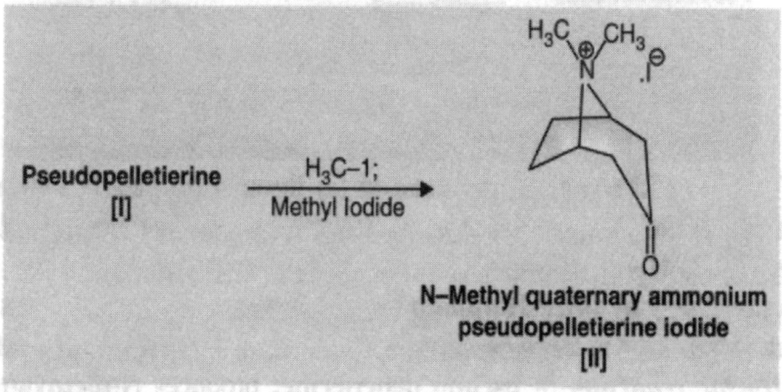

Pseudopelletierine [I] → H₃C–1; Methyl Iodide → N–Methyl quaternary ammonium pseudopelletierine iodide [II]

(3) **Presence of Keto (–C–) Moiety:** The *alkaloid* **pseudopelletierine (I)** readily forms an 'oxime' with *hydroxylamine,* and also yields a **secondary alcohol [–CH(OH)–]** upon reduction with H₂–Pt, as given below:

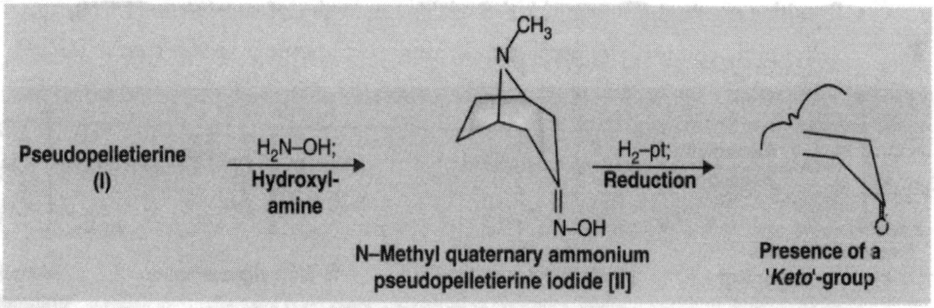

Pseudopelletierine (I) → H₂N–OH; Hydroxyl-amine → N–Methyl quaternary ammonium pseudopelletierine iodide [II] → H₂–pt; Reduction → Presence of a 'Keto'-group

(4) **Location of Keto [>C=O] moiety:** It has been duly observed that **pseudopelletierine** interacts with *two molecules of benzaldehyde* to yield an altogether *newer chemical entity* known as 'dibenzylidine derivative', as shown below:

(a)

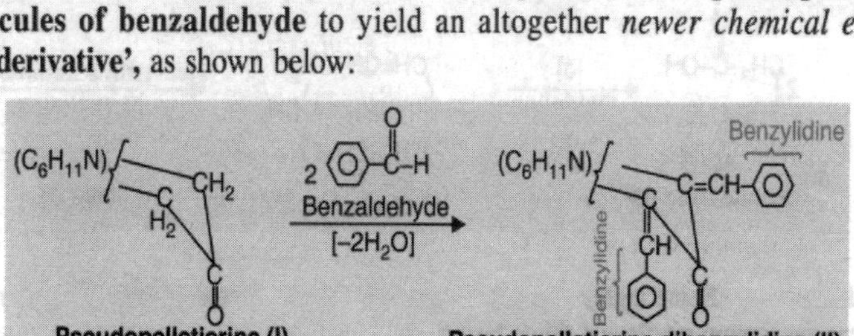

Pseudopelletierine (I) + 2 Benzaldehyde [–2H₂O] → Pseudopelletierine dibenzylidine (II)

Likewise, the **alkaloid (I)** on treatment with **nitrous acid [HONO]*** at **0–5°C** to yield **di-isonitroso**

(II) **derivative** as given under:

Pseudopelletierine (I) → HONO; 0–5°C → **Di-isonitroso pseudopelletierine** (II)

(b)

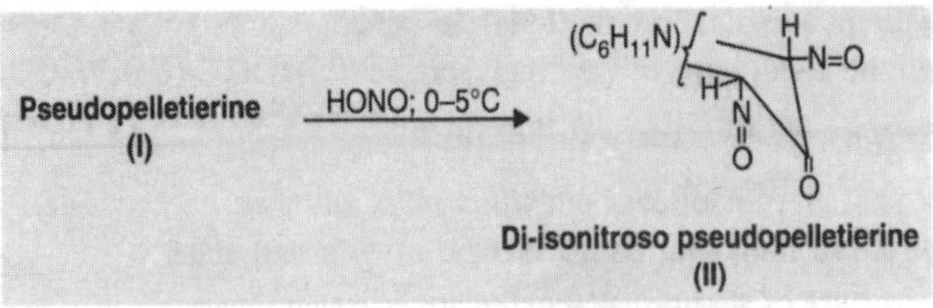

Similarly, interaction of **alkaloid (I)** with *two* moles of pure **piperidine** gives rise to the production of the corresponding **dipiperonylidine (III)** derivative as shown below:

Pseudopelletierine (I) → Piperidine (NH) → **Dipiperonylidine pseudopelletierine** (III)

(c)

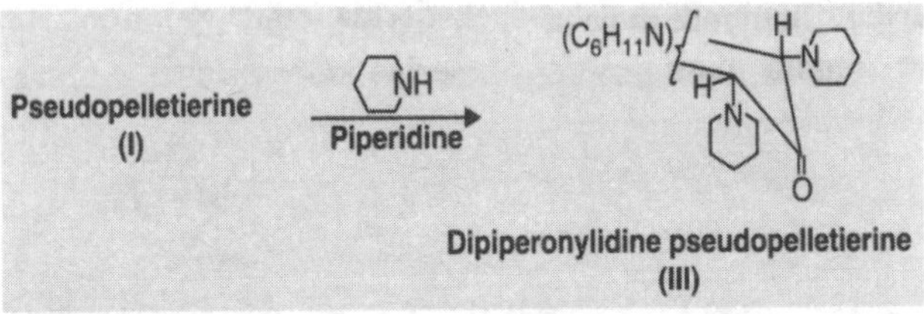

Inference: The aforesaid reactions described in (a) through (c) clearly indicate that the ketonic (>C=O) functional moiety duly present in **pseudopelletierine (I)** is definitely flanked by *two* **methylene** (–CH$_2$–) moieties *viz.,* –CH$_2$–CO–CH$_2$–.

(5) **Reduction followed by** *twice* **Hofmann Exhaustive Methylation Process:**

The various steps involved in these reaction are stated below in a chronological order:

- *Electrolytic reduction* of pseudo pelletierine: produces **N-methylgranatamine (I)** *via* **N-methylgranatoline** and **N–methylgranatenine** due to the reduction of **carbonyl (>C=O)** to **hydroxy methyl (–CH. OH) moiety,**
- **Product (I)** on being subjected *twice* to **Hofmann's exhaustive methylation** process gvies rise to the formation of **1, 5–cyclooctadiene (II)**–a *N–less product,* and
- Resulting **product (II)** on **oxidation** yields two moles of *succinic acid (III).*

Thus, we may have the following reactions:

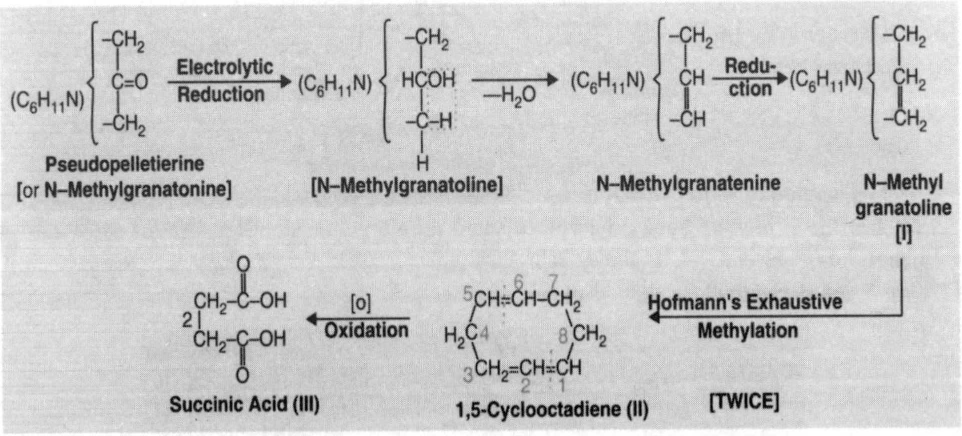

Pseudopelletierine [or N–Methylgranatonine] → Electrolytic Reduction → [N–Methylgranatoline] → –H$_2$O → N–Methylgranatenine → Reduction → N–Methyl granatoline [I]

Succinic Acid (III) ← [o] Oxidation ← 1,5-Cyclooctadiene (II) ← Hofmann's Exhaustive Methylation [TWICE]

---

* **Nitrous Acid [HONO] :** Being highly volatile and unstable **nhitrous acid [HONO]** is generated in the reaction mixture by reacting **sodium nitrite** and **dilute HCl** at 0–5°C.

(6) N-Atom Bonded to Cyclooctane Skeleton of Pseudopelletierine

Based on the remarkable, logical, and acceptable juggle of imaginative molecular structural design the **organic chemists** have put forward *three following* chemical structues for **pseudopelletierine**, namely:

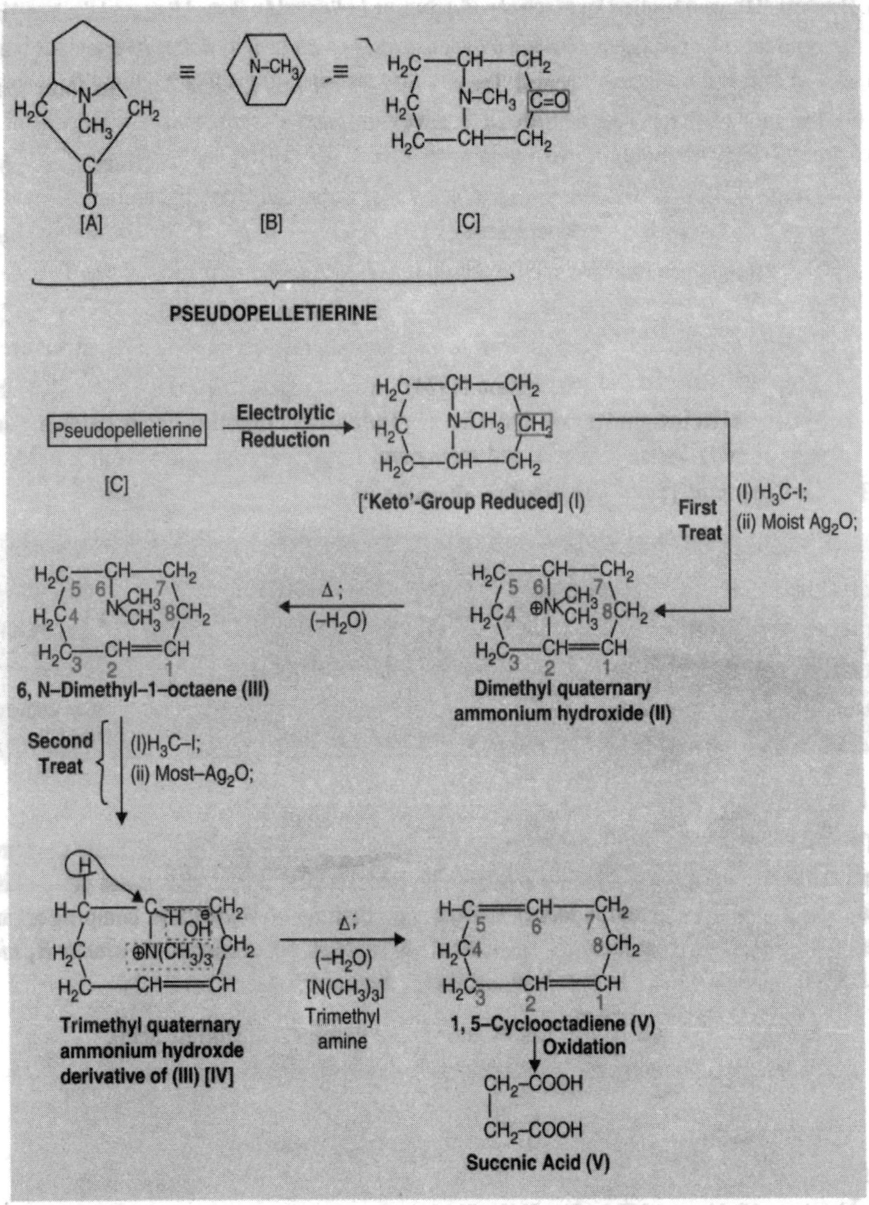

[A]   [B]   [C]

PSEUDOPELLETIERINE

Pseudopelletierine → (Electrolytic Reduction) → ['Keto'-Group Reduced] (I)

[C]

First Treat: (I) H$_3$C–I; (ii) Moist Ag$_2$O;

6, N–Dimethyl–1–octaene (III)   Dimethyl quaternary ammonium hydroxide (II)

Second Treat: (I) H$_3$C–I; (ii) Most–Ag$_2$O;

Trimethyl quaternary ammonium hydroxde derivative of (III) [IV]

Δ; (–H$_2$O) [N(CH$_3$)$_3$] Trimethyl amine

1, 5–Cyclooctadiene (V)
↓ Oxidation

CH$_2$–COOH
|
CH$_2$–COOH

Succnic Acid (V)

Explantions: The above course of reactions from **pseudopelletierine [C]** to **succinic acid** may be duly expatiated as under:

(1) **electrolytic reduction** of [C] reduces the **'Keto'–group** to **methylene –(CH₂–) group** to yield **product (I)**.

(2) **Product (I),** when subjected to the *'First Treat'* with *H₃ C-1 and moist Ag₂O* gives rise to the formation of **dimethyl quaternary ammonium hydroxide (II)**.

(3) **Product (II)** upon heating loses a mole of water, *as indicated by dotted line,* and the bond between **C-2 and N-atom** gets cleaved (shown by a *dotted line),* to yield **6, N–dimethyl–1–octane (III)**.

(4) **Product (III)** receives a **'Second Treat'** with *methyl iodide/most Ag₂O* to form **trimethyl quaternary ammonium hydroxide derivative of III (IV)**.

(5) **Product (IV)** upon **heating** loses a mole each of **water** and **trimethyl amine,** as shown by *dotted lines,* to produce **1, 5–cyclooctadiene (V)**.

(6) The resulting **product (V)** upon **oxidation** gives rise to the formation of **two moles of succinic acid.**

Inference: Based on the aforesaid evidential logical proof one may consider **structure [C]** as the **'proposed structure of pseudopelletierine'.**

(7) **Synthesis:** Both Menzies and Robinson (1924)*, and Schöpf (1937)** reported the synthesis of **pseudopelletierine** under the specific physiological condition from **calcium acetone dicarboxylate, glutardialdehyde,** and **methylamine.**

Besides, Cope *et al.* (1957) suggested an **alternate synthesis***** of **pseudopelletierine.**

Glutardialdehyde   Methyl   Calcium acetate   Pseudopelletie-
                   amine   dicarboxylate   rine [C]

3.3. Pyridine-Pyrrolidine Alkaloids

The survey of literature reveals that the **alkaloids** derived duly from the **nicotinic acid** are most commonly and collectively known as the **'Pyridine Alkaloids'******. In a broader perspective, the **alkaloids** that are usually found in *tobacco, Nicotiana tabacum* L., *Solanaceae,* comprise an array of vital and important plant constituents, namely: **nicotine, anabasine,** and **niacin** (*viz, vitamin B₃, nicotinic acid).* It is, however, pertinent to state here that the existence of:

- *pyridine nucleus* $\boxed{\langle O \rangle}$: has its genesis in **vitamin B₃ (nicotinic acid),**

- **pyridine-pyrrolidine** $\boxed{\langle O \rangle \langle N \rangle}$: gives rise to **'nicotine'.**

   combined nucleus

   * Menzies and Robinson : *J Chem Soc.,* 2163, 1924;

  ** Schöpf : *Angew Chem,* **50** : 799, 797, 1937;

 *** Cope *et al. : Org, Syn.,* 73m 1957.

**** Kar A : **Pharmacognosy and Pharmacobiotechnology,** New Age International, New Delhi, 2nd ed., 2007.

Nicotine* has proved to be **highly toxic alkaloid** belonging to this **'class of alkaloids'**; whereas, anabasine (or **neonicotine**), and **nornicotine** are solely responsible for attributing the typical characteristic **'aroma'** of the **tobacco smoke**. Following are the chemical structure of *three aforesaid* alkaloids:

Nicotine

Anabasine
[or Neonicotine]

Nornicotine

3.3.1. Isolation

The leaves and stems of the **tobacco plant** are dried in a **shade, powdered,** and **distilled** over **'milk of lime'**, when **'nicotine'** gets distilled. The **'distillate'** thus obtained is extracted with solvent ether successively. **Nicotine** appears as an **'oily colourless liquid'** after the evaporation of the solvent, and may be further purified by **fractional crystallization of its oxalate** (salt).

3.3.2. Characteristic Features

These essentially comprise:

(1) Freshly prepared **nicotine** is almost a colourless liquid.

(2) It is **hygroscopic** in nature.

(3) It is **levorotatory**, $[\alpha]^{20}_8 = -169.3$ (neat); $[a]_{5461} = -204.1°$

(4) Its bp_{745} is 246-247 °C.

(5) The natural (–)–**nicotine** are **dextrorotatory.**

(6) The natural (–)–**nicotine** shows the normal *physiolocial activity,* such as: **constriction of blood vessels, enhancement in heart-beat, increase in blood-pressure.**

(7) It gets darkened on exposure of **air** due to **auto-oxidation.**

(8) It is miscible with water in all proportions at <60°C.

3.3.3. Constitution of Nicotine

The structure of **nicoine** may be established based on a host of logical evidences and critical considerations pertaining to the most probable assigned **chemical structure:**

(1) **Molecular formula:** Based on the C, H, N analysis *vis-a-vis* **molecular weight determination** the molecular formula of **nicotine** is $C_{10}H_{14}N_2$.

(2) **Presence of Two Quaternary N-Atoms in Nicotine:** In fact, the very presence of the *two* **N-atoms in nicotine** are *quaternary in nature,* which may be easily ascertained due to the critical utilization of *two moles* of **methyl iodide** (H_3C-I) to yield a corresponding **dimethyl iodide derivative** as illustrated under:

---

* **Nicotine :** The name **'nicotine'** was duly assigned in the honour of **Sir J Nicot,** the ambassador of the King of France to Portugal from 1559–1561.

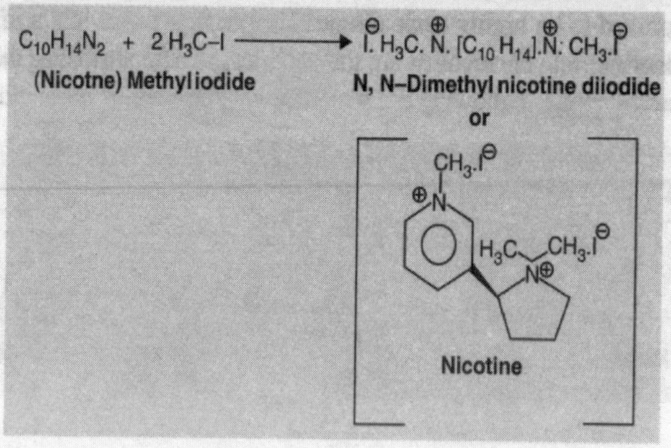

$$C_{10}H_{14}N_2 + 2 H_3C-I \longrightarrow I.\overset{\ominus}{}H_3C.\overset{\oplus}{N}.[C_{10}H_{14}].\overset{\oplus}{N}.CH_3.\overset{\ominus}{I}$$

(Nicotne) Methyl iodide   **N, N–Dimethyl nicotine diiodide**

or

Nicotine

Note: Interestingly, under *appropriate experimental parameters*, nicotine also gives rise to the formation of *two isomeric monomethyl-iodides*, of which one of the *tert*-N-atoms is having a N–methyl moiety.

(3) **Direct Oxidation with Chromic Acid [CrO₃] or Potassium Permanganate [KMnO₄]:** The direct oxidation of nicotine [A] with CrO_3 yields nicotinic acid [B], as shown under:

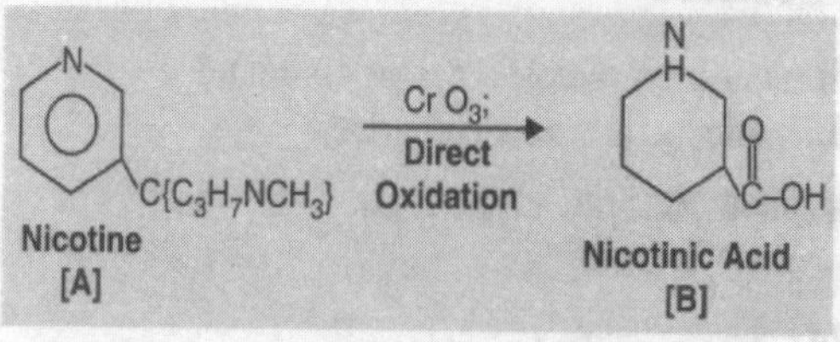

Nicotine
[A]   **Nicotinic Acid**
[B]

CrO_3;
Direct
Oxidation

Inference: The above reaction evidently indicated the actual presence of *six* of the *ten* C-atoms of nicotine [*Mol. Formula:* $C_{10}H_{14}N_2$]; and, therefore, suggests the placement of:

- *five* C-atoms in a 'Pyridine Ring',
- nicotine is a pyridine derivative having a *side-chain* at C-3 atom, and
- remaining *four* C-atoms in an adjoining 'Pyrrolidine Ring',

(4) **Presence of Pyridine Nucleus:** It is further substantiated and, ascertained the critical formation of a 'hexahydrate derivative, on being subjected to **reduction** with **Na and absolute ethanol.** Thus, we have the following reaction :

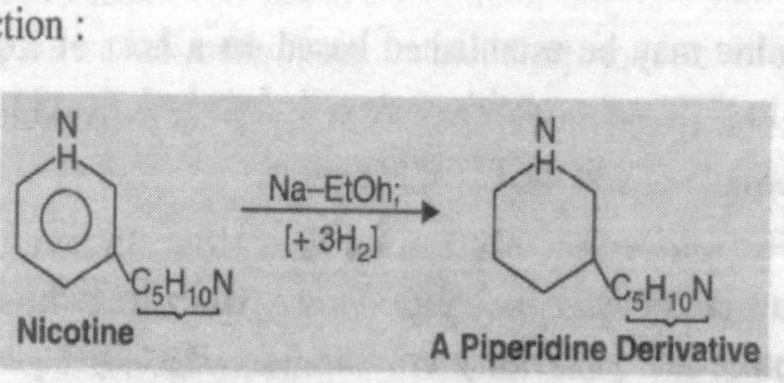

Nicotine   **A Piperidine Derivative**

Na–EtOh;
[+ 3H₂]

Conclusion: It may be concluded that *one* of the *two* N-atoms of the 'nicotine' is definitely present as a 'pyridine nucleus'.

(5) **Presence of an N-Methyl Moiety ; Herzing-Meyer analysis** *i.e,* the N-methylamine on heating with **hydroiodic acid (HI)** and **methyl iodide (H₃C–I)** formed in the reaction volatalizes from the

reaction mixtures to be *collected* and *assayed separately,* * helps to ascertain the presence of **N–methyl functional moiety** in nicotine.

(6) **Presence of 3 C–Fragment (C_3H_7):** The presence of a **3C–segment (C_3H_7) in 'nicotine'** remains to be accounted for logically. However, it may be duly demonstrated by virtue of the fact that:

- **'nicotine'** is absolutely devoid of any **'unsaturation',** and
- **'elemental composition'** reveals that the **[C_3H_7–NCH_3]** entity should be a **'cyclic structure'** by all means.

Therefore, the most **probable** and **obvious assumption** being that the said **3C–fragment** present in the **'nicotine'** must be a **'pyrrolidine ring',** thereby it could be assigned either structure **'X'** or **'Y'**:

Nicotine
['X']

Nicotine
['Y']

Importantly, one may observe that either of these *two* probable/assigned structures [*viz.,* **'X'** and **'Y'** would evidently account for the actual inherent overall activity activity of the **'alkaloid'**

Correctness of Assigned Structure ['X']: The correctness of this 'first approximate assumption' being adequately supported by the observation that when *'nicotine'* is heated vigorously with **zinc chloride ($ZnCl_2$) significant decomposition** comes into play with the respective production of **pyridine,** and **methylamine.** Indeed these facts fully supports the aforesaid *two structures* **'X'** and **'Y'**, it fails to make a **clear cut distinguishable** difference between them.

(7) **Attachment between the Pyridine (6–membered heterocyclic ring) and Pyrrolidine (5–membered heterocycle ring) in Nicotine:** The exact and precise attachment between the **pyridine** and **pyridine rings** was duly established by making use of the critical application of a **reaction characteristic** pertaining to the N–alkylpyridinium pyridines salts. Furthermore, the oxidation of an N–methylpyridinium salt by treatment with an **alkaline potassium ferricyanide [$K_3Fe(CN)_6$]** yield N–methyl–2–pyridone.

N–Methyl pyridinium
salt

2–Hydroxy–N
methyl pyridinium

N–Methyl–2
–pyridone

(8) **Nicotine Methiodide Produces Pyridone (I):** When Nicotine methiodide (I) was subjected to the treatment *as in (7) above,* it gives rise to the formation of the corresponding **pyridone (II).** The

---

* Reaction of the recovered **methyliodide** with excess of **alcoholic $AgNO_3$** gives rise to the formation of **silver iodide [AgI]** that may be estimated gravimetrically.

resulting **product (I)** was then duly oxidized with **chromic acid [CrO₃]** thereby causing complete destruction of the 'pyridine nucleus', leaving **L–hygrinic acid (III)** as the respective product.

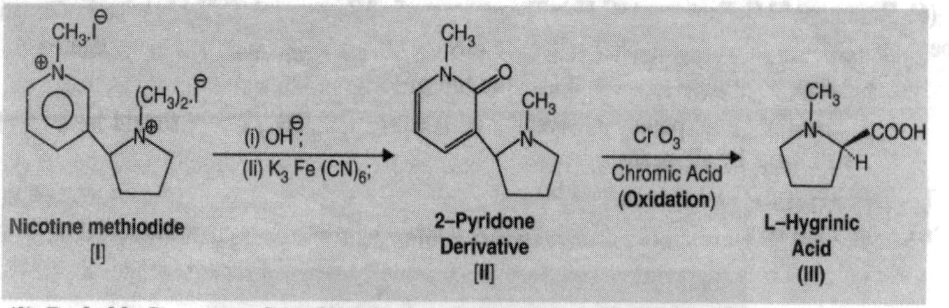

Nicotine methiodide
[I]

(i) OH⁻;
(ii) K₃Fe(CN)₆;

2–Pyridone
Derivative
[II]

Cr O₃
Chromic Acid
(Oxidation)

L–Hygrinic
Acid
(III)

(9) **Probable Structure of (–)–Nicotine:** As **L–hygrinic acid (III)**, in (8) above, has already an established known *'stereo chemistry,* the aforesaid net result obtained from (8) above categorically provides a complete definition with regard to structure of (–)–nicotine as given under:

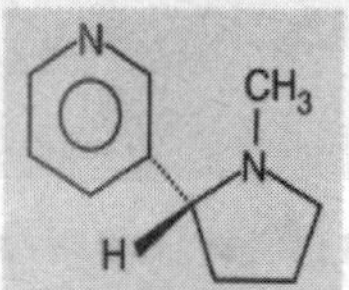

(–)-Nicotine

(10) **Synthesis:** In fact, the synthesis of **nicotine'** has been put forward by a number of *researchers* and *investigators* using a number of different routes, namely:

* **Pinner :** *Ber.* **26 :** 294, 1893;
* **Pictet and Rotschy :** *Ber.* **37 :** 1225, 1904.
* **Craig :** *J Am Chem Soc,* **55 :** 2854, 1933.
* **Nakane R and Hutchinson CR :** *J Org Chem,* **43 :** 3922, 1978.

However, the synthesis of Späth and Bretschneider (1938) is discussed in an elaborated manner as given under:

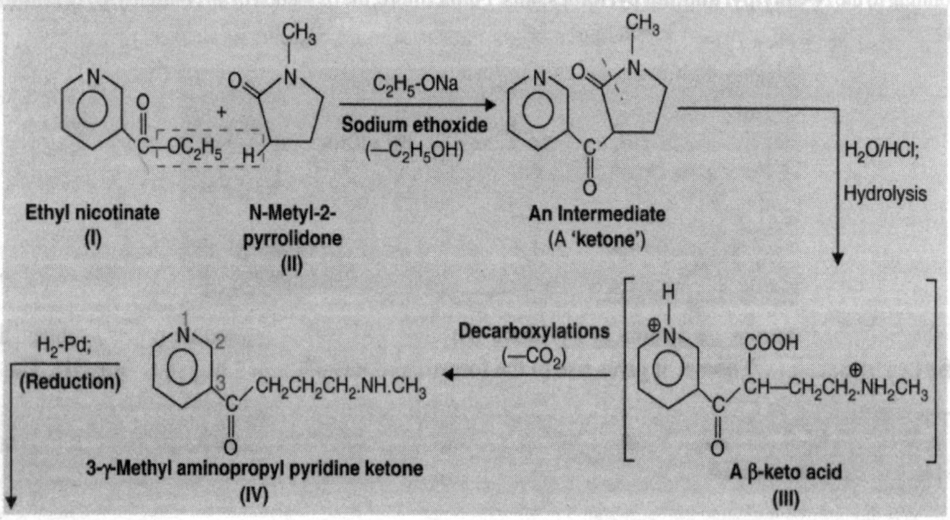

Ethyl nicotinate
(I)

N-Metyl-2-
pyrrolidone
(II)

C₂H₅-ONa
Sodium ethoxide
(—C₂H₅OH)

An Intermediate
(A 'ketone')

H₂O/HCl;
Hydrolysis

H₂-Pd;
(Reduction)

3-γ-Methyl aminopropyl pyridine ketone
(IV)

Decarboxylations
(—CO₂)

A β-keto acid
(III)

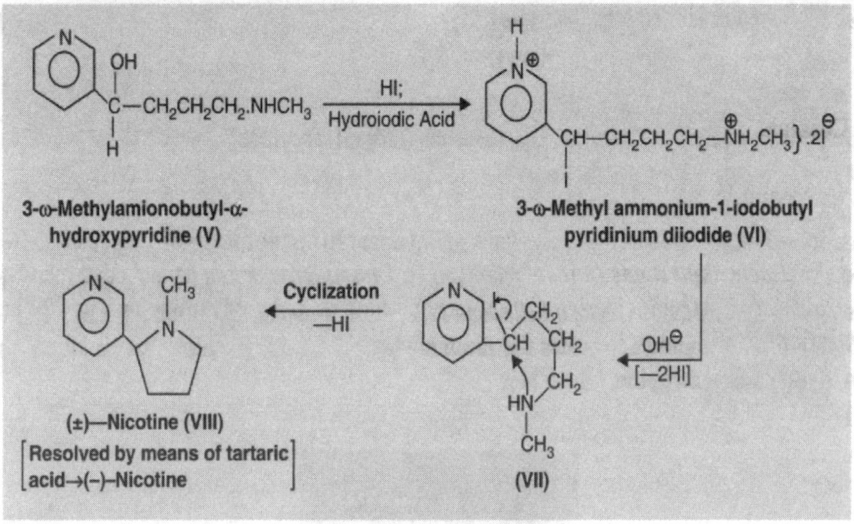

3-ω-Methylamionobutyl-α-
hydroxypyridine (V)

3-ω-Methyl ammonium-1-iodobutyl
pyridinium diiodide (VI)

(±)—Nicotine (VIII)

[Resolved by means of tartaric
acid→(–)-Nicotine]

(VII)

Explanation: The various steps involved in the synthesis of (–)–nicotine may be expatiated briefly as stated under:

(1) Interaction of **ethyl nicotinate (I)** and **N–Methyl–2–pyrrolidone (II)** in the presence of freshly prepared **sodium ethoxide (C_2H_5–ONa)** forms a **'ketone'** as an **intermediate (II)**, with the elimination of a mole of *ethanol.*

(2) The resulting **product (II)** upon hydrolysis in an **acidic medium** affords the cleavage of the **pyrrolidine ring** (as shown by **dotted line**) to form a **β–keto acid (III)**.

(3) **Product (III)** upon **decarboxylation (–CO_2)** yields 3–γ–methyl amino–propyl pyridine ketone **(IV)**, which on reduction with **H_2–Pd** forms 3–ω–methylamino butyl–α–hydroxy pyridine **(V)**.

(4) **Product (V)** when treated with **hydroiodic acid (HI)** gives rise to the formation of **3–ω–methyl ammonium–1–iodobutyl pyridinium diiodide (VI)**.

(5) The resulting **product (VI)** in an alkaline environment loses **two moles of HI** to yield **compound (VIII)**.

(6) **Product (VIII)** undergoes cyclization to lose a mole of **HI** to form (±)–nicotine.

Note: The racemic mixture of *'nicotine'* may be resolved by means of *tartaric acid* to yield ultimately the desired *compound (–)–nicotine.*

3.4. Tropane Alkaloids

Preamble: Tropane is a **'bicyclic compound'** usually formed by the *condensation* of a **pyrrolidine precursor** (*viz.,* 'ornithine') with *three* acetate–derived C–atoms predominantly. Importantly, both **pyrrolidine** as well as **piperidine ring systems** may be perceived with **'good judgement'** in the molecule.

In fact, the *3–hydroxy derivative of tropane* is invariably termed as **'tropine'**. It is, however, pertinent to state here that the careful *esterification* with (–)–**tropic acid** gives rise to the formation of **'hyoscyamine' (tropine tropate),** which eventually may be *racemized* to form **'atropine'.**

The following *two* potent and well-known *alkaloids* belonging to the class of '**Tropane Alkaloids**' shall be discussed in the sections that follows:

- **Atropine,** and
- **Cocaine**

3.4.1. Atropine [*Syn : Tropine Tropate; dl–Tropyl Tropate*]

3.4.1.1. Preamble

Atropine is recognized as the most vital and important '**tropane alkaloids**'. It occurs abundantly in the roots of *deadly night shade* (*Atropa belladona L., Datura Stramonium L.,* and other members of the family *Solanaceae,* **Atropine** designates the racemic form of **L–hyoscyamine** that rapidly undergoes **racemization*** to yield **atropine,** which being nothing

but ± – **hyoscyamine.**

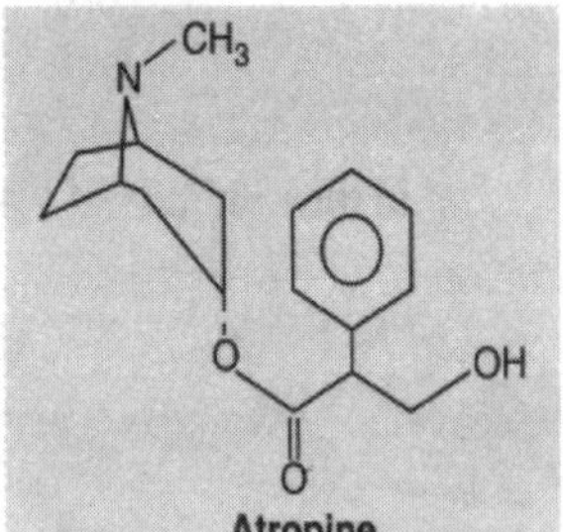

Atropine

3.4.1.2. Extraction

Chemnitius (1927)** carried out the extraction of '**atropine**' either from the air-dried **belladona roots** or from the **latex** (*juice*) of **datura plant.** Following steps are followed sequentially:

(1) The dried **belladona roots** or **latex from datura plant** is heated carefully with **potassium carbonate [K$_2$CO$_3$] solution** thereby rendering the *racemization of* **hyoscyamine** to **atropine.**

(2) The racemized **atropine** is now **extracted with chloroform,** preferably in a slightly *alkaline medium,* at least 3 to 4 times.

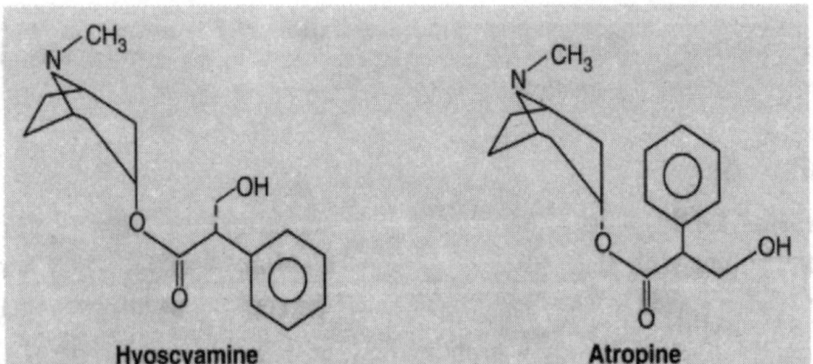

Hyoscyamine   **Atropine**

---

* **Racemization** takes place rapidly when **warmed** with an **ethanolic alkaline solution.**

** Chemnitius : *J Prakt Chem.,* **116 :** 276, 1927.

(3) The *combined chloroform layer* is filtered, solvent separated under vacuum, and finally the *crude alkaloid* (**atropine**) is carefully extracted with dilute H_2SO_4 [2.M] as its salt (**in an aqueous medium**).

(4) The resulting acidic solution is made alkaline with small amount of **potassium carbonate** [K_2CO_3] whereby the *alkaloid* **atropine** gets precipitated.

(5) Ultimately, the **precipitated atropine** is duly extracted with solvent ether successively, and subsequently purified by converting it into either an *oxalate* or a *sulphate*.

3.4.1.3. Characteristic Features

These essentially comprise:

(*i*) It is mostly obtained as **orthorhombic prisms** from **acetone**.

(*ii*) Its MP ranges between **114–116°C.**

(*iii*) **Atropine** *sublimes in high vacuum* at **93–110°C.**

(*iv*) Its **pK value** stands at **4.35 pH.**

(*v*) The pH of 0.0015 molar solution is found to be **10.0.**

(*vi*) **Solubility Profile:** Atropine 1g dissolves in 455 mL of H_2O at room temperature (20±2°C); 90 mL of H_2O at 80°C; 1.2mL ethanol at 60°C; 27mL glycerine; 25mL ether; 1mL chloroform; also soluble in benzene.

(*vii*) **Physiological Activity Profile :** These mostly consist of :

- **Atropine** causes dilation of the pupil of the **human eyes. (1 part in 40,000 parts of water).**
- Used most frequently in ophthalmological procedures.
- Reduces the activity of both **salivary glands** and **gastric glands.**
- Oral administration **first causes stimulation** and then followed by **depression of the central nervous system (CNS).**

> **Note:** *'Hyoscyamine'* closely resembles *'atropine'* in its physiological activity; however, the former turns out of be *'more active'* than the latter particularly upon the *peripheral nerves.*

3.4.1.4. Constitution of Atropine

The probable structure of 'atropine' molecule has been derived meticulously based upon the following critical evidences with logical supporting factual statements. However, the ultimate **'assigned structure'** was duly ascertained by its **'total synthesis'.**

(1) **Molecular Formula :** Atropine has the molecular formula $C_{17}H_{23}NO_3$.

(2) **Presence of an Ester Moiety :** The presence of an **'ester moiety'** in *atropine* may be clearly proved by carrying out its hydrolysis in the presence of **barium hydroxide [Ba(OH)$_2$] solution** and *heating* to obtain:

- a **racemic acid :** (±)–*tropic acid,* and
- an **alcohol** (**optically active**) : **tropine.**

Thus, we may have the following reaction:

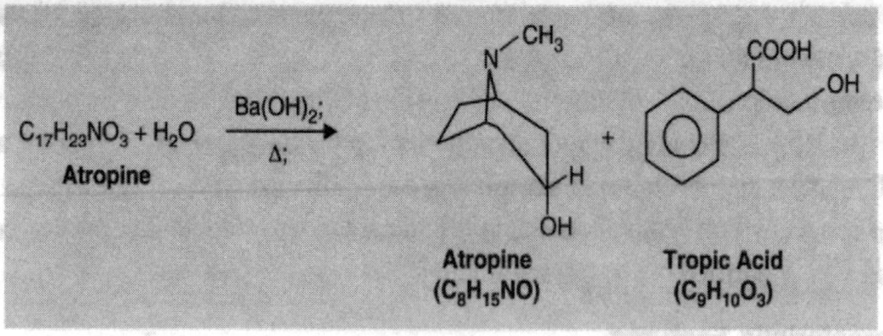

$$C_{17}H_{23}NO_3 + H_2O \xrightarrow[\Delta;]{Ba(OH)_2;}$$

Atropine

Atropine
$(C_8H_{15}NO)$

Tropic Acid
$(C_9H_{10}O_3)$

Inference : The aforesaid reaction evidently suggests that **atropine** is the *tropeine i.e.*, an **ester of tropic acid,** and it may also be called as **'Tropine Tropate'.**

Ladenburg (1883)* and Willstatter (1898)** further confirmed and ascertained that *'atropine'* is certainly a **tropine–tropate ester.** They, in fact, actually obtained *atropine* by evaporating an equimolar quantum of a mixture of **tropine** and **tropic acid** in the presence of **dilute hydrochloric acid (HCl).**

Besides, *'atropine'* may not be an **'amide'** [–C–NH–], which fact could be ascertained since the *product of hydrolysis,* **tropine,** happens to be a **'tertiary base'** [see section (2) above].

Interestingly, at this point in time the **constitution of atropine** may be resolved into *two* critical sections, namely:

- structure of **'Tropic Acid',** and
- structure of **'Tropine'.**

(3) **Structure of Tropic Acid:** Importantly, the structure of **tropic acid** is predominantly based upon the following evidences :

(a) **Molecular formula :** It is found to be $C_9H_{10}O_3$.

(b) **Saturated Monobasic Acid : Tropic acid** shows the following *two* vital and important reactions, such as:

- reacts with **one equivalent of alkali,** and
- **fails to add on a mole of bromine,**

thereby suggesting that it is a **saturated monobasic acid.**

(c) **Presence of one Hydroxyl Moiety : Tropic acid** on *acetylation* with acetic anhydride yields a *monoacetate* which indicates evidently that should contain **one hydroxyl moiety.**

(d) **Hydroxyl Moiety an 'Alcoholic Function':** In order to ascertain that the **hydroxyl (–OH) moiety** is an *alcoholic function* tropic acid is heated vigorously thereby losing a molecule of water and producing an *optically inactive unsaturated acid* known as **'atropic acid'** $[C_9H_8O_2]$. Therefore, it suggests explicitly that the **hydroxyl moiety** should be an **'alcoholic function'** only. The **hydroxyl moiety** should be an **'alcoholic function'** only. We may have the following reactions :

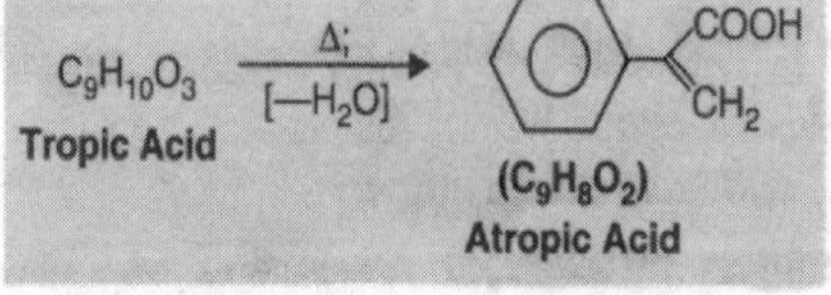

$$C_9H_{10}O_3 \xrightarrow[{[-H_2O]}]{\Delta;}$$

Tropic Acid

$(C_9H_8O_2)$
Atropic Acid

---

* Ladenburg, *Ann.*, **217 : 75,** 1883.

** Willstatter, *Ber,* **31 :**1537, 1898.

(e) **Oxidation :** **Atropic acid** upon oxidation yields **benzoic acid**, thereby indicating that **'atropic acid'**; and, therefore, **'tropic acid'** shall essentially comprise at least *one* **'benzene ring'** having a *side-chain* bearing a **carboxylic (–COOH)** moiety in their inherent **chemical structure.**

Thus, we may have :

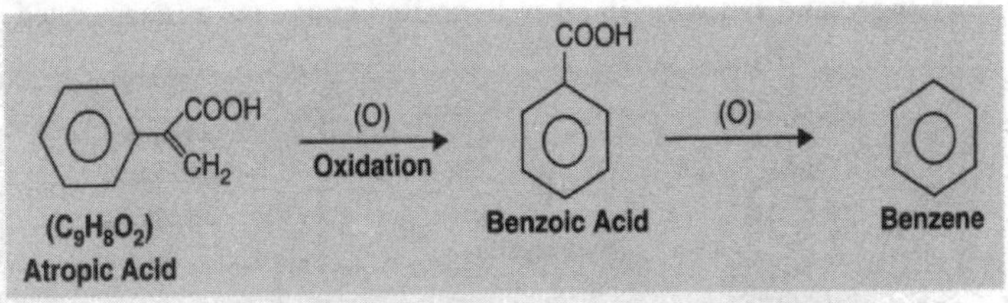

(C₉H₈O₂)
Atropic Acid

Benzoic Acid

Benzene

(f) **Isomeric Forms of Atropic Acid :** It has been observed that **atropic acid** is an **unsaturated acid** which suggests obviously that it may have the following *two* **isomeric structures,** namely:

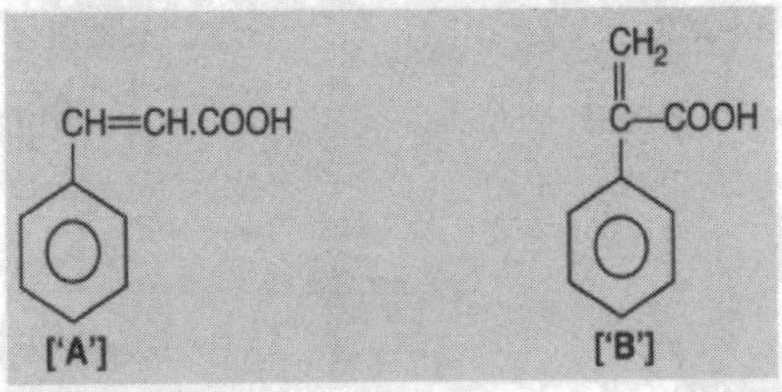

['A']

['B']

Inference : The above **structure ['A']** happens to be the structure of **cinnamic acid***, and hence, structure **['B']** should be the proposed (or designated) structure of **atropic acid.**

(g) **Confirmation of Atropic Acid Structure :** The confirmation of the structure of **atropic acid** may be accomplished by its **oxidation** with **potassium permanganate [KMnO₄]** that produces **phenylglyoxal,** as shown below:

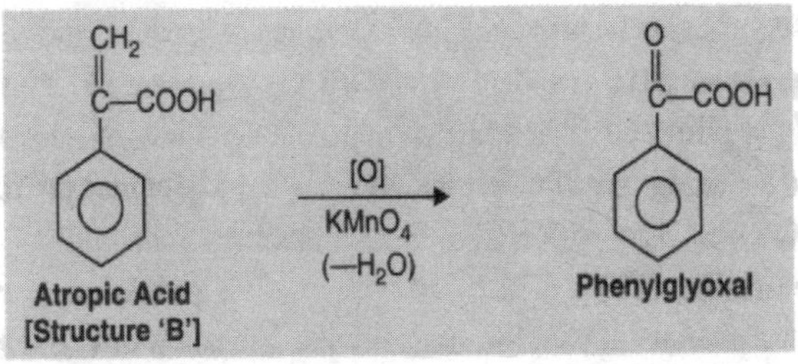

Atropic Acid
[Structure 'B']

Phenylglyoxal

(h) **Atropic Acid** *via* **Dehydration of Tropic Acid :** **Atropic acid** is duly obtained by the **dehydration of tropic acid,** as given under section (d), which evidently indicates that **'tropic acid'** could be either **'C'** or **'D'** :

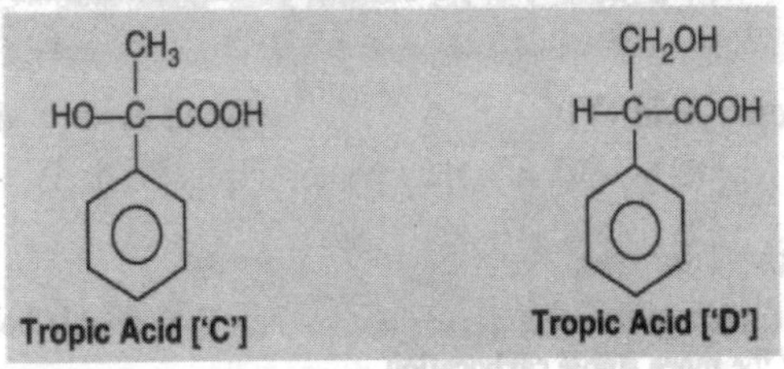

Tropic Acid ['C']

Tropic Acid ['D']

---

* Merck Index, Merck & Co., Inc., Whitehouse Station NJ (USA), 12th ed., 1996.

Nevertheless, the exact structure of **'tropic acid'** is proved and confirmed to be **structure ['D']** by its **'total synthesis'** put forward by Sletzinger and Paulsen (1945)* from **acetophenone**.

(*i*) **Synthesis of (±)–Tropic Acid :** The various steps involved in the synthesis of (±)–**tropic acid** suggested by **Sletzinger and Paulsen (1945)** from **acetophenone** are as enumerated under :

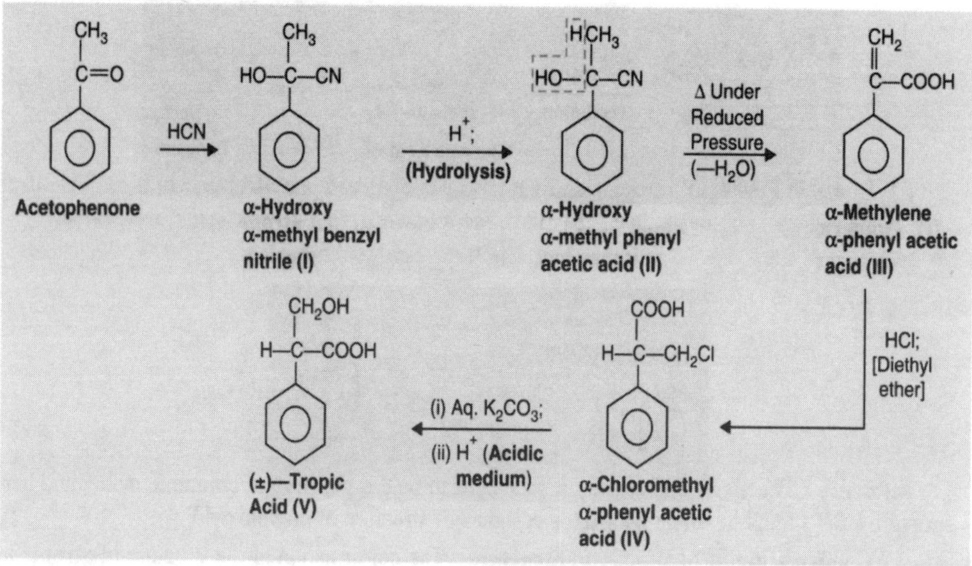

Explanation: The course of reaction in the synthesis of (±)–**tropic acid** may be explained as under :

(*i*) Acetophenone on treatment with HCN yields α–hydroxy–α–methyl benzyl nitrile (I).

(*ii*) **Product (I)** upon **hydrolysis** in an *acidic environment* produces the corresponding α–hydroxy–α–methyl phenyl acetic acid (II) (or **atrolactic acid**).

(*iii*) The resulting **product (II)** when heated under reduced pressure loses a mole of **water (H_2O)** (as *indicated* by *dotted-lines*) to yield α–methylene–α–phenyl acetic acid (III).

(*iv*) **Product (III)** on treating with HCl in diethyl ether (as a *medium*) produces α–chloromethyl–α–phenyl acetic acid (IV).

(*v*) Ultimately, the **product (IV)** when treated with aqueous **potassium carbonate (K_2CO_3)**; and subsequently reacting in an acidic medium gives rise to the formation of (±)–**tropic acid (V)**.

> **Note:** It is, however, pertinent to state here that the critical addition of HCl actually occurs quite contrary to the well-known *Markownikoff's Rule**. Obviously, if the said addition would have taken place just according to this 'rule', one might have again obtained another mole of *atrolactic acid (II)*.

Resolution of (±)–Tropic Acid (V): It may be accomplished easily and effectively by the aid of **quinine**.

---

* Sletzinger and Paulsen : US Patent 2390278 [1945 to Merck & Co.,].

** Makownikoff's Rule : It refers to the *electrophilic addition* to a **C–C double bond** that critically involves the *intermediate formation* of the **more stable carbocation**.

Synthesis by Blicke (1955)*: It essentially involves the treatment of **boiling phenylacetic acid** with a *Grignard's reagent :* **isopropyl– magnesium chloride** in an ethereal medium. The resulting product (i.e. a **Grignard derivative**) is finally treated with **formaldehyde (HCHO)**.

Thus, we may have the following reactions :

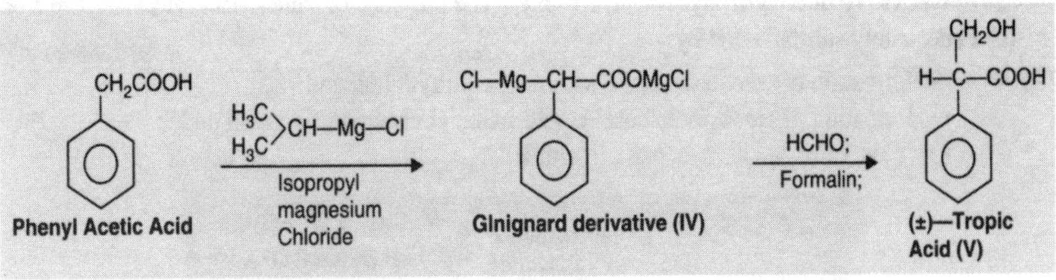

Based on the concrete evidential support to the structure of **Tropic Acid** by Sletzinger and Paulsen (1945) and Blicke (1955), two researchers, Fodor and Csepreghy (1961)* put forward the absolute configuration of the two possible isomers of **'tropic acid'**. However, it may be further ascertained by its careful *configuration* with a specific **amino acid (–)–alanine** as given under:

(4) **Structure of Tropine :** Likewise, the structure of **'tropine'** may be established based upon the following reactions supported by logistic explanations.

(a) **Molecular Formula :** Its molecular formula is found to be $C_8H_{15}NO$.

(b) **Addition Compound :** The interaction of **tropine** and **methyl iodide (CH$_3$–I)** produces a *crystalline* **addition compound** as shown below:

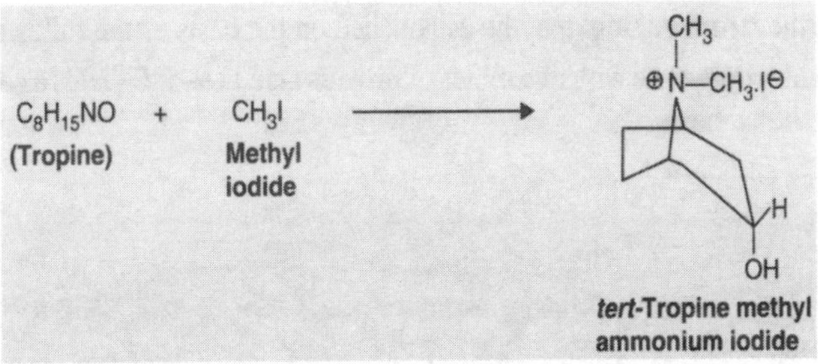

Inference : It suggests that the N–atom present in **'tropine'** is certainly **tertiary in nature**.

---

*   Blicke : US patent NO: 2716650 (1955) to Univ of Michigan, USA.

** Fodor and Cscpreghy : *J Chem Soc*, 3222, 1961.

(c) **Presence of N–Methyl Moiety :** **Tropine** on being fused with an alkali gives rise to the formation of **methyl amine,** thereby indicating that it should primarily contain a **N–methyl moiety** (=N–CH$_3$). However, the same aspect may also be duly ascertained by the fact that when **tropine** is heated with **HI (hydroiodic acid)** at 150°C *i.e.,* the well-known **Herzig-Meyer method,** it produces 'one molecule of methyl iodide (CH$_3$I),.

(d) **Presence of one Hydroxyl (–OH) Moiety :** The presence of 'one hydroxyl moiety' in *tropine* may be adequately substantiated by:

- formation of 'monoacetate' : with **acetic anhydride,** and
- formation of **'monobenzoate'** : with **benzoyl chloride,** as given under :

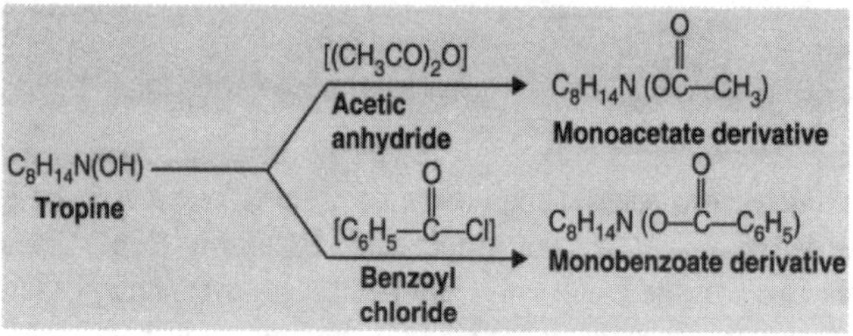

(e) **Presence of a *sec*–Alcoholic Function (–CHOH–) :** **Tropine** on being subjected to **oxidation** with *chromic acid (CrO$_3$)* yields a **ketone** termed as 'tropinone' [C$_8$H$_{15}$NO] as given below :

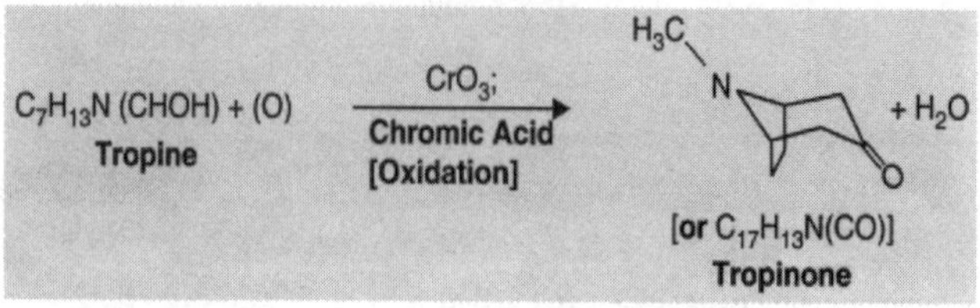

Inference : It clearly indicates that the **hydroxyl (–OH) moiety in tropine** should be a *secondary alcoholic function.*

(f) **Presence of a Secondary Alcoholic Moiety Flanked with Methylene (–CH$_2$–) Groups on its either Sides in Tropine Ring [-CH$_2$–CH(OH)–CH$_2$-]:**

Interestingly, the presence of a secondary alcoholic group flanked duly with methylene moieties on its either sides in the **'tropine ring'** may be established on the basis of the following *two* reactions:

(*i*) Treatment of **tropionone** with two moles of **nitrous acid** at *0–5°C* yields a corresponding **dioxime derivative (I)** as shown below :

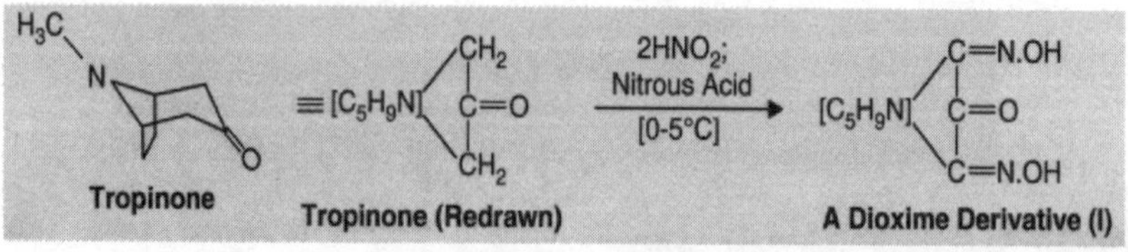

(*ii*) Treatment of **tropinone** with two moles of **benzaldehyde** give rise to the formation of **dibenzylidene derivative (II)** as given under:

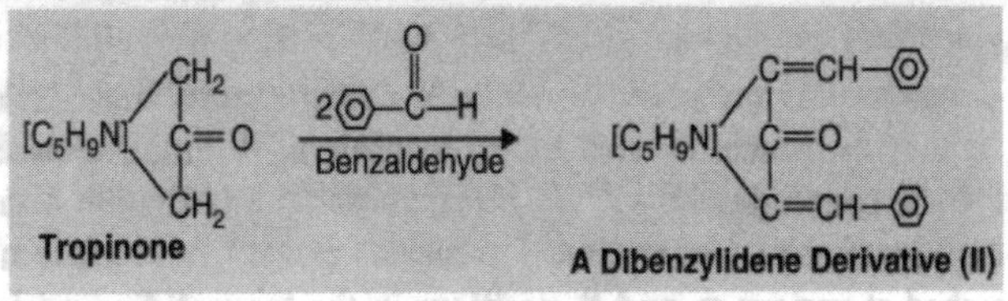

Tropinone → Benzaldehyde → A Dibenzylidene Derivative (II)

Inference : In fact, both these reactions (i) and (ii) above are the typical and characteristic reactions with regard to the **dimethylene carbonyl group** $\left[\begin{matrix}-CH_2 \\ -CH_2\end{matrix}\right>C=O\right]$

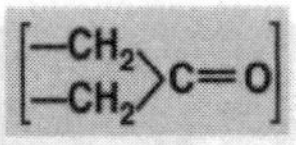

(*iii*) Besides, the careful **oxidation** of 'tropinone' produces a **dicarboxylic acid** known as 'tropinic acid, $[C_8 H_{13} NO_4]$ *without causing* any **loss of a C-atom**, thereby suggesting that the **dimethylene carbonyl moiety** $\left[\begin{matrix}-CH_2 \\ -CH_2\end{matrix}\right>C=O\right]$ is definitely present in the *ring system*, as shown below:

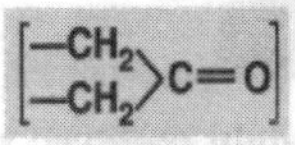

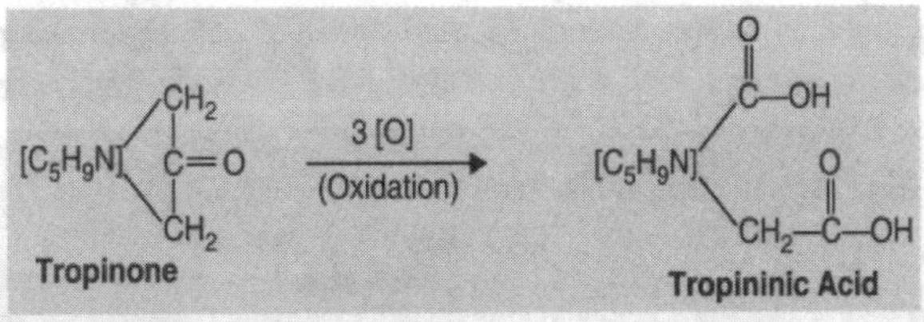

Tropinone → 3 [O] (Oxidation) → Tropininic Acid

Inference : Therefore, the above reaction suggests explicitly that the 'tropine' molecule essentially comprises the corresponding **secondary alcoholic moiety** [–CH$_2$–CH(OH)–CH$_2$–] in the **ring system.**

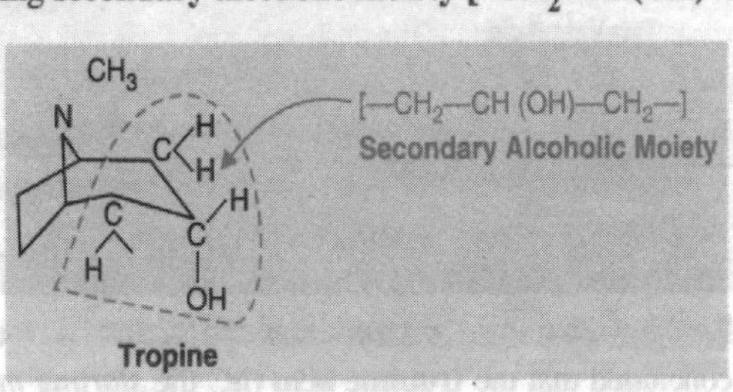

[—CH$_2$—CH (OH)—CH$_2$—]
Secondary Alcoholic Moiety

Tropine

(g) **Presence of N–Methylpyrrolidine Ring in Tropinone Tropine :** When **tropinic acid** is subjected to vigorous oxidation with **chromic acid and sulphuric acid (CrO_3/H_2SO_4)** it produces **N–methylsuccinimide** as shown under:

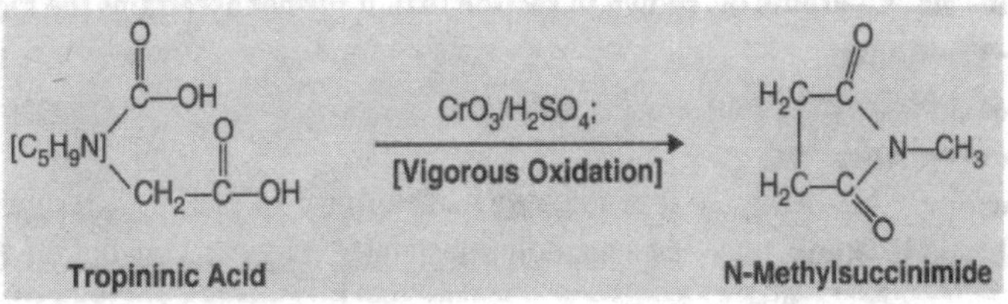

Tropininic Acid → CrO_3/H_2SO_4; [Vigorous Oxidation] → N-Methylsuccinimide

Inference: The above reaction clearly indicates the presence of a **N–methylpyrrolidine nucleus** in tropinone; and, therefore, in **tropine.**

Accountability of 5 Out of 8 C–Atoms in Tropine and Tropinic Acid :

Based upon the above candid inferences drawn the **N–methyl-pyrrolidine nucleus** actually account for **5 out of 8 C–atoms** duly present in **tropine** and **tropinic acid**. Since, **tropinic acid** designates a *'dibasic acid'*, it may be concluded that the rest of the **3C–atoms** should be present as **carboxyl (–COOH)** and **methylcarboxyl (=CH$_2$COOH)** moieties. Thus, the aforesaid *two* **functional groups** may be attached at various **α–and β–positions** in the **'pyrrolidine nucleus'**; and, therefore, these positions may be ascertained duly as given below:

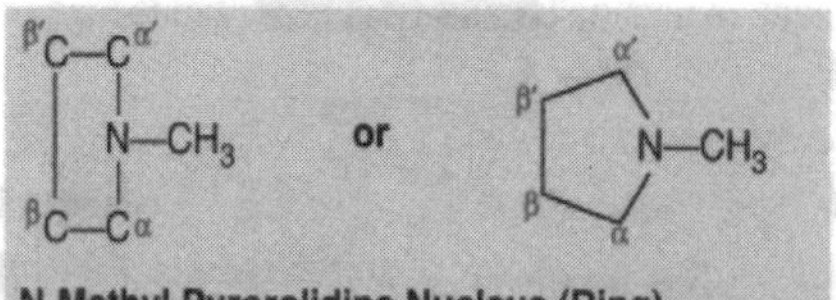

N-Methyl Pyrorolidine Nucleus (Ring)
[Showing Various α-and β-positions available]

(h) **Presence of Seven C–Atoms in Tropinic Acid :** It has been duly observed that **'tropinic acid'** [C$_8$H$_{13}$NO$_4$] on being subjected to **Hofmann's exhaustive methylation method produces** an unsaturated **dicarboxylic acid,** and thus subsequently on **reduction with H$_2$–Pt** to yield **'pimelic acid',** as given under:

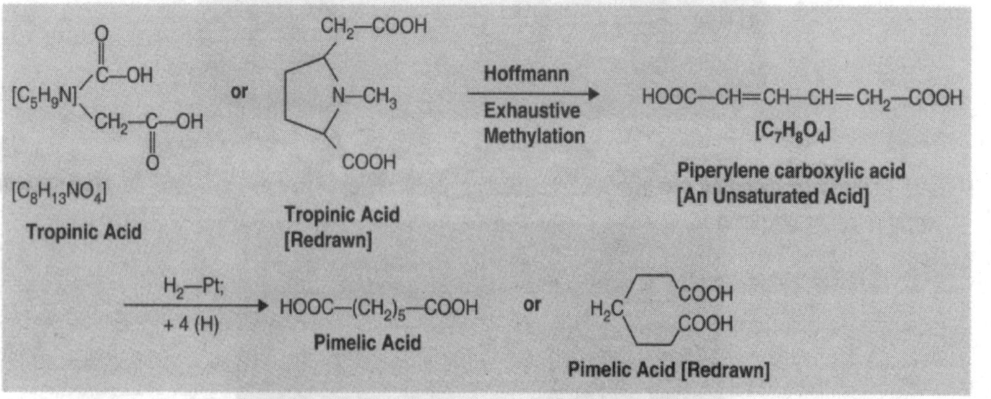

Explanation : Pimelic acid is a **dicarboxylic acid** which has **seven C-atoms** joined together in a series. Hence, it may be concluded that the **tropinic acid** (*ie.,* the *starting material*) should essentially comprise **seven C–atoms** connected to one another almost in the same manner; however, the glaring difference being that the *eighth C-atom* is duly attached to the (>N–CH$_3$) *i.e.,* **N–methyl moiety.**

Importantly, the aforesaid theoretical explanation only holds good if the *two* **prevailing carboxyl (–COOH) groups** *viz.,* C–atoms [as shown in section (g)]. It further ascertains the fact that the critical presence of a **7–membered C–ring system** is vehemently present in the following *three* structures:

- **Tropinone,**
- **Tropinic Acid, and**
- **Tropine.**

(i) **7–Membered C–Ring System Present in Tropine :** In fact, the presence of a **7–membered C–ring system** in **'tropine'** may be adequately substantiated by carrying out the following *three* steps in a *sequential manner;* namely :

- *Dehydration* of **'tropine'** with sulphuric acid (H$_2$SO$_4$) to lose a mole of water and yield **Tropidine (I),**

- *Hofmann's Degradation* (or *Hofmann's Exhaustive Methylation Method*) gives rise to the formation of 'tropilidene' (II) (or **cycloheptatriene**), and
- *Reduction* of the resulting **product (II)** with **H_2/Pt** takes up **6 H–atoms** to yield **cycloheptane (III).**

The aforesaid reactions may be expressed as under :

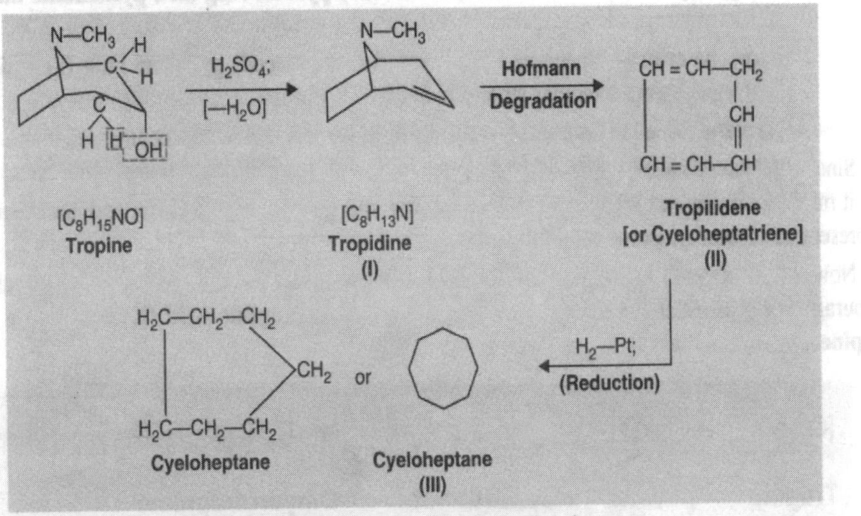

(j) Presence of a Reduced Pyridine Nucleus (Ring) in Tropine :

Importantly, the crucial presence of a **reduced pyridine nucleus** in the 'tropine' molecules may be duly established *via a series of reactions to arrive at 2–ethylpyridine* as the **end product,** as stated under:

- Treatment of **tropine** with **hydroiodic acid (HI)** at a temperature **below–150°C** to yield **tropine iodide (I),**
- *Reduction* of **Product (I)** gives **dihydrotropidine (II)** with the elimination of a mole of **HI,**
- Distillation with HCl produces **nordihydrotropidine (III)** and a mole of **methyl chloride,** and
- Finally, the resulting **product (III)** on being subjected to *reduction* with **Zn-dust/distillation** yields the desired product **2–elthylpyridine (IV)**

These reactions may be expatiated as under:

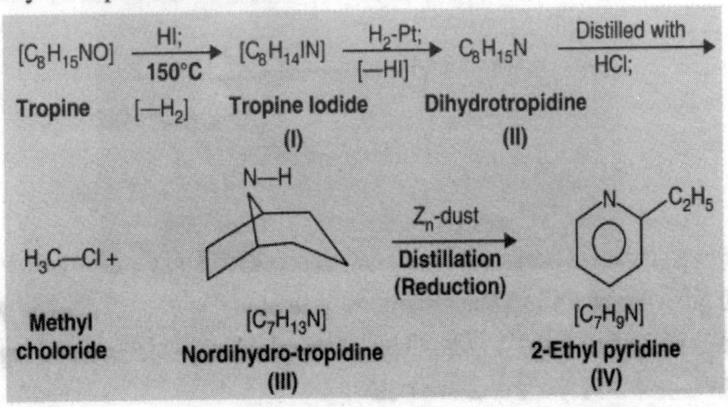

Conclusions : From the aforesaid approaches, reactions, and derivations one may summararily draw the following vital **conclusions,** namely :

- **Tropine** essentially has a **7–membered C–ring system** with a **secondary alcoholic group flanked by methylene moiety on its either side [–CH_2–CH(OH)–CH_2–],**
- **Tropine** possesses a **saturated (or reduced) pyrrole ring** (*i.e.*, **pyrrolidine nucleus)** in the structure,
- **Tropine** critically consists of **saturated (or reduced) pyridine ring** (*i.e.*, **piperidine nucleus)** in its molecular structure, and
- **Tropine** does have one **N–methyl (=N–CH_3) moiety** in its structure.

Since, **tropine** essentially possesses **only one N-atom in its molecular structure,** it clearly suggests that it must be **shared mutually between piperidine and pyrrolidine nucleus;** and, therefore, should be present as **N–methyl group (=N–CH_3)**

Now, based on the above supportive reasonings along with scientific evidences, it may be sufficiently encouraging to predict and assign the probable structures to **Tropinic Acid (X), Tropinone (Y),** and **Tropine (Z)** as given below:

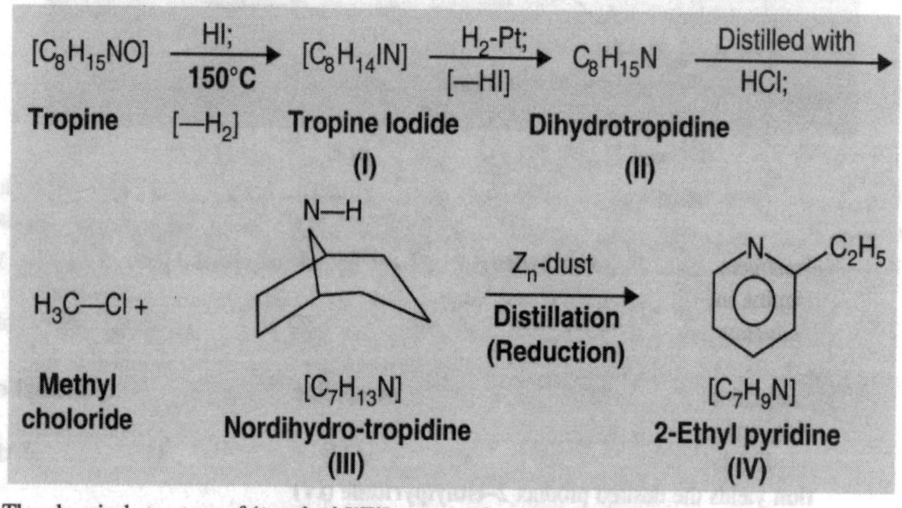
[C8H15NO]
Tropine
HI; 150°C
[—H2]
[C8H14IN]
Tropine Iodide
(I)
H2-Pt;
[—HI]
C8H15N
Dihydrotropidine
(II)
Distilled with
HCl;
H3C—Cl +
Methyl
choloride
N—H
[C7H13N]
Nordihydro-tropidine
(III)
Zn-dust
Distillation
(Reduction)
N
C2H5
[C7H9N]
2-Ethyl pyridine
(IV)

The chemical structure of **'tropine' ['Z']** can also be as given under:

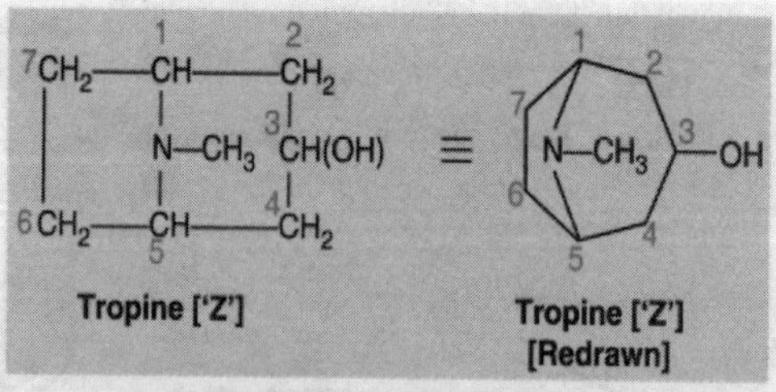
7CH2
CH
CH2
N—CH3
CH(OH)
6CH2
CH
CH2
≡
N—CH3
OH
Tropine ['Z']
Tropine ['Z']
[Redrawn]

(k) Logical Explanation of Various Reactions of *Tropine* Based on the Above Redrawn Structure ['Z']:

In fact, the following *four* important reactions of **'tropine'** may be expatiated logically based upon the **redrawn structure of tropine' ['Z'],** namely :.

- 2–Ethylpyridine from Tropine,
- Tropinone and Tropinic Acid from Tropine,
- Tropilidene (or Cycloheptatriene) from Tropine, and
- Pimelic acid from Tropinic Acid.

These aforesaid reactions shall now be treated individually in the sections that follows:

(a) 2–Ethylpyridine from Tropine :

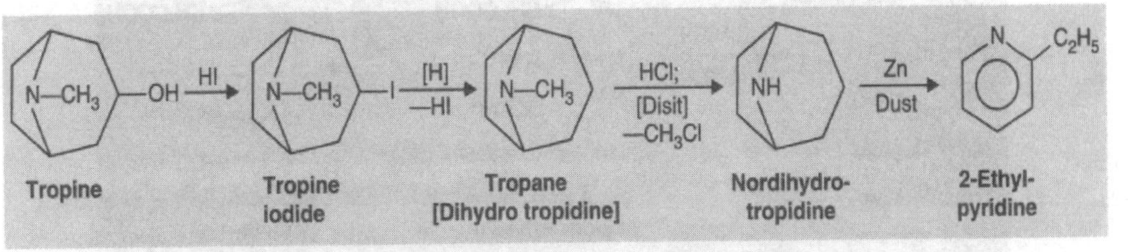

Explanation : Interaction of **tropine** with **HI** yields **tropine iodide** which upon reduction loses a mole of HI and form **tropane (dihydrotropidine)**. The resulting product when distilled with HCl loses a mole of **methyl chloride** producing **nordihydrotropidine**. The product thus obtained when reacted with **Zn–dust** gives rise to the formation of **2–ethyl pyridine**.

(b) Tropinone and Tropinic Acid from Tropine :

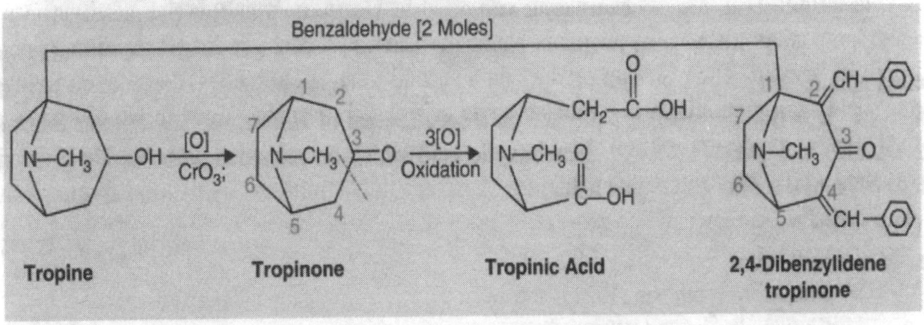

Explanation : **Tropine** on being subjected to oxidation with **chromic acid [CrO₃]** yields **tropinone**, which upon further oxidation using *three* nescent oxygen atom yield **tropinic acid** thereby affording cleavage between **C–3 and C–4** atoms. However, **tropinone** on treatment with *two* moles of **benzaldehyde** yields **2,4–dibenzylidene tropinone** directly by the loss of *two* moles of water.

(c) Tropilidine [or Cycloheptatriene] from Tropine :

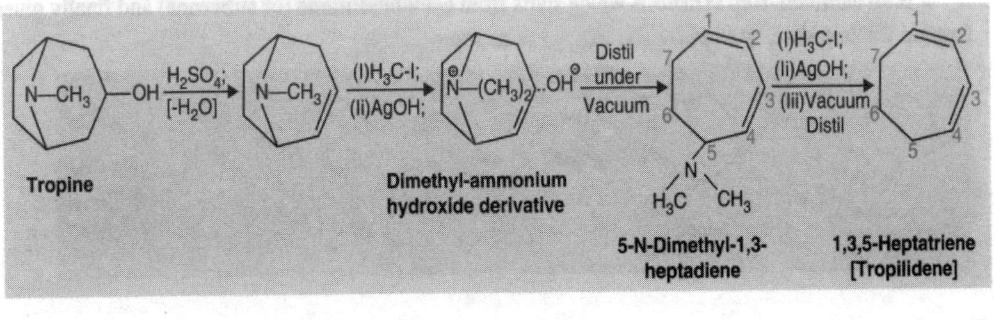

Explanation : Tropine when treated with sulphuric acid loses a mole of water to form an **unsaturated derivative.** The resulting product on being treated with **methyl iodide [H₃CI]** followed by **moist silver oxide [or AgOH]** the corresponding **ammonium hydroxide derivative.** The end product obtained above when distilled under vacuum yields **5N–dimethyl–1,3–heptadiene**, which on sequential treatment with H₃C–I, AgOH, and vacuum distillation yields **1,3,5–heptatriene [or tropilidene].**

(d) **Pimelic Acid From Tropinic Acid:**

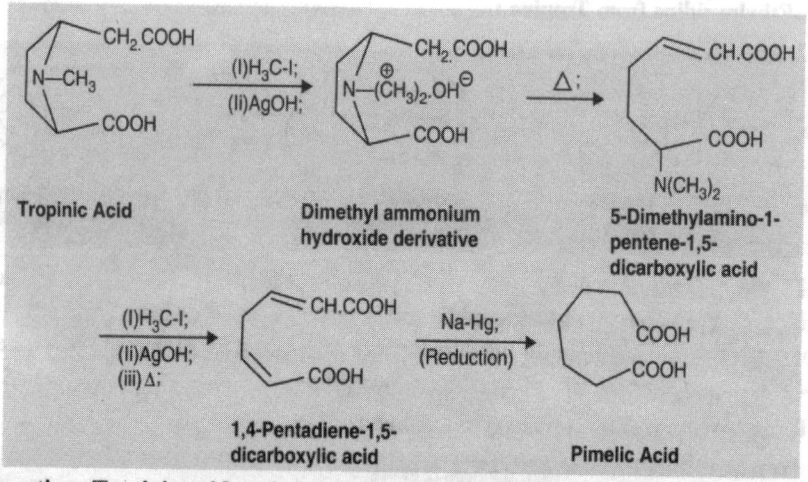

Tropinic Acid

Dimethyl ammonium hydroxide derivative

5-Dimethylamino-1-pentene-1,5-dicarboxylic acid

1,4-Pentadiene-1,5-dicarboxylic acid

Pimelic Acid

Explanation: **Tropinic acid** on being subjected to **Hofmann's exhaustive methylation** yields dimethyl ammonium hydroxide derivative, which on heating produces **5–dimethylamino–1–pentene–1, dicarboxylic acid.** The resulting product on a second treat of Hofmanns' degradation followed by heating yields **1, 4–pentadiene–1,5–dicarboxylic acid** which on reduction with **Na–Hg** gives rise to the formation of the desired **pimelic acid.**

(5) **Synthesis :** The structures of **'tropine'** has been duly confirmed by its **total synthesis in** *three* different manners, namely :

- Willstälter Synthesis (1900–1903),
- Robinson Synthesis (1921), and
- Elming Synthesis (1958).

These synthesis shall now be discussed at length with complete explanation individually in the sections that follows:

5.1. Willstälter Synthesis

It is an **eighteen–step synthesis** which starts from **cycloheptanone (or suberone)** and finally gives **tropine** *via* **tropidine, ψ-tropine,** and **tropinone** as stated under:

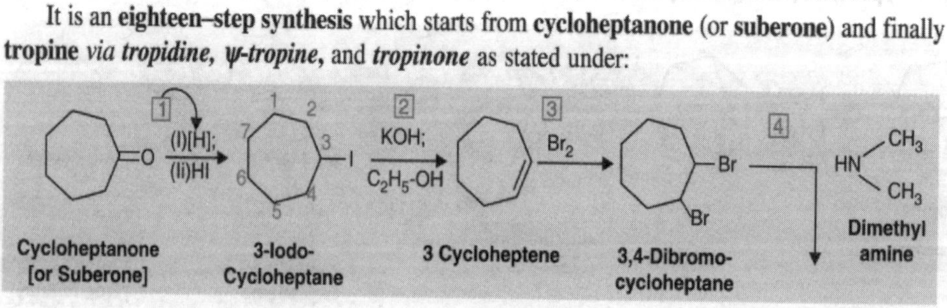

Cycloheptanone [or Suberone]

3-Iodo-Cycloheptane

3 Cycloheptene

3,4-Dibromo-cycloheptane

Dimethyl amine

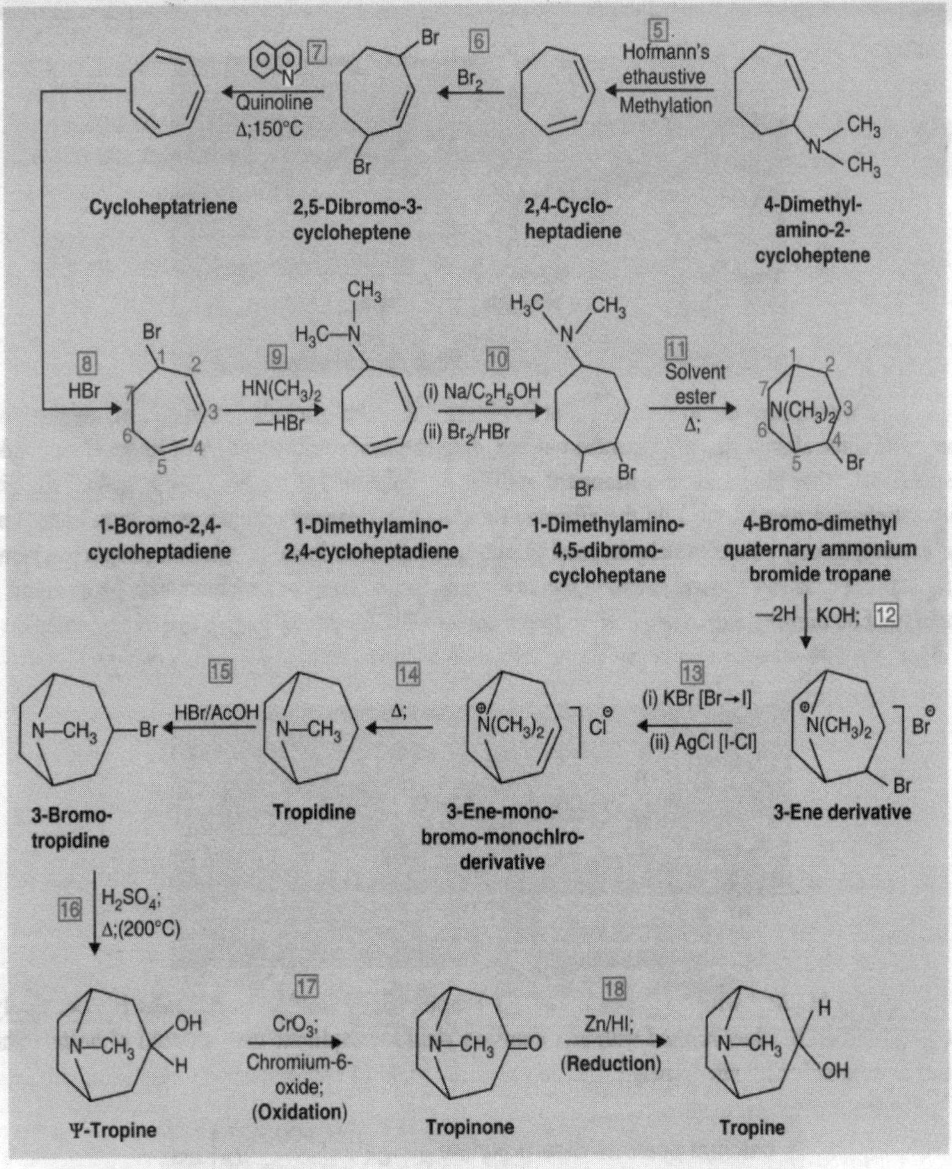

The reaction scheme shows the following compounds and transformations:

Cycloheptatriene ← [7] Quinoline Δ;150°C ← 2,5-Dibromo-3-cycloheptene ← [6] Br₂ ← 2,4-Cyclo-heptadiene ← [5] Hofmann's exhaustive Methylation ← 4-Dimethyl-amino-2-cycloheptene (N with CH₃, CH₃)

[8] HBr → 1-Boromo-2,4-cycloheptadiene → [9] HN(CH₃)₂ —HBr → 1-Dimethylamino-2,4-cycloheptadiene → [10] (i) Na/C₂H₅OH (ii) Br₂/HBr → 1-Dimethylamino-4,5-dibromo-cycloheptane → [11] Solvent ester Δ; → 4-Bromo-dimethyl quaternary ammonium bromide tropane

—2H | KOH; [12]

3-Bromo-tropidine (N—CH₃ —Br) ← [15] HBr/AcOH ← Tropidine (N—CH₃) ← [14] Δ; ← 3-Ene-mono-bromo-monochloro-derivative (N(CH₃)₂ Cl⁻) ← [13] (i) KBr [Br→I] (ii) AgCl [I-Cl] ← 3-Ene derivative (N(CH₃)₂ Br⁻ Br)

[16] H₂SO₄; Δ;(200°C)

Ψ-Tropine (N—CH₃ OH H) → [17] CrO₃; Chromium-6-oxide; (Oxidation) → Tropinone (N—CH₃ =O) → [18] Zn/HI; (Reduction) → Tropine (N—CH₃ H OH)

5.2. Robinson Synthesis

Robinson made a spectacular assumption that the meticulous and careful hydrolysis of **'tropinone'** causes a rupture in the **dihydro pyranotropane ring system** into the following *three* distinct chemical entities (or compounds), namely :

- **succinyldialdehyde,**
- **methylamine,** and
- **acetone.**

Thus, we may have the following expression:

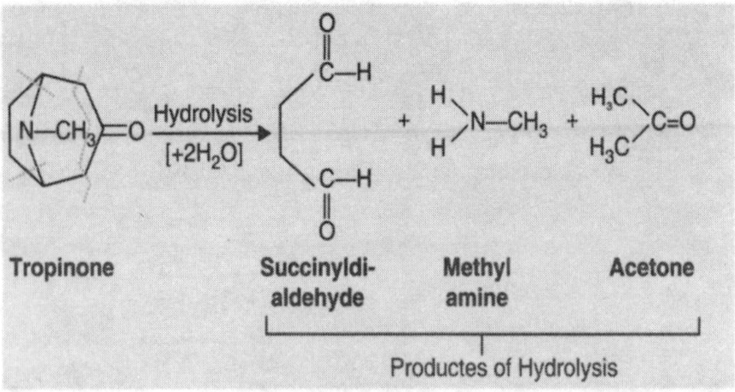

Tropinone	Succinyldi-aldehyde	Methyl amine	Acetone

Productes of Hydrolysis

It is, however, pertinent to state here that **Robinson** did advocate a rather strong belief, opinion and concept that the aforesaid *three* **chemical entities,** duly obtained as the *products of hydrolysis,* should be joined together by the help of **Mannich Reaction*** occurring *'twice'* to yield **'tropinone'** just in *'one-single-step only'.* Based on the aforesaid **ideology** and **concept** a mixture of *three* **chemical substances,** namely : **succinyldehyde, methylamine,** and **acetone** in **stoichiometric proportions** was duly exposed to an *aqueous environment* for a duration of **30 minutes** to give rise to the formation of a duration of **30 minutes** to give rise to the formation of **'tropinone'** in very *small quantum.*

Thus, we may have :

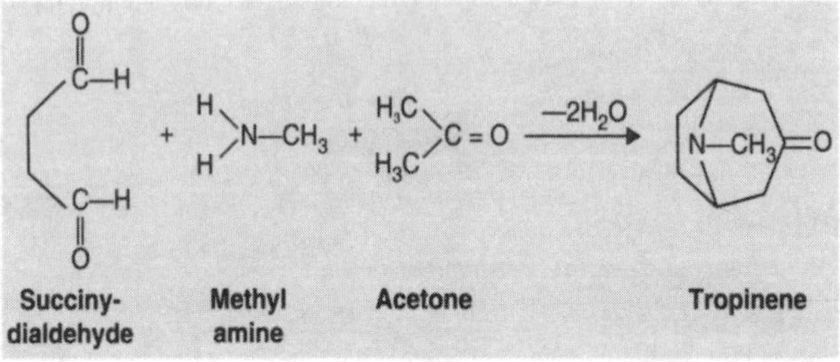

Succiny-dialdehyde	Methyl amine	Acetone	Tropinene

Modified Method with Enhanced Yield of Tropinone Upto 40%: An attempt has been duly made to modify the *older method* with an **enhanced yield of tropinone** using instead of *acetone* the following *two* **substituents,** namely :

- **calcium acetone dicarboxylate** $\left[O = C \begin{cases} CH_2-COO \\ CH_2-COO \end{cases} Ca \right]$,and

- **ethyl acetone dicarboxylate** $\left[O = C \begin{cases} CH_2.COOH \\ CH_2.COOH \end{cases} \right]$

Thus, we may have the following reactions :

---

* Mannich C and Krosche W : *Arch. Pharm.,* **250** : 647, 1912.

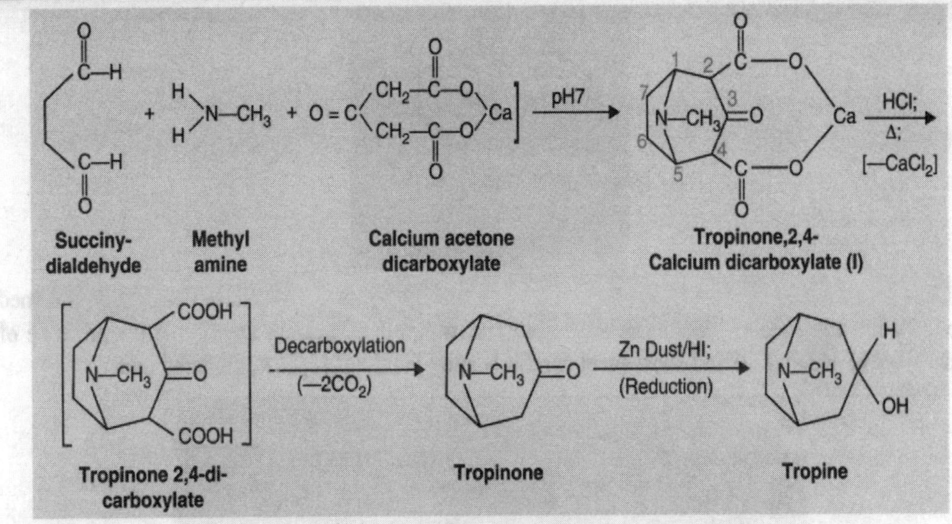

| Succiny-dialdehyde | Methyl amine | Calcium acetone dicarboxylate | | Tropinone,2,4-Calcium dicarboxylate (I) |

| Tropinone 2,4-di-carboxylate | Tropinone | Tropine |

Explanation : The interaction amongst **succinyl dialdehyde, methylamine,** and **calcium acetone dicarboxylate** at almost *neutral pH (7)* yields the corresponding **tropinone, 2,4–calcium dicarboxylate (I),** which on treatment with HCl and subsequent heating gives rise to the formation of an *intermediate i.e.,* **tropinone, 2,4–dicarboxylate** with the elimination of a mole of $CaCl_2$. The resulting intermediate on *decarboxylation* yields **tropinone (II)** which upon **reduction with Zn–dust** gives the desired product 'tropine'.

5.3. Elming Synthesis

Elming (1958) proposed a simple and convenient synthesis which only comprise *three* steps, namely :

- Conversion of **2, 5–dimethoxytetrahydrofuran (I)** with **hydronium ion (H_3O^+)** to produce **succinyldialdehyde (II),**
- treatment of the **product (II)** with *two* reactants, such as : **methylamine hydrochloride,** and **acetone dicarboxylate** (both being *water–soluble*) to yield the desired product 'tropinone (III) and
- reduction with Zn–dust/HI produces 'tropine' (IV).

We may have the following reactions :

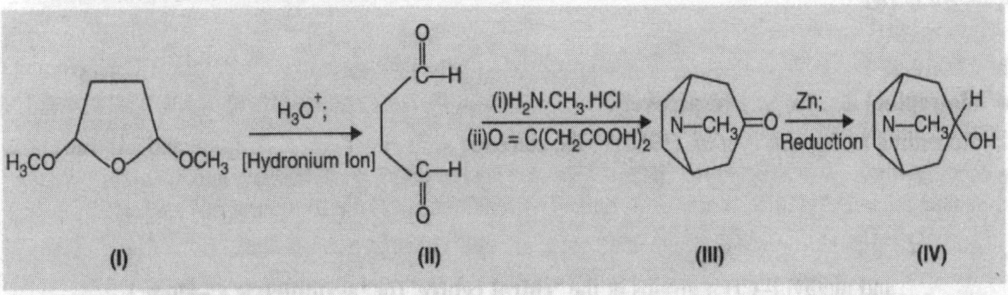

| (I) | (II) | (III) | (IV) |

(6) **Synthesis of Atropine :** It has already been proved and, established that 'atropine' is nothing but a 'tropine–tropate ester' [see section 3.4.1.4.(2)]. Based on the above logical findings and scientific evidences the following structure has been duly assigned for 'atropine' *viz.,*

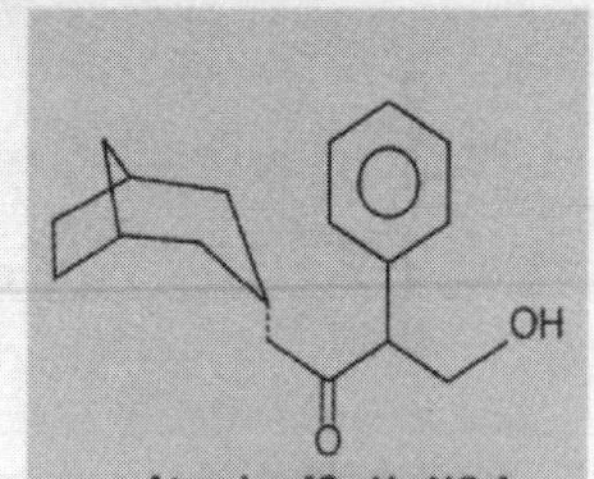

Atropine [$C_{17}H_{23}NO_3$]

Nevertheless, the **chemical structures** of both **tropic acid** and **tropine** have been duly confirmed explicitly in earlier sections by actual *total synthesis,* which upon heating together in the presence of **hydrogen chloride (HCl)** and subsequent careful heating yields '**atropine**', as stated under :

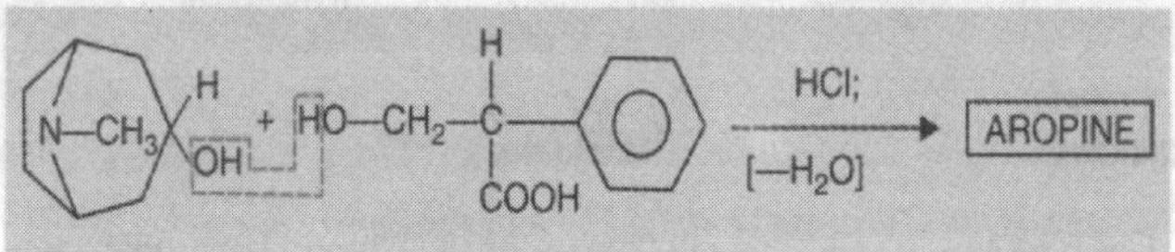

Tropine   Tropic acid

3.4.1.5. Conformation of *'Tropines'*

Both **pseudotropine** (or **ψ–Tropine**) and **tropine** do have the following conformations :

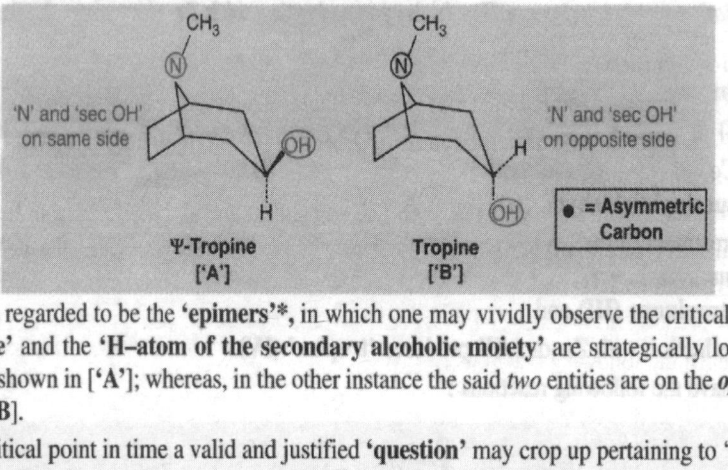

These are regarded to be the '**epimers**'*, in which one may vividly observe the critical positions of the '**N–bridge**' and the '**H–atom of the secondary alcoholic moiety**' are strategically located on the *same side,* as shown in ['**A**']; whereas, in the other instance the said *two* entities are on the *opposite side,* as shown in [**B**].

At this critical point in time a valid and justified '**question**' may crop up pertaining to 'which of the above two forms of geometrical isomers shall actually and logically represent **ψ–tropine** (or **pseudotropine**) and **tropine** respectively.

According to Beyerman *et al.* (1956)** the **stereochemistry** of **pseudotropine** and **tropine** could be visualized and explained more authoritatively if one considers the following *two* aspects:

- '**piperidine ring**' assumes a '**boat**' or '**chair**' *conformation,* and
- possibility of an '**equatorial orientation**' or an '**axial orientation**' of the hydroxyl (–OH) and methyl (–CH₃) groups at the '**chiral centre**' (or '**asymmetric C–atom**').

---

* **Epimer :** It refers to a molecule that differs from the configuration of another by one '**asymmetric carbon**'.
** Beyerman *et al.: Rec. Trav. Chim.* **75** : 1445, 1956.

It is, however, pertinent to state here that based on the analogy with *cyclohexane* and *pyranosides* the 'chair conformation' has been duly proposed for the **piperidine ring** present in **pseudotropine** and **tropine.** Beyerman *et al.* also suggested that :

- *secondary* alcoholic function in ψ–tropine is 'equitorial oriented', and
- *secondary* alcoholic moiety in **tropine** is 'axial oriented'.

Fodor and Csepreghy (1961)* articulately confirmed the aforesaid observations by performing :

- **X–ray Diffraction analysis** of 'tropine hydrobromide',
- Determining the **'Dipole Measurements'** and
- **¹H–NMR spectroscopic studies** of certain **'tropane structural analogues'.**

Nevertheless, most of these physico-chemical determinations reconfirmed and revealed that the critical **N–methyl moiety** in piperidine is predominantly having an *'equatorial chair conformation'.* Besides, the above spectacular findings were also duly supported overwhelmingly so as to reveal that in

the **'tropine molecule'** the inherent *'piperidine nucleus'* is strategically present as:

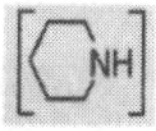

- ❑ **'chair conformation'** – *major quantum,* and
- ❑ **'boat conformation'** – *minor quantum.*

The following structures of both **pseudotropine** [ψ–*tropine*] and **tropine** are self–explanatory :

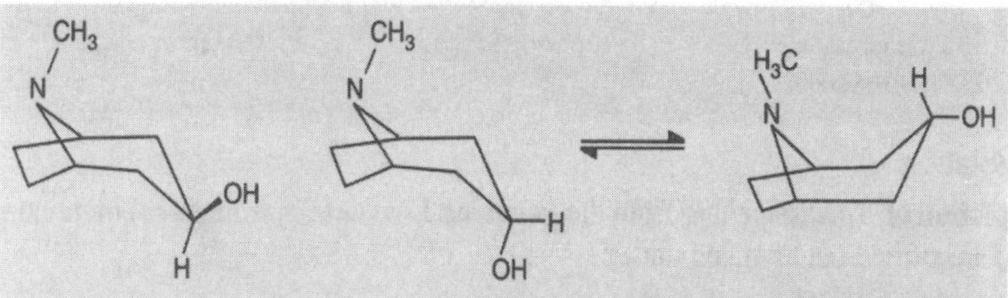

| Pseudotropine [Ψ-Tropine] | Tropine ['Chair' Conformation] | Tropine ['Boat' Conformation] |

Hyoscine [or Scopolamine] : Hyoscine, an *anticholinergic,* designates a **tropane alkaloid** duly isolated from : *Datura metal L., Scopola carniolica* Jacq., and other members of the family *Solanaceac.* Interestingly, **'homatropine'** represents a purely **synthetic alkaloid** duly prepared from tropine and **mandelic acid,** which acts as a **mydriatic** *i.e;* dilating the pupil of the eyes (like **atropine**).

The chemical structures of **hyoscine** (or **scopolamine**) and **homatropine** are as given under :

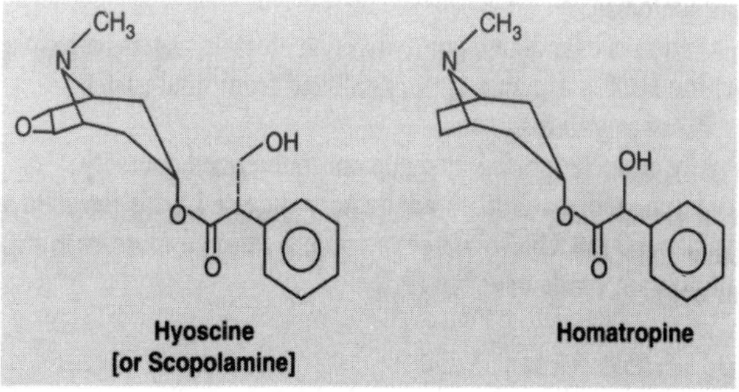

Hyoscine [or Scopolamine]   **Homatropine**

---

* Fodor and Csepreghy : *J Chem Soc.,* **115** : 3222, 1961

3.4.2. Cocaine [or β–Cocaine; Benzoylmethylecgonine]

3.4.2.1. Preamble

Squibb (1885)* was pioneer in the extraction and isolation of **'cocaine'** from the leaves of *Erythroxylon coca* Lam., and other species of **Erythroxylon** belonging to the natural order *Erythroxylaceae* or even by **synthesis** later on. It has been found that the leaves of *E. coca* may contain up to 1–2% **alkaloids** comprising mainly *'cocaine'* and also **cinnamyl cocaine** to certain extent only. In addition, it also contains several other vital and important constituents as :

- Tropacocaine (or Tropacaine),
- Benzoyl cocaine,
- Ecgonine,
- Benzoylecagonine,
- Cinnamoyl cocaine, and
- Hygrine.

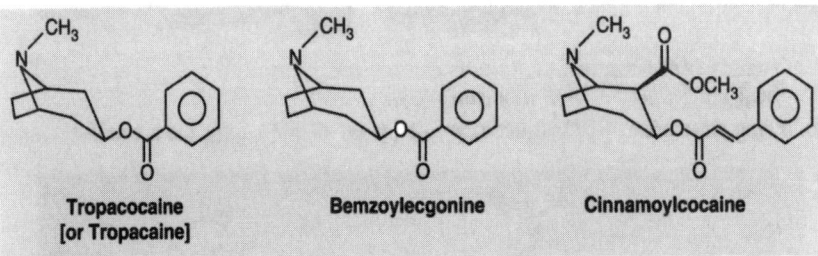

**Tropacocaine
[or Tropacaine]**   **Bemzoylecgonine**   **Cinnamoylcocaine**

3.4.2.2. Isolation

The **isolation of** *'cocaine'* either from the **Peruvian Leaves** (*i.e.*, from **Peru**) or **Javain Leaves** (*i.e.* from **Java**) are carried out as stated under:

S.No.   Isolation from Peruvian Leaves

1   Leaves are air–dried in shade, and powdered coarsely.

2   The powdered material is digested thoroughly either with **'lime powder'** or **'sodium carbonate'** *i.e., alkaline substances to allow the liberation of the* **'alkaloid'**) in a minimum quantum of dimineralized (DM) water.

3   Digested leaves is extracted with **light Petroleum Ether** (60–80°C) in a Soxhlet Extraction Assembly for **several hours**.**

4   The petroleum ether extract is acidified with **dilute HCl** to solubilize **cocaine** as its salt *viz.,* cocaine hydrochloride.

5   The acidic solution is evaporated gently over an **electric** water–bath (in a **fuming cup Board**), and the **cocaine HCl** is separated, recrystallized from dilute HCl.

S.No   Isolation From Javaian Leaves

1   Leaves are duly collected, dried in shade and pulverized coarsely.

2   Pulverized and sieved material is adequately digested with **lime/Na$_2$CO$_3$**–extraceted with Petroleum Ether (60–80°C)–Solvent evaporated–residual mass is acidified with dilute HCl – evaporated to obtain crude **cocaine HCl.**

---

* Squibb : *Pharm J*, [3], **15** : 755, 796; **16** : 67, 1985.

** Till a drop of the solvent collected fromthe lower end of the **'Soxhlet Thimble'** on a watch–glass upon evaporation fails to give a +ve test for alkaloids with **Dragendorff's Reagent.**

3   Recrystallization of the resulting product from dilute HCl yield **pure crystalline salt of cocaine.**

4   **Mother Liquor,** containing a mixture of other allied alkaloids is duly hydrolyzed to by boiling with dilute HCl for 60 mts to obtain **'ecgonine'.** The resulting solution when pow-ered into water (DM), the **truxillic acid*** gets knocked out. **Ecgonine HCl** crystallizes out, after evaporation of the **'filtrate'.** The hydrochloride gets decomposed with Na_2CO_3 to obtain free **'ecgonine',** which is duly extracted with **dilute alcohol.** Alcohol is evaporated off and **ecgonine** treated with **'benzoic anhydride'** to produce **benzoyl ecgonine.** The ex-cess of added *benzoic anhydride/benzoic acid* is duly removed by successive extraction with **ether.** Residue is washed with water to get rid of **'ecognine'.** The resulting **benzoyl ecgonine** is duly **methylated** with **CH_3I/Na–\MCOH** to obtain **'cocaine'** quantitatively.

3.4.3.3. Characteristic Features

These essentially include :

(1)   **Cocaine** is obtained as **monoclinic** colourless crystals mp 98°C.

(2)   **Interestingly, cocaine** is found to be volatile especially above 90°C, however, the **'subli-mate'** is not crystalline in nature.

(3)   It has a $bp_{0.1}$ 187–188°C.

(4)   It shows various specific rotations in different solvents:

$[\alpha]^{18}$ D : –35° [in 50% (*u/v*) ethanol];

$[\alpha]^{20}$ D : –16° [**c** = 4 in chloroform].

(5)   **pka/pkb. Values:** The aqueous solution of **cocaine** are found to be **alkaline to litmus :**
pka (15°) = **8.61**; and **pkb** (15°) = **5.59.**

(6)   **Solubility Pofile:** The **'solubility profile'** of **cocaine** are as given under:

1g **cocaine** dissolves in	=	600 mL water;
	=	270 mL water at 80° C;
	=	6.5 mL ethanol;
	=	3.5 mL diethyl ether;
	=	12 mL Olive oil; and Turpentine oil;
	=	30–50 mL Liquid Petrolatum.

CAUTION: *'Cocaine'* is mostly employed as a *local anaesthetic* in optical surgery and dentistry. Besides, 'cocaine' is a habit-forming drug; and hence, must be used with extreme *care* and *CAUTION.* When it is administered orally, it invariably enhances predominantly the *physical* as well as *mental* strength, power and alertness to such an extreme limit that the overall after–effect is apparently a 'deep depression'. Thus, individuals who are acutely suffering from *'chronic cocainism'* or *'habitual addiction'* usually suffer from ill health, wild/violent behaviour, hallucinations, and may ultimately become *in-sane.*

---

* **Truxillic Acid** : A cinnamic acid polymer(s) obtained from the **monor alkaloids** of *cocaine.*

3.4.3.4. Constitution of Cocaine

The 'constitution of cocaine' ultimately leading to its assigned structure and total synthesis are duly expatiated in the following section :

(1) **Molecular Formula** : Based on the analytical data of C, H, N, O analysis and molecular weight determination by mass spectroscopy, both the **emperical formula** and **molecular formula** remains the same and stands at $C_{17}H_{21}NO_4$.

(2) **Status and Nature of N-atom:** The exact and precise status/nature of **N–atom** present in *cocaine* having a relatively *strong tertiary base* (**pKa : 8.61**) may be ascertained by such methods as :

(a) **Formation of Methiodide:** A mole of **methyl iodide [H₃CI]** adds on to **cocaine** to give rise to the formation of **cocaine methiodide** as given under:

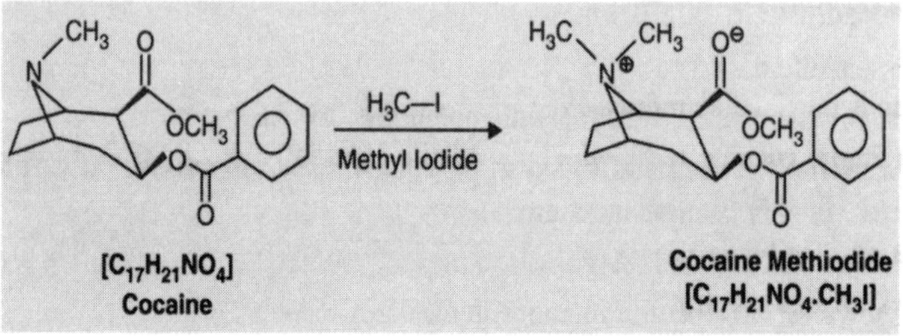

$[C_{17}H_{21}NO_4]$
Cocaine

Cocaine Methiodide
$[C_{17}H_{21}NO_4.CH_3I]$

(b) **Presence of N–Methyl Moiety :** Cocaine reacts with **cyanogen bromide (CNBr)** to yield *two* products of reaction, namely : **methyl bromide [H₃CBr]**, and **cyanonorcocaine,** as shown below :

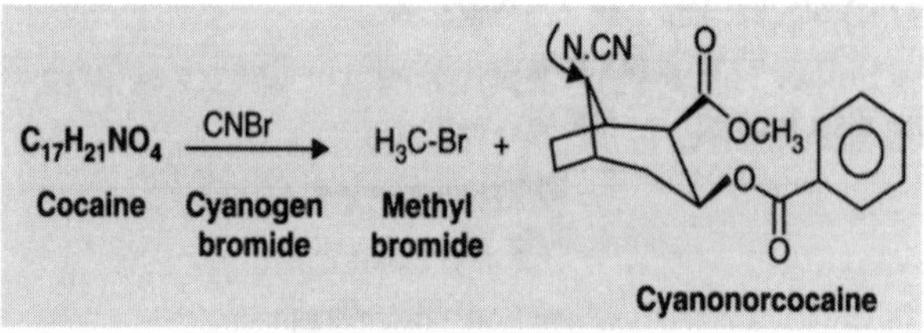

$C_{17}H_{21}NO_4$ $\xrightarrow{\text{CNBr}}$ $H_3C\text{-Br}$ +

Cocaine   **Cyanogen bromide**   **Methyl bromide**

Cyanonorcocaine

Thus, the **N–methyl** moiety **cocaine** and *bromide* from cyanogen bromide gets eliminated as **methyl bromide** there by yielding the corresponding **cyaonorcocaine.**

(3) **Hydrolysis :** Cocaine on being subjected to hydrolysis with gentle heating yields a mole each of **benzoylecgonine** and **methanol**

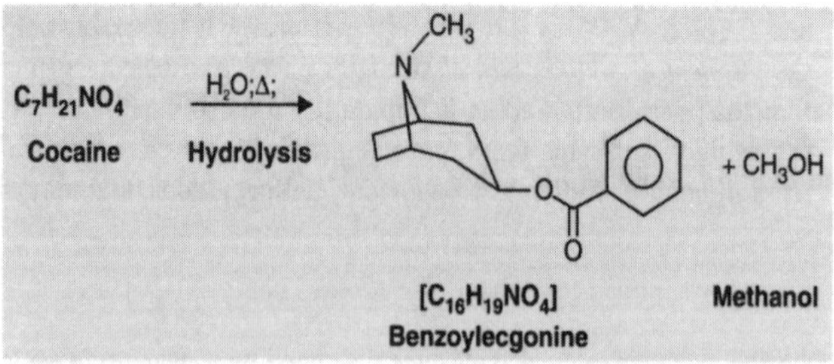

$C_7H_{21}NO_4$ $\xrightarrow{\text{H}_2\text{O};\Delta;}$

Cocaine   **Hydrolysis**

+ CH₃OH

$[C_{16}H_{19}NO_4]$ **Methanol**
Benzoylecgonine

Inference: Since, **benzoylecgonine** comprises a **carboxyl (–COOH) moiety** that critically suggests that *'cocaine'* must be the **'methyl ester of benzoylecgonine'**.

The above logical derivation may be further substantial by the fact that interaction between **methanol** and **benzoyl ecgonine** in HCl and heating yields *cocaine.*

Another Additional Supportive Evidence : **Benzoylecognine** when boiled with **barium hydroxide** **[Ba(OH)$_2$] solution** *i.e.,* in an *alkaline medium*, it crucially undergoes further hydrolysis to produce **ecgonine** and **benzoic acid** as stated under :

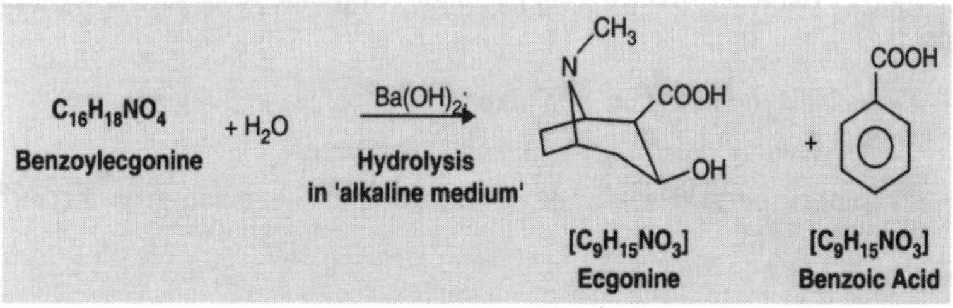

$$C_{16}H_{18}NO_4 \quad + H_2O \xrightarrow[\text{Hydrolysis}]{\text{Ba(OH)}_2;}$$

Benzoylecgonine

Hydrolysis in 'alkaline medium'

[C$_9$H$_{15}$NO$_3$]
Ecgonine

[C$_9$H$_{15}$NO$_3$]
Benzoic Acid

Inference : **Ecgonine** explicitly exhibits the typical reactions of an **'alcohol',** which strongly suggests that **'benzoylecgonine'** should be the **'benzoyl structural analogue'** of **ecgonine.**

(4) Constitution of Ecgonine : The various cardinal steps involved in the **constitution of ecgonine** are as enumerated under :

(a) **Molecular Formula :** Its molecular formula is C$_9$H$_{15}$NO$_3$.

(b) **Presence of *tert*–N–Atom :** **Ecgonine** has a *tertiary base,* since it forms the typical and characteristic crystalline **additive compound** *ecgonine methyl iodide* **[C$_9$H$_{15}$NO$_3$. CH$_3$I]** with **methyl iodide (CH$_3$–I)** as given under :

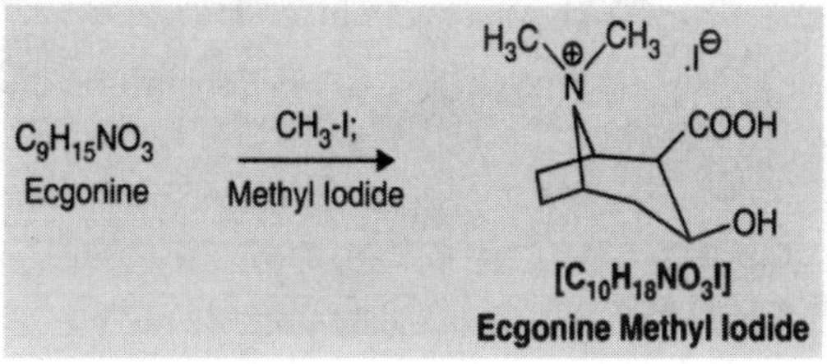

$$C_9H_{15}NO_3 \xrightarrow[\text{Methyl Iodide}]{\text{CH}_3\text{-I};}$$

Ecgonine

[C$_{10}$H$_{18}$NO$_3$I]
Ecgonine Methyl Iodide

Inference : The above reaction suggests that **ecgonine** molecule essentially consists of an inherent *tert*–N–atom.

(c) **Formation of an Ester and Salt :** **Ecgonine** gives rise to the formation of an **'ester'** with an *alcohol viz.,* **ethanol** thereby conforming that comprises **one carboxyl (–COOH) moiety.**

Besides, **ecgonine** also yields a corresponding salt with an **alkali** *viz.,* NaOH, thereby establishing that it possesses **one carboxyl (–COOH) group.**

(d) **Presence of a Hydroxyl (–OH) Moiety :** **Ecgonine** critically reacts with either an **acid chloride** *viz.,* **acetyl chloride [CH$_3$COCl]** or an **acid anhydride** *viz,* **acetic anhydride [(CH$_3$CO)$_2$O]** which suggests profusely that it has a **hydroxyl (–OH) moiety present in its structure.**

Inference : Because, the resulting **'acetyl derivative of ecgonine'** may be further duly *'esterified',and* it vividly ascertains the fact that **'ecgonine'** is both an **alcohol** and an **acid.**

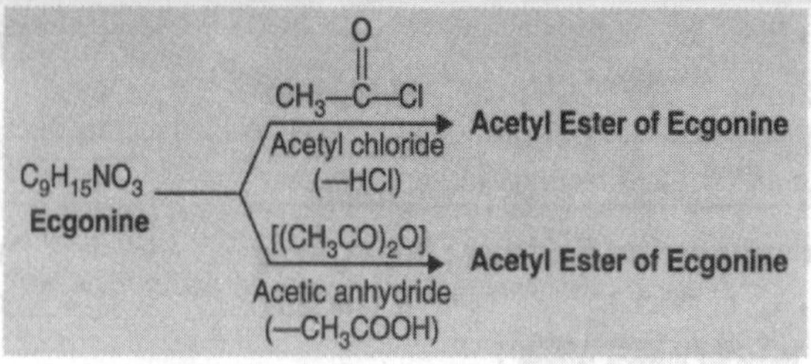

$C_9H_{15}NO_3$
Ecgonine

Acetyl chloride
(—HCl)
→ **Acetyl Ester of Ecgonine**

$[(CH_3CO)_2O]$
Acetic anhydride
(—CH$_3$COOH)
→ **Acetyl Ester of Ecgonine**

(e) **Degradative Products Obtained from Ecgonine : Ecgonine** yields several degradative products as given under :

- **Ecgonine** upon oxidation with chromium–6–oxide yield **ergoninone,**
- **Ergoninone** on decarboxylation yields **tropinone,**
- **Tropinone** on further oxidation with chromium–6–oxide gives **Ergoninic Acid** and **Tropinic Acid.**

All these *three* reaction may be summararily expressed as below:

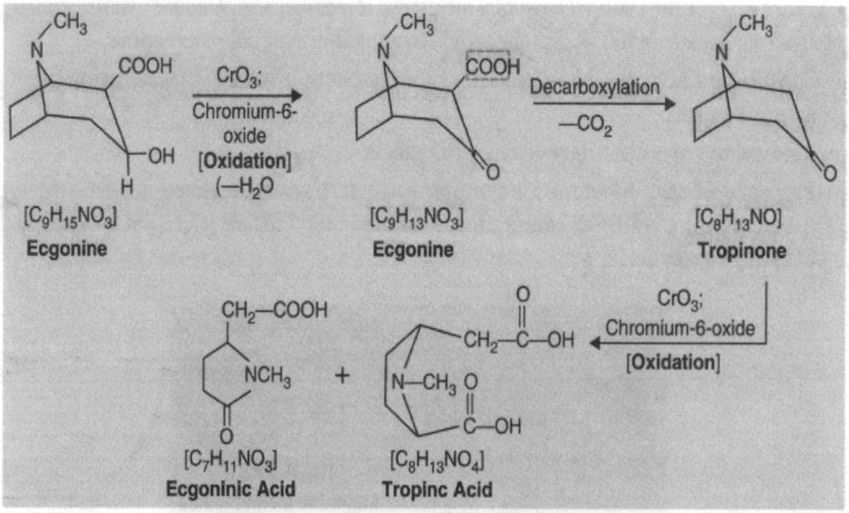

Based on the nature of *products of reaction* duly accomplished by **oxidation** of *ecgonine,* followed by **decarboxylation** and **oxidation,** as given above, one may arrive at the following *inferences,* and *conclusions,* namely :

(i) **Ecgonine possesses a 'Tropane' Skeleton and Position of the Secondary Alcoholic (–CHOH) Moiety remains akin to 'Tropine' :**

The course of reaction due to oxidation–decarboxylation of **'ecgonine'** first yields **'tropinone',** which upon further oxidation gives rise to the formation of **'tropinic acid'.**

Therefore, it evidently suggests that :

- **Ecgonine** essentially possesses a **'tropane skeleton',** and
- Position of the *secondary 'Alcoholic Moiety'* remains very much akin to **'Tropine',** as illustrated clearly under :

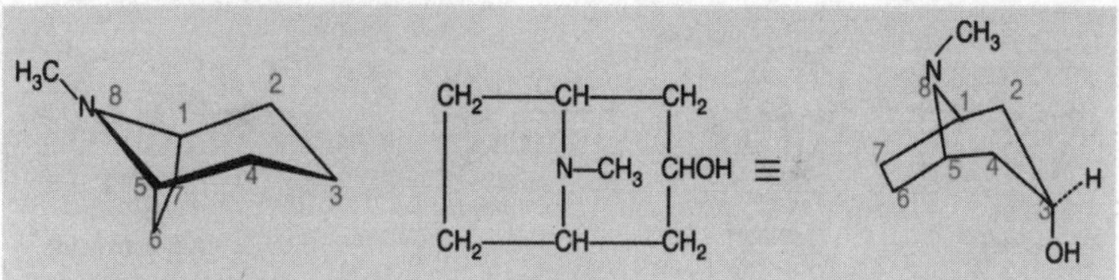

'Tropane' Skeleton   Tropine   Tropine

Close Resemblance to Structures of Ecgonine and Tropine : It has been duly observed that there exists a close similarity between the chemical structures of **'ecgonine'** and **'tropine'** based on the fact that :

- **Ecgonine** upon *dehydration* yields **'anhydrecgonine'**,
- **Anhydrocegonine** on *decarboxylation* produces **'tropidine'**, and
- **Tropidine** is also obtained by dehydration of **'Tropine'**.

Thus, we may have the following reactions :

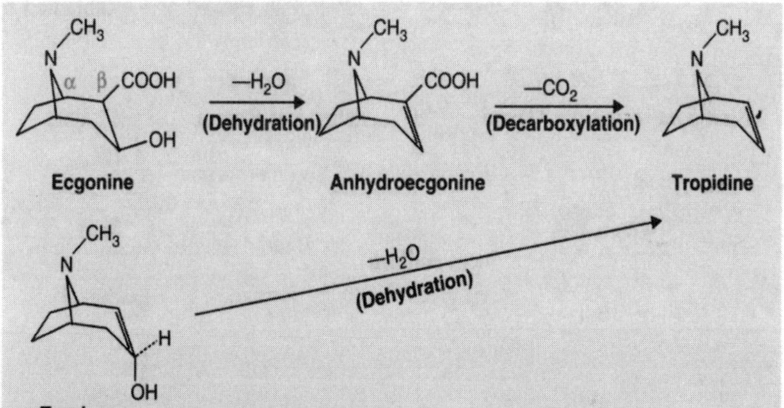

Ecgonine   Anhydroecgonine   Tropidine

Tropine

(*ii*) **Ecgonine** undergoes a rapid and fast decarboxylation phenomenon which suggests overwhelmingly that it must be a β–keto acid.

Willstatter *et al.* (1923) eventually confirmed the aforesaid **fact, observation,** and **interpretation** the actual critical formation of an **unstable ketonic acid** which ultimately is deprived of a molecule of **carbon dioxide (CO$_2$)** gives rise to the formation of **'tropinone'**.

Thus, we may have the structure of **'ecgonine'** as given under :

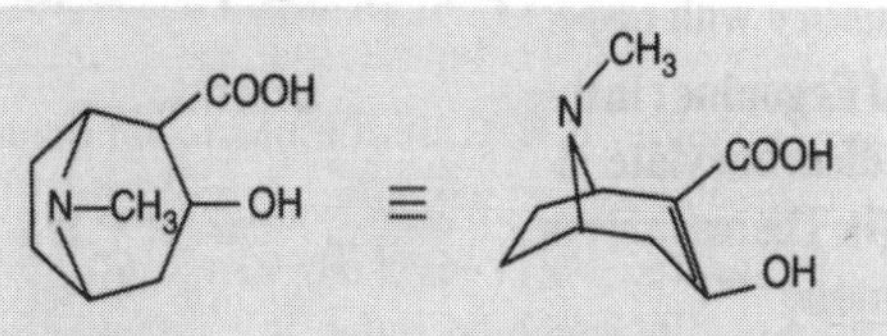

Ecgonine   Ecgonine

(f) It is, however, pertinent to state here that the above chemical structure of **'ecgonine'** obviously explains all its reactions :

---

    *   Willstalter *et al.: Ann.,* **434** : 111, 1923.

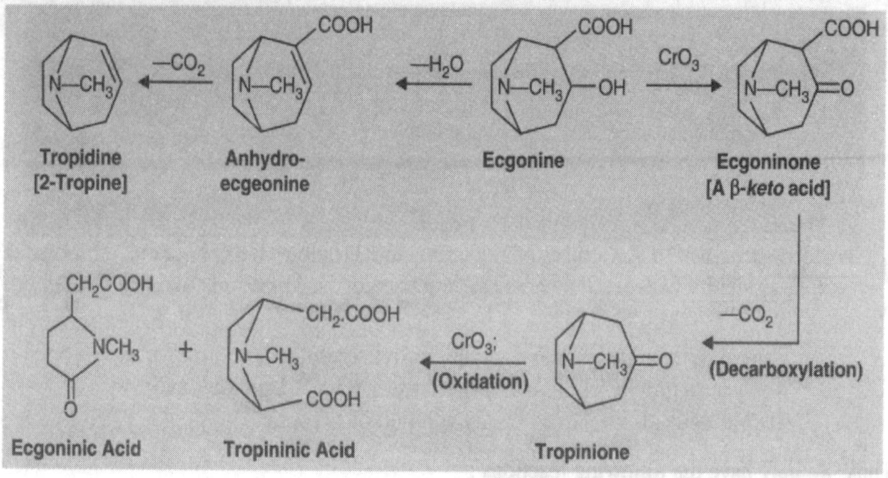

Tropidine
[2-Tropine]

Anhydro-
ecgeonine

Ecgonine

Ecgoninone
[A β-*keto* acid]

Ecgoninic Acid

Tropininic Acid

Tropinone

(g) **Synthesis** : The structure of 'ecgonine' is ultimately proved and established by its 'total synthesis'.

Willstätter put forward *two* separate synthesis of 'ecgonine', namely :

- *First Synthesis* from 'Tropinone' [Willstätter and Bommer, 1920]*, and
- *Second Synthesis* from 'Robinson's Method' [Willstatter *et al.* 1923]** and

(i) **First Synthesis of Ecgonine** : In this particular instance, the starting material is 'tropinone' (I), which on treatment with freshly cut **sodium (Na) metal** yields **tropinone sodium (III)**. The resulting **product (II)** on *carbonation* (CO_2) produces **sodium tropinone carboxylate (III)**, which finally upon reduction with **sodium amalgam (Na–Hg)** gives rise to the formation of **ecgonine (IV)**.

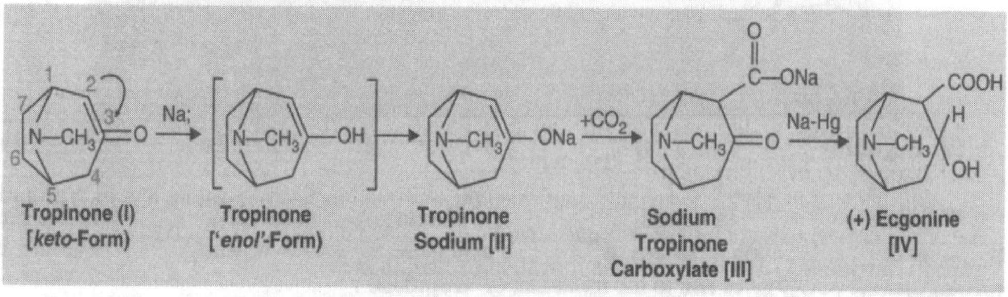

Tropinone (I)
[*keto*-Form]

Tropinone
['*enol*'-Form]

Tropinone
Sodium [II]

Sodium
Tropinone
Carboxylate [III]

(+) Ecgonine
[IV]

The resulting *racemic mixture* of (±)–**ecgonine** was found to be absolutely identical with (–)–**ecgonine** obtained from the *natural source* with respect to its **chemical properties**.

(ii) **Second Synthesis of Ecgonine :** In this case, the interaction amongst **succinyl dialdehyde, methyl amine,** and **ethyl acetone dicarboxylate** in the presence of *alkali* (KOH) yields **2–ethyl carboxylate tropinone–4–carboxylate (I).** The resulting **product (I)** undergoes **decarboxylation** on *heating* to produce **2–ethyl carboxylate tropinone (II)**, which on **reduction** first with **sodium amalgam (Na–Hg),** and subsequently upon **hydrolysis** gives rise to the formation of (±)–**ecgonine (III)**, as given under :

*   Willstätter *et al.: Ann.,* **434** : 111, 1923.

**   Willstätter and Bommer: *Ann.,* **422** :15,1920.

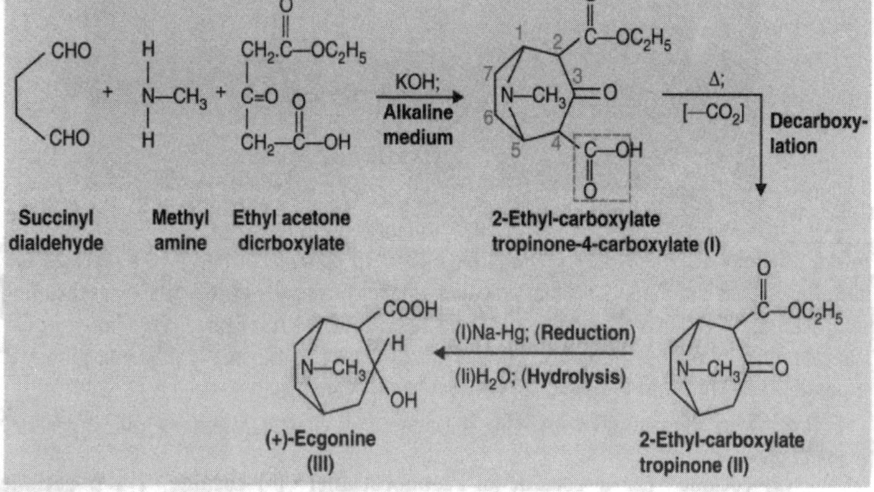

Succinyl dialdehyde + Methyl amine + Ethyl acetone dicrboxylate $\xrightarrow[\text{Alkaline medium}]{\text{KOH;}}$ 2-Ethyl-carboxylate tropinone-4-carboxylate (I) $\xrightarrow[\text{[—CO}_2\text{]}]{\Delta;}$ Decarboxylation

(I)Na-Hg; (Reduction); (Ii)H$_2$O; (Hydrolysis) → (+)-Ecgonine (III) ← 2-Ethyl-carboxylate tropinone (II)

Comments : It has been critically observed that instead of *only one racemic mixture* in the **final product** there were *three racemates,* namely :

- (±)–ecgonine,
- (±)–ψ–ecgonine, (or **pseudoecgonine**), and
- a '**third pair of entantiomers**'.

Nevertheless, the careful resolution of the racemic mixture of **ecgonine** [*i.e.,* (±)–ecgonine] yielded (–)–**ecgonine** which on being subjected to **esterification** with *methanol,* followed by **benzoylation** with **benzoyl chloride** produced (–)–**cocaine,** as shown under :

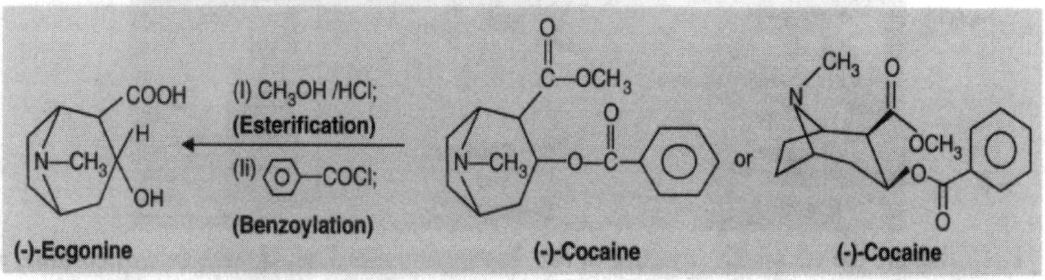

(-)-Ecgonine $\xrightarrow[\text{(Benzoylation)}]{\substack{\text{(I) CH}_3\text{OH /HCl;} \\ \text{(Esterification)} \\ \text{(Ii) } \bigcirc\text{—COCl;}}}$ (-)-Cocaine   or   (-)-Cocaine

(5) **Synthesis of Cocaine :** The above structure of '**cocaine**' has been justifiably proved and established by its **synthesis** that critically entails the meticulous resolution of the **racemic mixture of ecgonine** (±–ecgonine), immediately followed by benzoylation to yield '**cocaine**' which happens to be absolutely identical in every respect to the *naturally occurring* (–) **ecgonine** (*see above reaction*)..

Note: Likewise, (+) and (–)–ψ–cocaines (*i.e.,* pseudococaines) may be duly obtained from the racemic mixture of ψ–ecgonines [or (±)–ψ–ecgonines].

(6) **Stereochemistry of Cocaine :** It is, however, pertinent to state here that **ecgonine**; and, therefore, '*cocaine*' also possesses **four chiral centres** (or **four asymmetric C–atoms**) as indicated under :

---

* Woodward and Doering : *J Am. Chem. Soc,* **66** : 843, 1944.

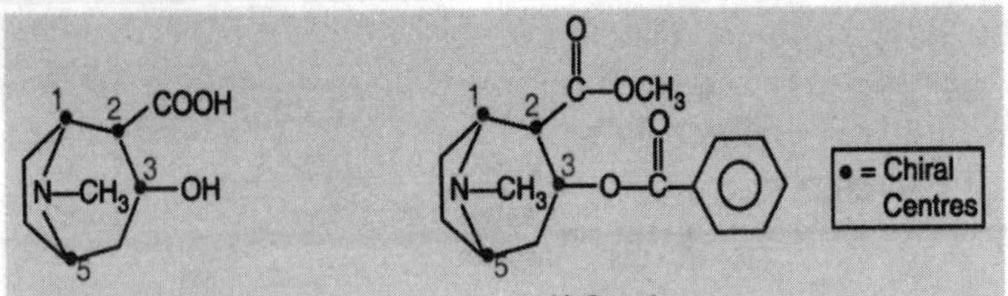

(-)-Ecgonine   (-)-Cocaine

Obviously, either of these *two* aforecited structures do have C–1, C–2, C–3 and C–5 as the 'chiral centres', thereby showing that they may have $(2^4=)$**16** *optically active forms* or **eight enantiomeric pairs**. Based on the preferential actual practices, one may critically observe the overwhelming feasibility and possibility of exclusively the so called '*cis–fusion*' of the *Nitrogen Bridge* between C–1 and C–5 yield categorically **only one configuration.** Thus, it gives rise to actually *four* **enantiomeric pairs** that are possible, of which, *three* have been duly prepared synthetically.

Following are the naturally occurring important and vital '**enatiomeric pairs**' generally found in the *coca leaves* :

- (±)–cocaine • (±)–ψ–cocaine (or **Pseudococaine**) • (–)–cocaine • (+)–ψ–cocaine.

(7) Conformational Variants of Cocaines and Ecgonines :

Findlay (1954)* and Kovacs *et al.* (1954)** explicitly proved and established that the following **conformational variants** do occur in :

- **Cocaine** and **Ecgonine**, and
- ψ–**Cocaine** [or **Pseudococaine**] and ψ–**Ecgonine** [or **Pseudococaine**],

as illustrated under :

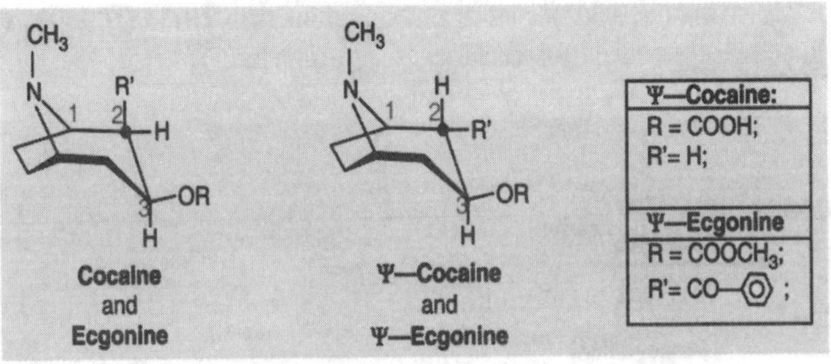

| Cocaine and Ecgonine | ψ—Cocaine and ψ—Ecgonine | ψ—Cocaine: R = COOH; R'= H;
ψ—Ecgonine R = COOCH₃; R'= CO—◯ ; |

Following are the **chemical structure** of *two* important **coca alkaloids** *viz.*, *benzoylcocaine*, and *cinnamoylcocaine :*

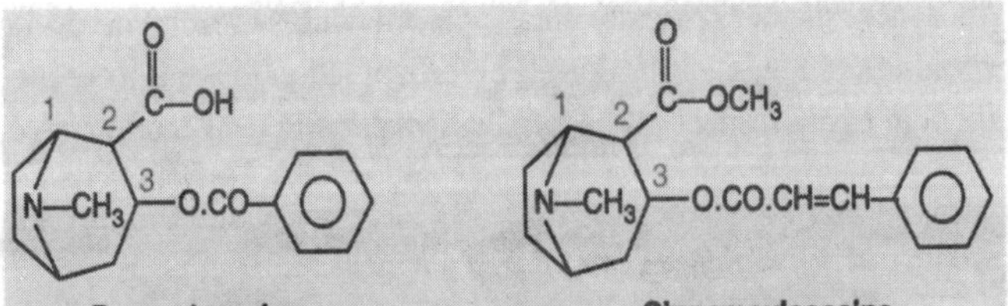

Benzoylcocaine   Cinnamoylcocaine

---

* Findlyay : *J Am Chem Soc.*, **76** : 2885, 1954.

** Kovacs O *et al.*, *Helv Chim Acta*, **37** : 892, 1954.

*** Kar, A : **Medicinal Chemistry,** New Age International, New Delhi, 5th ed., 2010.

(8) **Synthetic Substitutes of Cocaine :** The quantum leap forward with regard to the developments and progress in the ever expanding field of **'Medicinal Chemistry'** enabled the introduction of a plethora of **'synthetic substitutes of cocaine',** of which only *four* **important compounds** shall be discussed along with their *synthesis* and *explanation,* such as :

- Orthocaine,
- Procaine (or Novocaine),
- Lignocaine, and
- Phenacaine.

8.1. Orthocaine

Its **local anaesthetic** activity profile almost equals to *cocaine.*

Synthesis :

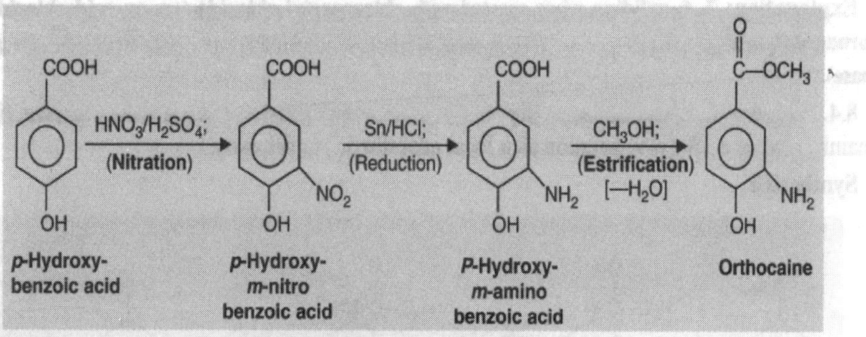

p-Hydroxy-benzoic acid	*p*-Hydroxy-*m*-nitro benzoic acid	*P*-Hydroxy-*m*-amino benzoic acid	Orthocaine

Explanation : The various aspects of the aforesaid synthesis are:

(1) *p*–**Hydroxy benoic acid** on nitration yields *p*–**hydroxy,** *m*–**nitro benzoic acid,** which upon reduction produces **p–hydroxy, *m*–amino benzoic acid.**

(2) The resulting product on esterification with methanol gives the desired product *'orthocaine'*.

8.2. Procaine [or Novocaine [R]]

Procaine is regarded to be the most potent and effective **local anesthetic,** which may be synthesized as stated under :

Synthesis : Kar* (2007) has described the synthesis of **'procaine'** by *three* methods, namely :

- **from 2–chloroethyl–*p*–amoni benzoate,**
- **from *p*–amino benoic acid [PABA], and**
- **from ethylene chlorohydrin.**

However, only on synthesis from **PABA** shall be discussed below :

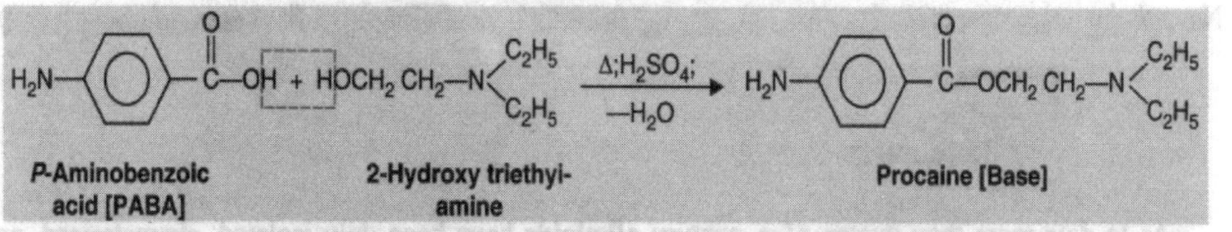

P-Aminobenzoic acid [PABA]	2-Hydroxy triethyi-amine	Procaine [Base]

Explanations : PABA interacts with **2–hydroxy triethyl amine** in the presence of sulphuric acid and gentle heating to lose a mole of water (as indicated above) and yields **'procaine'** which is a **base.**

---

* Kar, A : **Medicinal Chemistry**, New Age International, New Delhi, 4th ed., 2007.

8.3. Lignocaine

Lignocaine is found to be twice as potent as **procaine** and may be prepared as given under:

Synthesis :

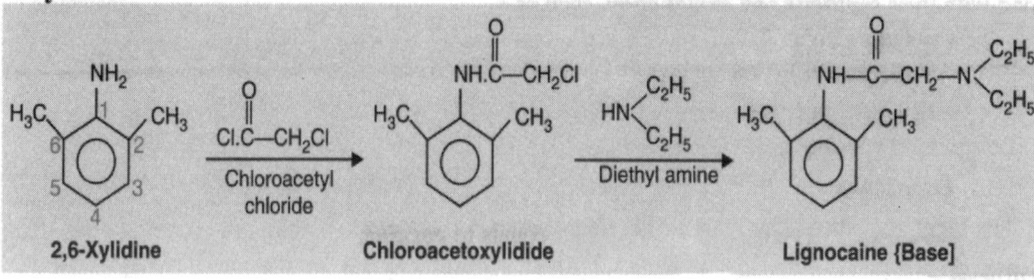

2,6-Xylidine   Chloroacetoxylidide   Lignocaine {Base}

Explanation: 2, 6–xylidine when reacted with **chloroacetyl chloride** (*i.e*, an **acid chloride**) yields **chloroacetoxylidide,** which upon treatment with diethylamine produces the desired compound **lignocaine** (a **base).**

8.4. Phenacaine : **Phenacaine** designates as one of the **oldest synthetic local anaesthetic,** which is mainly used as a 1% (*w/v*) solution as a local anaesthetic for the eyes.

Synthesis :

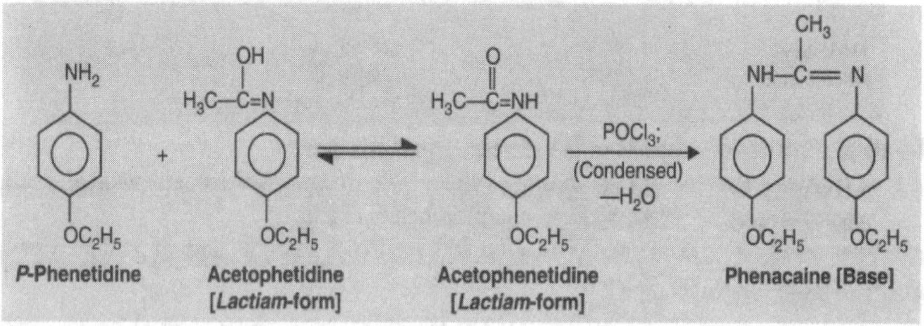

P-Phenetidine   Acetophetidine   Acetophenetidine   Phenacaine [Base]
                   [*Lactiam*-form]   [*Lactiam*-form]

Explanation: *p*–Phenetidine reacts with **acetophenetidine** (*lactam–form*) in the presence of POCl₃ (phosphorus oxychloride) to accomplish a **condensation reaction** (losing a mole of water) to yield **phenacaine** as a **base.**

3.5. Quinoline Alkaloids

Preamble : In a broader perspective, the alkaloids essentially comprising the **'quinoline nucleus'** do embrace a series of vital and important alkaloids that are solely derived from the **'cinchona bark'.** Nevertheless, the most critical and prominent members of this **specific class of compounds** are, namely :

- Quinine,
- Quinidine,
- Cinchonine, and
- Cinchonidine.

As to date more than quater–of–a–century **alkaloids** have been duly isolated, characterized, and studied exhaustively from the naturally occurring plant sources, such as :

- **Yellow Chinchona** [*Cinchona calisaya* Wedd.,], and *C. ledgeriana* Moens ex Trimen, and
- **Red Cinchona** [*Cinchona succirubra* Pavon ex Klotzsch belonging to the natural order *Rubiaceae.*

It has been duly reported that an *average commercial yield of the cinchona alkaloids* ranges between:

> **Quinine : 5.7 % ;**
> **Quinidine : 0.1–0.3%;**
> **Cinchonine : 0.2–0.4%; and**
> **Cinchonidine.**

However, the other closely related minor **alkaloids** do vary between **0.01–0.05 %** only.

Fundamental Structures of Cinchona Alkaloids : The well-known **'quinoline alkaloids'** do exhibit a host of vital therapeutic applications, namely : **quinine, quinidine, cinchonine,** and **cinchonidine.** Besides, having a closely related structure, they invariably possess almost similar medicinal values.

Importantly, these alkaloids essentially inherit the **'fundamental skeleton of 9'rubanol'** which is derived meticulously from the *'parent compound'* usually termed as **'ruban'.** In this manner, **'ruban'** is duly accomplished by the unique combination of *two* entirely distinct **'heterocyclic nuclei'***, namely :

- **4–Methyl quinoline residue,** and
- **Quinuclidine residue.**

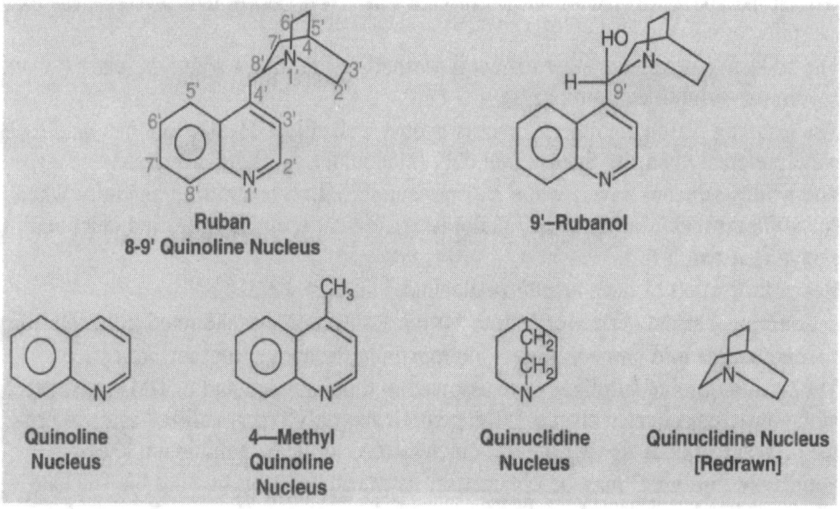

Ruban
8-9' Quinoline Nucleus

9'–Rubanol

Quinoline Nucleus

4—Methyl Quinoline Nucleus

Quinuclidine Nucleus

Quinuclidine Nucleus [Redrawn]

In the present context, the following *two* important **alkaloids** belonging to the class of **'quinoline alkaloids'** will be described at length in the sections that follows :

- Quinine, and
- Cinchonine.

3.5.1. Quinine

3.5.1.1.Preamble

Quinine designates the age-old, famous, wonderful, and extremely effective **'antimalarial drug'** obtained from the bark of *Cinchona callisaya* Wedd. It has the following **chemical structure.**

---

* In fact, this **nomenclature** was duly suggested by **Rabe** in order to simplify the naming of such compounds and also to signify its actual origin from the natural order *Rubiaceae.*

Quinine

3.5.1.2. Isolation

The various steps involved in the **isolation** of **'quinine'** are as stated under :

(1) The **cinchona bark** is carefully stripped and dried in the sun.

(2) The sun-dried bark is pulverized to a fine powder and sieved through a MS–mesh.

(3) The *uniform fine–powder* of the **cinchona bark** with a requisite quantum of **'lime'** and **'dilute NaOH solution'** so as to render it alkaline grossly. Allow the alkaline powdered mass to remain as such for nearly **16–20 hrs.**

(4) Extract the above mass with **hot petroleum ether (40=60°C)** and the solvent is drawn off carefully.

(5) The collective petroleum ether extract is washed several times with DM water (to get rid of any **water–soluble contaminants).**

(6) The resulting petroleum ether extract is treated with **dilute H_2SO_4** to a *just acidic pH level only* (preferably in a **Pb–lined** vessel duly fitted with a **mechanical stirrer.**

(7) The **acidic–aqueous layer,** while still remaining hot, is neutralized carefully when the so called **'neutral sulphates of the alkaloids** *e.g.,* **quinine, cinchonine, and cinchonidine'** got *crystallized out.*

(8) **Recrystallization** of the above *three* alkaloidal mixture yields :
   • **quinine** – as the *'first crop'*, since **'quinine sulphate'** has the **maximum solubility,** and
   • **cinchonine and cinchonidine** – do remain in the above **'mother liquor'.**

(9) The **'crude quinine sulphate'** is redissolved in minimum amount of DM water, decolorized with powdered activated charcoal (**Reagent Grade only**), recrystallized again to bring down the presence of both **cinchonine** and **cinchonidine** to the bare minimum level.

(10) Ultimately, **'quinine'** may be obtained in its purest form by treating the sulphate with an alkali, washing, and drying finally.

3.5.1.3. Characteristic Features

These predominantly comprise :
   • **Quinine** occurs as **orthorhombic needles** duly obtained from *absolute ethanol,* which are found to be **triboluminescent*.**
   • It has mp 177°C (with *decomposition* to a certain extent).
   • **Quinine** gets sublimed at high vacuum at 179–180°C.
   • It exhibits **specific rotation** in different medium as :
   Chloroform : $[\alpha]^{17}_D -117°$ (C = 1.5);
   Ethanol : $[\alpha]^{15}_D - 169°$ (C = 2 in 97% *v/v* ethanol)

---

* **Triboluminescent :** A **luminescence** (or *sparks*) duly prduced either by frictionor mechanical force applied to certain chemical crystals *viz.,* quinine.

$0.1 M H_2SO_4 : [\alpha]^{15}_D - 285° (C = 0.4 M in 1M H_2SO_4).$

- It exhibits *two* distinct **dissociation constants** *eg.*,
  $pk_1 = 5.07$ at 18°C; and
  $pk_2 = 9.7$ at 18°C.

3.5.1.4. Constitution of Quinine

Based on the various **chemical reactions, interpretations,** and **logical conclusions,** and above all the ultimate **'total synthesis'** go a long way to prove and establish the **constitution of quinine** as given in the sections that follows.

Pelletier and Caventau (1820)* were pioneer in the first and foremost isolation of both **'quinine'** and **'cinchonine'**. A survey of literature reveals that the **'cinchona bark'** from Java native trees almost yields **7–8 % quinine;** nevertheless, the latest teachinique of the **'grafted trees'**** showed an *exceptional 100% increase* (*i.e.,* **15–16% quinine**).

(a) **Molecular Formula :** Its molecular formula is $C_{20}H_{24}N_2O_2$.

(b) **Presence of Two *tert*–N–Atoms :** Quinine aptly takes on *two* **moles of methyl iodide (CH$_3$I)** to yield a **'diquaternary salt'** as given below

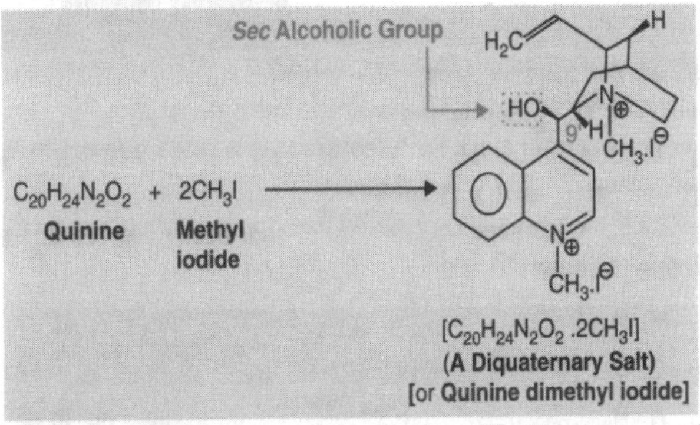

Sec Alcoholic Group

$C_{20}H_{24}N_2O_2$ + $2CH_3I$
Quinine   **Methyl iodide**

$[C_{20}H_{24}N_2O_2 .2CH_3I]$
(A Diquaternary Salt)
[or **Quinine dimethyl iodide**]

Inference : Quinine essentially possesses *two tertiary*–N–atoms, one each in the **'quinoline nucleus'**, and 'quinuclidine nucleus' as shown above clearly.

(c) **Presence of a *Secondary* Alcoholic Moiety :** Treatment of **'quinine'** with **acetic acid** and **benzoic acid** caused *esterification* to yield **monoacetate and monobenzoate derivatives** respectively as shown under:

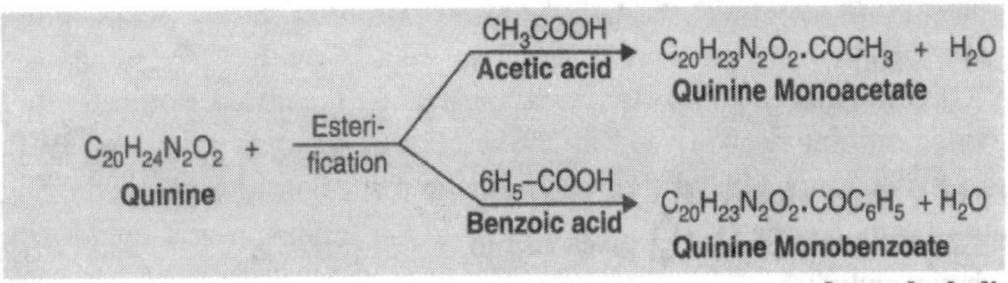

CH_3COOH
Acetic acid → $C_{20}H_{23}N_2O_2.COCH_3$ + H_2O
Quinine Monoacetate

$C_{20}H_{24}N_2O_2$ + Esterification
Quinine

$6H_5–COOH$
Benzoic acid → $C_{20}H_{23}N_2O_2.COC_6H_5$ + H_2O
Quinine Monobenzoate

Inference : The above reaction suggests that **'quinine'** has **one secondary alcoholic moiety.**

Besides, one may also show the presence of the **secondary alcoholic function** by the **oxidation** of **'quinine'** with **chromium–6–oxide (Cr$_2$O$_3$)** to produce the corresponding' quinone' (*i.e.,* a **ketone derivative**) as stated under :

---

* Pelletier and Caventau : *Ann Chem Phys,* **15** [2] : 291, 1820.
** **Graft** : A plant shoot fixed inot a cut in another plant to form a **'new growth'**.

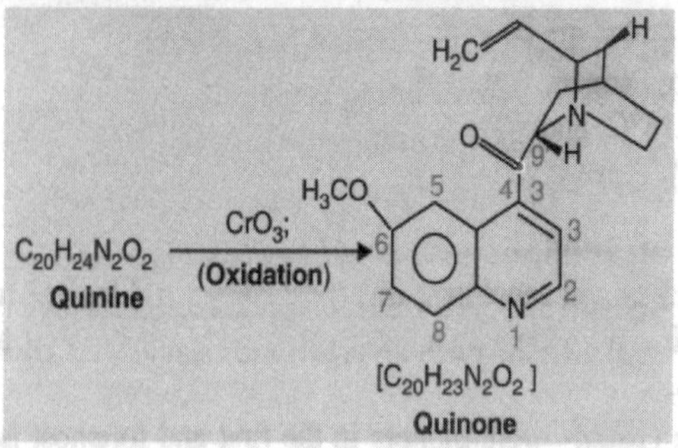

$$C_{20}H_{24}N_2O_2 \xrightarrow[\text{(Oxidation)}]{CrO_3;}$$

Quinine

$[C_{20}H_{23}N_2O_2]$
Quinone

i.e., replacement of the *sec*–alcoholic (–OH) function at C–9' with a **ketonic function** (O=C) at **C–9'**.

(d) **Presence of a Methoxyl (–OCH₃) Group at C–6 of Quinoline Nucleus of Quinine :**

Interestingly, **quinine** on being *heated with HCl,* duly eliminates one mole of **methyl chloride** **(CH₃I)** as indicated below :

$$C_{19}H_{21}N_2O.[OCH_3] \xrightarrow{HCl; \Delta;} C_{19}H_{21}N_2O.[OH] + CH_3\,Cl$$

Quinine Monoacetate   6—Hydroxy derivative
of Quinine

Inferences : There are *two* vital **inferences,** namely :

• a methoxyl (–OCH₃) group is present in **quinine,** and
• also suggests strongly that the **'second oxygen atom'** present in **'quinine'** is proved to be the one present in the **methoxy moiety.**

(e) **Presence of an Olefinic Linkage (–HC=CH–) :** It has been observed that **one mole of bromine** **[Br₂]** adds on to **'quinine'** as shown under :

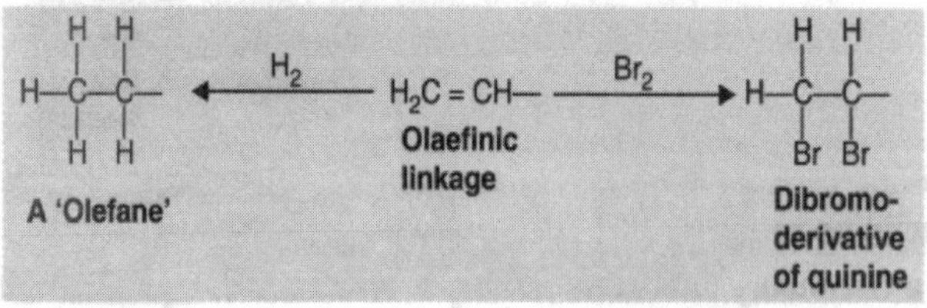

$$H-\overset{\overset{\displaystyle H}{|}}{\underset{\underset{\displaystyle H}{|}}{C}}-\overset{\overset{\displaystyle H}{|}}{\underset{\underset{\displaystyle H}{|}}{C}}- \xleftarrow{H_2} H_2C=CH- \xrightarrow{Br_2} H-\overset{\overset{\displaystyle H}{|}}{\underset{\underset{\displaystyle Br}{|}}{C}}-\overset{\overset{\displaystyle H}{|}}{\underset{\underset{\displaystyle Br}{|}}{C}}-$$

A 'Olefane'   Olaefinic
linkage   Dibromo-
derivative
of quinine

Inference : The above reaction suggests that **'quinine'** essentially has one olefinic (ethylenic) **double bond.** Nevertheless, the aforesaid analogy may also be substantiated duly by :

• absorption of a mole of **'hydrogen'** by **'quinine'** in the presence of a catalyst (*viz.,* **H₂–Pd),** and
• **'Olefinic linkage'** may also be established by the critical formation of **halogen substi- tuted quinine derivatives** with the reaction of *halogen acids* (*eg.,* HCl).

(f) **Presence of a Vinyl Moiety (H₂C=CH–):** The meticulously **controlled oxidation of quinine** with **potassium permangnate [KMnO₄]** gives rise to the formation of a **monocarboxylic acid** plus **formic acid** as shown under :

$$H_2C=CH- \xrightarrow[\text{(KMnO}_4)]{[O];} [-COOH + HCOOH$$

A Vinyl   A Mono— Formic Acid
Moiety in   carboxylic
Quinuclidine   acid
Nucleus of
Quinine

Inference : It evidently confirms the presence of a 'vinyl moiety' strategically located in the 'quinuclidine nucleus' of quinine.

(g) **Presence of a Quinoline Nucleus : Quinine** when duly fused with a **concentrated solution of KOH**, it produces an admixture of the following *two* products of reaction, namely :

- **6–Methoxyquinoline,** and
- **Lepidine [or 4–Methylquinoline],**

as given under :

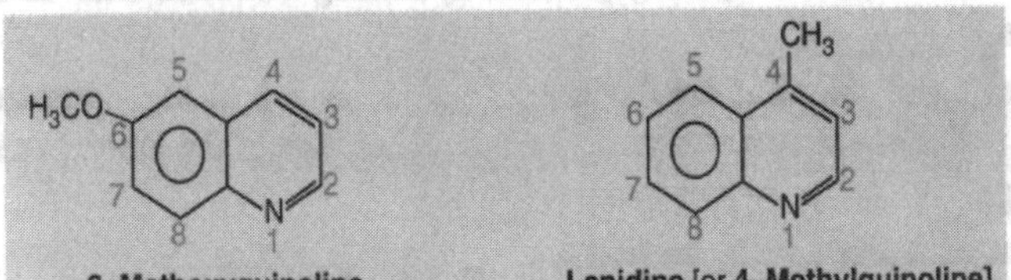

6–Methoxyquinoline   Lepidine [or 4–Methylquinoline]

Cinchonine when fused with KOH under exactly the *same experimental parameters* forms 'quinoline' and 'lepidine' as given under:

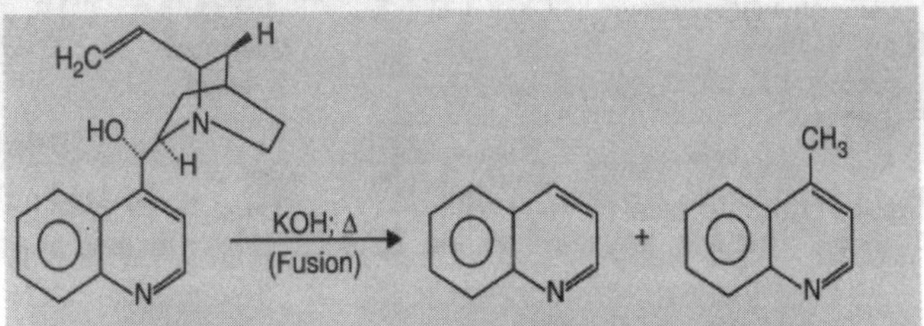

$[C_9H_{22}N_2O]$
Cinchonine

KOH; Δ
(Fusion)

Quinoline   +   Lepidine
[or 4–Methylquinoline]

(h) **Presence of Meroquinene [or Meroquinenine] : Quinine** on being subjected to vigorous oxidation with **chromic acid** gives rise to the formation of *two* **chemical** entities, namely :

- **quininic acid,** and
- **meroquinene*** [meroquinenine],

as shown under :

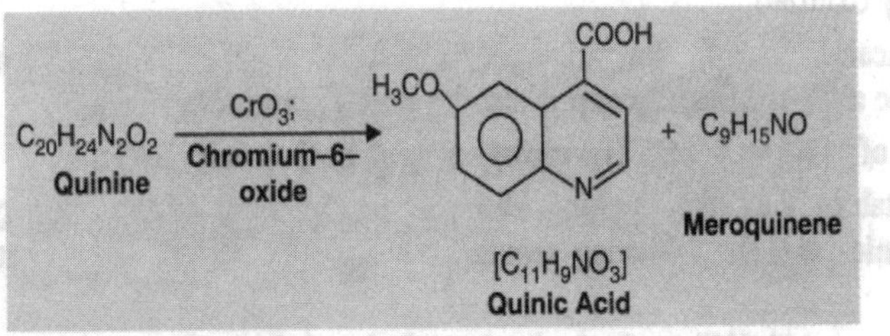

$C_{20}H_{24}N_2O_2$
Quinine

CrO_3;
Chromium–6–
oxide

$+ C_9H_{15}NO$

Meroquinene

$[C_{11}H_9NO_3]$
Quinic Acid

Remarks : In a solid evidential proof, one may get at the most preferred structure of **quinine'** provided one clearly ascertains the structures of :

---

* Earlier the second chemical component was commonly termed as *'second half' i.e.,* the segment having the'quinic-lidine nucleus$[C_9H_{15}NO]$.

- **Quininic acid,** and
- **Meroquinene.**

(*i*) **Structure of Quininic Acid :** The actual structure of **'quininic acid'** may be duly ascertained by adopting the following *four* steps in a sequential manner :

(a) **Presence of 6–Methoxyquinoline Nucleus :** The presence of **6–methoxy quinoline nucleus** in *'quininic acid'* (I) may be ascertained by :

- conversion of **(I)** to 6–hydroxyquinoline by heating with **soda lime*** to afford **decarboxylation ($-CO_2$),** and
- conversion of **'quininic acid'** to the corresponding **6–hydroxyquinoline** by heating carefully with HCl so as to confirm the *strategic position* of the inherent **methoxyl ($-OCH_3$) group.**

The above statement of facts may be seen in the following reactions :

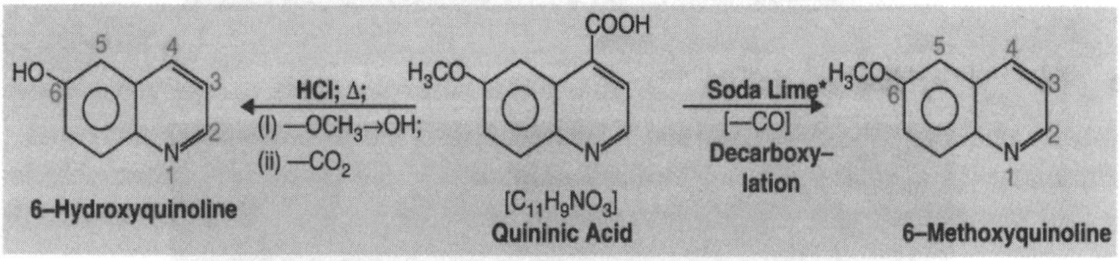

6–Hydroxyquinoline $[C_{11}H_9NO_3]$ 6–Methoxyquinoline
                            Quininic Acid

(b) **Oxidation of Quininic Acid forms Pyridine–2, 3,4,–Tricarboxylic Acid :** The *chromium–6–oxide* helps to oxidise **quininic acid** to yield **pyridine–2,3,4–tricarboxylic acid** as given under:

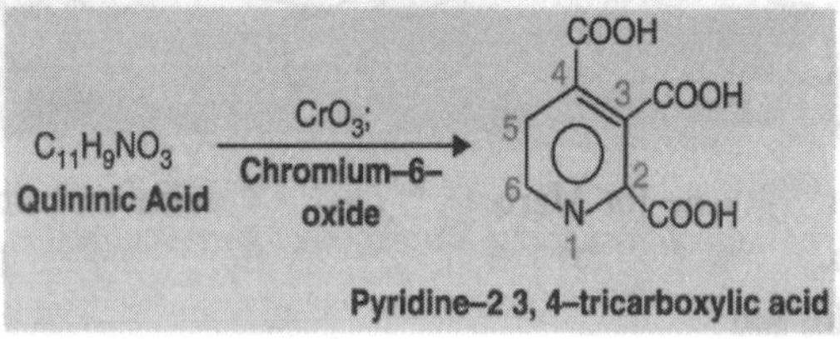

$C_{11}H_9NO_3$
Quininic Acid

CrO_3;
Chromium–6–
oxide

Pyridine–2 3, 4–tricarboxylic acid

Inferences : These essentially comprise :

(1) The adjacent **'phenyl ring'** present in *quininic acid* bearing the **methoxy ($-OCH_3$), moiety** gets duly oxidized.

(2) The critical presence of the **carboxyl ($-COOH$) moiety,** at **C–4** reveals emphatically that in **'quininic acid'** the **same group** is also located at position **C–4** only.

(c) **Presence of Methoxyl ($-OCH_3$) moiety at C–6 in Quininic Acid :**

In order to establish the exact location with regard to the position of the **methoxyl ($-OCH_3$) moiety** present in **'quininic acid'** the following sequential **reaction** were performed methodically as stated below :

- treatment with HCl to obtain the **demethylated derivative,** and
- **decarboxylation ($-CO_2$)** of the *resulting product* with *soda lime* yields **6–hydroxy quinoline.**

---

* **Soda Line :** It is made by fusing NaOH or a mixture of NaOH and KOH with **calcium hydroxide [Ca(OH)$_2$]** and finally making the **'fused mass'** into granules.

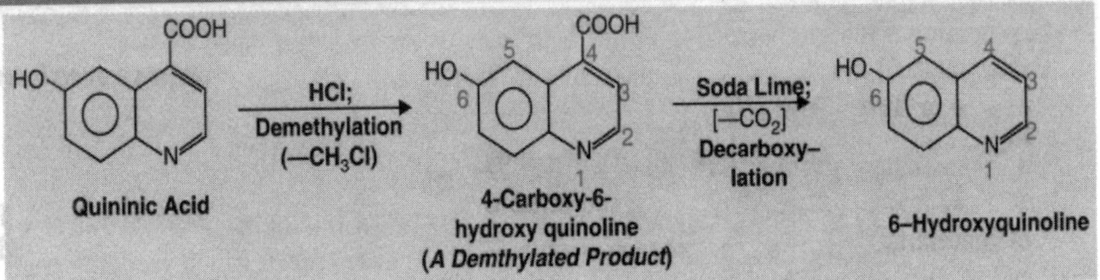

Quininic Acid → HCl; Demethylation (—CH₃Cl) → **4-Carboxy-6-hydroxy quinoline** (*A Demthylated Product*) → Soda Lime; [—CO₂] Decarboxylation → **6–Hydroxyquinoline**

Inference: The above **reactions** explicitly suggests that the **methoxyl moiety (–OCH₃)** in *quininic acid* is located strategically at **C–6 position.**

(d) **Synthesis :** The synthesis of '**quininic acid**' was duly put forward by Rabe and Kuliga (1909)* starting from *p*–**methoxy aniline (I).** as elaborated under:

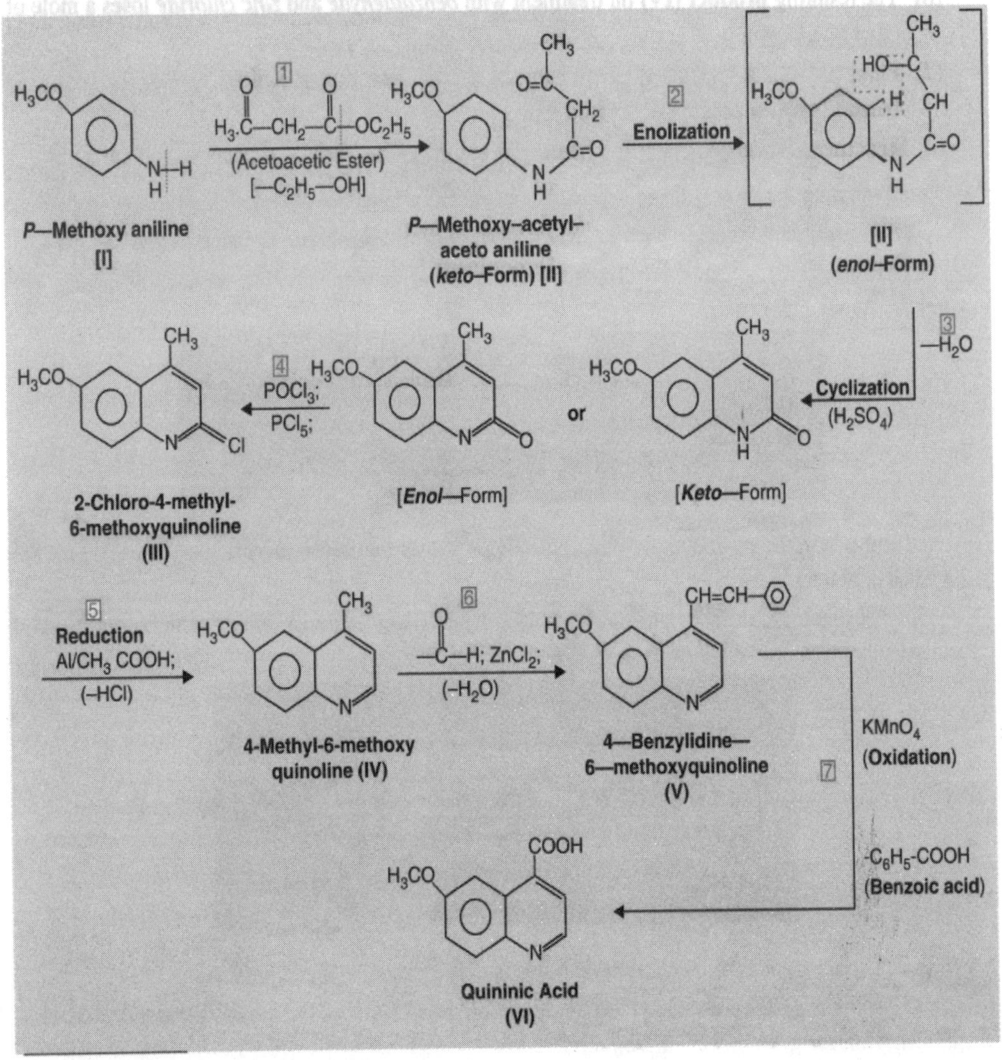

P—Methoxy aniline [I] → [1] Acetoacetic Ester [—C₂H₅—OH] → *P*—Methoxy–acetyl–aceto aniline (*keto*–Form) [II] → [2] Enolization → [II] (*enol*–Form)

2-Chloro-4-methyl-6-methoxyquinoline (III) ← [4] POCl₃; PCl₅; ← [Enol—Form] *or* [Keto—Form] ← [3] —H₂O Cyclization (H₂SO₄) ← [II] (*enol*–Form)

[5] Reduction Al/CH₃ COOH; (–HCl) → 4-Methyl-6-methoxy quinoline (IV) → [6] —C—H; ZnCl₂; (–H₂O) → 4—Benzylidine—6—methoxyquinoline (V) → [7] KMnO₄ (Oxidation) –C₆H₅-COOH (Benzoic acid) → Quininic Acid (VI)

* Rabe and Kuliga : *Ann.,* **346** : 346–349, 1909.

Explanation : The synthesis of **'quininic acid'** may be expatiated as stated under :

(1) Interaction of *p*–**methoxy aniline (I)** with **acetoacetic ester** yields *p–methoxy acetylaceto aniline* in *'keto'–form (II)*

(2) The resulting **product (II)** undergoes **'enolization'.**

(3) The **enolized product (II)** on being subjected to **'cyclization'** in the presence of a few drops of concentrated H_2SO_4 yields an *intermediate* in its *'keto'–*form, which gets duly **enolized.**

(4) The resulting enolized product when treated with $POCl_3/PCl_5$ produces **2–chloro–4–methyl–6–methoxy quinoline (III).**

(5) The **product (III)** upon reduction loses a mole of HCl to give **4–methyl–6–methoxy quinoline (IV).**

(6) The resulting **product (IV)** on treatment with *benzaldehyde* and *zinc chloride* loses a mole of water to yield **4–benzylidene–6–methoxy quinoline (V).**

(7) Finally, the **product (V)** on being subjected to **oxidation** with $KMnO_4$ eliminates a mole of **benzoic acid** to produce **'quininic acid'.**

(j) **Structure of Meroquinene [or Meroquinenine] :** The various aspects which categorically help in assigning the most probable structure of **'meroquinene'** are as stated under :

(1) **Molecular formula :** The *molecular formula* of **meroquinene** is found to be $C_9H_{15}NO_2$.

(2) **Presence of a Carboxyl Group :** **Meroquinene** produces both a **monosodium salt** and an **ester** as shown under :

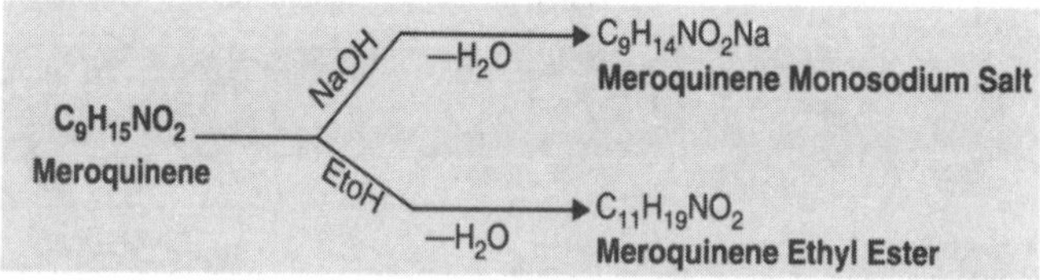

Inference : from the above *two* reactions one may infer that **meroquinone** possesses one **carboxyl (–COOH) moiety.**

3. **Presence of an Ethylenic Double Bond :** **Meroquinene** on being subjected to reduction catalytically takes up a mole of **hydrogen (H_2)** as shown under :

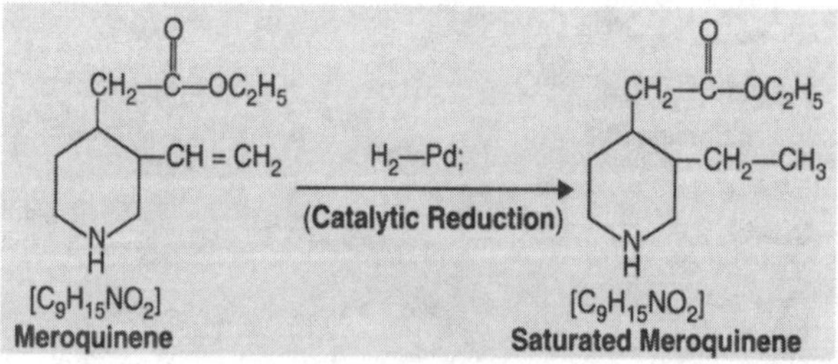

Inferences : (1) A **eythylenic double bond** is duly present in **meroquinene.**

(2) Furthermore, the presence of **ethylenic double bond** ($-CH=CH_2$) suggests that the **'side–chain'** still forms an *'integral part of meroquinene'.*

4. Presence of *Secondary* Amino (=NH) Function : Meroquinene may be duly **acetylated, benzoylated,** and **nitrosated** as given below :

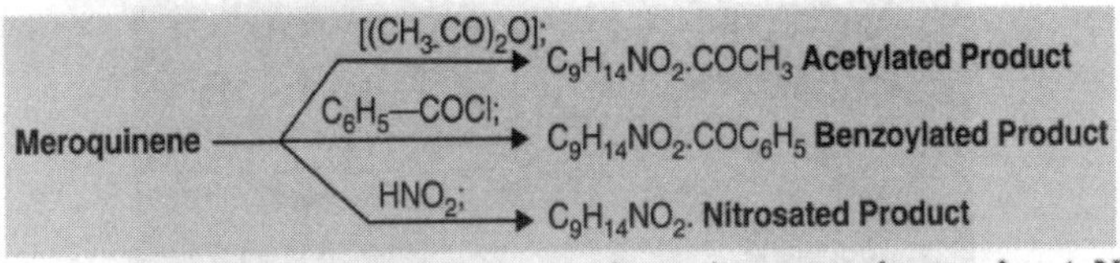

Inference : The amino moiety present in **meroquinene** is a *secondary* amino (=NH) function.

5. Oxidation : Meroquinene on being treated with **cold acidified potassium permanganate solution** yields a mole each of **formic acid** and **cincholoiponic acid** (*i.e.*, a **dicarboxylic acid**), which upon further **oxidation** gives rise to the formation of **'loiponic acid'** (also a **dicarboxylic acid**) as stated under :

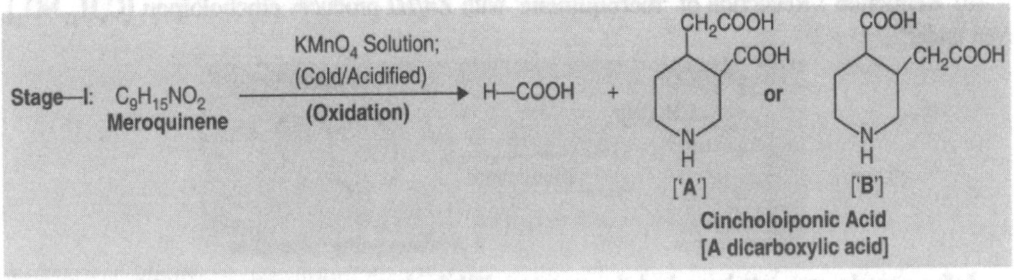

Inference : Formation of **formic acid** suggests that a **vinyl moiety** (–CH=CH$_2$) is present in the side–chain of **'meroquinene'**. The major product **cincholoiponic acid** thus obtained may be written either as **'A'** or **'B'**.

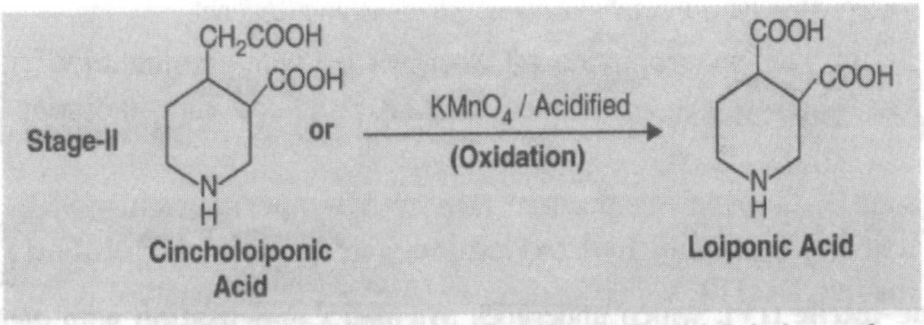

Explanation: The **cincholoiponic acid** gives **loiponic acid**, which being rather not-so-stable and gets isomerised to *more stable* **hexahydrocinchomeronic acid [C$_7$H$_{11}$NO$_4$]** (or piperidine–3,4–dicarboxylic acid) when treated duly with **KOH** at ~200°C. Therefore, the resulting *oxidized product* obtained from **cincholoiponic acid** must also be **'loiponic acid'** *i.e.*, another **dicarboxylic acid**.

Probable Structure of Cincholoiponic Acid : The most appropriate structure of **cincholoiponic acid** should be **'A'**, because upon heating with concentrated H$_2$SO$_4$ (36M) the former gives rise to the γ–picoline as given under :

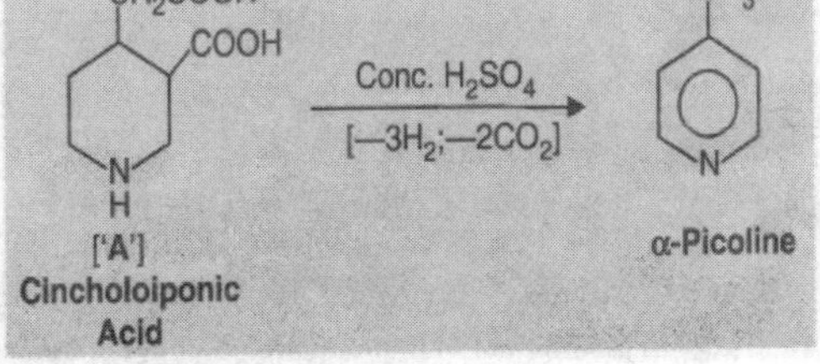

Other Exemplary Evidences for Structure 'A' of Meroquinene : (a) It has been duly observed that **meroquinene** upon heating with HCl at ~240°C produces **3-ethyl-4-methyl piperidine** as given under:

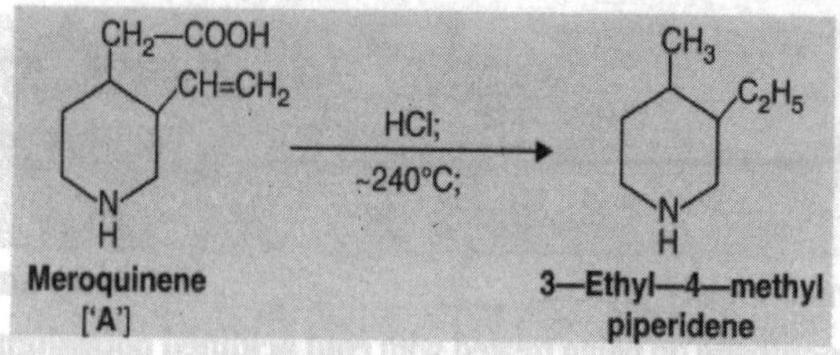

Meroquinene ['A']

HCl; ~240°C;

3—Ethyl—4—methyl piperidene

Inference: If **'meroquinene'** had the structure **'B'**, one would have gotten **4–propylpyridine** instead; and, therefore, **structure 'A'** is the assigned structure of *meroquinene*.

(b) **Reduction :** Reduction of **'meroquinene'** with **Zn/HI** produces **cincholoipon [$C_9H_{17}NO_2$]** as given under:

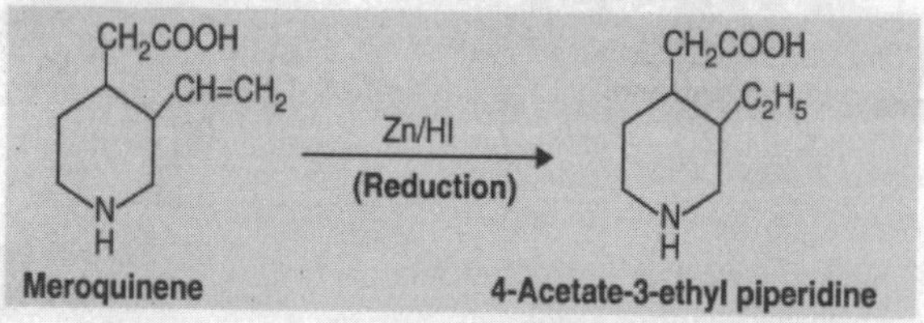

Meroquinene

Zn/HI (Reduction)

4-Acetate-3-ethyl piperidine

Inference: In case, we have had the structure **'B'** for **meroquinene,** we would have obtained a **'propyl group at C–4** instead. Hence, **structure 'A'** is the assigned structure of **'meroquinene'**.

(6) **Synthesis of Cincholoiponic Acid:** The structure **'A'** for **cincholoiponic acid** has been duly proved and confirmed by the synthesis put forward by Wohl *et al.* (1907).

The various steps involved in this **synthesis** are as given under :

(i) Two moles of β–chloropropionacetal when reacted with **ammonia** (NH$_3$) yields one mole of **imino dipropinacetal (I)**, which on treatment with HCl produces **dipropioiminoaldehyde (II)** as an **intermediate.**

(ii) Cyclization of the resulting **product (II)** yield **3–piperidine–ene–aldehyde (III)**, which on subsequent reactions : *first,* with **hydroxylamine,** and *secondly,* with **thionylchloride** produces **3–piperidine–ene–nitrile (IV)**, which undergoes **Michael Condensation** with *diethylmalonic ester* and freshly prepared *sodium ethoxide* to give rise to the formation of **piperidine–4–(diethylmalonic ester)–3–nitrile (V).**

(iii) Finally, the resulting **product (V)** when treated with barium hydroxide and then with HCl yields the racemic mixture of **cincholoiponic acid (VI).**

These reactions may be summarized as under :

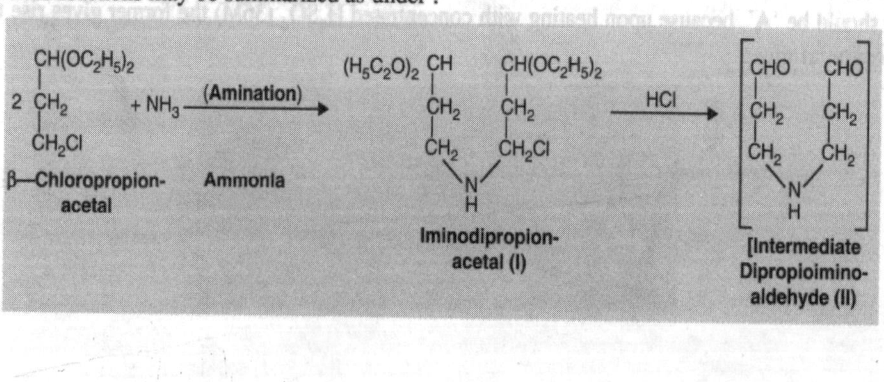

β—Chloropropion-acetal

Ammonia

(Amination)

Iminodipropion-acetal (I)

HCl

[Intermediate Diproploimino-aldehyde (II)]

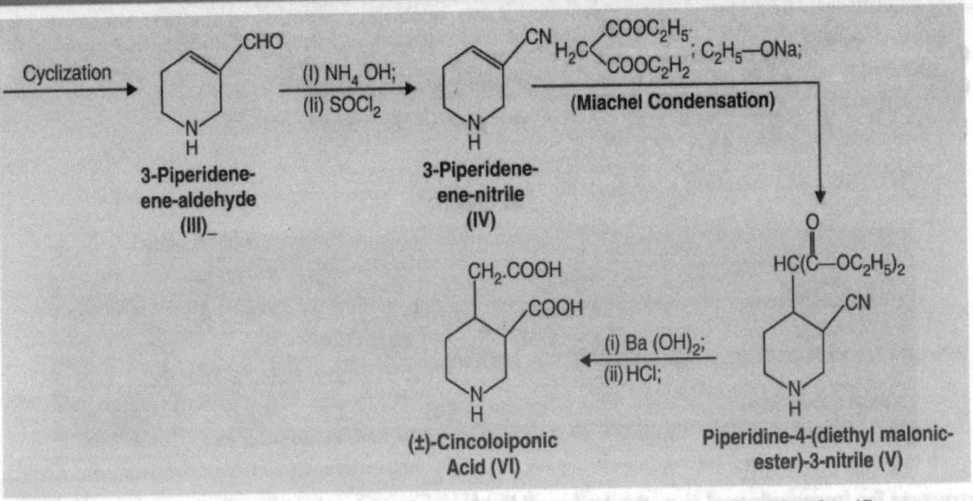

The most logical reaction scheme shows:

Cyclization → **3-Piperidene-ene-aldehyde (III)** (with CHO group) → (I) NH₄OH; (ii) SOCl₂ → **3-Piperidene-ene-nitrile (IV)** (with CN group) → H₂C(COOC₂H₅)₂; C₂H₅—ONa; **(Miachel Condensation)** → **Piperidine-4-(diethyl malonic-ester)-3-nitrile (V)** → (i) Ba(OH)₂; (ii) HCl; → **(±)-Cincoloiponic Acid (VI)** (with CH₂.COOH and COOH groups)

7. Exact Point of Linkage between 'Quininic Acid' and 'Meroquinene' in the 'Structure of Quinine': It is, however, pertinent to state here that the *'quinine molecule'* is completely devoid of any 'free carboxylic (–COOH) moiety'. Nevertheless, its **oxidation products** [see section 3.5.1.4.(c)], namely : **quininic acid** and **meroquinene**, do possess **free carboxylic (–COOH) moieties** predominantly, thereby suggesting explicitly that the *two* aforesaid 'oxidation products' are duly linked with each other by the help of the so called 'carboxylic C–atom' strategically present in the 'quinine residue'.

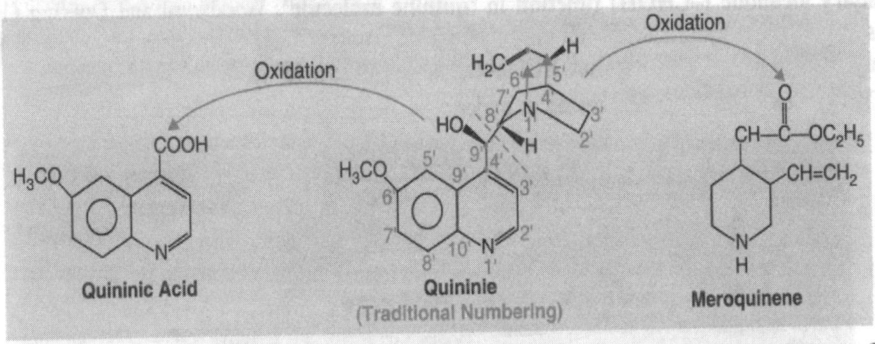

Quininic Acid   Quininie (Traditional Numbering)   Meroquinene

Besides, it is quite obvious that 'quinine molecule' does contain *two* tertiary nitrogen (basic) atoms; whereas, meroquinene essentially possesses a *single secondary N–atom* plus produces one single carboxyl (–COOH) moiety simultaneously. Evidently, such a typical behavioural pattern could only be possible provided the N–atom critically forms an integral part of the ensuing 'condensed ring system'.

Explanation : The most logical and acceptable explanation for the above theology may be suitably expatiated provided the, precursor of *'meroquinene'* has the following specific 'structure' (3–vinyl quinuclidine):

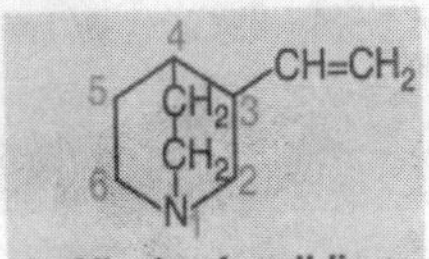

3—Vinyl quinuclidine

Oxidation of 3–Vinyl Quinuclidine : It has been duly established that *oxidation of 3–vinyl quinuclidine* with **chromium–6–oxide [CrO₃]** in the presence of *sulphuric acid [H₂SO₄]* yields an oxidative product that has essentially the *two* vital functional moieties:

- a *secondary N–atom* (from an initial **tertiary N–atom**) and
- a **carboxyl (–COOH) moiety.**

The **oxidation reaction** may be expressed as under :

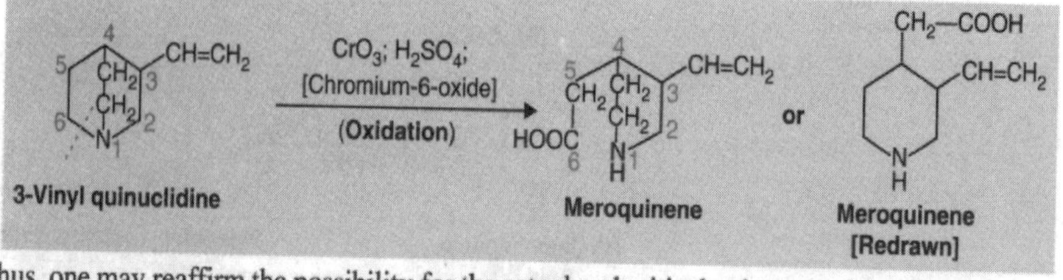

3-Vinyl quinuclidine   Meroquinene   Meroquinene
                                                       [Redrawn]

Thus, one may reaffirm the possibility for the actual and critical existence of the aforesaid *pattern of structure* for 'meroquinene' (*i.e.,* the *'quinuclidine'* **structure**) has been confirmed adequately by the synthesis of **3–ethylquinuclidine.**

Inference: From the above statement of facts and due logistics, the **'quinoline nucleus'** is critically linked at *position 4* with the corresponding **'quinuclidine nucleus'** at *position–8* (see the numbering of the two rings *viz.,* **isquinoline and quinuclidine**) in section (7) for **'structure of quinine'**

8. Exact Position of Secondary Alcoholic (–CHOH) Moiety : The exact status and position of the **secondary alcoholic (–CHOH) function** in 'quinine molecule'. Woodward and Doering (1944)* oxidised **quinine** to 'quinonone' with the help of **chromium–6–oxide (CrO₃).** The resulting **quinone** on being treated with *amyl nitrite* and *HCl* produces **'quininic acid'** as well as an **'oxime'.**

The various reactions are expressed as under :

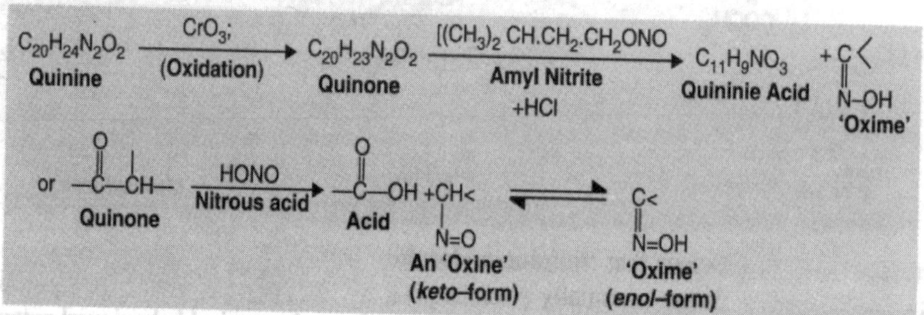

Importantly, the structure of the above *'oxime'* has been duly established by carrying out its *hydrolysis* to obtain one mole each of **meroquinene** and **hydroxylamine.**

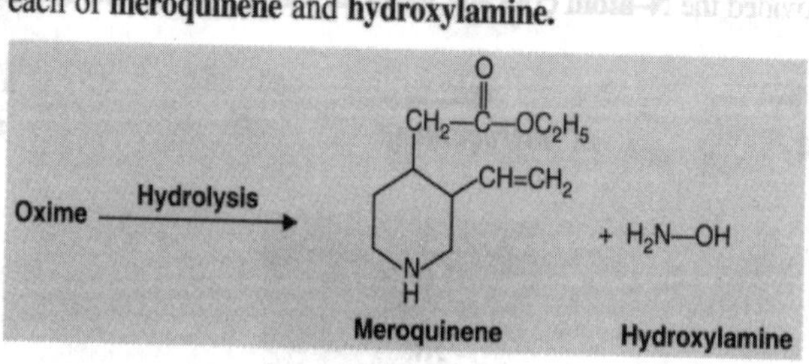

Oxime   **Hydrolysis**   + H₂N—OH

Meroquinene   Hydroxylamine

---

* Woodward and Doering: *J Am. Chem. Soc,* **66** : 84, 1944.

Inference : The above findings and observations reveal explicitly that the **'quinoline nucleus'** and the **'quinuclidine nucleus'** are joined together *via* the **secondary alcoholic (–CHOH) function.**

Therefore, based on the above facts, evidences, and logical explanations one may, assign the most feasible and possible structure of **'quinine'** as given under :

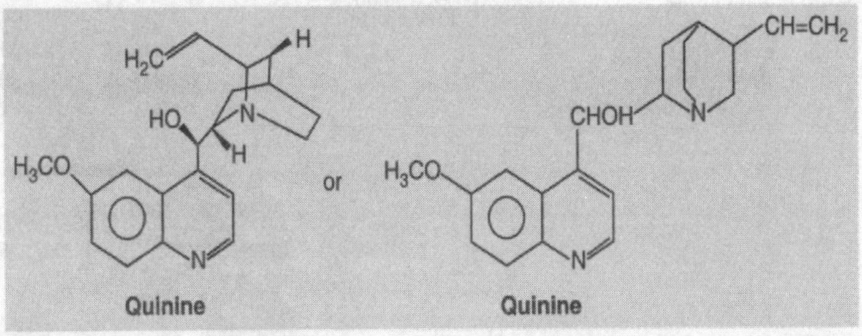

Quinine   or   Quinine

9. **Synthesis :** Ultimately, the structure of **'quinine'** is further confirmed by its **total synthesis** reported by Muhtadi *et al.* (1983)*; Woodward and Doering (1944)**. In fact, these dedicated researchers carried out the **total synthesis,** starting from *ab initio* up to the **racemic mixture of 'quinotoxine';** and from this point onward Rabe continued the synthesis till its completion.

The various steps involved in the **total synthesis** of **'quinine'** are as stated under :

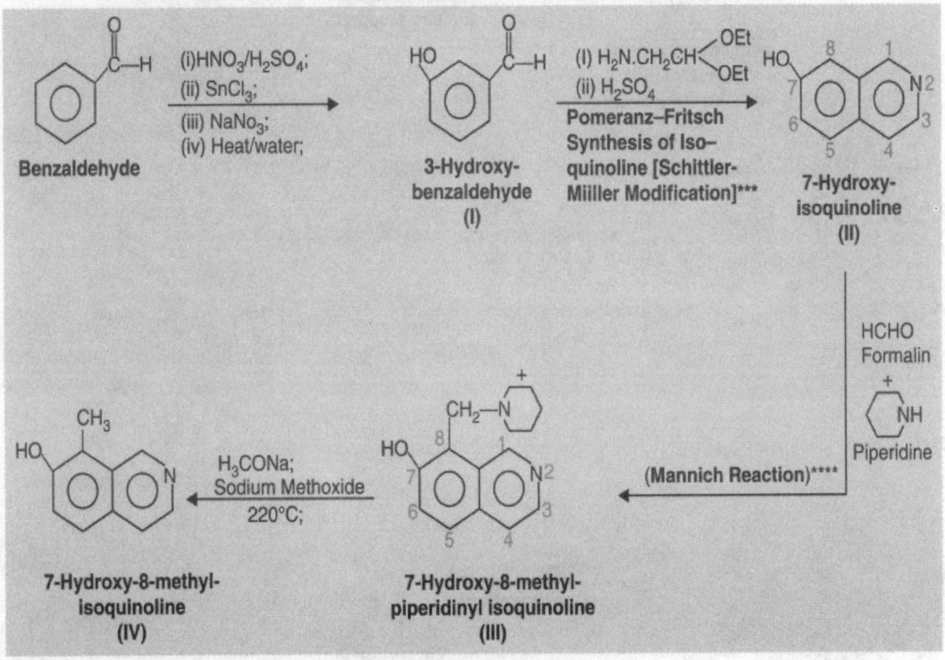

Benzaldehyde

(i) HNO$_3$/H$_2$SO$_4$;
(ii) SnCl$_3$;
(iii) NaNo$_3$;
(iv) Heat/water;

3-Hydroxy-benzaldehyde
(I)

(I) H$_2$N.CH$_2$CH$\begin{smallmatrix}OEt\\OEt\end{smallmatrix}$
(ii) H$_2$SO$_4$
Pomeranz–Fritsch Synthesis of Iso–quinoline [Schittler-Miiler Modification]**

7-Hydroxy-isoquinoline
(II)

HCHO
Formalin
+
Piperidine

7-Hydroxy-8-methyl-isoquinoline
(IV)

H$_3$CONa;
Sodium Methoxide
220°C;

7-Hydroxy-8-methyl-piperidinyl isoquinoline
(III)

(Mannich Reaction)****

---

 * Muhtadi FJ *et al.*: *Anal. Profile Drug Subs.*, **12** : 547-621, 1983.

 ** Woodward RB and Doering WE : *J Am Chem Soc.*, **66** : 849, 1944.

*** Pomeranz C : *Montash* : **14** : 116, 1893; Fritsch P : *Ber* : **26** : 419, 1893. Müller J : *Helvi Chim Acta*, : 914, 1119, 1948.

**** Mannich C and Krosche W : *Arc. Pharm.*, **250** : 647, 1912.

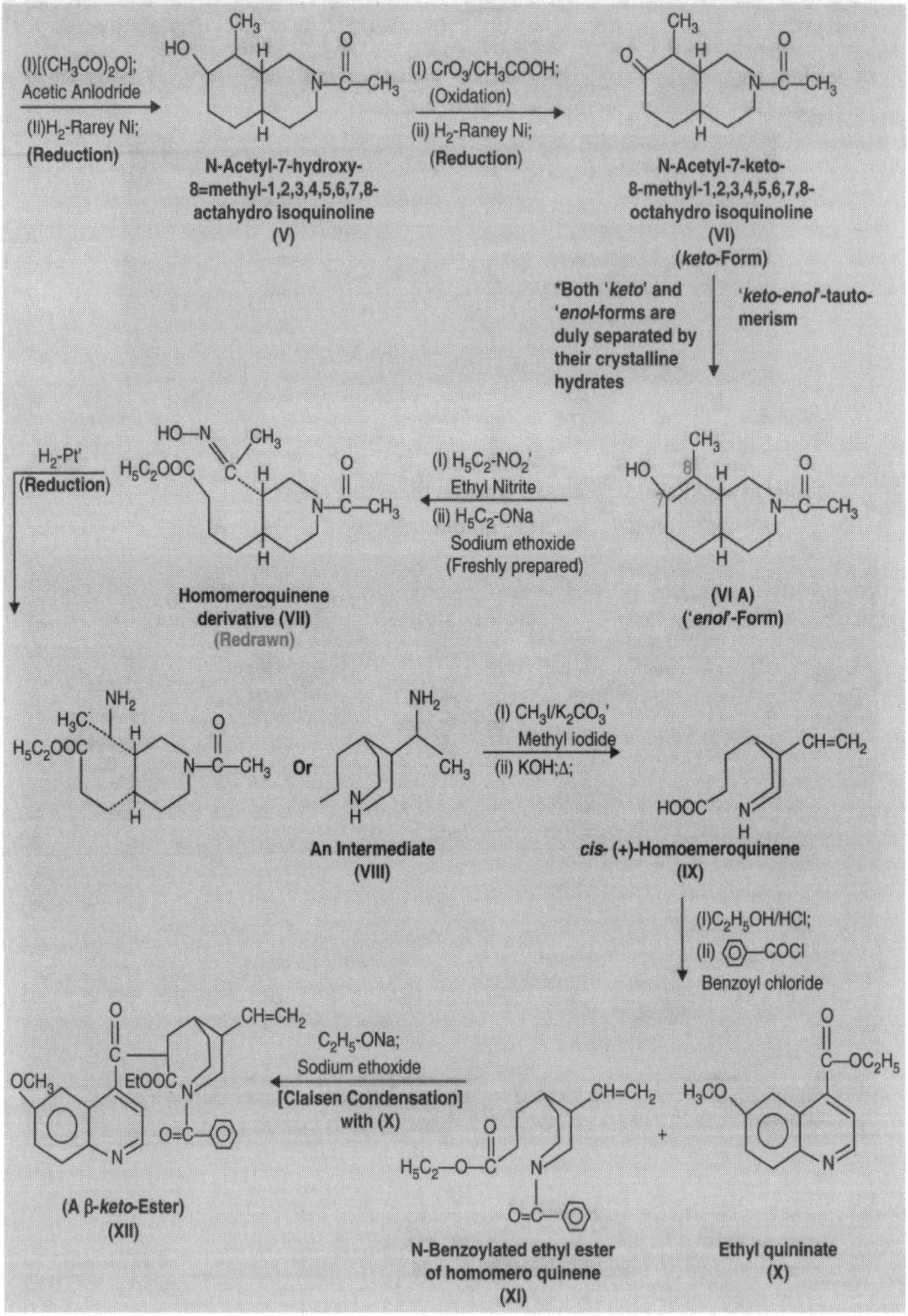

(I)[(CH₃CO)₂O];
Acetic Anlodride

(II)H₂-Rarey Ni;
(Reduction)

N-Acetyl-7-hydroxy-
8=methyl-1,2,3,4,5,6,7,8-
actahydro isoquinoline
(V)

(I) CrO₃/CH₃COOH;
(Oxidation)

(ii) H₂-Raney Ni;
(Reduction)

N-Acetyl-7-keto-
8-methyl-1,2,3,4,5,6,7,8-
octahydro isoquinoline
(VI)
(keto-Form)

*Both 'keto' and
'enol'-forms are
duly separated by
their crystalline
hydrates

'keto-enol'-tauto-
merism

H₂-Pt'
(Reduction)

Homomeroquinene
derivative (VII)
(Redrawn)

(I) H₅C₂-NO₂'
Ethyl Nitrite

(ii) H₅C₂-ONa
Sodium ethoxide
(Freshly prepared)

(VI A)
('enol'-Form)

An Intermediate
(VIII)

Or

(I) CH₃I/K₂CO₃'
Methyl iodide

(ii) KOH;Δ;

cis- (+)-Homoemerquinene
(IX)

(I)C₂H₅OH/HCl;

(Ii) ⬡—COCl
Benzoyl chloride

(A β-keto-Ester)
(XII)

C₂H₅-ONa;
Sodium ethoxide

[Claisen Condensation]
with (X)

N-Benzoylated ethyl ester
of homomero quinene
(XI)

Ethyl quininate
(X)

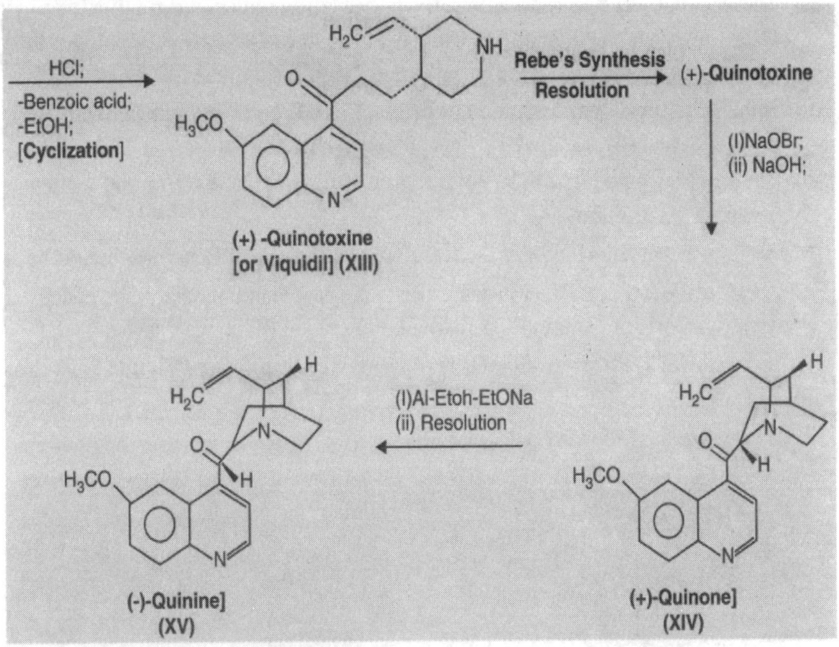

(+)-Quinotoxine
[or Viquidil] (XIII)

(-)-Quinine]
(XV)

(+)-Quinone]
(XIV)

Explanations : The various cardinal steps involved in the 'total synthesis' of 'quinine' by *three* researchers : **Muhtadi, Woodward,** and **Rabe** are explicitly enumerated as under ;

(1) **Benzaldehyde** when nitrated–reduced–diazotized–hydrolyzed yields **3–hydroxy benzaldehyde (I),** which upon **Pomeranz–Fritsch** synthesis using diethoxy ethyl amine and H_2SO_4 produces **7–hydroxy isoquinoline (II).**

(2) The resulting product on **Mannich Reaction** using *formalin* and *piperidine* yields **7–hydroxy–8–methyl piperidinyl isoquinoline (III),** which on treatment with freshly prepared **sodium methoxide at 220° C** produces **7–hydroxy–8–methyl–isoquinoline (IV).**

(3) The **product (IV)** when treated *first* with **acetic anhydride** and *secondly* with **Raney-Ni** (*i.e.,* reduction) yields **N–acetyl–7–hydroxy 8–methyl–1,2,3,4,5,6,7,8–octahydro–isoquinoline (V).**

(4) The resulting **product (V)** *first* with **oxidation** with **chromium–6–oxide,** and **reduction** with **Raney–Ni** yields **N–acetyl–7–keto–8–methyl–1,2,3,4,5,6,7,8–octahydro isoquinoline (VI).**

(5) The *keto*–form of product (VI) undergoes '*keto-enol*'–tantomerism to produce the corresponding 'enol' form (VI A).

(6) The resulting **product (VIA)** when treated *first* with **ethyl nitrite** and *secondly* with **sodium ethoxide** gives rise to the formation of **homomeroquinene derivative (VII)** due to the cleavage between **C–7** and **C–8,** which on subsequent *reduction* with H_2–Pt yields an **intermediate (VIII).**

(7) The *redrawn* intermediate (VIII) on *first* reaction with **methyl iodide** and K_2CO_3, and *secondly* with **KOH** and boiling produces *cis*–(±)–*homomeroquinene* (IX), which upon treatment with **EtOH/HCl** and **benzoyl chloride** yields *two* distinct *products* of reaction **ethyl quinindate (O)** and **N–benzoylated ethyl ester of homomeroquinene (XI).**

(8) The **product (X)** undergoes **Claisen Condensation** with **sodium ethoxide** to yield a β–keto ester **(XII)**, which on treatment with HCl undergoes **cyclization** to produce the racemic mixture of **quinotoxine (XIII)**, also known as **'viquidil'**.

(9) **Rabe's Synthesis** *i.e.,* resolution of **product (XIII)** gives **(+)–quinotoxine**, which upon treatment with **NaOBr** and **NaOH** yields **(+)–quinone (XIV)**.

(10) The resulting **product (XIV)** on reaction with **Al/EtOH/EtONa** and subsequent resolution produces **(–)–quinine (XV)**.

10. Mechanisms of Intermediary Conversions of Products of Reactions in Synthesis of Quinine:

Following are the *two* exemplary instances pertaining to the intermediary conversions of the products of reactions in synthesis of **'quinine'** [as discussed in section (9) above], such as:

(a) **Conversion of 7–Hydroxy–8–methyl piperidinyl isoquinoline (III) into 7–Hydroxy–8–methyl isoquinoline (IV) :**

Let us visualize the following reaction :

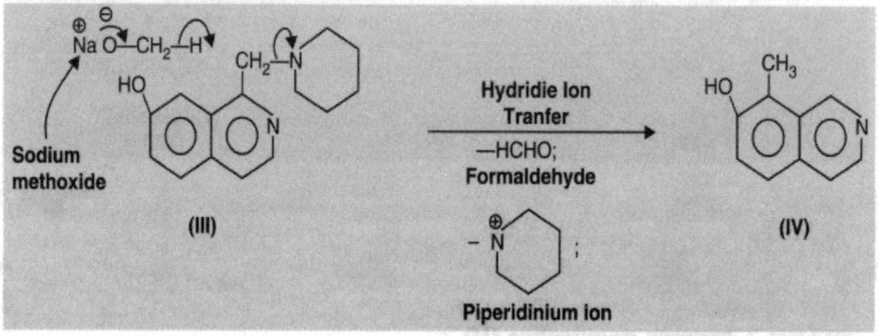

The exact and precise mechanism of the aforesaid reaction with the **methoxide ion (–CH$_3$O$^\ominus$) and compound (III)** most preferably comes into play by means of the critical **'hydride–ion transfer'**, thereby producing the resultant **compound (IV)** with the elimination of a mole each of **formaldehyde** and **piperidinium ion.**

(b) **Conversion of N–Acetyl–7–keto–8–methyl–1,2,3,4, 5, 6,7, 8–Oetahydro isoquinoline ['keto'– form] (VI) into Homomeroquinene derivative (VII):**

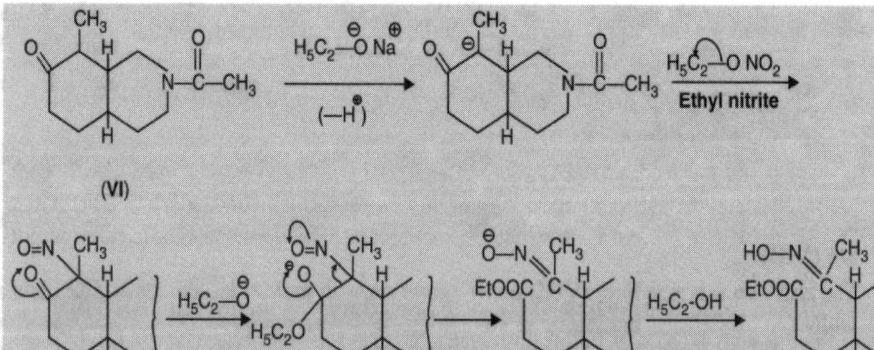

Explanation: The **mechanism** of conversion of **compound (VI)** into **compound (VIII)** may be expatiated as under :

(1) The *'keto'* form of N–acetyl–7–keto–8–methyl–1,2,3,4,5,6,7,8–octahydro isoquinoline (VI) when reacted with sodium ethoxide loses a **proton,** thereby C–8 retains a –ve charge on it.

(2) The resulting product on being treated with **ethyl nitrite** introduces a **nitroso (–N=O) moiety** strategically at **C–8,** which on further reaction with a **–vely charged ethoxide ion** helps to retain a **ethoxide (–OC$_2$H$_5$) residue at C–7.**

(3) The product thus obtained undergoes electronic transformations, thereby affording a distinct cleavage between the **C–7** and **C–8** atoms. Furthermore, C-7 retains a —COOEt group and C–8 retains a methyl plus a **nitroso moiety** bearing a –ve charge on the O–atom.

(4) Ultimately, the resulting product on treatment with *ethanol* yields the **compound (VIII).**

3.5.2. Cinchonine

Cinchonine is almost identical in structure with **'quinine',** except that it is devoid of the **methoxyl (H$_3$CO–) moiety** at **C–6** of the *quinoline nucleus.* Besides, like **quinine,** it is indeed a **'diacid tertiary base'. Cinchonine** invariably occurs in most varieties of *cinchona bark,* especially in the bark of *Cinchona micrantha* R. & P., *Rubiaceac.* However, it may be regarded to be the **'parent alkaloidal substance'** amongst the so called **'cinchona alkaloids'.**

Cinchonine

3.5.2.1. Isolation

In fact, the *'mother liquor'* obtained after the crystallization of **'quinine sulphate'** (refer to **'quinine'**) largely consists of **cinchonine** as well as other allied **alkaloids.** In usual practice, the **'mother liquor'** is subjected to :

- **neutralization,** and
- **alkalification** (careful *addition of NaOH solution*),

which allows the precipitation of **'cinchonine'.** Now, the precipitate thus obtained is filtered under suction and redissolved in a minimum volume of **solvent ether. Cinchonine** appears as solid deposits which may duly purified by recrystallization form **alcohol** in the form of either *needles* or *prisms.*

3.5.2.2. Characteristic Features

These essentially comprise:

(1) Prisms/needles of **cinchonine** obtained from *solvent ether* and *ethanol* show **mp 265°C.**

(2) **Cinchonine** usually sublimes at **220°C.**

(3) **Cinchonine** exhibits a **specific rotation in ethanol** as $[\alpha]_D$ + **229°.**

(4) **Solubility Profile:** 1g dissolves in 60 mL **ethanol;** 25 mL boiling **ethanol;** 110 mL **chloroform;** and 500 mL **solvent ether.**

3.5.2.3. Constitution of Cinchonine

The various **chemical reactions, inferences drawn,** and **logical explanations** in a concerted manner actually drive to a point so as to assign the most probable structure of **'cinchonine'**, namely:

(a) **Molecular Formula:** The **molecular formula** of *cinchonine* is determined to be $C_{19}H_{22}N_2O$.

(b) **Presence of Only One Hydroxyl (-CHOH) Moiety:** Cinchonine readily gives rise to the formation of either a **monoacetate** (with **acetic anhydride**) or a **monobenzoate** (with **benzoyl chloride**). Thus, we may have:

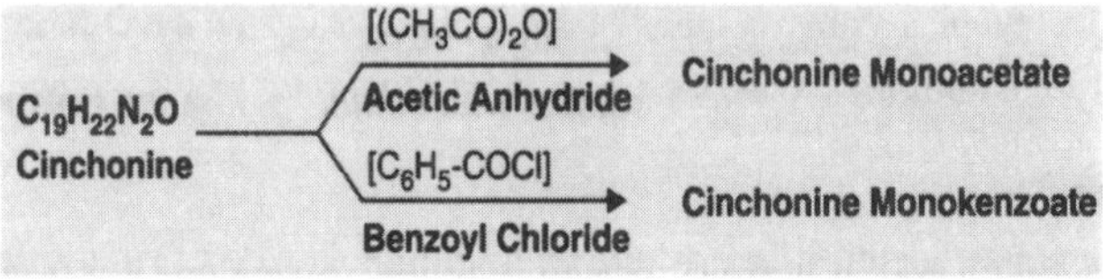

Inference: Cinchonine possesses only one essential **hydroxyl (-OH)** moiety in its molecule.

(c) **Nature of the Hydroxyl (-OH) Moiety:** Cinchonine on being subjected to **oxidation** yields the corresponding **'ketone'**, *cinchonine*, as given under:

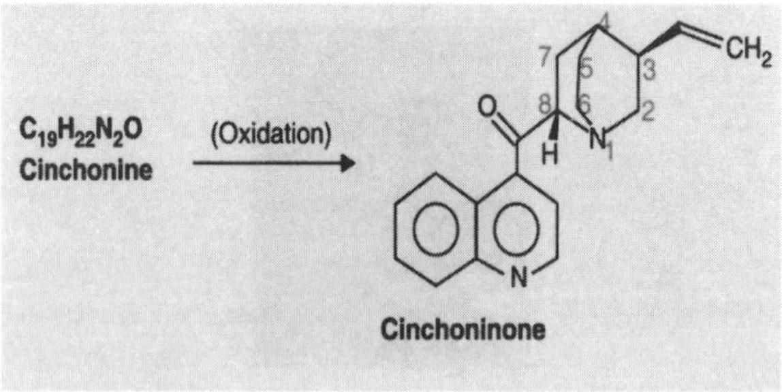

Inference: The aforesaid reaction suggests the **hydroxyl (-OH)** moiety present duly in **cinchonine** happens to be of **secondary alcoholic nature**.

(d) **Presence of Olephinic Double Bond [>C=C<]:** Cinchonine readily takes up one mole of **bromine** or **halogen acid** (viz., HI or HBr); and also gets easily *reduced (i.e., addition of hydrogen)* catalytically as given under:

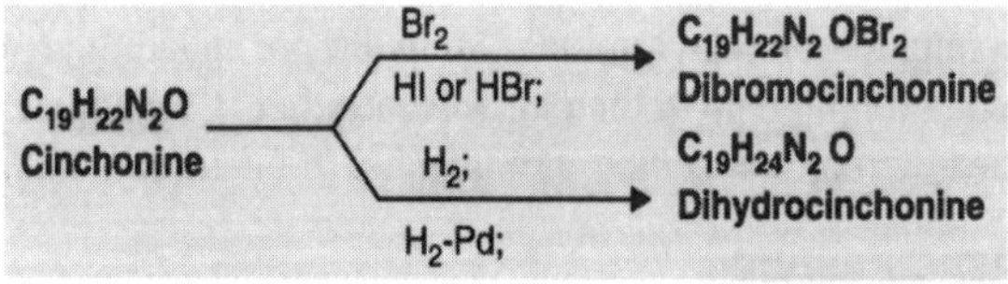

Inference: Thus, it may be ascertained that **cinchonine** essentially comprises one **olephinic double bond**.

(e) **Nature of N-Atom:** Cinchonine easily and conveniently adds on *two* moles of **methyl iodide (CH₃I)** as indicated under:

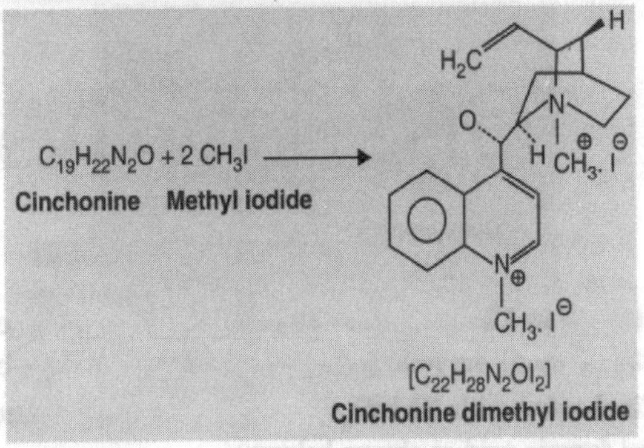

$C_{19}H_{22}N_2O + 2 CH_3I \longrightarrow$

Cinchonine   Methyl iodide

$[C_{22}H_{28}N_2OI_2]$
Cinchonine dimethyl iodide

Inference: It clearly indicates that 'cinchonine' essentially has a 'ditertiary base'.

(f) **Presence of Half-Quinoline Nucleus and Half-Quinuclidine Nucleus in Cinchonine:**

(1) **Quinoline Nucleus in Cinchonine:** The presence of 'quinoline nucleus in cinchonine may be ascertained by means of the following *two* steps as given under:

(i) **Fusion with KOH yields Lepidine: Cinchonine** on being fused with **potassium hydroxide (KOH)** yields **lepidine (I)** (or **4-methyl quinoline**) as depicted below:

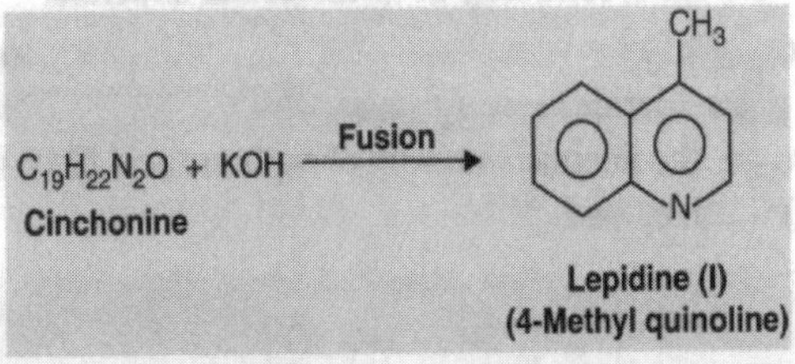

$C_{19}H_{22}N_2O + KOH \xrightarrow{\text{Fusion}}$

Cinchonine

Lepidine (I)
(4-Methyl quinoline)

Inference: From the above reaction it may be concluded that **cinchonine** has a **quinoline nucleus** with a side-chain at **C-4**.

(ii) **Vigorous Oxidation with Chromic Acid/H₂SO₄:** Cinchonine when subjected to vigorous oxidation with **chromic acid (CrO₃)** and **H₂SO₄** produces **cinchonine acid (II)** as given under:

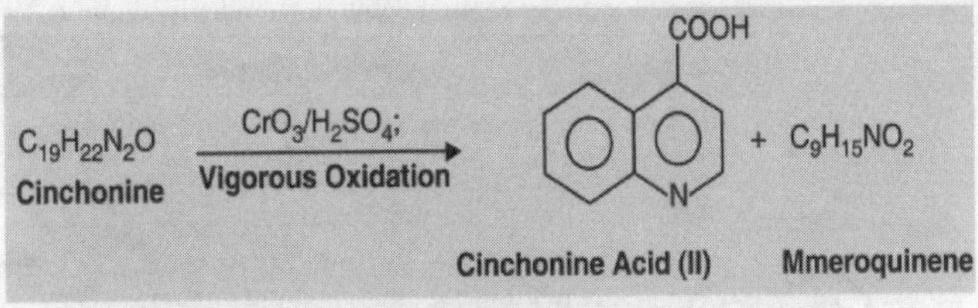

$C_{19}H_{22}N_2O \xrightarrow[\text{Vigorous Oxidation}]{\text{CrO}_3/\text{H}_2\text{SO}_4;} \quad + C_9H_{15}NO_2$

Cinchonine

Cinchonine Acid (II)   Mmeroquinene

Inference: The above vigorous reaction clearly, suggests that the presence of the **quinoline nucleus** in 'cinchonine' due to the formation of **cinchoninic acid [II]**.

> **NOTE:** Both steps (*i*) and (*ii*) vividly suggests that 'cinchonine' X may essentially comprise a *quinoline nucleus* with a side-chain strategically located at C-4 (of the *quinoline nucleus*) as shown below: Skraup suggested that the residue at C-1 is nothing but the 'second half', that ultimately proved to be the quinuclidine nucleus.

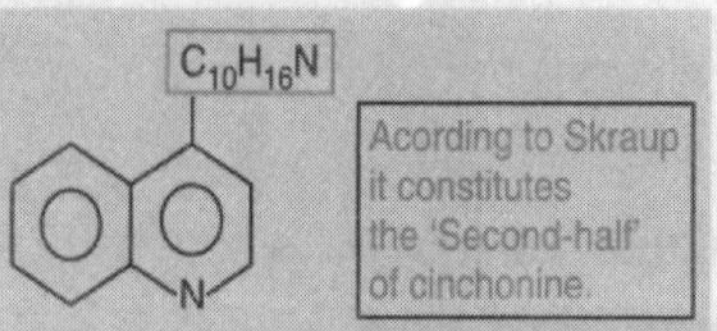

$C_{10}H_{16}N$

Acording to Skraup
it constitutes
the 'Second-half'
of cinchonine.

Cinchonine
(X)

(2) **Quinuclidine Nucleus in Cinchonine:** The **quinuclidine nucleus** or the so called **'second half'** of **cinchonine** is established by the following *two* steps duly supported by logistic explanations:

(*i*) **Oxidation of Cinchonine with KMnO₄:** Cinchonine when oxidized with $KMnO_4$ produces **cinchotenine** and **formic acid** as shown below:

$$C_{19}H_{22}N_2O + 4(0) \xrightarrow[\text{(Oxidation)}]{KM_nO_4} C_{18}H_{20}N_2O_3 + HCOOH$$

Cinchonine   Cinchotenine Formic
                                              Acid

Inference: The above oxidation reaction shows that **cinchonine** comprises a $-CH=CH_2$ **moiety** present duly in its side chain in the *second-half* of its molecular structure.

(*ii*) **Chlorination-Alcoholic KOH Treatment of Cinchonine:** Cinchonine on being chlorinated with phosphorus pentachloride (PCl_5) yields a chlorinated intermediate compound. Koenigs (1906)* further treated the intermediate with phosphoric acid (25%) to obtain **lepidine (I)** and **meroquinene (II)** as given under:

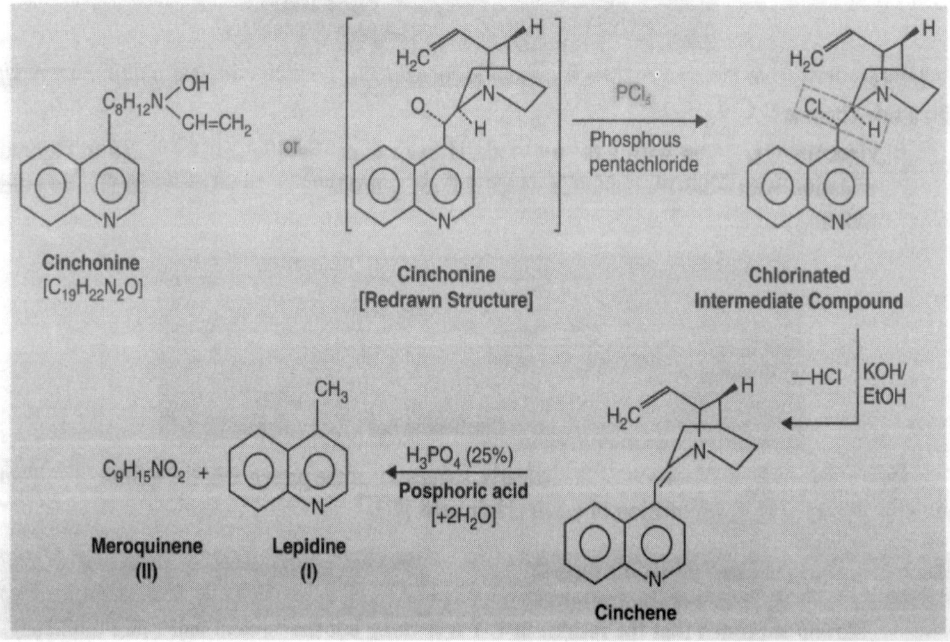

Cinchonine
[$C_{19}H_{22}N_2O$]

or

Cinchonine
[Redrawn Structure]

PCl₅
Phosphous
pentachloride

Chlorinated
Intermediate Compound

—HCl | KOH/
        EtOH

$C_9H_{15}NO_2$ +

Meroquinene   Lepidine
(II)   (I)

H_3PO_4 (25%)
Posphoric acid
[+2H₂O]

Cinchene

---

* Koenigs : *Ann*, **347** : 182, 1906.

Explanation: The various steps involved in the above reaction may be expatiated as stated under:

(1) **Cinchonine** on chlorination with PCl_5 replaces the *secondary alcoholic (-OH) group* to yield an **intermediate chlorinated compound**.

(2) The resulting product when treated with *ethanolic KOH solution* loses a mole of **HCl**, as shown by dotted line, to yield **cinchene-an unsaturated derivative**.

(3) The unsaturated product when reacted with H_3PO_4 (25%) takes up **two moles of water** to produce one mole each of **Lepidine (I)** and **Meroquinene (II)**.

However, ti has been duly observed that **cinchonine** when treated with CrO_3 (chromic acid) gives rise to the formation of **cinchoninic acid** and **meroquinene** as given under:

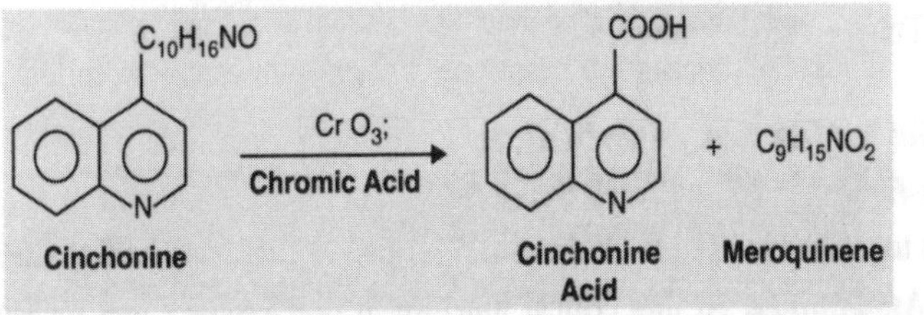

The structure of **meroquinene'** has already been proved and established under sections 2 (j) and 3, 4, 5 of **'Quinine'** earlier to be represented as:

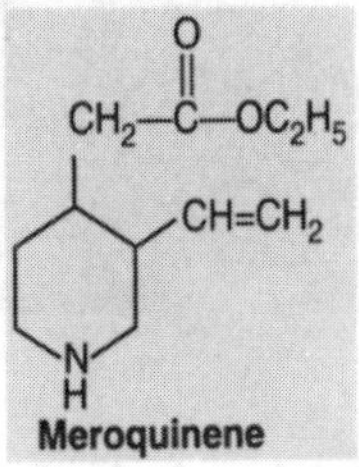

Meroquinene

3.5.2.4 Structure of Cinchonine

The structure of **meroquinene** has already been discussed *under section 3 and 4* of **quinine**. Hence, one may expect that the **C-atom** duly present in the **carboxyl (-COOH) moiety** in **'meroquinene'** should be the most probable point of **'attachment'** to the **'first-half'** (*i.e.*, the **'quinoline-half'**) at which the *fusion* of the **'second-half'** (*i.e.*, the **'quinuclidine-half'**) has occurred eventually.

As we know, that **'cinchonine'** happens to be a **'ditertiary base'**; hence, the **'second-half'** essentially comprises a **tertiary N-atom**.

It has also been duly established that **meroquinene** designates a *secondary base*; and, therefore, it is expected most logically that in the critical formation of **meroquinene** the following *two* cardinal changes come into being, namely:

- conversion of the *tertiary* **N-atom** [$\equiv N^{\oplus}$-] into the corresponding *secondary* **N-atom** [=N-], and

- formation of a **carboxyl (-COOH) moiety** simultaneously.

Explanation: The above observations and statement of facts may be argued reasonably and satisfactorily that the *tertiary* N-atom critically forms a specific segment pertaining to a **bridged-ring system**, thereby causing the obvious cessation of *one C-N bond* in **cinchonine** in the specific course of its **oxidation** by a mixture of chromium-6-oxide [CrO_3] and H_2SO_4 as shown under:

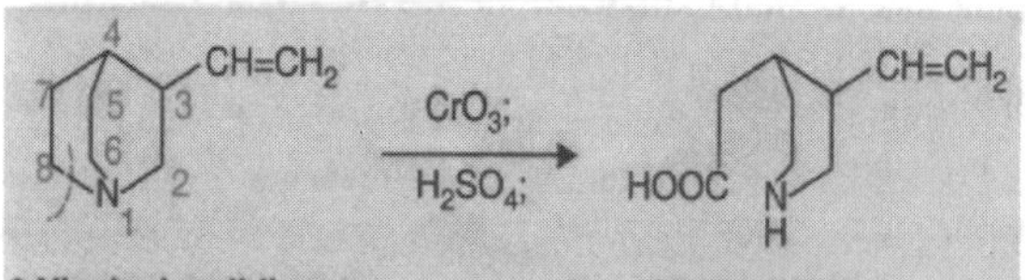

3-Vinylquinuclidine   Meroquinene

Inference: Thus, in **'cinchonine'**, one may critically observe the following vital and important aspects, namely:

- *First-Half i.e.,* the **'quinoline-half'**, and
- *Second-Half i.e.,* the **'quinuclidine-half'**,

should be linked together *via* its **side-chain** at **C-4** to the **quinuclidine-half** at **C-8**.

Important Assumption: At this critical juncture, if one assumes that the prevailing *secondary alcoholic moiety* form a linkage between the **'quinoline-half'** (*i.e.,* **First-Half**) to the **'quinuclidine-half'** (*i.e.,* the **Second-Half**), one may have the structure of **cinchonine** written as follows:

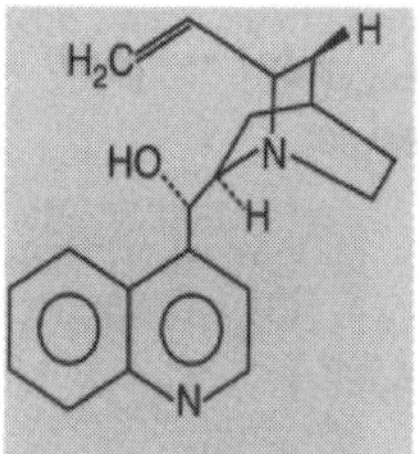

Cinchonine

3.5.2.5. Synthesis of Cinchonine

Rabe *et al.* (1908)* discovered the *partial synthesis of cinchonine* commencing from **cinchotoxine**, and the latter compound was duly prepared from **cinchonine** by means of the **Hydramine Fission** *i.e.,* a prolonged action of acetic acid with **cinchonine** as follows:

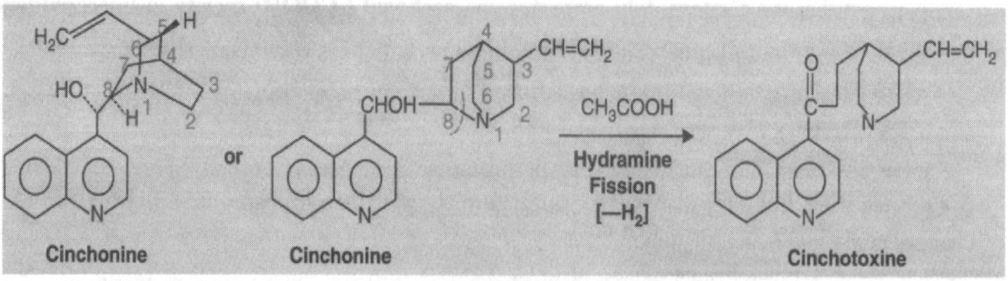

Cinchonine   Cinchonine   Cinchotoxine

---

* Rabe *et al.* : *Ber*, 41 : 63, 1908.

The 'hydramine fission' occurs due to the cleavage between N=1 and C-8 linkage, as shown by *dotted line*, thereby losing a **hydrogen atom** to yield **cinchotoxine**.

Importantly the above resulting product *i.e., cinchotoxine* undergoes *three* sequential steps to produce the desired product 'cinchonine':

- Reaction with sodium hypobromite (NaOBr) to obtain a **bromoderivative (I)**,
- Treatment with alkali (NaOH) to lose a mole of HBr by abstracting the *N-hydrogen atom* and *bromine atom* thereby causing 'cyclization' of the **quinuclidine nucleus** to yield 'cinchonine' **(II)**, and
- Reduction of **cinchonine (II)** with an ethanolic mixture of finely divided **Al-powder** and freshly prepared **sodium ethoxide (C_2H_5-ONa)** gives rise to the formation of a **racemic mixture of cinchonine (III)**.

These reactions may be summarized as given under:

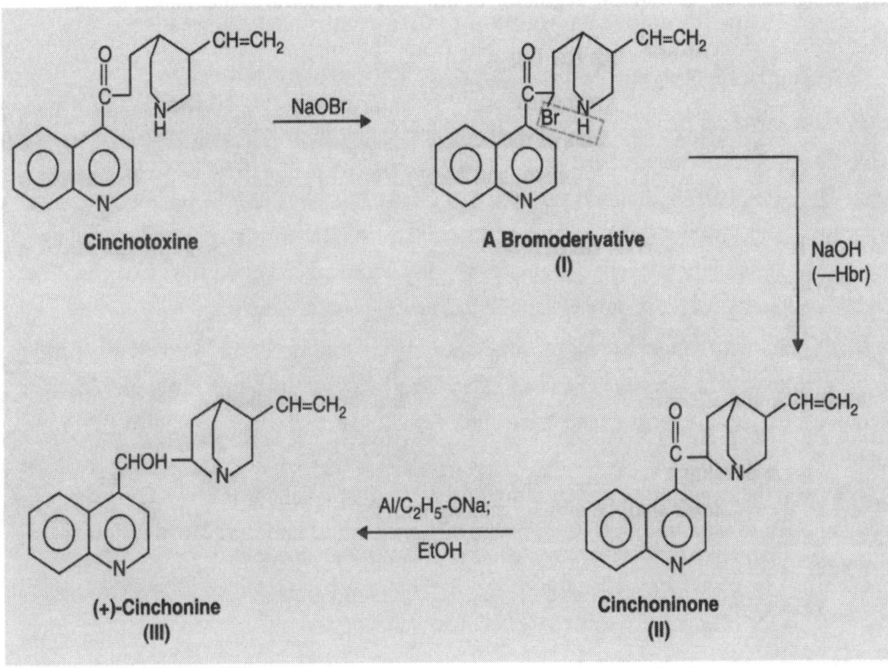

Cinchotoxine — A Bromoderivative (I) — Cinchoninone (II) — (+)-Cinchonine (III)

3.5.2.6. Absolute Configuration of Quinine and Cinchonine

Absolute configuration may be defined as the 'spatial arrangement' of *atoms* or *functional moieties* with regard to their relative position as well as status to one another.

Importantly the actual determination of the **absolute configuration** as, in fact, really a cumbersome phenomenon; and, therefore, may be accomplished genuinely provided the compound eventually forms **well-defined crystals**.

Besides, the *configuration* of a plethora of 'organic compounds', both derived from the **natural products** and the **synthetic products**, do invariably exhibit a remarkable correlation. *vis-a-vis* a **known compound**. In doing so, one may actually determine the so called *'relative configuration'*.

Following are the **'absolute configuration'** of *quinine* and *cinchonine*:

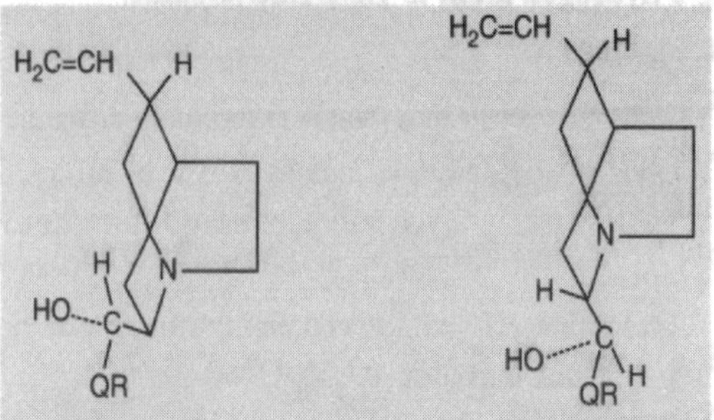

(+)-Cinchonine
(+)-Quinidine

(−)-Cinchonine
(−)-Quinidine

QR: designates the **'quinoloine segment'** of the various structrures mentioned above.

3.5.2.7. Synthetic Substitutes for Quinine

Preamble: Malaria has been recognized as a dreadful human disease since the 19th century and even before that era in several systems of medicine, such as: Indian, Chinese, Tibetan, Unani, African, Greek, Egyptian, Thai, Philipino, Brazilian, and Mexican to name a few. Without the advent of latest advancements and developments in science, knowledge, and technology the **antimalarial age-old drugs** were mostly based upon aqueous **decoctions*** duly prepared from the **barks of the cinchona tree** thereby heavily dependent upon the naturally occurring plant resources.

It has been widely acknowledged that–**'necessity is the mother of invention'**. During the well known **Second World War,** there was an exceptionally high demand for **'quinine'** (derived from the **cinchona bark**), and almost at the same time the ear-**marked cinchona-cultivated area in Java/ 'Sumatra'** duly occupied by the **Japanese**. The *time-framed challenge* as appropriately delivered by the researchers in the development of an array of the **'synthetic substitutes for Quinine'**, of which the following *three compounds* shall be discussed along with their **individual synthesis, explanation**, and **therapeutic uses:**

- **P maquin,**
- **Santoquin**, and
- **Hydroxychloroquine Sulphate**

3.5.2.7.1. Pamaquin

The synthesis of **'pamaquin'** may be accomplished by the following *three* steps in a **'sequential manner':**

(a) Preparation of Side-Chain,

(b) Preparation of Quinoline Nucleus, and

(c) Condensation of (a) and (b).

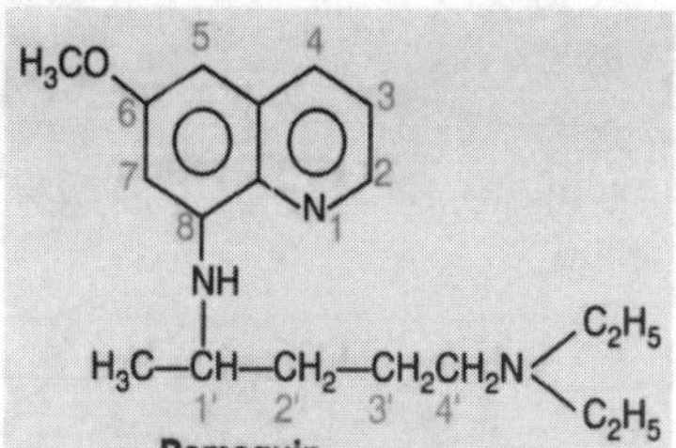

Pamaquin

[8-(4-Diethylamino-1-metylbutyl
amino)-6-methoxyquinoline]

(a) Preparation of Side-Chain [*i.e.*, 4-Amino-deithylamino pentane]:

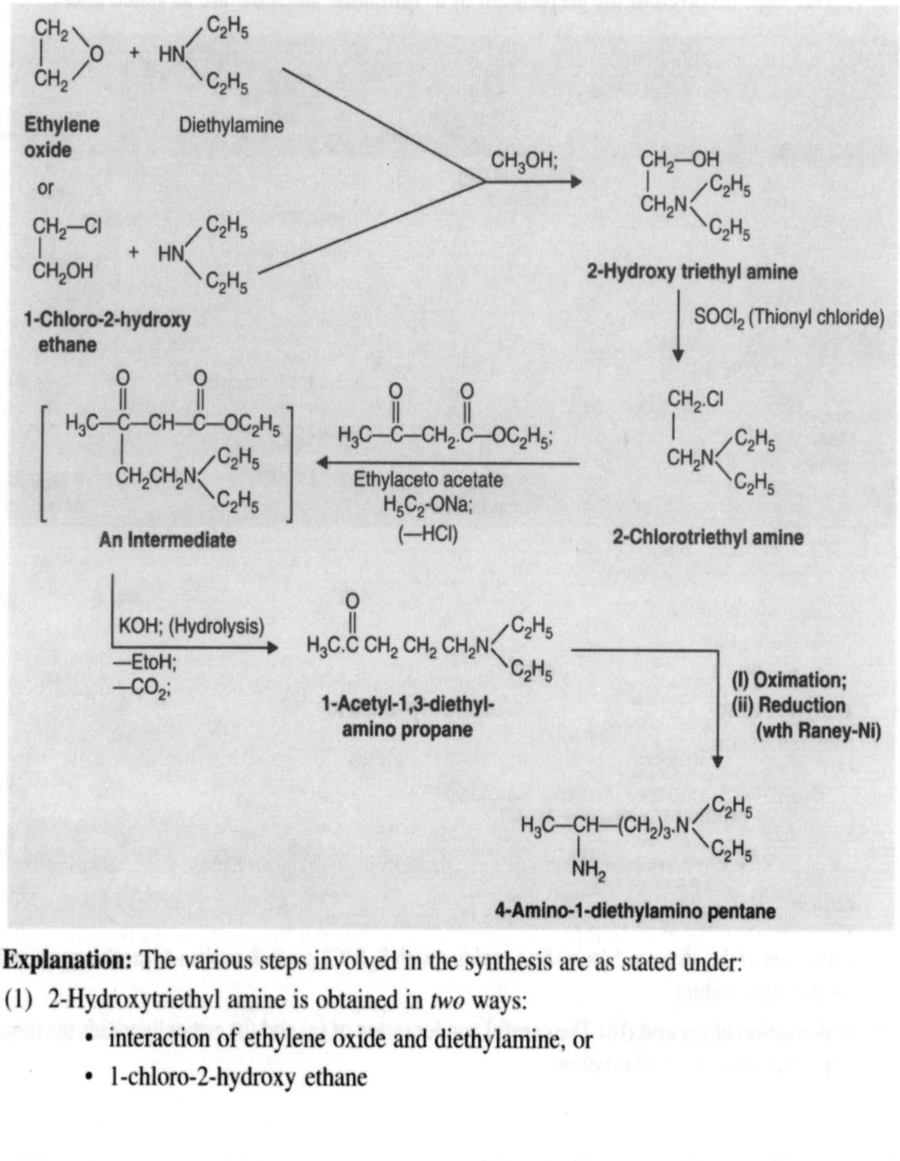

Explanation: The various steps involved in the synthesis are as stated under:

(1) 2-Hydroxytriethyl amine is obtained in *two* ways:

- interaction of ethylene oxide and diethylamine, or
- 1-chloro-2-hydroxy ethane

(2) The resulting product is duly chlorinated with thionyl chloride (SOCl$_2$) whereby the **2-hydroxy group** gets replaced with the **chloro group** to produce **2-chlorotriethyl amine**.

(3) The product thus obtained is treated with **ethylacetoacetate** (a **double-ester**) in the presence of freshly prepared **sodium ethoxide** (H$_5$C$_2$-ONa) loses a mole of HCl to form an 'intermediate compound'.

(4) The resulting 'intermediate' on being subjected to hydrolysis loses a mole each of **ethanol** and **carbon dioxide** to yield **1-acetyl-1,3-diethylamino propane**.

(5) Finally, the above resulting product upon **oximation** and subsequent *reduction* with **Raney-Ni** gives the desired *side-chain* **4-amino-1-diethylamino pentane**.

(b) **Preparation of Quinoline Nucleus [or 8-Amino-6-methoxy quinoline]:**

The various steps involved in the preparation of a 'quinoline nucleus' are as stated under:

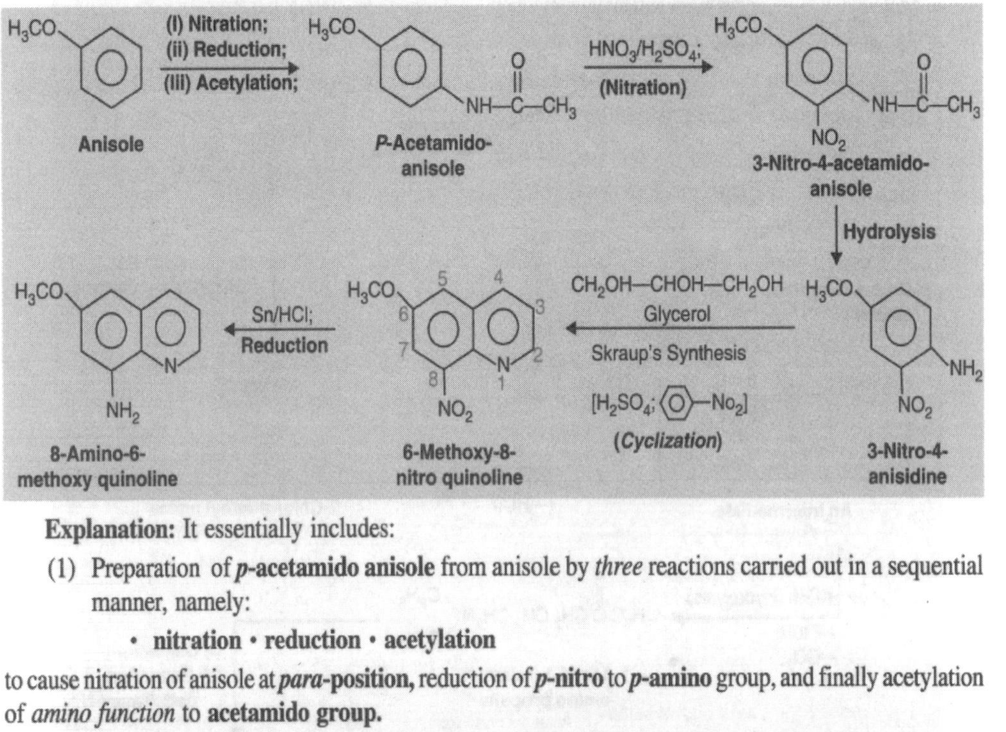

Explanation: It essentially includes:

(1) Preparation of *p*-acetamido anisole from anisole by *three* reactions carried out in a sequential manner, namely:

  • **nitration** • **reduction** • **acetylation**

to cause nitration of anisole at *para*-**position,** reduction of *p*-**nitro** to *p*-**amino** group, and finally acetylation of *amino function* to **acetamido group.**

(2) Further nitration of the resulting product yields **3-nitro-4-acetamidoanisole**, which upon **hydrolysis** produces **3-nitro-4-anisidine**.

(3) The resulting product on being subjected to **Skraup's Synthesis** in the presence of **glycerol/ H$_2$SO$_4$/p-nitrobenzene** undergoes cyclization to yield **6-methoxy-8-nitro-quinoline** *i.e.*, the **quinoline nucleus** gets introduced into the molecule.

(4) Reduction of the aforesaid *quinoline residue* with S$_4$/HCl gives **8-amino-6-methoxy quinoline** the desired product.

(c) **Condensation of (a) and (b):** The careful condensation of (a) and (b) critically yields the targeted compound 'pamaquine' as shown below:

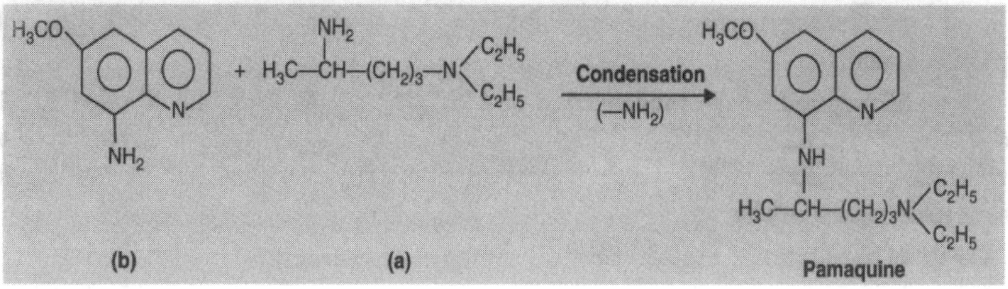

(b)   (a)   **Pamaquine**

3.5.2.7.2. Santoquin

Santoquin, another **quinoline synthetic substitute for quinine** may be synthesized by a **two-step reaction:**

- Interaction of *m*-chloroaniline and **diethylester of methylated oxosuccinate** to give **4,7-dichloro-3-methyl-quinoline**, and
- Treatment of resulting product with *3-amino-5-diethylamino pentane* to produce 'Satoquin' or **7-Chloro-4-{[4-(diethylamino)-1-methylbutyl]amino}-3-methyl-quinoline.**

The aforesaid reactions may be expressed as under:

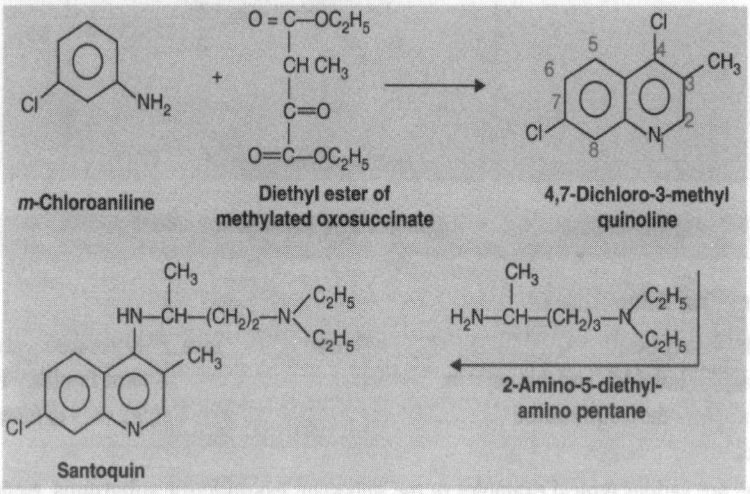

m-Chloroaniline   Diethyl ester of methylated oxosuccinate   4,7-Dichloro-3-methyl quinoline

2-Amino-5-diethyl-amino pentane

Santoquin

NOTE: *'Santoquin'* bears an additional methyl moiety at C-3 in the *'quinoline nucleus'* when compared to that of *chloroquine*. It is observed to be *less reactive* that *'chloroquine'*.

3.5.2.7.3. Hydroxychloroquine Sulphate

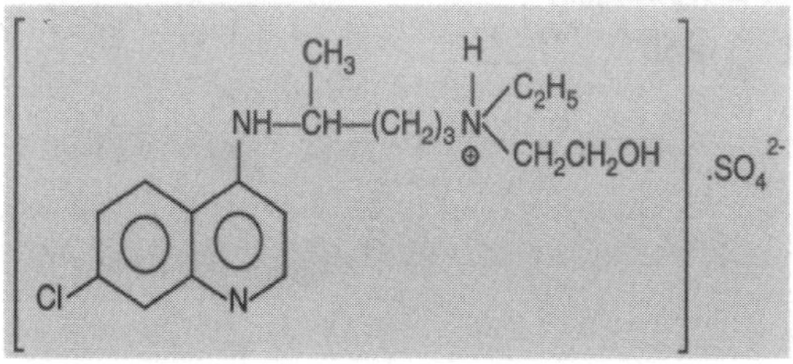

Hydroxychloroquine sulphate [or Ethanol-2-[[4-(7-chloro-4-quinolyl)amino]pentyl]ethylamino sulphate (1:1) salt].

It may be prepared by the **condensation** of **4,7-dichloro-quinoline** with **N'-ethyl-N''-(2-hydroxyethyl)-1,4-penetanediamine** when a mole of **HCl** is lost to yield the **hydroxychloroquine base,** which upon treatment with a **stoichiometric proportion** of H_2SO_4 produces the desired salt: **Hydroxychloroquine sulphate.**

The above reaction may be expressed as under:

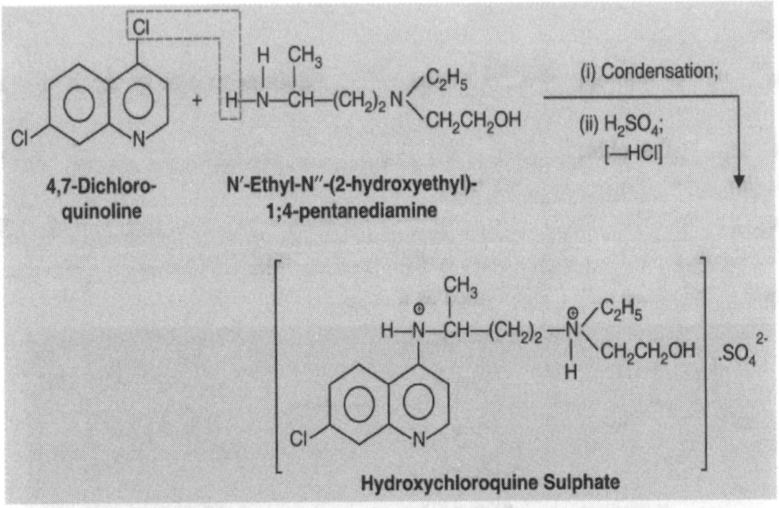

Hydroxychloroquine Sulphate

NOTE: Its therapeutic actions and applications are very much akin to 'chloroquine'.

3.6. Isoquinoline Alkaloids

Preamble: It may be observed that the 'isoquinoline structure' invariably occurs in an appreciable number of 'plant alkaloids', and that separated quite widely in a variety of plant families. However, in a broader perspective the isoquinoline alkaloids do represent the *largest single group of plant alkaloids*; and of course, showing an *enormous variation* in their actual respective structures.

Following are various typical examples of the important isoquinoline subgroups, namely:

(a) **Papaverine** *i.e.,* the *benzylisoquinolines,*

(b) **Sanguinarine** *i.e.,* the *benzophenathradines,*

(c) **Hydrastine** *i.e.,* the *phthalideisoquinolines,*

(d) **Morphine** *i.e.,* the *morphinans,*

(e) **Berberine** *i.e.,* the *protoberberines,*

(f) **Emetine,**

(g) **Narcotine,**

(h) **Ergotamine**

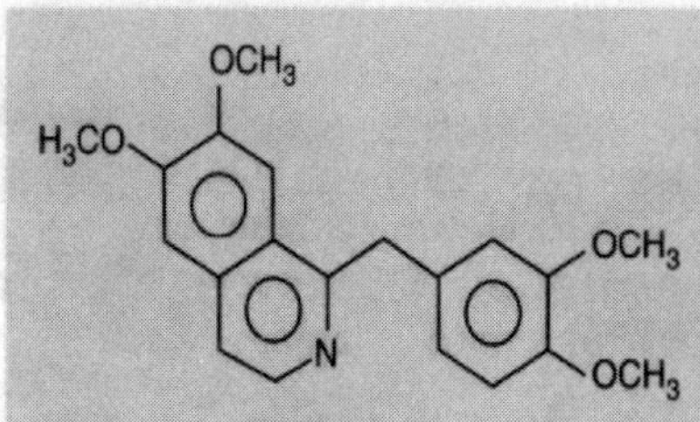

Papaverine

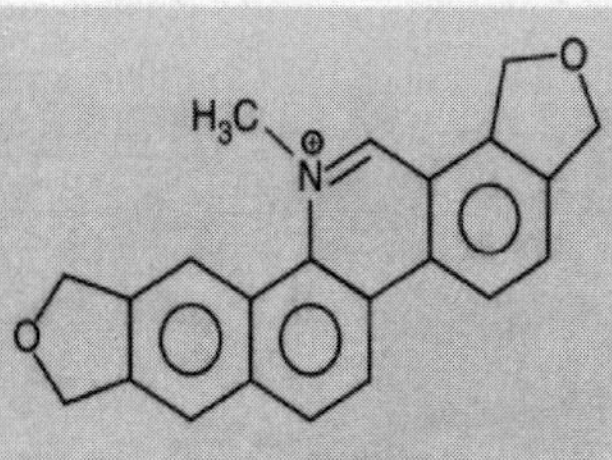

Sanguinarine

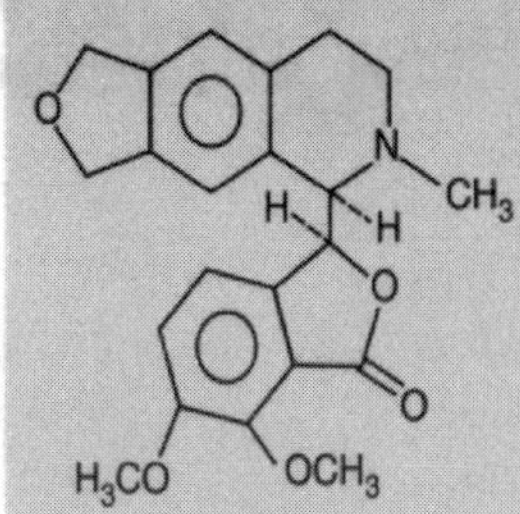

Hydrastine

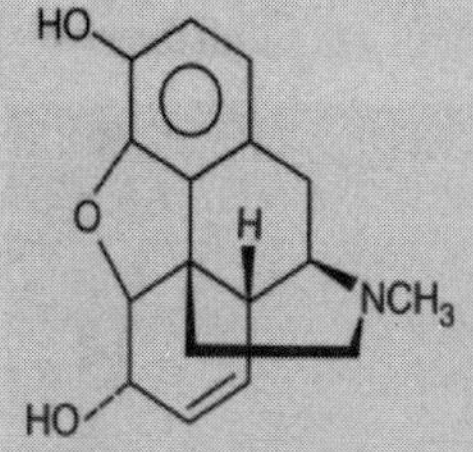

Morphine

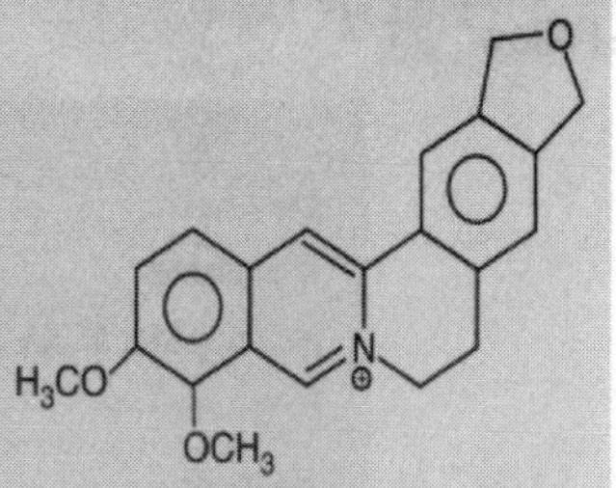

Berberine

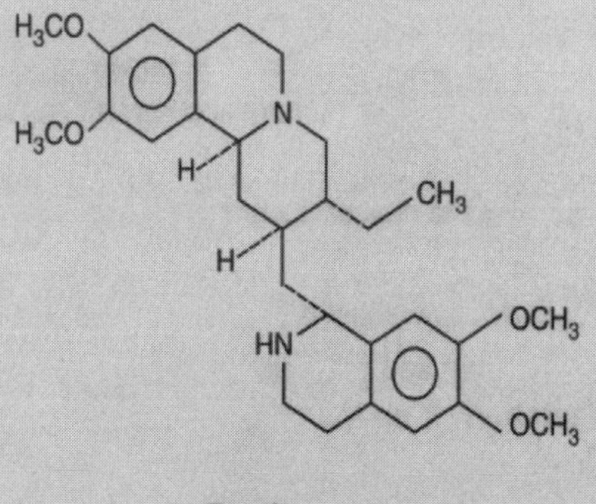

Emetine

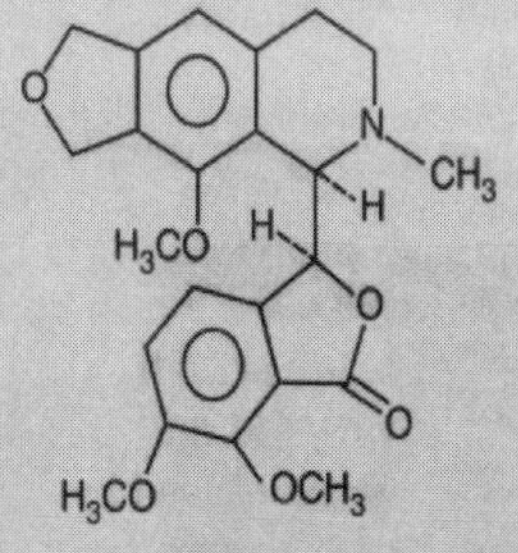

Narcotine

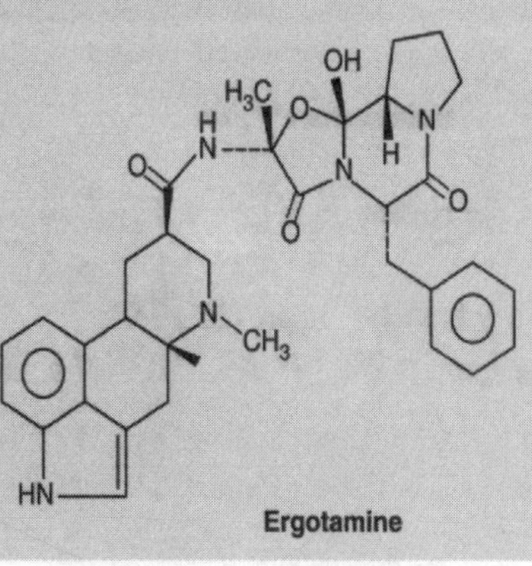

Ergotamine

Some of the important **isoquinoline alkaloid variants** shall now be treated individually in the section that follows:

3.6.1. Papaverine

A survey of literature would reveal most glaringly that the well-known *'Opium Alkaloids'* essentially comprise not only **'papaverine'** but also a cluster of almost *twenty-four* other alkaloids, of which a few vital and important members are namely: **codeine, laudanine, laudanosine, narceine, narcotine,** and **thebaine**, occurring predominantly in the **opium latex*** (**poppy plant**), *Papaver somniferum* Linne' (*family: Papaveraceae*); and, therefore, all of them are collectively termed as **'opium alkaloids'**.

Importantly, nearly altogether *thirty* **different alkaloids** have been duly **isolated, purified,** and **characterized** from *opium* as well as its *extracts*; howevere, a few of them are recognized duly as the **'alteration products'** of the ensuing **'alkaloids'** occurring *naturally* and *prominently* in the *'drug'* itself. Following are the **important alkaloids** along with their **exact and precise range in percentage(s)**:

Alkaloid		%
Codeine	:	0.8-2.5
Morphine	:	4-21
Noscapine (Narcotine)	:	4-8
Papaverine	:	0.5-2.5
Thebaine	:	0.5-2

3.6.1.1. Isolation of Papaverine

The dried **'milky juice'**, (or **'latex'**) duly obtained from the selected **'unripe poppy seed capsule'** is digested very carefully with freshly prepared **'milk of lime'** when one may accomplish the following products:

- **Morphine analogues:** usually *remain dissolved* in the **reaction mixture** (or **mother liquor**), and
- **Papaverine analogues:** invariably get *precipitated* and recovered subsequently.

From the latter *i.e.*, the **precipitated product** it is feasible and possible to obtain **'papaverine'** by means of meticulously carried out **'fractional precipitation'** from the ensuing mixture. Finally, pure **'papaverine'** may be accomplished by further purification as its respective **'hydrooxalate salt'**.

3.6.1.2. Characteristic Features

These essentially include:

(1) **Papaverine** is obtained as a **triboluminescence****, *orthorhombic prisms* from a mixture of *ethanol and solvent ether (1:1)*.

(2) It sublimes at 135-140°C at **11 mm pressure and 2 mm distance**.

(3) It has a density d_4^{20} 1.337.

(4) It is an **optically inactive tertiary base** having pK (25°) 8.07.

(5) **Papaverine** shows *uv max* (in ethanol): 239, 278, 280, 314, 327 nm (log ε : 4.83, 3.86, 3.60, 3.67).

(6) It is almost insoluble in water.

---

* That is, **'milky juice'**, and the hardened juice forms the **'opium'**.

* **Triboluminescence:** It refers to **'luminescence'** or **'sparks'** emnated by friction or mechanical force applied to certain **chemical crystals**.

(7) **Papaverine** is found to be soluble in **hot benzene, glacial acetic acid, acetone**; and slightly soluble in chloroform, carbon tetrachloride, petroleum ether.

> **NOTE:** The *'optimum pH'* for the critical storage of 'papaverine' in solution is ranges between 2.0-2.8.

3.6.1.3. Constitution of Papaverine

Goldschmidt *et al.* (1883-1885) meticulously carried out the **'oxidative degradation'** to determine and establish the **constitution of papaverine** and eventually the most probable structure of the **'alkaloids'**.

Thus, the following steps would ultimately lead to the nearest possible structure of **'papaverine'**:

(1) **Molecular Formula:** The molecular formula of **'papaverine'** is found to be $C_{20}H_{21}NO_4$, based on its analytical data/molecular weight determination.

(2) **Presence of Tertiary Base in Papaverine:** It has been duly proved and established that the critical nature of the N-atom in **'papaverine'** is having **'tertiary status'**, since it yields a **'quaternary salt'** on being treated with one mole of **methyl iodide** (H_3C-I) as given under:

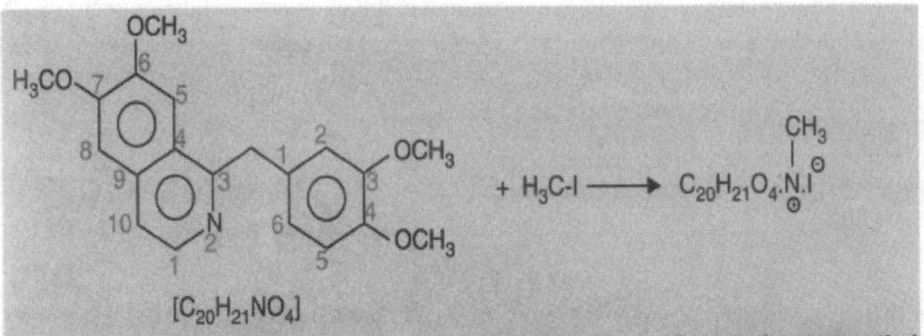

$$+ H_3C\text{-}I \longrightarrow C_{20}H_{21}O_4.N.I$$

| Papaverine | Methyl iodide | Papaverine-*tert*-methyl iodide |

$[C_{20}H_{21}NO_4]$

Inference: The N-atom present in **papaverine** is **tertiary in nature and status**.

(3) **Presence of Four Methoxyl (-OCH$_3$) Moieties: Zeisel Method*** helps to determine precisely the presence of *four* equivalents of methoxyl (-OCH$_3$) moieties in *papaverine* by subjecting it to vigorous constant boiling with **HI (hydroiodic acid)** as shown below:

$$C_{16}H_{12}N[OCH_3]_4 + 4HI \xrightarrow[\text{Vigorous Boiling}]{\Delta;} C_{16}H_9N[OH]_4 + {}_4H_3Cl$$

| Papareerine | Hydroiodic acid | Papaveroline [or Demethylated Papaverine] | Methyl iodide |

Inference: There are *two* most prevalent revelations, namely:

- 'papaverine' does contain *four* methoxyl groups, and
- 'all four O-atoms' in papaverine are duly present in *four* methoxyl (-OCH$_3$) moieties exclusively.

(4) **Presence of One Methylene (-CH$_2$=) Moiety: Papaverine** on being subjected to oxidation at *three* **different stages** yield the following *three* divergent products, namely:

---

* **Zeisel Method :** It refers to the fission by HI, and **argentometric determination** of the methyl iodide (H_3Cl) so formed.

(a) **Oxidation with cold dilute KMnO$_4$ solution:** It gives rise to the formation of **papaverinol** as shown below:

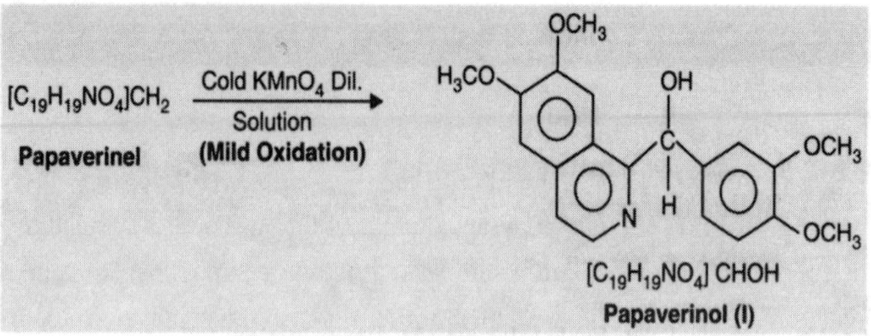

$$[C_{19}H_{19}NO_4]CH_2 \xrightarrow[\text{(Mild Oxidation)}]{\substack{\text{Cold KMnO}_4 \text{ Dil.} \\ \text{Solution}}}$$

Papaverinel

$[C_{19}H_{19}NO_4]$ CHOH
Papaverinol (I)

Inference: Papaverionol possesses a **secondary alcoholic function**.

(b) **More vigorous oxidation with hot KMnOS$_4$ Solution:** The resulting **product (I)** on vigorous oxidation with **hot KMnO$_4$ dilute solution** yields a **ketone** known as **papaveraldine (II)** as given under:

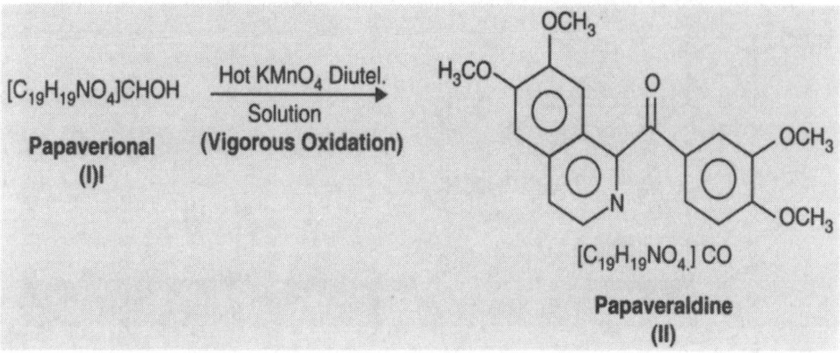

$$[C_{19}H_{19}NO_4]CHOH \xrightarrow[\text{(Vigorous Oxidation)}]{\substack{\text{Hot KMnO}_4 \text{ Diutel.} \\ \text{Solution}}}$$

Papaverional
(I)I

$[C_{19}H_{19}NO_4]$ CO
Papaveraldine
(II)

Inference: The critical formation of the **ketone, papaveraldine (II)**, reveals adequately that **papaverionol (I)** happens to be a *secondary* alcohol.

(c) **Prolonged oxidation with hot KMnO$_4$ solution:** Papaveraldine (II) undergoes prolonged oxidation with **hot KMnO$_4$ solution** produces **papaverenic acid (III)** as given under:

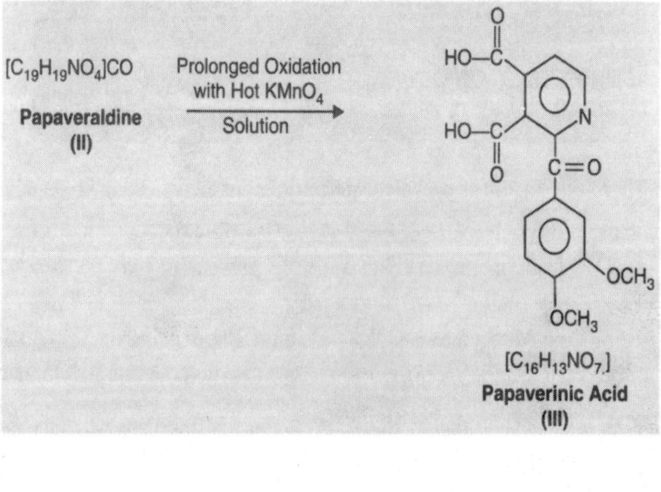

$$[C_{19}H_{19}NO_4]CO \xrightarrow[\text{Solution}]{\substack{\text{Prolonged Oxidation} \\ \text{with Hot KMnO}_4}}$$

Papaveraldine
(II)

$[C_{16}H_{13}NO_7]$
Papaverinic Acid
(III)

Inference: Papaverinic acid (III) is a **dibasic acid** and still holds the **keto moiety** in its **'precursor'**, since it critically forms an **'oxime'** *i.e.,* >C=N—OH. Besides, **papaverenic acid (III)** essentially comprises *two* **methoxyl functional moieties**. In short, all the evidences and inferences put forward from **section (a) through (c)** above suggests strongly that **papaverine** does contain a **methylene (-CH$_2$-) functional group**.

(5) **Fusion of Papaverine with KOH Yields Two Distinct Chemical Compounds:** These *two* chemical compounds do have the following molecular formula:

$$\text{Compound 'A' : } C_{11}H_{11}NO_2, \text{ and}$$
$$\text{Compound 'B' : } C_9H_{12}O_2,$$

together with a small quantum of **veratric acid**

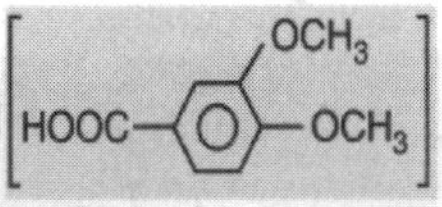

Interestingly, the **'combined molecular formulae'** of compounds **'A'** and **'B'** actually account for the **20 C-atoms,** which obviously suggests that these *two* **chemical entities** essentially constitute the *core molecule of papaverine.* Hence, it is quite important and pertinent to establish, determine, and peep into their **'actual relevant constitutions'**.

(a) **Structure of Compound 'A' [C$_{11}$H$_{11}$NO$_2$]:** There are *four* vital and important aspects to know the exact structure of the **compound 'A'**, namely:

(*i*) The N-atom in **compound 'A'** is found to be a **tertiary one**.

(*ii*) **Presence of** *two* **Methoxy (-OCH$_3$) Moieties**–has been duly determined by **Zeisel Method***.

(*iii*) **Demethylation** followed by **Zn-dust distillation** (*i.e.,* **reduction**) critically produces **isoquinoline residue** as given under:

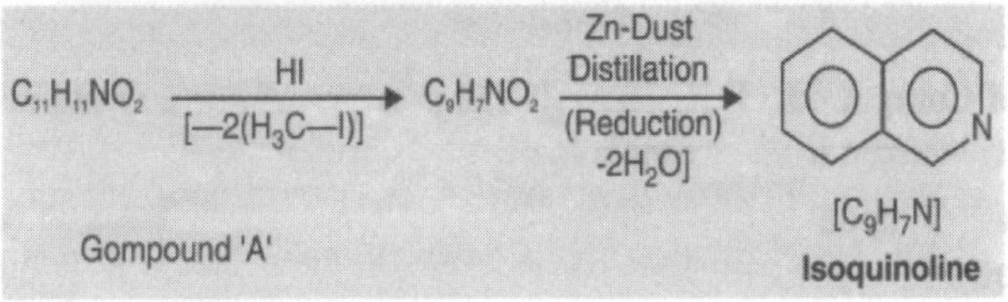

Inference: The above *two* sequential reactions explicitly suggests that **compound 'A'** is certainly a dimethoxyisoquinoline.

(iv) **Exact position of** *two* **methoxy (-OCH$_3$) moieties in Compound 'A':** One may easily determine the exact position of the *two* inherent methoxy (-OCH$_3$) moieties in compound

---

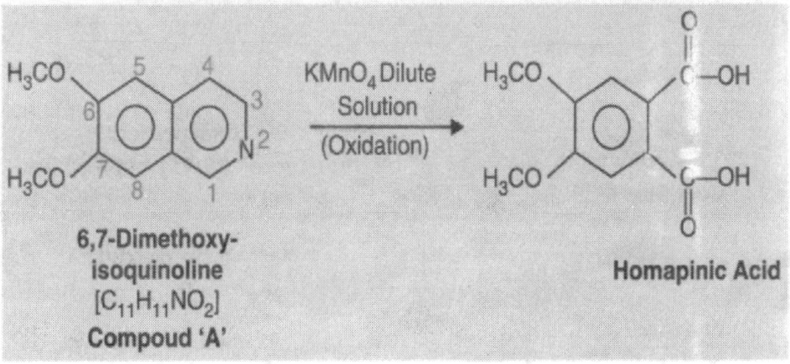

* **Zeisel Method :** It refers to the fission by hydroiodic acid (HI), and subsequent **'argentometric determination'** of the **methyl iodide (H$_3$CI)** so formed.

'A' by carrying out its *oxidation* (with dilute KMnO$_4$ solution) carefully to obtain **hemapinic acid** as shown below:

Inference: The aforesaid reaction suggests that **compound 'A'** is nothing but **6,7-dimethoxy isoquinoline**.

(v) **Further Evidence for the Structure of Compound 'A':** It is, however, possible to provide **further evidence for the structure of compound 'A'** (*i.e.*, **6,7-dimethoxy isoquinoline**) *via* its specific *synthesis* by the careful interaction of **veratric aldehyde** and **aminoacetal** to yield **4-ethoxy-6,7-dimethoxy isoquinoline (I)**. The resulting **product (I)** loses a mole of *ethanol* (H$_5$C$_2$OH) to give rise to the production of the desired **compound 'A'**, as expressed under:

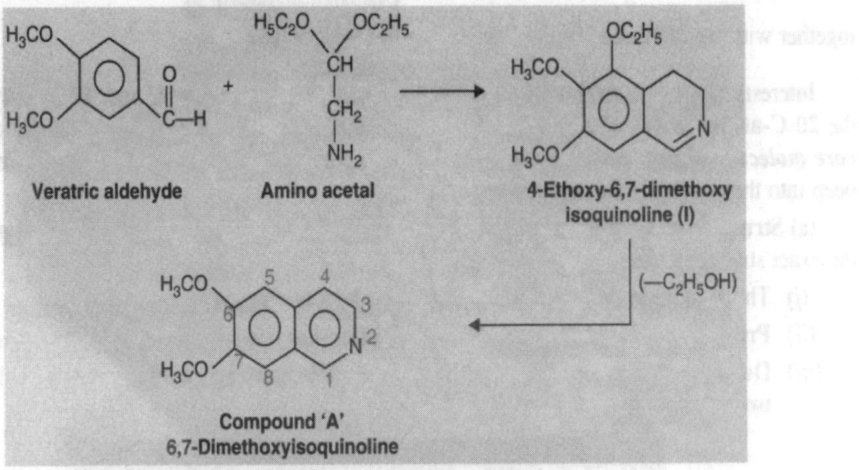

Veratric aldehyde   Amino acetal   4-Ethoxy-6,7-dimethoxy isoquinoline (I)

Compound 'A'
6,7-Dimethoxyisoquinoline

(b) **Structure of Compound 'B' [C$_9$H$_{12}$O$_2$]:** The presence of *two* 2 O-atoms in **compound 'B'** is solely due to the *two* inherent **methoxy moieties (-OCH$_3$)** present in it.

(*i*) **Compound 'B'** upon oxidation yields **'veratric acid'** (or **3,4-dimethoxy benzoic acid**) as stated below:

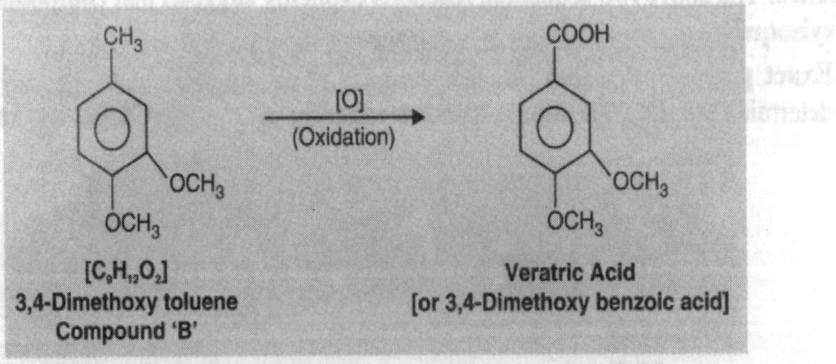

[C$_9$H$_{12}$O$_2$]
3,4-Dimethoxy toluene
Compound 'B'

Veratric Acid
[or 3,4-Dimethoxy benzoic acid]

(*ii*) **Compound 'B'** on *demethylation* with HI and followed by **oxidation** yields **protocatechuic acid** due to the conversions of:

- methyl group to carboxylic (—COOH) moiety, and
- methoxy moieties positioned at C-3 and C-4 by **phenolic (—OH) groups**.

Thus we may have:

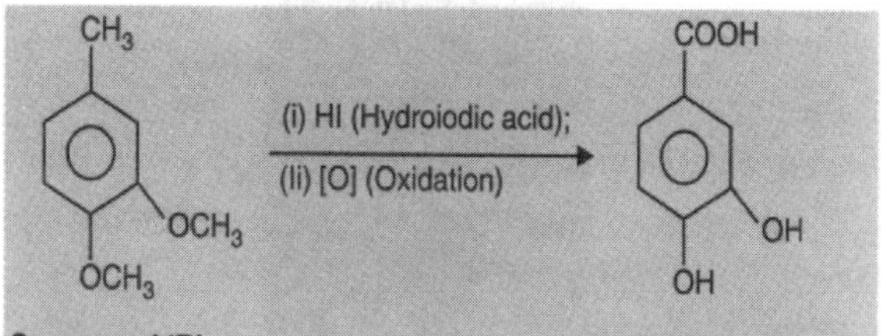

Compound 'B'   Protocatechuic Acid

Inference: From the aforesaid reactions (*i*) and (*ii*) one may infer that **compound 'B'** must be 3, 4-dimethoxytoluene or otherwise termed as **homoveratrole**.

> **NOTE:** The above logical explanation may also substantiate equally the specific and critical formation of some quantum of '*veratric acid*' in the *KOH fusion of papaverine* under section (5).

(6) **Point of Attachment Between Compounds 'A' and 'B':** It is, however, pertinent to state here that **papaverine** essentially possesses *four* distinct **methoxy (-OCH₃) moieties**, which further suggests and ascertains that the *two* **chemical entities (components)** critically holds **two-OCH₃ moieties** individually. Nevertheless, the point of attachment between **compounds 'A' and 'B'** cannot be *via* the **methoxy moieties**. Hence, there may be *two* viable and plausible options with respect to the aforesaid point of attachment, namely:

  - **Compound 'B'** must be linked to **Compound 'A'** *via* the **C-atom of the benzene nucleus,** or
  - **C-atom of the methyl (-CH₃) moiety in compound 'B'.**

Explanation: A rather more concrete evidence and acceptable observation being that the critical generation of '**veratric acid**' [see under section 5 (b) *i.e.*, during 'oxidation'] or during '**fusion**' predominantly reveals that the **compound 'B'** is duly attached *via* the **C-atom of the** *methyl moiety.* Hence, logically and scientifically the precise point of attachment of the '**isoquinoline nucleus**' may be decided by the following *two* cardinal factors, such as:

  - oxidation of '*papaverine*' to yield 6,7-dimethoxy isoquinoline-1-carboxylate [X], and
  - 2,3,4-pyridine tricarboxylate [Y] (or cinchomeronic acid).

The structures of the above *two* products are given as under:

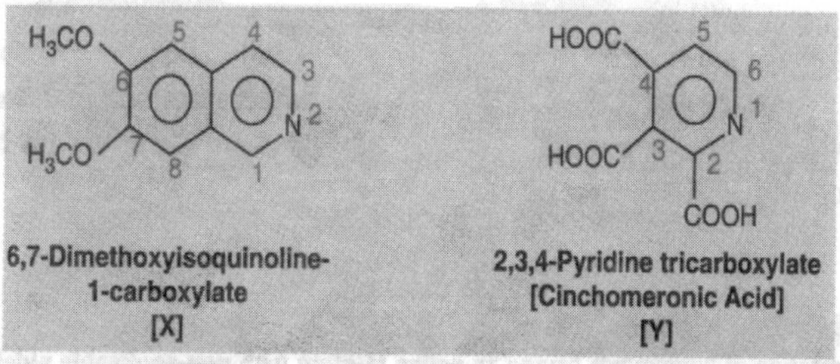

6,7-Dimethoxyisoquinoline-   2,3,4-Pyridine tricarboxylate
1-carboxylate   [Cinchomeronic Acid]
[X]   [Y]

(7) **Probable Structure of Papaverine:** Based on the various aforesaid scientific evidences one may assign the following as a **probable structure of** '*papaverine*'.

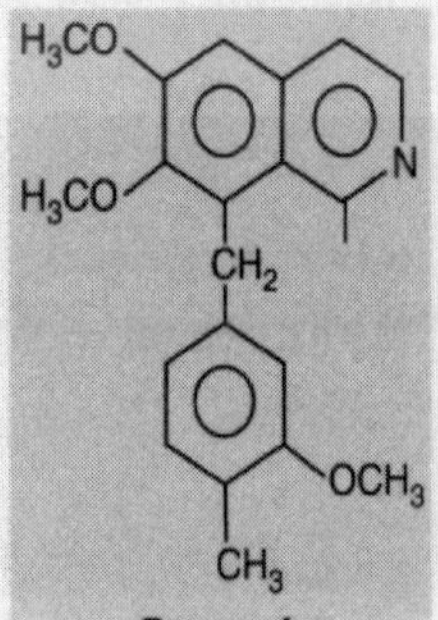

Papaverine

(8) **Synthesis of Papaverine:** A somewhat more diverse as well as complex class of the *'isoquinoline alkaloids' viz,* **papaverine** may be suitably exemplified by means of one of the simplest form **norladdanosoline**, also known as **tetrahydropapaveroline**, actually owe their origins in nature to the typical **Mannich Reactions*** of this kind. More elaborated meticulously designed experiment with the specific **isotopically labelled compounds** do indicate the under-mentioned generalized overall course of the *natural synthesis* being adopted appropriately:

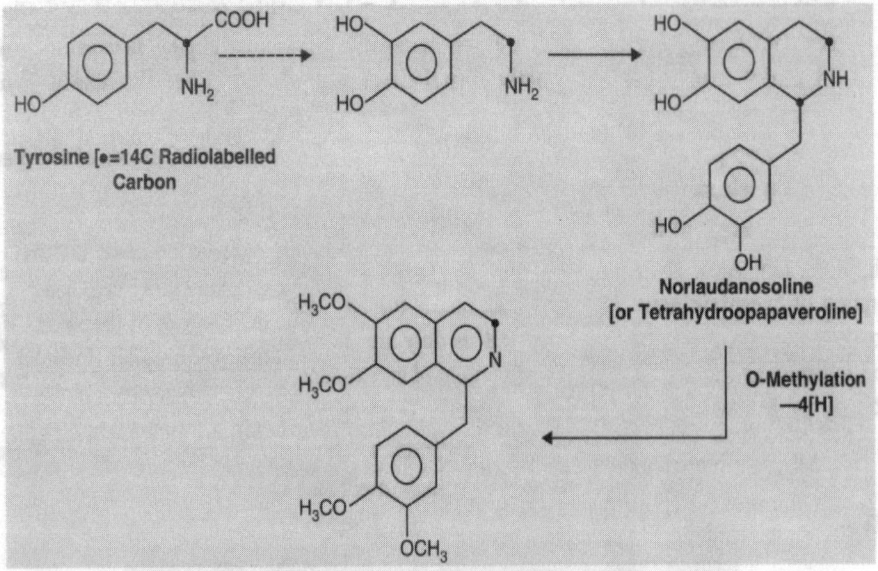

Explanation: The **radio-labelled tyrosine** (an *amino acid*) undergoes decarboxylation to yield a compound having a **primary amino function** which upon further reaction gives rise to the formation of **tetrahydropapaveroline** *i.e.,* **norlaudanosoline**. The resulting product on being, subjected to **O-methylation** loses **four H-atoms** to yield 'papaverine'.

---

* **Mannich C and Krosche W : *Arch Pharm*, 250 : 647, 1912.**

Ft refers to the reaction of compounds having an **active H-atom** with **non-enolizable aldehydes** and **ammonia** or **primary amines** to yield **aminomethyulated products** *i.e.,* **Maanich Bases.** $(CH_3)_2NH + HCHO + CH_3COCH_3 \rightarrow (CH_3)_2NCH_2OH_2COCH_3 + H_2O$

> **NOTE:** The 'broken arrow' (-- →) explicitly signifies the overall reactions normally taking place in the 'opium poppy', *Papaver somniferum*, which are not found to be relevant to the ring-closure reaction itself.

Ultimately, Pictet and Gams (1909)* put forward the synthesis of **'papaverine'**, which was later on improved by Braz and Chizhov (1958)**.

In usual practice, the **'synthesis of papaverine'** can be divided into *two* distinct individual steps, namely:

(a) Synthesis of Homoveratryl Amine and Homoveratroyl Chloride, and
(b) Condensatioin of the *two* Homoveratroyl derivatives.

8.1. Synthesis of Homoveratroyl Amine (I) and Homoveratroyl Chloride (II)

The various steps involved in the aforesaid synthesis are as stated under:

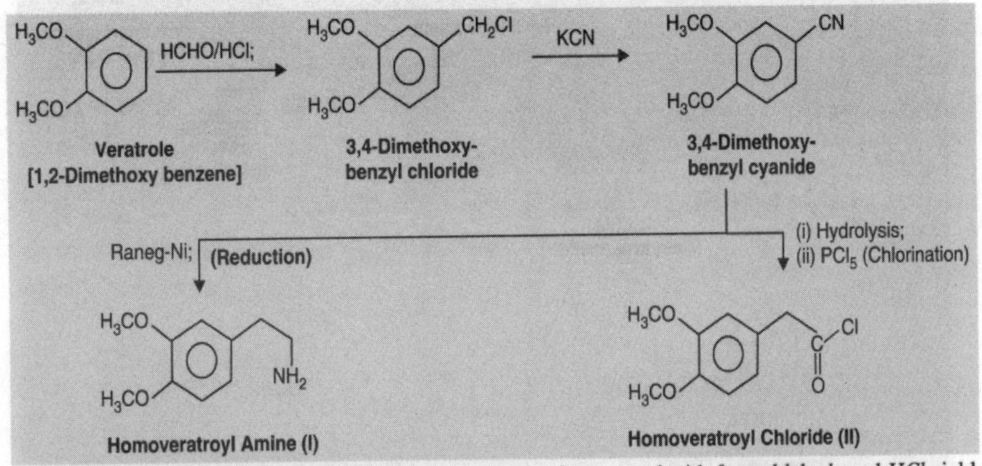

Explanation: Veratrole (or 1,2-dimethoxy benzene) when treated with *formaldehyde* and *HCl* yields **3,4-dimethoxy benzyl chloride**. The resulting product on treatment with potassium cyanide (KCN) forms **3,4-dimethoxybenzyl cyanide**. The **'cyanide' structural analogue** on being treated separately gives the *two* desired products:

- **homoveratroyl amine** – obtained by reduction with **Raney-Ni**, and
- **homoveratroyl chloride** – accomplished by *first* carrying out the **hydrolysis**; and *secondly* by **chlorination** with **PCl₅**.

8.2. Condensation of the Two Homoveratroyl Derivatives

The condensation of **homoveratryol amine (I)** and **homoveratroyl chloride (II)** takes place on heating to obtain an **open-chain amide** as an *intermediate* with the loss of one mole of **HCl** (as shown under). However, the resulting **open-chain amide** undergoes 'enolization' as the *first step*, which subsequently encounters with the well-known **Bischler-Napaieralsky Reaction***** as the *second step*, causing thereby the formation of **3, 4-dihydropapaverine (X)** (*i.e.*, **isoquinoline** nucleus gets generated). The resulting **compound (X)** is subjected to **dehydrogenation (—2H)** at 200°C to result into the

---

* Pictet and Gams : *Compt Rend*, **149** : 210, 1909; *Ber*, **42**: 2943, 1909.

** Braz and Chizov : *Soviet Pharm. Research*, **3** : 90-93 New York, 1958.

*** Bischler A and Napieralski B, *Ber*, **26** : 1903, 1893
It refers to the 'cyclodehydration of β-phenylamides' to **3, 4-dihydroisoquinoline** derivatives by means of such *condensing agents* as : **P₂O₅** or **ZnCl₂**.

The resulting **compound (X)** is subjected to **dehydrogenation (—2H) at 200°C** to result into the formation of the desired compound **papaverine.** The various steps involved are as indicated below:

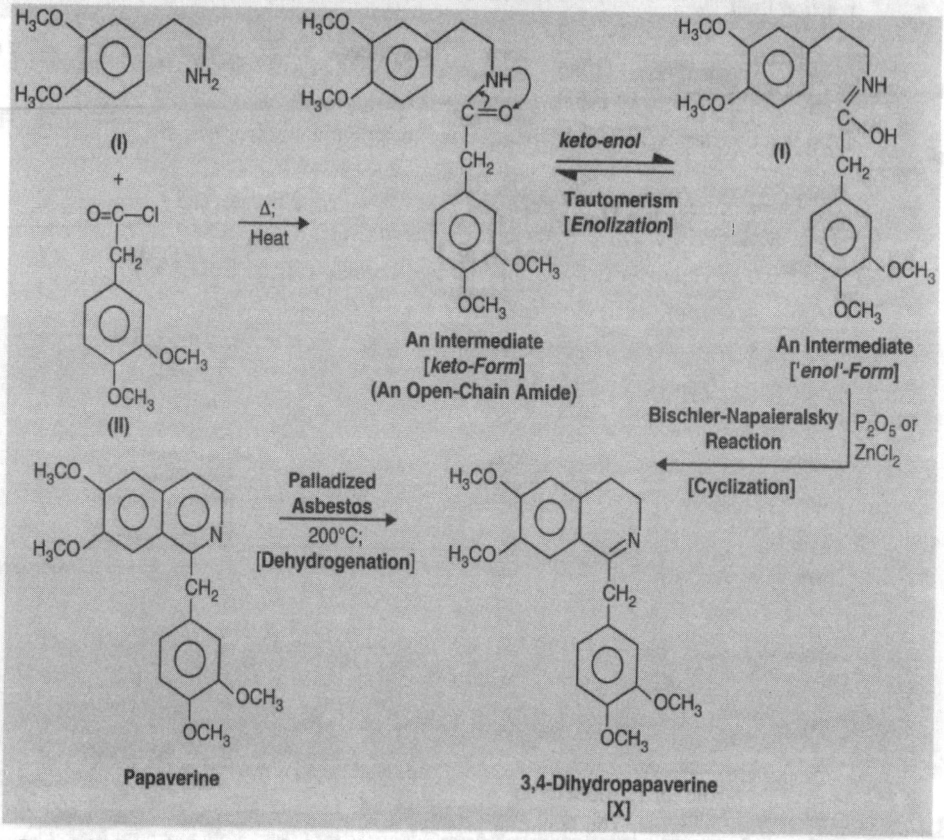
H3CO
H3CO
NH2
(I)
+
O=C—Cl
CH2
OCH3
OCH3
(II)
Δ;
Heat
NH
C=O
CH2
OCH3
OCH3
An Intermediate
[keto-Form]
(An Open-Chain Amide)
keto-enol
Tautomerism
[Enolization]
(I)
NH
C
OH
CH2
An Intermediate
['enol'-Form]
Bischler-Napaieralsky
Reaction
P2O5 or
ZnCl2
[Cyclization]
Palladized
Asbestos
200°C;
[Dehydrogenation]
N
Papaverine
3,4-Dihydropapaverine
[X]

Interestingly, one may also prepare the aforesaid **'open-chain amide'** starting from the more accessible **veratric acid** *via* the ***Arndt-Eistert Reaction*** as suggested by Arndt and Eistert (1935)*

The various steps involved in the **Arndt-Eistert Reaction** are given as under:

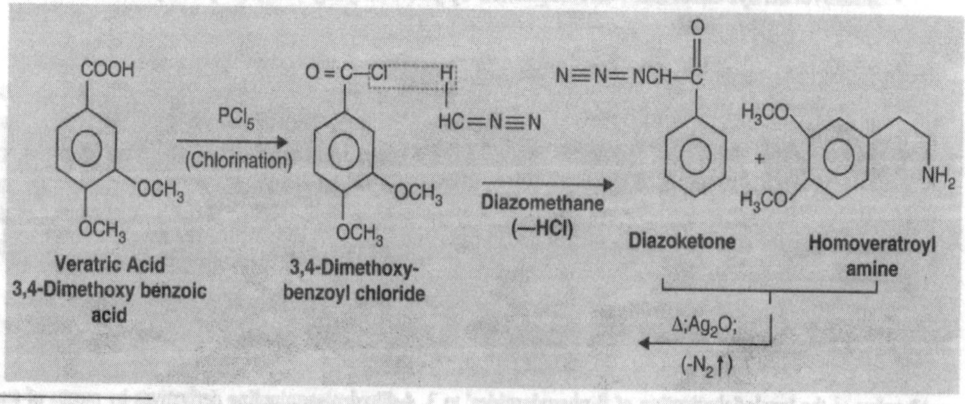
COOH
OCH3
OCH3
Veratric Acid
3,4-Dimethoxy benzoic
acid
PCl5
(Chlorination)
O = C—Cl H
HC═N≡N
3,4-Dimethoxy-
benzoyl chloride
Diazomethane
(—HCl)
N≡N═NCH—C
O
Diazoketone
+
H3CO
H3CO
NH2
Homoveratroyl
amine
Δ;Ag2O;
(-N2↑)

(Contd....)

*Ardnt F and Eistent B, *Ber*, **68** : 200, 1935.

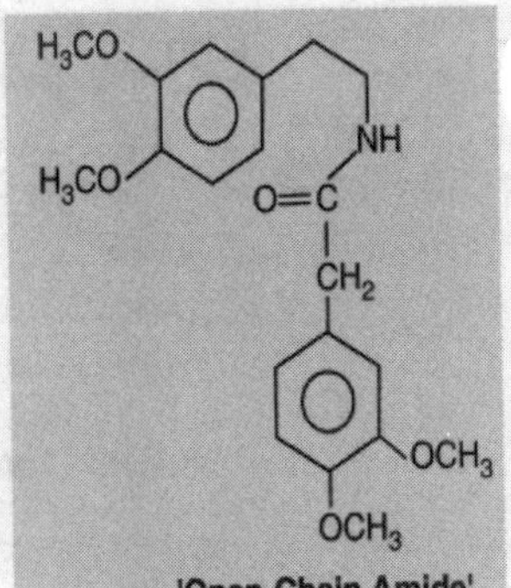

'Open-Chain Amide'
['*keto*'-Form]

Formation of Norlaudanosoline (or Tetrahydropapaveroline) and Pavine *via* Reduction of 'Papaverine': The meticulous and careful **reduction** of **'papaverine'** with *Sn and dilute HCl* yields norlaudanosoline and *pavine* as given under:

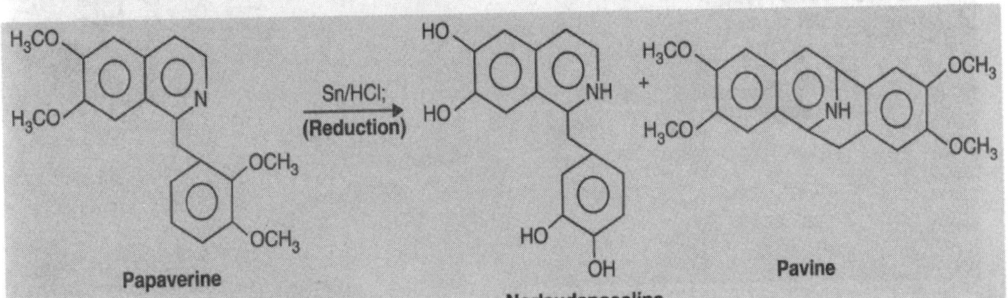

Mechanism: Importantly, the critical formation of **norlaudanosoline** is presumed to be the most probable product; whereas, the distinct formation of **pavine** essentially causes to participate certain extent of remarkable **'intramolecular rearrangement'**. In all, there are *four* sequential steps as stated under:

Step 1: Hydrogenation of **papaverine** to produce **1, 2-dihydropapaverine (I)**,

Step 2: The resulting **product (I)** upon protonation gives rise to the production of an **'iminium salt' (II)**,

Step 3: **Product (II)** undergoes specific **'intramolecular nucleophilic cyclization'** to *benzenoid type* **intermediate product (III)**,

Step 4: Finally the **intermediate product** undergoes **'deprotonation'** to yield **pavine (IV)**. The aforesaid *three* steps are expressed as under:

3.6.2. Sanguinarine

Sanguinarine is an *isoquinoline alkaloid* obtained from the dried rhizome of **Sanguinaria** or **Bloodroot,** *Sanguinaria canadensis* Linne' (*Family: Papaveraceae*). Interestingly, the **'generic name'** of the *alkaloid* has been duly derived from *sanguinarius* which means **'bloody'**; and actually signifies the inherent *blood red colour* of the *latex* (or *juice*). In Canada, the plant *canadensis* has its **native habitat.**

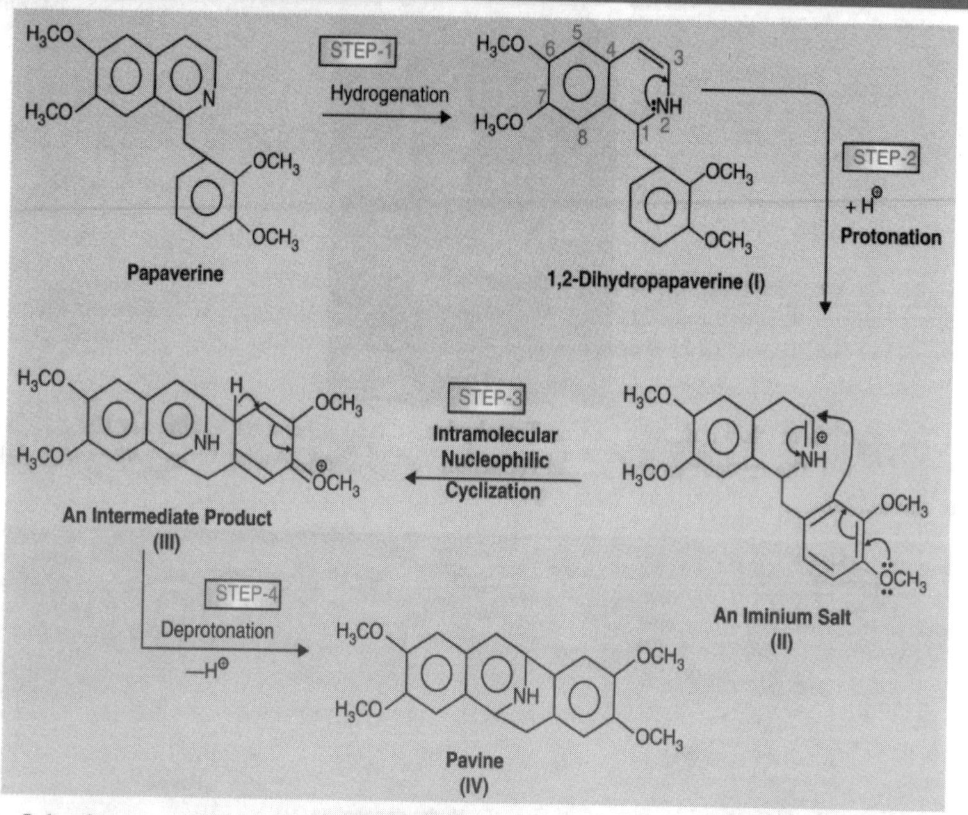

Papaverine

STEP-1
Hydrogenation

1,2-Dihydropapaverine (I)

STEP-2
+ H⊕
Protonation

An Intermediate Product
(III)

STEP-3
Intramolecular
Nucleophilic
Cyclization

An Iminium Salt
(II)

STEP-4
Deprotonation
—H⊕

Pavine
(IV)

It has been reported that the plant grows abundantly in rich-open woodland in **North America** located strategically to the east of the **Mississippi (USA)**. The major collection occurs predominantly, in the eastern States of USA.

The '**Indians**' profusely made use of '**Bloodroot**' in staining their *facial outlook*. It was also employed as an '**acrid emetic**'. Evidence from the literature reveals that the so called '**early settlers**' used **Blood root** in the preparation of '*homemade cough remedies*'.

The major **alkaloids** present in '**Bloodroot**' are as stated under:

Name of Alkaloid		Percentage (%)
Sanguinarine	:	~1
Allocryptopine	:	<0.5
Chelerythrine	:	<0.5
Protopine	:	<0.5

Characteristic Features: These essentially include:

(1) The '**esanguinaria alkaloids**' are usually colourless initially, but have a tendency to form '**coloured salts**'.

(2) **Sanguinarine** gives rise to the formation of typical and characteristic *reddish salts* with sulphuric acid (H_2SO_4) or nitric acid (HNO_3); *yellowish salts* with '**chelerythrine alkaloid**'.

(3) **Sanguinarine** designates a **benzophenathradine class of isoquinoline alkaloids**.

(4) Almost all alkaloids of **sanguinaria** are invariably found in also other members of the natural order **Papaveraceae**.

Therapeutic uses largely include: emetic properties, expectorant, antiseptic in **'Toothpastes'**, **'Mouthwashes'**.

The structure of **sanguinarine** is duly provided under section 3.6.

3.6.3. Hydrastine

Hydrastine is a **phthalideisquinoline** alkaloid which is obtained from *Hydrastis canadanesis* L., (**Family**: *Ranunculaceae*). It is found to be **readily soluble** in *ethanol, chloroform*, and *ether* but almost **insoluble** in *water*. Its structure has been provided in section 3.6.

Importantly, the **'protoberberine'** skeleton present duly in **scoulerine** may be duly subjected to appropriate modification, as given under:

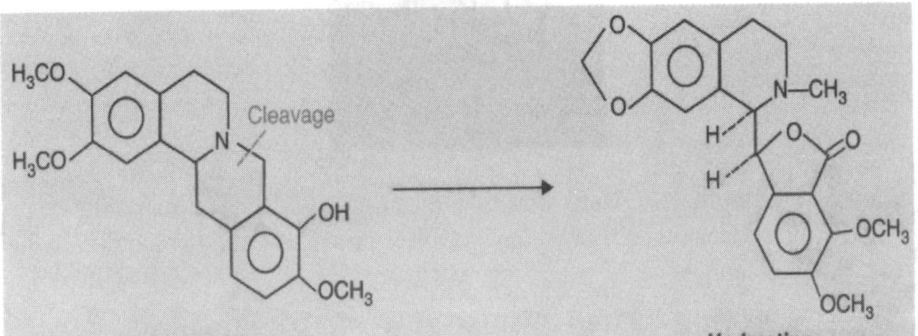

Scole-erine
[Protoberberine Type]

Hydrastine
[Phthalideisoquinoline-type]

It is, however, pertinent to observe the critical **cleavage** of the ensuing **'heterocyclic ring systems'** located strategically *adjacent to the N-atom*, as shown above with dotted line, thereby giving rise to altogether **'newer skeletal entities'**, such as: *'hydrastine'* from *Hydrastis canadensis* (*Ranunculaceae*).

Interestingly, **'hydrastine'** is found to be beneficial as a popular traditional remedy in the critical control of the **'uterine bleeding'**.

3.6.4. Morphine

Preamble: Morphine belongs to the class of **Phenanthrene Alkaloids** or the relatively older nomenclature **'Opium Alkaloids'**. It is regarded to be the most important amongst the **phenanthrene alkaloids**. It is, however, worthwhile to mention at this point in time that both **morphine** and its **allied alkaloids** categorically belong to the so called **morphinan isoquinoline structural analogues**.

Morphine is regarded to be the **'prototype of opiate analgesics'** that eventually exerts its action at several sites in the **central nervous system (CNS)** for causing **analgesia** (*i.e.*, relief from **pain sensation**). In general, the so called *structural features*, such as:

- presence of **'quaternary C-atom'** *i.e.*, a C-atom having no **H-substitution**,
- critical presence of a **phenyl moiety** or an **'isostere'** intimately linked to this **quaternary C-atom**,
- presence of a **'tertiary N-atom'**, and
- inclusion of a **2C-bridge** separating the **'tertiary N-atom'** and the **'central C-atom'**.

Thus, **morphine**, see *structure with numbering* below, plus the allied **opium alkaloids**, which do exhibit predominant **analgesic activity profile** essentially possess these aforesaid *vital, important* and *critical* **structural basic requirements**.

Morphine possesses the following *structural features*:

C-13 : central C-atom;

Ph-Ring : attach to C-13 comprises essentially **C-atoms** 1→4 and 11, and 12;

tert **N-atom** : linked duly *via* a **2C-bridge** (*viz.,* **C-15** and **C-16**) to the **central C-atom** (*i.e.,* **C-13**).

Morphine

Isolation of Phenanthrene Alkaloids (or Opium Alkaloids): The following steps need to be followed rigidly for the isolation of **morphine** and **other important alkaloids** *viz.,* **codeine, narcotine, papaverine, thebaine** etc.

(1) The '**raw opium**' *i.e.,* **dried latex** (or *milky juice*), is exhaustively extracted with *cold* **dichloromethane [CH$_2$Cl$_2$]**. Thus, we may have *two* residual segments:

- **Solvent Layer (CH$_2$Cl$_2$)** – contains solely **narcotine**, gum, and **papaverine**,
- **Insoluble Residue** – consists of **morphine** and other components (*e.g.,* **codeine**, thebaine etc.).

The **solvent layer** is separated instantly.

(2) **Dichloromethane Layer:** It is accomplished from **step (1)** from which the solvent is separated under vacuum so as to obtain a '**dry residual mass**', which is duly extracted with **hot dilute HCl** treated with activated charcoal powder (for *decolourization*)–and filtered ultimately under suction. The perfectly clear **filtrate** is neutralized caustiously with NH$_4$OH, when the following *two* **opium alkaloids** get separated:

- **papaverine**, and
- **narcotine**.

Papaverine may be separated from the aforesaid mixture of alkaloids by extracting it with **ethanol** successively. The *ethanolic extract* when treated with '**acid oxalate**' salts out the '**papaverine**' which may be collected-purified by recrystallization. **Narcotine** present in its '**crude form**' in the residual portion obtained from the alcohol extract, is subjected to purification on the same lines.

(3) **Residue from Dichloromethane Extraction [Step-1]:** It is shaken up carefully with freshly prepared '**lime water**' at room temperature (~20°C) only. In this way, the *three* **major alkaloids**, namely: **Morphine, Codeine** and **Thebaine** remain very much in the '**lime-water**'. The resulting lime-water extractive mixture when extracted repeatedly with the *non-polar solvent* '**benzene**' specifically takes up both **codeine** and **thebaine** in the ensuing **benzene layer**.

Now, the pH of the **benzene layer** is duly raised to **pH 8** (with NH_4OH) when the **crude morphine** gets precipitated. However, the **'filtrate'** still contains **morphine**, which on being evaporated under vacuum - extracted with amyl alcohol-produces another **'crop of crude morphine'**. Thus, the *two* aforesaid samples of **'crude morphine'** are mixed duly.

(4) The resulting **'crude morphine'**, obtained in **step-3**, is carefully dissolved in minimum quantity of **dilute HCl**; and subsequently, filtered. The clear acidified filtrate (containing *'morphine hydrochloride'*), is cautiously neutralized with NH_4OH and followed immediately by the incorporation of *ethanol*, which enables **'morphine'** to emerge as a **precipitate**. The precipitated **'morphine'** is duly dissolved in a minimum quantum of dilute HCl. Finally, the resulting solution so obtained, upon concentration and refrigeration (cooling) gives out the beautiful crystals of **'morphine hydrochloride'**.

(5) **Treatment of Benzene Extract Obtained in Step-3:** The **benzene extract** duly obtained from **step-3** is evaporated under reduced pressure to yield a residue, which is dissolved in ethanol-decolourized-chilled, and filtered. The clear **filtrate** on treatment with dilute H_2SO_4 produces a precipitate of **'codeine sulphate'**, which is collected under suction. The resulting filtrate is duly treated with **tartaric acid** solution carefully to obtain a precipitate containing **'thebaine'** and **'thebaine-acid tartrate'**.

Characteristic Features: The various **characteristic features** of **'morphine'** are as stated under:

(1) **Morphine** occurs as short, orthorhombic, columnar prism from *anisole*.

(2) It gets decomposed at 254°C which also represents a **metastable phase**.

(3) It has mp 197°C; however, the **high-melting form** sublimes at 190-200°C (0.2 mm pressure at 2 mm distance).

(4) It is very *sparingly soluble* in solvent ether, chloroform, and water; whereas, it is *extremely soluble* in ethanol, and alkaline solution(s).

(5) **Morphine** usually, acts as a **monoacid base**; and, therefore, forms well defined **'salts'** with various mineral acids *viz.*, **morphine hydrochloride, morphine sulphate**, etc.

3.6.4.1 CONSTITUTION OF MORPHINE

The most probable structure of **'morphine'** may be obtained on scientific and logical basis by means of the following steps adopted in a sequential manner:

(I) **Molecular Formulae:** The molecular formula of **morphine** is determined to be $C_{17}H_{19}NO_3$.

(II) **Nature of N-Atom: Morphine** takes up one mole of **methyl iodide** [H_3C-I] to form a **quaternary methyl iodide** (a salt) as given under:

$C_{17}H_{19}NO_3$ $\xrightarrow{[CH_3-I];}$

Morphine

$[C_{18}H_{22}NO_3I]$
Morphine methyl iodide
[A Quatennary Ammonium Salt]

Besides, the *tertiary* **nature of N-atom** may be further duly ascertained by the following *two* named reactions separately:

(i) **Hofmann Degration Method:** The tertiary nature of **N-atom** is further confirmed by the **Hofmann degradation of codeine derivative** that eventually shows the presence of **N-atom** in the ring itself as given under:

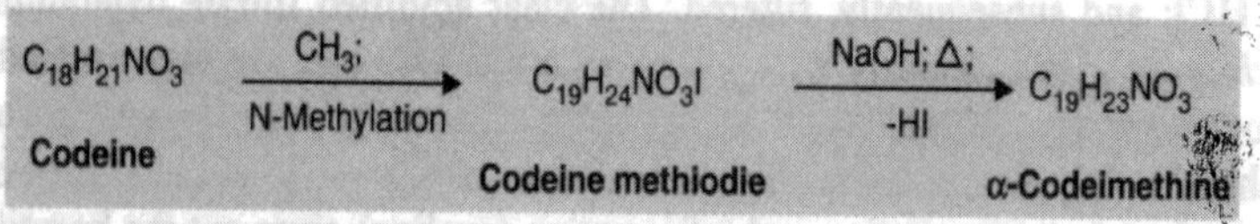

$$C_{18}H_{21}NO_3 \xrightarrow[\text{N-Methylation}]{CH_3;} C_{19}H_{24}NO_3I \xrightarrow[-HI]{NaOH;\ \Delta;} C_{19}H_{23}NO_3$$

Codeine   **Codeine methiodie**   **α-Codeimethine**

Thus, **Hofmann degradation** of **codeine** *via* N-methylation and alkaline treatment gives the end-product known as **α-codeimethine.**

From the aforesaid **chemical transformations** one may correlate these observed changes corresponding to the **Hofmann degradation** as applicable to **N-methylpiperidine**, which suggests vehemently that the **N-atom** should by all means be present in a **ring system**, as depicted below:

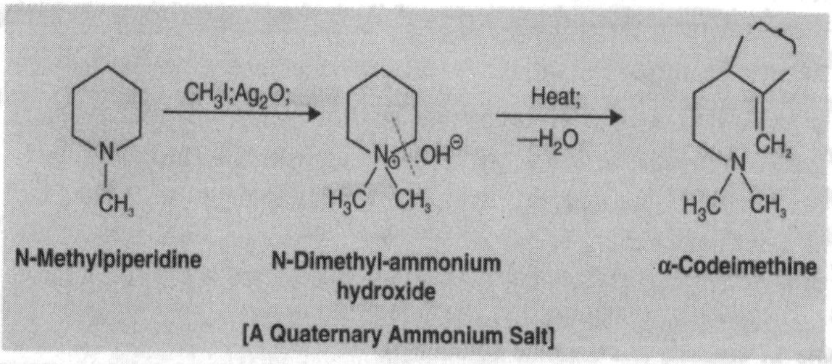

N-Methylpiperidine   **N-Dimethyl-ammonium hydroxide**   **α-Codeimethine**

[A Quaternary Ammonium Salt]

(ii) **Herzig-Meyer Method:** It indicates explicitly the presence of **N-methyl [=N-CH₃] moiety** in *morphine*. The critical formation of **dimethylaminoethanol** $[(CH_3)_2NCH_2CH_2OH]$ from α-methylmorphimethine $C_6H_{15}O \begin{bmatrix} =NCH_3 \\ -OCH_3 \\ -CHOH \end{bmatrix}$ generously reveals that *codeine*, which being a methylated morphine, essentially comprises a N-methyl [=N–CH₃] moiety.

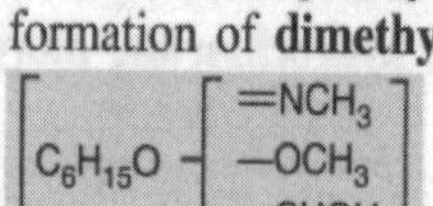

Note: Importantly, the above revelations may be further substantiated by the fact that *codeine* on being subjected to *von Braun Degradation* eventually takes up *one N-atom* and in turn loses 3 H-atoms. Nevertheless, the *results* thus arrived at may be readily interpreted by drawing the following logical conclusions, namely:

• actual conversion of N-methyl (=N-CH₃) moiety into N-nitrile [=N-CN] moiety, and
• ascertain strongly that both 'codeine' and 'morphine' predominantly contain one N-methyl group.

(III) **Natura of O-Atoms:** **Morphine** contains *three* **O-atoms** the presence of which may be categorically proved and confirmed by means of the following *four* **specific reactions,** namely:

(i) **Presence of *two* Hydroxyl (-OH) Moieties:** It has been observed that **morphine** when carefully **acetylated** or **benzoylated** produces the corresponding **diacetyl** or **dibenioyl** derivatives as given below:

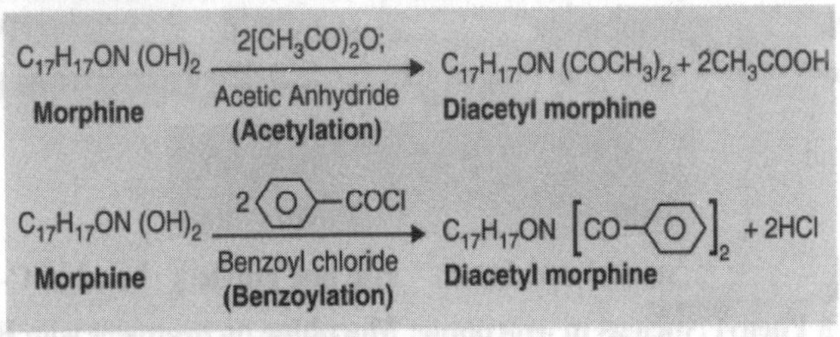

$$C_{17}H_{17}ON(OH)_2 \xrightarrow[\substack{\text{Acetic Anhydride}\\ \text{(Acetylation)}}]{2[CH_3CO)_2O;} C_{17}H_{17}ON(COCH_3)_2 + 2CH_3COOH$$

Morphine   **Diacetyl morphine**

$$C_{17}H_{17}ON(OH)_2 \xrightarrow[\substack{\text{Benzoyl chloride}\\ \text{(Benzoylation)}}]{2 \langle O \rangle - COCl} C_{17}H_{17}ON\left[CO-\langle O \rangle\right]_2 + 2HCl$$

Morphine   **Diacetyl morphine**

Inference: Both '**acetylation**' and '**benzoylation**' reactions indicate that **morphine** essentially contains *two* hydroxyl (-OH) moieties.

(ii) **Phenolic Nature of One Hydroxyl Group:** It can be proved by the following *two* reactions:

First Reaction: Morphine with **ferric chloride (FeCl$_3$) Solution** gives a typical characteristic colouration.

Second Reaction: Monosodium morphine (salt) is obtained when **morphine** is reacted with NaOH solution; and from this solution **morphine** is duly **recovered** by passing CO$_2$/gas through it.

Inference: The above *two* specific reactions reveal that one of the two hydroxyl moieties is definitely of a **phenolic** nature.

(iii) **Alcoholic Nature of Second Hydroxyl Group: Morphine** on being treated with **halogen acids** (*viz.,* HCl, HBr, HI) produces a corresponding **monohalogeno structural analogue** whereby **one hydroxyl group is duly replaced by a halogen atom.** Obviously, this particular reaction is **characteristic of alcohols only.**

Inference: It may be concluded that the *second hydroxyl moiety* is of an **alcoholic nature** positively.

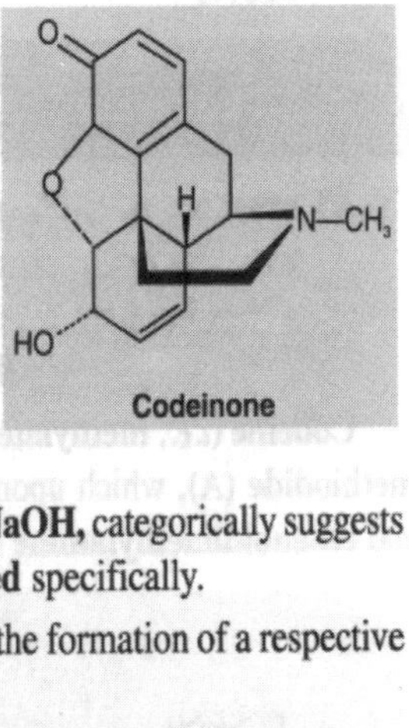

Alternatively, the above statement of facts may be further substantiated with the following **chemical reactions:**

- **Morphine** when heated with **methyl iodide (H$_3$CI)** in the presence of *aqueous KOH* forms a **methylated product** known as '**codeine**' [C$_{18}$H$_{21}$NO$_3$].

Codeinone

Inference: Codeine fails to produce any colouration with ferric chloride solution; besides, remains absolutely insoluble in **aqueous NaOH,** categorically suggests that the **second hydroxyl group** in **morphine** has been **methylated** specifically.

- **Codeine** on being oxidized with **chromic acid [CrO$_3$]** gives rise to the formation of a respective ketone, termed a **codeinone.**

Inference: The aforesaid reaction confirms that the '**second hydroxyl group**' in **codeine** is of a **secondary alcoholic nature (-CHOH-).** In addition, the above reactions further substantiates that **codeine** happens to be the **monomethyl (phenolic) ether of morphine.**

(iv) **The Third O-Atom Forming a Lactone Ring between C-4 and C-5:** Based on the critical observations that the **third O-atom** is quite **unreactive in nature,** in addition to the various **degradation products of morphine** one may infer logically that the said **O-atom in morphine** is duly present as an '**ether linkage**' or farming a '**lactone ring**'.

(IV) **Presence of Ethylenic Bond at C-7 and C-8:** It has been duly observed that when **codeine** (*i.e.*, **methylated morphine**) is reduced catalytically in the presence of **Pd (palladium)**, it eventually adds on **one mole of hydrogen (H$_2$)** to yield **dihydrocodeine** as given under:

$$C_{18}H_{21}NO_3 \xrightarrow{\text{H}_2\text{-Pd;}} C_{18}H_{21}NO_3$$

Codeine   Dihydro codeine

Inference: Both **morphine** and **codeine** comprise **one ethylenic bond at C-7 and C-8.**

(V) **Presence of a Phenyl Nucleus in Morphine: Morphine** on treatment with **Br$_2$** (bromination) produces a **mono-bromo derivative** with the evolution of **hydrogen bromide** (gas), which suggests obviously that it predominantly contains a **phenyl (benzene) nucleus.**

(VI) **Presence of a Cyclic *tertiary* Amine System in Morphine: Codeine** (*i.e.*, the **methylated morphine**) when subjected to **exhaustive methylation** gives rise to the formation of **α-codeimethine** that essentially contains **one additional methylene (-CH$_2$-) moiety than codeine**; besides, the N-atom remains very much at its **original position** (*i.e.*, it never gets eliminated at all). Furthermore, it may be derived that, in case, **codeine molecule** does comprise an **acyclic tertiaryamine system,** the resulting product (under the aforesaid experimental parameters) obtained should have predominantly contained **lesser number of C-atoms** amalgamated with the definite **loss of a N-atom.**

Now, based upon the fact that **codeine** contains a **cyclic *tertiary* amine system**, the above cited results may be duly expatiated and expressed by the reactions given already under **section (b)(i) above.**

Note: Importantly, the structures of both *α-codeimethine* and its respective isomer *β-codeimethine* are very much identical with those of *α-* and *β-methylmorphimethine* respectively.

(VII) **Presence of a Phenanthrene Nucleus in Morphine: Morphine** upon vigorous distillation with **Zn-dust** produces **phenanthrene** and a few *basic components* that vividly suggests that **morphine** may essentially comprise a **phenanthrene nucleus.** However, these observations and findings are further supported and confirmed by the following evidences:

Codeine (*i.e.*, **methylated morphine**) on being treated with methyl iodide (H$_3$Cl) produces **codeine methiodide (A)**, which upon boiling with **acetic anhydride** yields a mixture of **methyl morphol (C)** and **ethanoldimethylamine (D).** These aforesaid reactions may be expressed as under:

Codeine methiodide   α-Methylmorph-methine   Methyl morphol Ethanoldimethylamine
        (A)   (B)   (C)   (D)

At this juncture, it becomes worthwhile to establish the following *two* cardinal aspects related to the very structure of **'morphine':**

- **Structure of Methyl Morphol [C]**, and
- **Presence of N-Methyl (N-CH$_3$) Group.**

(i) **Structure of Methyl Morphol [C]**: It may be established by heating **methyl morphol (C)** with HCl at 180° under pressure to form a mole each of **methyl chloride (CH$_3$Cl)** and **morphole (E)** (*i.e.*, **dihydroxyphenanthrene**).

Subsequently, the *acetylated product i.e.*, **diacetylmorphol** on being oxidized yields the corresponding **diacetylphenanthraquinone**, thereby ascertaining the fact that the respective **C-9**, and **C-10** positions in **(C)** are absolutely **free**.

Diacetylphenanthraquinone when oxidised more vigorously with **KMnO$_4$** gives rise to **phthalic acid** thereby showing that the 2 hydroxyl moieties are present in the same ring.

Besides, the **2 hydroxyl moieties** in **methyl morphol (C)** are strategically located at the *ortho-position*, since **morphine** when fused with *alkali* produces **photocatechuic acid.**

The above observations may be summarized as given under:

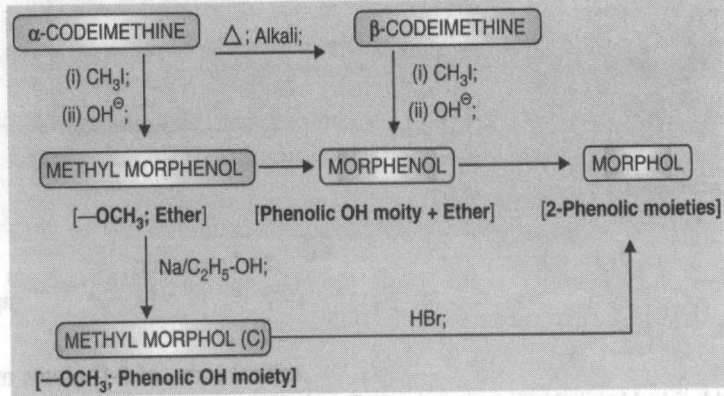

(ii) **Synthesis of Methyl Morphol (C)**: Knorr and Horlein (1907)* synthesized **methyl morphol (C)** (*i.e.*, **4-hydroxy-3-methoxyphenanthrene**) starting from **3-acetoxy-4-methoxy-2-nitrobenzaldehyde (X)**.

The **compound (X)** was duly prepared by the interaction of **3,4-dimethoxy-2-nitrobenzaldehyde (I)** and the **sodium salt of phenyl acetic acid (II)** in the presence of **acetic anhydride** to obtain an intermediate product **3,4-dimethoxy-2-nitro-α-phenylcinnamic acid (III)**. The **product (III)** is sequentially subjected to treatment with an **acid**, **NaNO$_2$/H$_2$SO$_4$**, and **C$_4$-powder** gives the penultimate **product (X)**, which upon heating yields the final desired **product (C)**. These reactions are as given under:

(iii) **Presence of N-Methyl [>N-CH$_3$] Moiety in Morphine Molecule**: The critical formation of **ethanoldimethylamine (D)** from **α-methylmorphimethine (B)**, under section (g), distinctly reveals that *'codeine'* (*i.e.*, **methylated morphine**) does contain one **N-methyl function (>N-CH$_3$)**. However, the above findings may be further ascertained and substantiated by the obvious fact that

---

*Knorr and Horlein : *Ber.*, **40**: 2032, 3341, 4889 (1907).

codeine on being subjected to the well-known **von Braun degradation** helps to **add on one N-atom** at the cost of a **loss of 3 H-atoms.**

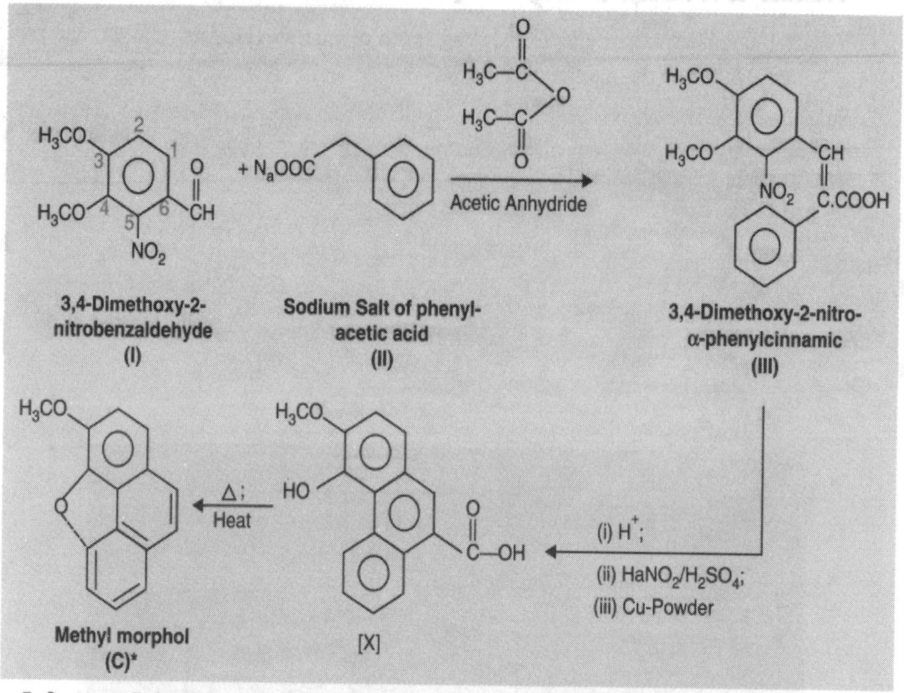
H3CO
H3CO
NO2
CH
+ NaOOC
H3C—C
H3C—C
Acetic Anhydride
H3CO
H3CO
NO2
CH
C.COOH
3,4-Dimethoxy-2-nitrobenzaldehyde (I)
Sodium Salt of phenyl-acetic acid (II)
3,4-Dimethoxy-2-nitro-α-phenylcinnamic (III)
(i) H+;
(ii) HaNO2/H2SO4;
(iii) Cu-Powder
H3CO
HO
C—OH
[X]
Δ; Heat
H3CO
Methyl morphol (C)*

Inference: It is, therefore, may be inferred that the actual conversion of **>N-CH_3 to >N-C≡N** does take place thereby suggesting strongly that both ***codeine*** and ***morphine*** comprises an **N-methyl moiety.**

(VIII) **Position of 3-Oxygen Atoms in Morphine:** The critical status of **2-O-atoms** may be fully justified by elucidating the structure of **morphenol (Y).**

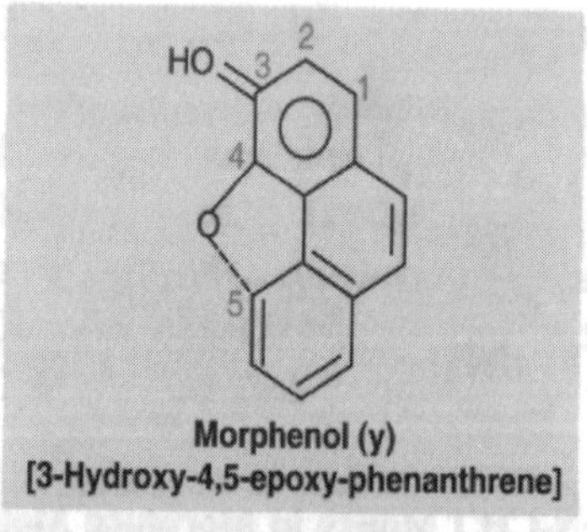
HO
1
2
3
4
5

Morphenol (y)
[3-Hydroxy-4,5-epoxy-phenanthrene]

Structure of Morphenol: It may be accomplished by adopting the following steps:

(1) The starting material **β-methylmorphimethine (1)** on being heated in an aqueous medium gives rise to the formation of a mixture of *three* **products of reaction** viz., **trimethylamine, ethylene,** and **methylmorphenol.**

(2) **Methylmorphenol** on demethylation with HCl yields **morphenol (Y)** which essentially contains one **'phenolic hydroxyl (-OH) moiety'** and an inert **epoxy O-atom** in it.

β-Methylmorphimethine (1) | Trimethylamine | Ethylene | Methyl morphenol

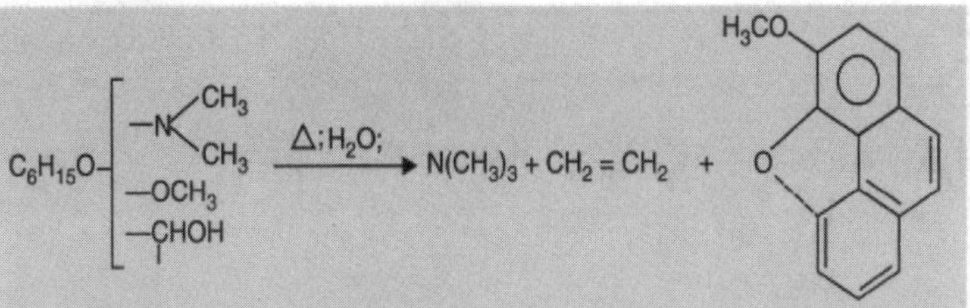

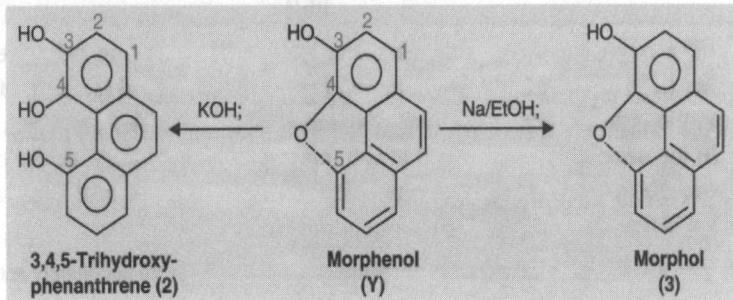

(3) **Morphenol (Y)** when fused with KOH, it gives rise to the formation of 3,4,5-trihydroxy-**phenathrene (2)**.

[2]

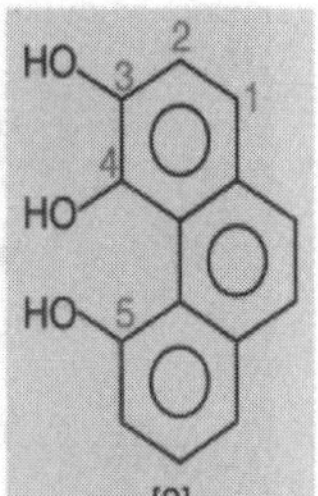

Interestingly, the **chemical structure of compound (2)** was duly proved by:

- its synthesis (*i.e.* **methylating the trihydroxyphenanthrene obtained from morphenon,** and
- **morphenol** (Y) on being subjected to reduction with **Na/Et-OH** yields **morphol (3).**

The aforesaid reactions and their respective products may be expatiated if one assigns the following structure to **morphenol (Y),** that prominently holds on **epoxy bridge** between **C-4** and **C-5** in the ***phenanthrene nucleus*** as given under:

3,4,5-Trihydroxy-phenanthrene (2) | Morphenol (Y) | Morphol (3)

Accountability of 3 O-Atoms: Of the ***three*** **O-atoms** in ***morphine two*** of them are duly accounted for as stated under:

- **the basic structure of morphenol (Y),** and
- **the production of (Y) from codeine,**

first-located at C-3, and *second*-forms the ***epoxy-bridge*** between **C-4** and **C-5** in the **phenanthrene nucleus.** However, the exact position of the ***third*** **O-atom** yet remains to be fully justified and accounted for.

Presence of *Third* O-Atom at C-6 in Phenanthrene Nucleus: The interaction of **codeine methiodide (A)** and **codeinone methiodide (B)** separately with an admixture of **acetic anhydride [$(CH_3CO)_2O$],** and **sodium acetate [CH_3COONa]** followed by gentle heating yields **3-methoxy-4-acetoxyphenanthrene (A-1)** and **3-methoxy-4,6-diacetoxyphenanthrene (B-1)** repectively plus a mole each of **dimethylaminoethanol** as given under:

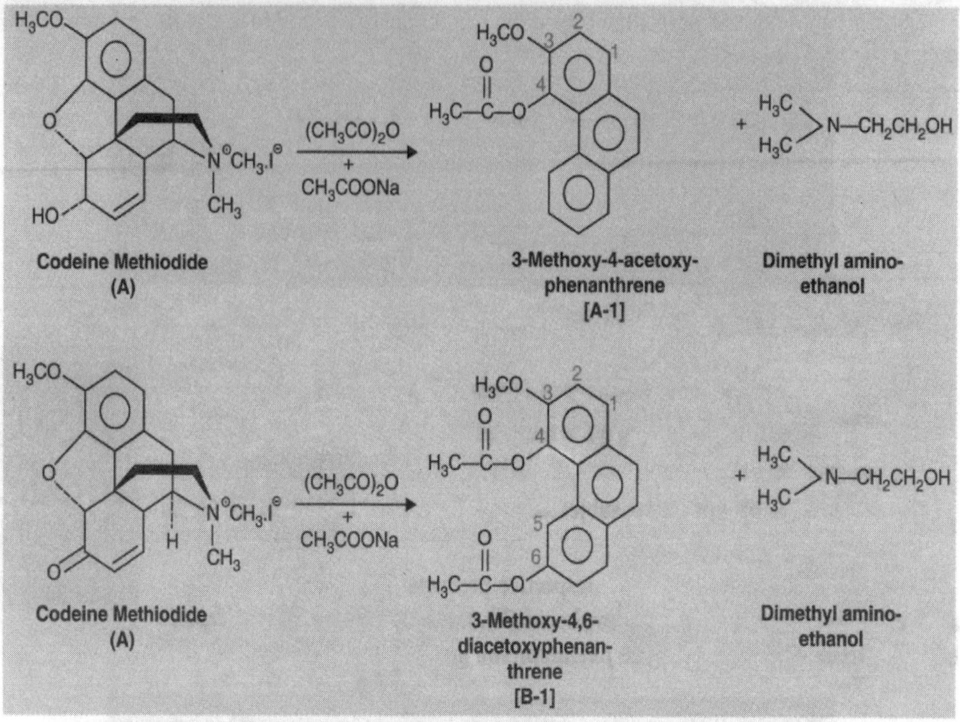

Codeine Methiodide
(A)

3-Methoxy-4-acetoxy-
phenanthrene
[A-1]

Dimethyl amino-
ethanol

Codeine Methiodide
(A)

3-Methoxy-4,6-
diacetoxyphenan-
threne
[B-1]

Dimethyl amino-
ethanol

Inference: The very presence of an **additional acetoxyl moiety** at **C-6** in **B-1** evidently suggests that in the **former (A)** the secondary **alcoholic moiety** (-OH) gets eliminated as a mole of water during the **dehydrogenation phenomena** to the respective *aromatic product* **[A-1]**; whereas, in the **latter** instance[*i.e.*, **B**] the **ketonic function** (>CO) at C-6 (instead of the *secondary alcoholic moiety*) gets duly **enolized** enroute to the corresponding **aromatic product (B-1)**; and, therefore, it appears as an 'acetoxy' moiety' in the **final product B-1**.

In this manner, all the *three* O-atoms in **morphine molecule** have been adequately accounted for as follows:

- *first* O-atom at C-3,
- *second* O-atom as epoxy bridge between C-5 and C-6, and
- *third* O-atom as secondary alcoholic function at C-6 in the phenanthrene nucleus.

(IX) **Structure of Morphine:** In order to assign a proper **structure of morphine** the following important steps need to be followed rigidly:

(1) **Assigning Partial Structure to Morphine:** The following observations need to be considered:

- **morphine** yields *monobromo derivative* with **Br$_2$** and a **monosodium salt,** that suggests it essentially comparises **one benzenoid structure,**
- **ethanoldimethylamine (D)** is obtained as one of the products during the exhaustive methylation of **codeimethines** [under section (g)],
- **ethylene** is also formed [under section (h)], and
- presence of a **double bond** and a **tertiary N-atom** in it.

Based on the aforesaid distinguished facts one may assign a *partial structure to morphine* as given under:

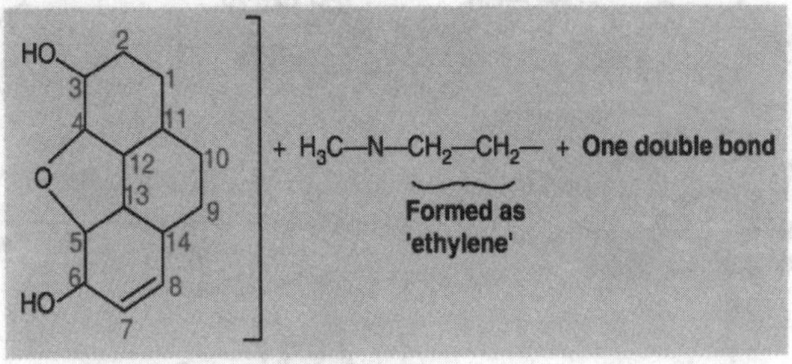

+ $H_3C-N-CH_2-CH_2-$ + **One double bond**

Formed as 'ethylene'

Remarks: At this point in time, the *two* critical problems need to be addressed, namely:

- **exact positions of 'double bond',** and
- point of attachment of side-chain **[$H_3C-N-CH_2-CH_2$]** in morphine,

in order to give a proper explanation to **all the reactions of morphine.**

(2) **Point of Attahment to the Side-chain [$H_3C-N-CH_2-CH_2-$]:** It has been duly observed that when:

- **Codeine*** is gently oxidized with **chromic acid (CrO$_3$)** and results into the formation of **hydroxy codeine** plus **codeinone.**
- **Hydroxy codeine** upon **Hofmann's exhaustive methylation** forms **ketocodeimethine..**
- **Ketocodeimethine** - when subjected to *acetolysis* with **acetic anhydride [(CH$_3$CO)$_2$O]** gives rise to the production of **methoxy-acetoxyphenanthrene.**
- **Methoxyacetoxyphenanthrene** on *oxidation* yields the product **methoxymonoacetoxy phenanthraquinone.**

The above **sequence of reactions** may be summarized as under:

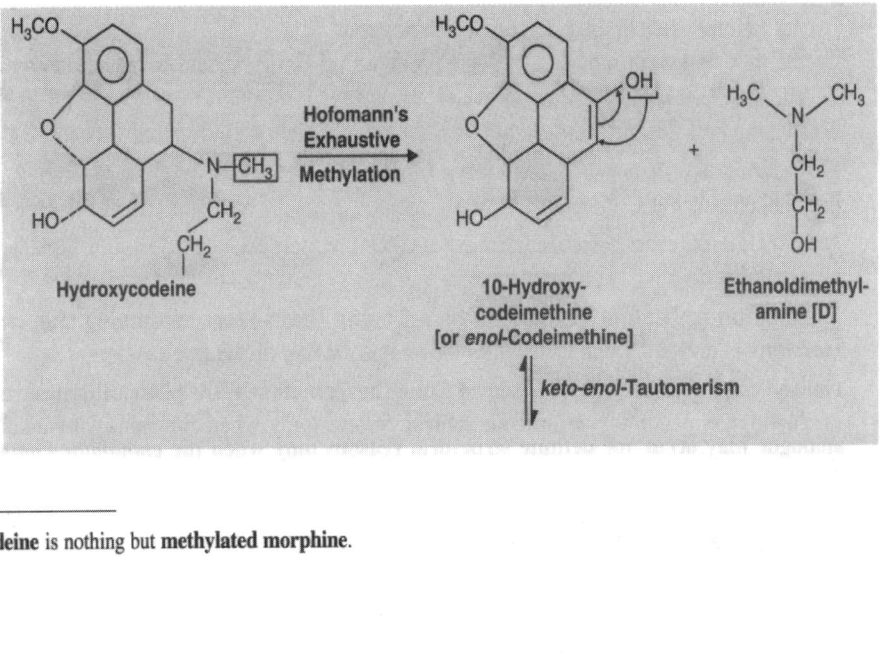

Hydroxycodeine

Hofomann's Exhaustive Methylation

10-Hydroxy-codeimethine [or *enol*-Codeimethine]

keto-enol-Tautomerism

Ethanoldimethyl-amine [D]

---

* **Codeine** is nothing but **methylated morphine.**

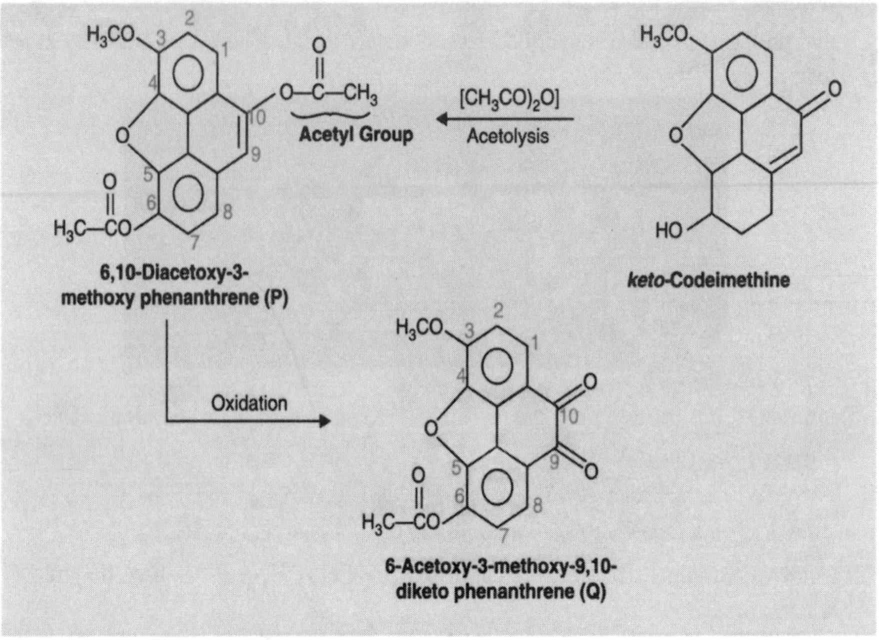

6,10-Diacetoxy-3-methoxy phenanthrene (P)

keto-Codeimethine

6-Acetoxy-3-methoxy-9,10-diketo phenanthrene (Q)

Inferences: These essentially include:

(*i*) The critical loss of **one acetyl group** from '**P**' (*i.e.*, **6, 10-diacetoxy-3-methoxy phenanthrene**) to '**Q**' (*i.e.*, **6-acetoxy-3-methoxy-9, 10-diketophenanthrene**) in the course of the *last oxidation reaction* clearly confirms that one of the *two* **acetoxyl moieties** in '**P**' should be strategically positioned at **C-9** or **C-10**.

(*ii*) Besides, the aforesaid the **acetoxyl moiety** gets duly introduced *via* the **ketonic function** during the **aromatization process** (*i.e.*, **acetolysis with acetic anhydride**). This obviously suggests that:

- the **ketonic moiety** in *keto*-**codeimethine**, and
- the **new hydroxyl moiety in hydroxycodeine** (at **C-10**), should by all means be located either at **C-9** or **C-10**.

(iii) Finally, the **new double bond** should be most appropriately be introduced between **C-9 and C-10** during the '*fusion of N-containing ring system*'; and, therefore, the **N-atom** should be duly linked either at **C-9** or **C-10**.

(iv) Nevertheless, the exact point of attachment at **C-9** is precisely established after the **total synthesis of morphine** discursed at a latter stage.

(3) **Elimination of N-atom Containing Side-Chain:** The **N-atom containing side chain** in *morphine* is invariably eliminated with the **aromatization of the nucleus.**

Gulland and Robinson (1925)* observed that - '**the formation of the phenanthrene structure-analogue may occur for definite structural reasons only when the** *enthamine-chain*' gets duly displaced'.

---

* Gulland, Robison: *J. Chem Soe.,* **123;** 1980, 1928.

Based on the relevant fact that since the **N-end of the side-chain** has already been shown as **linked to C-9**, thereby suggesting that **C-end of the ensuing side-chain** should be critically positioned at an *'angular position'*; and hence, its subsequent **extrusion fro C-9 position** actually becomes essential and feasible for the **aromatization phenomena.**

Importantly, of the *two* possible deemed to be positions *e.g.*, **C-13** and **C-14**, the former is specifically chosen based on the fact this *pattern of structure* justifiably and reasonably explains the acceptable rearrangement of *thebaine* to *thebainone*; and, therefore, the **'partial structure of morphine'** may be given as under:

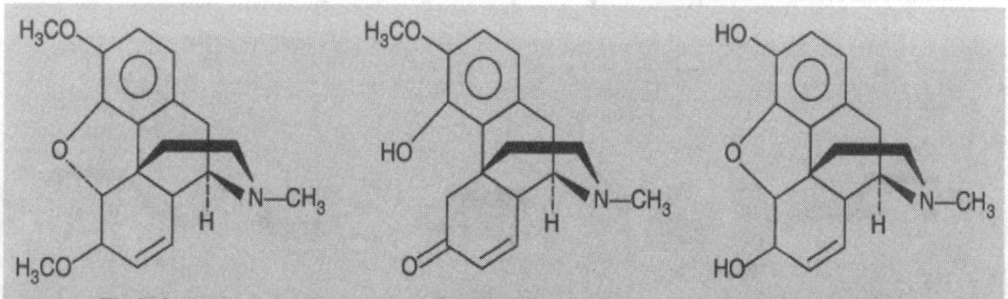

Thebaine   Thebainone   Partial Structure of
                                           Morphine

(4) **Position of the Double Bond:** Chlorination of **codeine** with *PCl₅* gives rise to the corresponding α-chloroiodide. The resulting product on being subjected to hydrolysis with **aqueous CH₃COOH** affords a mixture of : **codeine, isocodeine, psendocodeine,** and **allopseudocodeine** (*i.e.,* the respective **'positional isomers'**).

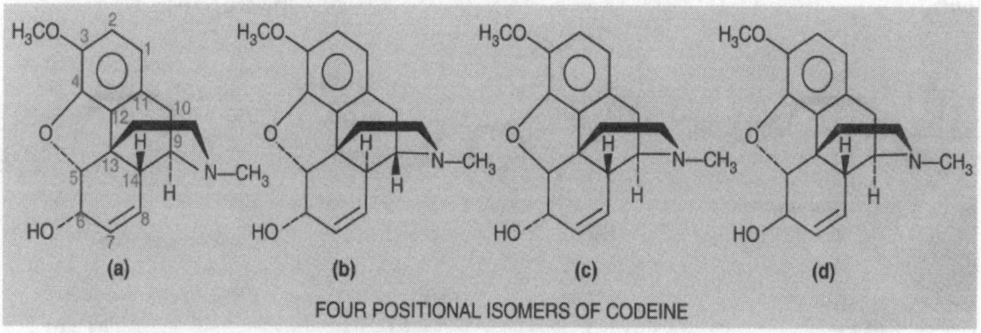

FOUR POSITIONAL ISOMERS OF CODEINE

Inferences: They essentially comprise:

(i) Both **(a)** and **(d)** yield the *same ketone* (**codeinone**) upon careful **oxidation** suggesting thereby that they distinctly differ **only with regard to the position of the -OH moiety at C-6.**

(ii) The remaining *two* isomers **(b)** and **(c)** usually give the **same ketone** upon *oxidation* thereby indicating that these *two entities* do differ in the critical position of the **-OH moiety** which remains at **C-8.**

Thus, one may explain all these aforesaid changes provided it is assumed perceptively that the **'double bond'** exists between **C-7** and **C-8.** Therefore, the tentative and probable structures of both **morphine** and **codeine** may be as given under:

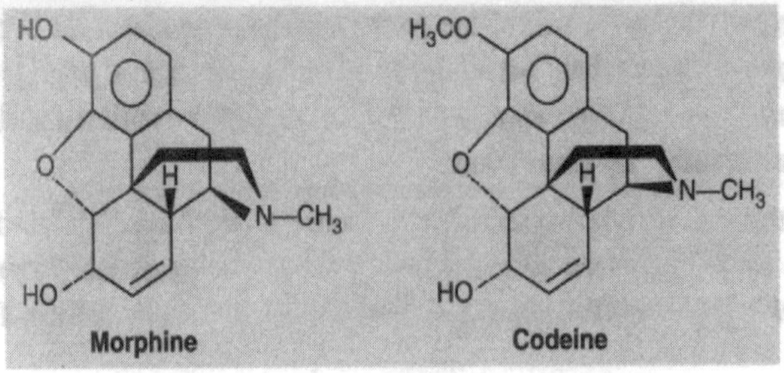
HO
H
N—CH3
HO
Morphine
H3CO
N—CH3
HO
Codeine

The various reactions of **morphine** may be explained justifiably on the basis of the aforesaid facts:

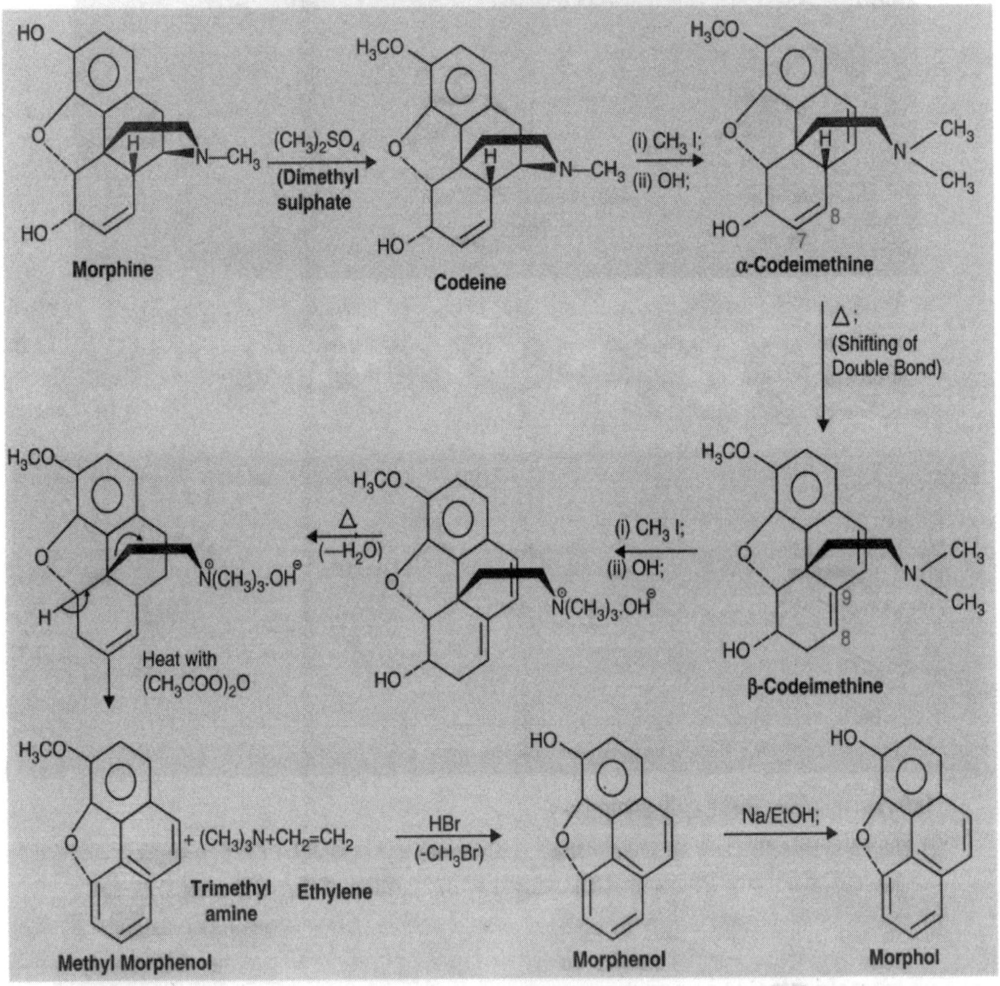
HO
N—CH3
HO
Morphine
(CH3)2SO4
(Dimethyl sulphate
H3CO
Codeine
(i) CH3 I;
(ii) OH;
CH3
N
CH3
α-Codeimethine
Δ;
(Shifting of Double Bond)
β-Codeimethine
(i) CH3 I;
(ii) OH;
N(CH3)3.OH
Δ
(—H2O)
Heat with
(CH3COO)2O
+ (CH3)3N+CH2=CH2
Trimethyl amine
Ethylene
Methyl Morphenol
HBr
(-CH3Br)
Morphenol
Na/EtOH;
Morphol

(X) **Synthesis of Morphine**: The total synthesis of **morphine** may be accomplished by the methods put forward by:

(a) **Gates *et al.* (1956)** [Gates *et. al.*: **J Am Chem. Soc.** 78:1380, 1956.

(b) **Morrison *et al.* (1967)** [Morrison *et al.*: **Tetrahadron letteres**,41: 4055, 1967

[A] Gate's Synthesis:

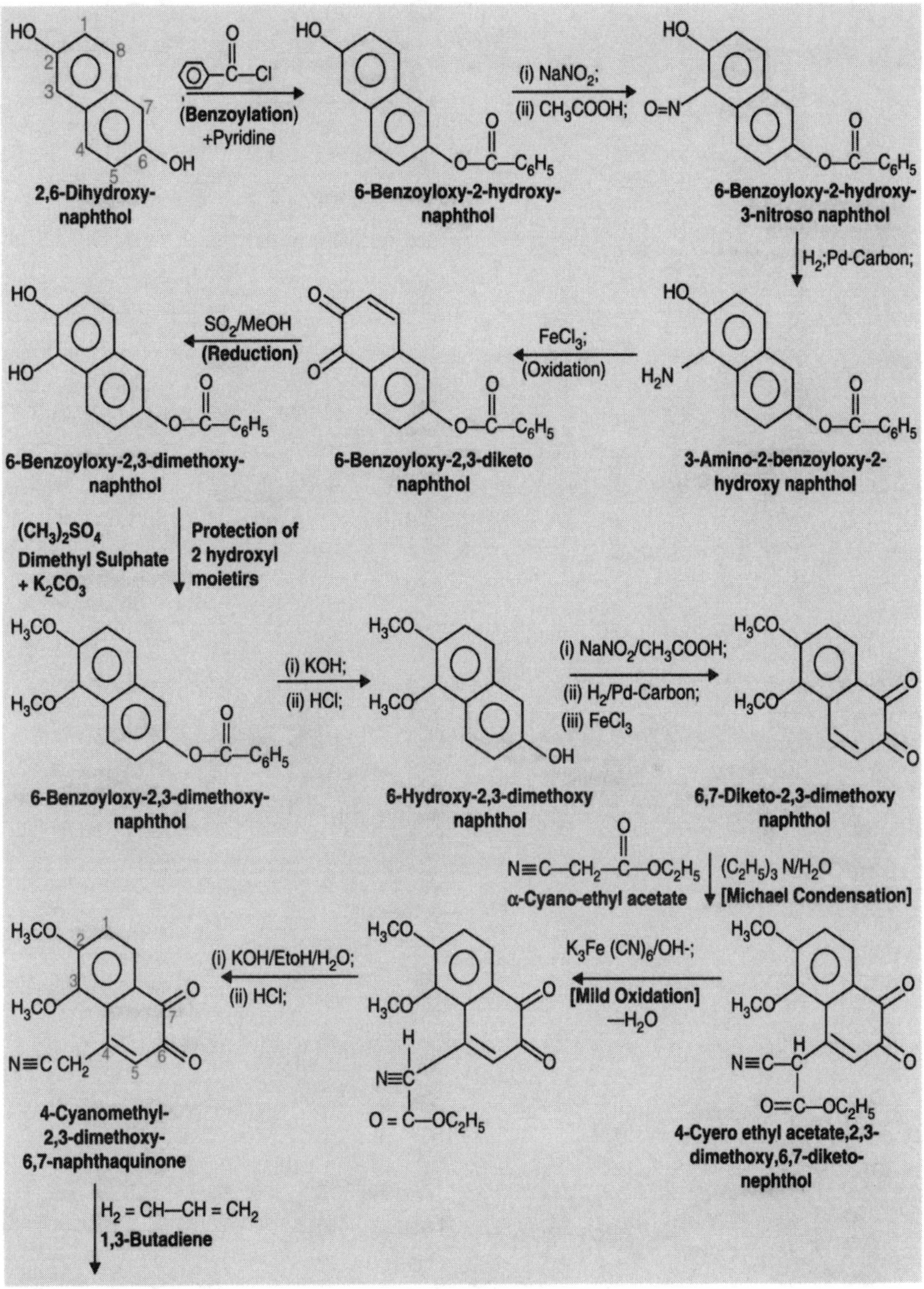

2,6-Dihydroxy-naphthol → (Benzoylation) +Pyridine → **6-Benzoyloxy-2-hydroxy-naphthol** → (i) NaNO₂; (ii) CH₃COOH; → **6-Benzoyloxy-2-hydroxy-3-nitroso naphthol** → H₂;Pd-Carbon; → **3-Amino-2-benzoyloxy-2-hydroxy naphthol** → FeCl₃; (Oxidation) → **6-Benzoyloxy-2,3-diketo naphthol** → SO₂/MeOH (Reduction) → **6-Benzoyloxy-2,3-dimethoxy-naphthol** → (CH₃)₂SO₄ Dimethyl Sulphate + K₂CO₃ | Protection of 2 hydroxyl moietirs → **6-Benzoyloxy-2,3-dimethoxy-naphthol** → (i) KOH; (ii) HCl; → **6-Hydroxy-2,3-dimethoxy naphthol** → (i) NaNO₂/CH₃COOH; (ii) H₂/Pd-Carbon; (iii) FeCl₃ → **6,7-Diketo-2,3-dimethoxy naphthol**

N≡C—CH₂—C—OC₂H₅ α-Cyano-ethyl acetate | (C₂H₅)₃ N/H₂O [Michael Condensation]

→ **4-Cyero ethyl acetate,2,3-dimethoxy,6,7-diketo-nephthol** → K₃Fe (CN)₆/OH⁻; [Mild Oxidation] —H₂O → **4-Cyanomethyl-2,3-dimethoxy-6,7-naphthaquinone** → (i) KOH/EtOH/H₂O; (ii) HCl;

H₂ = CH—CH = CH₂ 1,3-Butadiene

---

* Cates *et al.* : *J Am Chem Soc.*, **78** : 1380, 1956.

** Morrison *et al.* : *Tetrahedron Letters*, (41), 4055, 1967.

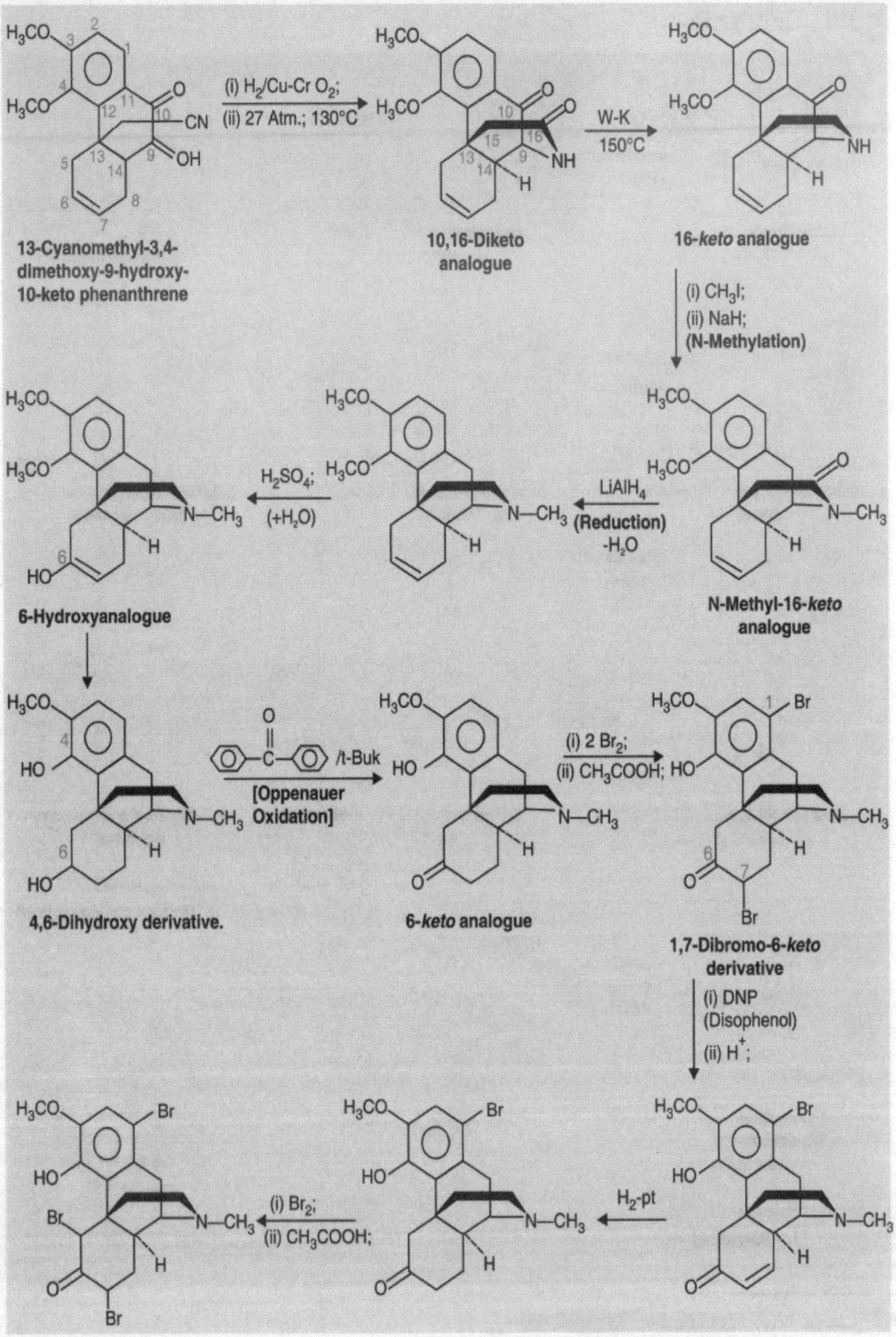

13-Cyanomethyl-3,4-
dimethoxy-9-hydroxy-
10-keto phenanthrene

(i) H$_2$/Cu-Cr O$_2$;
(ii) 27 Atm.; 130°C

10,16-Diketo
analogue

W-K
150°C

16-*keto* analogue

(i) CH$_3$I;
(ii) NaH;
(N-Methylation)

6-Hydroxyanalogue

H$_2$SO$_4$;
(+H$_2$O)

LiAlH$_4$
(Reduction)
-H$_2$O

N-Methyl-16-*keto*
analogue

4,6-Dihydroxy derivative.

[Oppenauer
Oxidation]

/t-Buk

6-*keto* analogue

(i) 2 Br$_2$;
(ii) CH$_3$COOH;

1,7-Dibromo-6-*keto*
derivative

(i) DNP
(Disophenol)
(ii) H$^+$;

(i) Br$_2$;
(ii) CH$_3$COOH;

H$_2$-pt

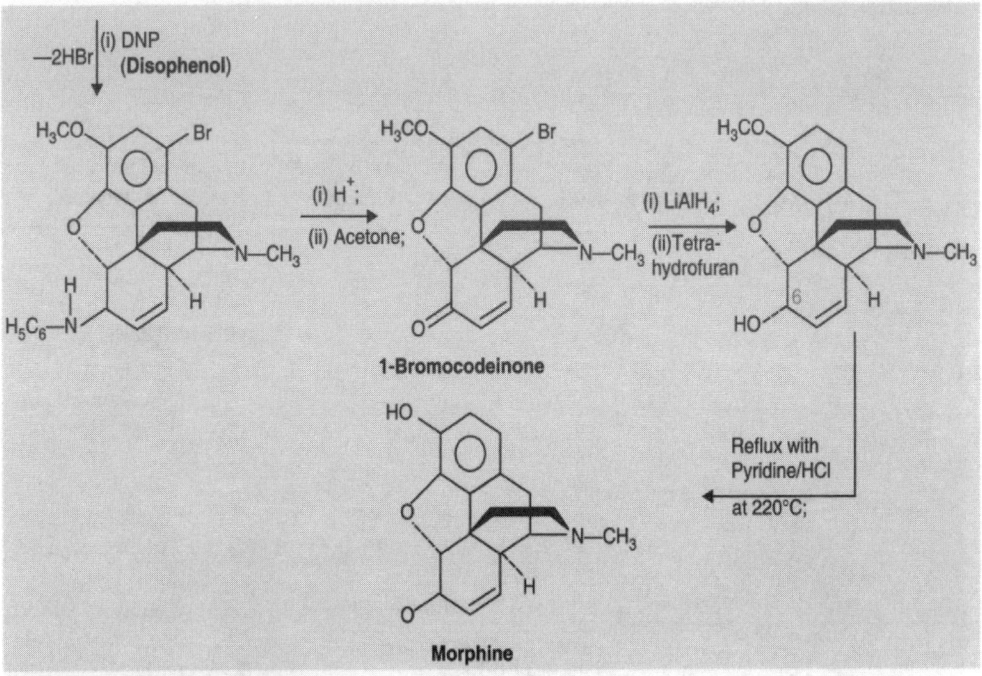
(i) DNP
(Disophenol)
—2HBr
H3CO
Br
O
N—CH3
H
H5C6—N
H
(i) H+;
(ii) Acetone;
1-Bromocodeinone
(i) LiAlH4;
(ii)Tetra-
hydrofuran
6
HO
Reflux with
Pyridine/HCl
at 220°C;
HO
Morphine

[B] Morrison *et al.* Synthesis:

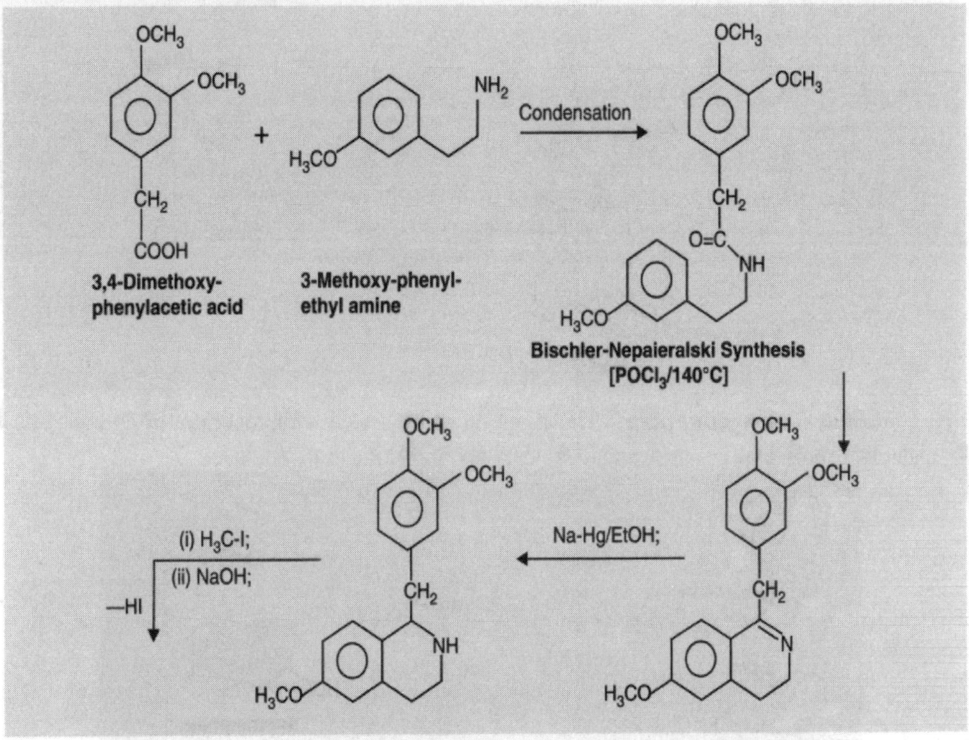
OCH3
OCH3
CH2
COOH
3,4-Dimethoxy-
phenylacetic acid
+
NH2
H3CO
3-Methoxy-phenyl-
ethyl amine
Condensation
O=C
NH
Bischler-Nepaieralski Synthesis
[POCl3/140°C]
Na-Hg/EtOH;
N
(i) H3C-I;
(ii) NaOH;
—HI

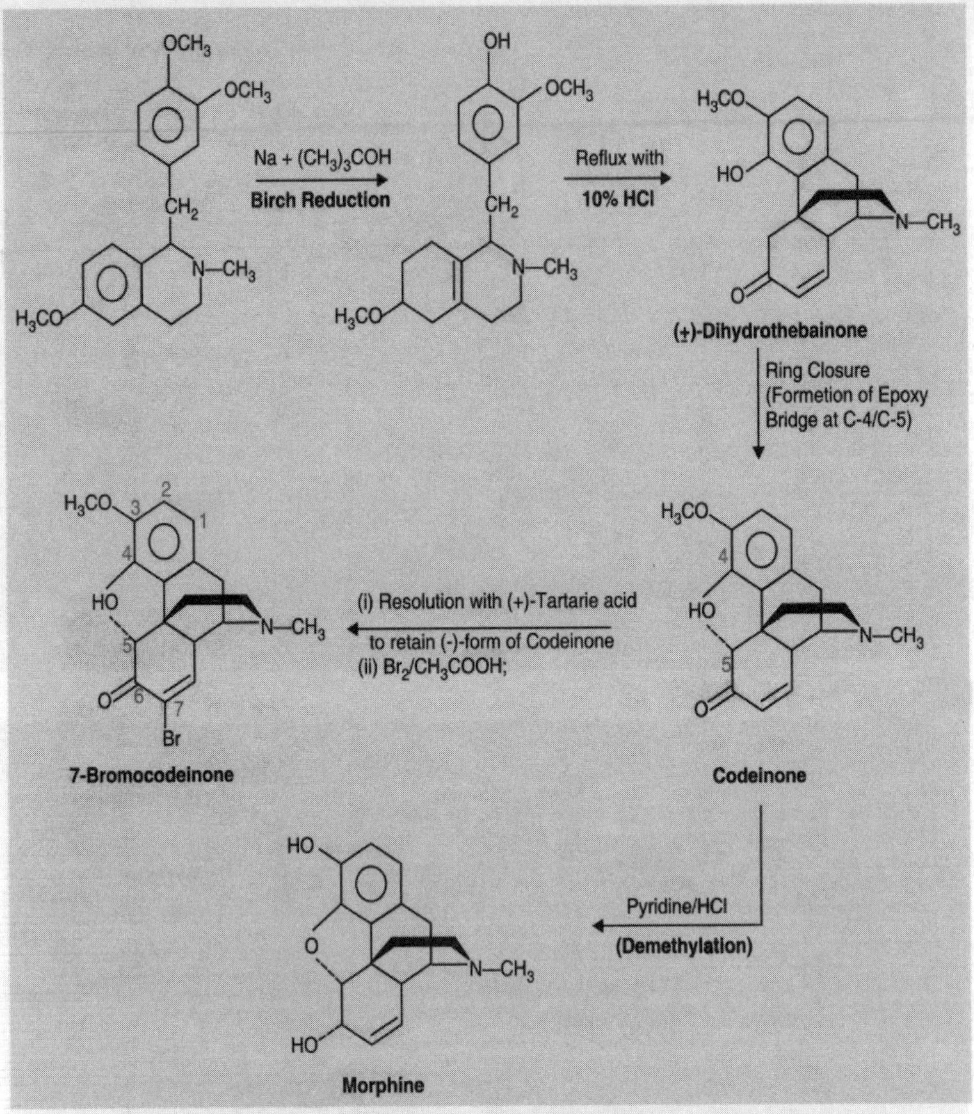
OCH3
OCH3
CH2
N–CH3
H3CO
Na + (CH3)3COH
Birch Reduction
OH
OCH3
CH2
N–CH3
H3CO
Reflux with
10% HCl
H3CO
HO
N–CH3
O
(±)-Dihydrothebainone
Ring Closure
(Formetion of Epoxy
Bridge at C-4/C-5)
H3CO
HO
N–CH3
O
Codeinone
(i) Resolution with (+)-Tartarie acid
to retain (-)-form of Codeinone
(ii) Br2/CH3COOH;
H3CO
HO
N–CH3
O
Br
7-Bromocodeinone
Pyridine/HCl
(Demethylation)
HO
O
N–CH3
HO
Morphine

Formaton of Apomorphine: **Morphine** on being heated with concentrated hydrochloric acid, undergoes **molecular rearrangement** to yield **apomorphine** as given under:

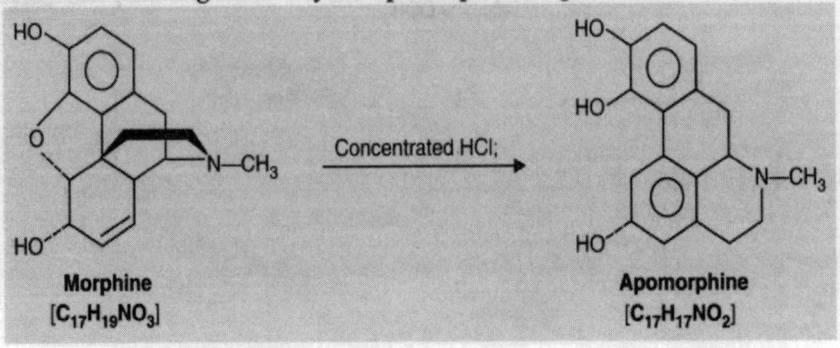
HO
O
N–CH3
HO
Morphine
[C17H19NO3]
Concentrated HCl;
HO
HO
N–CH3
HO
Apomorphine
[C17H17NO2]

The above rearrangement proceeds with loss of water residues.

Synthetic Substitutes for Morphine:* Due to the addictive tendencies of **morphine**, it has been duly replaced by a few tested and tried **syntheutic substitutes for morphine**, having comparatively simpler **chemical structure**, but showing therapetic actions quantitatively comparable to **morphine**,

Examples: A few typical **examples** are given below:

(a) **Methadone Hydrochloride:**

Methadone Hydrochloride

Its narcotic analgesic actions are fairly comparable to **morphine**.

(b) **Pethidine Hydrochloride:**

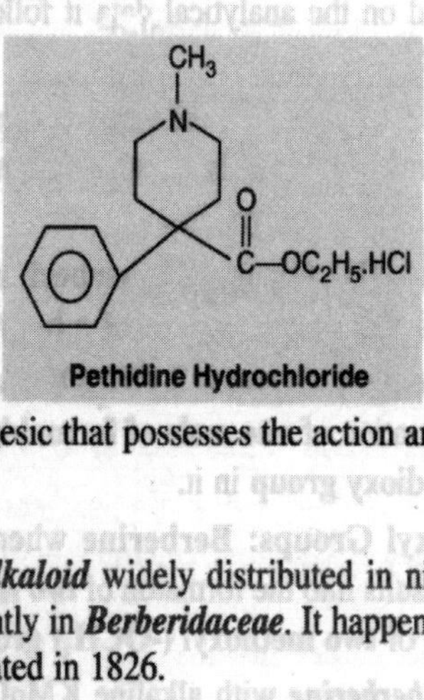

Pethidine Hydrochloride

It is a synthetic nercotic analgesic that possesses the action and uses of **morphine**.

3.6.5 Berberine

Berberine is a *tautomeric alkaloid* widely distributed in nine or more botanical families, but occurs most abundantly and frequently in *Berberidaceae*. It happens to be a vital component of several traditional medicines, and first isolated in 1826.

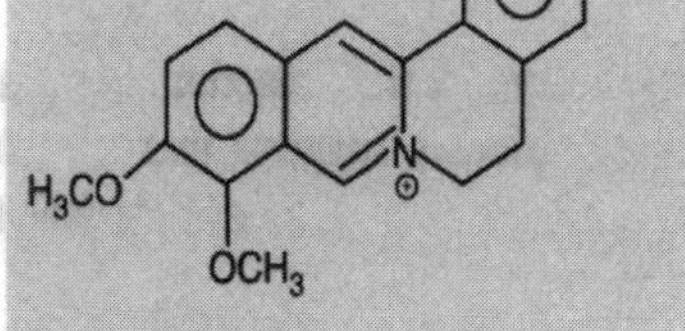

Berberine 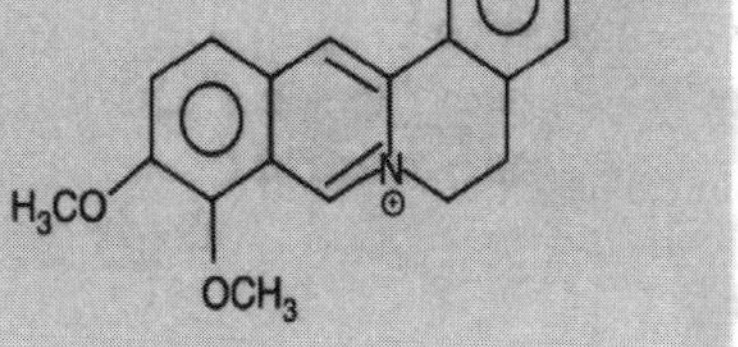   **Palmatine**

---

* Kar A: **Medicinal Chemistry**, New Age International, New Delhi, 5th edn., 2010.

Interestingly, **beriberine** and its related alkaloids (*vz.*, **Palmatine**) are of frequent occurrence in the order **Rhodales** that essentially includes such families as:

Papaveraceae and *Fumariaceae*.

Characteristic Features: These essentially include:

(1) it appears as yellow needles from ether.

(2) It melts at 145°C

(3) Its UV_{max} : 265, 343 nm.

(4) Its pK value is 2.47.

(5) It has a bitter taste.

(6) Its pKa value is 11.8.

(7) It is fairly soluble in ethanol and boiling water; and is sparingly soluble in chloroform, cold water, and benzene.

3.6.5.1 Constitution of Berberine

The following steps need to be followed in a sequential manner so as to arrive at the probable structure assigned to **berberine**.

(1) **Molecular Formula**: Based on the analytical data it follows that the **molecular formula of berberine** is found to be $C_{20}H_{18}NO_5$.

(2) **Nature of N-Atom:** Since, it gives rise to the formation of **methyl iodide** by interaction with $CH_3.I$ (methyl iodide), it arcertains that the N-atem is tertiary in nature; and, therefore, must form the **integral segment of a cyclic system.**

(3) **Presence of a Hydroxyl (-OH) Group:** As **berberine** yields both the **monoacetyl** and **monobenzoyl** derivatives, it shows the presence of a **hydroxyl group**.

(4) **Presence of one Methylenedioxy Group [-O-CH₂-O]:** Berberne on being heated with dilute H_2SO_4 gives rise to the formation of **one mole of formaldehyde (HCHO)** thereby showng the presence of one **methylenedioxy group** in it.

(5) **Presence of Two Methoxyl Groups:** Berberine when heated gently with an excess of **hydroioidic acid (HI),** it results into the formaton of *two* **moles of methyl iodide (H_3C-I)** there by confirming the presence of **two methoxyl (-OCH₃) groups** in it.

(6) **Oxidation: Oxidation** of *berberine* with alkaline $KMnO_4$ usually gives *three* **products of reaction**, such as:

- **Berberillic acid** : $C_{20}H_{19}NO_9$;
- **Berberal** : $C_{20}H_{17}NO_7$; and
- **Oxyberberine** : $C_{20}H_{17}NO_5$.

It has been duly observed that in all the aforesaid *three* products of reaction essentially do **retain twenty (20) carbon atoms** ineach of them. It, therefore, obviously suggests that the critical inherent structures of these **three products** shall offer the probable idea with regard to the **basic skeleton structure of berberine**.

It is, therefore, pertinent to state at this material time that one should know and establish the **basic structures** of these *three* **products** before assigning the probable structure to **'berberine'**.

(7) **Structure of Berberilic Acid [$C_{20}H_{19}NO_9$]:** The various steps followed in establishing the structure of berberilic acid are as stated under:

(i) Its molecular formula is found to be $C_{20}H_{19}NO_9$.

(ii) By means of standard prescribed tests one may easily ascertain that berberilic acid contains:

- **two** - carboxyl (-COOH) groups,
- **two** - methoxyl (-OCH$_3$) groups, and
- **one** - methylenedioxy (-O-CH$_2$-O-) group,

which clearly shows the presence of one amide function $(-\overset{\overset{\displaystyle O}{\|}}{C}-NH_2)$ present in **berberilic acid**. Thus, we may have:

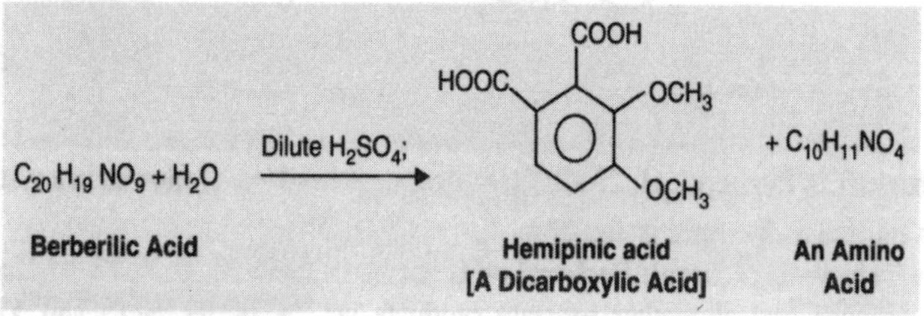

$$C_{20}H_{19}NO_9 + H_2O \xrightarrow{\text{Dilute } H_2SO_4;} \text{Hemipinic acid} + C_{10}H_{11}NO_4$$

Berberilic Acid   **Hemipinic acid**   **An Amino**
                           [A Dicarboxylic Acid]   **Acid**

(iii) ω-**Aminoethyl piperonylic acid:**

The generated amino acid is found to be **ω-aminoethyl piperonylic acid (X)** by means of the following sequential reactions:

- Hofmann's exhaustive methylation,
- mild oxidation, and
- preduction,

in order to remove the **methylenedioxy moiety** specifically to obtain the resulting product as **4, 5-dihydroxy-phthalic acid.**

(Y). These aforesaid reactions may be summarized as given under:

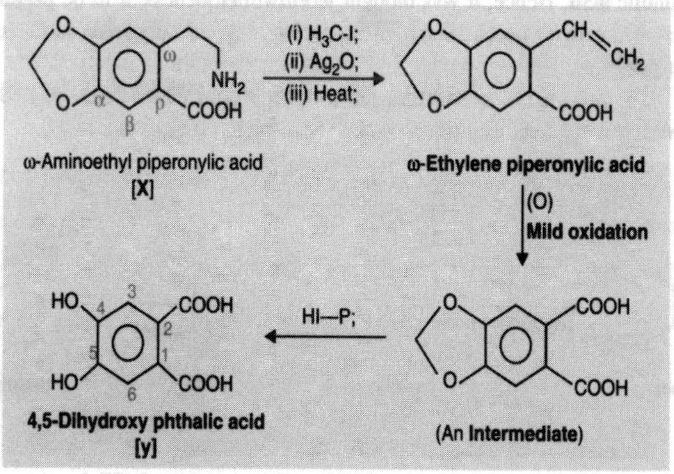

ω-Aminoethyl piperonylic acid
[X]

(i) H$_3$C-I;
(ii) Ag$_2$O;
(iii) Heat;

ω-Ethylene piperonylic acid

(O)
Mild oxidation

(An Intermediate)

HI—P;

4,5-Dihydroxy phthalic acid
[y]

(IV) Proposed Structure of Berberific Acid: Based on the structure of **hemipinic acid** [as in (ii) above] and the amino acid i.e., **w-aminoethyl piperonylic acid (X)** [as in (iii) above], which eventually constitute the products of hydrolysis obtained from berberilic acid, the latter may be assigned the following **proposed structure:**

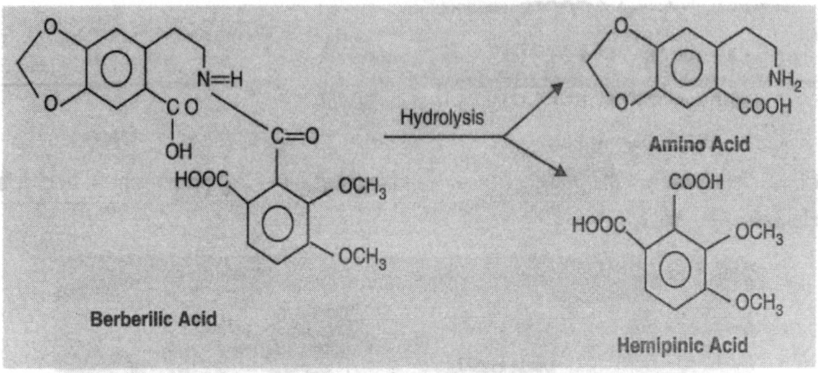

Berberilic Acid

Hemipinic Acid

(8) **Structure of Berberal [C$_{20}$H$_{17}$NO$_7$]:** Following steps need to be followed in a systematic manner:

   (i) Its molecular formula is found to be C$_{20}$H$_{17}$NO$_7$.

   (ii) **Hydrolysis of Berberal: Berberal** on being heated with dilute H$_2$SO$_4$ undergoes hydrolysis to produce two altogether separate products viz., **X** [C$_{10}$H$_{10}$NO$_5$] and Y[C$_{10}$H$_9$NO$_3$], as stated under

$$\underset{\text{Berberal}}{C_{20}H_{17}NO_7} \xrightarrow[\text{(Hydrolysis)}]{\text{Dilute H}_2\text{So}_4} \underset{(X)}{C_{10}H_{10}NO_5} + \underset{(Y)}{C_{10}H_9NO_3}$$

   (iii) **Compound 'X' as y-Opianic Acid (or Pseudo-opianic Acid):** The compound X (C$_{10}$H$_{10}$NO$_5$) has been shown to possess **two methoxy, one aldehydic,** and **one carboxylic** groups. Besides, when X is duly oxdized produces **hemipinic acid** (i.e., an organic molecule with having an establshed structure) thereby suggesting that **compound X** bears a close resemblance to pseudo-opianic acid. Obviously, X fails to resemble 100% with y-opianic acid. Hence, it was thought worthwhiletobelieve it to be **pseudo-opnianic acid** wherein the two functional moieties: *carboxylic* and *aldehydic* have been interchanged completely.

Finally, the **compound X** is ascertained to be ψ-opianic acid which could be further confirmed by its careful **reduction** to meconin (or meconinic acid lactone) as given under:

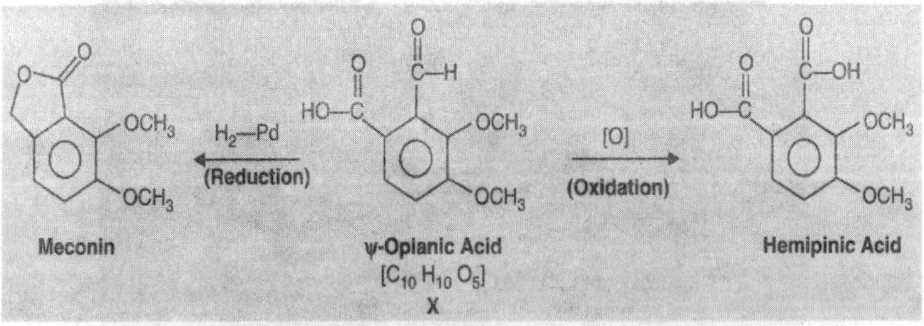

Meconin

ψ-Opianic Acid
[C$_{10}$H$_{10}$O$_5$]
X

Hemipinic Acid

   (iv) **Compound 'Y' as Noroxyhydrostininie:** The compound **'Y'** has been critically shown to posses one group each of : *keto, secondary N-atom,* and *methylene dioxy.* Following **two** steps of reactions take place using noroxyhydrastinne as the starting material, namely:

- **Hydrolysis** of **Y** affords an amino acid; which suggests that it should be a '*cyclic amide*' and
- Treatment with an alkali followed by **Hofmann's exhaustive methylation** and oxidation yields a compound duly identified as **4, 5-methylenedioxy phthalic acid**.

Thus, we may have the following reactions:

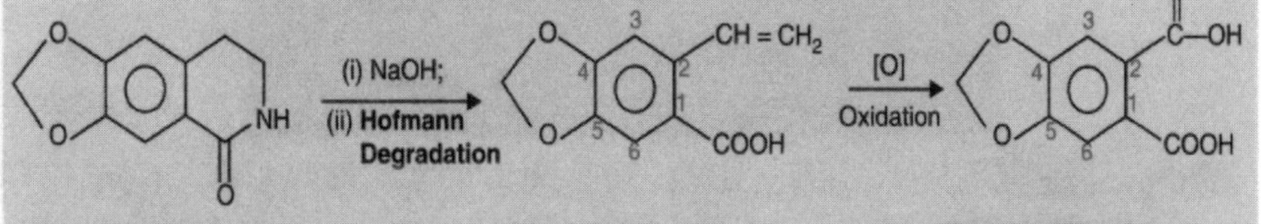

| Noroxyhydrastinine [Y] | 4,5-Methylene dioxy-2-ethylene benzoic acid | 4,5-Methylene dioxy-phthalic acid |

(9) **Proposed Structure of Berberal:** Based on the structure of ψ-opianic acid (**X**) [see section (ii) above] and **noroxyhydrastimine (Y)** [see section **(iv)** above] one may elegantly propose the structure of **berberal** as I; however, it was intelligently replaced by structure **II**. The above statement rests on the glaring fact that ψ-opianic acid (**X**) on being treated upon with amiline, gets duly linked to the **N-atom** of the 'amiline' *via* the **aldehydic C-atom**. Thus, we may have the following structures of **I** and **II**:

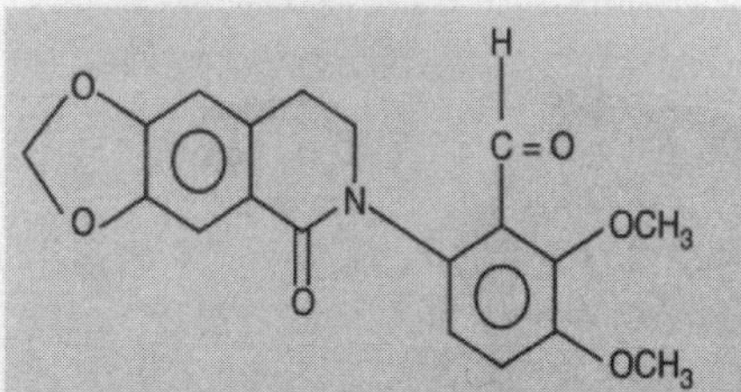

[I] 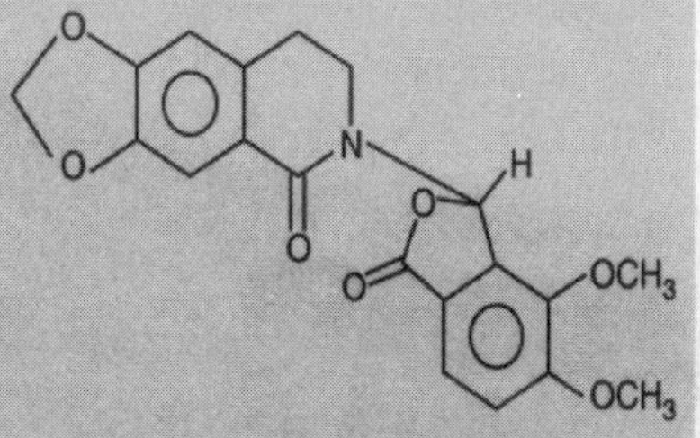   [II]

(10) **Structure of Berberine:** So far we have keenly taken due cognizance of the fact that both **berberal** and **berberilic acid** are obtained by the oxidation of '**berberine**'; and, therefore, the structure of the latter compound may be designated as given below in structure (III):

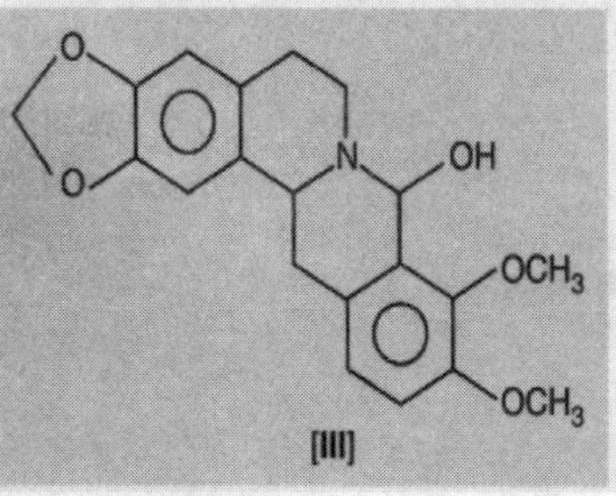

[III]

(11) **Another Structure of Berberine [IV] (as a Strong Basic Hydroxide):** Berberine, when isolated meticulously from the respective sulphate by barium hydroxide [Ba(OH)$_2$] hydrolysis, it usually appears as a **strong basic hydroxide**, as structure **[IV]**.

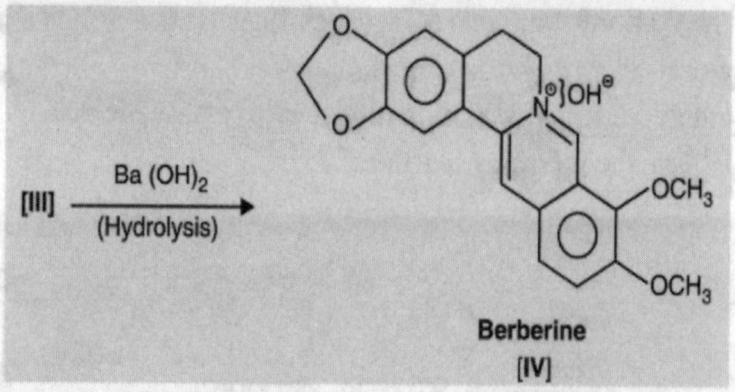

[III] $\xrightarrow[\text{(Hydrolysis)}]{\text{Ba (OH)}_2}$

Berberine
[IV]

(12) **Synthesis of Berberine:** The proposed structure **(IV)** for berberine was adequately proved by its synthesis (Haworth, 1927).*

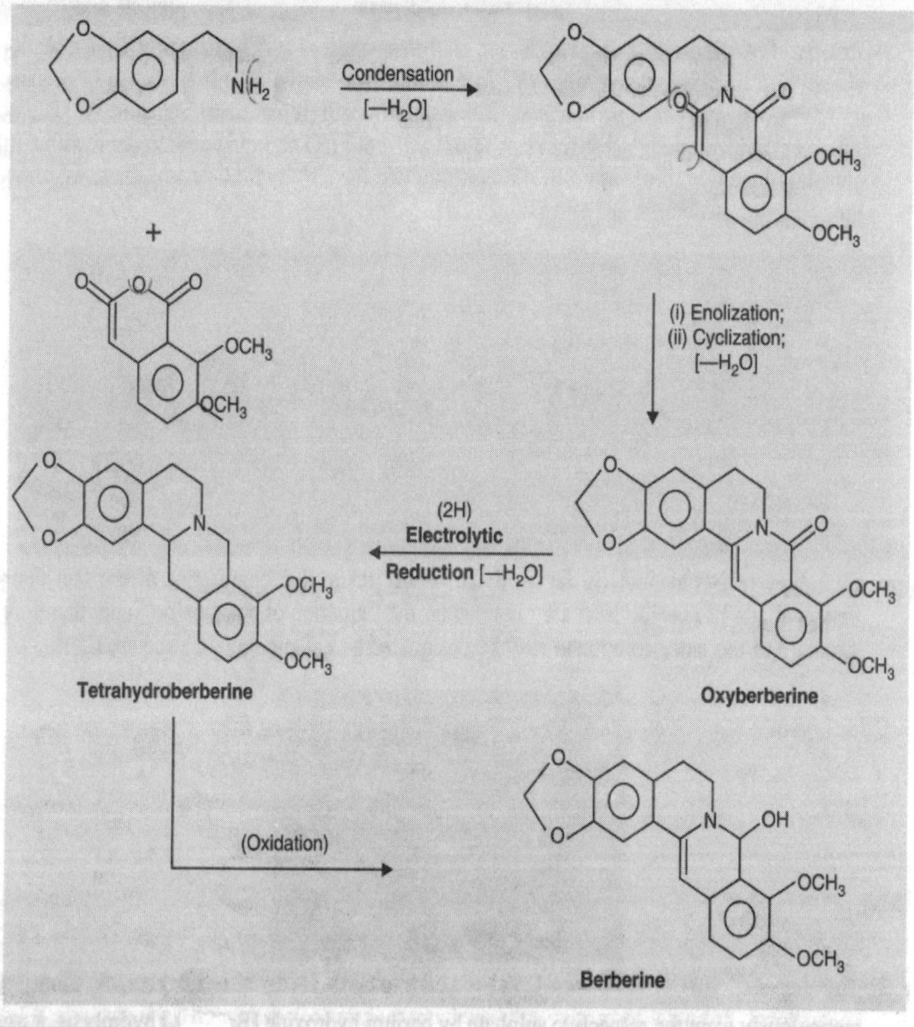

Tetrahydroberberine

Oxyberberine

Berberine

Condensation [—H₂O]

(i) Enolization;
(ii) Cyclization;
[—H₂O]

(2H)
Electrolytic
Reduction [—H₂O]

(Oxidation)

* Haworth: J Chem Soc., 548, 1927.

It is a simple **four-step synthesis** which exclusively ascertains and confirms the proposed structure of berberine (IV) as stated earlier.

3.6.6. Emetine

Emetine is the principal alkaloid of **ipecac**, the ground roots of **Uragoga ipecacuanha** (Brot.) Baill. **Rubeaceae**, and is found to occur naturally as the (-)-form.

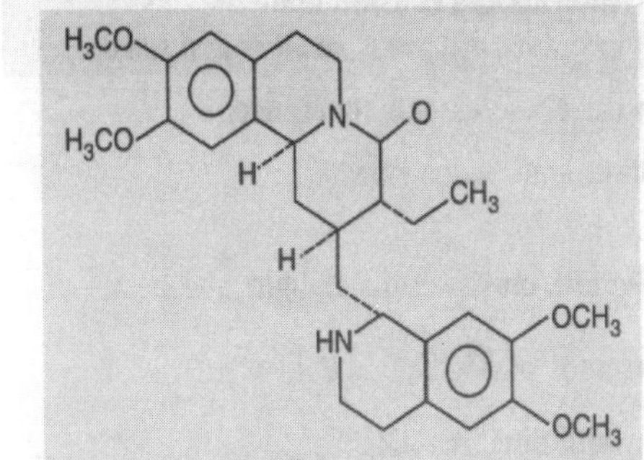

Emetine

Isolation: Pelletier and Magendie (1817) were first to report the isolation of emetine from the roots of **Caphaelis ipecacuanha**. In addition to **emetine**, ipecae contains **four** other alkaloids, namely: **Cephaeline, psychotrine, O-methylpsychotrine**, and **emetamine** (collectively known as **Ipecacuanha alkaloids**), as given under:

Modus Operandi: The powdered roots of *U. ipecacuanha* are duly agitated with ethanol/amyl alcohol/a **mixture of benzene and light petroleum ether (bp 60-80°C)**, when the alkaloids are dissolved in organic solvent. The resulting solvent is first removed by **decantation,** and **scondly** treated with dilute HCl (to cause complete dissolution of the alkaloid (s) as their hydrochloride salts), which upon **alkalization** with **ammonia**, and **thirdly** treated with **ether** to obtain two distinct layers:

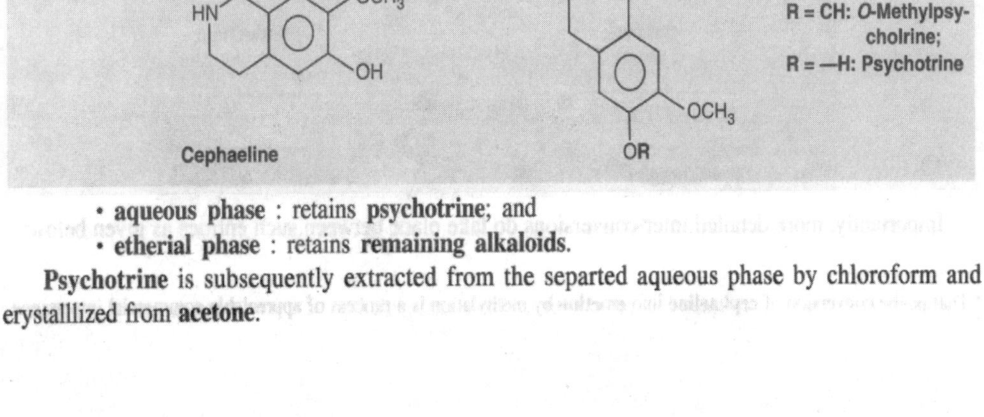

Cephaeline

R = CH: O-Methylpsy-
cholrine;
R = —H: Psychotrine

• **aqueous phase** : retains **psychotrine**; and
• **etherial phase** : retains **remaining alkaloids**.

Psychotrine is subsequently extracted from the separted aqueous phase by chloroform and erystalllized from **acetone**.

Ethereal solution when concentrated under a reduced pressure (or vacuum) - alkaloids are duly converted into the hydrochloride salt thereby salting out **"emetine"** as a **crystalline product**; whereas, the **remaining alkaloids** (*viz.*, **cephaeline** and *O*-methyl psychotrine ether) are converted into the **oxalates** in the ethanolic solution to help in the subsequent separation of O-methyl psychotrine ether and **emetine.**

> **Note:** In short, one may actually come across *two distinct types* of crystals being separated:
> * one : for psychotrine and emetine, and
> * *secondaly:* for *O-methyl psychotrine ether* and *emetine.*

Characteristic Features: These essentially comprise:

(1) Emetine is a white amorphous powder.

(2) Its m.p. is 74°C

(3) Emetine twins yellow on exposure to light and heat.

(4) its $[\alpha]_D^{20}$ value stands at $-50°$ (C = 2 in $CHCl_3$).

(5) It gives a **strong alkaline reaction**:
  $pK_1 = 5.77$; and $pK_2 = 6.64$.

(6) It is found to be freely soluble in CH_3OH; $C_2H_5\text{-}OH$; CH_3COCH_3; $CH_3\text{-}COOC_2H_5$; $C_2H_5OC_2H_5$; $CHCl_3$.

(7) **Emetine** is sparingly soluble in petroleum ether and water; moderately soluble **in dilute NH_4OH,** but sparingly soluble in solutions of **NaOH** and **KOH.**

Uses: Ipecaeuanha roots, containing **emetine** finds its abundant use as a therapeutic agent for the treatment and cure of such **human ailments** as: **emetic, expectorant,** and **amoebic dysentry.**

Structural Inter-relationship Amongst Ipecae Alkaloids: The structures of the *four alkaloid* variants of present in **ipecae alkaloids** *viz.*, **cephaeline (I), emetine (II),** *O-*methyl psychotrine ether **(III),** and psychotrine **(IV)** have been duly established as a result of the metiecullous research of **Pyman, Spath,** and **Robinson,** and their respective collaborators. The relationship between some of them may be explicitly expressed as under*:

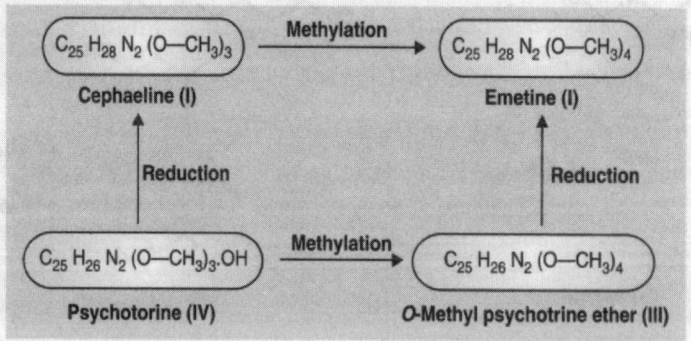

Importantly, more detailed inter-conversions do take place between such entities as given below:

---

* That is, the conversion of **cephaeline** into **emetine** by methylation is a process of appreciable commercial importance.

- Psychotrine (IV) to *O*-Methyl psychotrine ether (III);
- Psychotrine (IV) to *iso*-cephaeline (V);
- Cephaeline (I) to Emetine (II);
- Emetine (II) to Emetamine (VI);
- Emetamine (VI) to *iso*-Emetine (VII)
- *O*-Methyl psychotrine ether (III) to Emetine (II);
- *O*-Methyl psychotrine ether (III) to *iso*-Emetine (VIII).

All these **inter-relationships** (or **inter conversions**) may be summarized as stated under:

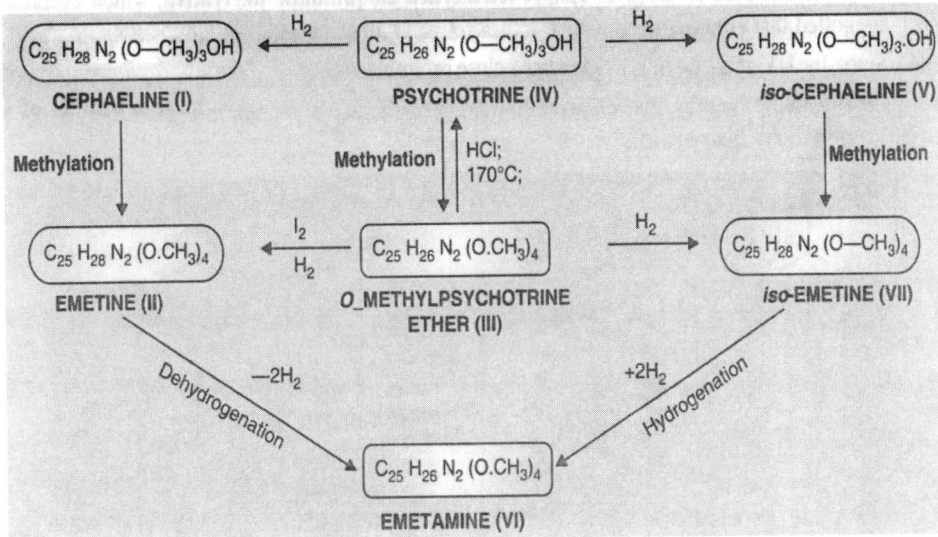

3.6.6.1. Constitution of Emetine

The underlying steps may be followed sequentially in order to obtain the probable structure assigned logically with scientific evidences to **emetine:**

(1) **Molecular Formula:** Its molecular formula is found to be $C_{29}H_{40}N_2O_4$.

(2) N-methyl moieties are totally absent in *emetine*. However, the **two N-atoms** in it usually exist as:

- a **secondary N-atom** (°N-), and
- a **tertiary N-atom** (°N),

and we have: $pK_{a1} = 8.5$, and $pK_{a2} = 7.7$.

(3) **Zeisel's method:** Emetine when heated with HI (hydroiodic acid) gives *four* moles of methyl iodide (H_3C-I), thereby suggesting the presence of four methoxyl (-OCH$_3$) moieties present in it.

(4) **Oxidation of Emetine:** The oxidation of **emetine** with KMnO$_4$/acetone gives rise to the formation of 3-hemipinic acid (A), plus a small quantum of **6,7-dimethoxyisoquinolive-1-carboxylic acid (B)**. Likewise, **emetine** when duly oxidized with chromic acid (CrO$_3$) is yield **4,5-dimethoxyphthalonimide (C).**

Evidently, the crucial production of these oxidation products reveals appropriately that emetine comprises one unit of **6,7-dimethoxyisoquinoline.**

Thus, we may have:

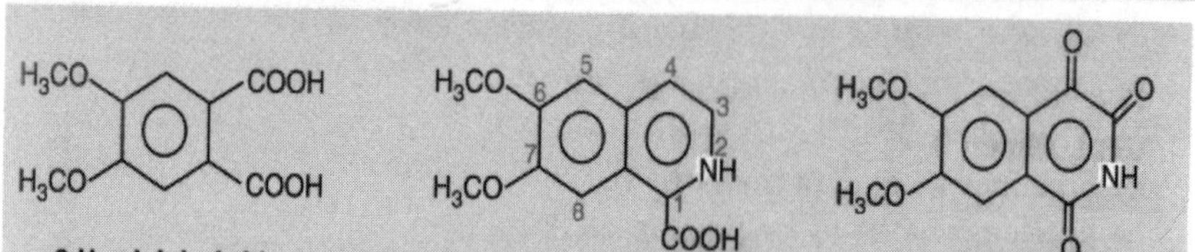

| 3-Hemipinic Acid (A) | 6,7-Dimethoxy-isoquinoline-1-Carboxylic acid (B) | 4,5-Dimethoxyphthalonimide (C) |

(5) Dobbie and Fox (1954) observed that the UV-speetra of **emetine** and **cephaeline** do show a close resmkblance to those of **1,2,3,4-tetrahydro isoquinoline derivative**, which eventually suggested that **emetine** should have **tetrahydroisoquinoline** as its **integral component**.

(6) Since, the **UV-spectrum** of *emetine* has a close resemblance to that of the **tetrahydro-papaverine** suggesting thereby that **emetine** should predominantly comprise at least **two units of O-dimethoxy benzene** in it.

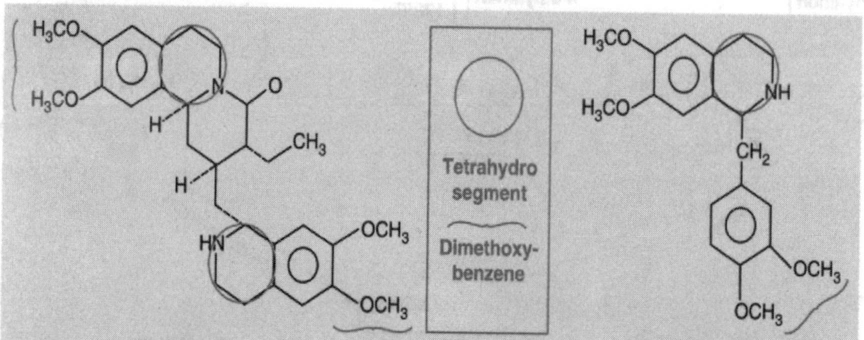

| Emetine | Tetrahydro segment / Dimethoxy-benzene | Tetrahydropapaverine |

(7) **Alkaline Oxidation of Emetine:** Emetine upon gentle oxidation with alkaline $KMnO_4$ yields two distinct products *viz.*, **copydaldiue (D)**, a **tetrahydro isoquinoline derivative**, and 3-**hemipinic acid (A)** as

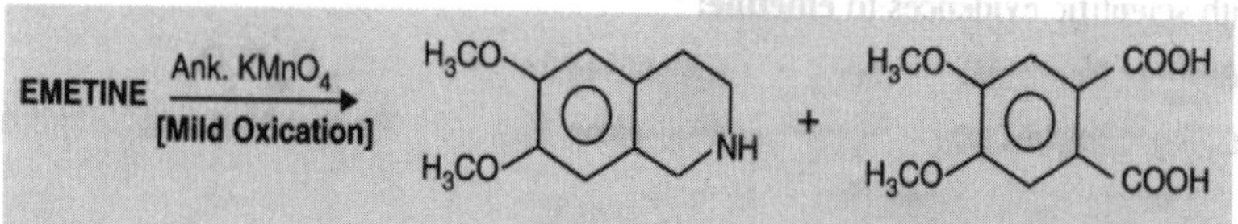

| Corydaldine | [A] |

Inferences: These essentially include:

(a) The combined total yield of both **[D]** and **[A]** as the definite bye products obtained by mild oxidation of **emetine** amounts to almost **0.96 mole per mole of emetine** strongly suggests that **emetine** contains *two* **units of 6,7-dimethroxy-isoquinoline.**

(b) The alkaloid **papaverine**, with only *one* such unit (*i.e.*, **6,7-dimethoxy isoquinoline**), yields almost **50% of the two aforesaid products [D] and [A]** thereby indicating the presence of only one unit of **6,7-dimethoxyisoquninoline.**

(8) **Presence of two units of 6,7-Dimethoxyisoquinoline in Emetine Further Substantiated:**

In fact, the very critical presence of **two units of 6,7-dimethoxyisoquinoline** is further substantiated by means of the following two critical observations, namely:

(a) Cephaeline on *ethylation* gives **ethyl ether of cephaline** which upon *oxidation* yields a mixture of two products, namely:

- corydadline (**D**), and
- 6-elthroxy-7-methroxy tetrahydro-isoquinoline-1-one (**E**).

H_5C_2O — 6 5 4 3 2 NH

H_3CO — 7 8 1

O

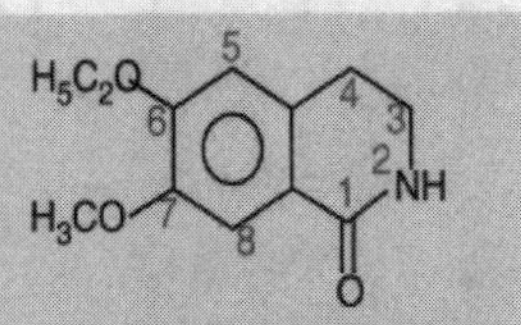

6-Ethoxy-7-nethoxy-tetrahydro-
isoquinoline-1-one (E)

The mixture of (**D**) and (**E**) could not be separated, which on oxidation gives **3-hemipinic acid (A)** and **4-ethoxy-5-methroxyphthalic acid (F)** as expressed under:

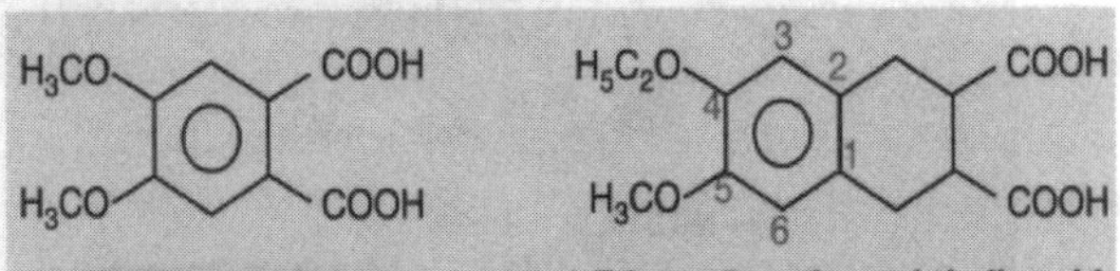

H_3CO — COOH

H_3CO — COOH

**3-Hemipinic Acid
[A]**

H_5C_2O — 3 2 COOH

H_3CO — 4 5 6 1 COOH

**4-Ethoxy-5-methoxyphthalic acid
[F]**

(b) It may be observed specifically that upon oxidation:

Emetine: gives 75% of compound (A) (*i.e.*, 3-hemipinic acid),

Papaverine: with exclusively *one* 6, 7-dimethroxyisoquinoline unit yields hardly up to 25% of compound (A) under identical experimental conditions.

Therefore, it virtually conforms the critical existence of two isoquinoline units in emetine.

(9) **Point of Linkage of Two Units of Tetrahydroiso quinoline to Rest of Emetine Molecule:**

The above conceptualized analogy may be acscertained from the exact position of the OXO or carboxyl (-COOH) moiety duly present in compounds:

[B] : **6, 7-Dimethoxy isoquinoline - 1- carboxylic acid,**

[C] : **4, 5-Dimethoxyphthalonimide**, and

[D] : **Corydaldine**.

As we know that **one of the two N-atoms present in *emetine*** is of **tertiary status**; whereas, the **second N-atom** must be linked to the **remaining portion of the molecule**. Hence, it is now logically feasible and possible to assign a probable **partial-structure of emetine** as given under:

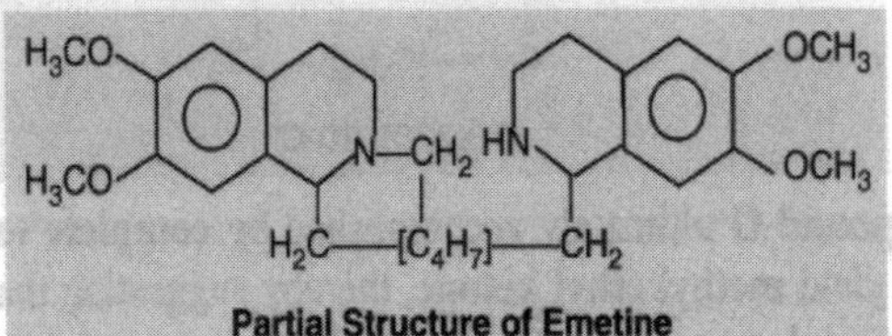

H_3CO — OCH$_3$

H_3CO — OCH$_3$

N—CH$_2$ HN

H_2C — [C$_4$H$_7$] — CH$_2$

Partial Structure of Emetine

(10) **Nature of C_4H_7 Fragment: Hofmann's degradation** (*i.e,* Hofmann's exhaustive methylation method) of *emetine* is carried out carefully in order to **eliminate the N-atoms completely.** Following are the various sequential degradation steps of *emetine* by the aforesaid Hofmann's exhaustive methylation method:

Emetine

COMPOUND-G

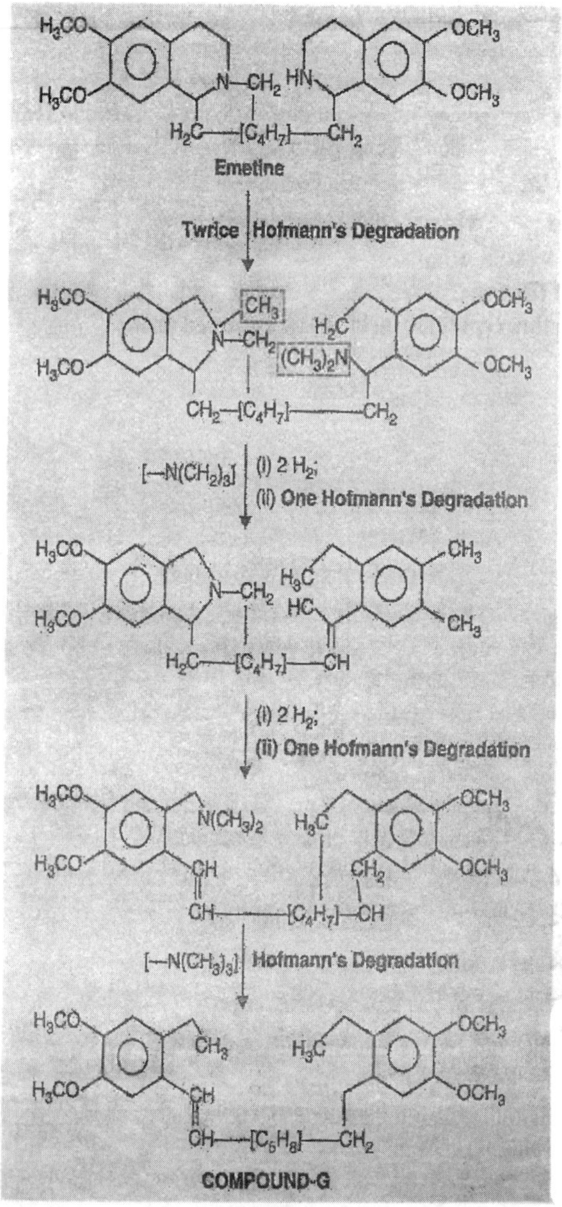

Interestingly, the **compound-G** ultimately accomplished by **complete exhaustive methylation is** subjected to **ozonolysis** to yield **methyl ethyl ketone,** thereby suggesting that the ensuring **bracketed** component C_5H_8 obtained as the respective **Hofmann degraded product from emetine contains a** *typical functional grouping* as shown under:

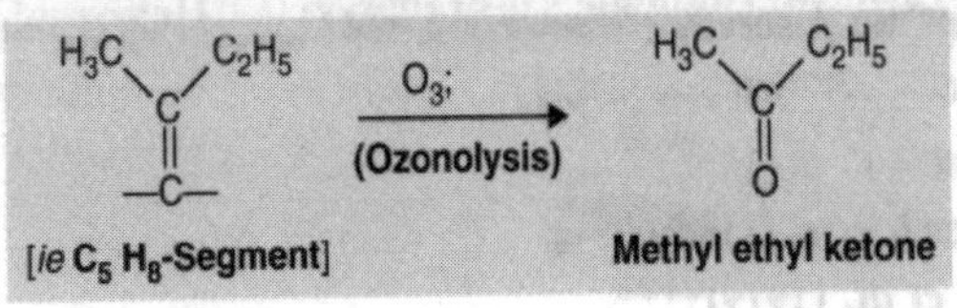

$$H_3C-\underset{\underset{|}{C}}{C}-C_2H_5 \xrightarrow[\text{(Ozonolysis)}]{O_3;} H_3C-\underset{\underset{\parallel}{O}}{C}-C_2H_5$$

[*ie* $C_5 H_8$-Segment]   **Methyl ethyl ketone**

Based on the above statement of facts, it may now be possible to provide a '**complete structure**' of the *degraded produc*t as given below:

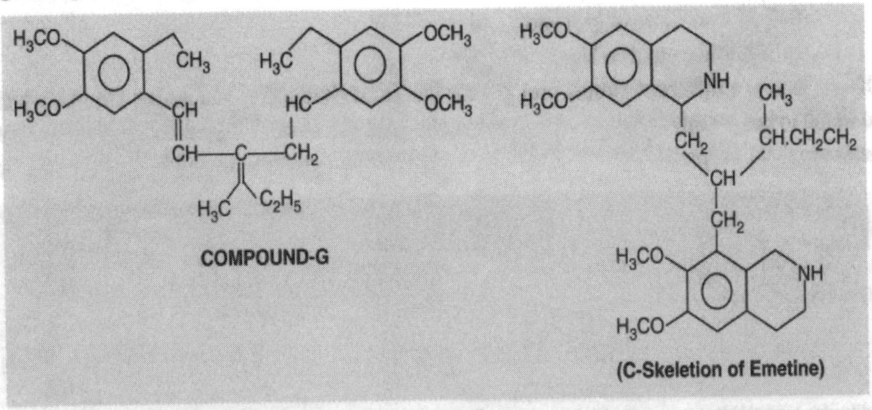

COMPOUND-G

(C-Skeletion of Emetine)

(11) **Full C-skeleton of Emetine:** If one move *backward* from **compound-G,** one may arrive at the **full C-skeleton of emetine** as shown above. However, the only problem still remains to satisfy the *third valency of one of the N-atoms* in order to render it to be a *tertiary* N-atom.

In order to justify the above statement with definite logical possibilities to assign the following *three* altogether **different molecular structures** for 'emetine' as **H, J,** and **K.**

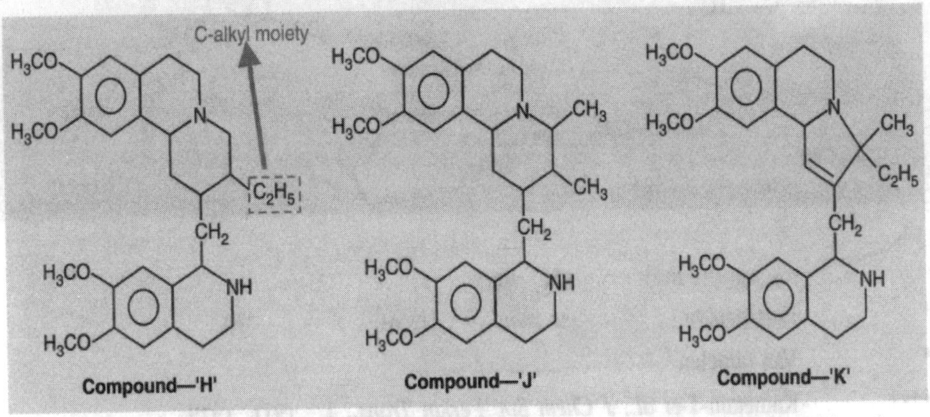

Compound—'H'   **Compound—'J'**   **Compound—'K'**

(12) **Compound-'H' : Proposed Structure of Emetine Based on Biogenetic Considerations:** Robinson (1948)* proposed the structure of *emetine* similar to **compound-'H'** as above which is exclusively based upon the **biogenetic considerations.** Besides, *emetine* essentially possesses only **one C-alkyl moiety** that suggests strongly that the remaining two structures viz., compound - 'J' and **compound 'K',** may be ignored or discarded totally.

(13) **Synthesis of Emetine:** A survey of literature would reveal that there are seven different routes of the total synthesis of emetine, namely:

   (a) Evstigneeva *et al.* (1950)**,

   (b) Van Tamelen *et al.* (1969)***,

   (c) Kametani *et al.* (1979)****,

   (d) Burgstahler and Bithos (1960)*****

   (e) Takano *et al.* (1978),******

   (f) Fujii and Yoshifuji, (1980), and *******

   (g) Openshow and Whiltaker (1963).

Of all the **seven different routes** propsed for the **Total Synthesis of Emetine** the one which was indeed found to be not only efficient, but also proved to be a viable commercial method for the **synthesis of emetine** is that of **Openshaw and Whittaker (1963)**; and, therefore, it would be discussed as under:

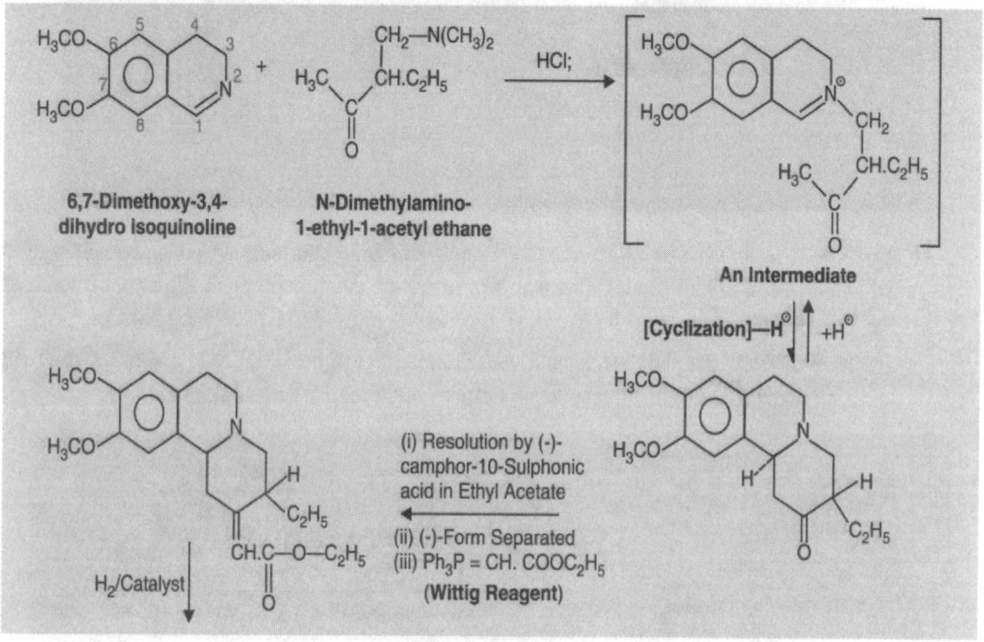

6,7-Dimethoxy-3,4-dihydro Isoquinoline   **N-Dimethylamino-1-ethyl-1-acetyl ethane**

An Intermediate

[Cyclization]—H⊕ ↑↓ +H⊕

(i) Resolution by (-)-camphor-10-Sulphonic acid in Ethyl Acetate

(ii) (-)-Form Separated

(iii) Ph₃P = CH. COOC₂H₅

(Wittig Reagent)

H₂/Catalyst

*   Robinson R : *Nature*, 162: 524, 1948.

**   Enstigneeva *et al.* : *Proc Aead Sci.* USSR, **75**: 539, 1950.

***   Van tamelen et. al. : *J Am shem Soc.*, **91**, 7359, 1969.

*****   Kametani-T-*et al.*, *J Chem Soc Perkin Trans.*, **1** : 1911, 1979.

******   Burgstahler and Bithos, *J. Am Chem Soc.*, **82**: 5446, 1960.

*******   Tanano S *et al. J Org Chem.*, **43**: 4169, 1978.

******   Fujii T and Yoshi fuji S : *Tetrahedron*, **36**: 1539, 1980.

*******   Openshaw and Whittaper, *J Chem Soc.*, 1461, 1963.

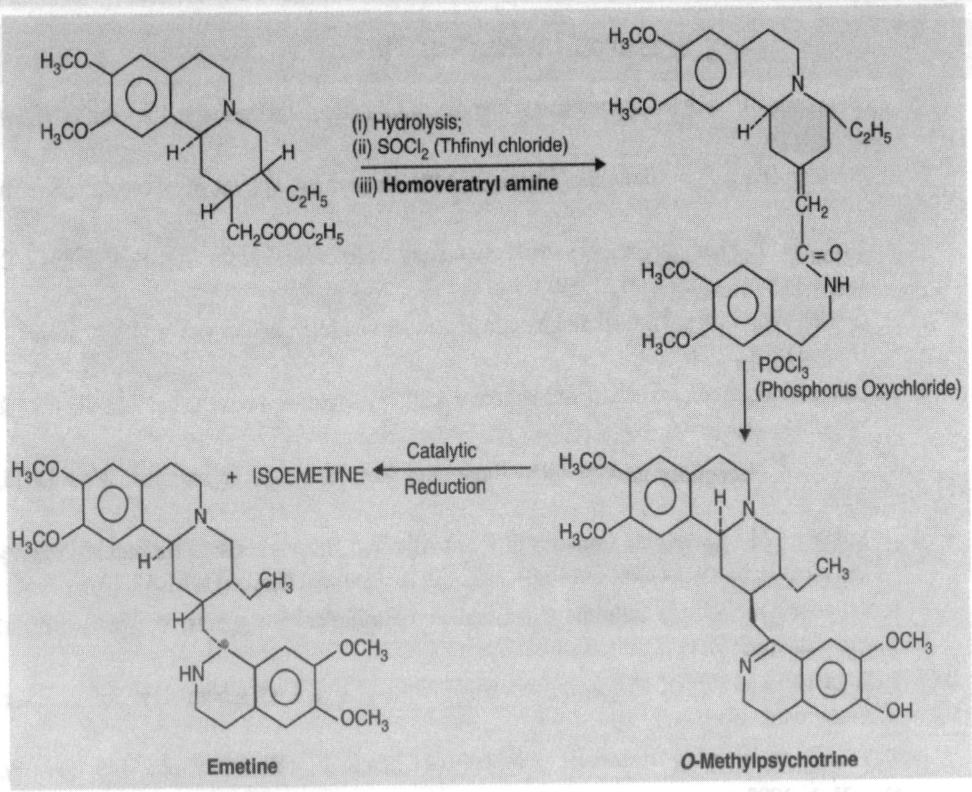

Isoemetine: A stereo isomer of 'emetine'.

Observations: These essentially consist of:

(1) Both **isoemetine** and *emetine* happen to be the **stereoisomers**, whereby the two structures only differ in their specific configuration at **C(1′)** indicated above.

(2) Basesd on the fact that 'isoemetine' bears hardly any *amoebecidal activity*; and, therefore, it is subjected to conversion into corresponding 'emetin' by means of the following three steps, namely:

    (a) formation of *N-chloro derivative* by **sodium hypochlorite**

    (b) *Dehydrochlorination* by a *alkali* to form *O-methyl psychotrine*, and

    (c) Catalytic reduction of the latter to yield a mixture of **emetine** and **isoemetine**.

(3) Nevertheless, the **end-product** thus obtained is again treated carefully to obtain **emetine** which eventually increases the overall yield of **emetine** appreciably.

RECOMMENDED FURTHER READINGS

1. Bernstein BB : **Kirk-Othmer-Encyclopedia of Chemical Technology**, 4th, ed., Vol. 2, Wiley, New York, 1992.

2. Brossi A (Ed.): **The Alkaloids, Chemistry, and Pharmacology**, Vol. 39, Academic, San Diego, 1990.

3. Bruneton J: **Pharmacognosy: Phytochemsitry Medicinal Plants**, 2nd Edn, TEC & DOC, Intercept LTD., New York, 1999.

4. Cordell GA (Ed.) : **The Alkaloids, Chemistry, and Pharmaeology**, Vol. 46, Academic, San Diego (USA), 1998.

5. Dewick PM: **Medicinal Natural Prodcuts: A Biosynthetic Approach**, 2nd Edn. John Wiley & Sons Ltd., Chichester (UK), 2001.

6. Kar A : **Pharmacognosy and Biotechnology**, 2nd Edn. New Age International, New Delhi, 2008.

7. Kutchan TM : **Molecular Genetics of Plant Alkaloid Biosysthesis**. *The Alkaloids, Chemistry* and Pharmacology, Cordell GA (Ed.) : Vol. 50, Academic, San Diego (USA), 1998.

8. Pelletier SW (Ed.): **Alkaloids, Chemical and Biological Perspectives**, Vol. 10, Elsevier, Amsterdam, 1996.

9. Stockigt J and Rupert M : **Comprehensive Natural Products Chemistry**, Vol. 4, Elsevier, Amsterdam, 1999.

10. Van Arnum SD: **Kirk-Othmer Encyclopedia of Chemical Technology**, Vol. 25, 4th Edn., Wiley, New York, 1998.

11. Verpoorte R et al. : **The Alkaloids, Chemistry, and Pharmacology** (ed Cordell GA), Vol. 50, Academic, San Diego, 1998.

REVIEW QUESTIONS

1. What are **alkaloids**? Discuss briefly the **alkaloids** under following categories giving suitable examples:

    (i) **Protoalkaloids**, and

    (ii) **Pseudoalkaloids**.

2. Discuss the '**Isolation of Alkaloids**' with reference to:

    (a) **Specific alkaloidal tests**, and

    (b) **Extraction of Alkaloids**.

3. Explain the general methods of **Structure Elucidation of Alkaloids**. Give suitable examples in support of your answer.

4. How will you determine the following in an '**alkaloid**'?

    (a) **Functional Nature of N-atom**

    (b) **Estimation of C-Methyl Moieties**.

(c) **Hofmann's Exhaustive Degradation Method**

(d) **Limitations of Hofmann's Exhaustive Degradation Method**.

5. Write Short notes on the following:

   (i) **Emde's Degradation Method**

   (ii) **von Braun's Method for Tertiary Cyclic Amines**

   (iii) **Reductive Degradation**

   (iv) **Dehydrogenation**

6. Desribe the folloiwng **Physico-Chemical Methods of Analysis**

   (a) **UV-Spectrophotometry**,

   (b) **IR-Spectroscopy**,

   (c) **^{1}H-NMR - Spectroscopy**, and

   (d) **Mass spectroscopy**

   Explain the above aspects with some typical examples of **alkaloids** that you have studied.

7. How do the following Analytical Techniques help largely in determining the structure of an "**unknown alkaloid**".:

   (i) **Circular Dichroism (CD)**,

   (ii) **Optical Rotary Dispersion (ORD)**,

   (iii) **X-Ray Diffraction Analysis**,

   (iv) **Emission Spectroscopy**,

   (v) **Conformational Analysis**.

8. How would you classify the '**Alkaloids**'? Give the chemical strcuture, plant source, and uses of at least ONE well-known example from each class of compound.

9. Discuss the **CONSTIUTION** of any **ONE** *alkaloid* given under:

   (a) **Hygrine**,

   (b) **Cuscohygrine**,

   (c) **Ricimine**,

   (d) **piperine**.

10. Give a comprehensive account with regard to any **ONE** of the **CONSTIUTION** of the following 'Alkaloids':

    (i) **Pelletierine**,

    (ii) **Isopelletierine**

    (iii) **Pseudopelletierine**

11. Describe the constitution of any one of the following '**alkaloids**':

    (a) **Nicotine**

    (b) **Atropine**

    (c) **Cocaine**

12. What are the 'Quinine Alkaloids' you have studied? Elaborate the fundamental structure of the **Cinchona Alkaloids**.

13. Discuss comprehensively the **Constitution of Quinine**. Give all relevant reactions in this particular example.

14. Give a detailed account on any one of the following:
    (i) **Constitution of Cinchonine**
    (ii) **Constitution of Papaverine**.

15. Discuss the '**Constitution of Morphine**' giving an elaborated account of all the important steps and chemical reactions involved.

16. Write short notes on any **one** of the following:
    (a) **Berberine**
    (b) **Emetine**.

Index

A

H

T

U

V

W

X

Y

Z